Wide Bandgap Semiconductor Based Electronics

Wide Bandgap Semiconductor-Based Electronics

Edited by
Fan Ren and Stephen J Pearton
University of Florida, Gainesville FL, USA

IOP Publishing, Bristol, UK

ISBN   978-0-7503-2516-5 (ebook)
ISBN   978-0-7503-2514-1 (print)
ISBN   978-0-7503-2517-2 (myPrint)
ISBN   978-0-7503-2515-8 (mobi)

DOI   10.1088/978-0-7503-2516-5

Version: 20200901

IOP ebooks

British Library Cataloguing-in-Publication Data: A catalogue record for this book is available from the British Library.

Published by IOP Publishing, wholly owned by The Institute of Physics, London

IOP Publishing, Temple Circus, Temple Way, Bristol, BS1 6HG, UK

US Office: IOP Publishing, Inc., 190 North Independence Mall West, Suite 601, Philadelphia, PA 19106, USA

Contents

6 Breakdown in Ga_2O_3 rectifiers—the role of edge termination and impact ionization

Yu-Te Liao, Minghan Xian, Randy Elhassani, Patrick Carey IV, Chaker Fares, Fan Ren, Marko Tadjer and S J Pearton

7 Radiation damage in Ga_2O_3 materials and devices

S J Pearton, Fan Ren, Jihyun Kim, Michael Stavola and Alexander Y Polyakov

Part II   Gallium Nitride/Aluminum Nitride

10   The effect of growth parameters on the residual carbon   10-1
concentration in GaN high electron mobility transistors:
theory, modeling, and experiments
Indraneel Sanyal and Jen-Inn Chyi

11   High Al-content AlGaN-based HEMTs   11-1
Albert G Baca, B A Klein, A M Armstrong, A A Allerman,
E A Douglas and R J Kaplar

18 Irradiation effects on high aluminum content AlGaN channel devices   18-1

Patrick Carey, Fan Ren, Jinho Bae, Jihyun Kim and Stephen Pearton

Part III Zinc Oxide

19 BeMgZnO wide bandgap quaternary alloy semiconductor   19-1

*Kai Ding, Vitaliy Avrutin, Natalia Izyumskaya, Ümit Özgür and
Hadis Morkoç*

Part IV Boron Nitride

Part V Diamond

Preface

There is increasing interest in the development of wide bandgap and ultra-wide bandgap semiconductors for use in more efficient power switching devices and solar-blind UV detection [1–3]. As just one example, electricity currently accounts for ~38% of primary energy consumption in the US and is the fastest growing form of end-use energy [1]. Achieving high-power conversion efficiency requires low-loss power semiconductor switches. Today's incumbent power, Si-based switch technology, includes metal–oxide field effect transistors (MOSFETs), insulated gate bipolar transistors (IGBTs), and thyristors [1]. However, silicon power semiconductor devices have several important limitations, including high switching losses and low switching frequencies. The relatively low Si bandgap of 1.1 eV and the low critical electric field (approximately 30 V per micron) require high voltage devices to achieve a substantial critical thickness. The large thickness translates to devices with high resistance and associated conduction losses and low switching frequency [1, 2]. Silicon high voltage power MOSFETs require large die areas to keep the conduction losses low. Power electronics play a significant role in the delivery of electricity as they control and convert electrical power to provide optimal conditions for transmission, distribution, and load-side consumption. The forecast for the fraction of electricity processed through some form of power electronics is estimated to increase to 80% by 2030, roughly double that of today. The relatively low silicon bandgap also contributes to high intrinsic carrier concentrations in silicon-based devices, resulting in a high leakage current at elevated temperatures. Temperature variation of the bipolar gain in IGBTs amplifies the leakage and limits the maximum junction temperature of many IGBTs to 125 °C. This upper limit of the device operating temperature can be significantly increased by the use of wider bandgap semiconductors.

Therefore, advances in power electronics have the potential for enormous energy efficiency and a range of operating condition improvements. The development of advanced power electronic devices with exceptional efficiency, reliability, functionality, and form factors will provide a competitive commercial advantage in the deployment of advanced energy technologies [3]. Power semiconductor devices based on wide bandgap materials, such as SiC and GaN, offer the potential of breakthrough performance for a wide range of applications. Advances in wide bandgap (WBG) semiconductor materials are enabling a new generation of power semiconductor devices that far exceed the performance of silicon-based devices. The higher critical electric fields in these WBG materials ($\geqslant$2 MV cm^{-1}) enable thinner, more highly doped voltage-blocking layers, which can reduce on-resistance by two orders of magnitude in the majority carrier architectures, such as MOSFETs, relative to an equivalent Si device. Moreover, the high breakdown electric field and low conduction losses mean that WBG materials can achieve the same blocking voltage and on-resistance with a smaller form factor. This reduced capacitance allows higher frequency operation compared to a Si device.

In the automotive, aerospace, and defense industries, where power electronics devices and related packaging play a significant role in the power conversion and transmission process, SiC and to a lesser extent GaN devices have established themselves in important market niches where they can supplement or replace Si devices. Renewable energy sources must also be switched into the existing power grid. In many parts of the world, power systems are commonly experiencing instantaneous penetration levels of inverter-based power sources such as wind, solar photovoltaics (PV), and battery storage in excess of 50%–60% relative to system demands.

Ultra-wide bandgap semiconductor devices present a potential new advance for the power electronics industry, motivated by the potential for improved performance, higher switching efficiency, and higher power density compared to existing technologies. These materials are generally defined as having bandgaps exceeding that of GaN (3.4 eV) and include gallium oxide (Ga_2O_3), diamond (C), $AlN/Al_xGa_{1-x}N$ and cubic boron nitride (c-BN). Because many figures of merit for device performance scale nonlinearly with the bandgap, these semiconductors have long been known to have potential compelling advantages over their narrower bandgap cousins in high-power and RF electronics, as well as in deep-UV optoelectronics, quantum information, and extreme-environment applications. The benefits of these materials for high-power devices stems from the increase in critical electric field (E_C) with increasing bandgap (E_G). A higher critical field permits the use of a thinner drift region. This results in a lower resistance to the flow of electrons when the junction is not depleted (i.e. when the device is in the conducting on-state). Even more efficient switching could be expected from the semiconductors with larger bandgaps listed above. Ga_2O_3 has attracted increasing attention in recent years. The high theoretical breakdown electrical field (8 MV cm^{-1}), ultra-wide bandgap (~ 4.8 eV), and large Baliga's figure of merit (BFOM) of Ga_2O_3 make it a potential candidate material for next-generation high-power electronics. Hexagonal boron nitride (h-BN) has a bandgap of about 6.5 eV and possesses novel physical properties. For example, freestanding h-BN epilayers can be sliced into varying shapes and tiled together to form arrays that are flexible with good conformability and transferability. Devices based on freestanding h-BN epilayers can be attached to rigid, flat, or curved surfaces. As such, these materials potentially offer a disruptive platform to design a wide range of photonic and electronic devices with flexible form factors. The transverse electric (TE) mode is generally preferred for optoelectronic device applications because this mode allows for surface emitting light-emitting diodes and is associated with a lower threshold and higher optical gain, and higher lasing beam quality for laser diodes. Due to the combination of the TE mode, a large exciton binding energy, and its two-dimensional (2D) nature, h-BN light-emitting devices are expected to be very efficient and, in principle, h-BN based deep-UV lasers should perform better than AlN-based deep-UV lasers. Other oxide systems, such as the BeMgZnO material systems, are also attracting attention for applications in transparent thin film transistors, photodetectors, and solar cells, owing to their tunable bandgap (over the range 3.37 eV (ZnO) to 10.6 eV (BeO)) and high electron saturation velocity.

However, even with the considerable advantages of these materials, a number of challenges are preventing the widespread adoption of ultra-wide bandgap semiconductors and only SiC has a significant market penetration for power electronics based on wide bandgap materials. GaN does play an important role in military electronics and is promising for electric vehicle infrastructure. Significant work remains to overcome these barriers and realize the full potential of all these materials in power electronics.

The purpose of this book is to provide a detailed overview of the challenges and progress in wide and ultra-wide bandgap semiconductors. Authors from around the world have provided reviews on the status of the development of materials and devices in these systems. This collection of chapters is therefore a resource for those entering or already in this rapidly expanding field.

S J Pearton and Fan Ren
University of Florida, Gainesville, FL, USA

References

[1] Kizilyalli I C, Carlson E P, Cunningham D W, Manser J S, Xu Y and Liu A Y 2018 *Wide Band-Gap Semiconductor Based Power Electronics for Energy Efficiency* (Washington, DC: United States Department of Energy) 20585
[2] Tsao J Y *et al* 2018 Ultrawide-bandgap semiconductors: research opportunities and challenges *Adv. Electron. Mater.* **4** 1600501
[3] Pearton S J, Jiancheng Yang P, Cary H IV, Ren F, Kim J, Tadjer M J and Mastro M A 2018 A review of Ga_2O_3 materials, processing, and devices *Appl. Phys. Rev.* **5** 011301

Acknowledgements

The editors would like to acknowledge support from the HDTRA12020002 cooperative research agreement 'Interaction of Ionizing Radiation with Matter University Research Alliance' funded by the Defense Threat Reduction Agency (DTRA), particularly Jacob Calkins.

Editor biography

Fan Ren

Fan Ren is Distinguished Professor of Chemical Engineering at the University of Florida, Gainesville, FL, USA. He joined UF in 1997 after 12 years as a Member of Technical Staff at AT&T Bell Laboratories, where he was responsible for high speed compound semiconductor device development. He is a Fellow of AIChE, ECS, IEEE, APS, MRS and AVS. He was the recipient of the Gordon Moore award from ECS and the Albert Nerken award from AVS.

S J Pearton

Steve Pearton is Distinguished Professor and Alumni Chair of Materials Science and Engineering at the University of Florida, Gainesville, FL, USA. He has a PhD in Physics from the University of Tasmania and was a postdoc at UC Berkeley prior to working at AT&T Bell Laboratories in 1994-2004 .His interests are in the electronic and optical properties of semiconductors. He is a Fellow of the IEEE, AVS, ECS, TMS, MRS, SPIE and APS. He was the recipient of the J.J. Ebers Award from IEEE, Gordon Moore Award from ECS, John Thornton award from AVS, Adler award from APS and the Bardeen award from TMS.

Contributor list

Elaheh Ahmadi
Department of Electrical Engineering and Computer Science, University of Michigan, Ann Arbor, MI, USA

A A Allerman
Sandia National Laboratories, PO Box 5800, Albuquerque, NM, 87185-1085, USA

Manuel Alonso-Orts
Departamento Física de Materiales, Facultad de CC Físicas, Universidad Complutense Madrid, Spain

A M Armstrong
Sandia National Laboratories, PO Box 5800, Albuquerque, NM, 87185-1085, USA

Vitaliy Avrutin
Department of Electrical and Computer Engineering, Virginia Commonwealth University, Richmond, VA, USA

Albert G Baca
Sandia National Laboratories, PO Box 5800, Albuquerque, NM, 87185-1085, USA

Jinho Bae
Department of Chemical and Biological Engineering, College of Engineering, Korea University, Anam-dong 5-1, Sungbuk-gu, Seoul 02841, Republic of Korea

Kwang Hyeon Baik
Department of Materials Science and Engineering, Hongik University, Jochiwon, Sejong 30016, Republic of Korea

Mahitosh Biswas
Department of Electrical Engineering and Computer Science, University of Michigan, Ann Arbor, MI, USA

Patrick Carey
Department of Chemical Engineering, University of Florida, Gainesville, FL, 32611, USA

Jen-Inn Chyi
National Central University, Taiwan, Republic of China

Kai Ding
Department of Electrical and Computer Engineering, Virginia Commonwealth University, Richmond, VA, USA

Hang Dong
Key Laboratory of Microelectronics Devices and Integration Technology, Institute of Microelectronics, Chinese Academy of Sciences, Beijing, People's Republic of China

E A Douglas
Sandia National Laboratories, PO Box 5800, Albuquerque, NM, 87185-1085, USA

Randy Elhassani
Department of Chemical Engineering, University of Florida, Gainesville, FL, 32611, USA

Madeline Esposito
Sandia National Laboratories

Chaker Fares
Department of Chemical Engineering, University of Florida, Gainesville, FL, 32611, USA

Philip X-L Feng
Department of Electrical and Computer Engineering, Herbert Wertheim College of Engineering, University of Florida, Gainesville, FL, 32611, USA

Houqiang Fu
School of Electrical, Computer, and Energy Engineering, Arizona State University, Tempe, AZ, 85287, USA

Kai Fu
School of Electrical, Computer, and Energy Engineering, Arizona State University, Tempe, AZ, 85287, USA

Nicholas Glavin
Air Force Research Laboratory, 2941 Hobson Way, Wright-Patterson AFB, OH, 45433, USA

Marius Grundmann
Universität Leipzig, Felix-Bloch-Institut für Festkörperphysik, 04103 Leipzig, Germany

Weibing Hao
School of Microelectronics, University of Science and Technology of China, Hefei, People's Republic of China

Aman Haque
Department of Mechanical Engineering, The Pennsylvania State University, University Park, PA, 16802, USA

Jennifer Hite
US Naval Research Laboratory, Washington, DC, 20375, USA

Zahabul Islam
Department of Mechanical Engineering, The Pennsylvania State University, University Park, PA 16802, USA

Natalia Izyumskaya
Department of Electrical and Computer Engineering, Virginia Commonwealth University, Richmond, VA, USA

Soohwan Jang
Department of Chemical Engineering, Dankook University, Yongin 16890, Republic of Korea

H X Jiang
Department of Electrical and Computer Engineering, Texas Tech University, Lubbock, TX, 79409, USA

R J Kaplar
Sandia National Laboratories, PO Box 5800, Albuquerque, NM, 87185-1085, USA

Janghyuk Kim
Department of Chemical and Biological Engineering, College of Engineering, Korea University, Anam-dong 5-1, Sungbuk-gu, Seoul 02841, Republic of Korea

Jihyun Kim
Department of Chemical and Biological Engineering, College of Engineering, Korea University, Anam-dong 5-1, Sungbuk-gu, Seoul 02841, Republic of Korea

Suhyun Kim
Department of Chemical and Biological Engineering, College of Engineering, Korea University, Anam-dong 5-1, Sungbuk-gu, Seoul 02841, Republic of Korea

B A Klein
Sandia National Laboratories, PO Box 5800, Albuquerque, NM, 87185-1085, USA

F Max Kneiß
Universität Leipzig, Felix-Bloch-Institut für Festkörperphysik, 04103 Leipzig, Germany

Mark E Law
Electrical and Computer Engineering, University of Florida, Gainesville, FL, 32611, USA

J Li
Department of Electrical and Computer Engineering, Texas Tech University, Lubbock, TX, 79409, USA

Yu-Te Liao
Department of Electrical and Computer Engineering, National Chiao Tung University, Hsinchu, Taiwan

J Y Lin
Department of Electrical and Computer Engineering, Texas Tech University, Lubbock, TX, 79409, USA

Ming Liu
Key Laboratory of Microelectronics Devices and Integration Technology, Institute of Microelectronics, Chinese Academy of Sciences, Beijing, People's Republic of China

Shibing Long
School of Microelectronics, University of Science and Technology of China, Hefei, People's Republic of China

Michael A Mastro
US Naval Research Laboratory, Washington, DC, 20375, USA

Joana C Mendes
Instituto de Telecomunicações, Campus Universitário de Santiago, 3810-193 Aveiro, Portugal

Bianchi Méndez
Departamento Física de Materiales, Facultad de CC Físicas, Universidad Complutense Madrid, Spain

Hadis Morkoç
Department of Electrical and Computer Engineering, Virginia Commonwealth University, Richmond, VA, USA

Debarati Mukherjee
Instituto de Telecomunicações, Campus Universitário de Santiago, 3810-193 Aveiro, Portugal
Department of Electronics, Telecommunications and Informatics, University of Aveiro, 3810-193, Aveiro, Portugal

Miguel Neto
CICECO, Department of Materials and Ceramic Engineering, University of Aveiro, 3810-193 Aveiro, Portugal

Emilio Nogales
Departamento Física de Materiales, Facultad de CC Físicas, Universidad Complutense Madrid, Spain

Filipe J Oliveira
CICECO, Department of Materials and Ceramic Engineering, University of Aveiro, 3810-193 Aveiro, Portugal

Ümit Özgür
Department of Electrical and Computer Engineering, Virginia Commonwealth University, Richmond, VA, USA

Erin Patrick
Electrical and Computer Engineering, University of Florida, Gainesville, FL, 32611, USA

S J Pearton
Department of Material Science and Engineering, University of Florida, Gainesville, FL, 32611, USA

Luiz Pereira
Department of Physics and I3N—Institute for Nanostructures, Nanomodulation and Nanofabrication, University of Aveiro, 3810-193 Aveiro, Portugal

A Y Polyakov
National University of Science and Technology MISiS, Moscow, 119049, 4 Leninsky Ave, Russian Federation

Fan Ren
Department of Chemical Engineering, University of Florida, Gainesville, FL, 32611, USA

Shlomo Rotter
Smart Diamond Technologies Lda, Aveiro, Portugal

Indraneel Sanyal
National Central University, Taiwan, People's Republic of China

Rui F Silva
CICECO, Department of Materials and Ceramic Engineering, University of Aveiro, 3810-193 Aveiro, Portugal

Michael Stavola
Physics Department, Lehigh University, Bethlehem, PA, 18015, USA

Marko Tadjer
US Naval Research Laboratories, Washington DC, 20375, USA

Q W Wang
Department of Electrical and Computer Engineering, Texas Tech University, Lubbock, TX, 79409, USA

Yu-Lin Wang
Institute of Nanoengineering and Microsystems, Department of Power Mechanical Engineering, National Tsing Hua University, Hsinchu, 30013, Taiwan, People's Republic of China

Holger von Wenckstern
Universität Leipzig, Felix-Bloch-Institut für Festkörperphysik, 04103 Leipzig, Germany

Chang-Run Wu
Institute of Nanoengineering and Microsystems, Department of Power Mechanical Engineering, National Tsing Hua University, Hsinchu, 30013, Taiwan, People's Republic of China

Minghan Xian
Department of Chemical Engineering, University of Florida, Gainesville, FL, 32611, USA

Xueqiang Xiang
School of Microelectronics, University of Science and Technology of China, Hefei, People's Republic of China

Wenhao Xiong
School of Microelectronics, University of Science and Technology of China, Hefei, People's Republic of China

Guangwei Xu
School of Microelectronics, University of Science and Technology of China, Hefei, People's Republic of China

E B Yakimov
Institute of Microelectronics Technology and High Purity Materials, Russian Academy of Sciences, 6 Academician Ossipyan str., Chernogolovka, Moscow Region 142432, Russian Federation
National University of Science and Technology MISiS, Moscow, 119049, 4 Leninsky Ave, Russian Federation

Yuji Zhao
School of Electrical, Computer, and Energy Engineering, Arizona State University, Tempe, AZ, 85287, USA

Xu-Qian Zheng
Department of Electrical and Computer Engineering, Herbert Wertheim College of Engineering, University of Florida, Gainesville, FL, 32611, USA

Xuanze Zhou
School of Microelectronics, University of Science and Technology of China, Hefei, People's Republic of China

Part I

Gallium oxide

IOP Publishing

Wide Bandgap Semiconductor-Based Electronics

Fan Ren and Stephen J Pearton

Chapter 1

Low-dimensional β-Ga$_2$O$_3$ semiconductor devices

Suhyun Kim, Jinho Bae, Janghyuk Kim and Jihyun Kim

1.1 Introduction

Certain properties of wide bandgap semiconductors have received significant attention and have led to pivotal breakthroughs in (opto)electronics. Their light absorption and emission in the ultraviolet (UV) and visible regions are attractive in optoelectronics, while their high breakdown voltage makes them suitable for high-power electronics. Furthermore, ultra-wide bandgap semiconductors, which have higher bandgap energies than those of conventional wide bandgap materials such as SiC or GaN, can exhibit superior device performance as the relationship between the bandgap and the figures of merit is nonlinear; an increase in the bandgap causes a much more significant increase in the figures of merit. Gallium oxide (Ga$_2$O$_3$) is the newest and least mature among the major materials that have paved the way for the advancement of ultra-wide bandgap semiconductors. Ga$_2$O$_3$ has five polymorphs, among which the monoclinic β phase is the most thermodynamically stable crystal. Owing to its bandgap energy of 4.6–4.9 eV, β-Ga$_2$O$_3$ is suitable for deep-UV solar-blind photodetectors [1, 2]. Moreover, the superior Baliga's figure of merit for β-Ga$_2$O$_3$ shows great potential for applications in high-power devices and its high chemical and thermal stabilities allow device operation under harsh conditions [3, 4].

Low-dimensional β-Ga$_2$O$_3$ nanostructures, including nanowires, nanobelts, nanorods, nanosheets, and nanomembranes, provide further advantages in addition to all the attractive properties of a single-crystal substrate. Therefore, these nanostructures, which can be fabricated by either top-down or bottom-up methods, have been widely applied in electronic and optoelectronic devices in the form of an individual nanobelt, an array, or a bridged network of nanostructures. Using substrates with higher thermal conductivity when fabricating devices based on β-Ga$_2$O$_3$ nanostructures could be a solution to the low thermal conductivity, which is a critical drawback in applying β-Ga$_2$O$_3$ in high-power electronics. Furthermore,

doi:10.1088/978-0-7503-2516-5ch1

the nanostructures have fewer defects and less strain when they form heterostructures, while the deposition of bulk material on a substrate usually induces a lattice mismatch between the material and the substrate. Through van der Waals integration with other two-dimensional (2D) materials, the unique properties of β-Ga$_2$O$_3$ nanostructures can be further enhanced. After a brief discussion regarding the preparation methods and the contact properties of β-Ga$_2$O$_3$ nanostructures, the fabrication and characterization of various transistor structures are demonstrated in this chapter.

1.1.1 Preparation of low-dimensional Ga$_2$O$_3$ nanostructures

Nanostructures can be prepared through either bottom-up or top-down methods. The bottom-up methods for growing β-Ga$_2$O$_3$ nanostructures include physical evaporation, arc discharge, laser ablation, carbothermal reduction, microwave plasma chemical vapor deposition (CVD), and metal–organic CVD. The CVD methods are most widely used as they allow the reproducible synthesis of high-purity β-Ga$_2$O$_3$ nanostructures at a high deposition rate. Nanostructures with different properties and structures can be achieved by controlling the growth parameters, such as the precursors, catalysts, growth temperature, growth time, distance between the source and the substrate, and flow rates of the source gases. Gallium and oxygen are commonly used as precursors when growing Ga$_2$O$_3$ nanostructures using CVD equipment. Auer *et al* also used these materials as the source and grew different types of nanostructures including nanorods, nanoribbons, nanowires, and cones of monoclinic β-Ga$_2$O$_3$ depending on the growth temperature and the presence of Au catalysts in the CVD process. The nanoribbons and nanorods were synthesized with and without catalysts, respectively, through the vapor–solid growth mechanism, while the nanowires were grown in the presence of catalysts through the vapor–liquid–solid (VLS) mechanism [5]. Kumar *et al* used a different catalyst, Fe, to synthesize crystalline β-Ga$_2$O$_3$ nanowires through the VLS mechanism and obtained structural, morphological, and optical properties comparable to those grown using Au catalysts [6]. Similarly, the diameter of the β-Ga$_2$O$_3$ nanowires can also be effectively controlled by the growth parameters; the obtained diameters are usually larger for higher values of temperature, time, and gas flow rate in the presence of catalysts, as shown in figure 1.1 [7]. Furthermore, a larger catalyst size or smaller distance between the metal source and the substrate produces nanowires with larger diameters [7].

Monoclinic β-Ga$_2$O$_3$ can also be combined with other materials to produce functional nanostructures for high-performance device applications. Hsieh *et al* synthesized Au/Ga$_2$O$_3$ core–shell nanowires through the VLS mechanism and nitridized them at a relatively low temperature of 600 °C to form Au/Ga$_2$O$_3$/GaN nanowires (figure 1.2). The nanowires had a metal–oxide–semiconductor (MOS) structure that could be applied to vertical high-power nanoelectronics [8]. Similarly, Kumar *et al* investigated the influence of ammonification on the synthesis of coaxial GaN/Ga$_2$O$_3$ [9]. They reported that as the ammonification temperature increased, the decomposition rate of the β-Ga$_2$O$_3$ nanowires increased, leading to increased GaN conversion. Therefore, coaxial GaN/Ga$_2$O$_3$ was formed at relatively low

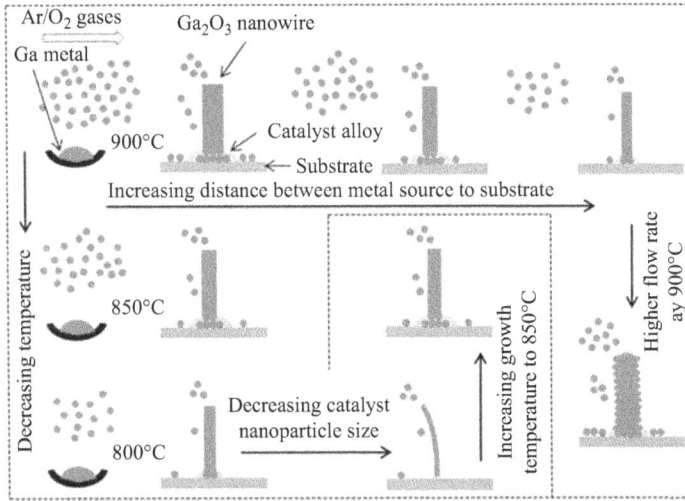

Figure 1.1. Schematic of diameter tuning of β-Ga$_2$O$_3$ nanowires by controlling the growth parameters. Reproduced with permission from [7]. Copyright 2017 Springer Nature.

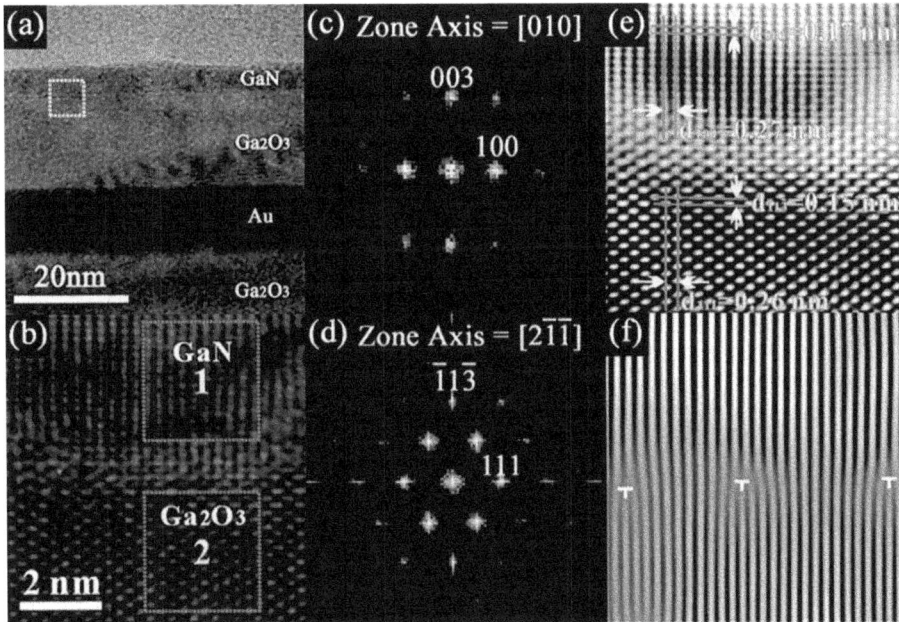

Figure 1.2. Structural characterization of Au/Ga$_2$O$_3$/GaN nanowires with MOS structure. Reproduced with permission from [8]. Copyright 2008 the American Chemical Society.

ammonification temperatures while the complete conversion of GaN from β-Ga$_2$O$_3$ was observed at a high temperature of 1050 °C (figure 1.2). Furthermore, Au-decorated β-Ga$_2$O$_3$ nanowires were also demonstrated to modify the optical properties of the nanostructures and enhance their photocatalytic effect [10].

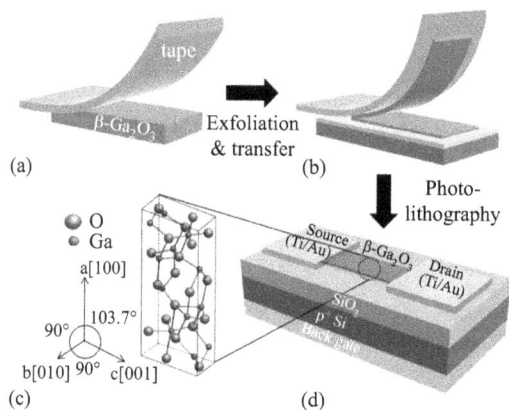

Figure 1.3. Schematic of the fabrication process for field-effect transistors based on mechanically exfoliated β-Ga$_2$O$_3$ flakes. Reproduced with permission from [11]. Copyright 2016 PCCP Owner Societies.

The top-down approach for forming nanostructures usually involves the etching of bulk materials. For example, an inductively coupled plasma etching of a GaN-based planar LED covered with Ni nanoclusters and a SiO$_2$ layer transformed the device into a nanorod LED and improved its light extraction efficiency and output power. Although there has been no reported research regarding the formation of β-Ga$_2$O$_3$ nanostructures through etching, another top-down approach is available for β-Ga$_2$O$_3$ owing to its crystal structure. The exceptionally high lattice constant along one direction ($a = 1.2$ nm, $b = 0.3$ nm, and $c = 5.8$ nm) in monoclinic β-Ga$_2$O$_3$ enables mechanical exfoliation using the convenient Scotch tape method [11–13]. Mechanical exfoliation is commonly applied to 2D materials formed by the van der Waals interaction between each layer. The cleavage planes parallel to the (100) and (001) planes allow mechanical cleavage of β-Ga$_2$O$_3$ substrates into thin flakes without the need for etching, as shown in figure 1.3 [11]. Because the flakes are exfoliated from a single-crystal substrate, high crystallinity is maintained in the nanobelts. The exfoliated flakes show low surface roughness and are ~20–400 nm in thickness [11, 12, 14–18]. As the exfoliation and the transfer processes are all performed under dry conditions and the exfoliated β-Ga$_2$O$_3$ flakes form van der Waals heterostructures with other materials, a clean interface is observed between the layers [13, 16].

1.1.2 Contact properties of β-Ga$_2$O$_3$ nanodevices

The study of contact properties is essential to fully characterize β-Ga$_2$O$_3$ nanobelt-based devices, which have a smaller contact area than the bulk-substrate-based devices. This is because the contact properties have a significant influence on the electrical characteristics of the device. Therefore, optimal contact metal selection and process optimization are essential to avoid unintentional device degradation due to poor contact.

Generally, a Schottky contact is formed when a metal is in contact with a wide bandgap semiconductor. Therefore, the contact resistance between β-Ga$_2$O$_3$, which is an n-type wide bandgap semiconductor, and the contact metal depends on the Schottky barrier height (SBH), Φ_B. The SBH for n-type semiconductors satisfies the following equation:

$$q\Phi_B = q\Phi m - E_{EA},$$

where q is the electron charge, Φ_m is the work function of the contact metal, and E_{EA} is the electron affinity of the n-type semiconductor. In the case of devices consisting of ohmic contacts such as metal–oxide–semiconductor field-effect transistors (MOSFETs), high contact resistance can be responsible for the degradation of the electrical properties. On the other hand, the Schottky contact properties determine the performance of devices such as metal–semiconductor field-effect transistors (MESFETs) or Schottky barrier diodes (SBDs). Hence, to fabricate and characterize β-Ga$_2$O$_3$ nanobelt devices that suit the purpose of a device, the contact properties between various metals and β-Ga$_2$O$_3$ nanobelts were investigated.

1.1.3 The ohmic contacts of β-Ga$_2$O$_3$ nanobelt devices

The first approach to form ohmic contacts is to select a metal electrode with the smallest SBH, considering the electron affinity of β-Ga$_2$O$_3$ and the work function of the contact metal. This led to the study of various metal electrodes that can achieve the lowest contact resistance and excellent ohmic contact properties in β-Ga$_2$O$_3$ nanobelt-based devices. Yao et al analyzed the electrical properties of nine different contact metals, namely Ti, In, Ag, Sn, W, Mo, Sc, Zn, and Zr, and evaluated the contact properties of each metal electrode [19]. The study showed that Au-capped Ti metal electrodes have the best ohmic properties among the aforementioned nine metal electrodes. Considering the existence of the Schottky barrier, which is calculated by the work function of Ti and the electron affinity of the unintentionally doped β-Ga$_2$O$_3$ (4.33 eV and ~4.00, respectively), an underlying mechanism for forming the ohmic contact is observed, which will be described later.

Ohmic contact in the bulk-substrate β-Ga$_2$O$_3$ devices was achieved through pre-treatments such as ion implantation, reactive ion etching, and plasma bombardment in the defined region before the metal electrode deposition [20]. This pre-treatment process is not commonly used for nanobelt-based devices because it causes unpredictable damage to the β-Ga$_2$O$_3$ and the underlying substrate, moreover its ohmic properties are not reproducible. Instead, post-treatments such as annealing after metal electrode deposition are mainly used for ohmic contact confirmation. The annealing process is expected to reduce the damage before and after the fabrication process.

Bae et al investigated the annealing conditions of β-Ga$_2$O$_3$ nanobelt-based devices and their influence on the contact properties of the contact metal (figure 1.4) [13]. The results confirmed that the electrical characteristics of the Ti/Au contacts were significantly improved, compared to the as-deposited device, when the annealing process was performed at 500 °C. Research has also shown that out-diffusion of oxygen from β-Ga$_2$O$_3$ to Ti, leading to the formation of oxygen vacancies as donors, significantly improves the electrical properties of devices. In addition, Ti$_x$O$_y$, which

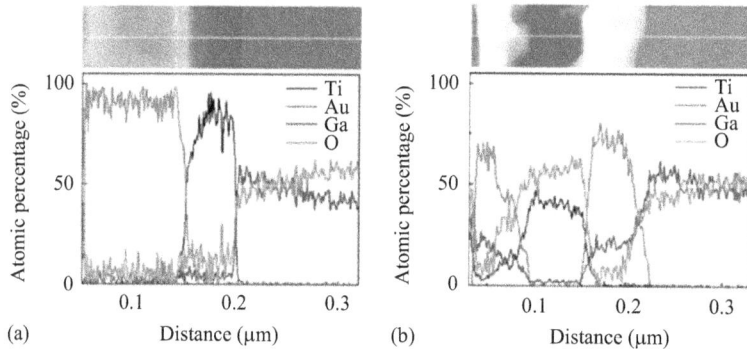

Figure 1.4. Atomic percentage profiles of contact metal and β-Ga$_2$O$_3$ obtained using energy dispersive x-ray spectrometry (a) before and (b) after annealing at a temperature of 500 °C. Reproduced with permission from [13]. Copyright 2017 the Electrochemical Society.

has a lower work function due to the reaction between the out-diffused oxygen and Ti, has a significant influence on the formation of the ohmic contact.

This result is consistent with that reported by Zhen *et al*, who performed Ar-ambient annealing on a β-Ga$_2$O$_3$ nanobelt FET and confirmed that the contact resistance decreased from ~430 to ~0.387 Ω mm^{-1} before and after annealing, respectively [15]. Thus, it was confirmed that most Ti/Au contacts can form an ohmic contact through annealing at ~450 °C–500 °C. However, Bae *et al* confirmed that oxygen diffusion degrades the metal electrode at temperatures at or above 700 °C [13]. Therefore, multilayer metal deposition such as Ti/Al/Au and Ti/Al/Ni/Au is being investigated to achieve stable operation at high temperatures.

1.1.4 β-Ga$_2$O$_3$ nanobelt Schottky contacts

The Schottky contact is essential for the fabrication of structures such as MESFETs and SBDs, which are commonly used in power devices. Since β-Ga$_2$O$_3$ has been attracting attention as a high-voltage device due to its wide bandgap and ultrahigh breakdown field, studies of the Schottky contact for bulk substrates have been conducted. Farzana *et al* proposed guidelines for selecting a Schottky contact metal by calculating the SBH for four metals—Pd, Ni, Pt, and Au—using three independent methods (*I–V*, *C–V*, and internal photoemission) (figure 1.5) [21]. The suitable contact metal for a Schottky barrier device was determined by analyzing the current transport mechanism of different contact materials. However, there have been no specific studies regarding the Schottky contact for β-Ga$_2$O$_3$ nanobelt devices, although Ni or Pt is used as a Schottky contact for nanobelt devices based on the existing research on bulk substrates. Bae *et al* proposed the use of Au-capped Ni contacts in β-Ga$_2$O$_3$ nanobelt devices as the top-gate electrode for MESFET devices and measured their electrical properties [22, 23]. Swinnich *et al* fabricated a flexible substrate-based SBD. They deposited a Pt Schottky contact on the β-Ga$_2$O$_3$ nanobelt to fabricate an SBD that showed an excellent breakdown voltage even when bent [24]. However, Kim *et al* demonstrated

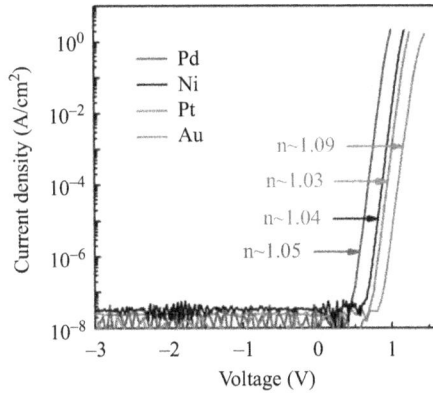

Figure 1.5. Schottky diode properties of β-Ga$_2$O$_3$ nanobelt FETs with various Schottky contact metals. Reproduced with permission from [21]. Copyright 2017 AIP Publishing.

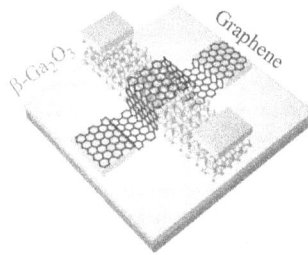

Figure 1.6. Schematic of β-Ga$_2$O$_3$/graphene heterostructure photodetector devices. Graphene is used as the transparent top-gate electrode of the photodetector. Reproduced with permission from [27]. Copyright 2019 the American Chemical Society.

Pt/Au electrode degradation during high-temperature operations. Therefore, further studies are required to confirm stable operation under harsh conditions [25].

The greatest advantage of the β-Ga$_2$O$_3$ nanobelts is that they can be integrated with other 2D materials to fabricate devices that can utilize the advantages of both materials. Therefore, there have been attempts toward the integration and bandgap engineering of β-Ga$_2$O$_3$ nanobelts with mechanically exfoliated 2D materials. For example, Yan *et al* fabricated a graphene barristor device utilizing the Schottky contact between exfoliated graphene and a β-Ga$_2$O$_3$ nanobelt to achieve a high breakdown field of 5.2 MV cm^{-1} [26]. Kim *et al* used graphene as a transparent top-gate electrode for β-Ga$_2$O$_3$ nanobelts, minimizing the dark current and shadow effect under the electrode (figure 1.6) [27]. Further studies are underway to form multifunctional Schottky contacts that cannot be applied to metal electrodes, such as transparent electrodes or junction FETs, through strain-free van der Waals bonding between the β-Ga$_2$O$_3$ nanobelt and 2D materials.

1.2 β-Ga$_2$O$_3$-based nanoelectronic devices

To explore β-Ga$_2$O$_3$-based nanoelectronic devices, there have been several studies on the fabrication and characterization of various types of FETs and SBDs using β-Ga$_2$O$_3$ nanobelts. The β-Ga$_2$O$_3$ nanobelt, which can be easily exfoliated from the bulk crystal and inherits the quality of the bulk crystal, is more suitable for nanoelectronic devices because of its superior quality compared to the other β-Ga$_2$O$_3$ nanostructures [28]. Generally, the β-Ga$_2$O$_3$ nanobelts used for fabricating devices have thicknesses in the range of tens to hundreds of nanometers and are exfoliated along the (100) or (001) directions [29]. Moreover, β-Ga$_2$O$_3$ exhibits low thermal conductivity of 11–27 W m^{-1} K^{-1}, which is notably lower compared to competing materials such as SiC (360–490 W m^{-1} K^{-1}) and GaN (150–200 W m^{-1} K^{-1}) [4, 30]. Therefore, research on improving the thermal conductivity of β-Ga$_2$O$_3$-based devices by integrating with other substrates or materials is essential for the development of β-Ga$_2$O$_3$-based power electronic devices. As β-Ga$_2$O$_3$ nanobelts are easily transferred to various substrates and materials, they can be applied in research on the electrical and thermal properties of β-Ga$_2$O$_3$-based nanoelectronic devices on various substrates in preference to conventional bulk crystals. In addition, β-Ga$_2$O$_3$ nanobelts can be combined with other semiconductor materials to fabricate various types of new devices.

1.2.1 Single β-Ga$_2$O$_3$ nanobelt-based field-effect transistors

Hwang *et al* successfully demonstrated the possibility of fabricating MOSFETs using β-Ga$_2$O$_3$ nanobelts for the first time [12]. Since then, there have been various studies regarding the fabrication of β-Ga$_2$O$_3$-based nanoelectronic devices using nanobelts.

The thickness of the nanobelt was in the range of 20–100 nm and TEM observations showed that it is mainly exfoliated along the (100) direction. The energy-dispersive x-ray spectroscopy and the absorption spectra of the (100) β-Ga$_2$O$_3$ nanobelt confirmed that the properties of the bulk crystal are maintained in the nanobelt. The fabricated β-Ga$_2$O$_3$ nanobelt MOSFET was fabricated by depositing a Ti/Au metal electrode followed by transferring a 100 nm thick β-Ga$_2$O$_3$ nanobelt onto an SiO$_2$/Si substrate, as shown in figure 1.7.

The β-Ga$_2$O$_3$ nanobelt MOSFET exhibits μ_{FE} and SS values of ~70 cm^2 V^{-1} s^{-1} and ~200 mV dec^{-1}, respectively, as shown in figure 1.8. These values were relatively low compared to those of bulk β-Ga$_2$O$_3$ devices, indicating that the metal contact and interfaces have not been optimized. The device exhibited n-type semiconductor behavior originating from atomic defects or impurities in the β-Ga$_2$O$_3$. The fabricated β-Ga$_2$O$_3$ nanobelt MOSFET exhibited a high gate modulation of ~10^7 even under a high drain voltage of 20 V.

Kim *et al* fabricated back-gated MOSFETs using ~200 nm thick β-Ga$_2$O$_3$ nanobelts that were mechanically exfoliated from unintentionally n-type doped (~3 × 10^{17} cm^{-3}) commercial β-Ga$_2$O$_3$ substrates and studied their stability and electrical characteristics under various operating temperatures in the range of 25 °C–250 °C [11]. They reported that the electrical conductance of the β-Ga$_2$O$_3$ MOSFET

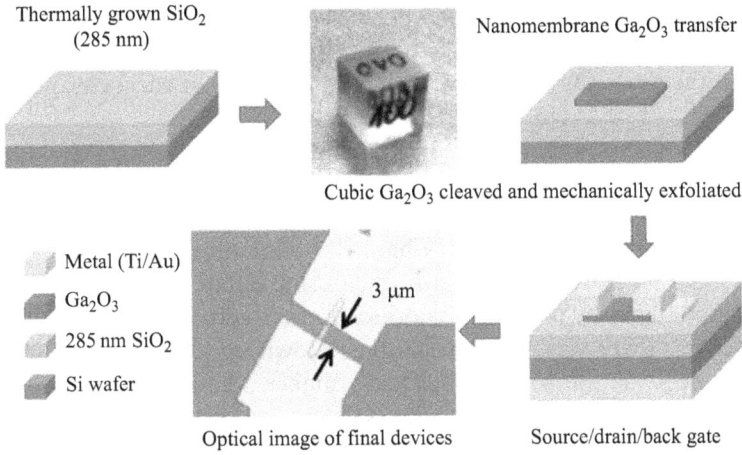

Figure 1.7. Schematic process flow for the fabrication of β-Ga$_2$O$_3$ nanobelt field-effect transistors. Reproduced with permission from [12]. Copyright 2014 AIP Publishing.

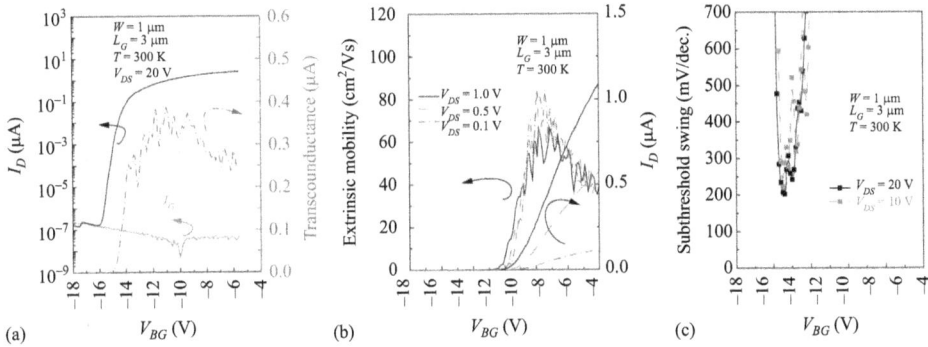

Figure 1.8. (a) Drain current (I_D) versus back-gate-to-source voltage (V_{BG}), showing an on–off current ratio of ~10^7 and n-type semiconductor behavior. (b) Field-effect mobility (μ_{FE}) and (c) subthreshold swing (SS) versus V_{BG}. Reproduced with permission from [12]. Copyright 2014 AIP Publishing.

increased with temperature and the activation energy was approximately 0.25 eV, implying that the increased conductivity of the β-Ga$_2$O$_3$ nanobelts was caused by activation of deep donor level oxygen vacancies at higher temperatures. Furthermore, an electrical breakdown was not observed in these measurements for a V_{DS} of up to +40 V and a V_{GS} of −60 V between 25 °C and 250 °C.

Tadjer *et al* reported the fabrication of enhancement-mode (E-mode) FETs based on β-Ga$_2$O$_3$ nanobelts using a high-*k* HfO$_2$ gate dielectric [31]. The threshold voltage was +2.9 V at a V_{DS} of 100 mV, which, according to the authors, originated from a larger conduction band offset between HfO$_2$ and β-Ga$_2$O$_3$. A maximum I_{DS} of about 11.1 mA mm^{-1} and an on-resistance (R_{ON}) of about 818 Ω mm were measured at V_{GS} = 18 V, and at V_G < 0 a large off-state current leakage was observed, which originated from the traps at the HfO$_2$–β-Ga$_2$O$_3$ or β-Ga$_2$O$_3$–SiO$_2$ interfaces. Electron spin resonance (ESR) measurements were performed at room temperature to investigate the origin of the electrons in the β-Ga$_2$O$_3$ nanobelts and the results indicated that the conductivity originated from the presence of oxygen vacancies in the β-Ga$_2$O$_3$ nanobelts: the density of oxygen vacancies was measured to be 2.3×10^{17} ($\pm 50\%$) cm^{-3}.

Ahn *et al* demonstrated back- and top-gated MOSFETs using Al$_2$O and SiO$_2$ as the dielectrics of the top and back gates, respectively [32]. The I_D–V_{DS} output characteristics of the β-Ga$_2$O$_3$ nanobelt FETs operating at 25 °C with either or both front and back gates were analyzed. Channel modulation was observed to improve when both gates were used rather than when using either of the gates. The I_{DS} was effectively modulated by V_{GS} with good saturation and sharp pinch-off character-istics. Moreover, there was no electrical breakdown up to biases of $V_{DS} = +100$ V and $V_{GS} = -100$ V, showing stable performance. A saturation mobility (μ) of ~1.35 cm^2 V^{-1} s^{-1} was achieved, which was determined using the following equation:

$$\mu = \frac{2L}{(W)} \times m^2 \times C_G,$$

where $I_{D,SAT}$ is the drain saturation current, m is the slope of a regression fit to the straight-line portion of the $(I_{D,SAT})^{1/2}$–V_G transfer curve, W is the gate width, L is the gate length, and C_G is the gate capacitance. The authors noted that the mobility in depletion-mode (D-mode) structures was often overestimated as the β-Ga$_2$O$_3$ nanobelt FET is a bulk conduction channel device.

Zhou *et al* demonstrated the fabrication of E-mode and D-mode β-Ga$_2$O$_3$ nanobelt-based MOSFETs by optimizing the thickness of a β-Ga$_2$O$_3$ nanobelt [33]. They used Sn-doped β-Ga$_2$O$_3$ nanobelts with a doping concentration of 2.7×10^{18} cm^{-3} as the channel layer. Ar plasma treatment was implemented to decrease the contact resistance. As a result, a relatively high maximum drain current density ($I_{D,MAX}$) of 600 mA mm^{-1} was achieved by a D-mode MOSFET using Sn-doped 94 nm thick β-Ga$_2$O$_3$ nanobelt. The authors reported that the threshold voltage (V_{th}) shifted from the negative to positive direction as the thickness was gradually reduced. The dependence of V_{th} on the thickness of the β-Ga$_2$O$_3$ nanobelt was attributed to the surface depletion effect caused by the dangling bond on the β-Ga$_2$O$_3$ nanobelt surface. Meanwhile, the E-mode β-Ga$_2$O$_3$ nanobelt FETs demonstrated a high breakdown voltage of 185 V and an average electrical field (E_{av}) of 2 MV cm^{-1}, demonstrating the great potential of β-Ga$_2$O$_3$ nanobelt-based FETs in future power devices.

β-Ga$_2$O$_3$ nanobelt-based high-power devices are attracting attention due to their advantages, which include minimizing power loss due to their downsizing and economizing power system through efficiency maximization. Among the available fabrication techniques, the field plate (FP) technique is widely used owing to its ease of fabrication and efficiency. The ability to disperse concentrated electric fields at specific locations through a gate FP, source-connected FP, or multiple FPs has already been confirmed in devices such as GaN and AlGaN, which are commonly used for the fabrication of conventional power device materials. A study for increasing the breakdown voltage of β-Ga$_2$O$_3$ devices by applying the FP technique

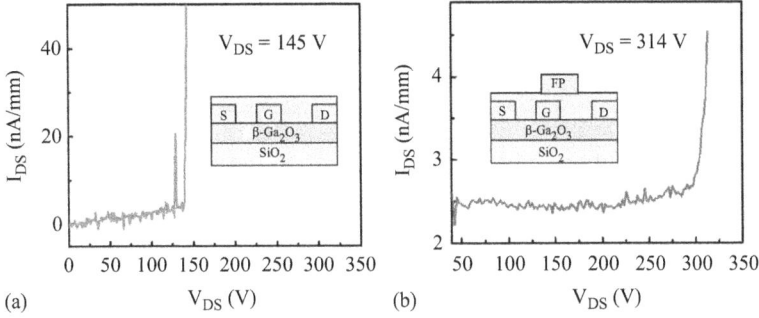

Figure 1.9. Off-state three-terminal hard breakdown results of the fabricated β-Ga$_2$O$_3$ nanoFET (a) without and (b) with the source-connected field-modulating plate. The insets show the schematic of each device. Reproduced with permission from [23]. Copyright 2019 The Royal Society of Chemistry.

was also conducted on a nanobelt-based device. Bae *et al* conducted a study to maximize the breakdown voltage of a quasi-2D power device by introducing a source-connected FP to a β-Ga$_2$O$_3$ nanobelt MESFET (figure 1.9) [23]. The maximum value of the electric field on the β-Ga$_2$O$_3$ surface was decreased from ~11 to ~6 MV cm^{-1} by introducing the source FP, which is smaller than the breakdown field of β-Ga$_2$O$_3$. The three-terminal off-state breakdown voltage of the fabricated devices was improved to 314 V, which is twice that of the conventional devices, confirming the possibility of a high-power nanodevice.

A high $I_{D,MAX}$ of 1.5/1.0 A mm^{-1} for D/E-mode β-Ga$_2$O$_3$ nanobelt FETs was achieved by increasing the doping concentration of the β-Ga$_2$O$_3$ nanobelts from 3.0×10^{18} to 8.0×10^{18} cm^{-3}, respectively [34]. Further, a lower contact resistance of 0.75 Ω mm and a higher electric field velocity of 7.3×10^6 cm s^{-1} were achieved from the high doping channel compared with those of the low doping channel. In addition, the self-heating effect was studied using thermo-reflectance measurements [35]. The results showed that even at a low bias power regime ($P = V_{DS} \times I_D = 1.2$ W mm^{-1}), or in unbiased devices, the temperature of the device was 35 °C higher than room temperature. Moreover, the device temperature increased more significantly in higher power bias regimes, leading to reduced electron mobility, reliability, and breakdown voltage.

Furthermore, the authors reported thermodynamic investigations of a β-Ga$_2$O$_3$ nanobelt FET on a sapphire substrate, which has half the ΔT compared to the SiO$_2$/Si substrate [36]. The thermal resistances were measured to be 4.6×10^{-2} and 1.47×10^{-1} mm^2 K W^{-1} for the sapphire and SiO$_2$/Si substrates, respectively. As a result of the reduced self-heating, an $I_{D,MAX}$ of 535 mA mm^{-1} was achieved on the sapphire substrate, which is 2.5× higher than that on the SiO$_2$/Si substrate. These results show that incorporating β-Ga$_2$O$_3$ channels in substrates with high thermal conductivity can help solve the low thermal conductivity problems of β-Ga$_2$O$_3$ in power electronics applications.

Ma *et al* reported abnormal positive threshold voltage (V_{th}) shifts under negative bias stress conditions while operating β-Ga$_2$O$_3$ nanobelt MOSFETs [37]. This is attributed to the surface depletion effects originating from the surface state of the molecules absorbed on the β-Ga$_2$O$_3$ nanobelt surface. The surface depletion effects were moderated by passivating with an ALD–Al$_2$O$_3$ layer. These results reveal the importance of proper passivation on the β-Ga$_2$O$_3$ surface.

1.2.2 β-Ga$_2$O$_3$ nanobelt-based heterostructured transistors

The 2D β-Ga$_2$O$_3$ nanobelts have been attracting increased interest as potential nanoscale building blocks for future high-power (opto)electronic devices. The combination of 2D materials and β-Ga$_2$O$_3$ nanobelts is expected to produce a great synergy. The concept of 2D heterostructures was first demonstrated in FETs by integrating mechanically exfoliated β-Ga$_2$O$_3$ nanobelts and h-BN flakes [16].

As h-BN has an extraordinarily flat and clean surface, the h-BN dielectric provides a minimal density of charged impurities on its interface with the β-Ga$_2$O$_3$ nanobelts. Deformations or faults in each layer and in the interface between the β-Ga$_2$O$_3$ nanobelts and h-BN flakes were not observed from the cross-sectional TEM images, indicating the formation of a van der Waals heterostructure. The fabricated β-Ga$_2$O$_3$/h-BN heterostructured transistors demonstrated low gate leakage as well as a high I_{DS} on–off ratio of $\sim 10^7$. Furthermore, the author demonstrated that the top-gate threshold voltage can be linearly controlled by a back-gate voltage using a dual gate operation.

Bae et al introduced a gate FP structure by integrating a 2D material on a β-Ga$_2$O$_3$ nanobelt MESFET (figure 1.10) [22]. After the precise transfer of h-BN on the β-Ga$_2$O$_3$ channel, half of the gate electrode was defined on h-BN to introduce a gate FP structure. The h-BN gate FP, like the source FP, mitigated the electric field concentrated at the hot gate edge, preventing premature breakdown of the device, securing the reliability of the device, and thereby improving the breakdown voltage significantly. The three-terminal off-state breakdown voltage of the fabricated h-BN gate FP β-Ga$_2$O$_3$ nanobelt MESFET was 344 V, which exhibited stability under high-voltage operation conditions.

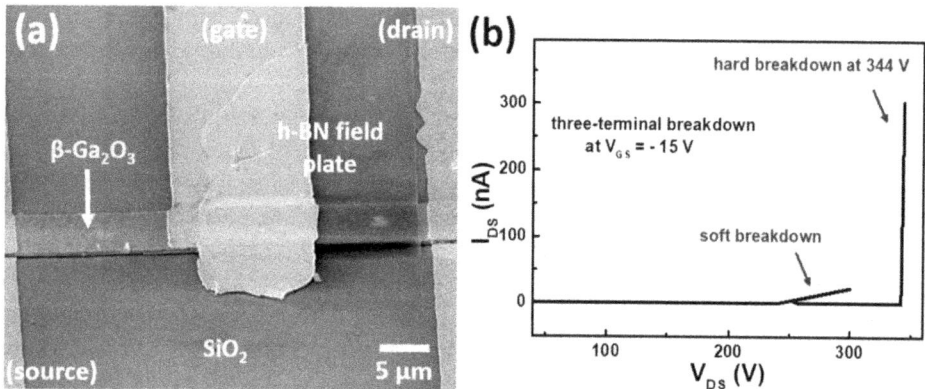

Figure 1.10. (a) SEM image of the fabricated β-Ga$_2$O$_3$ FET with a h-BN gate FP device. (b) Three-terminal off-state breakdown voltage of the fabricated device. Reproduced with permission from [22]. Copyright 2018 AIP Publishing.

Furthermore, a heterostructure n-channel depletion-mode β-Ga$_2$O$_3$ junction field-effect transistor (JFET) was fabricated by integrating a β-Ga$_2$O$_3$ nanobelt with a WSe$_2$ flake as shown in figure 1.11 [38]. The p-type WSe$_2$ flakes were used as the p-gate for the JFET instead of Ga$_2$O$_3$, which has not yet been implemented. The fabricated WSe$_2$–Ga$_2$O$_3$ heterojunction p–n diode displayed proper rectifying behavior with a high rectifying ratio of $\sim10^5$. The fabricated JFET exhibited excellent transistor characteristics with a high I_{DS} on–off ratio of $\sim10^8$, a low subthreshold swing of 133 mV dec^{-1}, and a three-terminal breakdown voltage of +144 V. This synergetic integration of 2D materials and β-Ga$_2$O$_3$ nanobelts introduces β-Ga$_2$O$_3$ as a nanoscale building block for future high-power devices and opens the possibility of more diverse forms of β-Ga$_2$O$_3$ nanobelt-based heterostructures.

1.2.3 β-Ga$_2$O$_3$ nanobelt-based Schottky barrier diode

An SBD is a widely used structure in high-power or high-frequency devices because of low voltage loss in forward connections and easy driving in high-frequency environments. Using β-Ga$_2$O$_3$, which has been attracting attention as a potential near-future high-power material, various SBD devices have been fabricated. For example, Yang $et\ al$ fabricated a Schottky rectifier using a Si-doped β-Ga$_2$O$_3$ epitaxial layer and analyzed its electrical properties and breakdown voltage [39]. The results showed an excellent on–off ratio of 3×10^7 that was not affected by temperature. Moreover, the device exhibited a breakdown voltage of more than

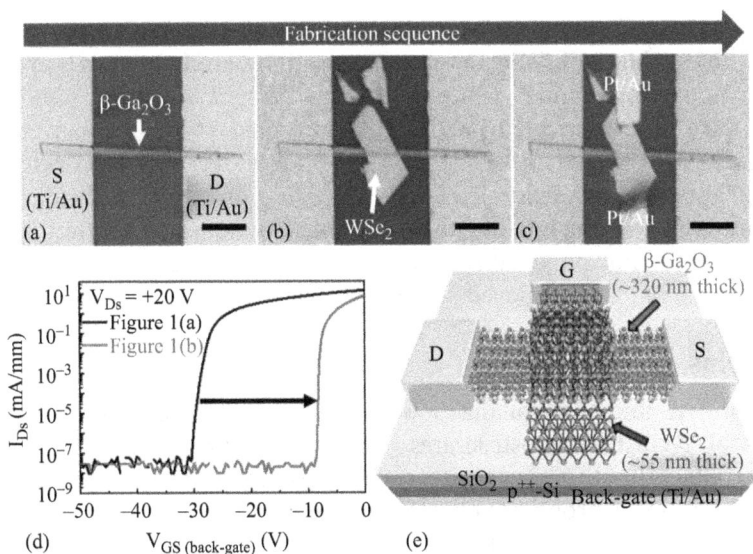

Figure 1.11. (a)–(c) Optical images of the fabrication sequence of the WSe$_2$/Ga$_2$O$_3$ heterojunction JFET. The scale bars represent 10 μm. (d) I_{DS}–V_{GS} transfer characteristics of the device at V_{DS} = +20 V (a) before and (b) after the multilayer WSe$_2$ was transferred. (e) Schematic illustration of the WSe$_2$/Ga$_2$O$_3$ heterojunction JFET on the SiO$_2$/Si substrate. Reproduced with permission from [38]. Copyright 2018 the American Chemical Society.

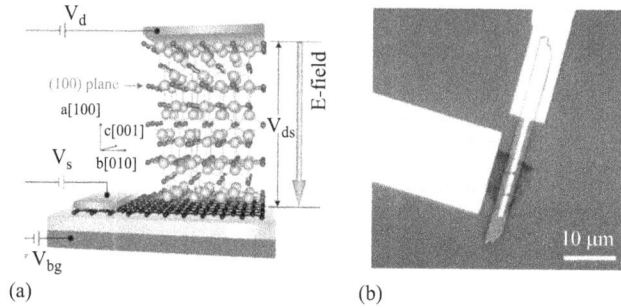

Figure 1.12. (a) Schematic of a β-Ga$_2$O$_3$/graphene vertical barristor heterostructure. (b) Optical microscope image of a fabricated device. Reproduced with permission from [26]. Copyright 2018 AIP Publishing.

1600 V, which demonstrates the rapid progress in this material system, exhibiting impressive power switching applications.

There have also been studies on the use of SBDs in β-Ga$_2$O$_3$ nanobelt devices. For instance, Swinnich *et al* fabricated a high-power flexible SBD by transferring β-Ga$_2$O$_3$ nanobelts onto a polyimide substrate using the Scotch tape method and depositing ohmic and Schottky electrodes [24]. A breakdown voltage of -119 V and a critical breakdown field of 1.2 MV cm^{-1} were measured, which demonstrated superior stability in high-power operations even after the bending test. This study has opened the possibility of various applications of β-Ga$_2$O$_3$ by implementing β-Ga$_2$O$_3$ nanobelt SBDs fabricated on versatile flexible substrates.

Yan *et al* fabricated a β-Ga$_2$O$_3$/graphene vertical barristor heterostructure by integrating a β-Ga$_2$O$_3$ nanobelt with graphene (figure 1.12) [26]. Taking advantage of the easy integrability of β-Ga$_2$O$_3$ nanobelts with other 2D materials, Yan's group fabricated a heterostructure device by overlapping a β-Ga$_2$O$_3$ nanobelt on a graphene flake. The fabricated barristor device could switch the current through the gate bias and exhibited an excellent on–off ratio of over 10^4. The device achieved a remarkable breakdown field of 5.2 MV cm^{-1} in a direction perpendicular to the (100) plane, which is much better than any other previously reported lateral FET device. This study not only utilized a β-Ga$_2$O$_3$ nanobelt-based device to fabricate heterojunctions, but also achieved stability in high-voltage driving.

1.3 Conclusion

Different growth and functionalization techniques have been used to control the properties of β-Ga$_2$O$_3$ nanostructures. For device applications, the ohmic and Schottky contacts to the nanostructures were investigated by introducing a 2D material with metallic properties or different metals that have been commonly used for wide bandgap materials. Although the research on β-Ga$_2$O$_3$ nanoelectronics is still in its early stages, there have been several studies on the fabrication and characterization of devices based on β-Ga$_2$O$_3$ nanostructures. Most β-Ga$_2$O$_3$-based nanoelectronics devices have been fabricated using β-Ga$_2$O$_3$ nanobelts, which inherit the characteristics of single-crystal β-Ga$_2$O$_3$. Using the advantages of β-Ga$_2$O$_3$ nanobelts, such as their large bandgap, high breakdown field, and thermal and chemical stabilities, various types of high-performance electronics including MOSFETs, MESFETs, SBDs, and JFETs have been demonstrated. β-Ga$_2$O$_3$ nanobelts are considered potential candidates for establishing nanoscale device platforms for heat management and future novel nanoelectronic applications.

References

[1] Tsao J Y *et al* 2018 *Adv. Electron. Mater.* **4** 1600501

[2] Pearton S J, Yang J, Cary P H, Ren F, Kim J, Tadjer M J and Mastro M A 2018 *Appl. Phys. Rev.* **5** 011301

[3] Higashiwaki M, Sasaki K, Kuramata A, Masui T and Yamakoshi S 2014 *Phys. Status Solidi Appl. Mater. Sci.* **211** 21

[4] Higashiwaki M, Sasaki K, Murakami H, Kumagai Y, Koukitu A, Kuramata A, Masui T and Yamakoshi S 2016 *Semicond. Sci. Technol.* **31** 34001

[5] Auer E, Lugstein A, Löffler S, Hyun Y J, Brezna W, Bertagnolli E and Pongratz P 2009 *Nanotechnology* **20** 434017

[6] Kumar S, Sarau G, Tessarek C, Bashouti M Y, Hähnel A, Christiansen S and Singh R 2014 *J. Phys. D: Appl. Phys.* **47** 435101

[7] Kumar M, Kumar V and Singh R 2017 *Nanoscale Res. Lett.* **12** 184

[8] Hsieh C-H, Chang M-T, Chien Y-J, Chou L-J, Chen L-J and Chen C-D 2008 *Nano Lett.* **8** 3288

[9] Kumar M, Sarau G, Heilmann M, Christiansen S, Kumar V and Singh R 2017 *J. Phys. D: Appl. Phys.* **50** 035302

[10] Lu J, Xing J, Chen D, Xu H, Han X and Li D 2019 *J. Mater. Sci.* **54** 6530

[11] Kim J J, Kim J J, Oh S, Mastro M A and Kim J J 2016 *Phys. Chem. Chem. Phys.* **18** 15760

[12] Hwang W S *et al* 2014 *Appl. Phys. Lett.* **104** 203111

[13] Bae J, Kim H-Y and Kim J 2017 *ECS J. Solid State Sci. Technol.* **6** Q3045

[14] Oh S, Mastro M A, Tadjer M J and Kim J 2017 *ECS J. Solid State Sci. Technol.* **6** Q79

[15] Li Z, Liu Y, Zhang A, Liu Q, Shen C, Wu F, Xu C, Chen M, Fu H and Zhou C 2018 *Nano Res.* **12** 143

[16] Kim J, Mastro M A, Tadjer M J and Kim J 2017 *ACS Appl. Mater. Interfaces* **9** 21322

[17] Kwon Y, Lee G, Oh S, Kim J, Pearton S J and Ren F 2017 *Appl. Phys. Lett.* **110** 131901

[18] Liu Y, Du L, Liang G, Mu W, Jia Z, Xu M, Xin Q, Tao X and Song A 2018 *IEEE Electron Device Lett.* **39** 1696

[19] Yao Y, Davis R F and Porter L M 2017 *J. Electron. Mater.* **46** 2053

[20] Sasaki K, Higashiwaki M, Kuramata A, Masui T and Yamakoshi S 2013 *Appl. Phys. Express* **6** 086502

[21] Farzana E, Zhang Z, Paul P K, Arehart A R and Ringel S A 2017 *Appl. Phys. Lett.* **110** 202102

[22] Bae J, Kim H W, Kang I H, Yang G and Kim J 2018 *Appl. Phys. Lett.* **112** 122102

[23] Bae J, Kim H W, Kang I H and Kim J 2019 *RSC Adv.* **9** 9678

[24] Swinnich E, Hasan M N, Zeng K, Dove Y, Singisetti U, Mazumder B and Seo J H 2019 *Adv. Electron. Mater.* **5** 1800714

[25] Kim S and Kim J 2019 *ECS J. Solid State Sci. Technol.* **8** Q3122

[26] Yan X, Esqueda I S, Ma J, Tice J and Wang H 2018 *Appl. Phys. Lett.* **112** 032101

[27] Kim S, Oh S and Kim J 2019 *ACS Photonics* **6** 1026

[28] Kim M, Seo J-H, Singisetti U and Ma Z 2017 *J. Mater. Chem. C* **5** 8338

[29] Pearton S J, Ren F, Tadjer M and Kim J 2018 *J. Appl. Phys.* **124** 220901

[30] Higashiwaki M, Murakami H, Kumagai Y and Kuramata A 2016 *Jpn J. Appl. Phys.* **55** 1202A1

[31] Tadjer M J, Mahadik N A, Wheeler V D, Glaser E R, Ruppalt L, Koehler A D, Hobart K D, Eddy C R and Kub F J 2016 *ECS J. Solid State Sci. Technol.* **5** P468

[32] Ahn S, Ren F, Kim J J, Oh S, Kim J J, Mastro M A and Pearton S J 2016 *Appl. Phys. Lett.* **109** 62102

[33] Zhou H, Si M, Alghamdi S, Qiu G, Yang L and Ye P D 2017 *IEEE Electron Device Lett.* **38** 103

[34] Zhou H, Maize K, Qiu G, Shakouri A and Ye P D 2017 *Appl. Phys. Lett.* **111** 92102

[35] Maize K, Ziabari A, French W D, Lindorfer P, Oconnell B and Shakouri A 2014 *IEEE Trans. Electron Devices* **61** 3047

[36] Zhou H, Maize K, Noh J, Shakouri A and Ye P D 2017 *ACS Omega* **2** 7723

[37] Ma J and Yoo G 2019 *Jpn J. Appl. Phys.* **58** SBBD01

[38] Kim J, Mastro M A, Tadjer M J and Kim J 2018 *ACS Appl. Mater. Interfaces* **10** 29724

[39] Yang J, Ren F, Tadjer M, Pearton S J and Kuramata A 2018 *ECS J. Solid State Sci. Technol.* **7** Q92

IOP Publishing

Wide Bandgap Semiconductor-Based Electronics

Fan Ren and Stephen J Pearton

Chapter 2

β-Ga$_2$O$_3$ power field-effect transistors

Hang Dong, Guangwei Xu, Xuanze Zhou, Wenhao Xiong, Xueqiang Xiang, Weibing Hao, Shibing Long and Ming Liu

2.1 Introduction

Energy consumption has been increasing dramatically with the development of society, in particular since the industrial revolution. Indeed, the application of electric energy has supported the social and technological developments of recent decades. In this process, the invention and application of semiconductor power devices have been notable milestones, because such devices enormously increase the power capacity of transmission and production systems, and also reduce wastage and save energy. Power devices are the semiconductor components that operate under conditions of high voltage, large current, high frequency, and high temperature. Therefore, they are largely different from information processing devices in terms of device structure, performance parameters, and basic functions. Various special structures have been designed to redistribute the internal electrical field of devices under a high voltage bias, such as field plates, buried-channels, and super junctions. On-resistance (R_{on}) is an important parameter for power devices, due to the fact that even a tiny conduction resistance will produce tremendous energy consumption and joule heat due to the high on-state current. Rectification and switching are the main roles played by power devices in modern high voltage and large current electrical systems.

As can be expected from the above, the developmental path of power devices certainly separates them from other electronic devices. Excellent performance is the first aim for a power device. In addition, most power devices are discrete due to their large current capability and cooling demands, therefore, the feature sizes of power devices tend to be far larger than those of integrated devices. The development of fabrication processes and the use of new materials for power devices are described in the following.

Silicon (Si) is the first choice for power devices due to its mature processing platform and cheap fabrication cost. Another reason is that new types of devices and structure designs have given silicon based power devices sufficient performance to meet specific performance requirements (high voltage, large current, and fast switching speed), such as vertical double-diffused metal–oxide field-effect transistors (VDMOSFETs), thyristors, insulated gate bipolar transistors (IGBTs), and so on.

Concurrently with the development of silicon based power devices, various so-called wide bandgap semiconductors, such as GaN, SiC, Ga_2O_3, and diamond, have also been considered to overcome the theoretical intrinsic material limitations of silicon. Figure 2.1 shows the different bandgap semiconductors and table 2.1 the intrinsic properties of several power device materials. In order to judge the potential of materials for power devices, many figures of merit have been put forward. Baliga's figure of merit $\varepsilon \mu E_c^3$ (BFOM) is used to evaluate the specific on-resistance for high voltage devices.

Although wide bandgap semiconductors seem to be completely superior to traditional materials, in particular in breakdown field strength (E_c), many technical difficulties have slowed their development, such as issues in the high quality substrate, epitaxial growth, and doping technology. These problems have been solved for GaN and SiC in recent decades. Specifically, the high mobility two-dimensional electron gas (2DEG) at the heterojunction of GaN/AlGaN was discovered and gives GaN based high electron mobility transistors (HEMTs) both a larger breakdown voltage and a faster switching speed than Si. In addition, silicon carbon (SiC) based power devices have also been developed, including Schottky diodes (SBDs) and metal–oxide field-effect transistors (MOSFETs) with superior breakdown and temperature characteristics. After decades of investigation, both GaN and SiC based power devices have become commercially available in recent years, as shown in figure 2.2. For example, GaN based HEMTs have been used to increase the charging power of new-generation adapters, and SiC based MOSFETs have also been applied in the converters of electric vehicles.

Si, Ge   GaAs, InP   SiC, GaN Ga_2O_3, Diamond

Common bandgap   Wide bandgap   Ultra wide bandgap

Figure 2.1. Comparison of wide bandgap semiconductors. Reprinted from [1] with permission of AIP publishing.

Table 2.1. Intrinsic properties of several power materials.

Material	E_g(eV)	ε	μ_n(Cm^{2}V^{-1}s^{-1})	E_c(MV/Cm)	BFOM
Si	1.1	11.8	1480	0.3	1
4H-SiC	3.25	9.7	1000	2.5	317
GaN	3.4	9	1250	3.3	846
Diamond	5.5	5.5	2000	10	24660
β-Ga_2O_3	4.8	10	300	8	3214

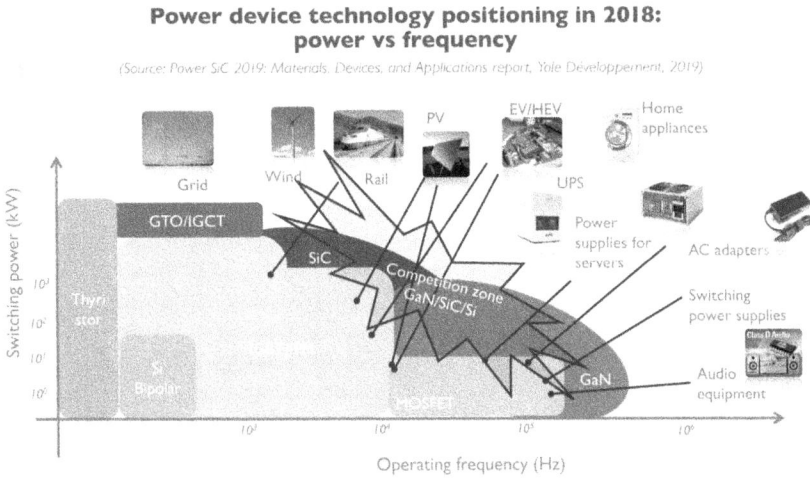

Figure 2.2. The application fields of various power devices (Si, GaN, and SiC) [2]. Reprinted with permission of Yole Développement.

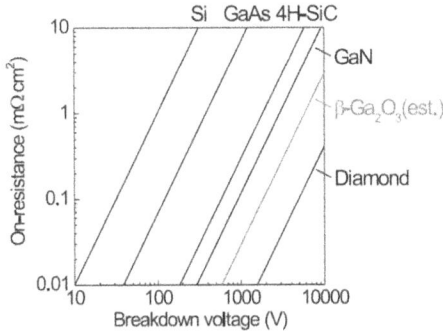

Figure 2.3. The theoretical limits ($V_{BR} \sim R_{ON}$) of β-Ga$_2$O$_3$ relative to other materials. Reprinted from [1] with permission of AIP publishing.

Gallium oxide (Ga$_2$O$_3$) is another promising wide bandgap semiconductor with a larger bandgap ($E_g \sim 4.8$ eV) and breakdown field strength ($E_{br} \sim 8$ MV cm^{-1}) than GaN or SiC. Due to Ga$_2$O$_3$'s high BFOM value of 3214, an increasing number of researchers hope that Ga$_2$O$_3$-based power electronics may be a candidate for ultrahigh voltage applications with lower on-state loss, as shown in figure 2.3.

After the successful fabrication of large area and high-quality single-crystal Ga$_2$O$_3$ using an edge-defined film-fed growth (EFG) method [3], various high-quality homoepitaxial growth techniques have been reported in recent years, including molecular beam epitaxy (MBE) [4–8], metal–organic chemical vapor deposition (MOCVD) [9–11], and hydride vapor phase epitaxy (HVPE) [12–14]. The EFG method is a kind of modified melting technique, and it has the advantages of low loss and high output. Therefore, Ga$_2$O$_3$ has the potential of a lower cost than even SiC or GaN, as shown in figure 2.4.

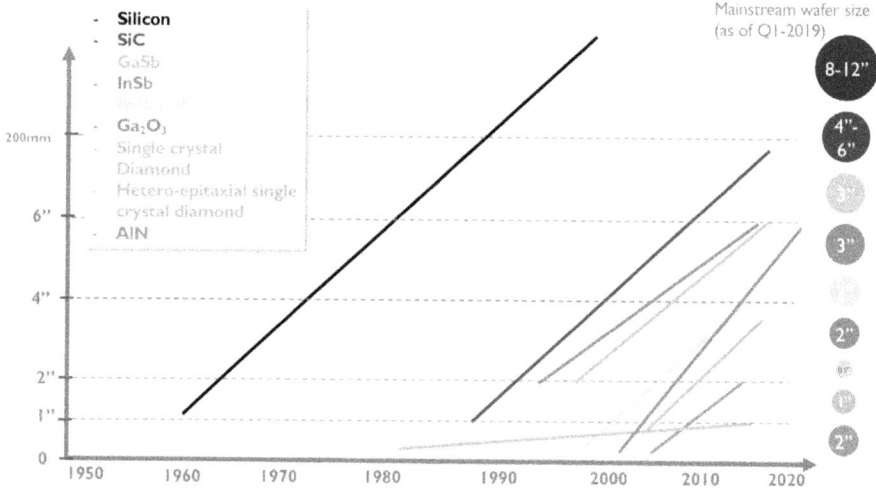

(a)

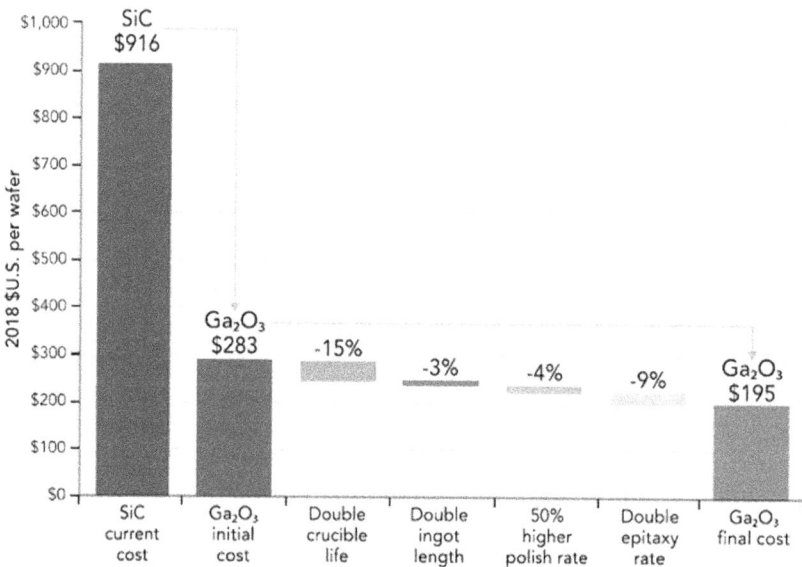

(b)

Figure 2.4. Development of the fabrication of Ga_2O_3 crystal substrate (top) [15], reprinted with permission of Yole Développement, and the Ga_2O_3 cost reduction potential relative to SiC (bottom) [16], copyright 2019 reprinted with permission from Elsevier.

This chapter will mainly introduce the recent developments in β-Ga_2O_3-based power transistors. The investigation of β-Ga_2O_3 based transistors began with a planar metal–semiconductor transistor (MESFET) in 2012 [1]. Depletion-mode (d-mode) MOSFETs were successfully fabricated in 2013 [17]. In order to realize enhancement-mode (e-mode) devices without p-type doping, a low doping channel, fin channel, and gate trench structure, as well as nitrogen doping techniques and gate channel thermal oxidation, have been employed by researchers [18–23]. In addition, vertical transistors with fin channels have also been developed to reach higher blocking voltages (>1 kV) and lower on-state resistance [24], and the source field plate further increased the voltage to 1.6 kV [25]. In addition to high voltage, switching speed and radio frequency (RF) gain have also been focused on. A turn-on time of 28.6 ns and a turn-off time of 94.0 ns were tested in a trench-gate transistor [22], and a T-gate was used to realize a current cut-off frequency of 27 GHz [26]. Moreover, researchers have tried to fabricate 2DEG transistors through a δ-doping technique and constructing a Ga_2O_3/$(AlGa)_2O_3$ heterojunction [26–36].

To make a direct comparison with various types of transistors and obtain a correct understanding of the development level of β-Ga_2O_3 based transistors, we review and discuss some key parameters of the state-of-the-art β-Ga_2O_3-based transistors in the next section.

2.2 Key parameters of a β-Ga_2O_3 power field-effect transistor

In simple terms, power transistors are three-terminal (gate, drain, and source electrodes) devices that can switch quickly between the off and on states controlled by gate bias and can sustain a high off-state blocking voltage with small on-state resistance. Transistors are unipolar devices, therefore, they have a faster switching speed due to the absence of the minority carrier charge-storage effect. In addition, it is easily driven with low loss due to the gate voltage control and a lower gate insulator leakage current.

For high voltage transistors, when the transistors turn off, the drift region, a major part of the conduction path, will be depleted to sustain the high voltage. However, when transistors turn on, a large current will pass through the same conduction path and may produce a lot of heat and destroy the device if the on-resistance is too big. Figure 2.5 shows a diagrammatic sketch of vertical transistors,

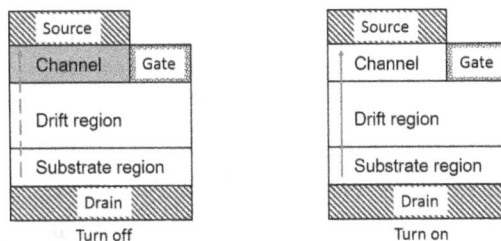

Figure 2.5. Schematic diagram of turn-off (left) and turn-on (right) MOSFETs.

where the channel is the region controlled by the gate and will be protected due to its vulnerable gate electrics, the substrate region is the supporting layer and tends to be highly doped to decease the on-resistance, and the drift region is the major part of the breakdown voltage and on-resistance. As shown in the following simplified equation, a large breakdown voltage ($V_{br,dr}$) and small on-resistance ($R_{on,dr}$) of the drift region is actually a relationship of interaction, in particular doping concentration ($N_{D,dr}$). The specific derivation process can be found in [37]:

$$R_{on,dr} = \frac{2V_{br,dr}}{q\mu E_c N_{D,dr}}.$$

A large breakdown field strength (E_c) and high carrier mobility (μ) can optimize the trade-off relationship between the breakdown voltage (V_{br}) and on-resistance (R_{on}). This is the reason why wide bandgap semiconductors with high breakdown field strength (E_c) were introduced into the power electronics field.

Several performance indices are particularly important for practical power transistor applications and should be the main considerations when developing and designing such devices. The following is a summary of these key parameters:

(a) *Breakdown voltage.*

The breakdown voltage (V_{br}) is the critical drain–source voltage of a power transistor when it turns off. As we know, a high drain–source voltage will induce a large electric field distribution in the whole conduction path (the channel, drift, and substrate regions). Therefore, the drain–source voltage is the integration of the electrical field in the whole conduction path. However, impact ionization will take place and destroy the material's crystal structure under a critical electrical field, which is the so-called breakdown field strength (E_c) that evaluates the electrical resistance ability of an insulation material (dielectrics). Simply increasing the integration path will also make the on-resistance increase. Therefore, materials with excellent breakdown characteristics are a better choice.

Except for the above conditions, an inhomogeneous distribution of the electrical field will also limit the breakdown voltage due to high electric field point, which leads to degradation before other conditions. Therefore, a series of measurements are taken to adjust the distribution of the electric field or protect the weak points, such as the field plate (FP), guide ring (GR), buried channel, termination protection, super-junction (SJ), and so on.

(b) *On-resistance.*

Power transistors actually work in the linear region when they turn on, therefore, the whole conduction on-resistance (R_{on}) is closely related to system efficiency and thermal loss. On-resistance is composed of several parts, as shown by the following equation:

$$R_{on} = R_{co} + R_{ch} + R_{dr} + R_{sub},$$

where R_{co} is the contact resistance, R_{ch} is the channel resistance, R_{dr} is the drift region resistance, and R_{sub} is the substrate resistance. We do not discuss R_{ch} and R_{sub} here. For R_{co}, metal function matching, stacking alloys, high doping, and post-annealing are the common methods to decrease the contact resistance. In addition, drift region resistance accounts for the main part, and an SJ is an effective structure to optimize the restrictive relationship between on-state resistance and off-state breakdown voltage. In short, an SJ is a structure in which p-type and n-type doping regions are distributed alternately and SJs are used widely in Si-based power devices. In recent years, the SJ structure has also been reported to have been applied in commercial SiC based MOSFETs.

(c) *Threshold voltage.*

The threshold voltage is the critical voltage at which transistors switch between the on and off states, and it is a common parameter for both transistors in the integrated circuit and power transistors. Many factors determine the threshold voltage, such as the gate dielectric capacitance, gate metal work function, channel doping concentration, and so on.

For power transistors, the threshold voltage determines its operation mode, driving, and safety performance. Enhancement-mode transistors are more user-friendly because they are easily driven by a simple driven circuit with lower losses, and can protect the system when there is an emergency power-down. Therefore, an e-mode transistor is a better choice than d-mode one. For example, d-mode GaN HEMTs are usually used in an integration cascode circuit to realize the e-mode function. In addition, the threshold voltage should be appropriate due to the instability of the driven voltage. A small turn-on voltage is easily driven and the probability of erroneous turn-on will also be larger. Conversely, if the threshold voltage is too large, the lifetime of the gate dielectrics will decline and also a large surge gate voltage will destroy the device. Based on the above, the threshold voltage should be designed carefully for practical applications.

(d) *Switching time.*

The switching time is the time that transistors switch between the on and off states, including turn-on time (t_{on}) and turn-off time (t_{off}). Transistors are voltage-controlled devices and the switching speed is determined by the charging and discharging speeds of the parasitic capacitance between the electrodes. Due to their unipolarity, power transistors tend to operate as fast switching elements. In fact, the channel carrier mobility and driven circuit will also have an influence on the switching time.

Switching time characteristics are very significant for practical low-loss switching applications. As shown in the following, where $t_{d(on)}$ and $t_{d(off)}$ are the so-called delay turn-on and turn-off times, which are the device's response times after the driven gate voltage operates, and t_r and t_f are the drain–source current rising and falling times after turning on or off:

$$t_{on} = t_{d(on)} + t_r$$
$$t_{off} = t_{d(off)} + t_f.$$

(e) *Gate charge.*

Gate charge (Q_g) is the necessary quantity of gate charge that drives transistors to turn on and is one of the key dynamic parameters. For driven circuit designers, gate charge is also a key parameter to determine the driven current and loss. In addition, the gate charge versus gate–source voltage curve (Q_g–V_{gs}) can be used to evaluate the parasitic capacitance and measure driven loss.

(f) *Parasitic capacitance.*

Parasitic capacitances are the inter-electrode capacitances due to the device structure and fabrication process, which include gate–source capacitance (C_{gs}), gate–drain capacitance (C_{gd}), and drain–source capacitance (C_{ds}). Parasitic inter-electrode capacitances will inevitably deteriorate the switching time, in particular C_{gd} because of the Miller effect. However, they will work through three termination capacitances, including the input capacitance (C_{iss}), the output capacitance (C_{oss}), and the reverse transfer capacitance (C_{rss}). The relationship between termination capacitances and inter-electrode capacitances is shown by

$$C_{iss} = C_{gd} + C_{gs}$$
$$C_{oss} = C_{gd} + C_{ds}$$
$$C_{rss} = C_{gd}.$$

The driven circuit and C_{iss} determine the delay time ($t_{d(on)}$ and $t_{d(off)}$). For soft switching applications, C_{oss} will be considered to avoid system resonance. In addition, C_{rss} is an important parameter for the switching time and turn-off delay time.

(g) *RF gain cut-off frequency.*

The ratio frequency (RF) gain cut-off frequency is the highest frequency at which a transistor has a signal current gain when operating in the RF power amplification circuit of high voltage or large current applications. Therefore, a higher gain cut-off frequency is the goal of RF devices.

In the following, we will introduce the development of β-Ga$_2$O$_3$ transistors in the following order: planar depletion-mode transistors, planar enhancement-mode transistors, vertical depletion-mode transistors, vertical enhancement-mode transistors, homojunction HEMTs, heterojunction HEMTs, and film transistors.

2.3 Planar depletion-mode transistors

Depletion-mode transistors are normally-on devices with a threshold voltage smaller than zero for n-type devices.

Planar depletion-mode transistors based on β-Ga$_2$O$_3$ were first demonstrated in 2012, when a single-crystal (010) oriented gallium oxide (Ga$_2$O$_3$) MESFETs was fabricated on an MBE layer with Sn doping ($N_D = 7.0 \times 10^{17}$ cm^{-3}) by Higashiwaki *et al* [1], as shown in figures 2.6(a) and (b). A breakdown voltage of about 250 V was

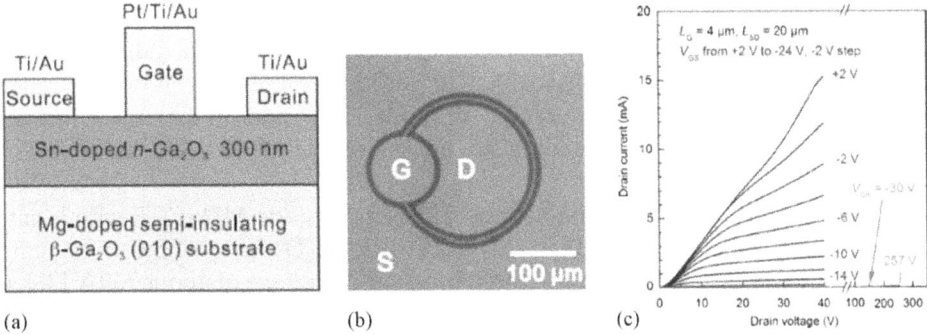

Figure 2.6. (a) Cross-sectional schematic, (b) optical microscope micrograph, and (c) DC output characteristics of the Ga_2O_3 MESFET. Reprinted from [1] with the permission of AIP Publishing.

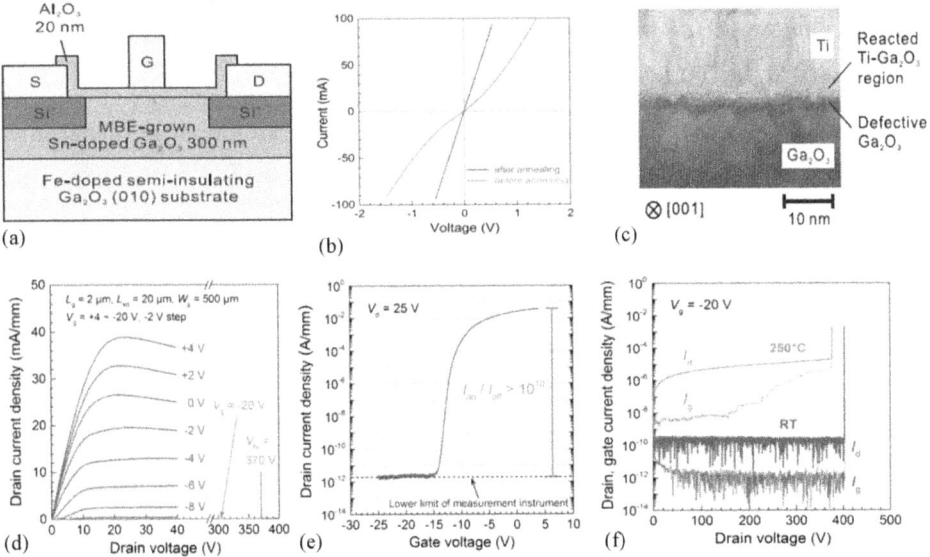

Figure 2.7. (a) Schematic cross-section of the MOSFET. (b) I–V characteristics of the implanted Ga_2O_3 film and (c) a cross-sectional TEM micrograph of the Ti–Ga_2O_3 interface after Ti/Au electrode annealing. (d) DC output curves, (e) transfer curves at RT, and (f) off-state breakdown characteristics of the MOSFET. Reprinted from [1] with the permission of AIP Publishing.

obtained with Pt/Ti/Au deposited as Schottky gates, as shown in figure 2.6(c). The device exhibited a comparable performance to standard structures and preliminarily proved the potential of Ga_2O_3 power transistors.

In order to further decrease the leakage current and increase the on–off ratio, in 2013 they demonstrated a β-Ga_2O_3 MOSFET with 20 nm Al_2O_3 as the gate dielectrics made using atomic layer deposition (ALD), as shown in figure 2.7(a) [17]. By adopting Si-ion implantation doping in the source and drain regions (5.0×10^{19} cm^{-3}) and 470 °C annealing in N_2 for 1 min, the specific contact resistance was about 8.1×10^{-6} Ω cm^2, as shown in figure 2.7(b). Defective Ga_2O_3 and the reacted Ti–Ga_2O_3 region at the Ti–Ga_2O_3 interface after annealing were likely to account for the excellent contact, as shown in figure 2.7(c). From this, classical methods to form ohmic contacts have been identified, including Ti as the contact metal and 470 °C N_2 annealing. In addition, the device showed excellent performance with a breakdown voltage of about 370 V and an extremely low off-state drain leakage current, leading to an on–off ratio of 10^{10}, as shown in figures 2.7(d)–(f).

In 2016, Wong *et al* first manufactured a depletion-mode MOSFET with a gate-connected field plate on an MBE epitaxial (010) β-Ga_2O_3 layer of 3.0×10^{17} cm^{-3} with implanted Si$^+$, achieving an enhanced breakdown voltage of about 750 V [38], as shown in figure 2.8(a). In addition, the source and drain regions are highly doped (5.0×10^{19} cm^{-3}) with Si to improve the ohmic contact. The off-state drain and gate leakage current were reduced effectively by implementing highly resistive unintentionally doped (UID) Ga_2O_3 and effective surface passivation with SiO_2, as shown figure 2.8(b). The on–off ratio decreases from 10^9 to 10^3 with increasing temperature, as shown in figure 2.8(c), which indicates that the transistor can operate stably against thermal stress up to at least 300 °C. In addition, the current collapse phenomenon caused by the self-heating effect is observed in the output curves measured under the DC and pulse test conditions, as shown in figure 2.8(d). Therefore, heat management

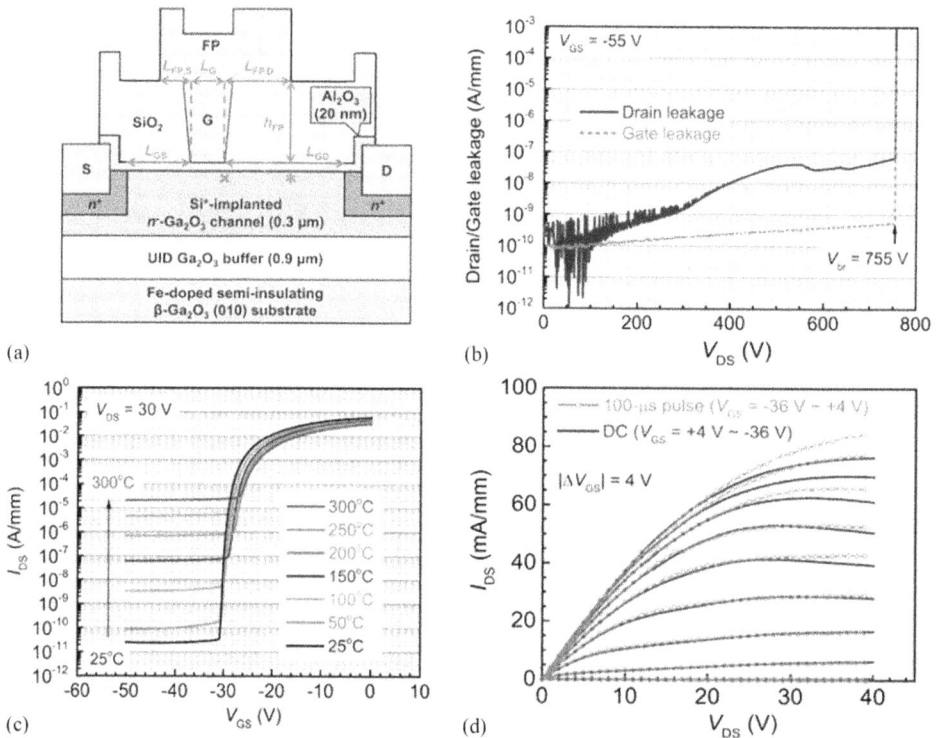

Figure 2.8. (a) Schematic cross-section, (b) the off-state drain/gate leakage and breakdown curves, (c) temperature-dependent transfer characteristics at $V_{ds} = 30$ V, and (d) DC and pulsed output curves of the β-Ga_2O_3 FP-MOSFET. Copyright 2016 IEEE. Reprinted with permission from [38].

is significant for devices on Ga_2O_3 with relatively low thermal conductivity. The characterization of channel temperature and radiation hardness against gamma ray irradiation in similar devices was also tested and analyzed in [39].

In 2016, Green et al demonstrated a depletion-mode MOSFET on a Sn-doped Ga_2O_3 (1.7×10^{18} cm^{-3}) film grown on a Mg-doped (100) β-Ga_2O_3 single-crystal substrate using metal–organic vapor phase epitaxy (MOVPE) [40]. Figure 2.9(a) shows the representative two-finger MOSFET with a 2 μm gate length. They reported a breakdown voltage of 230 V with a small gate–drain distance of 0.6 μm. The experimentally measured breakdown electrical field of 3.8 MV cm^{-1} is the highest reported value for Ga_2O_3 transistors at that time. Although this value is far smaller than Ga_2O_3's theoretical value of 8 MV cm^{-1}, it surpasses the theoretical limits of SiC and GaN. As shown in figures 2.9(b) and (c), the device exhibits great performance with an on–off ratio of 10^7 and a threshold voltage of about −25 V.

In 2017, they demonstrated another RF Ga_2O_3 transistor with a modulation channel achieved with an incompletely depleted sub-micro gate recess [41], as shown in figure 2.9(d). The transistor exhibited an on-resistance ($R_{on,sp}$) of 50.4 mΩ mm, a threshold voltage (V_{TH}) of −10.1 V, a maximum current density (I_{DS}) of 150 mA mm^{-1}, an on–off ratio of 10^6, and a transconductance (g_m) of 21 mS mm^{-1}, as shown in figure 2.9(e). The extrinsic cut-off frequency (f_T) and maximum oscillating frequency (f_{max}) of the transistor were 3.3 GHz and 12.9 GHz, respectively. In addition to a P_{out} of 0.23 W mm^{-1}, other RF performance results, including a power gain of 5.1 dB and a power-added efficiency of 6.3%, were confirmed in detail with a passive source and load

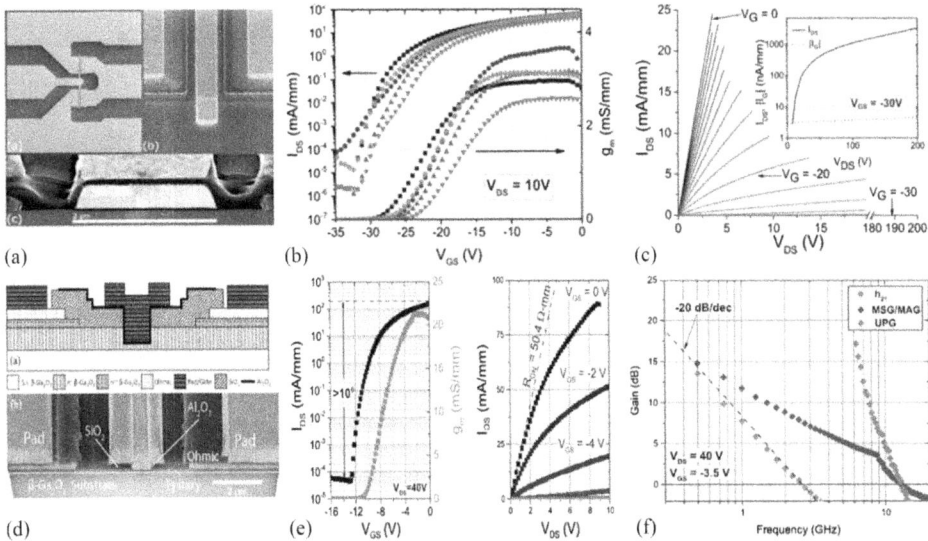

Figure 2.9. (a) Top-down scanning electron microscopy (SEM) image of the two-finger MOSFET on β-Ga_2O_3, (b) transfer characteristics at $V_{ds} = 10$ V, and (c) DC output curves for a 2×50 μm MOSFET. Copyright 2016 IEEE. Reprinted with permission from [40]. (d) Cross-section schematic and focused ion beam (FIB) cross-sectional image of the RF transistor. (e) DC transfer, output characteristics and (f) extrinsic small signal RF gain performance of the device. Copyright 2017 IEEE. Reprinted with permission from [41].

tuning at 800 MHz. A high current density of 150 mA mm^{-1} was also obtained. All results preliminarily implied that β-Ga$_2$O$_3$ has a great potential in applications for power switching and RF electronics.

In 2017, Zeng et al first reported a β-Ga$_2$O$_3$ MOSFET with spin-on-glass (SOG) doping technology, realizing a source–drain ohmic contact and a breakdown voltage of 382 V [42], as shown in figures 2.10(a)–(c). In 2018, a planar field-plate device was fabricated on an MBE grown 200 nm Sn-doped epitaxial layer with a concentration of 2×10^{17} cm^{-3} and a 200 nm UID buffer layer on an Fe-doped semi-insulating Ga$_2$O$_3$ substrate [43], as shown in figure 2.10(d). It is worth mentioning that highly doped source and drain regions have also been realized using SOG, and a composite field plate with denser high-quality ALD SiO$_2$ at the top close to the gate edge and a thick PECVD SiO$_2$ layer at the bottom are used to increase the breakdown voltage. A device with L_{gb} = 20 µm measured a breakdown voltage of 1.85 kV in Fluorinert, as shown, and an average electric field strength of 4.2 ± 0.2 MV cm^{-1} was obtained for a device with L_{gd} = 0.8 µm, as shown in figure 2.10(e). The electric field strength is more than half the theoretically predicted 8 MV cm^{-1} limit. In 2019, a highly doped capping layer was used to form the ohmic contact [44], as shown in figure 2.10(f).

In 2018, Lv et al first reported depletion-mode MOSFETs with source field-plated protection. The devices were fabricated on an MOCVD grown 240 nm Si-doped epitaxial layer with a concentration of 8×10^{17} cm^{-3} and a 400 nm UID buffer on an Fe-doped semi-insulating (010) β-Ga$_2$O$_3$ substrate [45]. As shown in figure 2.11(a),

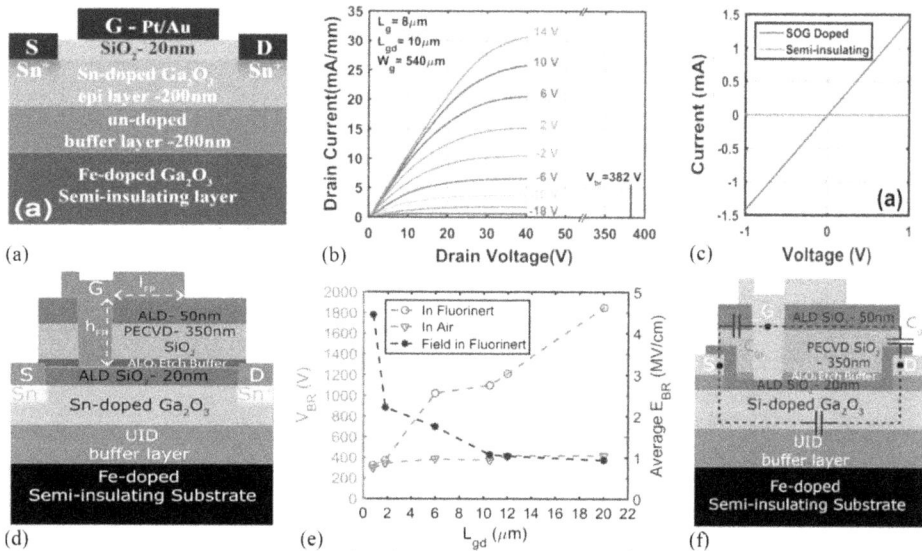

Figure 2.10. (a) Schematic cross-section of an MOSFET with SOG doping, (b) the output characteristics, and (c) a comparison of the conducting current between the semi-insulating layer and the SOG doped layer. Copyright 2017 IEEE. Reprinted with permission from [42]. (d) A field-plate MOSFET with SOG S/D doping and (e) the breakdown voltage V_{br} and averaged breakdown electric field E_{br} versus gate–drain separation L_{gd} in the measured devices. Copyright 2018 IEEE. Reprinted with permission from [42]. (f) A field-plate MOSFET with a highly doped capping layer. Copyright 2019 IEEE. Reprinted with permission from [38].

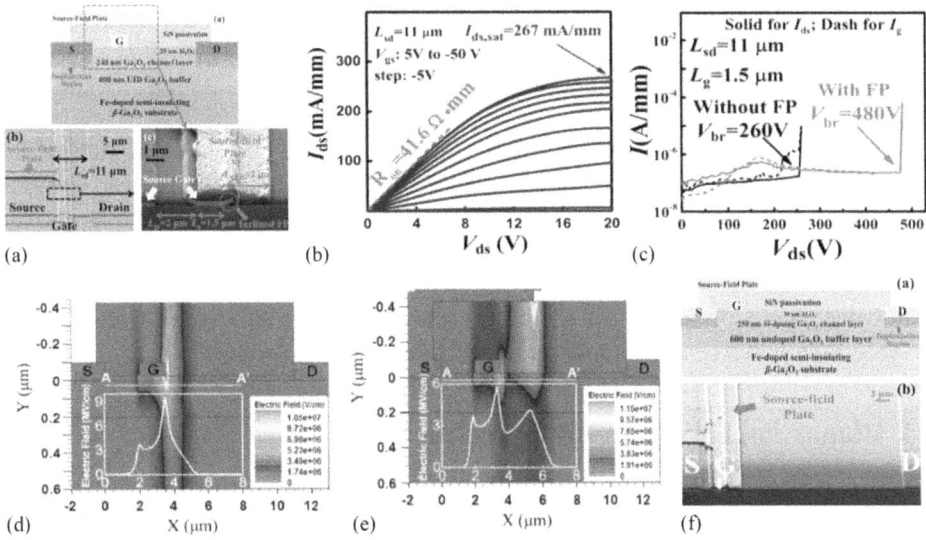

Figure 2.11. (a) Schematic cross-section and SEM image, (b) output characteristics, and (c) three-terminal off-state breakdown characteristics of the source field-plate MOSFET. Simulation of the electrostatic field (d) without and (e) with the source field plate. Copyright 2019 IEEE. Reprinted with permission from [45]. (f) Schematic cross-section and SEM image of the 2360 V source field-plate MOSFET [46], copyright IOP Publishing, reproduced with permission, all rights reserved.

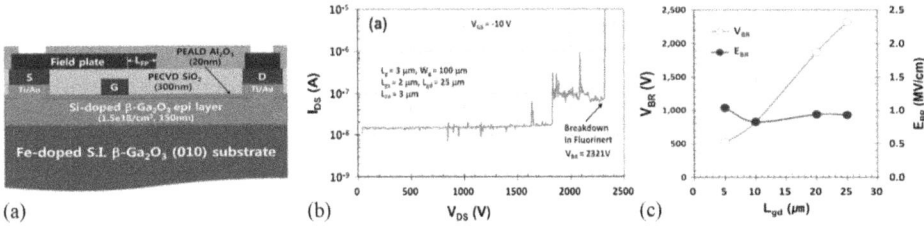

Figure 2.12. (a) Cross-section view of a source-connected FP β-Ga$_2$O$_3$ MOSFET, (b) the breakdown voltage (V_{BR}) and (c) breakdown electric field (E_{BR}) versus gate–drain spacing (L_{gb}) [47], copyright IOP Publishing, reproduced with permission, all rights reserved.

the source field plate is realigned to the source contact with a drain extension ($L_{fp,d}$) of 2 μm. Figures 2.11(b) and (c) show that devices with $L_{gd} = 11$ μm have a saturation drain current ($I_{d,sat}$) of 267 mA mm^{-1}, specific on-resistance ($R_{on,sp}$) of 4.57 mΩ cm^2, and a destructive voltage (V_{br}) of 480 V. In addition, the power figure of merit ($V_{br}^2/R_{on,sp}$) reaches 50.4 MW cm^{-1} for a device with $L_{gd} = 11$ μm. The simulations of the electrostatic field in figures 2.11(d) and (e) show that the source field plate effectively modulates the electric field in the channel. In 2019 they reported that the device with a larger $L_{gd} = 23$ μm and lower channel doping has a concentration of higher breakdown voltage of 2360 V, as shown in figure 2.11(f) [46].

In 2019, Mun *et al* reported a high voltage source field-plated MOSFET on a highly Si-doped (1.5×10^{18} cm^{-3}) epitaxial layer which was grown on an Fe-doped semi-insulating (010) substrate [47], as shown in figure 2.12(a). Figure 2.12(b) shows the linear relationship between the breakdown voltage (V_{br}) and the distance of the gate to-drain (L_{gb}). We can see that V_{br} increases with increasing L_{gb}, and the maximum V_{br} of 2321 V was obtained for devices with $L_{gb} = 25$ μm, as shown in figures 2.12(b) and (c).

As seen in the above, depletion-mode transistors have been investigated widely and researchers have focussed on increasing the breakdown voltage using a gate-connected or source-connected field plate and improving the field-plate oxide property. The field plate is an effective structure to modulate the electric field distribution and protect the weak breakdown points. In addition, a large drain–source distance is of benefit for increasing the breakdown voltage, however, it will increase the conduction path and thus degrade the on-resistance.

2.4 Planar enhancement-mode transistors

Enhancement-mode transistors are normally off and only turn on when gate bias voltage is operated. Due to the absence of p-type doping, researchers have tried a variety of methods to realize a normally-off conduction channel for Ga_2O_3 transistors.

In 2016, Chabak *et al* demonstrated enhancement-mode Ga_2O_3 wrap-gate fin-array field-effect transistors (FinFETs) with a high breakdown voltage [18]. As shown in figure 2.13(a), fin channels are etched using inductively couple plasma (ICP) with Cr as the mask, and then ohmic contacts consisting of Ti/Al/Ni/Au are grown in the drain and source regions, followed by rapidly annealing for 1 min at 470 °C in nitrogen. The gate dielectrics and metal electrodes are deposited using the conventional process. Cross-sectional views of the triangular-shaped fins are shown in figure 2.13(b), and a channel doping density of $N_D \sim 2.3 \times 10^{17}$ cm^{-3} is extracted

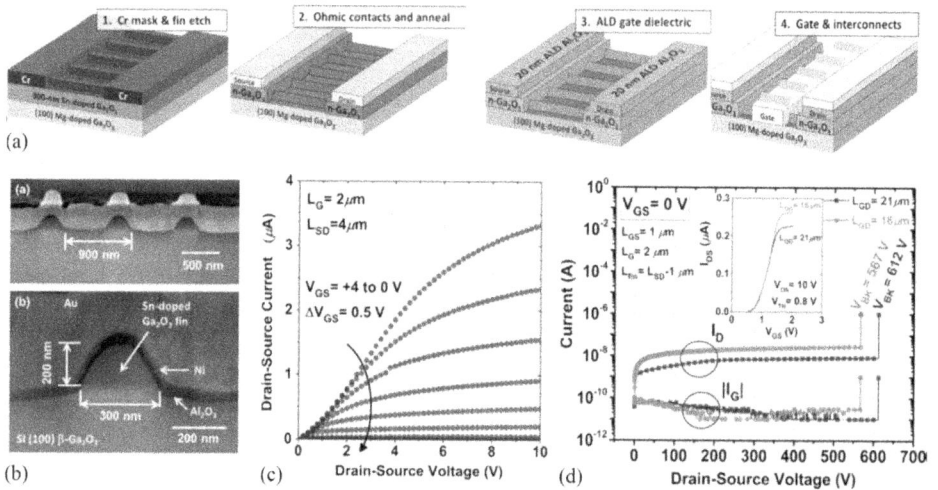

Figure 2.13. (a) Fabrication process of Ga_2O_3 FinFETs. (b) Cross-sectional SEM images of the fin channel. (c) The output curves from $V_G = +4$–0 V. (d) The breakdown voltages of devices with $L_G = 2$ μm and $L_{GD} = 16$ and 21 μm. Reprinted from [18], with the permission of AIP Publishing.

by $1/C^2$–V_{GS}. As a result, an output current of 3.5 μA @ V_{GS} = +4 V, V_{DS} = 10 V, and a threshold voltage (V_{TH}) larger than zero are obtained, which indicates its enhancement-mode operation, as shown in figure 2.13(c). In addition, a breakdown voltage (V_{BR}) of 612 V is found for devices with L_{GD} = 21 μm.

In 2017, enhancement-mode Ga_2O_3 MOSFETs employing an UID Ga_2O_3 channel layer were fabricated by Wong *et al* [19], as shown in figure 2.14(a). The ultra-low channel background carrier concentration of 4×10^{14} cm^{-3} leads to the threshold voltage of ~+6 V, as shown in figure 2.14(b). In addition, selective-area Si+ implantation of 5×10^{19} cm^{-3} was implemented to improve the ohmic contact in the source and drain regions. The DC output characteristics indicate that the maximum saturation current reaches 1.4 mA mm^{-1} when V_{GS} = 38 V and V_{DS} = 15 V, as shown in figure 2.14(c). Fermi level pinning limits the drain–source current (I_{DS}) modulation of V_{GS}, and the peak extrinsic transconductance of 0.38 mS mm^{-1} accounts for the charge trapping of an unoptimized dielectrics interface. The off-state gate leakage current of 10 pA mm^{-1} and the on–off ratio near 10^6 are presented in figures 2.14(b) and (d), respectively.

In 2018, Chabak *et al* demonstrated an enhancement-mode Ga_2O_3 MOSFET on a 200 nm Si-doped (5×10^{17} cm^{-3}) homoepitaxial layer grown using MBE. The recess process was first adopted for the gate region, which was depleted by gate interface states [21], as shown in figure 2.15(a). The threshold voltage (V_{th}) of +2 V indicates that the device is normally off, and the on-resistance (R_{on}) reaches

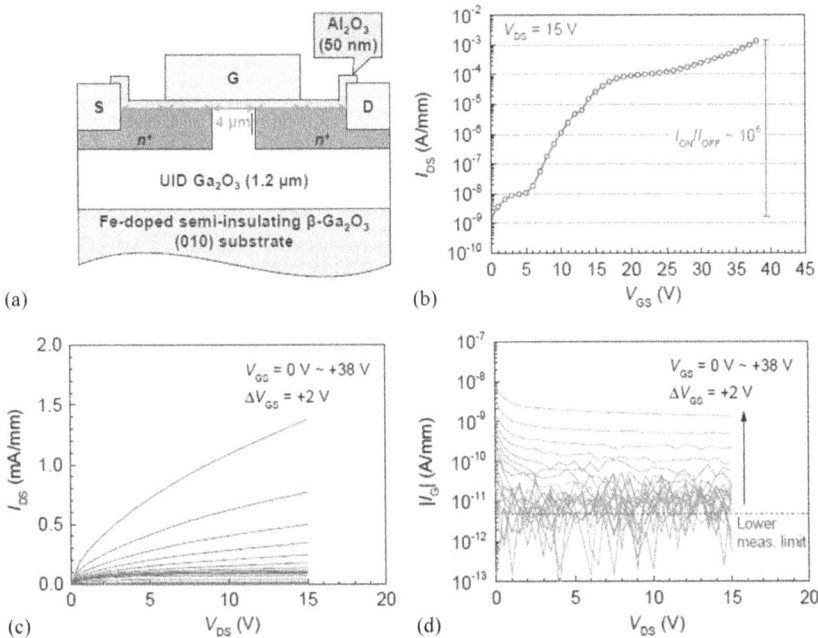

Figure 2.14. (a) Schematic cross-section, (b) I_{DS}–V_{GS} characteristics plotted in semi-logarithmic scales, (c) DC I_{DS}–V_{DS} characteristics, and (d) I_G–V_{DS} characteristics of the enhancement-mode Ga_2O_3 MOSFET [19]. Copyright IOP Publishing, reproduced with permission, all rights reserved.

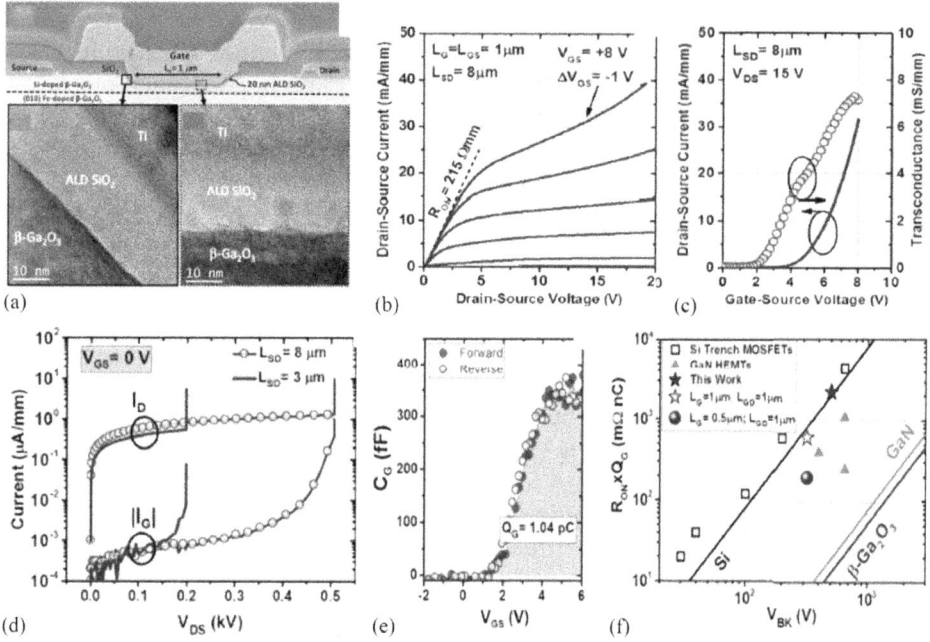

Figure 2.15. (a) SEM false-colored cross-section view and high-resolution transmission electron microscopy (HR-TEM) of the recess-gate β-Ga$_2$O$_3$ MOSFETs. (b) Family of I–V curves with up to +8 V forward gate bias (V_{GS}), (c) linear transfer characteristics at $V_{DS} = 15$ V, and (d) its gate–source and drain–source breakdown curves. (e) C_G–V_{GS} curve to estimate the total gate charge (Q_G) and (f) the device's dynamic switch loss figure of merit, $Q_G R_{on}$, compared to the devices based on Si and GaN. Copyright 2018 IEEE. Reprinted, with permission, from [21].

215 Ω mm and the drain–source current density (I_{DS}) reaches 40 mA mm^{-1} due to the highly doping channel layer, as shown in figures 2.15(b) and (c). The on-resistance components are calculated and analyzed in detail, and ungated access resistances ($R_{SH,S}$ and $R_{SH,D}$) are the major part (56%). In addition, the breakdown voltage reaches 505 V for devices with a drain–source distance (L_{DS}) of 8 μm, as shown in figure 2.15(d). The total gate charge (Q_G) is estimated by the C_G–V_{GS} curve, as shown in figure 2.15(e), and dynamic switch losses are first discussed using the product of total gate charge (Q_G) and on-resistance (R_{on}), as shown in figure 2.15(f). It is worth noting that the dynamic switch losses of the fabricated Ga$_2$O$_3$ transistors have surpassed Si technology and will compete with GaN power switches when the optimization steps are applied in the future.

In 2019, Kamimura *et al* implemented normally-off MOSFETs via unintentionally co-doped (N and Si) β-Ga$_2$O$_3$ grown by plasma-assisted molecular beam epitaxy (PAMBE) [23], as shown in figure 2.16(a). The concentration of Si and N can be controlled by the O–Ga flux ratio during growth, as shown figure 2.16(b). The output and transfer characteristics of the MOSFET with N and Si concentrations of 1×10^{18} cm^{-3} and 2×10^{17} cm^{-3}, respectively, are shown in figures 2.16(c) and (d). The co-doped β-Ga$_2$O$_3$ has shown a trend to form an inversion channel and will turn on when a larger gate voltage (V_g) than +8 V is applied. In addition, the on–off ratio

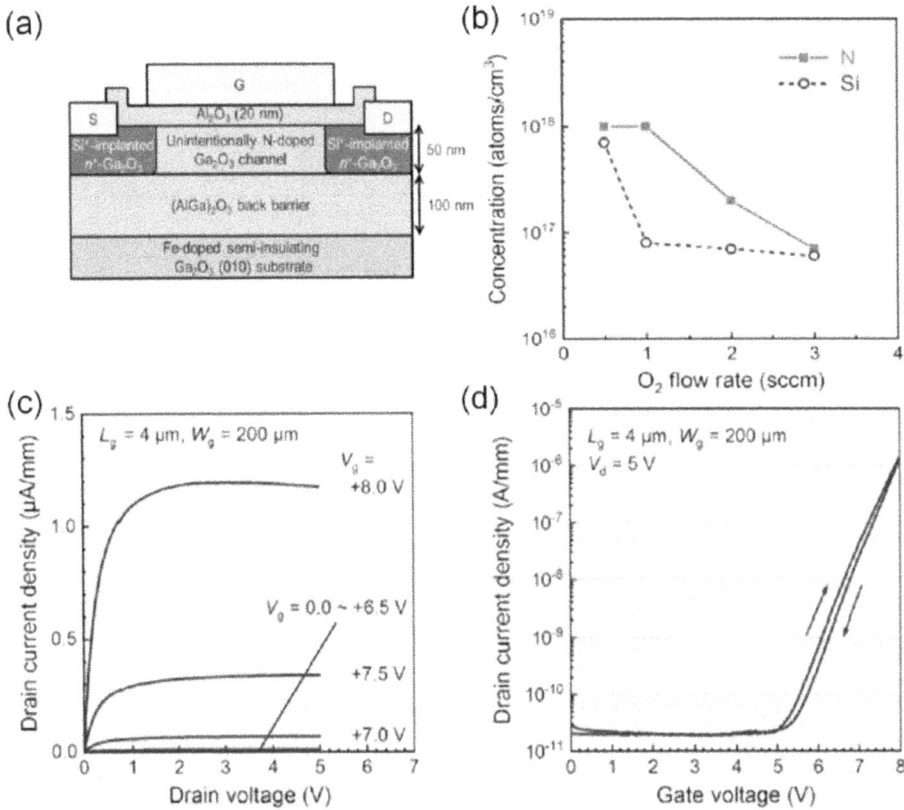

Figure 2.16. (a) Schematic cross-section of the normally-off β-Ga$_2$O$_3$ MOSFET with an unintentionally N-doped channel. (b) O$_2$ flow rate dependences of N and Si concentrations. The DC (c) I_d–V_d and (d) I_d–V_g characteristics of the device. Reprinted from [23], with the permission of AIP Publishing.

has exceeded five orders of magnitude, indicating that the leakage current is ultra-low. Unintentional N-doping is an attempt to realize normally-off devices, and the developments of intentional N-doping technologies are expected to pave the way for β-Ga$_2$O$_3$ power switching transistors in the future.

In 2019, Dong *et al* reported an enhancement-mode MOSFET with a trench-gate structure [22], as shown in figure 2.17(a). The gate isolation layer is 20 nm Al$_2$O$_3$ (ALD) and the passivation layer is 100 nm SiO$_2$ (PECVD). The device shows a decent static performance, including an on-resistance (R_{on}) of 364 Ω mm, a threshold voltage (V_{th}) of +4.2 V, and an on–off ratio of 10^7. In addition, the dynamic switching characteristics are tested with a resistive load and analyzed first for β-Ga$_2$O$_3$ transistors. As shown in figure 2.17(b), the device under test (DUT) is driven by the pulse generator and the load resistance (R_L) is connected between the drain and source electrodes in series. The fast switching time, including the turn-on time (t_{on}) of 28.6 ns and turn-off time (t_{off}) of 94.0 ns, are obtained for the first time and provide preliminary results for β-Ga$_2$O$_3$ high-speed switching applications. In addition, off-state parasitisc capacitances have also been tested at 100 kHz,

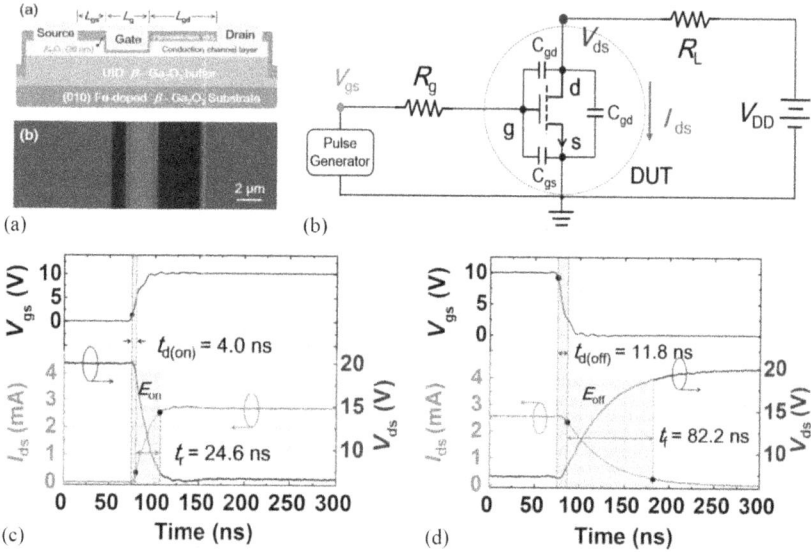

Figure 2.17. (a) Schematic cross-section of a trench-gate β-Ga$_2$O$_3$ MOSFET. (b) Diagram of the dynamic test circuit with resistive load. (c) Switching-on and (d) switching-off curves of the device [22]. Copyright IOP Publishing. Reproduced with permission. All rights reserved.

including an input capacitance (C_{iss}) of 37 pF mm^{-1}, output capacitance (C_{oss}) of 42 pF mm^{-1}, and reverse transfer capacitance (C_{rss}) of 14 pF mm^{-1}. It is necessary to mention that the switching times of devices with smaller parasitic capacitances will be reduced further.

In the above, best efforts have been made to realize normally-off operation without an effective p-type dopant. Until now, the low doping sub-conduction channel has also limited the on-state current, and the co-doping (N and Si) epitaxial layer meets the same difficulty of forming a conduction path under the effect of gate voltage. Therefore, the process optimization of N-doping will be necessary to solve this problem in the future. In addition, physical isolation, such as gate trench structure by etching, seems to be an effective method with the depletion effect of interfacial states. However, the stability of devices needs to be taken into consideration carefully. In fact, a heterogeneous p-type oxide may be a candidate to form a conduction layer or depletion of the channel in the future.

2.5 Vertical depletion-mode transistors

Vertical-type power devices have the advantage of high breakdown voltage due to the large electrical field blocking distance between the drain and source electrodes.

In 2017, the first vertical β-Ga$_2$O$_3$ depletion-mode MOSFET was demonstrated by Sasaki *et al* [20]. The trench-gate structure was realized using etching through (001) Ga$_2$O$_3$ halide vapor phase epitaxy (HVPE), and the mesa width and gate length were approximately 2 μm and 1 μm. Cu and HfO$_2$ were used as the gate metal and dielectric layer, while SiO$_2$ was the isolation layer between the gate and source electrodes, as shown in figures 2.18(a) and (b). Based on the DC output and transfer

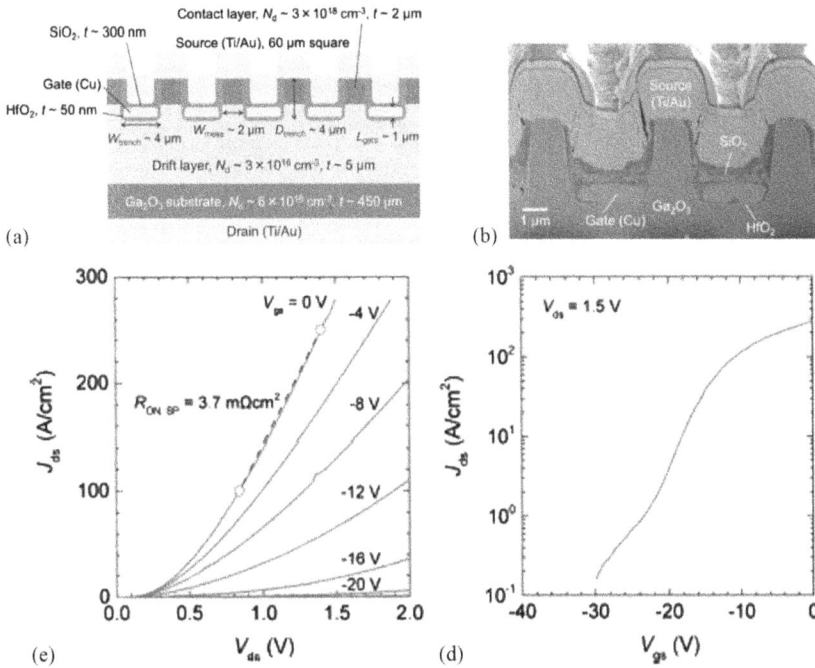

Figure 2.18. (a) Schematic illustration and (b) SEM image of the vertical β-Ga$_2$O$_3$ trench MOSFET. (c) DC output characteristics and (d) transfer characteristics of the device [20]. Copyright IOP Publishing. Reproduced with permission. All rights reserved.

curves shown in figures 2.18(c) and (d), a specific on-resistance ($R_{ON,SP}$) of 3.7 mΩ cm^2, a threshold voltage (V_{th}) of <-30 V, and an on–off ratio of 10^3 were obtained. The gate–source leakage current limited the on–off ratio and the low breakdown voltage of 10–20 V was accounted for by the unoptimized fabrication process. In addition, decreasing the mesa width and/or the donor concentration (N_D) of the mesa are expected to develop a normally-off vertical device.

In 2018, a vertical β-Ga$_2$O$_3$ depletion-mode MOSFET with a planar-gate architecture was fabricated by Wong *et al* [48]. They used a buried current blocking layer (CBL) formed by Mg ion (Mg^{++}) implantation doping to electrically isolate the source and drain electrodes, as shown in figure 2.19(a). As shown in figure 2.19(b), the transistors with an aperture opening length (L_{ap}) of 15 μm had an on-state current (I_{DS}) of ~1 kA cm^{-2} and an estimated specific on-resistance ($R_{DS,sp}$) of ~5 mΩ cm^2. However, due to the poor inter-device isolation as a result of residual n-type conductivity in the Mg-doped CBL, the device had a large source–drain leakage current, leading to a small on–off ratio, as shown in figure 2.19(c). In addition, the phenomenon that thermal annealing at 1000 °C for 30 min in N$_2$ causes Mg diffusion to compensate the background Si dopant was observed, as shown in figure 2.19(d).

In order to further develop this kind of transistor, Wong *et al* studied acceptor doping by Mg and N implantations in detail [49]. As shown in figures 2.20(a) and

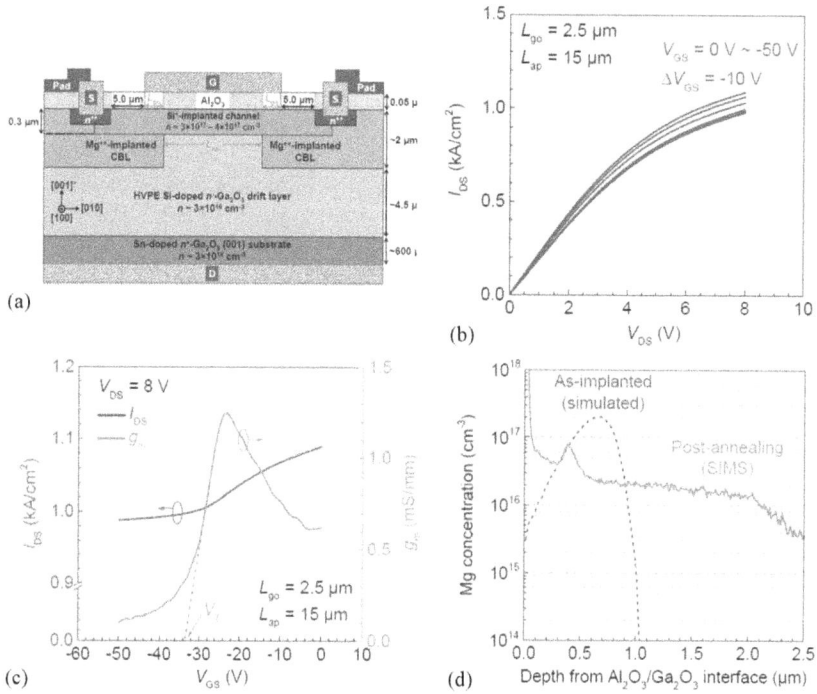

Figure 2.19. (a) Schematic cross-section of the current vertical β-Ga$_2$O$_3$ MOSFET. (b) The DC I_{DS}–V_{DS} output and (c) DC I_{DS}–V_{GS} transfer characteristics of the device. (d) Mg profiles obtained before (simulated profile) and after (SIMS profile) annealing at 1000 °C for 30 min in N$_2$ [48]. Copyright IOP Publishing. Reproduced with permission. All rights reserved.

(b), significant Mg diffusion will take place when the annealing temperature is higher than 900 °C, while the N dopant will remain stable until 1100 °C. In addition, the I–V characteristics of the CBL implanted by the Mg and N dopant are tested. The leakage current increases with higher temperature annealing (600 °C $< T_a <$ 900 °C) recovering from the implantation damage, behaving as deep traps in Mg^{++} implanted Ga$_2$O$_3$. Activation of the Mg ion will effectively compensate n-type doping after 1000 °C annealing, leading to a leakage current decrease, as shown in figure 2.20(c). However, the electron barrier was not dominated by the effect of lattice defects in N^{++} implanted Ga$_2$O$_3$, therefore, the leakage current will show monotonic negative dependence on T_a only due to the recovery from damage, as shown in figure 2.20(d).

Thus, in 2019, Wong *et al* reported new current aperture vertical β-Ga$_2$O$_3$ MOSFETs with a N^{++} implantation CBL [50], as shown in figure 2.21(a). The SIMS depth profiles of N and Si indicate the formation of a conduction channel and CBL structure due to the stability of N implantation, as shown in figure 2.21(b). In addition, N^{++}-implanted CBL can block the leakage current more effectively than Mg^{++}-implanted CBL, as shown in figure 2.21(c). The electrical parameters of the device are extracted based on the figures 2.21(d) and (e), including a specific on-resistance (R$_{on,sp}$) of 31.5 mΩ cm^2, an on–off ratio of 10^8, and a threshold voltage

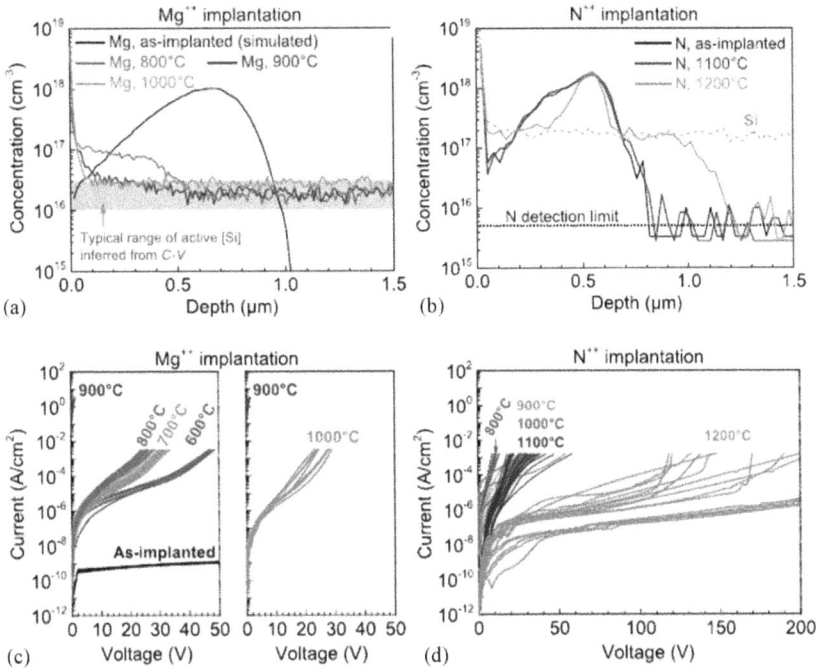

Figure 2.20. SIMS depth profiles of (a) Mg in Mg^{++}-implanted β-Ga_2O_3 with light Si-doping and (b) N in N^{++}-implanted UID Ga_2O_3 with light Si-doping. Curves for as-implanted and post-annealing at different temperatures are shown. Significant Mg diffusion is observed. (c) *I–V* characteristics of Mg^{++} implantation and (d) N^{++} implantation Ga_2O_3 after different temperature annealing. Reprinted from [20] with the permission of AIP Publishing.

(V_{th}) of -56 V. In addition, the breakdown voltage is limited to less than 30 V by the failure of Al_2O_3, as shown in figure 2.21(f). Improvements to the dielectric and doping processes are necessary to avoid premature device failure.

In addition, in 2017 Hu *et al* reported a vertical depletion-mode fin transistor with an on–off ratio $>10^9$ [51], as shown in figure 2.22(a). A 400 nm width fin structure on the UID $N_D = \sim 1 \times 10^{17}$ cm^{-3} (-201) β-Ga_2O_3 is defined by a metal hard mask patterned by electron beam lithography (EBL), and implemented using a BCl_3/Ar based dry etch process. The excellent morphology of the etched fin structure is shown in figure 2.22(b), which provides a precondition for the good performance of the device. The output current density reaches 1 kA cm^{-2}, and the threshold voltage is about -17 V, as shown in figure 2.22(c). In addition, a breakdown voltage (V_{br}) of >185 V for a 200 nm width fin device is obtained for devices with a fin width of 200 nm, as shown in figure 2.22(d). The failure of dielectrics near the bottom of the fin is responsible for the premature breakdown due to the field crowding effect. Therefore, field plates are expected to be implemented to improve the breakdown characteristics.

For vertical depletion-mode transistors, the current blocking effect of ion implantation and thermal diffusion are the key points to develop further. In addition, although the fin-channel structure is also used to realize the depletion-mode, the transistors will be normally-off with the fin width decreasing further.

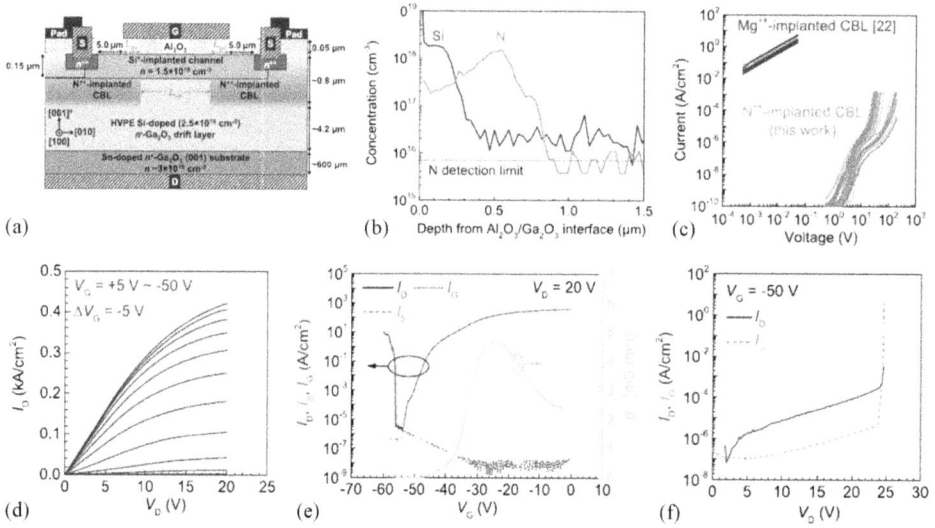

(a) (b) (c)

(d) (e) (f)

Figure 2.21. (a) A cross-sectional schematic of the current aperture vertical β-Ga$_2$O$_3$ MOSFET. (b) SIMI depth profiles of N and Si in the gate–CBL overlap region of the completed device. (c) Comparison of vertical leakage through N^{++}-implanted and Mg^{++}-implanted CBLs. (d) DC output, (e) DC transfer, and (f) off-state breakdown characteristics of the device. Copyright 2019 IEEE. Reprinted, with permission, from [50].

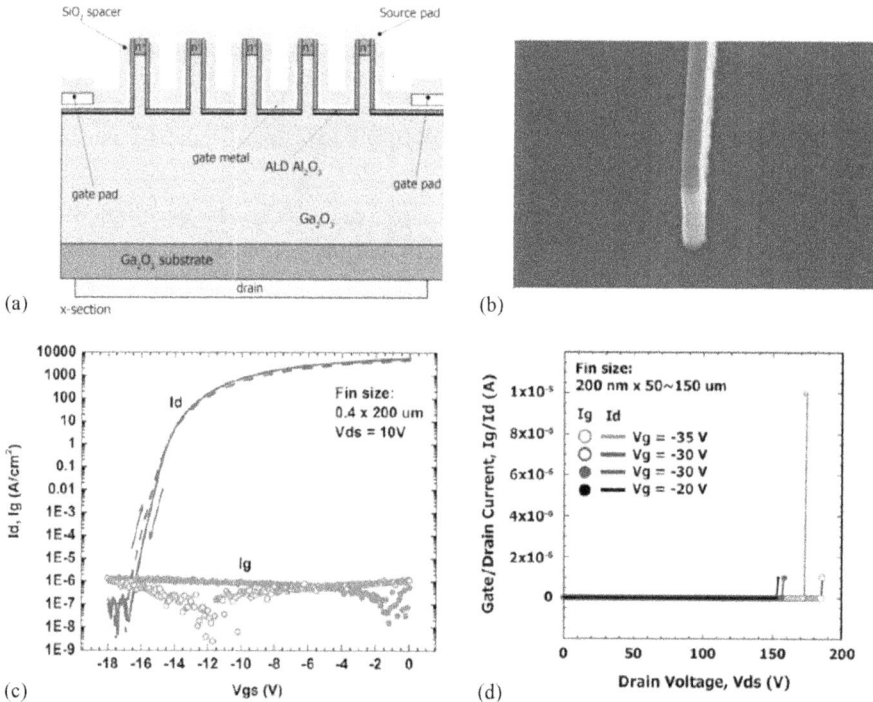

(a) (b)

(c) (d)

Figure 2.22. (a) Schematic cross-section of the β-Ga$_2$O$_3$ vertical FinFET and (b) an SEM image of a fin structure on the (−201) UID Ga$_2$O$_3$ substrate after dry etching. (b) SIMI depth profiles of N and Si in the gate–CBL overlap region of the completed device. (c) DC transfer characteristics of the device with a 400 nm wide fin and (f) off-state breakdown characteristics of the device with a 200 nm wide fin. Copyright 2017 IEEE. Reprinted with permission from [51].

2.6 Vertical enhancement-mode transistors

Vertical enhancement-mode transistors have the advantages of a high breakdown voltage and being easily driven, therefore, the investigation of this type of transistor is quite significant.

In 2018, Hu *et al* demonstrated an enhancement-mode Ga_2O_3 vertical MISFET with a high breakdown voltage (1057 V). The MISFET was fabricated on a bulk β-Ga_2O_3 (001) substrate ($N_D = 2 \times 10^{18}$ cm^{-3}) with a 10 μm low doping ($<2 \times 10^{16}$ cm^{-3}) HVPE layer [24]. To realize enhancement-mode operation, 330 nm wide and 795 nm long vertical fin channels were etched in an ICP system using Pt metal masks, patterned using EBL. More specific process steps can be found in [24] and the outstanding structure of the completed MISFET is shown in figure 2.23(a).

The enhancement-mode vertical Fin-MISFET showed admirable performance, including a drain–source current density (I_d) of ~350 A cm^{-2} @ $V_{gs} = 3$ V, and $V_{ds} = 10$ V, differential on-resistance ($R_{on,sp}$) of ~18 mΩ cm^{-2}, and a threshold voltage (V_{th}) of ~2.2 V, as well as the subthreshold slope (SS) of ~85 mV dec^{-1} [24]. In addition, the hysteresis is less than 0.2 V, which indicates a good gate interfacial property, and there is no obvious short-channel effect due to favorable electrostatic control of the gate electrode, as shown in figures 2.23(b) and (c). The three-terminal off-state breakdown curve of the device is shown in figure 2.23(d), indicating that

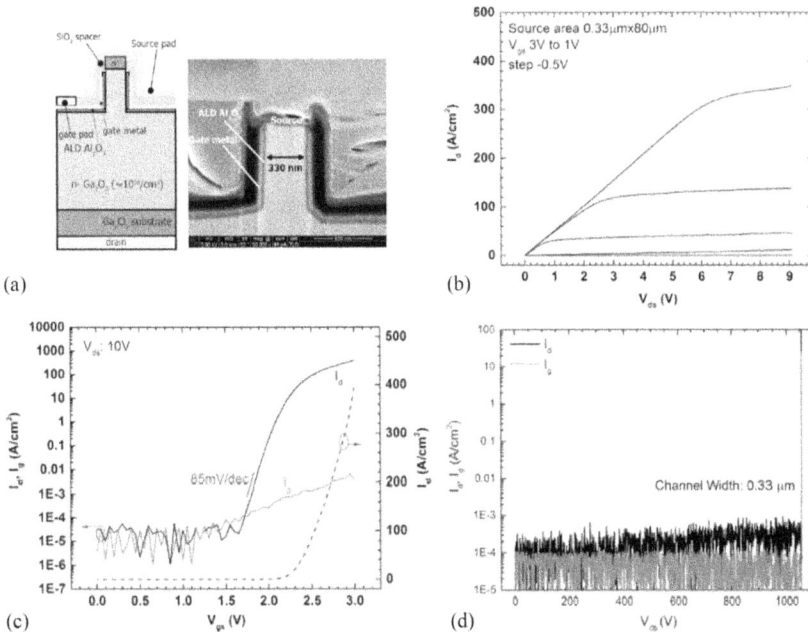

Figure 2.23. (a) Schematic cross-section and scanning electron microscopy cross-section image of a β-Ga_2O_3 vertical Fin-MISFET. (b) Representative pulsed I_{ds}–V_{ds} output characteristics, (c) I_d/I_g–V_{gs} transfer characteristics, and (d) three-terminal off-state I_d/I_g–V_{ds} of a Ga_2O_3 vertical power MISFET. Copyright 2018 IEEE. Reprinted with permission from [24].

both the drain (I_d) and gate leakage currents (I_g) will remain low at $V_{gs} = 0$ V until a hard breakdown takes place near 1057 V. Moreover, two-dimensional device simulations revealed that the values of electric field peaks in this device were much higher than 1.44 MV cm^{-1}.

To conclude, this is the first reported enhancement-mode vertical transistor with high breakdown voltage and this is sufficient for it to be a milestone in the application of β-Ga$_2$O$_3$ based power transistors. In addition, edge termination techniques are expected to further increase the breakdown voltage, such as a field plate or ion implantation.

Thus, Hu *et al* reported a new-generation vertical source-plate FinFET with a higher breakdown voltage of 1.6 kV at the 31st International Symposium on Power Semiconductor Devices and ICs (ISPSD) in 2019 [25]. The superior electrostatic design based on reduced surface field (RESURF) principles and a novel fabrication process flow can account for the performance improvement compared to previous works. To be specific, the size of the fin channel is 1.3 μm high and 480−560 nm wide on the 10 μm low doping ($N_D = \sim 1.2 \times 10^{16}$ cm^{-3}) HVPE epitaxial layer. Acid/base wet chemical treatments are used to remove the etching plasma damage, resulting in a the bottleneck structure, as shown in figure 2.24(a). In addition, a reduced gate area and an extended source pad outside the gate edge (source field plate) are implemented without complicating the processing steps compared to previous works [24]. Other similar processing steps can be found in [25].

Finally, the vertical β-Ga$_2$O$_3$ FinFETs are tested with a drain–source current density (I_d) of ~ 600 A cm^{-2} @ $V_{gs} = 5$ V and $V_{ds} = 10$ V, a specific on-resistance ($R_{on,sp}$) of 5.5 mΩ cm^2, a threshold voltage (V_{th}) of ~ 3.8 V, and an on–off ratio of

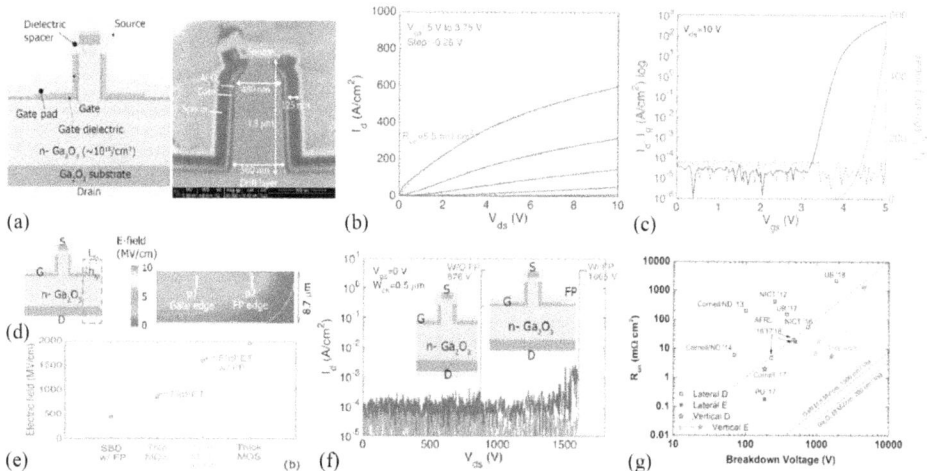

Figure 2.24. (a) Schematic structure and SEM cross-section image of a vertical β-Ga$_2$O$_3$ (001) FinFET. (b) Representative DC I_{ds}–V_{ds} output characteristics and (c) I_d/I_g–V_{gs} transfer characteristics in log/linear scales. (d) Simulation of the off-state electric field distribution at $V_{ds} = 1600$ V. (e) Measured highest reverse BVs of each equivalent structures of the transistors. (f) Off-state I_d/I_g–V_{ds} of Ga$_2$O$_3$ vertical FinFETs with/without source FPs. (g) Performance comparison of all lateral and vertical Ga$_2$O$_3$ power transistors. Copyright 2019 IEEE. Reprinted with permission from [25].

2-24

10^8, as shown in figures 2.24(b) and (c). It is necessary to state that a higher V_{th} than expected may be attributed to a negative interface charge and a short-channel effect is observed without sensible cause.

In addition, 2D simulation of the electric field indicates that the electric field peaks appear at both the gate (P1, as shown in figure 2.24(d)) and source FP edges (P2, as shown in figure 2.24(d)). Both the thickness of the supporting dielectric (h_{fp}) of 125 nm and the FP extension outside the gate (L_{fp}) of 10 μm are chosen due to the practical process and simulation results. Some equivalent structures are prepared at the same time with vertical FinFET fabrication and are tested. The results reveal that an FP can significantly reduce the electric field peak at the gate edge (P1) comparing thin MOS and thin MOS w/FP, as shown in figure 2.24(e). The source FP can increase the breakdown voltage of the vertical β-Ga$_2$O$_3$ FinFET from 876 V to 1605 V, as shown in figure 2.24(f). The source-connected FP FinFETs have shown excellent performance and have become a benchmark for Ga$_2$O$_3$ power transistors, as shown in figure 2.24(g). In addition, multiple field plates, resistive ion implantation, and floating guard rings are expected to further improve the blocking characteristics.

In conclusion, the fin channels have enabled the realization of vertical Ga$_2$O$_3$ transistors with the enhancement-mode without p-type doping, and the promising performance has made researchers focus on realizing the application of Ga$_2$O$_3$ power transistors. In addition, a further decrease of the on-resistance is necessary for applications with lower loss.

2.7 Homojunction HEMT

High electron mobility transistors (HEMTs) have faster switching frequency and realize higher system transfer efficiency. Ga$_2$O$_3$ HEMTs are investigated first by using an advanced technology, namely δ-doping to realize a 2DEG.

In 2017 a multi-layer δ-doped β-Ga$_2$O$_3$ MOSFET with a sheet charge density of 2.4×10^{14} cm^{-2} with a high mobility of 82 cm^2 V^{-1} s^{-1} and low sheet resistance of 320 Ω sq^{-1} was manufactured by Krishnamoorthy *et al* [27]. Silicon δ-doped layers are separated by 4 nm undoped Ga$_2$O$_3$ layers grown by plasma-assisted MBE, as shown in figure 2.25(a), and the solid Si source will be heated to 1290 °C to desorb the surficial oxide growth between the layers, as shown in red in figure 2.25(b). The other processing details can be obtained from [27]. As shown in figure 2.25(c), the equilibrium band diagram leads to a pulsed charge distribution. As a result, the drain current (I_{DS}) reached 236 mA mm^{-1} @ V_G = +2 V and V_{DS} = 7 V. A transconductance (g_m) of 26 mS mm^{-1} and a threshold voltage (V_{th}) of -14 V are obtained. However, the off-state current is 215 μA mm^{-1} at V_G = -14 V and V_{DS} = 7 V, leading to an on–off ratio of $\sim10^3$, as shown in figures 2.25(d) and (e). The breakdown voltage of 51 V is limited, as shown in figure 2.25(f).

In 2018, Xia *et al* fabricated Si δ-doped MESFETs with low source and drain contact resistances (1.5 Ω mm^{-1}), which are achieved by a regrown high doping (N_D = 2×10^{20} cm^{-3}) β-Ga$_2$O$_3$ layer and patterned using PECVD with SiO$_2$ [30], as shown in figure 2.26(a). The δ-doped channel is shown in figure 2.26(b), and the

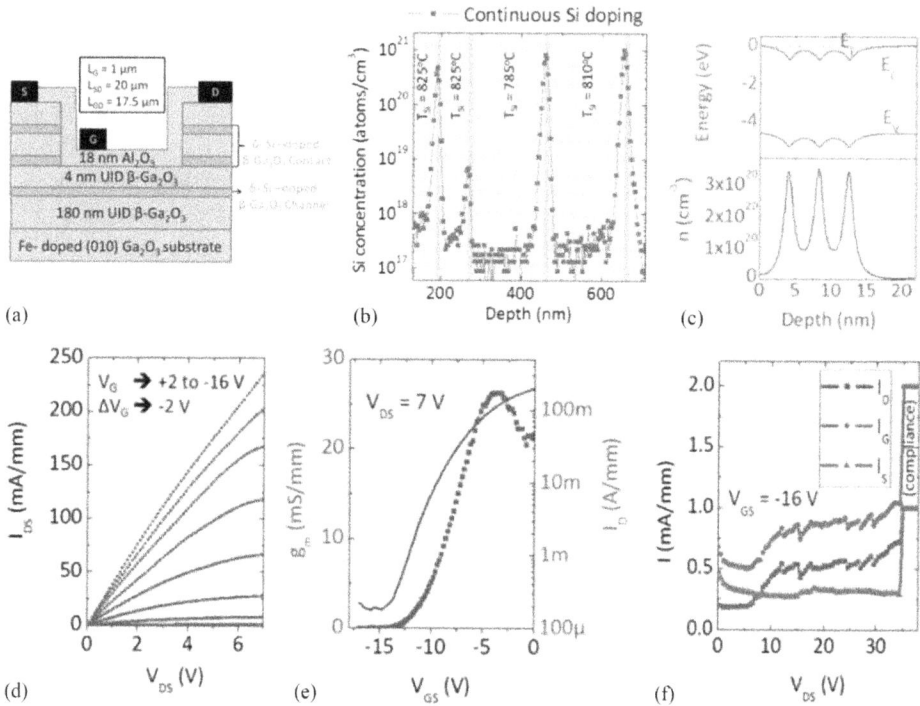

Figure 2.25. (a) Schematic of a δ-doped β-Ga$_2$O$_3$ FET. (b) SIMS profile of Si dopants. (c) Equilibrium band diagram and charge profile. (d) Output I_{DS}-V_{DS} and (e) transfer I_D-V_{GS} characteristics in log/linear scales, and (f) three-terminal off-state breakdown characteristics of the device [27]. Copyright IOP Publishing. Reproduced with permission. All rights reserved.

drain–source current density (I_{DS}) of 140 mA mm^{-1} indicates the well channel conduction, as shown in figure 2.26(c). Figures 2.26(d) and (e) show a threshold voltage (V_{th}) of -3.5 V, a transconductance (g_m) of 34 mS mm^{-1}, and a breakdown voltage of 170 V. In addition, the peak electron mobility reaches 95 cm^2 V^{-1} s^{-1} at a density of 3.5×10^{12} cm^{-2}, as shown in figure 2.26(f). Although the RF performance of the device is not reported in the literature, the results imply that δ-doping can be a potential process to realize Ga$_2$O$_3$ high frequency FETs.

Furthermore, Joishi *et al* investigated the effect of buffer iron doping on a δ-doped β-Ga$_2$O$_3$ MESFET [28], as shown in figure 2.27(a). Based on the diffusion profile shown in figure 2.27(b), a >200 nm UID buffer layer can prevent the effect of Fe diffusion at least. In addition, the 2DEG will possess a larger mobility and density with increasing thickness of the UID buffer layer, as shown in figures 2.27(c) and (d). Therefore, MESFETs with a thicker buffer layer will have a larger pinch-off voltage, as shown in figure 2.27(e). In addition, the presence of Fe in the substrate will limit the saturated current density (I_{DS}) of the MESFET, as shown in figure 2.27(f). To sum up, this work establishes the effect of the buffer layer design in δ-doped transistors.

In addition, the trapping effects in Si δ-doped Ga$_2$O$_3$ MESFET were investigated by McGlone *et al* in 2018 [29], as shown in figure 2.28(a). The traps filled with

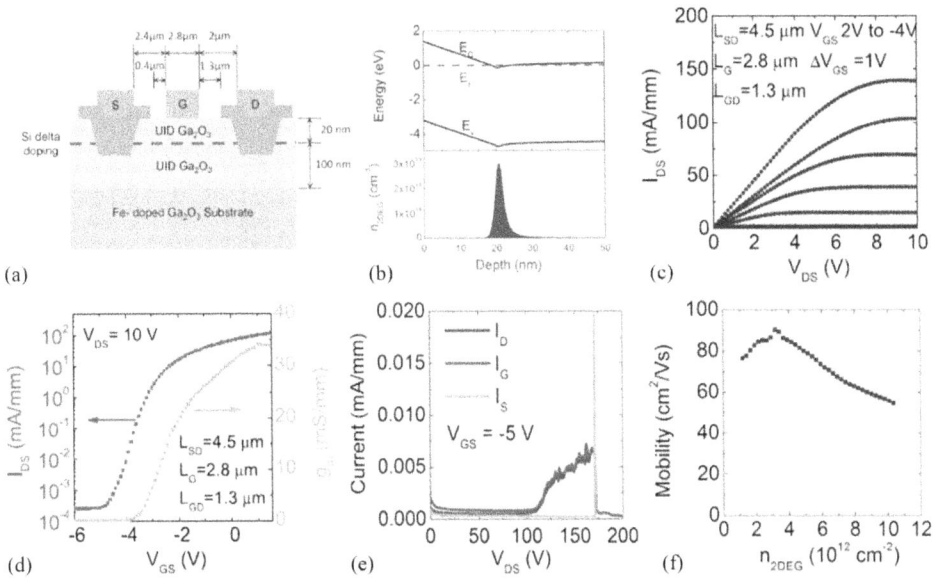

(a) (b) (c)

(d) (e) (f)

Figure 2.26. (a) Schematic of the δ-doped β-Ga$_2$O$_3$ MESFET with a regrown ohmic contact layer. (b) Equilibrium energy band diagram and 2DEG charge profile. (c) Output, (d) transfer, and (e) breakdown characteristics of the device. (f) Density-dependent field-effect mobility extracted from the MESFET. Copyright 2018 IEEE. Reprinted with permission from [30].

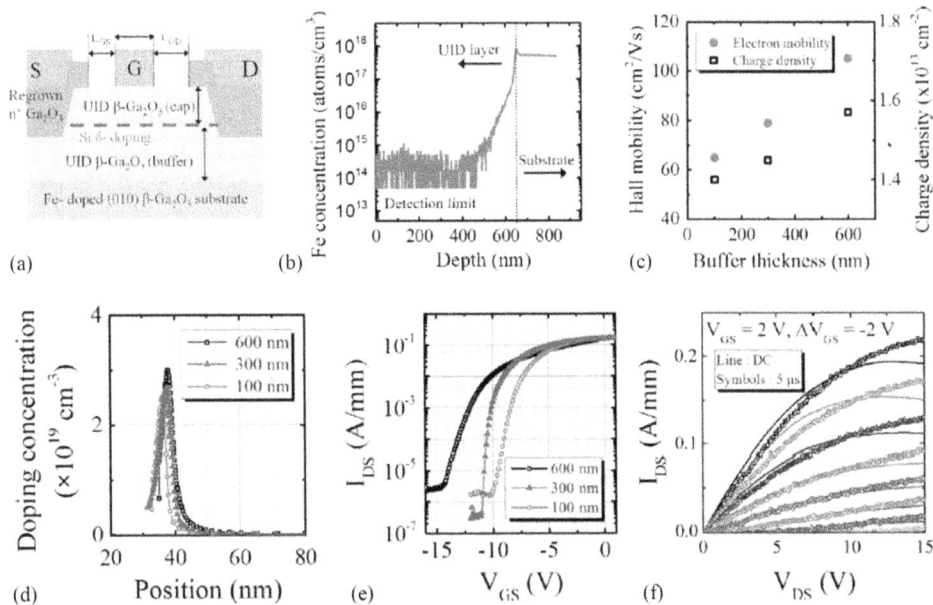

(a) (b) (c)

(d) (e) (f)

Figure 2.27. (a) Two-dimensional schematic of the Si δ-doped MESFET. (b) Fe depth profile of the as-grown UID Ga$_2$O$_3$ buffer using SIMS. (c) Hall mobility and their corresponding charge density dependence with UID buffer thickness. (d) The 2DEG profile extracted from the C–V curve. (e) Transfer curves (I_{DS}–V_{GS}) and (f) DC and pulsed I_{DS}–V_{DS} curves for the MESFET with a 600 nm buffer. Reprinted from [28] with permission from AIP Publishing.

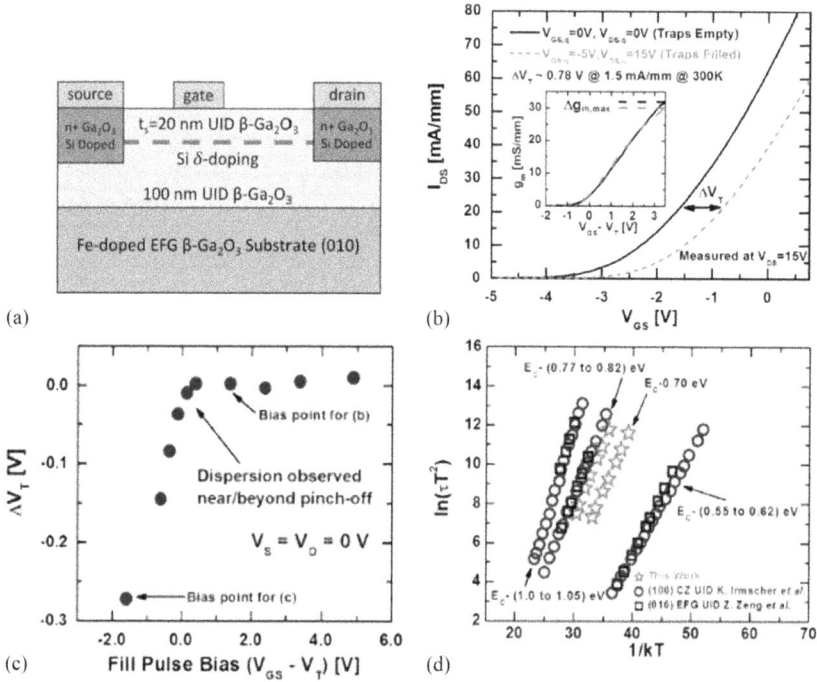

Figure 2.28. (a) Cross-sectional diagram of the grown and processed Si δ-doped Ga_2O_3 MESFET. (b) Double pulsed transfer characteristics and 0.78 V threshold voltage shift between empty ($V_{GS} = 0$ V) and filled ($V_{GS} = -5$ V) traps. (c) Trap-induced V_T shift dependence with the gate bias ($V_{GS}-V_T$). (d) The trap energy and capture cross-section σ_n extracted from the Arrhenius plot. Copyright 2018 IEEE. Reprinted with permission from [29].

electron make the threshold voltage shift positively ($\Delta V_T = +0.78$ V), thereby reducing the channel charge, as shown in figure 2.28(b). In addition, the V_T only shifts dramatically when $V_{GS} < V_T$, as shown in figure 2.28(c). We can explain that it is likely to be buffer traps rather than surface traps because the channel will screen the electric field below the channel when $V_{GS} \geqslant V_T$. Two trap levels of $E_C - 0.70$ eV and $E_C - 0.77$ eV in MBE grown Ga_2O_3 are revealed by deep level transient spectroscopy (DLTS) measurement, as shown in figure 2.28(d).

In 2019, a T-shaped gate δ-doped β-Ga_2O_3 MESFET with a cut-off frequency of 27 GHz was demonstrated by Xia *et al* [26], as shown in figure 2.29(a). The T-shaped Schottky gate was defined by electron beam lithography using a PMMA/MMA/PMMA resist stack, as shown in figure 2.29(b). An integrated sheet carrier density of 9.9×10^{12} cm^{-2} was extracted by $C–V$ measurement, as shown in figure 2.29(c). A current density (I_{DS}) of 260 mA mm^{-1}, a transconductance (g_m) of 44 mS mm^{-1}, and a breakdown voltage (V_{BR}) of 150 V were obtained, as shown in figures 2.29(d) and (e). Finally, a current gain cut-off frequency of 27 GHz was tested, as shown in figure 2.29(f). This work represents an important step towards the application of high frequency β-Ga_2O_3 transistors.

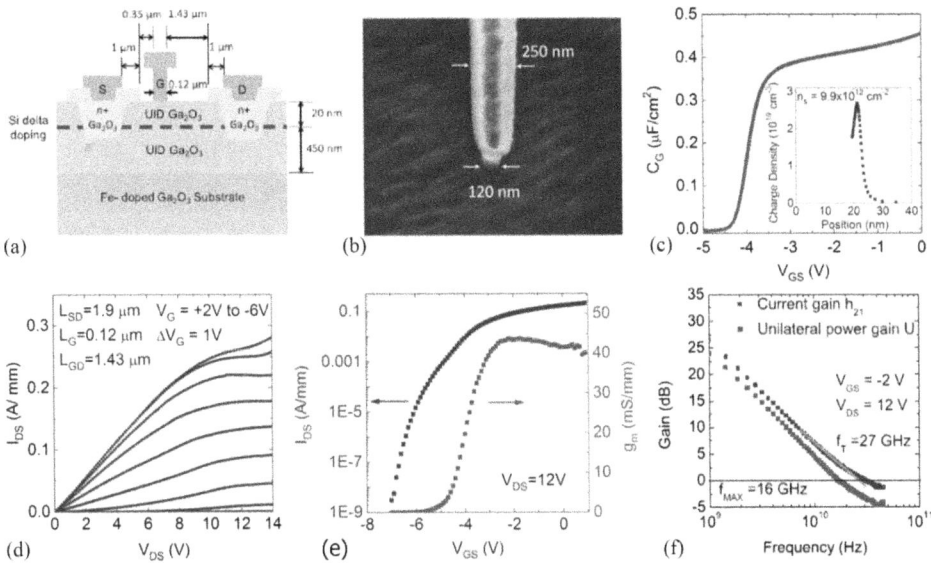

Figure 2.29. (a) Device schematics of the T-shaped gate MESFET. (b) Side-view SEM image of a 120 nm T-shaped gate. (c) $C-V$ characteristics and (inset) electron concentration profile of 2DEG. (d) Output and (e) transfer characteristics of δ-doped MESFET. (f) Measured small signal performance and cut-off frequency (f_T) of the transistor. Copyright 2019 IEEE. Reprinted with permission from [26].

Homojunction HEMTs have exhibited a large on-state current due to high electrical field mobility and a decent homojunction interfacial property due to the absence of lattice mismatch. The improvement of epitaxial growth has promise to be applied for high frequency switches. In addition, a larger potential well is necessary to form a higher concentration 2DEG, therefore the heterojunction is developed to pursue larger current densities.

2.8 Heterojunction HEMT

In addition to homo-interfacial Si δ-doped β-Ga$_2$O$_3$ HEMTs, the investigation of β-Ga$_2$O$_3$ based heterojunction HEMTs has also been carried out extensively.

In 2017, Krishnamoorthy et al reported the formation of a modulation-doped 2DEG at the β-(Al$_x$Ga$_{1-x}$)$_2$O$_3$/Ga$_2$O$_3$ (AGO/GO) heterojunction using silicon δ-doping [33]. An AGO/GO epitaxial stack grown on the Fe-doped (010) β-Ga$_2$O$_3$ semi-insulating substrate is shown in figure 2.30(a), and an RMS roughness of 0.7 nm indicates a smooth surface morphology as shown in figure 2.30(b). However, the crystalline quality of the AGO layer should be improved further due to its higher full width half maximum (FWHM) value of 0.24° than the 0.2° FWHM for AlGaN. The STEM image clearly shows the presence of the AGO/GO heterojunction, as shown in figure 2.30(c). In addition, through a Schrödinger–Poisson simulation and CV measurement, an Al$_2$O$_3$ mole fraction of 20% for the AGO barrier, an AGO dielectric constant of 13, an AGO/GO conduction band offset of 0.6 eV, and a donor energy level (E_C-E_D) of 135 meV are obtained. It is

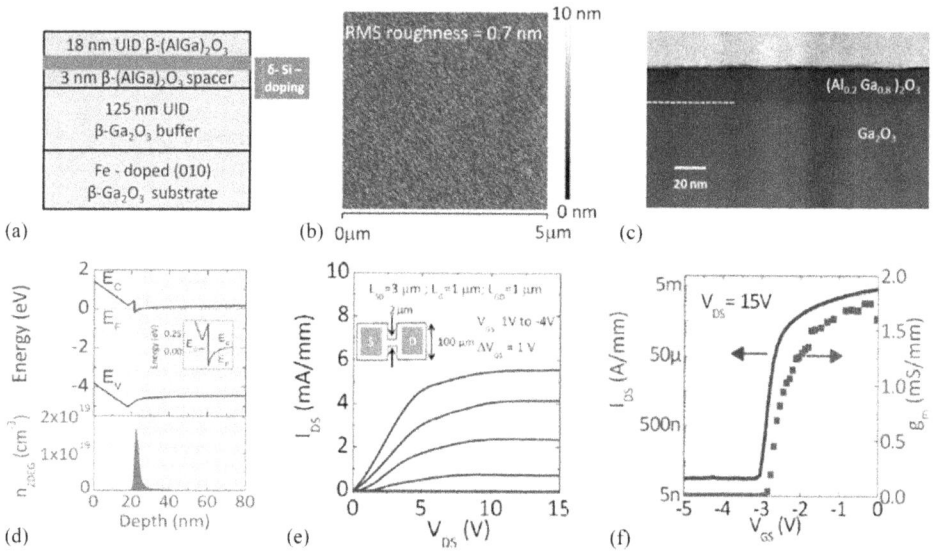

Figure 2.30. (a) Epitaxial structure, (b) atomic force microscopy (AFM) image, and (c) cross-sectional STEM image of a modulation-doped (MOD) AGO/GO heterojunction. (d) Equilibrium energy band diagram and 2DEG charge profile of the AGO/GO heterojunction. (d) Output characteristics and (e) transfer characteristics of the MODFETs. Reprinted from [33] with permission from AIP Publishing.

necessary to mention that the bandgap difference appears completely as the conduction band offset at the AGO/GO heterojunction, as shown in figure 2.30(d). The electrical characteristics of the AGO/GO MODFET without ohmic source/drain contacts are shown in figures 2.30(e) and (f). A maximum current density of 5.5 mA mm^{-1}, a pinch-off voltage of -3 V, and a peak transconductance of 1.75 mS mm^{-1} were measured.

A Ge-doped β-(Al$_x$Ga$_{1-x}$)$_2$O$_3$/Ga$_2$O$_3$ heterojunction has also been applied to fabricate MODFETs. We do not introduce it in detail due to its similar effects, but the parameters can be obtained from [32].

In 2018, HMETs with 2DEG at the interface of the β-(Al$_x$Ga$_{1-x}$)$_2$O$_3$/Ga$_2$O$_3$ heterostructure and a Si δ-doped layer were reported and investigated by Zhang *et al* [31], as shown in figure 2.31(a). A larger sheet carrier density of $\sim 2 \times 10^{12}$ cm^{-2} was measured for sample B with a 360 nm UID Ga$_2$O$_3$ buffer compared to sample A with a 130 nm buffer, as shown in figure 2.31(b). The polar optical phonon scattering mainly limits the electron mobility at high temperatures, although 2790 cm^2 V^{-1} s^{-1} for sample B can be measured at 50 K, as shown in figure 2.31(c). A maximum drain current (I_{DS}) of 46 mA mm^{-1}, a peak transconductance (g_m) of 39 mS mm^{-1}, and a threshold voltage (V_{th}) of $+0.5$ V are obtained, as shown in figures 2.31(d) and (e). High frequency small signal measurements show a cut-off frequency of 3.1 GHz for this device, as shown in figure 2.31(f).

After that, β-(Al$_x$Ga$_{1-x}$)$_2$O$_3$/Ga$_2$O$_3$ double heterostructure FETs were demonstrated by Zhang *et al*, and electrons can be transferred from below and above the Ga$_2$O$_3$ quantum well (QW), leading to a higher 2DEG charge density of 3.85×10^{12} cm^{-2}

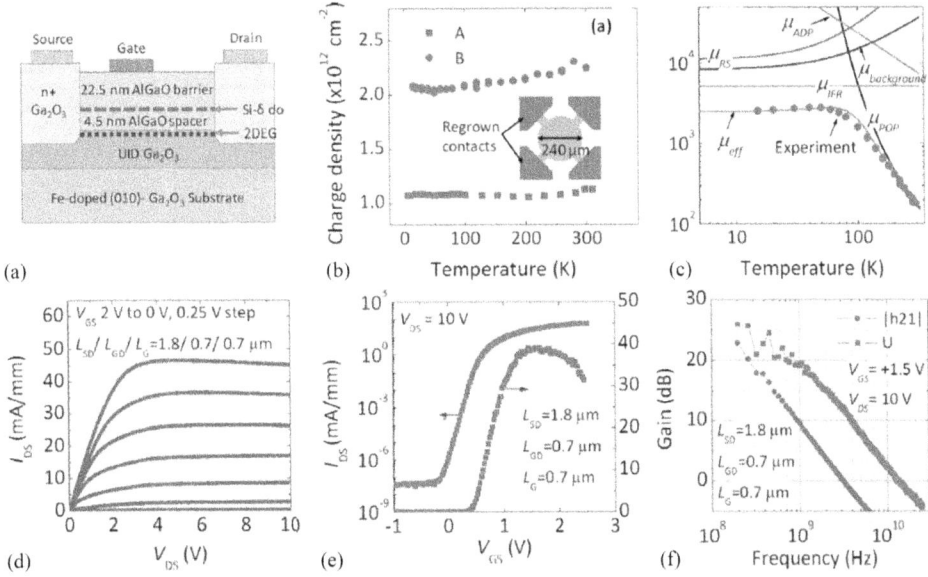

Figure 2.31. (a) Schematic epitaxial stack of the MODFET structure. (b) Temperature-dependence of charge density for sample A (130 nm buffer) and sample B (360 nm buffer). (c) Experimental and calculated electron mobility for sample B (360 nm buffer). (d) Output characteristics, (e) transfer characteristics, and (f) RF characteristics of the MODFET on sample B (360 nm buffer). Reprinted from [31] with permission from AIP Publishing.

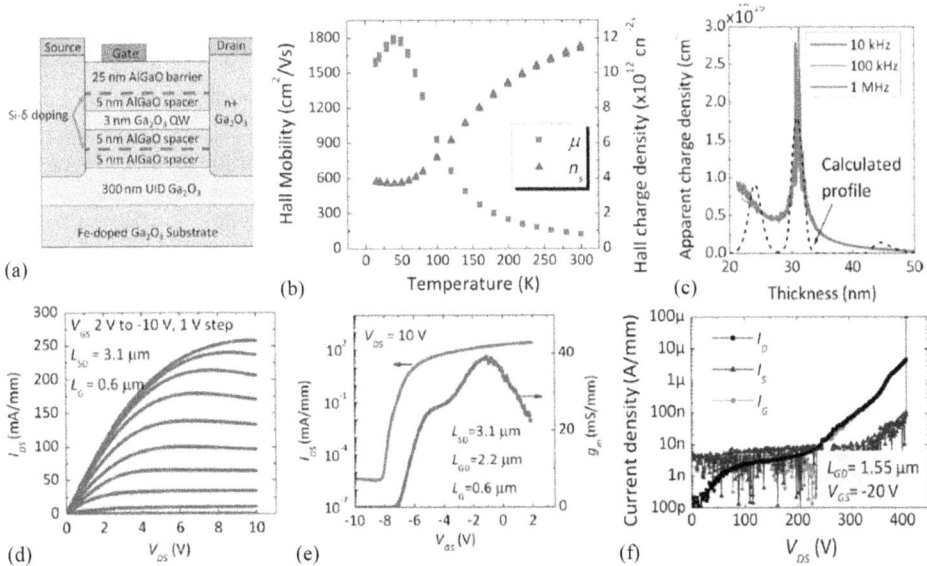

Figure 2.32. (a) Schematic epitaxial stack of the double heterostructure MODFET. (b) Temperature-dependence of Hall mobility and charge density using a van der Pauw configuration. (c) The apparent charge distribution extracted from CV measurements. (d) Output characteristics, (e) transfer characteristics, and (f) three-terminal off-state breakdown characteristics of the MODFET with a gate–drain spacing of 1.55 μm. Reprinted from [34] with permission from AIP Publishing.

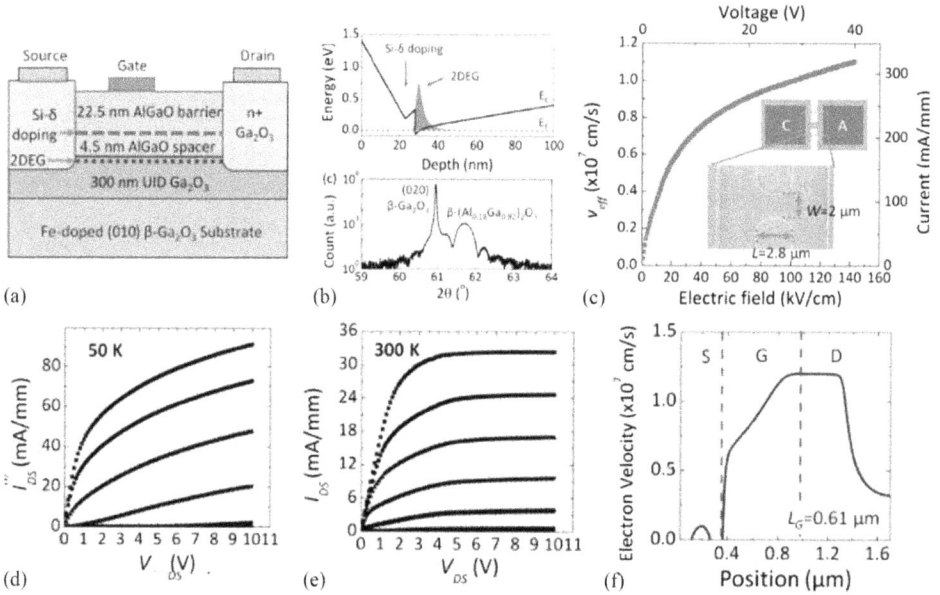

Figure 2.33. (a) Schematic of the MODFET device. (b) Equilibrium energy band diagram and charge distribution profile (top), and an x-ray diffraction ω–2θ diffraction pattern for (020) plane reflections (bottom). (c) Measured velocity-field profile at 50 K using the device structure in the inset. Output characteristics measured (d) at 50 K and (e) at 300 K. (f) Simulated electron velocity profile inside the channel at $V_{DS} = 10$ V and $V_{GS} = +4$ V. Copyright 2019 IEEE. Reprinted with permission from [36].

than for a single heterostructure [34], as shown in figure 2.32(a). The Hall mobility of 123 cm^2 V^{-1} s^{-1} and carrier density of 1.14×10^{13} cm^{-2} are measured at room temperature using a van der Pauw configuration, as shown in figure 2.32(b). The 2DEG channel at 31.0 nm below the top surface is extracted by CV measurement, as shown in figure 2.32(c). The devices showed a maximum drain current (I_{DS}) of 257 mA mm^{-1}, a peak transconductance (g_m) of 39 mS mm^{-1}, a pinch-off voltage of −7.0 V, and a three-terminal breakdown voltage around 400 V for HEMTs with $L_{GD} = 1.5$ μm, as shown in figures 2.32(d) and (f).

The low temperature performance of β-(Al$_x$Ga$_{1-x}$)$_2$O$_3$/Ga$_2$O$_3$ MODFETs was also investigated by Zhan *et al* in 2019 [36]. As shown in figures 2.33(a) and (b), both the device structure and film properties are as before. An effective velocity of 1.1×10^7 cm s^{-1} is obtained at 143 kV cm^{-1} and 50 K, as shown in figure 2.33(c). In addition, the HEMT reaches a larger saturation current of 87 mA mm^{-1} at 50 K, as shown in figures 2.33(d) and (e). The low electron velocity in the gate channel accounts for the low 2DEG charge density and long gate length, as the simulated electron velocity profile shown in figure 2.33(f). Therefore, increasing the 2DEG charge density and scaling the gate length are better choices for RF applications of β-(Al$_x$Ga$_{1-x}$)$_2$O$_3$/Ga$_2$O$_3$ MODFETs.

In 2019 high breakdown voltage β-(Al$_x$Ga$_{1-x}$)$_2$O$_3$/Ga$_2$O$_3$ MODFETs with a gate field plate were demonstrated by Joishi *et al* [35], as shown in figure 2.34(a). The unpassivated Schottky gate leakage current limits the breakdown voltage to ~350 V,

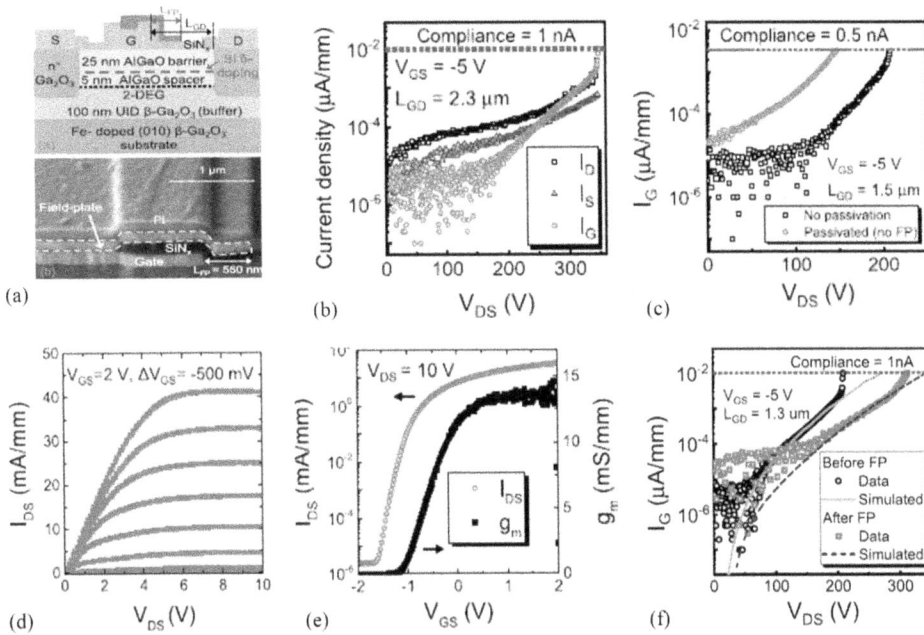

Figure 2.34. (a) 2D schematic of the β-$(Al_{0.22}Ga_{0.78})O_3$/Ga_2O_3 MODFET with a gate-connected field plate (top), and cross-sectional SEM of the device (bottom). (b) Three-terminal off-state breakdown voltage of ~350 V. (c) Comparison of gate leakage current (I_G) before and after SiN_x passivation without a field plate. (d) Output (I_{DS}–V_{DS}) and (e) transfer (I_{DS}–V_{GS}) characteristics of the device. (f) The Schottky gate leakage current (I_G) with and without field plate and Fowler–Nordheim tunneling simulation. Copyright 2019 IEEE. Reprinted with permission from [35].

as shown in figure 2.34(b). In fact, passivation with SiN_x results in an increased gate leakage instead, because such superficial states (damage during the PECVD SiN_x deposition) causes an increase of the peak electric field at the gate edge, as shown in figure 2.34(c). The MODFETs with a source to drain distance (L_{SD}) of 2 μm, a gate length (L_{GD}) of 1.3 μm, and a gate to drain distance (L_{GD}) of 0.32 μm show a maximum current density of 42 mA mm^{-1}, a pinch-off voltage of -1.2 V, and an on–off ratio of ~10^8, as shown in figures 2.34(d) and (e). The gate leakage current is fitted with Fowler–Nordheim tunneling, as shown in figure 2.34(f). In addition, a breakdown voltage (V_{BR}) of 1.37 kV and an R_{on} of 120.1 mΩ cm^2 ms obtained for transistors with L_{GD} = 16 μm.

To sum up, the growth of AGO is significant for the performance of β-Ga_2O_3 base heterojunction HEMTs, and they still have room for improvement for an on-state current with high blocking voltage.

2.9 Nanomembrane transistors

Nano-membranes of β-Ga_2O_3 can be acquired mechanically in the same way as 2D materials, due to the monoclinic crystal structure of β-Ga_2O_3 with the lattice of the a-axis being longer than the other two axes. Therefore, a lot of research on β-Ga_2O_3 film transistors has been reported for power applications in recent years.

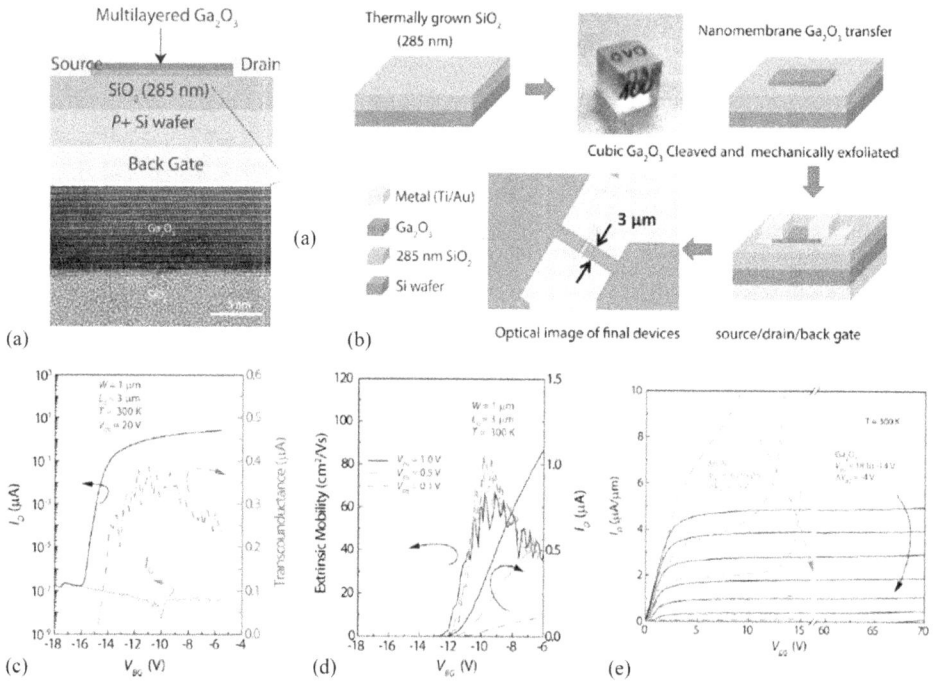

Figure 2.35. (a) Cross-sectional schematic of a nanomembrane β-Ga$_2$O$_3$ FET and TEM image of a flat interface between β-Ga$_2$O$_3$ and the SiO$_2$ dielectrics. (b) Schematic process flow for nanomembrane β-Ga$_2$O$_3$ FET. (c) Transfer properties, (d) field-effect mobility, and (e) output characteristics of β-Ga$_2$O$_3$ devices with a 100 nm thick nanomembrane. Reprinted from [52] with permission from AIP Publishing.

In 2014, Hwang *et al* first demonstrated high voltage nanomembrane β-Ga$_2$O$_3$ FETs by exfoliating films from bulk crystals [52]. The detailed process flow for nanomembrane FETs is shown in figure 2.35(b). The cross-section image of the completed device and the atomically smooth interface between Ga$_2$O$_3$ and SiO$_2$ are clearly shown in figure 2.35(a). The transistors with a 100 nm thick β-Ga$_2$O$_3$ nanomembrane ($N_D = 5.5 \times 10^{17}$ cm^{-3}) show a high gate modulation of $\sim 10^7$ and a threshold voltage of ~ -16 V, as shown in figure 2.35(c). The uncorrected field-effect mobility of ~ 70 cm^2 V^{-1} s^{-1} is extracted due to source and drain contact resistances (~ 55 Ω mm), as shown in figure 2.35(d). Compared to MoS$_2$ FETs, Ga$_2$O$_3$ FETs do not show the signatures of avalanche or impact-ionization breakdown when implementing a high voltage (~ 70 V), as shown in figure 2.35(e). This work indicates that nanomembrane β-Ga$_2$O$_3$ transistors have the potential for integration on foreign substrates and to realize high voltage switching operation.

In 2016 Ahn *et al* applied both SiO$_2$ (300 nm) and Al$_2$O$_3$ (15 nm) as the back and front gate dielectrics of β-Ga$_2$O$_3$ ($N_D = \sim 3 \times 10^{17}$ cm^{-3}, ~ 200 nm) flake based FET [53], as shown in figure 2.36(a). Both-gates operation with a smaller threshold voltage ($V_{TH,both}$) of ~ -20 V and a larger transconductance ($G_{M,both}$) of ~ 4.4 mS mm^{-1} improves the channel modulation relative to individual back gating or front gating ($V_{TH, front} = \sim -25$ V, $G_{M, front} = \sim 3.7$ mS mm^{-1}), and the transistors exhibit a

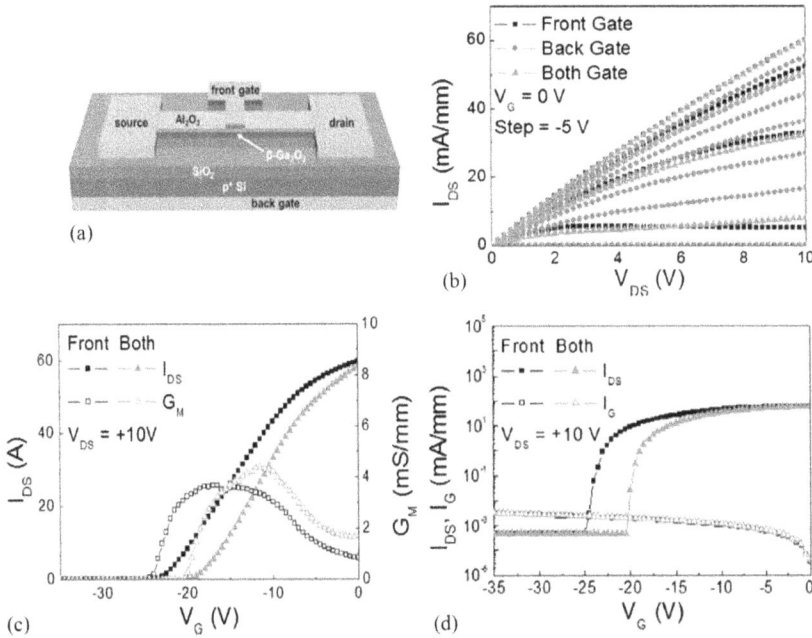

Figure 2.36. (a) The schematic of the double-gate β-Ga$_2$O$_3$ flake based FET. (b) Drain I–V with front, back, and both gates, (c) typical transfer characteristics with front or both gates, and (d) the gate voltage (V_G) dependent drain current (I_{DS}) and gate leakage current (I_{GS}) of the devices. Reprinted from [53] with permission from AIP Publishing.

maximum source–drain current density of 60 mA mm^{-1} and an on–off ratio of ~10^5 due to the smaller gate leakage current, as shown in figure 2.36(b) and (c). In addition, the field-effect mobility of ~1.35 cm^2 V^{-1} s^{-1} of the nano-belt FET is calculated.

In 2017 Zhou et al has fabricated a back-gate transistor with a high record drain current density of 600/450 mA mm^{-1} and a high on–off ratio of 10^{10} in both the depletion and enhancement modes [54], as shown in figures 2.37(b)–(e). It is necessary to mention that the nanomembrane of the D-mode GOOI-FET is 94 nm thick β-Ga$_2$O$_3$ (N$_D$ = 2.7 × 10^{18} cm^{-3}), while the thickness is 79 nm for the e-mode device. Figure 2.37(f) shows the determined thickness dependent V_T, indicating that the operation mode of transistors can be transferred though controlling the film thickness. In addition, a breakdown voltage of 185 V and a low subthreshold slope of 140 mV dec^{-1} are achieved in e-mode devices with an average breakdown electric field of 2 MV cm^{-1}, which shows the great potential of GOOI-FETs for power electronics.

In 2018 Ma et al demonstrated a high-performance nanomembrane β-Ga$_2$O$_3$ (N$_D$ = 4.8 × 10^{17} cm^{-3}) ~250 nm FET with a field-effect mobility (μ_{FE}) of 60.9 cm^2 V^{-1} s^{-1}, an on–off ratio of 10^9, and a subthreshold slope (SS) of 210 mV dec^{-1} [55], as shown in figure 2.38(a). The temperature-dependent performance was measured in the temperature range of 300 K to 475 K. As shown in figures 2.38(b) and (c), μ_{FE} decreased from 61.3 cm^2 V^{-1} s^{-1} to 38.5 cm^2 V^{-1} s^{-1} due to phonon scattering, and threshold voltage (V_{TH}) increased due to the thermally activated donor induced

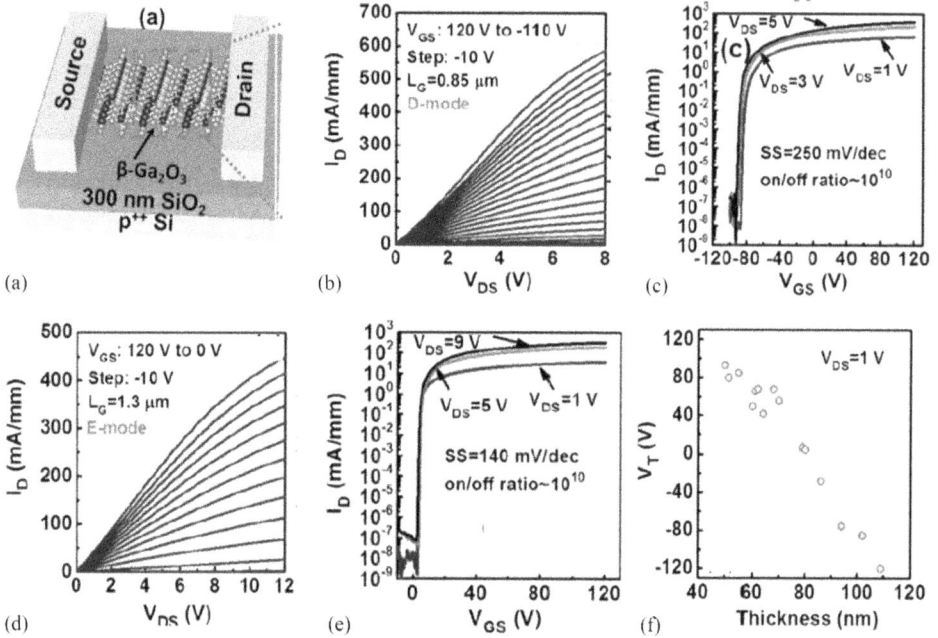

Figure 2.37. (a) Schematic view of GOOI-FETs. (b) I_D–V_{DS} output characteristics and (c) log-scale I_D–V_{GS} transfer characteristics of a D-mode GOOI-FET with a 94 nm thick nanomembrane. (d) and (e) Electric characteristics of an e-mode GOOI-FET with a 79 nm thick nanomembrane. (f) V_T versus nanomembrane thickness. Copyright 2017 IEEE. Reprinted with permission from [54].

carriers in the channel. Normal positive V_{TH} shift under PBTS is shown in figure 2.38(e), and abnormal positive V_{TH} shift under NBTS is shown in figure 2.38(d), which is attributable to significant surface depletion effects. In addition, an ALD–Al_2O_3 passivation layer can induce a negative shift, as shown in figure 2.38(f). Therefore, the high-quality passivation is necessary to ensure the electrical performance of nanomembrane β-Ga_2O_3 FETs.

In 2018, Noh *et al* reported a top gate β-Ga_2O_3 ($N_D = 2.7 \times 10^{18}$ cm^{-3}, 50–100 nm) field-effect transistor on diamond with a record high drain current of 980 mA mm^{-1} [56], as shown in figures 2.39(a)–(c). Compared to similar devices on a sapphire or SiO_2/Si substrate, β-Ga_2O_3 FETs on a diamond substrate exhibited a 63% larger drain current (I_D) than sapphire (132% larger than the SiO_2/Si substrate) due to a smaller self-heating effect (SHE) and better transfer properties, and similar on-resistance (R_{ON}) and threshold voltage (V_{TH}) due to their independence from the substrate. However, the transconductance (g_m) of FETs on a diamond substrate reaches 35 mS mm^{-1} due to the higher field-effect mobility at a lower channel temperature, compared to 21 mS mm^{-1} for sapphire and 14 mS mm^{-1} for the substrate.

In 2018 Bae *et al* demonstrated a 344 V quasi-two-dimensional β-Ga_2O_3 (4×10^{17} cm^{-3}, 200 nm–400 nm) field-effect transistors with a boron nitride (BN) field plate [57], as shown in figure 2.40(a). The transistors exhibited an obvious saturation drain

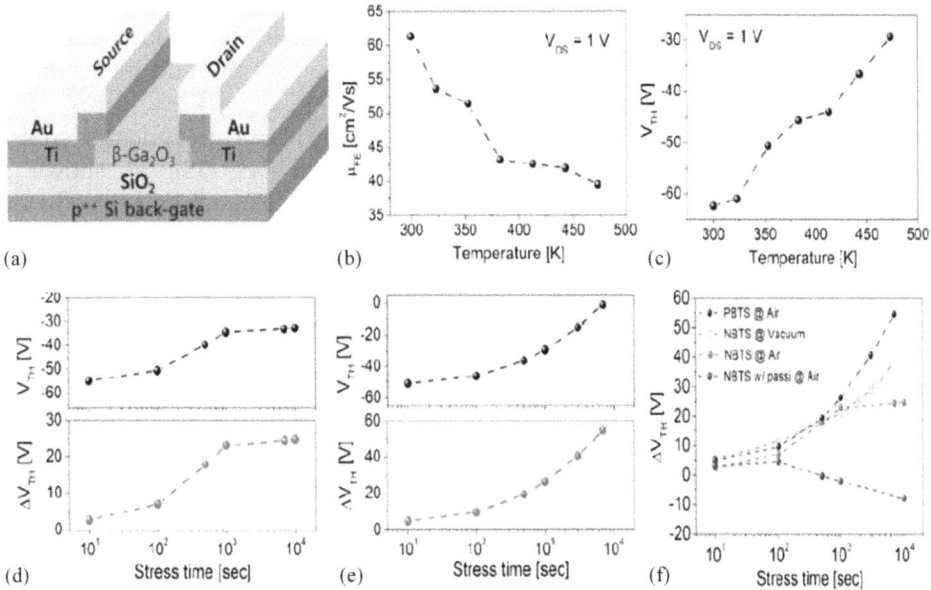

Figure 2.38. (a) Schematic illustration of the β-Ga$_2$O$_3$ FET with a channel thickness of ~250 nm and the bottom-gate configuration. (b) The extracted field-effect mobility (μ_{FE}) and (c) threshold voltage shift (ΔV_{TH}) dependence on the operation temperature. Threshold voltage shift (ΔV_{TH}) versus (d) stress time under negative bias-temperature stress (NBTS) of $V_{GS} - V_{TH} < 0$ V, $T_{STR} = 80$ °C, (e) stress time under positive bias-temperature stress (PBTS) of $V_{GS} - V_{TH} > 0$ V, $T_{STR} = 80$ °C at $V_{DS} = 1$ V. (f) Comparison of ΔV_{TH} as a function of stress under PBTS (ambient), NBTS (ambient and vacuum), and NBTS (ambient) of the device with Al$_2$O$_3$ passivation. Copyright 2018 IEEE. Reprinted with permission from [55].

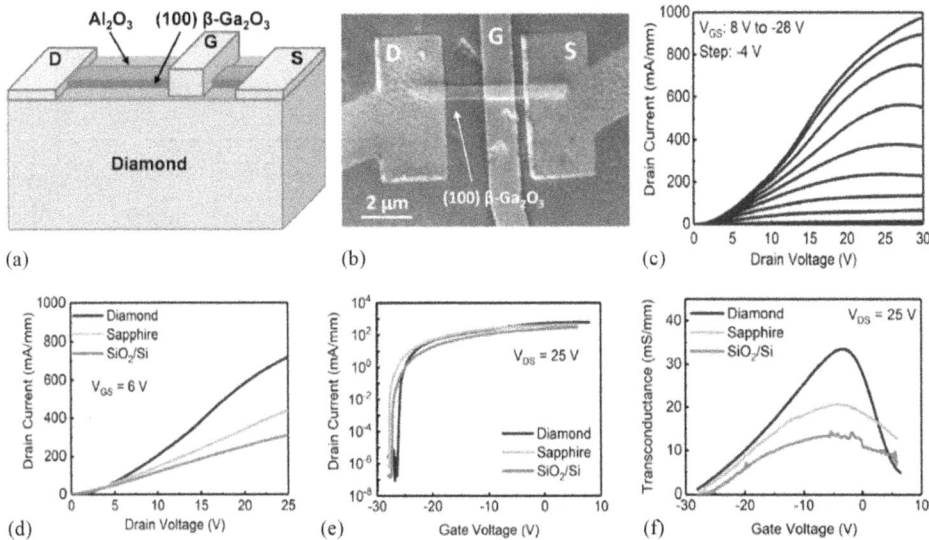

Figure 2.39. (a) Schematic view and (b) SEM image of a top-gate β-Ga$_2$O$_3$ FET on a diamond substrate. (c) The record high maximum drain current density (I_D) of 980 mA mm^{-1} among all top-gate β-Ga$_2$O$_3$ FETs. Comparison of the (d) I_D–V_D, (e) I_D–V_{GS}, and (f) g_m–V_{GS} characteristics of FETs on a diamond, sapphire, and SiO$_2$/Si substrate. Copyright 2018 IEEE. Reprinted with permission from [56].

current density of $\sim$2.0 mA mm^{-1}, as shown in figure 2.40(b). In addition, the BN field plate decreased the drain–gate leakage current effectively and increased the breakdown voltage to 344 V, as shown in figures 2.40(c) and (d).

In 2019 Kim *et al* proposed the use of remote fluorine plasma treatment to control the threshold voltage of β-Ga$_2$O$_3$ field-effect transistors [58], as shown in figure 2.41(a). The XPS spectrum suggests the existence of Ga–F and the effect of plasma treatment, as shown in figure 2.41(b). Compared to the transfer curves of the double-gated pristine and fluorinated β-Ga$_2$O$_3$ MESFET shown in figures 2.41(c) and (d), fluorine plasma treatment introduces a positive shift of threshold voltage (V_{TH}) from −1.1 to +2.2 V.

To sum up, β-Ga$_2$O$_3$ nanomembrane transistors have realized both depletion and enhancement modes through controlling the thickness of the film and have shown a larger current density and higher voltage than common 2D materials. This indicates that they have potential applications for integrated smart power electronics in the future.

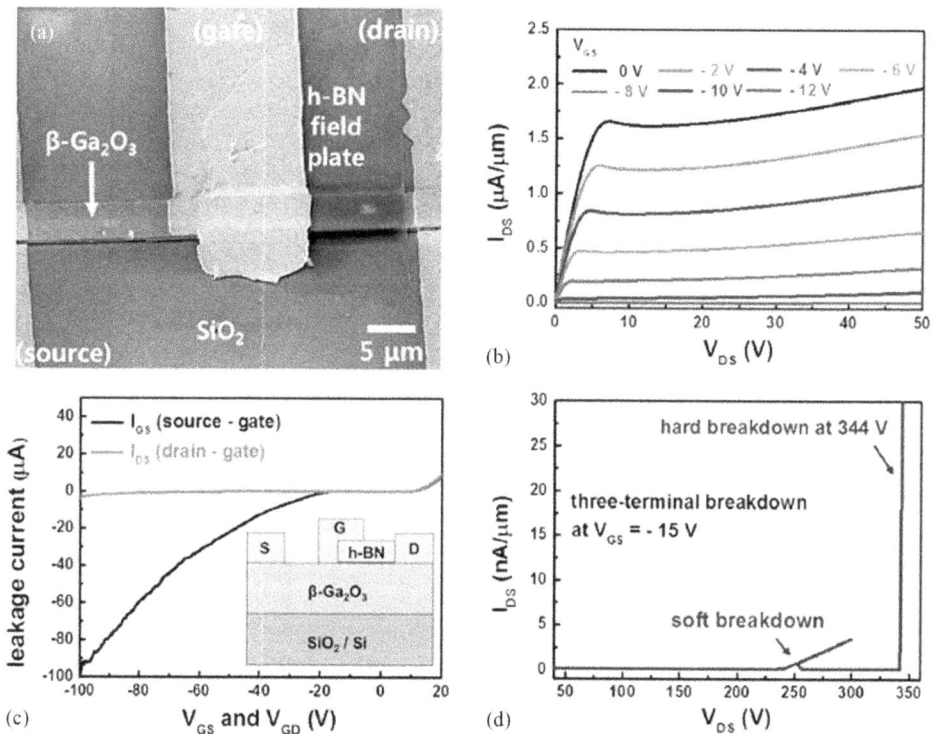

Figure 2.40. (a) SEM image of the β-Ga$_2$O$_3$ (200–400 nm thick) MESFET with a BN field plate ($\sim$70 nm thick). (b) DC output characteristics, (c) two-terminal gate leakage current (I_g) in the source-gate (V_{GS}) and drain–gate (V_{GD}) regions, and (d) three-terminal breakdown curve of the devices. Reprinted from [57] with permission from AIP Publishing.

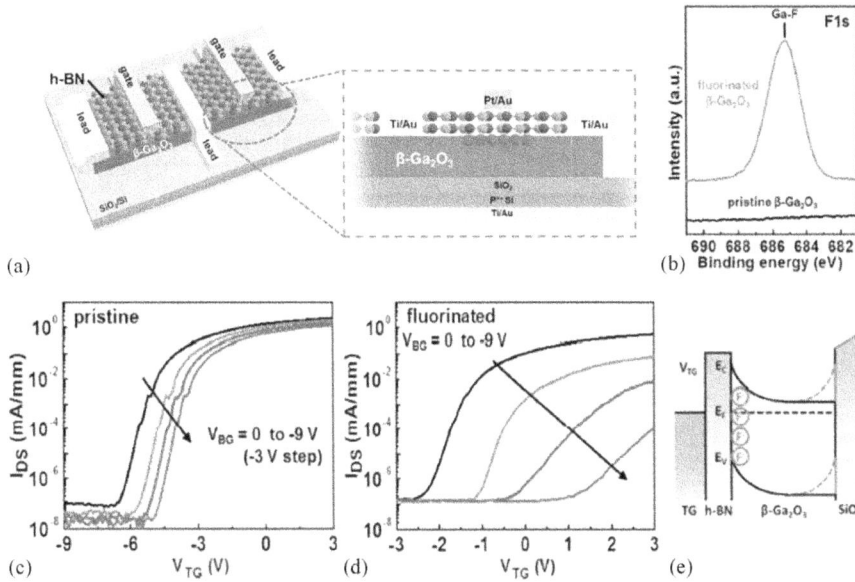

Figure 2.41. (a) Schematic of the series of pristine and fluorinated β-Ga$_2$O$_3$ MISFETs with front and back-gate electrodes. (b) XPS spectrum of F1s for pristine and fluorinate β-Ga$_2$O$_3$. Transfer curve of (c) pristine and (d) fluorinated β-Ga$_2$O$_3$ MISFETs. (e) The energy band diagram of the fluorinated channel under $V_{BG} < 0$ V and $V_{TG} = 0$ V. Reproduced from [58] with permission of The Royal Society of Chemisty.

2.10 Conclusion

Due to the breakthroughs in substrate growth and epitaxial techniques, β-Ga$_2$O$_3$ has triggered active research for various transistors. The experimental results have preliminarily shown the outstanding properties of this ultra-wide bandgap semiconductor, in particular in the breakdown electrical field strength. In the future, more efforts should be taken to solve the difficulty of p-type doping and thermal issues. In addition, 2DEG, realized by the δ-doped doping technique, is a promising project for high frequency applications.

References

[1] Higashiwaki M, Sasaki K, Kuramata A, Masui T and Yamakoshi S 2012 Gallium oxide (Ga$_2$O$_3$) metal–semiconductor field-effect transistors on single-crystal β-Ga$_2$O$_3$ (010) substrates *Appl. Phys. Lett.* **100** 013504

[2] Lin H, Dogmus E and Villamor A 2019 *Power SiC 2019: Materials, Devices, and Applications* ed J Palmour (Oxford: Elsevier)

[3] Kuramata A, Koshi K, Watanabe S, Yamaoka Y, Masui T and Yamakoshi S 2016 High-quality β-Ga$_2$O$_3$ single crystals grown by edge-defined film-fed growth *Jpn J. Appl. Phys.* **55** 1202A2

[4] Villora E G, Shimamura K, Kitamura K and Aoki K 2006 RF-plasma-assisted molecular-beam epitaxy of β-Ga$_2$O$_3$ *Appl. Phys. Lett.* **88** 031105

[5] Oshima T and Fujita S 2007 Ga$_2$O$_3$ thin film growth on *c*-plane sapphire substrates by molecular beam epitaxy for deep-ultraviolet photodetectors *Japan. J. Appl. Phys.* **46** 110091

[6] Sasaki K, Kuramata A, Masui T, Víllora E G, Shimamura K and Yamakoshi S 2012 Device-quality β-Ga$_2$O$_3$ epitaxial films fabricated by ozone molecular beam epitaxy *Appl. Phys. Express* **5** 035502

[7] Okumura H, Kita M, Sasaki K, Kuramata A, Higashiwaki M and Speck J S 2014 Systematic investigation of the growth rate of β-Ga$_2$O$_3$ (010) by plasma-assisted molecular beam epitaxy *Appl. Phys. Express* **7** 095501

[8] Sasaki K, Higashiwaki M, Kuramata A, Masui T and Yamakoshi S 2014 Growth temperature dependences of structural and electrical properties of Ga$_2$O$_3$ epitaxial films grown on β-Ga$_2$O$_3$ (010) substrates by molecular beam epitaxy *J. Cryst. Growth* **392** 30–3

[9] Huang C-Y, Horng R-H, Wuu D-S, Tu L-W and Kao H-S 2013 Thermal annealing effect on material characterizations of β-Ga$_2$O$_3$ epilayer grown by metal organic chemical vapor deposition *Appl. Phys. Lett.* **102** 011119

[10] Gogova D, Wagner G, Baldini M, Schmidbauer M, Irmscher K, Schewski R, Galazka Z, Albrecht M and Fornari R 2014 Structural properties of Si-doped β-Ga$_2$O$_3$ layers grown by MOVPE *J. Cryst. Growth* **401** 665–9

[11] Wagner G, Baldini M, Gogova D, Schmidbauer M, Schewski R, Albrecht M, Galazka Z, Klimm D and Fornari R 2014 Homoepitaxial growth of β-Ga$_2$O$_3$ layers by metal–organic vapor phase epitaxy *Phys. Status Solidi* A **211** 27–33

[12] Nomura K, Goto K, Togashi R, Murakami H, Kumagai Y, Kuramata A, Yamakoshi S and Koukitu A 2014 Thermodynamic study of β-Ga$_2$O$_3$ growth by halide vapor phase epitaxy *J. Cryst. Growth* **405** 19–22

[13] Murakami H *et al* 2015 Homoepitaxial growth of β-Ga$_2$O$_3$ layers by halide vapor phase epitaxy *Appl. Phys. Express* **8** 015503

[14] Oshima Y, Víllora E G and Shimamura K 2015 Quasi-heteroepitaxial growth of β-Ga$_2$O$_3$ on off-angled sapphire (0 0 0 1) substrates by halide vapor phase epitaxy *J. Cryst. Growth* **410** 53–8

[15] Lin H and Dogmus E 2019 *Emerging Semiconductor Substrate: Market and Technology Trends* Yole Développement (France) http://www.yole.fr/EmergingSemiconductorSubstrates_Overview.aspx#.Xz_NrchKiUk

[16] Reese S B, Remo T, Green J and Zakutayev A 2019 How much will gallium oxide power electronics cost? *Joule* **3** 1–5

[17] Higashiwaki M, Sasaki K, Kamimura T, Hoi Wong M, Krishnamurthy D, Kuramata A, Masui T and Yamakoshi S 2013 Depletion-mode Ga$_2$O$_3$ metal–oxide–semiconductor field effect transistors on β-Ga$_2$O$_3$ (010) substrates and temperature dependence of their device characteristics *Appl. Phys. Lett.* **103** 123511

[18] Chabak K D *et al* 2016 Enhancement-mode Ga$_2$O$_3$ wrap-gate fin field-effect transistors on native (100) β-Ga$_2$O$_3$ substrate with high breakdown voltage *Appl. Phys. Lett.* **109** 213501

[19] Wong Y N M H, Kuramata A, Yamakoshi S and Higashiwaki M 2017 Enhancement-mode Ga$_2$O$_3$ MOSFETs with Si-ion-implanted source and drain *Appl. Phys. Express* **10** 041101

[20] Sasaki K, Thieu Q T, Wakimoto D, Koishikawa Y, Kuramata A and Yamakoshi S 2017 Depletion-mode vertical Ga$_2$O$_3$ trench MOSFETs fabricated using Ga$_2$O$_3$ homoepitaxial films grown by halide vapor phase epitaxy *Appl. Phys. Express* **10** 124201

[21] Chabak K D *et al* 2018 Recessed-gate enhancement-mode β-Ga$_2$O$_3$ MOSFETs *IEEE Electron Device Lett.* **39** 67–70

[22] Dong H *et al* 2019 Fast switching β-Ga$_2$O$_3$ power MOSFET with a trench-gate structure *IEEE Electron Device Lett.* **40** 1385–8

[23] Kamimura T, Nakata Y, Wong M H and Higashiwaki M 2018 Normally-off Ga_2O_3 MOSFETs with unintentionally nitrogen-doped channel layer grown by plasma-assisted molecular beam epitaxy *IEEE Electron Device Lett.* **39** 814

[24] Hu Z, Nomoto K, Li W, Tanen N, Sasaki K, Kuramata A, Nakamura T, Jena D and Xing H G 2018 Enhancement-mode Ga_2O_3 vertical transistors with breakdown voltage >1 kV *IEEE Electron Device Lett.* **39** 869–72

[25] Hu K N Z, Li W, Jinno R, Nakamura T, Jena D and Xing H G 2019 1.6 kV vertical Ga_2O_3 FinFETs with source-connected field plates and normally-off operation *31st Int. Symp. on Power Semiconductor Devices and ICs (ISPSD)*

[26] Xia Z *et al* 2019 β-Ga_2O_3 delta-doped field effect transistors with current gain cutoff frequency of 27 GHz *IEEE Electron Device Lett.* **40** 1052–5

[27] Krishnamoorthy S, Xia Z, Bajaj S, Brenner M and Rajan S 2017 Delta-doped β-gallium oxide field-effect transistor *Appl. Phys. Express* **10** 051102

[28] Joishi C, Xia Z, McGlone J, Zhang Y, Arehart A R, Ringel S, Lodha S and Rajan S 2018 Effect of buffer iron doping on delta-doped β-Ga_2O_3 metal semiconductor field effect transistors *Appl. Phys. Lett.* **113** 123501

[29] McGlone J F, Xia Z, Zhang Y, Joishi C, Lodha S, Rajan S, Ringel S A and Arehart A R 2018 Trapping effects in Si-doped-Ga_2O_3 MESFETs on an Fe-doped-Ga_2O_3 Substrate *IEEE Electron Device Lett.* **39** 1042–5

[30] Xia Z, Joishi C, Krishnamoorthy S, Bajaj S, Zhang Y, Brenner M, Lodha S and Rajan S 2018 Delta doped beta-Ga_2O_3 field effect transistors with regrown ohmic contacts *IEEE Electron Device Lett.* **39** 568–71

[31] Zhang Y *et al* 2018 Demonstration of high mobility and quantum transport in modulation-doped β-$(Al_xGa_{1-x})_2O_3$/Ga_2O_3 heterostructures *Appl. Phys. Lett.* **112** 173502

[32] Ahmadi E, Koksaldi O S, Zheng X, Mates T, Oshima Y, Mishra U K and Speck J S 2017 Demonstration of β-$(Al_xGa_{1-x})_2O_3$/Ga_2O_3 modulation doped field-effect transistors with Ge as dopant grown via plasma-assisted molecular beam epitaxy *Appl. Phys. Express* **10** 071101

[33] Krishnamoorthy S *et al* 2017 Modulation-doped β-$(Al_{0.2}Ga_{0.8})_2O_3$/Ga_2O_3 field-effect transistor *Appl. Phys. Lett.* **111** 171501

[34] Zhang Y, Joishi C, Xia Z, Brenner M, Lodha S and Rajan S 2018 Demonstration of β-$(Al_xGa_{1-x})_2O_3$/Ga_2O_3 double heterostructure field effect transistors *Appl. Phys. Lett.* **112** 233503

[35] Joishi C, Zhang Y, Xia Z, Sun W, Arehart A R, Ringel S, Lodha S and Rajan S 2019 Breakdown characteristics of β-$(Al_{0.22}Ga_{0.78})_2O_3$/$Ga_2O_3$ field-plated modulation doped field effect transistors with SiN_x passivation *IEEE Electron Device Lett.* **40** 1241–5

[36] Zhang Y, Xia Z, McGlone J, Sun W, Joishi C, Arehart A R, Ringel S A and Rajan S 2019 Evaluation of low-temperature saturation velocity in $(Al_xGa_{1-x})_2O_3$/Ga_2O_3 modulation-doped field-effect transistors *IEEE Trans. Electron Devices* **66** 1574–8

[37] Baliga B J 2010 *Fundamentals of Power Semiconductor Devices* (Berlin: Springer)

[38] Wong M H, Sasaki K, Kuramata A, Yamakoshi S and Higashiwaki M 2016 Field-plated Ga_2O_3 MOSFETs with a breakdown voltage of over 750 V *IEEE Electron Device Lett.* **37** 212–5

[39] Wong A T M H, Makino T, Ohshima T, Sasaki K, Kuramata A, Yamakoshi S and Higashiwaki M 2017 Radiation hardness of Ga_2O_3 MOSFETs against gamma-ray irradiation *75th Annual Device Research Conf. (DRC), South Bend, IN, 2017* pp 1–2

[40] Green A J *et al* 2016 3.8-MV cm^{-1} breakdown strength of MOVPE-grown Sn-doped β-Ga_2O_3 MOSFETs *IEEE Electron Device Lett.* **37** 902–5

[41] Green A J et al 2017 β-Ga$_2$O$_3$ MOSFETs for radio frequency operation IEEE Electron Device Lett. **38** 790–3

[42] Zeng K, Wallace J S, Heimburger C, Sasaki K, Kuramata A, Masui T, Gardella J A and Singisetti U 2017 Ga$_2$O$_3$ MOSFETs using spin-on-glass source/drain doping technology IEEE Electron Device Lett. **38** 513–6

[43] Zeng K, Vaidya A and Singisetti U 2018 1.85 kV breakdown voltage in lateral field-plated Ga$_2$O$_3$ MOSFETs IEEE Electron Device Lett. **39** 1385–8

[44] Zeng K, Vaidya A and Singisetti U 2019 A field-plated Ga$_2$O$_3$ MOSFET with near 2-kV breakdown voltage and 520 m$\Omega \cdot$ cm^2 on-resistance Appl. Phys. Express **12** 081003

[45] Lv Y et al 2019 Source-field-plated β-Ga$_2$O$_3$ MOSFET with record power figure of merit of 50.4 MW cm^{-2} IEEE Electron Device Lett. **40** 83–6

[46] Lv Y, Zhou X, Long S, Liang S, Song X, Zhou X, Dong H, Wang Y, Feng Z and Cai S 2019 Lateral source field-plated β-Ga$_2$O$_3$ MOSFET with recorded breakdown voltage of 2360 V and low specific on-resistance of 560 mΩ cm^2 Semicond. Sci. Technol. **34** 11LT02

[47] Jae Kyoung Mun Z K C, Chang W, Chang H-W, Jung and Do J 2019 2.32 kV breakdown voltage lateral β-Ga$_2$O$_3$ MOSFETs with source-connected field plate ECS J. Solid State Sci. Technol. **8** Q3079

[48] Wong M H, Goto K, Morikawa Y, Kuramata A, Yamakoshi S, Murakami H, Kumagai Y and Higashiwaki M 2018 All-ion-implanted planar-gate current aperture vertical Ga$_2$O$_3$ MOSFETs with Mg-doped blocking layer Appl. Phys. Express **11** 064102

[49] Wong M H, Lin C-H, Kuramata A, Yamakoshi S, Murakami H, Kumagai Y and Higashiwaki M 2018 Acceptor doping of β-Ga$_2$O$_3$ by Mg and N ion implantations Appl. Phys. Lett. **113** 102103

[50] Wong M H, Goto K, Murakami H, Kumagai Y and Higashiwaki M 2019 Current aperture vertical beta-Ga$_2$O$_3$ MOSFETs fabricated by N- and Si-ion implantation doping IEEE Electron Device Lett. **40** 431–4

[51] Song B, Verma A K, Nomoto K, Zhu M, Jena D and Xing H G 2016 Vertical Ga$_2$O$_3$ Schottky barrier diodes on single-crystal β-Ga$_2$O$_3$ (−201) substrates 74th Annual Device Research Conf. (DRC), Newark, DE pp 1–2

[52] Hwang W S et al 2014 High-voltage field effect transistors with wide-bandgap β-Ga$_2$O$_3$ nanomembranes Appl. Phys. Lett. **104** 203111

[53] Ahn S, Ren F, Kim J, Oh S, Kim J, Mastro M A and Pearton S J 2016 Effect of front and back gates on β-Ga$_2$O$_3$ nano-belt field-effect transistors Appl. Phys. Lett. **109** 062102

[54] Zhou H, Si M, Alghamdi S, Qiu G, Yang L and Ye P 2017 High performance depletion/enhancement-mode β-Ga$_2$O$_3$ on insulator (GOOI) field-effect transistors with record drain currents of 600/450 mA mm^{-1} IEEE Electron Device Lett. **38** 201

[55] Ma J, Lee O and Yoo G 2018 Abnormal bias-temperature stress and thermal instability of Ga$_2$O$_3$ nanomembrane field-effect transistor IEEE J. Electron Devices Soc. **6** 1124–8

[56] Noh M S J, Zhou H, Tadjer M J and Ye P D 2018 The impact of substrates on the performance of top-gate p-Ga$_2$O$_3$ field-effect transistors—record high drain current of 980 mA-mm on diamond 76th Device Research Conf. (DRC), IEEE

[57] Bae J, Kim H W, Kang I H, Yang G and Kim J 2018 High breakdown voltage quasi-two-dimensional β-Ga$_2$O$_3$ field-effect transistors with a boron nitride field plate Appl. Phys. Lett. **112** 122102

[58] Kim J, Tadjer M J, Mastro M A and Kim J 2019 Controlling the threshold voltage of β-Ga$_2$O$_3$ field-effect transistors via remote fluorine plasma treatment J. Mater. Chem. C **7** 8855–60

Chapter 3

Beta gallium oxide (β-Ga$_2$O$_3$) nanomechanical transducers: fundamentals, devices, and applications

Xu-Qian Zheng and Philip X-L Feng

Beta gallium oxide (β-Ga$_2$O$_3$) is an emerging ultra-wide bandgap (UWBG, ~4.8 eV) semiconductor with attractive properties for future power electronics, optoelectronics, and sensors for detecting gases and ultraviolet radiation. In addition to such potential, the β-Ga$_2$O$_3$ crystal possesses excellent mechanical properties, making it pertinent as a material for resonant nanoelectromechanical systems (NEMS). These devices can act as a significant addition to the rapidly emerging β-Ga$_2$O$_3$ electronics. Single-crystal β-Ga$_2$O$_3$ nanomechanical resonators have been demonstrated experimentally with the material's Young's modulus characterized as $E_Y = 261$ GPa. Oscillators based on such β-Ga$_2$O$_3$ nanomechanical resonators have been applied to solar-blind UV (SBUV) sensing applications. Further, by integrating a conventional photocurrent-based detection scheme into the β-Ga$_2$O$_3$ resonator, dual-modality SBUV sensing can be achieved in the same transducer.

3.1 Introduction

Wide bandgap (WBG) semiconductors, such as silicon carbide (SiC) and gallium nitride (GaN), are widely considered as emerging candidates with great promise for developing semiconductor power electronics that offer better miniaturization and scaling perspectives, as well as devices and integrated systems for applications in harsh environments. Such expectations are derived largely from the WBG nature and superior electrical properties they offer over those provided by conventional silicon (Si) devices [1]. In addition to their outstanding electrical properties, WBG semiconductors also offer advantages for enabling functional components in micro/nanoelectromechanical systems (MEMS/NEMS), including higher Young's modulus engineerable electromechanical and optomechanical coupling effects, and a wide range of insulating, semiconducting, and conducting characteristics based on doping [2].

Beta gallium oxide (β-Ga$_2$O$_3$), an emerging ultra-wide bandgap (UWBG) semiconductor, has recently attracted increasing interest for future generations of power electronics [3–5] and ultraviolet (UV) optoelectronics [6, 7] due to its UWBG ($E_{g,\beta\text{-Ga}_2\text{O}_3} \approx 4.5$–4.9 eV) that is significantly wider than those of GaN and SiC [8, 9]. It offers a very high critical field strength ($\varepsilon_{br,\beta\text{-Ga}_2\text{O}_3} = 8$ MV cm^{-1} predicted and $\varepsilon_{br,\beta\text{-Ga}_2\text{O}_3} = 5.7$ MV cm^{-1} measured) and electron mobility up to $\mu_n = 300$ cm^2 V^{-1} s^{-1} at room temperature [3, 4, 10, 11]. These contribute to an excellent Baliga's figure of merit (BFOM), making β-Ga$_2$O$_3$ a promising contender for future-generation power devices [4, 5].

Beyond its electrical properties, β-Ga$_2$O$_3$ also exhibits promising mechanical strength (Young's modulus $E_{Y,\beta\text{-Ga}_2\text{O}_3} = 261$ GPa) along with a mass density of $\rho = 5950$ kg m^{-3}, which contributes to a promising acoustic velocity of $c = 6623$ m s^{-1}, comparable to those of Si (8415 m s^{-1}) and GaN (8044 m s^{-1}) [12, 13]. Importantly, the material also possesses extraordinary chemical and thermal (melting point at 1820 °C) stability [14, 15]. The excellent ensemble of attributes in β-Ga$_2$O$_3$ makes it suitable for new UWBG M/NEMS beyond SiC and GaN M/NEMS, as future electromechanically coupled and tunable β-Ga$_2$O$_3$ electronic, optoelectronic, and physical sensing devices. Further, β-Ga$_2$O$_3$ is sensitive to solar-blind UV (SBUV) light [6, 7] and offers a reversible response to oxidation and reduction gases [16], enabling UV light and gas sensing applications, toward future integration of critical mechanical functions into electronics. Table 3.1 summarizes the important electrical and mechanical properties that are beneficial for making promising β-Ga$_2$O$_3$ M/NEMS devices.

Importantly, the low thermal expansion coefficient, moderate Young's modulus, and high specific heat capacity of the β-Ga$_2$O$_3$ crystal contributes to an extremely high thermoelastic damping (TED) limited quality factor (Q) for β-Ga$_2$O$_3$ resonant mechanical devices. One theoretical study [17] shows that the TED-limited Q of β-Ga$_2$O$_3$ can be on the order of 10^8, and the Akhiezer limit of frequency–quality

Table 3.1. The intrinsic properties of different materials for MEMS/NEMS applications [4, 13, 17].

Material	Si	GaAs	4H-SiC	GaN	β-Ga$_2$O$_3$
Bandgap, E_g (eV)	1.1	1.4	3.3	3.4	**4.8**
Dielectric constant, ε	11.8	12.9	9.7	9	**10**
Breakdown field, E_{br} (MV cm^{-1})	0.3	0.4	2.5	3.3	**8**
Electron mobility, μ (cm^2 (V s)$^{-1}$)	1480	8400	1000	1250	**300**
Thermal conductivity, κ (W (cm K)$^{-1}$)	1.5	0.5	4.9	2.3	**0.1–0.3**
BFOM, $\varepsilon\mu E_{br}^3$	1	15	317	846	**3214**
Young's modulus, E_Y (GPa)	165	118	605	398	**261**
Acoustic velocity, c (m s^{-1})	8415	2470	13 100	8044	**6623**
Akhiezer limit of f–Q product, $f \times Q$ ($\times 10^{13}$ Hz)	2.5	NA	25	35	**330**

factor product ($f \times Q$) of resonators based on β-Ga$_2$O$_3$ is extremely high at 3.3×10^{15} Hz, higher than that of Si, SiC, and GaN (table 3.1). In other words, β-Ga$_2$O$_3$ may enable among the highest TED-limited Q for M/NEMS resonators using conductive materials, including metals and semiconductors. The other dissipation pathways, including anchor loss and surface dissipation, can be mitigated through a special geometrical design and an optimized fabrication process, respectively. Thus, the Q of β-Ga$_2$O$_3$ M/NEMS resonators can potentially be higher than that of resonators made of Si, SiC, and GaN, making them promising for both RF signal processing and physical sensing applications.

Additionally, in contrast to other demonstrated WBG M/NEMS materials, including SiC and GaN, bulk β-Ga$_2$O$_3$ crystals can be produced using homoepitaxy (thus preventing threading dislocations) and other more cost-effective melting growth methods, such as the Czochralski (CZ) [18, 19], floating zone (FZ) [20–22], and edge-defined film-fed growth (EFG) [23] techniques. The variety of synthesis methods has already led to the realization of various β-Ga$_2$O$_3$ nanostructures, including nanowires [24], nanobelts [25], nanorods [26], and nanosheets [27]. Using high-density plasma etch techniques with BCl$_3$ as the etchant, β-Ga$_2$O$_3$ crystal can be effectively attacked and removed in mask-defined areas [28]. Thus equipped with various effective synthesis and etching methods, there is great potential to realize β-Ga$_2$O$_3$ devices using micromachining techniques.

In this chapter, we focus on reviewing recent advances in the experimental demonstration of β-Ga$_2$O$_3$ resonant NEMS devices, from basic nanomechanical resonators to functional transducers. We first review single-crystal β-Ga$_2$O$_3$ nanomechanical resonators made using β-Ga$_2$O$_3$ nanoflakes grown via low-pressure chemical vapor deposition (LPCVD). From the measurements, multimode resonances and spatial visualization of the multimode motion are resolved to extract the mechanical properties, i.e. the material's Young's modulus, $E_Y = 261$ GPa, and a device's built-in stress. Then, the device platform is further applied to SBUV sensing applications. Both resonators and oscillators using suspended β-Ga$_2$O$_3$ nanostructures are constructed and exposed to SBUV irradiation for the investigation of SBUV detection based on resonance frequency shift. Further, a conventional photocurrent-based detection scheme is integrated into the β-Ga$_2$O$_3$ resonator, achieving dual-modality SBUV sensing in the same transducer. Finally, we propose the all-electrical transduction (including both actuation and detection) of the mechanical motion of a β-Ga$_2$O$_3$ vibrating channel transistor (VCT).

3.2 β-Ga$_2$O$_3$ circular drumhead resonators

The first investigations of β-Ga$_2$O$_3$ resonant NEMS utilized β-Ga$_2$O$_3$ nanomechanical resonators in a circular drumhead structure and investigated the mechanical properties of such devices [12, 29]. Figure 3.1 illustrates the first demonstration of β-Ga$_2$O$_3$ resonant NEMS in the form of circular drumhead nanomechanical resonators. β-Ga$_2$O$_3$ is the most stable polymorph of Ga$_2$O$_3$ which has a monoclinic crystalline structure (figure 3.1(a)) [30]. The β-Ga$_2$O$_3$ drumhead resonators are constructed by single-crystal β-Ga$_2$O$_3$ nanoflakes grown using the LPCVD method

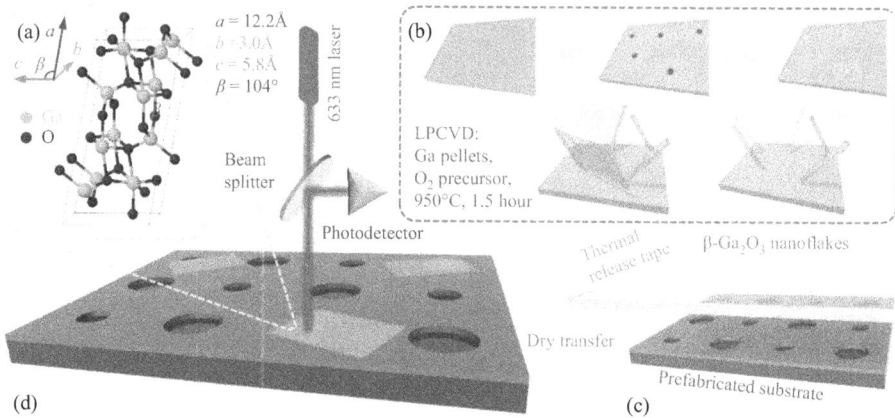

Figure 3.1. Schematic of experimental demonstration of single-crystal β-Ga_2O_3 nanomechanical resonators. (a) β-Ga_2O_3 crystal structure. (b) Formation of β-Ga_2O_3 nanoflakes in low-pressure chemical vapor deposition (LPCVD). (c) Illustration of all-dry transfer of β-Ga_2O_3 nanoflakes by using thermal release tape as a stamp to create suspended β-Ga_2O_3 devices on pre-defined microtrenches and arrays. (d) Illustration of fabricated β-Ga_2O_3 resonators under study by using the scanning laser interferometry motion detection and spatial mapping system. Adapted with permission from [12]. Copyright 2017 American Chemical Society.

(figure 3.1(b)) [31]. The as-grown nanoflakes have widths of ~2–30 μm and thicknesses of ~20–140 nm. Using the synthesized β-Ga_2O_3 nanoflakes, suspended β-Ga_2O_3 nanostructures are fabricated by employing an all-dry transfer technique (figure 3.1(c)). The fabricated β-Ga_2O_3 circular drumhead resonators have thicknesses of ~20 to ~80 nm and diameters of ~3.2 and ~5.2 μm.

The fabricated free-standing β-Ga_2O_3 nanomechanical resonators are characterized by using an ultrasensitive laser interferometry system (figure 3.1(d)). Figure 3.2 shows the measured multimode resonances of a suspended β-Ga_2O_3 diaphragm with a thickness of 23 nm and a diameter of 5.7 μm. A total of six resonance peaks are found in the frequency range of ~9–36 MHz with Q of ~220–280.

After resonance measurements, a thermal annealing process (250 °C in N_2 environment at ~500 mTorr for 1.5 h) is performed on the β-Ga_2O_3 nanomechanical resonators to remove residues and adsorbates on the device surface. After annealing, the resonance frequencies exhibit large upshifts of the resonance frequencies. The annealing effects on the resonance characteristics can be attributed to the elimination of added mass and surface loss pathways arising from the thermal release tape residues and gas adsorbates (e.g. O_2 gas molecules) on the device surface. Later, after storing the device for over 1 month in a moderate vacuum (~70 Torr) at room temperature, the multimode resonances are measured again, showing considerable frequency downshifts from ~3% to 15%. This observation indicates that β-Ga_2O_3 can adsorb gas molecules efficiently even in a moderate vacuum. A second annealing is conducted under the same conditions. The resonance frequencies are slightly higher than those after the first annealing, indicating refreshing of the device from the surface adsorbates and further removal of the remaining tape residues. Figure 3.2 shows the resonance performance of a β-Ga_2O_3 nanomechanical

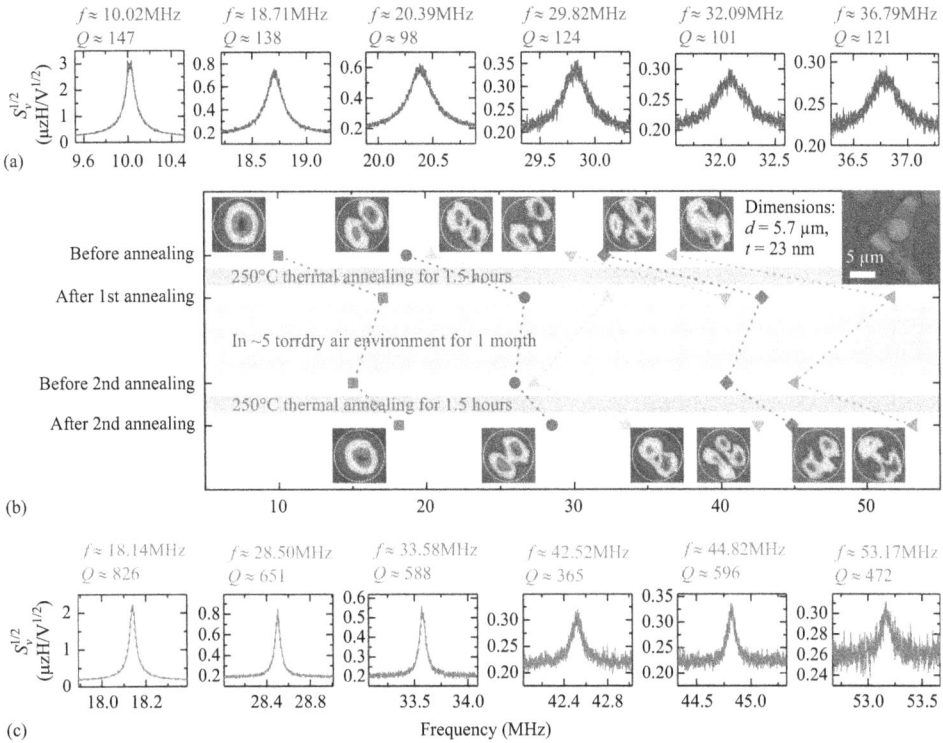

Figure 3.2. Resonance frequencies and effects of annealing on a β-Ga$_2$O$_3$ nanomechanical resonator. (a) Undriven thermomechanical mode resonance spectra of the first six modes before annealing with corresponding frequencies and Q. (b) Resonance frequencies of the device before annealing, after the first annealing, after ~1 month storage in a ~70 Torr desiccator environment, and after the second thermal annealing. (c) The undriven thermomechanical mode's resonance spectra of the first six modes immediately after the second annealing with corresponding resonance frequencies and Q. Optical images of the device, and spatial mapping and visualization of the resonance mode shapes are shown in the insets of (b). Adapted with permission from [29]. Copyright 2017 Springer Nature.

resonator at different stages of the annealing process. The post-annealing measured upshifts in f and enhanced Q can be qualitatively explained by the annealing enabled cleaning of possible adsorbates and residues, and the alleviation or elimination of their associated dissipation and energy loss processes.

To quantitatively understand the measured resonances, analysis of multimode resonances and frequency scaling for these circular β-Ga$_2$O$_3$ diaphragms is performed. In any given circular drumhead device, both flexural rigidity (dominated by thickness and elastic modulus) and built-in tension (stress) can be important, thus we have [32]

$$\omega_m = (k_m a) \sqrt{\frac{D}{\rho \cdot b \cdot a^4} \left[(k_m a)^2 + \frac{\gamma \cdot a^2}{D} \right]}, \qquad (3.1)$$

where m denotes the mode number, $\omega_m = 2\pi f_m$ is the mth mode angular resonance frequency, $(k_m a)^2$ is the eigenvalue, a is the radius of circular resonator, D is the flexural rigidity $D = E_Y b^3 / [12(1 - \nu^2)]$ with b being the thickness of the device, ν is Poisson's ratio, ρ is the volume mass density of β-Ga$_2$O$_3$, and γ is the surface pretension evenly distributed in the plane.

Equation (3.1) yields a 'mixed elasticity' model that captures both the 'disk' and 'membrane' limits of frequency scaling for such devices, as well as the transition between these two regimes. In the 'disk' regime, the resonance frequencies are positively dependent on resonator thickness, $\omega_m \propto t$. In the 'membrane' regime, the resonance frequencies scale with resonator thickness as $\omega_m \propto t^{-1/2}$. For circular drumhead resonators in the 'disk' regime, the built-in tension has a negligible effect on determining the resonance frequency. Therefore, if the device is in the 'disk' regime, the Young's modulus can be revealed from the measured fundamental-mode resonance frequency ω_0 by

$$E_Y = \frac{12a^4\rho(1 - \nu^2)}{[(k_0 a)^2]^2 b^2}\omega_0^2, \qquad (3.2)$$

with $(k_0 a)^2 = 10.215$ for the fundamental mode [32]. Using equation (3.2) and the measured fundamental-mode resonances of some other devices in this study, an average Young's modulus, $E_Y \approx 261$ GPa, is extracted for the β-Ga$_2$O$_3$ nanoflakes.

By applying the measured Young's modulus to the resonator model, figure 3.3 shows the frequency scaling of β-Ga$_2$O$_3$ nanomechanical resonators from the 'membrane' regime to the 'disk' regime. The scaling can serve as a guideline for the future design of β-Ga$_2$O$_3$ NEMS resonators with desired size and operating frequencies.

To further investigate the mechanical properties of the β-Ga$_2$O$_3$ resonators, this study [12] also conducted spatial mapping of the multimode resonances of the β-Ga$_2$O$_3$ resonators. As shown in figure 3.4, the multimode device behaves in the transition regime, where both built-in tension and flexural rigidity are important in determining its resonance frequencies. In FEM simulations, a uniform stress of $\sigma = 70$ MPa (1.61 N m^{-1} surface tension) yields a result that matches the fundamental-mode resonance frequency. By sweeping the anisotropic, biaxial built-in tension in modeling, one can find that when the biaxial built-in tensions are 105 and 35 MPa (corresponding to 2.42 and 0.81 N m^{-1} of surface tension, respectively), the simulated resonances are in good agreement with the measured multimode responses for both the frequencies and the mode shapes.

3.3 Resonant solar-blind ultraviolet (SBUV) transducers

Based on the establishment of β-Ga$_2$O$_3$ nanomechanical resonators, further investigations can be carried out on utilizing these resonant NEMS devices as transducers. A resonator is an outstanding platform for ultrasensitive detection owing to its high responsiveness and fast responses in resonance frequency shifts to external

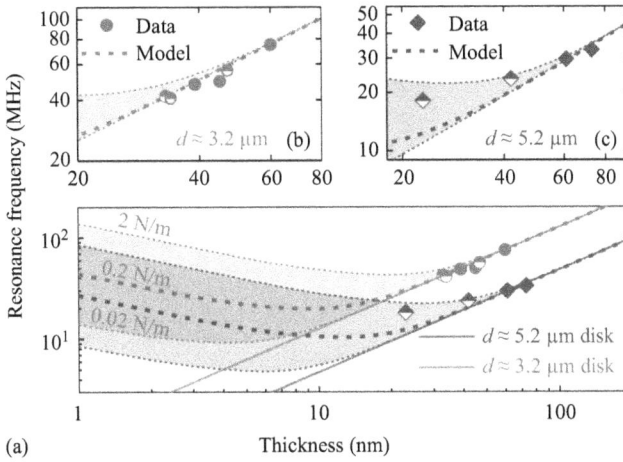

Figure 3.3. Modeling and elucidating the mixed elastic behavior and transition from the 'disk' regime to the 'membrane' regime in β-Ga$_2$O$_3$ diaphragm nanoresonators. The green circles and blue diamonds represent devices with diameters of $d \approx 3.2$ μm and $d \approx 5.2$ μm, respectively. The scattered symbols represent experimental data from optical interferometry measurements. The curved lines show the calculated resonance frequency versus device thickness, each with 0.02, 0.2, and 2 N m^{-1} surface tensions. (a) Frequency scaling of β-Ga$_2$O$_3$ diaphragm nanoresonators with thicknesses in the 1–100 nm range. The dashed lines indicate the relation for ideal 'disk' resonators. (b) and (c) Zoomed-in scaling of the measured devices' thickness ranges for resonators with diameters of $d \approx 3.2$ μm and $d \approx 5.2$ μm, respectively. Reproduced with permission from [12]. Copyright 2017 American Chemical Society.

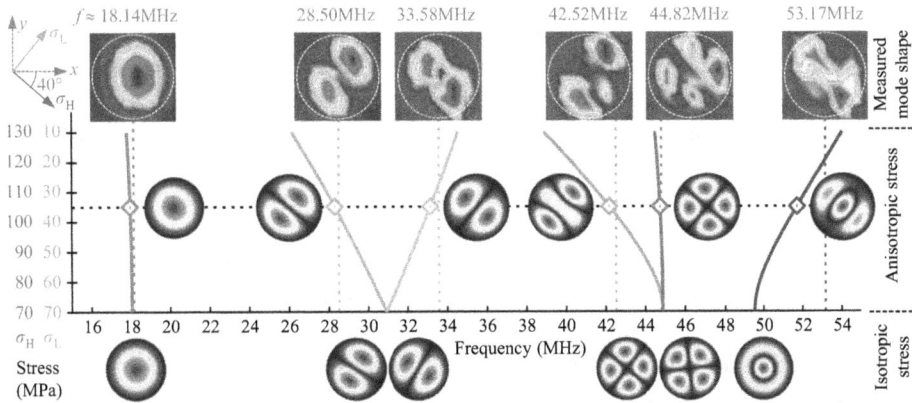

Figure 3.4. Determination of anisotropic tension in the circular β-Ga$_2$O$_3$ diaphragm nanoresonator by matching the simulated mode shapes and frequencies with the spatially resolved and visualized multimode resonances. Reproduced with permission from [12]. Copyright 2017 American Chemical Society.

perturbations. In addition, the wide bandgap ($E_g \approx 4.5$–4.9 eV) of the β-Ga$_2$O$_3$ crystal makes it an excellent candidate for SBUV sensing with a cut-off wavelength below ~250–260 nm. Naturally, β-Ga$_2$O$_3$ nanomechanical resonators should exhibit promising UV sensing functionality, utilizing its optical and mechanical properties.

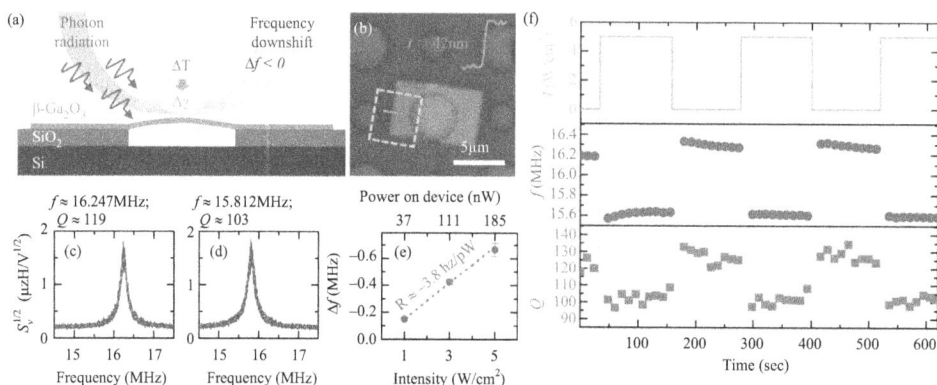

Figure 3.5. Middle ultraviolet (MUV) light detection using a ~5.2 μm diameter β-Ga$_2$O$_3$ nanomechanical drumhead resonator. (a) Illustration of the MUV light sensing mechanism. (b) Optical image of the device and AFM trace with measured thickness. The device's fundamental-mode thermomechanical resonance (c) without light radiation and (d) with 3 W cm^{-2} (111 nW on device) light radiation. (e) Frequency shifts under different source light intensities. (f) Fundamental thermomechanical resonance frequency and Q with a cycling illumination of 5 W cm^{-2} light intensity (185 nW of power on device). Adapted with permission from [33]. Copyright 2017 IEEE.

3.3.1 Resonator

A follow-up study [33] on β-Ga$_2$O$_3$ circular drumhead nanomechanical resonators investigated their potential as middle ultraviolet (MUV, 200–300 nm) light detectors. Figure 3.5(a) illustrates the light sensing mechanism of the β-Ga$_2$O$_3$ nanomechanical resonator. The photothermal effect induced by the incident MUV light elevates the temperature and expands the suspended β-Ga$_2$O$_3$ crystal, leading to a resonance frequency downshift. By probing the resonance frequency shift of the β-Ga$_2$O$_3$ device, the intensity of the incident MUV light can be resolved. The laser interferometry system is used to measure both the undriven thermomechanical resonance and the photothermally driven response of the β-Ga$_2$O$_3$ resonator in a moderate vacuum (~20 mTorr).

A continuous UV light source of 200–600 nm in wavelength applies photon radiation on β-Ga$_2$O$_3$ resonators down to the MUV regime. Figures 3.5(c) and (d) show the detection of MUV light (111 nW of power on device) for a β-Ga$_2$O$_3$ drumhead resonator ($d \approx 5.2$ μm) with a thickness of ~42 nm. By periodically switching the MUV light on and off (185 nW of power on device), the study measures the frequency responses of the resonator due to MUV light illumination (figure 3.5(f)). The results show a linear response between frequency shift and incident light power on the device, with an average responsivity of −3.8 Hz pW^{-1} upon radiation down to the MUV wavelength (figure 3.5(e)).

3.3.2 Oscillator

To improve the sensing capability, a closed-loop self-sustaining oscillator is constructed by adding a feedback circuit to the experimental system (figure 3.6(a)) [34].

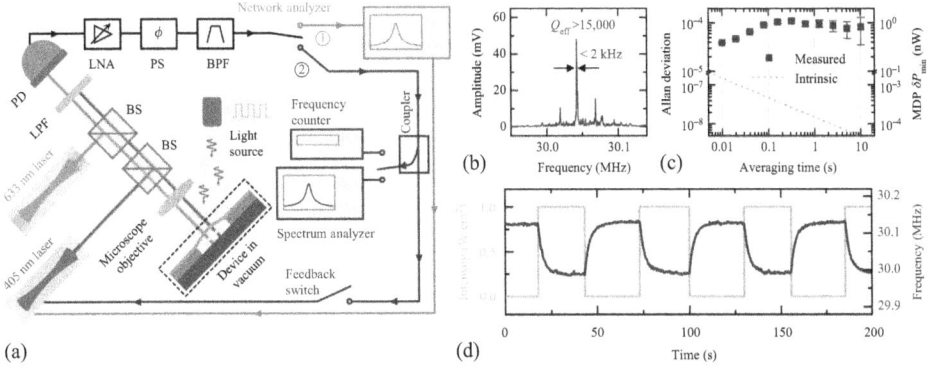

Figure 3.6. Performance of the β-Ga$_2$O$_3$ feedback oscillator. (a) Schematic of the measurement. (b) The closed-loop oscillation spectrum. (c) Measured and intrinsic Allan deviation. (d) Responses of the β-Ga$_2$O$_3$ oscillator to cyclic photon irradiation. Adapted with permission from [34]. Copyright 2018 IEEE.

A self-sustaining oscillator can generate an oscillating signal by extracting energy from a non-periodic source. The β-Ga$_2$O$_3$ resonator is acting as a frequency-determining element in the circuit. The construction of the closed-loop oscillator can significantly narrow the resonance spectral linewidth of the resonator, achieving a resonant based sensor with much higher precision. The oscillator also enables a much faster detection speed by eliminating the need for the much slower frequency sweeping type measurement.

The oscillator is first characterized in the frequency domain by a spectrum analyzer. With the oscillator, the Q of the ~30 MHz resonance is boosted from ~200 to an effective Q_{eff} >15 000 (figure 3.6(b)), which is a >70-fold enhancement. In addition, the oscillator frequency fluctuations are measured, and the Allan deviation is computed to evaluate the frequency stability of the oscillator (figure 3.6(c)), which yields an Allan deviation of $\sigma_A(\tau)$ ~3.9×10^{-5} for $\tau = 10$ ms.

When cyclic MUV light with an incident power level of ~49 nW is applied on the suspended device area, the oscillator exhibits real-time tracking of incident light in frequency. Figure 3.6(d) shows clear frequency downshifts upon cyclic illumination. Using $\mathfrak{R} = \Delta f/P_D$, a responsivity of $\mathfrak{R} = -3.1$ Hz pW^{-1} is obtained. Given the Allan deviation, the oscillator has a frequency fluctuation of $\delta f(\tau = 10$ ms$) = (2)^{1/2}\sigma_A f_0 \approx$ 1.66 kHz. Thus, the minimum detectable power (MDP) of the MUV sensing oscillator is $\delta P_{min} = \delta f/\mathfrak{R} \approx 1.4$ nW at $\tau = 1$ s, and $\delta P_{min} \approx 0.53$ nW at $\tau = 10$ ms, i.e. a better resolution in higher speed detection. This clearly demonstrates an intrinsic advantage of feedback oscillators for real-time, high speed sensing.

Further, to analyze the fundamental limit of the β-Ga$_2$O$_3$ resonator and oscillator for MUV detection, the thermomechanical noise limited Allan deviation is calculated, $\sigma_{A,th}(\tau) = (\pi k_B T/(P_c \tau Q^2))^{1/2}$, where k_B is the Boltzmann constant, T is the temperature, and P_c is the operating power of the resonator [35]. Therefore, for $\tau = 10$ ms, the Allan deviation limited by thermomechanical noise is $\sigma_A = 1.50 \times 10^{-7}$ which is more than two orders of magnitude lower than the measured Allan deviation. The discrepancy could be attributed to added noise from the electrical–optical feedback circuit. The noise in the circuit includes the noise of the amplifier, the noise of the laser, the noise of the photodetector (PD), and the noise of the other circuit components. By optimizing the feedback circuit design and utilizing low noise components, for example, replacing the optical–electrical feedback scheme with a pure electrical scheme, the noise performance of the oscillator can be improved significantly in the future, in particular in integrated versions.

In addition, one can also calculate the noise equivalent power (NEP) limited by the thermomechanical noise. Using the thermomechanical noise limited Allan deviation, the thermomechanical frequency noise spectral density is $S_\omega^{1/2}(\omega) = \omega_0\sqrt{\tau/\pi}\,\sigma_A$. Given $\tau = 10$ ms and $\omega_0 = 2\pi \times 30.13$ MHz, the calculated noise spectral density is $S_\omega^{1/2}(\omega_0) = 1.60$ Hz$^{-1/2}$. Thus, one can obtain the thermomechanical noise limited noise equivalent power (NEP) of the device NEP$_{th} = S_\omega^{1/2}(\omega)/2\pi|\Re| = 8.2 \times 10^{-14}$ W Hz$^{-1/2}$, which is comparable to the NEP (limited by shot noise) of the state-of-the-art β-Ga$_2$O$_3$ PDs. Table 3.2 summarizes the performance comparison between this study and a few other β-Ga$_2$O$_3$ photodetectors in the literature.

3.3.3 Dual-modality transducer

Another study demonstrates a single NEMS transducer where SBUV light can be detected in two sensing modalities, optoelectronically and photo-thermomechani-cally [38]. The demonstration facilitates the direct comparison between two sensing mechanisms and establishes a platform that can take advantage of both modalities. Figure 3.7(a) illustrates the device structure and optical and electrical circuitries. For the optoelectronic modality (modality I), the electrical conductivity of the β-Ga$_2$O$_3$ channel could be enhanced by photons in two different processes: modulation of channel conductance and lowering of the Schottky barrier height at the β-Ga$_2$O$_3$–metal contact. In the resonant NEMS modality (modality II), the SBUV photon radiation will photothermally heat up the suspended nanostructure. The elevated temperature will expand the doubly clamped beam, lower the stress in the device, and downshift the mechanical resonance frequency of the resonator. Therefore, by monitoring the resonance frequency of the NEMS transducer, the incident SBUV irradiation can be resolved.

Table 3.2. Performance comparison with β-Ga$_2$O$_3$ PDs.

Active area	Responsivity $|\Re|$	MDP[a] δP_{min}	NEP[b]	References
21.2 µm^2	3.1 Hz pW^{-1}	0.53 nW	8.2×10^{-14} W Hz$^{-1/2}$	This work
0.8 cm^2	39.3 A W^{-1}	28.0 nW	1.5×10^{-14} W Hz$^{-1/2}$	[7]
~7 mm^2	0.07 A W^{-1}	1–10 nW	~8×10^{-14} W Hz$^{-1/2}$	[36]
~0.8 mm^2	8.7 A W^{-1}	1–10 nW	~7×10^{-15} W Hz$^{-1/2}$	[37]

[a] MDPs for PDs are calculated here, using $\delta P_{min} = I_D/\Re$; I_D is the dark current.
[b] NEPs for PDs are calculated here, using NEP $= (2qI_D)^{1/2}/\Re$, where q is the electronic charge, $\Re$ is the responsivity, and $2qI_D$ is the shot noise of the PD.

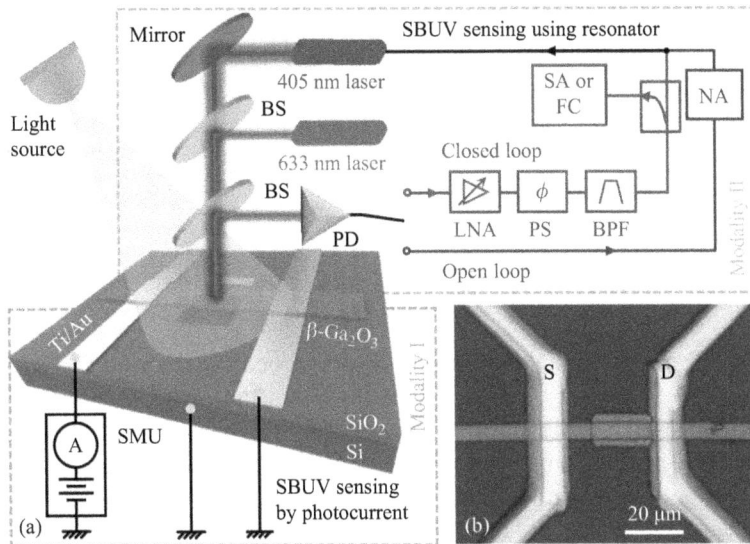

Figure 3.7. (a) Schematic illustration of a β-Ga$_2$O$_3$ transducer in the form of a suspended doubly clamped beam with electrical contacts for two modalities of SBUV detection, modality I and modality II. (b) Optical image of the β-Ga$_2$O$_3$ dual-modality transducer. Adapted with permission from [38]. Creative Commons Attribution 4.0 International License.

The β-Ga$_2$O$_3$ transducer with dual-modality SBUV sensing is fabricated using β-Ga$_2$O$_3$ flakes mechanically cleaved from bulk crystal synthesized using the edge-defined film-fed growth (EFG) method [39]. Thanks to the high-quality bulk β-Ga$_2$O$_3$ with a controllable doping level and thus controllable electrical properties, these flakes can be used to form high-performance β-Ga$_2$O$_3$ electronics [5, 40]. Although β-Ga$_2$O$_3$ is not what is conventionally regarded as a layered material, such as 2D materials where strong in-plane covalent bonds and much weaker interlayer van der Waals bonds co-exist, it has been reported that β-Ga$_2$O$_3$ has two strong cleavage planes {100} and {001}, with the {100} plane easier to cleave than the {001} plane [19, 40]. Density functional theory calculations also suggest that the {100} plane is easier to break with the {100} surface covalent bonds being up to 6 eV per bond stronger than the bonds perpendicular to the {100} plane [41]. Therefore, β-Ga$_2$O$_3$ bulk crystal could be mechanically cleaved even through an exfoliation method. In addition, mechanically cleaved β-Ga$_2$O$_3$ flakes show a well-defined crystal orientation. They are usually in belt shapes, where the large surfaces are coincident with the {100} plane and the long sides are parallel with the [010] axis of the crystal thanks to the two strong cleavage planes.

After identifying a β-Ga$_2$O$_3$ flake with desired size and thickness, the flake prepared on the polydimethylsiloxane (PDMS) stamp is transferred to a 290 nm SiO$_2$-on-Si substrate with pre-defined microtrenches using a dry transfer technique. The transferred flake is suspended over a 900 nm deep microtrench, forming a doubly clamped beam structure. Subsequently, deposition of metal electrodes (40 nm Au on 150 nm Ti) on both sides of the microtrench to electrically access the β-Ga$_2$O$_3$ device is performed through a high-precision stencil mask (shadow mask) without using photoresists or wet chemicals, and thus the pristine nature of the fabricated β-Ga$_2$O$_3$ device is ensured. The device has a 5 μm wide, 32 μm long, and 310 nm thick β-Ga$_2$O$_3$ channel defined by two metal electrodes with a 20 μm long portion of the channel suspended over the microtrench (figure 3.7(b)).

To analyze the optoelectronic sensing modality, the responsivity can be expressed using [42]

$$\Re_I = \frac{G\eta e}{(hc/\lambda)},\tag{3.3}$$

where G is the photoconductive gain, η is the optical absorbance of the β-Ga$_2$O$_3$ material, e is the elementary charge, h is Plank's constant, c is the speed of light, and λ is the wavelength of the photon. Therefore, for a single wavelength of light irradiation, the optoelectronic responsivity of the device only relates to G. The G can also be expressed by the ratio between carrier recombination lifetime (τ_{life}) and carrier transit time (τ_{trans}) in the device channel, $G = \tau_{\text{life}}/\tau_{\text{trans}}$. Thus, G can be enhanced by decreasing τ_{trans} or increasing τ_{life}. The transit time can be further determined by $\tau_{\text{trans}} = L^2/(\mu V)$, where L is the channel length, μ is the electron mobility, and V is the applied voltage across the channel. To reduce τ_{trans}, one can either reduce the channel length by making electrodes closer or optimize the Schottky barrier at the semiconductor–metal interface to modulate the contact resistance. We can also enhance the responsivity by manipulating the conduction channel in the β-Ga$_2$O$_3$ crystal to increase the carrier lifetime τ_{life}. By engineering the parameters of the device so that it is at the edge of opening a conductive channel, weak SBUV photon illumination can significantly modulate the channel conductance and induce a high current response. The methods to achieve this include (i) selecting a β-Ga$_2$O$_3$ nanoflake with an ideal thickness and doping level, and (ii) band modulation of the channel using gating.

To experimentally demonstrate the optoelectronic sensing modality, the photo-current response (modality I) of the β-Ga$_2$O$_3$ PD is measured using a source measure unit (SMU) as illustrated in figure 3.7(a). Figure 3.8(a) shows the channel current of the β-Ga$_2$O$_3$ under variable bias under different illumination conditions. While illumination by blue (460 nm), green (518 nm), and red (646 nm) light emitting diodes (LEDs) only induces a negligible change in current, the device is sensitive to SBUV LED illumination with a wavelength at about 255 nm. Upon 12 nW of SBUV light irradiated on the active area of the β-Ga$_2$O$_3$ device (a 5×32 μm^2 area of the flake between electrodes), the device exhibits a current increase from 20.53 pA to 67.71 pA at 10 V bias, which represents a responsivity $\Re_I$ of ~4 mA W^{-1}. With a dark current of $I_{\text{dark}} \approx 21$ pA, the PD has a noise equivalent power NEP$_I = (2eI_{\text{dark}})^{1/2}/\Re_I = 6.3 \times 10^{-13}$ W Hz$^{-1/2}$.

Further, the sensing modality utilizing resonance frequency shift induced by the photothermal effect is demonstrated using the same device (figure 3.7(a), modality II). Upon SBUV illumination, the suspended β-Ga$_2$O$_3$ flake absorbs photons with an energy higher than the bandgap. The photon energy transfers to the crystal lattice

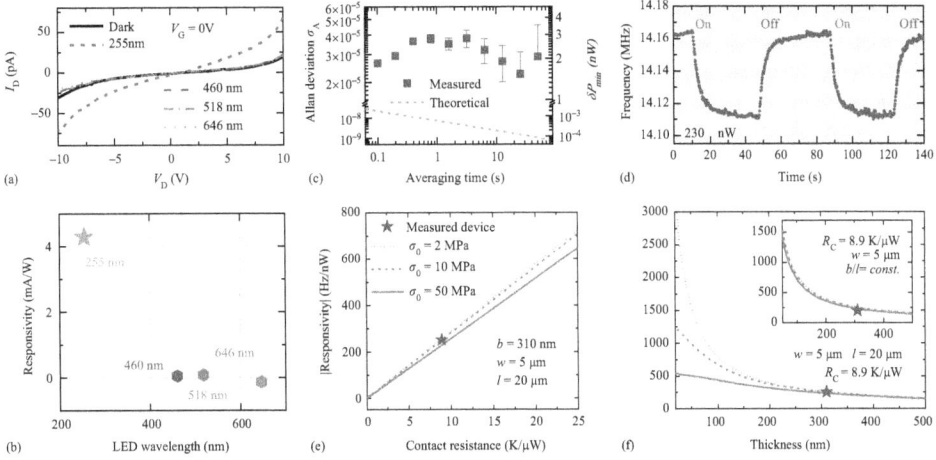

Figure 3.8. Performance of the dual-modality β-Ga$_2$O$_3$ transducer for SBUV detection. (a) Transport characteristics of the PD in the dark and upon illumination by SBUV (255 nm), blue (460 nm), green (518 nm), and red (646 nm) LEDs. (b) Photocurrent responsivity of the β-Ga$_2$O$_3$ transducer at different wavelengths. (c) Allan deviation of the feedback oscillator. The magenta curve shows the Allan deviation limited by thermomechanical noise. (d) Frequency response of the oscillator upon cyclic mercury lamp light irradiation (230 nW on device). (e) Responsivity scaling with respect to the contact resistance. (f) Responsivity scaling for β-Ga$_2$O$_3$ resonators with thicknesses in the 20–500 nm range. Inset: responsivity scaling for thicknesses with constant b/l. Adapted with permission from [38]. Creative Commons Attribution 4.0 International License.

and generates heat; thus, the temperature of the suspended structure increases. The generated heat diffuses along the suspended β-Ga$_2$O$_3$ flake and then dissipates to the substrate. Accordingly, the thermal resistance of the β-Ga$_2$O$_3$ doubly clamped beam and the contact thermal resistance between β-Ga$_2$O$_3$ and the substrate dictate the device's temperature distribution. With an elevated temperature, the β-Ga$_2$O$_3$ flake will expand because of the positive thermal expansion coefficient of β-Ga$_2$O$_3$, and it lowers the built-in stress of the device. The stress change can be determined by

$$\Delta\sigma = -\alpha E_{\mathrm{Y}}\frac{\int_0^l \Delta T(x)\mathrm{d}x}{l} = -\alpha E_{\mathrm{Y}}\eta P_{\mathrm{i}}\left(\frac{l}{12\kappa wb} + R_{\mathrm{C}}\right), \tag{3.4}$$

where $\Delta T(x)$ is the elevated temperature distribution, x denotes the location along the beam, $\alpha = 3.37 \times 10^{-6}$ K^{-1} is the thermal expansion coefficient in the [010] direction of the β-Ga$_2$O$_3$ crystal, [43] E_{Y} is the Young's modulus of β-Ga$_2$O$_3$ crystal, $\kappa = 27$ W (m K)$^{-1}$ is the thermal conductivity of β-Ga$_2$O$_3$ crystal in the [010] direction (along the beam), w and b are the width and thickness of the beam, respectively, and R_{C} is the contact thermal resistance. The total stress in the device is the combination of the initial stress σ_0 and the stress change induced by the thermal expansion $\sigma = \sigma_0 + \Delta\sigma$. The resonance frequency of a tensioned, doubly clamped beam is governed by the following equation [44]:

$$f_n = A_n \frac{b}{l^2} \sqrt{\frac{E_Y}{\rho}} \sqrt{1 + \frac{\sigma l^2}{3.4 E_Y b^2}}, \tag{3.5}$$

where A_n is the eigenvalue for the resonance mode and $\rho = 5950$ kg m^{-3} is the volume mass density of the β-Ga$_2$O$_3$ crystal.

Since the light sensing responsivity of the resonator $\mathfrak{R}_f$ is the differential ratio between the resonance frequency shift and the incident irradiation power, $\mathfrak{R}_f = df/dP_i$, by combining equation (3.4) and (3.5), one can express the responsivity of the SBUV sensing resonator using

$$\mathfrak{R}_f = \frac{df_n}{dP_i} = -\frac{\alpha \eta A_n}{6.8b} \sqrt{\frac{E_Y}{\rho}} \left(\frac{l}{12 \kappa w b} + R_C \right) \left(1 + \frac{\sigma l^2}{3.4 E_Y b^2} \right)^{-1/2}. \tag{3.6}$$

Although the responsivity is related to multiple variables, equation (3.6) suggests that a higher contact thermal resistance R_C can improve the responsivity of the resonance based SBUV sensor.

The detection of the UV light irradiation is resolved by measuring the resonance frequency shifts of the β-Ga$_2$O$_3$ doubly clamped beam (figure 3.7(b)) using an ultrasensitive laser interferometry system (figure 3.7(a), modality II). The motion of mechanical structure is driven by an amplitude modulated blue (405 nm) laser. The driven motion of the device is interferometrically picked up by a 633 nm laser light and detected by a low noise PD. For the open loop measurement (the dark blue path in figure 3.7(a), modality II) and using the third resonance mode ($f \approx 14.320$ MHz with $Q \approx 1170$) of the device, after shining light from the mercury lamp onto the resonator, the resonator exhibits a 55 kHz frequency downshift upon 230 nW illumination on the device and a 32 kHz downshift upon 115 nW illumination.

Then, a self-sustained oscillator is built based on the β-Ga$_2$O$_3$ doubly clamped beam resonator by adding a feedback loop to the laser interferometry system (the dark green path in figure 3.7(a), modality II). By measuring the spectrum of the feedback oscillator using a spectrum analyzer, the oscillator provides an ~40-fold improvement in Q, with an effective Q of the oscillator, $Q_{osc,eff} \approx 48\,000$.

The frequency stability of the β-Ga$_2$O$_3$ oscillator can be calibrated by measuring its Allan deviation σ_A. As shown in figure 3.8(c), the oscillator achieves an Allan deviation of $\sigma_A = 2.6 \times 10^{-5}$ for an averaging time of $\tau = 0.1$ s and $\sigma_A < 4 \times 10^{-5}$ for an averaging time of up to $\tau = 50$ s. The Allan deviation represents the frequency stability of the oscillator and can be translated into the frequency fluctuation δf of the oscillator $\delta f (\tau = 0.1 \text{ s}) = \sqrt{2} \sigma_A f_c = 530$ Hz.

The oscillator is then used to detect light irradiation in real time. By illuminating periodic light from the mercury lamp to the β-Ga$_2$O$_3$ resonator, the real-time tracking of the oscillator's frequency shows frequency downshifts upon illumination (figure 3.8(d)). The 230 nW mercury lamp illumination on the suspended device exhibits a -52 kHz frequency shift of the oscillator. With the results of the closed-loop feedback oscillation, one can extract an average frequency responsivity of the β-Ga$_2$O$_3$ resonator of $\mathfrak{R}_f = \Delta f / P_i \approx 250$ Hz nW^{-1}, and the minimum detectable power (MDP) of the UV sensing oscillator $\delta P_{min} = \delta f / \mathfrak{R}_f \approx 2$ nW.

Similar to the calculation introduced in the last section, the Allan deviation limited by thermomechanical noise at an averaging time of 0.1 s is $\sigma_{A,th}(\tau = 0.1 \text{ s}) = 2.29 \times 10^{-8}$. The thermomechanical frequency noise spectral density can be extracted by [35] $S_{f,th}^{1/2}(f_c) = 2f_c \sqrt{\pi\tau} \sigma_{A, th}$, and with an averaging time of 0.1 s, the thermomechanically limited frequency noise of the device is 0.38 Hz Hz$^{-1/2}$. Thus, the thermomechanical noise limited NEP of the device can be calculated, where NEP$_{th} = S_{f,th}^{1/2}(f_c)/(2\pi|\mathfrak{R}_f|) = 2.4 \times 10^{-13}$ W Hz$^{-1/2}$.

In order to fully understand the NEMS sensing modality and give guidelines for designing resonators with higher SBUV sensing responsivity, equation (3.6) is used to scale the responsivity with respect to different device parameters (figures 3.8(e) and (f)). From the results, one can tell that the responsivity improves linearly with respect to the thermal contact resistance. It can also be noted that the thermal resistance of the beam plays a minimal role in the responsivity since the responsivity is approaching zero when the contact resistance is zero. In addition, by shrinking the thickness of the β-Ga$_2$O$_3$ crystal, the responsivity improves. Figure 3.8(f) shows the responsivity of the resonator with different device thicknesses with a thermal contact resistance of 8.9 K µW^{-1}. From the plot, the responsivity can be much improved by making the device thinner. Further, by reducing the thickness and the length of the resonator together at constant b/l (figure 3.8(f) inset), the responsivity of the device is improved, and the scaling shows much less dependence on the initial stress in the device. To design a better SBUV sensing device using the NEMS modality in the future, two methods could be used to improve the responsivity, including (i) increasing the thermal contact resistance from the β-Ga$_2$O$_3$ flake to the substrate and (ii) reducing the size (length and thickness) of the resonator.

3.4 Conclusions and future perspectives

In this chapter, we have reviewed the demonstration of a new type of UWBG nanomechanical resonator based on LPCVD synthesized single-crystal β-Ga$_2$O$_3$ nanoflakes. These β-Ga$_2$O$_3$ nanoresonators demonstrate robust multimode resonances in the HF and VHF bands ranging from 18 to 75 MHz. From devices operating in the 'disk' regime, the measured resonances determine a Young's modulus of $E_Y \approx 261$ GPa. Multimode spatial mapping discloses new mode splitting features, which in combination with parametric modeling reveal anisotropic built-in tension in devices that operate in the transition regime. The effects of thermal annealing on resonance characteristics have been measured, revealing that β-Ga$_2$O$_3$ can adsorb gas molecules in air efficiently. In terms of practical applications, we then present the first demonstration of a β-Ga$_2$O$_3$ nanomechanical resonator for MUV sensing. The feedback oscillator is further employed to demonstrate real-time sensing of cyclic MUV light irradiation onto the β-Ga$_2$O$_3$ resonator. To further improve and validate the UV sensing using β-Ga$_2$O$_3$ resonant NEMS devices, we review the demonstration of the first β-Ga$_2$O$_3$ SBUV transducer with dual sensing modalities, including photocurrent modulation caused by the photoelectric effect, and resonance frequency shift induced by the photothermal effect.

To drive β-Ga_2O_3 resonant MEMS and NEMS toward future applications in sensing, RF, and in particular in the platforms of integrated microsystems, an all-electrical readout of β-Ga_2O_3 resonators is highly desirable. To this goal, lately we have made the initial effort to demonstrate all-electrical excitation and detection of β-Ga_2O_3 NEMS resonators based on β-Ga_2O_3 vibrating channel transistors (VCTs) [45], where the device mechanical vibration modulates the transistor's channel conductance, which is exploited for decoding the device motion through a frequency modulation (FM) down-mixing scheme. Looking toward the future, first a direct readout of the device motion without using the FM mixing technique is highly desired for integration of β-Ga_2O_3 M/NEMS with an RF circuit. To achieve this, a higher aspect ratio (suspended-area-to-thickness ratio) is needed to achieve a more pronounced mechanical motion, higher doping of β-Ga_2O_3 material and better metal–β-Ga_2O_3 contact is desired for improved transistor conductivity, and a shallower gate-to-suspended-flake distance is desired to enhance the gating effect. In addition, more effort should be focused on applying the VCTs to practical applications, e.g. SBUV sensing and RF signal processing. Finally, the means to mass-produce β-Ga_2O_3 resonant MEMS and NEMS devices should be developed to guide the technology toward industry.

References

[1] Shenai K, Dudley M and Davis R F 2013 Current status and emerging trends in wide bandgap (WBG) semiconductor power switching devices *ECS J. Solid State Sci. Technol.* **2** N3055

[2] Cimalla V, Pezoldt J and Ambacher O 2007 Group III nitride and SiC based MEMS and NEMS: materials properties, technology and applications *J. Phys. D: Appl. Phys.* **40** 6386

[3] Green A J *et al* 2016 3.8-MV/cm breakdown strength of MOVPE-grown Sn-doped β-Ga_2O_3 MOSFETs *IEEE Electron Device Lett.* **37** 902–5

[4] Higashiwaki M, Sasaki K, Murakami H, Kumagai Y, Koukitu A, Kuramata A, Masui T and Yamakoshi S 2016 Recent progress in Ga_2O_3 power devices *Semicond. Sci. Technol.* **31** 034001

[5] Zhou H, Si M, Alghamdi S, Qiu G, Yang L and Ye P D 2017 High-performance depletion/enhancement-mode β-Ga_2O_3 on insulator (GOOI) field-effect transistors with record drain currents of 600/450 mA/mm *IEEE Electron Device Lett.* **38** 103–6

[6] Zou R, Zhang Z, Liu Q, Hu J, Sang L, Liao M and Zhang W 2014 High detectivity solar-blind high-temperature deep-ultraviolet photodetector based on multi-layered (l00) facet-oriented β-Ga_2O_3 nanobelts *Small* **10** 1848–56

[7] Kong W-Y, Wu G-A, Wang K-Y, Zhang T-F, Zou Y-F, Wang D-D and Luo L-B 2016 Graphene–β-Ga_2O_3 heterojunction for highly sensitive deep UV photodetector application *Adv. Mater.* **28** 10725

[8] Lorenz M R, Woods J F and Gambino R J 1967 Some electrical properties of the semiconductor β-Ga_2O_3 *J. Phys. Chem. Solids* **28** 403–4

[9] Ueda N, Hosono H, Waseda R and Kawazoe H 1997 Anisotropy of electrical and optical properties in β-Ga_2O_3 single crystals *Appl. Phys. Lett.* **71** 933–5

[10] Xia Z *et al* 2019 Metal/$BaTiO_3$/β-Ga_2O_3 dielectric heterojunction diode with 5.7 MV/cm breakdown field *Appl. Phys. Lett.* **115** 252104

[11] Ma N, Tanen N, Verma A, Guo Z, Luo T, Xing H G and Jena D 2016 Intrinsic electron mobility limits in β-Ga$_2$O$_3$ *Appl. Phys. Lett.* **109** 212101

[12] Zheng X-Q, Lee J, Rafique S, Han L, Zorman C A, Zhao H and Feng P X-L 2017 Ultrawide band gap β-Ga$_2$O$_3$ nanomechanical resonators with spatially visualized multimode motion *ACS Appl. Mater. Interfaces* **9** 43090–7

[13] Rais-Zadeh M, Gokhale V J, Ansari A, Faucher M, Théron D, Cordier Y and Buchaillot L 2014 Gallium nitride as an electromechanical material *J. Microelectromech. Sys.* **23** 1252–71

[14] Yu M-F, Atashbar M Z and Chen X 2005 Mechanical and electrical characterization of β-Ga$_2$O$_3$ nanostructures for sensing applications *IEEE Sens. J.* **5** 20–5

[15] Nikolaev V I, Maslov V, Stepanov S I, Pechnikov A I, Krymov V, Nikitina I P, Guzilova L I, Bougrov V E and Romanov A E 2017 Growth and characterization of β-Ga$_2$O$_3$ crystals *J. Cryst. Growth* **457** 132–6

[16] Jangir R, Porwal S, Tiwari P, Mondal P, Rai S K, Ganguli T, Oak S M and Deb S K 2012 Photoluminescence study of β-Ga$_2$O$_3$ nanostructures annealed in different environments *J. Appl. Phys.* **112** 034307

[17] Chen Y-J, Chung C-J and Li J V 2020 Theoretical potential of extremely high quality factors of β-Ga$_2$O$_3$ based MEMS resonators *Jpn J. Appl. Phys.* **59** 011002

[18] Tomm Y, Reiche P, Klimm D and Fukuda T 2000 Czochralski grown Ga$_2$O$_3$ crystals *J. Cryst. Growth* **220** 510–4

[19] Galazka Z, Uecker R, Irmscher K, Albrecht M, Klimm D, Pietsch M, Brützam M, Bertram R, Ganschow S and Fornari R 2010 Czochralski growth and characterization of beta-Ga$_2$O$_3$ single crystals *Cryst. Res. Technol.* **45** 1229–36

[20] Ueda N, Hosono H, Waseda R and Kawazoe H 1997 Synthesis and control of conductivity of ultraviolet transmitting beta-Ga$_2$O$_3$ single crystals *Appl. Phys. Lett.* **70** 3561–3

[21] Tomm Y, Ko J M, Yoshikawa A and Fukuda T 2001 Floating zone growth of β-Ga$_2$O$_3$: a new window material for optoelectronic device applications *Sol. Energy Mater. Sol. Cells* **66** 369–74

[22] Villora E G, Shimamura K, Yoshikawa Y, Aoki K and Ichinose N 2004 Large-size β-Ga$_2$O$_3$ single crystals and wafers *J. Cryst. Growth* **270** 420–6

[23] Kuramata A, Koshi K, Watanabe S, Yamaoka Y, Masui T and Yamakoshi S 2016 High-quality β-Ga$_2$O$_3$ single crystals grown by edge-defined film-fed growth *Jpn J. Appl. Phys.* **55** 1202A2

[24] Kumar S, Sarau G, Tessarek C, Bashouti M Y, Hähnel A, Christiansen S and Singh R 2014 Study of iron-catalysed growth of β-Ga$_2$O$_3$ nanowires and their detailed characterization using TEM, Raman and cathodoluminescence techniques *J. Phys. D: Appl. Phys.* **47** 435101

[25] Zhang J, Jiang F, Yang Y and Li J 2005 Catalyst-assisted vapor–liquid–solid growth of single-crystal Ga$_2$O$_3$ nanobelts *J. Phys. Chem.* B **109** 13143–7

[26] Rafique S, Han L, Zorman C A and Zhao H 2016 Synthesis of wide bandgap β-Ga$_2$O$_3$ rods on 3C-SiC-on-Si *Cryst. Growth Des.* **16** 511–7

[27] Ohira S, Sugawara T, Nakajima K and Shishido T 2005 Synthesis and structural investigation of β-Ga$_2$O$_3$ nanosheets and nanobelts *J. Alloys Compd.* **402** 204–7

[28] Hogan J E, Kaun S W, Ahmadi E, Oshima Y and Speck J S 2016 Chlorine-based dry etching of β-Ga$_2$O$_3$ *Semicond. Sci. Technol.* **31** 065006

[29] Zheng X-Q, Lee J, Rafique S, Han L, Zorman C A, Zhao H and Feng P X-L 2018 Free-standing β-Ga$_2$O$_3$ thin diaphragms *J. Electron. Mater.* **47** 973–81

[30] Yoshioka S, Hayashi H, Kuwabara A, Oba F, Matsunaga K and Tanaka I 2007 Structures and energetics of Ga_2O_3 polymorphs *J. Phys. Condens. Matter* **19** 346211

[31] Rafique S, Han L, Lee J, Zheng X-Q, Zorman C A, Feng P X-L and Zhao H 2017 Synthesis and characterization of Ga_2O_3 nanosheets on 3C-SiC-on-Si by low pressure chemical vapor deposition *J. Vac. Sci. Technol.* B **35** 011208

[32] Suzuki H, Yamaguchi N and Izumi H 2009 Theoretical and experimental studies on the resonance frequencies of a stretched circular plate: application to Japanese drum diaphragms *Acoust. Sci. Technol.* **30** 348–54

[33] Zheng X-Q, Lee J, Rafique S, Han L, Zorman C A, Zhao H and Feng P X-L 2017 Wide bandgap β-Ga_2O_3 nanomechanical resonators for detection of middle-ultraviolet (MUV) photon radiation *Proc. 30th IEEE Int. Conf. on Micro Electro Mechanical Systems (MEMS'17) (Las Vegas, NV, USA, 22–26 January 2017)* (New York: IEEE) 209–12

[34] Zheng X-Q, Lee J, Rafique S, Karim M R, Han L, Zhao H, Zorman C A and Feng P X-L 2018 β-Ga_2O_3 NEMS oscillator for real-time middle ultraviolet (MUV) light detection *IEEE Electron Device Lett.* **39** 1230–3

[35] Cleland A N and Roukes M L 2002 Noise processes in nanomechanical resonators *J. Appl. Phys.* **92** 2758–69

[36] Nakagomi S, Momo T, Takahashi S and Kokubun Y 2013 Deep ultraviolet photodiodes based on β-Ga_2O_3/SiC heterojunction *Appl. Phys. Lett.* **103** 072105

[37] Oshima T, Okuno T, Arai N, Suzuki N, Ohira S and Fujita S 2008 Vertical solar-blind deep-ultraviolet Schottky photodetectors based on β-Ga_2O_3 substrates *Appl. Phys. Express* **1** 011202

[38] Zheng X-Q, Xie Y, Lee J, Jia Z, Tao X and Feng P X-L 2019 Beta gallium oxide (β-Ga_2O_3) nanoelectromechanical transducer for dual-modality solar-blind ultraviolet light detection *APL Mater.* **7** 022523

[39] Mu W, Jia Z, Yin Y, Hu Q, Zhang J, Feng Q, Hao Y and Tao X 2017 One-step exfoliation of ultra-smooth β-Ga_2O_3 wafers from bulk crystal for photodetectors *Cryst. Eng. Comm.* **19** 5122–7

[40] Hwang W S *et al* 2014 High-voltage field effect transistors with wide-bandgap β-Ga_2O_3 nanomembranes *Appl. Phys. Lett.* **104** 203111

[41] Barman S K and Huda M N 2019 Mechanism behind the easy exfoliation of Ga_2O_3 ultra-thin film along (100) surface *Phys. Status Solidi Rapid Res. Lett.* **13** 1800554

[42] Zheng W, Lin R, Zhu Y, Zhang Z, Ji X and Huang F 2018 Vacuum ultraviolet photodetection in two-dimensional oxides *ACS Appl. Mater. Interfaces* **10** 20696–702

[43] Orlandi F, Mezzadri F, Calestani G, Boschi F and Fornari R 2015 Thermal expansion coefficients of β-Ga_2O_3 single crystals *Appl. Phys. Express* **8** 111101

[44] Jun S C, Huang X M H, Manolidis M, Zorman C A, Mehregany M and Hone J 2006 Electrothermal tuning of Al–SiC nanomechanical resonators *Nanotechnology* **17** 1506–11

[45] Zheng X-Q, Lee J and Feng P X-L 2020 Beta gallium oxide (β-Ga_2O_3) vibrating channel transistor *Proc. 33rd IEEE Int. Conf. on Micro Electro Mechanical Systems (MEMS'20) (Vancouver, BC, Canada, 18–22 January 2020)* (New York: IEEE) pp 31–5

IOP Publishing

Wide Bandgap Semiconductor-Based Electronics

Fan Ren and Stephen J Pearton

Chapter 4

Epitaxial growth of monoclinic gallium oxide using molecular beam epitaxy

Mahitosh Biswas and Elaheh Ahmadi

Gallium oxide (Ga_2O_3), which is referred to as an ultra-wide bandgap semiconductor, has attracted a great deal of technological interest in the domain of power electronic devices and solar-blind ultra-violet (UV) detectors due to its wide bandgap of 4.5–5.3 eV [1, 2] (depending on the crystal structure) and high breakdown electric field of ~8 MeV cm^{-1}. Ga_2O_3 exists in five different polymorphs among which β-Ga_2O_3, with a monoclinic crystal structure, is the most thermodynamically stable and this material is the focus of this chapter.

Single-crystal β-Ga_2O_3 substrates are grown by a variety of melt-growth methods, such as the floating-zone [3, 4], edge-defined film-fed growth [5], and Czochralski [6, 7] methods. Epitaxial β-Ga_2O_3 films have been also successfully demonstrated using various vapor phase techniques, including pulsed-laser deposition [8], halide vapor phase epitaxy [9], MIST epitaxy [10], metal–organic chemical vapor deposition (MOCVD) [11], and molecular beam epitaxy (MBE) [2, 12].

This chapter briefly covers the basic properties of β-Ga_2O_3 and is mainly focused on the MBE growth of β-Ga_2O_3 films and their heterostructures. Modulation-doped field effect transistors (MODFETs) are then discussed as an example of the application of MBE to fabricate high-power transistors. For more discussion of β-Ga_2O_3-based electronic devices, we refer the reader to the review papers published recently on this topic [13, 14]

4.1 The properties of Ga_2O_3

4.1.1 Polymorphs

As reported by Roy *et al* [15], Ga_2O_3 exists in five different polymorphs, namely α- (corundum), β- (monoclinic), γ- (defective spinel), δ- (cubic), and ε- (hexagonal) Ga_2O_3. Although α-Ga_2O_3 has higher bandgap (5.3 eV) compared to β-Ga_2O_3 (4.9 eV), β-Ga_2O_3 is the most thermodynamically stable phase (with a melting point

doi:10.1088/978-0-7503-2516-5ch4

of 1800 °C). The other polymorphs are meta-stable and can be transformed into β-Ga$_2$O$_3$ in the temperature range of 750 °C–900 °C. It has been shown that δ-Ga$_2$O$_3$ is a nano-crystalline form of ε-Ga$_2$O$_3$. More information about the five different polymorphs can be found in table 4.1. The tendency of their formation free energies is $\beta < \varepsilon < \alpha < \delta < \gamma$ at low temperatures. Figure 4.1 summarizes the synthesis and interconversion of different polymorphs of Ga$_2$O$_3$ [16].

4.1.2 Crystal structure, electronics, and thermal properties of β-Ga$_2$O$_3$

As mentioned earlier, the most stable polymorph is β-Ga$_2$O$_3$ which possesses a body-centered monoclinic structure. The unit cell of β-Ga$_2$O$_3$ is shown in figure 4.2.

Table 4.1. The properties of polymorphs of Ga$_2$O$_3$. The reported bandgaps in this table are all direct bandgaps. Indirect bandgaps for γ- and ε-Ga$_2$O$_3$ were also reported, with values of 4.4 and 4.5 eV, respectively [22, 24].

Polymorphs	Structures	Space group	Lattice parameters (Å) (experimental)	Bandgap (eV)
α	Corundum	$R\bar{3}c$	$a = 4.982$, $c = 13.433$ [17]	5.3 [18]
β	Monoclinic	C2/m	$a = 12.23$, $b = 3.04$, and $c = 5.8$ [19]	4.9 [20]
γ	Defective spinel	Fd$\bar{3}$m	$a = 8.24$ [21]	5.0 [22]
δ	Cubic	Ia$\bar{3}$	$a = 10.0$ [3]	—
ε	Hexagonal	P6$_3$mc	$a = 2.90$, $c = 9.25$ [23]	5.0 [24]

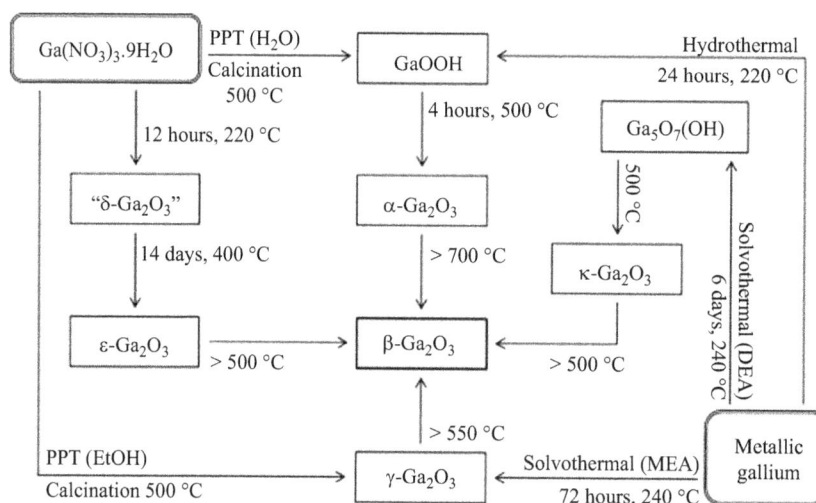

Figure 4.1. The synthesis and interconversion of six different polymorphs of Ga$_2$O$_3$. Upon heating, gallium oxyhydroxide transforms into a transient phase called κ-Ga$_2$O$_3$. Reproduced with permission from [16]. Copyright 2013 John Wiley and Sons.

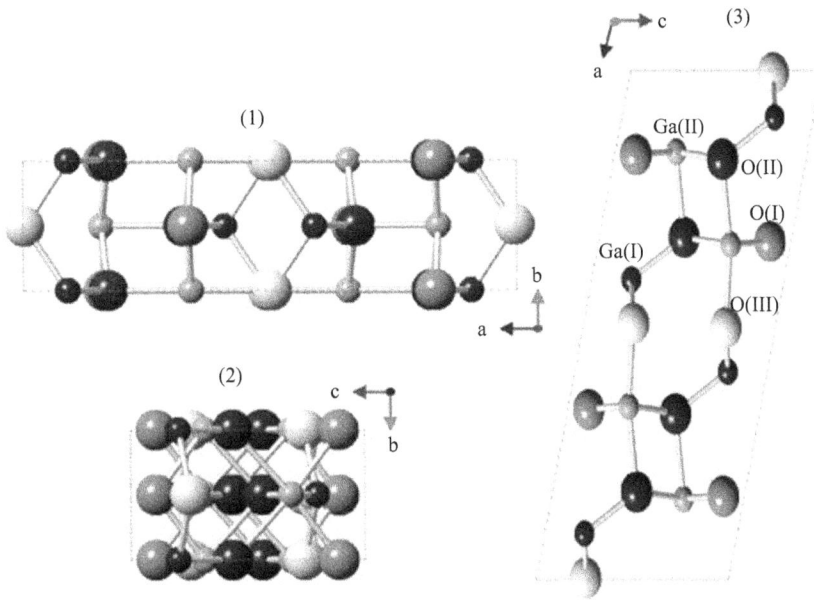

Figure 4.2. Unit cell of β-Ga$_2$O$_3$ along the c- (1), a- (2), and b-axes (3). Reproduced with permission from [25]. Copyright IOP Publishing and Deutsche Physikalische Gesellschaft. CC BY-NC-SA.

Table 4.2. Comparison of the material properties between Si, 4H-SiC, GaN, diamond, and β-Ga$_2$O$_3$.

Materials properties	Si	4H-SiC	GaN	Diamond	β-Ga$_2$O$_3$
Bandgap, E_g (eV)	1.1	3.3	3.4	5.5	4.9
Breakdown field, E_c (MV cm^{-1})	0.3	2.5	3.3	10	6.8
Electron mobility, μ (cm^2 V^{-1} s^{-1})	1480	1000	1250	2000	300
Dielectric constant, ε	11.8	9.7	9	5.5	10
BFOM, $\mu\varepsilon E_c^3$	1	317	846	24 660	3214
Thermal conductivity, λ (W cm^{-1} K^{-1})	1.5	2.7	2.1	20	0.13 (101), 0.21 (010)

The unit cell has two crystallographically inequivalent Ga sites, one is in tetrahedral geometry Ga(I) and the other in octahedral geometry Ga(II). There are three inequivalent O-sites; one appears tetrahedrally O(I) and the other two are coordinated trigonally, O(II) and O(III). The average interionic distances are 1.83 Å (tetrahedral Ga–O), 2 Å (octahedral Ga–O), 3.02 Å (tetrahedron edge O–O), and 2.84 Å (octahedron edge O–O) [26]. The lattice parameters and structure in the space group C2/m of β-Ga$_2$O$_3$ were reported first by Kohn *et al* [19] and Geller [26], respectively. The material properties of β-Ga$_2$O$_3$ in comparison to Si, 4H-SiC, GaN, and diamond are summarized in table 4.2.

Density functional theory (DFT) [27, 28] was previously used to understand the band structure of β-Ga$_2$O$_3$. However, since DFT uses ground state theory, it underestimates the exchange-correlation potential among the excited electrons and hence the results are not accurate. Hybrid DFT [29, 30], on the other hand, provides results for structures, energetics, and bandgaps which are closer to the experimental values determined by angle-resolved photoemission [31]. The electronic band structure of β-Ga_2O_3 is shown in figure 4.3.

Calculations predict that the conduction band minimum appears at the Γ k-point and the valence band is almost flat. The valence band maximum is almost degenerate at the Γ and M k-points. Thus, there exists a direct bandgap of 4.69 eV at Γ and an indirect gap of 4.66 eV at the M point. Therefore, the indirect bandgap of β-Ga_2O_3 is only ~30 meV smaller than the direct bandgap. The calculated value of the electron effective mass ($m_e{}^*$) is 0.27–0.28 m_0 [30, 32] and it is isotropic, which agrees well with the experimental value of 0.28 m_0 [31]. Since the valance band is almost flat, it is expected to have a large effective mass for holes ($m_h{}^*$). The value of

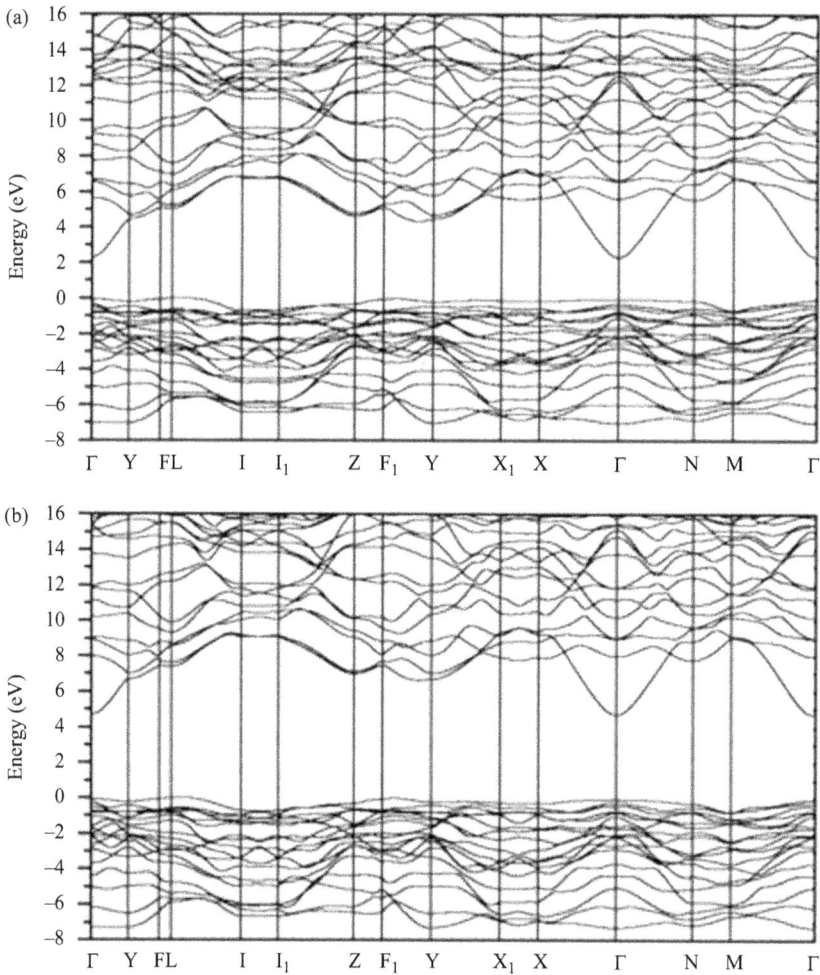

Figure 4.3. Electronic band structure of β-Ga_2O_3. Reproduced with permission from [33]. Copyright 2017 the American Physical Society.

m_h^* is estimated to be 40 m_0 and 0.40 m_0 along the Γ–Z and the Γ–A directions, respectively [29].

Due to crystalline anisotropy, thermal conductivity (λ) in β-Ga$_2$O$_3$ is very different in different crystallographic directions. By using the laser flash method, the value of λ was estimated to be 13 W m K^{-1} along the [001] direction [34] and 21 W m K^{-1} along the [010] direction [35], which are 2–3 orders of magnitude lower than those reported for other wide bandgap semiconductors [36]. Very recently, Mengle *et al* [37] studied thermal conductivity in β-Ga$_2$O$_3$ using DFT and density functional perturbation theory and showed that it is strongly influenced by anharmonic phonon–phonon interactions, which are quantified by the optical mode Gruneisen parameters. Poor thermal conductivity was attributed to large values of Gruneisen parameters. However, an ordered alloy was suggested to provide improved lattice thermal conductivity. Sai *et al* [38] theoretically showed that at 50% Al concentration, a highly stable ordered alloy of AlGaO$_3$ is formed, in which Al atoms occupy octahedral sites [39], leading to enhanced components of the thermal conductivity tensor by 70%–100%. This was attributed to higher group velocities and a decrease in scattering.

4.1.3 Optical properties of MBE-grown β-Ga$_2$O$_3$

The absorption spectrum of β-Ga$_2$O$_3$ exhibits a power-law expressed as [41]

$$(\alpha h v) = B\left(h v - E_g\right)^{1/2}, \tag{4.1}$$

where $h v$ is the incident photon energy, α is the absorption coefficient, B is the absorption edge width parameter, and E_g is the bandgap. As can be seen in figure 4.4, $(\alpha h v)^2$ versus $h v$ resulted in a linear plot in the high energy region which suggests that direct allowed transitions occur across the bandgap of Ga$_2$O$_3$ [40].

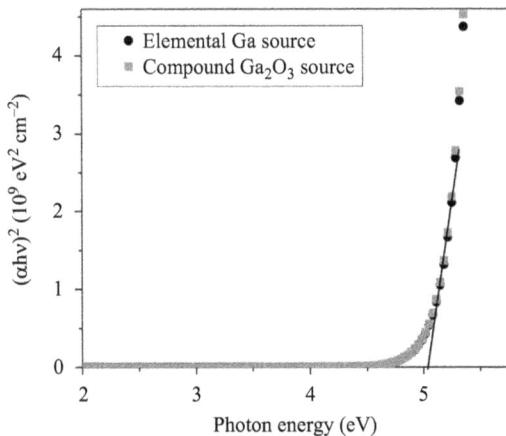

Figure 4.4. $(\alpha h v)^2$ versus photon energy, $h v$ for the β-Ga$_2$O$_3$ films grown on c-sapphire by PAMBE. The optical bandgap is estimated by extrapolation to $h v = 0$. Reproduced from [41] with permission of AIP Publishing.

The value of E_g is estimated by extrapolating the linear plot to $h\nu = 0$. Using this technique, the bandgap of the β-Ga$_2$O$_3$ films grown using plasma-assisted MBE (PAMBE) on c-sapphire at 750 °C was estimated to be ~5.02 eV [41]. This suggests excellent optical transparency of the gown films in the UV and visible regions. Ueda *et al* [42] reported that absorption is sensitive to the polarization of the incident light and found the fundamental absorption edges for $E\|b$ and $E\|c$ at 4.9 and 4.7 eV, respectively.

Most photoluminescence (PL) studies performed on β-Ga$_2$O$_3$ bulk or thin films do not show near-band-edge emission in the deep UV (4.4–5.0 eV), but only emission in the visible to the UV-A range (~2.0–3.5 eV). These PL spectra typically consist of UV (3.4 eV), blue (2.95 eV), and green (2.48 eV) bands which are ascribed to the localization of a self-trapped hole polaron [43], donor–acceptor (V_{Ga} or V_{Ga} +V_O) recombination [44], and hole trapped O$_i$ (interstitial oxygen) [45], respectively. In addition, red emission has also been observed at 2.15 eV because of a neutral oxygen vacancy [46]. Near-band-edge emission has only been reported in β-Ga$_2$O$_3$ nanowires, which may be due to a reduced density of defects [47]. The different emission bands are schematically shown in figure 4.5. Based on first-principle calculations, Mengle *et al* [48] predicted that near-band-edge emission can be achieved at a sufficiently high carrier concentration and temperature.

4.2 Molecular beam epitaxy

Due to the ultra-high vacuum environment, high purity source materials, and computer-controlled shutters, MBE has been a widely accepted tool for the epitaxial growth of semiconductor films and their heterostructures. In particular, the growth of various wide bandgap oxides (WBGOs) has also been successfully realized using

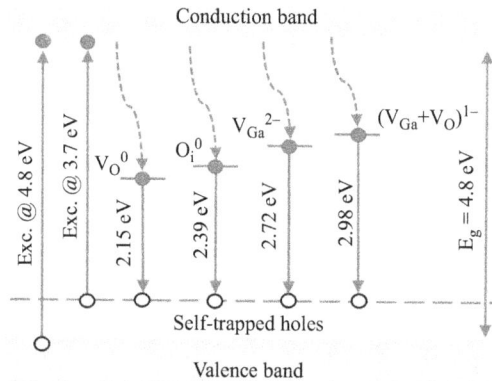

Figure 4.5. Sketch of the band diagram involving the PL mechanism associated with the different emission bands in crystalline β-Ga$_2$O$_3$ thin films under 3.7 and 4.8 eV excitation. The filled circles denote electrons and the empty circles denote holes. The curved arrows correspond to the non-radiative process of an electron from the conduction band to the specific deep level. Reproduced with permission from [46]. Copyright 2019 IOP Publishing Ltd and Deutsche Physikalische Gesellschaft.

MBE [49–51]. Perovskite stannates (an emerging class of WBGOs), e.g. BaSnO$_3$ and SrSnO$_3$, which are of interest for future use in RF and high-power devices, have also been grown using MBE [52, 53].

MBE is a physical vapor phase deposition technique and takes advantage of an ultra-high vacuum of typically 10^{-10} Torr and high purity (6–7 N) source materials and hence reduces impurity incorporation during film growth. This technique was invented to deposit GaAs and GaAs/AlGaAs compound semiconductor materials in 1960 at the Bell Telephone Laboratories by Arthur and Cho [54]. The schematic of an MBE system is shown in figure 4.6. Due to the ultra-high vacuum, the mean free path (~100 m) of the impinging atoms or molecules from the effusion cell exceeds the distance (~30 cm) between the cell and substrate inside the growth chamber. Hence, the evaporated atoms or molecules from single or multiple cells do not interact with each other in the vapor phase. They impinge on the single crystalline substrate heated to a particular temperature and result in the formation of the desired epitaxial films.

To monitor the growth *in situ*, the system is equipped with a reflection high energy electron diffraction (RHEED) which typically employs a high energy (~15 keV) electron beam directed towards the growing sample surface at a grazing angle (2°–3°). Because of the grazing angle incidence of an electron beam, it is scattered within the first few atomic layers giving rise to a surface-sensitive diffraction pattern.

If the electron interacts only with the first atomic layer with a perfectly flat and ordered surface, three-dimensional lattice points should form parallel infinite rods theoretically. However, due to divergence and dispersion of the electron beam, the rod in the Ewald sphere is restricted to having a finite thickness. Therefore, the diffraction pattern from a flat surface and an ordered surface consists of streaks with modulated intensity. However, if the surface is not flat, many electrons are

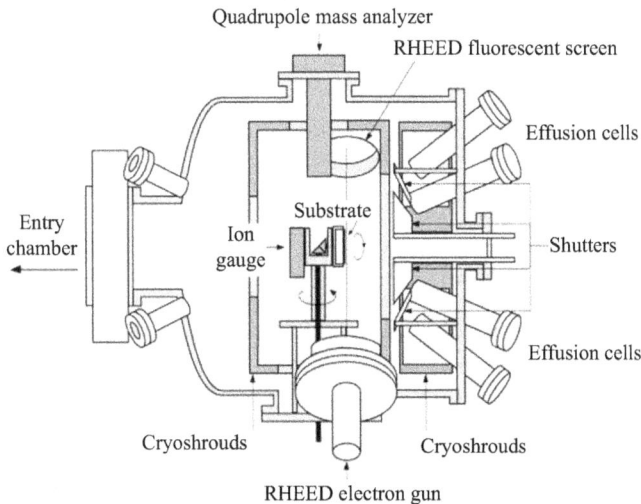

Figure 4.6. Schematic of a typical solid source MBE growth chamber. Reproduced with permission from [55]. Copyright 2003 Elsevier.

transmitted through rough edges and scattered in different directions, leading to a RHEED pattern with spotty features. For thin film growth, obtaining a streaky RHEED pattern is desired. Another *in situ* diagnostic tool named residual gas analyzer helps us determine impurities in the growth chamber.

The ultra-high vacuum environment of an MBE system makes it attractive for applications in which precise control of the heterointerface, alloy composition, and doping is required. On the other hand, the operation and maintenance of MBE is significantly more expensive and much slower growth rates (typically 200 nm h^{-1}–500 nm h^{-1}) can be achieved via this system compared to other epitaxial growth techniques such as MOCVD and halide vapor phase epitaxy (HVPE). The combination of low growth rate and costly maintenance of MBE makes this system unsuitable for device applications in which a thick drift layer is required (e.g. vertical power devices).

4.3 Growth modes

For understanding the growth mechanism on a substrate, it is essential to illustrate the physics governing nucleation and growth. The change in surface free energy (ΔG) during the nucleation of heteroepitaxy (the films and substrate are off different materials) can be written as

$$\Delta G = \gamma_{LV} - \gamma_{SV} + \gamma_{LS}, \qquad (4.2)$$

where γ_{LV} is the change in free energy between the layer (L) and total volume (V), γ_{SV} is the change in free energy between the substrate (S) and V, and γ_{LS} is the change in free energy between L and S. Based on ΔG, the growth mechanism can be classified into three modes, namely, layer-by-layer growth (Frank van der Merwe growth), layer growth followed by island growth (Stranski–Krastanov (S–K) growth), and island growth mode (Volmer–Weber growth).

Layer-by-layer growth. In this growth mode, atoms impinging on the surface diffuse and nucleate in two-dimensional (2D) islands then attract more atoms until the layer is complete. Atoms or molecules are more strongly bonded to the substrate than to each other. The epitaxial growth of semiconductor films employs a layer-by-layer mode. For this mode to occur, $\gamma_{SV} > \gamma_{LV} + \gamma_{LS}$ must be satisfied. If the substrate is cut slightly misoriented from the low index plane, the step terrace width reduces and hence the distance that impinging atoms have to defuse on the surface decreases. This leads to a step-flow growth which is typically the desired growth regime for epitaxy of high-quality nitrides and Ga$_2$O$_3$ films.

S–K growth. Heteroepitaxial growth of InAs on GaAs is an example of the S–K growth mode. This involves the growth of an initial 2D layer (called a wetting layer) followed by a three-dimensional layer (an island). When the lattice mismatch between the substrate and the films is in the range of 3%–7%, the S–K growth mode occurs [56]. Immediately after the growth of two monolayers, the wetting layer becomes energetically unfavorable and the misfit strain is relieved via the formation of islands.

Island growth mode. In this growth mode, the smallest group of atoms or clusters nucleates on the substrate surface in three dimensions. The atoms are more closely bonded to each other than to the substrate. Metal-on-insulator is a typical example of this growth mode. For the island growth mode to occur, $\gamma_{SV} < \gamma_{LV} + \gamma_{LS}$ must be satisfied.

4.4 Epitaxial growth of β-Ga$_2$O$_3$ thin films by MBE

The growth of Ga$_2$O$_3$ thin films typically employs a source of gallium and a source of oxygen to completely oxidize Ga atoms on the surface. In an MBE system, the source of gallium is elemental gallium evaporated from an effusion cell, whereas the source of oxygen (O$_2$) could be a reactive oxygen plasma or ozone (O$_3$) or an O$_3$/O$_2$ mixture. It was observed that Ga$_2$O$_3$ cannot be grown by supplying only molecular oxygen (O$_2$). Among the O$_2$ sources, O$_3$ is highly toxic and flammable and hence needs to be handled with special care. Although elemental Ga is the most common source of gallium, a Ga$_2$O$_3$ compound source has also been used to grow Ga$_2$O$_3$ films. In a conventional PAMBE system, a highly active oxygen beam of atoms or molecules from an RF source operated at 13.6 MHz is used to oxidize the evaporated Ga atoms on a heated substrate to form a layer of Ga$_2$O$_3$. To have stoichiometric control during the growth of Ga$_2$O$_3$, an oxygen-rich condition is typically used to maintain oxygen overpressure to the growth surface. On the other hand, a Ga-rich growth condition is used for reducing the density of Ga vacancies which serve as compensating acceptors in β-Ga$_2$O$_3$ [57].

Competition between the formation of Ga$_2$O$_3$ on a substrate surface and desorption of volatile Ga$_2$O from the surface determines the growth rate of the films [58, 59]. To reduce the formation and desorption of Ga$_2$O, the oxygen flux needs to to be increased. An alternative source containing more reactive oxygen species, such as ozone, can reduce the desorption of Ga$_2$O and hence increase the growth rate. However, there is a threshold for the partial pressure of oxygen to be used in an MBE chamber due to the risk of oxidation of the ion gauge, RHEED, and heater filaments as well as the metallic source materials in the effusion cells. To resolve this issue, a Ga$_2$O$_3$ compound source has been suggested to be used which would give rise to Ga$_2$O species, instead of Ga species, and avoid the requirement for high oxygen partial pressure to the growing surface. As a result, Ga$_2$O molecules and oxygen would lead to the formation of crystalline Ga$_2$O$_3$ growth at a much lower substrate temperature. The Ga$_2$O$_3$ compound source also helps us obtain a relatively higher growth rate compared to the elemental Ga source. A similar approach was attempted for the growth of BaSnO$_3$ thick films using MBE where a SnO$_2$ source was employed, instead of an elemental Sn source, to overcome the high overpressure of Sn and the difficulty of incorporation into the tertiary films [50]. Additionally, to avoid oxidation of the source material in the oxide environment, hybrid MBE [60, 61] has been used previously for the growth of SrTiO$_3$, where titanium tetra isopropoxide was used as the titanium source, whereas a conventional effusion cell was utilized for Sr.

4.4.1 Growth of β-Ga$_2$O$_3$ from an elemental Ga source using PAMBE

In the PAMBE system, to grow the stoichiometric Ga$_2$O$_3$ layer reactive oxygen species are introduced from an RF plasma source. Both the homoepitaxial and heteroepitaxial growth of β-Ga$_2$O$_3$ films have been studied using PAMBE. However, one of the distinguishable advantages of β-Ga$_2$O$_3$ compared to other wide bandgap materials (e.g. GaN, SiC, AlN, and diamond) is that high-quality, large-scale, and cost-effective β-Ga$_2$O$_3$ substrates are achievable via melt-growth techniques. Therefore, in our opinion, there is no compelling reason to justify the heteroepitaxial growth of β-Ga$_2$O$_3$ on foreign substrates. Nevertheless, the hetero-epitaxial growth of β-Ga$_2$O$_3$ on c-plane sapphire (0001) and MgO (100) [62, 63] has been investigated.

Homoepitaxial growth of β-Ga$_2$O$_3$ has been studied on various β-Ga$_2$O$_3$ planes including (100), (010), ($\bar{2}$01), and (001) [61, 64–66]. Since large-scale (4′) β-Ga$_2$O$_3$ (001) substrates are available, this plane is of particular interest for the fabrication of cost-effective power devices. However, the epitaxial growth of Ga$_2$O$_3$ films by MBE on this plane has been challenging due to low growth rates and rough surface morphologies.

It has been observed that the growth rate on the (100) plane is much lower than for the other orientations. The lower growth rate on this plane is due to the lower adhesion energy on (100) terraces and, therefore, higher desorption of atoms supplied to the (100) terraces. This is the most stable surface and is one of the cleavage planes due to the weak bonds on the (100) surface and leads to enhance-ment of the step-flow growth mode [59]. Villora *et al* [63] monitored β-Ga$_2$O$_3$ film growth on a β-Ga$_2$O$_3$ (100) substrate at a substrate temperature of 820 °C and under a chamber pressure of 3.3×10^{-3} Pa, *in situ* through RHEED, as can be seen in figure 4.7. During the growth, they observed that the streak lines became sharper, while the RHEED intensity oscillated intensively, suggesting a layer-by-layer growth mode.

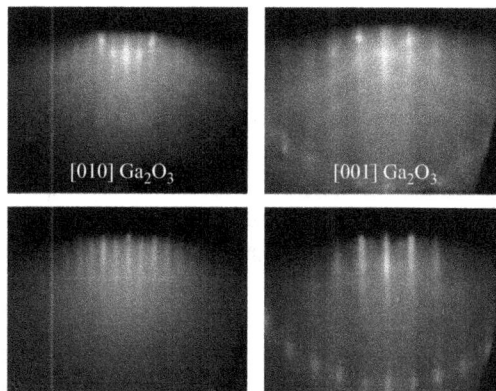

Figure 4.7. RHEED images along the [010] and [001] azimuths. Top: β-Ga$_2$O$_3$ (100) substrate. Bottom: 0.75 μm epilayer. Reproduced with permission from [63]. Copyright 2006 AIP Publishing.

Oshima *et al* [64] investigated the homoepitaxial growth and etching character-istics of (001) β-Ga$_2$O$_3$ using PAMBE. They showed that the growth rate of β-Ga$_2$O$_3$ increased with increasing Ga flux and reached a plateau of 56 nm h^{-1} and then decreased at a higher Ga flux. The reduction in growth rate by further increasing the Ga flux was attributed to the formation of volatile Ga$_2$O, as will be discussed in more detail in the following. The maximum achievable growth rate was decreased when the substrate temperature was increased from 750 °C to 800 °C. It was also shown that the growth rate was negative (net etching) when only Ga flux was supplied and the etch rate would increase by increasing Ga flux and substrate temperature. It was found that Ga-etching of (001) β-Ga$_2$O$_3$ substrates prior to the homoepitaxial growth markedly improved the surface roughness of the film. The growth of β-Ga$_2$O$_3$ thin films on ($\bar{2}$01)-oriented β-Ga$_2$O$_3$ substrate using PAMBE has been also demonstrated by Hao *et al* [67]. However, the surface roughness was relatively high (an RMS surface roughness of ~7 nm).

Homoepitaxial growth of Ga$_2$O$_3$ using PAMBE has been mostly studied on β-Ga$_2$O$_3$(010) [67, 68] due to the higher-quality films and smoother surface morphology as well as higher growth rates achievable on this plane. Okumura *et al* [69] grew 60 nm β-Ga$_2$O$_3$(010) layers on a chemical mechanical polishing-treated Sn-doped n-type β-Ga$_2$O$_3$(010) substrate and investigated the film quality as a function of growth temperature. The smooth surface of the β-Ga$_2$O$_3$(010) films could be attributed to the step-flow growth mode [70] due to the large miscut angle (>2°) of the substrate. Kaun *et al* [65] also studied the growth of β-Ga$_2$O$_3$ (010) as a function of substrate temperature. Their investigations indicated a suitable substrate temperature range of 600 °C–675 °C for the homoepitaxial growth of β-Ga$_2$O$_3$ (010). At a substrate temperature below 600 °C, layer-by-layer growth occurred, whereas at a substrate temperature above 675 °C, multi-steps were formed.

In PAMBE, the growth of Ga$_2$O$_3$ is typically performed under oxygen-rich conditions by a supply of more O than Ga atoms to the growth surface. This contrasts with the growth of other Ga- or O-based semiconductor films, such as GaN [71] and In$_2$O$_3$ films [72], where metal-rich growth conditions result in improved structural properties. The growth rate (Γ) of Ga$_2$O$_3$ is limited by Ga flux (Φ_{Ga}) under O-rich conditions and thus increases linearly with Φ_{Ga} up to the stoichiometric Ga flux (Φ_{SF}), which consumes all available oxygen atoms for Ga$_2$O$_3$ formation. Higher Φ_{Ga} fluxes ($\Phi_{Ga} > \Phi_{SF}$) lead to a Ga-rich growth regime, where Γ decreases with increasing Φ_{Ga}. This is attributable to the oxygen-deficiency-induced formation of volatile suboxide, Ga$_2$O, which desorbs from the growth surface rather than contributing to the formation of Ga$_2$O$_3$ [73]. Nevertheless, Ga-rich growth conditions could reduce the formation of Ga vacancies that act as compensating acceptors in Ga$_2$O$_3$ [57].

According to Vogt *et al* [58], the reactions involved in the Ga$_2$O$_3$ growth and suboxide formation on c-sapphire substrate by PAMBE are expressed as

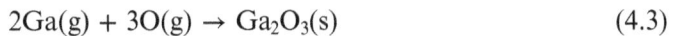

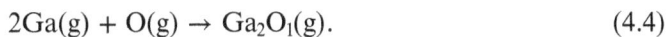

$$2Ga(g) + 3O(g) \rightarrow Ga_2O_3(s) \tag{4.3}$$

$$2Ga(g) + O(g) \rightarrow Ga_2O_1(g). \tag{4.4}$$

The letters, 'g' and 's', inside the first bracket denote the gas and solid phase, respectively. They studied the growth dynamics of Ga_2O_3 by MBE using a quadruple mass spectrometer. The Φ_{Ga}, Ga–O ratio (r_{Ga}), and growth temperature (T_G) help us understand the reaction kinetics of Ga_2O_3 growth. At a low T_G of 500 °C in the O-rich regime, i.e. at $r_{Ga} \leqslant 1$, Γ increases linearly with Φ_{Ga}. All of the Ga atoms are fully incorporated into the films and hence the sticking coefficient is unity in this case. After reaching a maximum at $r_{Ga} = 1$, the growth regime switches to Ga-rich at 660 °C and Γ shows a linear decrease due to suboxide formation. As can be seen in the growth diagram in figure 4.8, there are three distinctive growth regimes: (i) complete Ga incorporation (i.e. Ga transport-limited growth regime (O-rich)); (ii) plateau of growth rate at increasing Φ_{Ga} (i.e. Ga_2O-desorption-limited growth regime (O-rich)); and (iii) a decreasing growth rate at increasing Φ_{Ga} (i.e. O-transport-limited growth regime (Ga-rich)). No Ga desorption was observed at the investigated temperatures.

4.4.2 Growth of β-Ga_2O_3 using a Ga_2O_3 compound source

A Ga_2O_3 compound source was used previously to deposit polycrystalline Ga_2O_3 films [74, 75] as gate-insulating dielectrics. Monoclinic β-Ga_2O_3 films have also been heteroepitaxially grown on a c-plane sapphire substrate using MBE with a poly-crystalline Ga_2O_3 compound source. The growth of Ga_2O_3 films using the compound source follows the reactions [76]

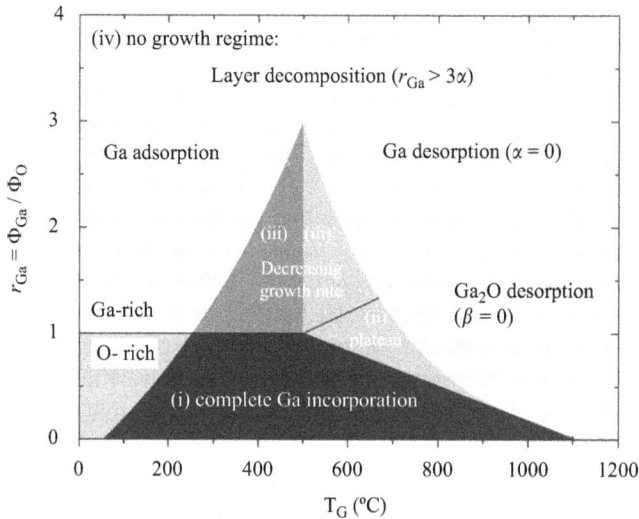

$$Ga_2O_3 \text{ (heat to 1750 °C)} \rightarrow Ga_2O + O_2 \qquad (4.5)$$

Figure 4.8. A growth diagram (GD) for the Ga_2O_3 MBE growth including all experimental parameters (Φ_{Ga}, Φ_O) and T_G. It is divided into two major regimes: O-rich and Ga-rich. The GD illustrates regimes of complete (i), partial (ii, iii, iii$'$), and no Ga incorporation (iv) as a function of T_G and r_{Ga}. Reproduced with permission from [58]. Copyright 2016 AIP Publishing.

$$Ga_2O + 2O \rightarrow Ga_2O_3 \text{ or } Ga_2O + O_2 \rightarrow Ga_2O_3. \tag{4.6}$$

The Ga_2O_3 compound source is thermally decomposed to Ga_2O and oxygen molecules. The oxygen deficiency found in the Ga_2O_3 films using the elemental source is expected to be reduced via this technique since there is no requirement of oxidizing the metallic Ga atoms. Ghose *et al* [76] compared the quality of the films obtained with the use of a Ga_2O_3 compound source alone and a combination of the compound source and an RF plasma source, and obtained no clear difference. Thus, this suggests that the compound source provides enough oxygen for the nucleation of a crystalline Ga_2O_3 film. The authors also observed that the films grown at 500 °C showed amorphous behavior, while the films grown at a substrate temperature $\geqslant$600 °C exhibited $(\bar{2}01)$-oriented Ga_2O_3 films on a c-sapphire substrate. The crystal quality of the films grown at a substrate temperature of 700 °C was analyzed using high angle annular dark field (HAADF), as can be seen in figure 4.9. The analysis revealed the orientation relationship of $(\bar{2}01)$-β-$Ga_2O_3\|(0001)$ Al_2O_3, a sharp Ga_2O_3/Al_2O_3 interface, and no noticeable phase separation.

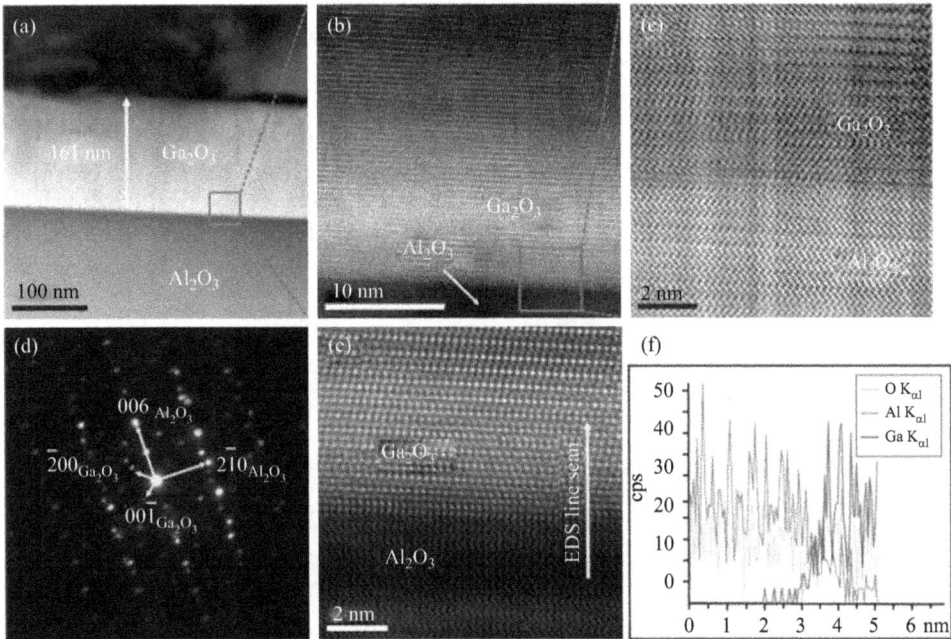

Figure 4.9. (a) Low magnification and (b) high magnification of an HAADF image of a Ga_2O_3/sapphire heterointerface. (c) Atomic-resolution of the HAADF image of the heterointerface taken from the small red square marked in (b). (d) selected-area electron diffraction pattern of Ga_2O_3 films. (e) HAADF image of the interface where the energy dispersive spectroscopy (EDS) line scan direction is marked in yellow. (f) EDS line scan profile of the yellow arrow shown in (e). Reproduced with permission from [76]. Copyright 2017 AIP Publishing.

4.4.3 Growth of β-Ga$_2$O$_3$ using ozone-MBE

Using ozone-MBE, Sasaki *et al* [77] reported the growth of β-Ga$_2$O$_3$ films on a Si-doped n-type β-Ga$_2$O$_3$ substrate and Mg-doped semi-insulating β-Ga$_2$O$_3$ substrates. As a source of oxygen, a mixture of ozone (5%) and oxygen (95%) was used. An undoped β-Ga$_2$O$_3$ film growth was carried out on a Si-doped n-type β-Ga$_2$O$_3$ (010) substrate at 600 °C, under a Ga BEP of 2×10^{-4} Torr and with a mixed (ozone (5%) and oxygen (95%)) flow of 5 sccm. The grown films exhibited an RMS roughness of 0.7 nm. Similar to β-Ga$_2$O$_3$ growth using PAMBE, it was found that the growth rate on the (010) and (310) planes ($\sim$125 nm h^{-1}) was much higher than that on the cleavage planes, such as the (100) plane ($\sim$10 nm h^{-1}). This growth technique has been used by the Hiashiwaki group to fabricate metal–oxide semiconductor FETs (MOSFETs) [78, 79, 90].

4.5 Investigation of deep level defects and traps in MBE-grown β-Ga$_2$O$_3$

Both shallow and deep level defects in semiconductors can greatly influence device performance, such as threshold voltage, carrier mobility, on-resistance, and DC-to-RF dispersion through trapping effects [80, 81]. Hence, it is important to study and better understand the properties of various defects in β-Ga$_2$O$_3$ films and their impact on device performance. Farzana *et al* [82] investigated deep level defects generated in Ge-doped β-Ga$_2$O$_3$ films, grown using PAMBE, through deep level transient spectroscopy and deep level optical spectroscopy (DLOS) techniques, and compared the results to traps in bulk unintentionally doped (UID) β-Ga$_2$O$_3$ grown using edge-defined film-fed growth (FEG).

They also compared the energy levels and concentrations of the deep level defects generated in EFG-grown UID β-Ga$_2$O$_3$ (010) and PAMBE-grown Ge-doped β-Ga$_2$O$_3$ (010), as shown schematically in figure 4.10. The trap states near $E_c - 4.37$ eV,

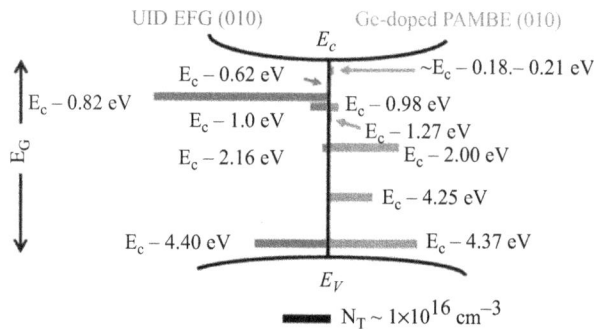

Figure 4.10. Summary of the distribution (energy levels and concentrations) of deep level defect states detected for an EFG-grown UID (Si background limited n-type) β-Ga$_2$O$_3$ (010) substrate and Ge-doped β-Ga$_2$O$_3$ (010) PAMBE epitaxial layers. The energy values of the DLOS detected levels in the figure are extracted from the Pässler model. The horizontal bars revealing the concentration of each individual trap state in the figure have been drawn to the scale of 1×10^{16} cm^{-3} as depicted by the black line ($N_T \sim 1 \times 10^{16}$ cm^{-3}). Reprinted from [82] with the permission of AIP Publishing.

E_c − 2.0 eV, and E_c − 0.98 eV exist with similar concentrations in both samples grown using PAMBE or EFG. The E_c − 2.0 eV trap state is very important as it lies near the midgap and has high concentrations, and hence can serve as an effective center in the recombination-generation process. According to Varley et al [83], oxygen vacancies are predicted to serve as deep donors with (0/2) transition states at −1.6, 2.2, and 1.3 eV below E_c for V_O^I, V_O^{II}, and V_O^{III}, respectively. A self-trapped hole in a small polaron state, h + 1, with transition states (+/0) was estimated to be at 4.61 eV [84]. Thus the DLOS detected energy levels at −2.0 and −4.4 eV can be attributed to V_O^{II} and h^+1, respectively [84]. The trap states at E_c − 3.25, E_c − 1.7 eV, and E_c − 0.18–0.21 eV measured on the PAMBE-grown sample were absent in the FEG-grown sample. On the other hand, the trap state at E_c − 0.82 eV measured on the EFG-grown sample was absent in the PAMBE-grown sample. This level was found to be dominant in all β-Ga_2O_3 substrates synthesized either by FEG or melt-based bulk crystal growth methods [85, 86]. The calculations performed by Deak et al [84] suggest that Ga vacancies with (−2/−3) transition states appear at 0.67 and 1.16 eV below E_c for V_{Ga}^I and V_{Ga}^{II}, respectively. Therefore, the trap levels at E_c − 0.82 eV measured on the FEG-grown sample and at E_c − 1.0 eV measured on both samples could be attributed to V_{Ga}^I and V_{Ga}^{II}, respectively.

Neal et al [87] studied the electron density and mobility of samples grown by a variety of methods, including EFG, Czochralski, MBE, and low pressure chemical vapor deposition via temperature-dependent van der Pauw and Hall effect measurements. Through simultaneous, self-consistent fitting of the temperature-dependent carrier density and mobility, they estimated a similar ionization energy of ~30 meV for Si and Ge in β-Ga_2O_3. They also observed an order of magnitude higher density of compensating centers in MBE-grown samples compared to those in all the other samples. This could be attributed to the diffusion of Fe from the insulating substrate into the thin film n-type doped Ga_2O_3 (~300 nm) grown using PAMBE [88].

4.6 The status of dopants in MBE-grown β-Ga_2O_3

PAMBE has been used successfully to fabricate devices such as deep ultra-violet photodetectors [89], MOSFETs [90], and MODFETs [70, 91]. To enhance the device performance further, controllable doping is essential. It is widely believed in the community that it is not possible to achieve p-type doping in β-Ga_2O_3. Regardless, magnesium (Mg) [92, 93] and iron (Fe) [94, 95] have been identified as deep acceptors which can be utilized to achieve insulating Ga_2O_3 substrates.

Si, Ge, and Sn have been considered as promising n-type dopants in Ga_2O_3 and all of them substitute in Ga sites. Sn^{4+} is predicted to favor an octahedrally coordinated Ga (II) site, whereas Ge^{4+} preferentially incorporates in a tetrahedrally coordinated Ga (I) site [84, 96]. This makes Ge^{4+} a more effective dopant than Sn^{4+}. The estimated activation energies are 16–50 meV for Si [97, 98], 7.4–60 meV for Sn [99, 100], and 17.5 meV for Ge [96], but as mentioned earlier an activation energy of 30 meV has been measured by Neal et al [87] for both Si and Ge in β-Ga_2O_3.

Krishnamoorthy et al [101] studied the Si-doping of β-Ga_2O_3 using PAMBE and systematically investigated the effect of Si oxidation in the plasma environment as a function of the time/thickness of the films and Si cell temperature and found strong source oxidation and reduction of Si flux during growth, which led to a non-uniform Si-doping profile. To avoid this issue, a shutter pulsing scheme was introduced. The Si cell was pulsed for 1 s (with a duty cycle of 1 min) while maintaining the Si source at a relatively high temperature of 850 °C–900 °C to decompose the oxide. This is referred to as the Si delta doping approach, using which a nearly flat doping profile was achieved. Very recently, the same group studied the origin of high Si flux during the Si-doping of Ga_2O_3 using PAMBE [102]. They concluded that Si flux is not limited by the vapor pressure of Si. In the oxide environment volatile SiO forms, which has a low sublimation energy and, therefore, its flux has weak dependence on Si cell temperature and, instead, a strong dependence on the background oxygen pressure. However, extended exposure to activated oxygen leads to the reduction of the SiO flux because of the formation of SiO_2 on the Si surface. Therefore, at a relatively low Si effusion cell temperature (lower than that typically used in nitrides), a very high doping concentration can be achieved and the Si-doping concentration cannot be controlled in a wide range by changing the cell temperature.

Sasaki *et al* [77] reported a Sn-doped n-type Ga_2O_3 layer on Mg-doped semi-insulating Ga_2O_3 (010) substrates at 700 °C using ozone-MBE with SnO_2 powder as the Sn source and estimated an Sn dopant density of 7×10^{17} cm^{-3} from secondary ion mass spectroscopy (SIMS) measurements. To achieve a wide range of Sn concentration in β-Ga_2O_3 through ozone-MBE, the substrate temperature was reduced to 600 °C. As shown in figure 4.11, a carrier concentration ranging from 10^{16} to 10^{19} cm^{-3} was obtained. This was the first report where the authors claimed to have achieved a controllable carrier concentration in thick homoepitaxially grown Ga_2O_3 films. Sn doping has also been demonstrated using PAMBE in Ga_2O_3 films grown on b-plane and c-plane β-Ga_2O_3 substrates [66, 68] and Sn concentrations ranging from 5×10^{16} to 1×10^{20} cm^{-3} have been achieved via this technique. The doping concentration was shown to be independent of the growth

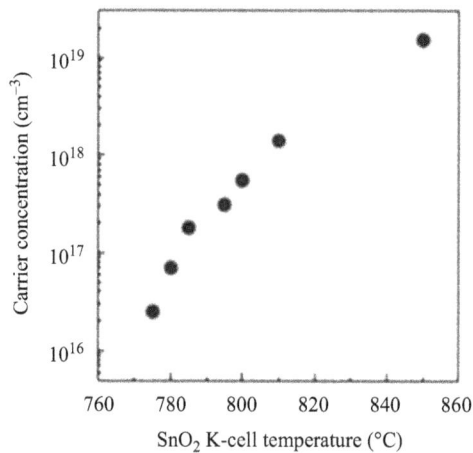

Figure 4.11. Carrier concentration as a function of SnO_2 K-cell temperature. Reproduced with permission from [77]. Copyright IOP Publishing, reproduced with permission, all rights reserved.

temperature in the range studied (600 °C–800 °C). Nonetheless, due to oxidation of Sn over time (elemental Sn was used as the Sn source in these studies), achieving reproducible Sn concentrations for a specific Sn cell temperature has been challenging.

The Ge doping of Ga_2O_3 films on b-plane and c-plane β-Ga_2O_3 substrates has been also demonstrated [68]. It was shown that O-rich growth conditions drastically improved Ge incorporation. It was also shown that, in contrast to Sn/Si-doping, the Ge concentration in the Ga_2O_3 films is strongly dependent on the substrate temperature and by increasing the substrate temperature, the Ge concentration is reduced significantly. Additionally, an increasing Ge cell temperature initially increased Ge incorporation but subsequently decreased it. Figure 4.12 shows Ge and Sn concentrations in Ga_2O_3 films grown on a β-Ga_2O_3 (001) substrate using PAMBE.

4.7 β-$(Al_xGa_{1-x})_2O_3$/Ga_2O_3 heterostructures and superlattices

Similar to the (Al, Ga, In)N and (Al, Ga, In)As material systems, alloying β-Ga_2O_3 with In and Al to form β-(Al, Ga, In)$_2O_3$ films and their heterostructures is desirable as these materials enable the design and fabrication of novel device structures with superior characteristics. For example, β-$(Al_xGa_{1-x})_2O_3$/Ga_2O_3 heterostructures could be used for fabrication of MODFETs which can be potentially very important for high-power and high-frequency applications. However, alloying β-Ga_2O_3 with Al or In is challenging due to their limited solubility in Ga_2O_3 [103, 104]. Additionally, Al_2O_3 ($E_g = 8.7$ eV [105]) and In_2O_3 ($E_g = 2.9$ eV [106]) are thermally stable in corundum [107] and bixbyite [108] crystal structures, respectively. Nevertheless, β-$(Al_xGa_{1-x})_2O_3$ films with Al content as high as 20% and 40% have been demonstrated using PAMBE [65] and MOCVD [109], respectively. Oshima *et al* [110] has also demonstrated the epitaxial growth of $(In_xGa_{1-x})_2O_3$/Ga_2O_3 hetero- structures on a c-sapphire substrate using PAMBE. When a $(In_xGa_{1-x})_2O_3$ layer was

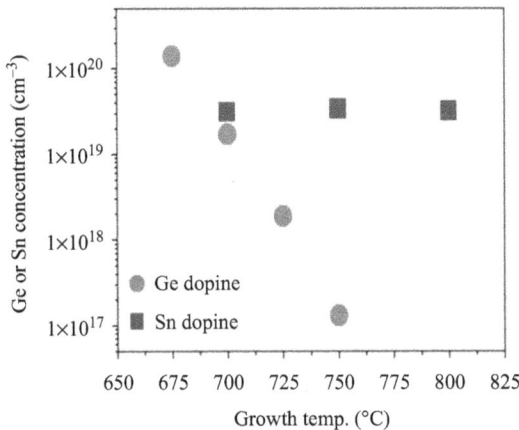

Figure 4.12. Ge and Sn concentrations as a function of growth temperature as studied using SIMS. Reproduced with permission from [66]. Copyright 2018 IOP Publishing Ltd and Deutsche Physikalische Gesellschaft.

grown at a high substrate temperature of 700 °C–800 °C, In segregation and phase separation were observed for In content as low as 8%. On the other hand, it was shown that a low growth temperature (~600 °C) was necessary to suppress the phase separation. They showed $(In_xGa_{1-x})_2O_3$ films of In composition of up to 35% with ($\bar{2}01$)-oriented monoclinic structure and sharp absorption edges with clear fringes.

According to the Ga_2O_3–Al_2O_3 phase-diagram constructed by Hill *et al* [111], the solubility limit of Al_2O_3 in Ga_2O_3 in the temperature range of 850 °C–1950 °C was reported to be 65%. However, MBE typically uses a range of temperatures of 600 °C–800 °C which greatly reduces the solubility limit of Al_2O_3 in Ga_2O_3 due to the formation of an $AlGaO_3$ intermediate compound.

The PAMBE growth of $(Al_xGa_{1-x})_2O_3$/Ga_2O_3 heterostructures was first systemically investigated by Kuan *et al* [65] on Fe-doped β-Ga_2O_3 (010) substrates in a substrate temperature range of 600 °C–750 °C. In this study, since at the time a model was not yet developed to determine the composition from the HRXRD profile for monoclinic crystal structure, the compositions were determined through Al:Ga flux ratios. At a substrate temperature of 600 °C, β-$(Al_xGa_{1-x})_2O_3$ (010) films with Al content ranging from 10% to 18% were demonstrated (figure 4.13). The HRXRD profile for films with 15% and lower Al content showed Pendellosung fringes, indicating films with high quality. In contrast, the layer peak corresponding to the β-$(Al_{0.18}Ga_{0.82})_2O_3$ film was broad and barely noticeable and no Pendellosung fringes could be observed in the HRXRD profile measured on this sample, indicating poor film quality as the Al content in the film increased. Increasing the substrate temperature to 650 °C improved the film quality.

Since then, modulation-doped β-$(Al_xGa_{1-x})_2O_3$/Ga_2O_3 heterostructures grown on Fe-doped β-Ga_2O_3(010) substrates using PAMBE have been used for the fabrication of MODFETs by several groups [91, 111, 113]. Ahmadi *et al* [112] and Zhang *et al* [114] have separately reported $(Al_xGa_{1-x})_2O_3$–Ga_2O_3 modulation-doped FETs using Ge and Si as n-type dopants in the barrier, respectively. As discussed earlier, Ge incorporation in β-Ga_2O_3 films reduces as the substrate temperature increases [73]. Therefore, there is a trade-off between achieving a

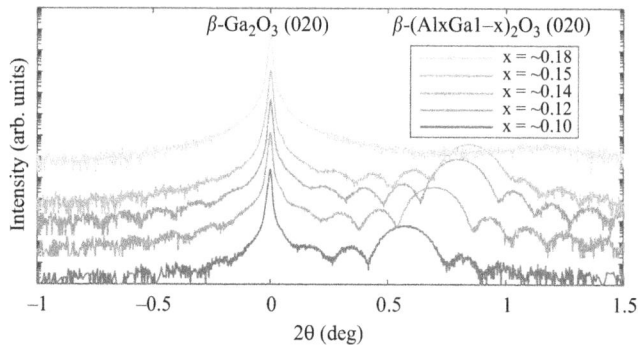

Figure 4.13. Symmetric HRXRD ω–2θ scans around the β-Ga_2O_3 (020) reflection for the Al flux series at 600 °C, which included ~60 nm thick β-$(Al_xGa_{1-x})_2O_3$ (010) layers on a β-Ga_2O_3 (010) substrate. Reproduced from [65] with the permission of AIP Publishing.

high concentration of n-type doping in the barrier and growing high-quality $(Al_xGa_{1-x})_2O_3$ film, when Ge is used as the dopant. In contrast, Si incorporation does not depend on the substrate temperature and allows high growth temperatures for $(Al_xGa_{1-x})_2O_3$ films, and consequently films with higher Al content are achievable [68].

High Al content in $(Al_xGa_{1-x})_2O_3$ is, in particular, important to realize high-density two-dimensional electron gas (2DEG) in these heterostructures. 2DEG density depends on the δ-doping concentration in the barrier, the distance between the δ-doped layer and the channel, and the conduction band discontinuity (ΔE_C) between the Ga_2O_3 and $(Al_xGa_{1-x})_2O_3$ (figure 4.14). Higher Al content leads to larger ΔE_C which not only results in higher 2DEG density, but also allows for a larger doping concentration in the barrier before a parasitic channel can be formed. Nonetheless, as mentioned earlier, increasing Al content beyond 20% in $(Al_xGa_{1-x})_2O_3$ grown using PAMBE has been challenging as it leads to poor quality films and phase separation [115]. Metal–oxide catalyzed epitaxy (MOCATAXY) was recently developed by the Speck group at UCSB to enhance the quality of $(Al_xGa_{1-x})_2O_3$ films with high Al content [116]. This technique has been previously used to expand the growth regime of β-Ga_2O_3 using PAMBE, allowing the growth to be performed at higher temperatures and with higher Ga flux to achieve higher growth rates. In this technique, indium is used as a catalyst and involves an exchange mechanism between Ga and In in In_2O_3 formed at the growth surface through the reactions shown below [117]:

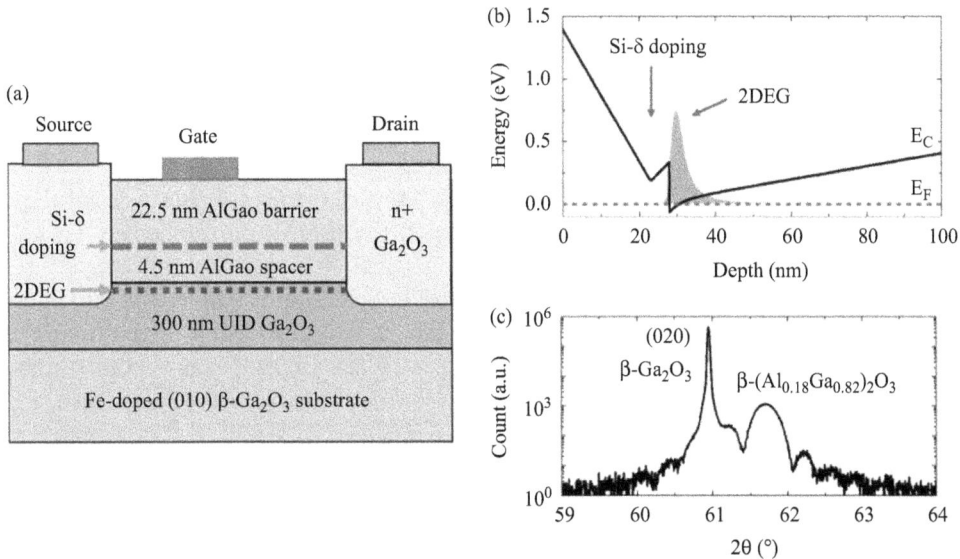

$$2In(a) + 3O(a) \rightarrow In_2O_3(s) \tag{4.7}$$

Figure 4.14. (a) Schematic of the MODFET device. (b) Equilibrium energy band diagram and charge distribution profile under a free surface. (c) XRD ω–2θ diffraction pattern for (020) plane reflections. Reproduced with permission from [120]. Copyright 2019 IEEE.

$$2Ga(a) + In_2O_3(s) \rightarrow Ga_2O_3(s) + 2In(a). \qquad (4.8)$$

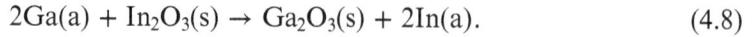

It was expected that, similar to AlGaAs–GaAs or AlGaN–GaN heterostructures, higher electron mobilities can be achieved by forming a 2DEG in modulation-doped β-$(Al_xGa_{1-x})_2O_3$/Ga_2O_3 heterostructures beyond what can be achieved in n-type doped β-Ga_2O_3 films. However, the highest room-temperature electron mobility reported so far in β-$(Al_xGa_{1-x})_2O_3$–Ga_2O_3 double-heterostructures has been only 180 cm^2 V^{-1} s^{-1} for a 2DEG density of ~2×10^{12} cm^{-2} limited by phonon scattering [49, 118, 119]. The electron mobility increased to 2790 cm^2 V^{-1} s^{-1} at 50 K. This high electron mobility allowed for observation of Shubnikov–de-Haas oscillations from which an electron effective mass of 0.33m_e was extracted. Achieving high electron mobility at low temperatures also allowed this group to measure the saturation velocity for this material system [120]. They reported a saturation velocity of above 1.1×10^7 cm s^{-1} at 50 K. Very recently, the same group achieved a 2DEG sheet charge density up to 6.1×10^{12} cm^{-2} with a mobility of 147 cm^2 V^{-1} s^{-1} by reducing the spacer thickness between the Si δ-doping and heterojunction interface in β-$(AlGa)_2O_3$/Ga_2O_3 modulation-doped structures. This is the highest 2DEG density achieved in these heterostructures so far.

A high breakdown voltage of about 1370 V has been measured on a β-$(AlGa)_2O_3$/Ga_2O_3 MODFET with a gate-to-drain separation of 16 μm [121]. These breakdown characteristics were shown to be limited by the Schottky gate, suggesting that developing a reliable dielectric with a high breakdown voltage for Ga_2O_3 is of crucial importance for high-power applications.

4.8 Summary

There has been rapid progress in the bulk and epitaxial growth of Ga_2O_3 as well as the fabrication of Ga_2O_3-based electronic and optoelectronic devices. In particular, MBE has played and will continue to play an important role in expanding our understanding of fundamental properties, including doping, defects, deep/shallow trap levels, and heterostructures in this material system. Additionally, the first demonstration of various β-Ga_2O_3-based FETs was made possible using MBE-grown epi-structures. There remain many open questions and challenges for the community to explore. Some of these challenges include achieving high Al content β-$(Al_xGa_{1-x})_2O_3$ films, increasing 2DEG density and mobility in β-$(Al_xGa_{1-x})_2O_3$–Ga_2O_3 modulation-doped heterostructures, and realizing an n-type dopant source which enables reliable and consistent doping with a wide range of concentrations and effective p-type dopants.

References

[1] Tomm Y, Reiche P, Klimm D and Fukuda T 2000 *J. Cryst. Growth* **220** 510–4
[2] Higashiwaki M, Sasaki K, Kuramata A, Masui T and Yamakoshi S 2012 *Appl. Phys. Lett.* **100** 013504
[3] Víllora E G, Shimamura K, Yoshikawa Y, Aoki K and Ichinose N 2004 *J. Cryst. Growth* **270** 420–6

[4] Suzuki N, Ohira S, Tanaka M, Sugawara T, Nakajima K and Shishido T 2007 *Phys. Status Solidi* C **4** 2310–3

[5] Aida H, Nishiguchi K, Takeda H, Aota N, Sunakawa K and Yaguchi Y 2008 *Jpn J. Appl. Phys.* **47** 8506

[6] Galazka Z *et al* 2010 *Cryst. Res. Technol.* **45** 1229–36

[7] Tomm Y, Ko J, Yoshikawa A and Fukuda T 2001 *Sol. Energy Mater. Sol. Cells* **66** 369–74

[8] Orita M, Hiramatsu H, Ohta H, Hirano M and Hosono H 2002 *Thin Solid Films* **411** 134–9

[9] Murakami H *et al* 2014 *Appl. Phys. Express* **8** 015503

[10] Oshima Y, Víllora E G and Shimamura K 2015 *J. Cryst. Growth* **410** 53–8

[11] Kim H W and Kim N H 2004 *Mater. Sci. Eng.* B **110** 34–7

[12] Sasaki K, Higashiwaki M, Kuramata A, Masui T and Yamakoshi S 2013 *J. Cryst. Growth* **378** 591–5

[13] Jian A Z, Khan K and Ahmadi E 2019 *Int. J. High Speed Electron. Syst.* **28** 1940006

[14] Ahmadi E and Oshima Y 2019 *J. Appl. Phys.* **126** 160901

[15] Roy R, Hill V G and Osborn E F 1952 *J. Am. Chem. Soc.* **74** 719–22

[16] Playford H Y, Hannon A C, Barney E R and Walton R I 2013 *Chem. Eur. J.* **19** 2803

[17] Marezio M and Remeika J P 1967 *J. Chem. Phys.* **46** 1862

[18] Shinohara D and Fujita S 2008 *Jpn J. Appl. Phys.* **47** 7311–3

[19] Kohn J, Katz G and Broder J D 1956 *Am. Minerol* **42** 398

[20] Matsuzaki K, Hiramatsu H, Nomura K, Yanagi H, Kamiya T, Hirano M and Hosono H 2006 *Thin Solid Films* **496** 37

[21] Zinkevich M, Morales F M, Nitsche H, Ahrens M, Ruhle M and Aldinger F 2004 *Z. Met. Res. Adv. Tech* **95** 756

[22] Oshima T, Nakazono T, Mukai A and Ohtomo A 2012 *J. Cryst. Growth* **359** 60–3

[23] Playford H Y, Hannon A C, Barney E R and Walton R I 2013 *Chemistry* **19** 2803–13

[24] Nishinaka H, Tahara D and Yoshimoto M 2016 *Jpn J. Appl. Phys.* **55** 1202bc

[25] Janowitz C *et al* 2011 *New J. Phys.* **13** 085014

[26] Geller S 1960 *J. Chem. Phys.* **33** 676–84

[27] Yamaguchi K 2004 *Solid State Commun.* **131** 739

[28] Zhang L, Yan J, Zhang Y, Li T and Ding X 2012 *Sci. China Phys. Mech. Astron* **55** 19

[29] Varley J B, Weber J R, Janotti A and Van de Walle C G 2010 *Appl. Phys. Lett.* **97** 142106

[30] He H, Orlando R, Blanco M, Pandey R, Amzallag E, Baraille I and Rerat M 2006 *Phys. Rev.* B **74** 195123

[31] Janowitz C *et al* 2011 *New J. Phys.* **13** 085014

[32] Peelaers H and Van de Walle C G 2015 *Phys. Status Solidi* B **252** 82

[33] Mock A, Korlacki R, Briley C, Darakchieva V, Monemar B, Kumagai Y, Goto K, Higashiwaki M and Schubert M 2017 *Phys. Rev.* B **96** 245205

[34] Villora E G, Shimamura K, Ujiie T and Aoki K 2008 *Appl. Phys. Lett.* **92** 2

[35] Galazka Z *et al* 2014 *J. Cryst. Growth* **404** 184

[36] Higashiwaki M, Sasaki K, Murakami H, Kumagai Y, Koukitu A, Kuramata A, Masui T and Yamakoshi S 2016 *Semicond. Sci. Technol.* **31** 34001

[37] Mengle K A and Kioupakis E 2019 *AIP Adv.* **9** 015313

[38] Mu S, Peelaers H and Van de Walle C G 2019 *Appl. Phys. Lett.* **115** 242103

[39] Peelaers H, Varley J B, Speck J S and Van de Walle C G 2018 *Appl. Phys. Lett.* **112** 242101

[40] Rebien M, Henrion W, Hong M, Mannaerts J P and Fleischer M 2002 *Appl. Phys. Lett.* **81** 250

[41] Ghose S, Rahman S, Rojas-Ramirez S, Caro M, Droopad R, Arias A and Nedev N 2016 *J. Vac. Sci. Technol.* B **34** 02L109

[42] Ueda N, Hosono H, Waseda R and Kawazoe H 1997 *Appl. Phys. Lett.* **71** 933

[43] Kuznetsov A I, Abramov V N and Uibo T V 1985 *Opt. Spectrosc. (USSR)* **58** 368

[44] Binet L and Gourier D 1998 *J. Phys. Chem. Solids* **59** 1241

[45] Ho Q D, Frauenheim T and Deák P 2018 *Phys. Rev.* B **97** 115163

[46] Berencén Y, Xie Y, Wang M, Prucnal S, Rebohle L and Zhou S 2019 *Semicond. Sci. Technol.* **34** 035001

[47] Li Y, Tokizono T, Liao M, Zhong M, Koide Y, Yamada I and Delaunay J J 2010 *Adv. Funct. Mater.* **20** 3972

[48] Mengle K A, Shi G, Bayerl D and Kioupakis E 2016 *Appl. Phys. Lett.* **109** 212104

[49] Fujita M, Kawamoto N, Sasajima M and Horikoshi Y 2004 *J. Vac. Sci. Technol.* B **22** 1484

[50] Raghavan S, Schumann T, Kim H, Zhang J Y, Cain T A and Stemmer S 2016 *APL Mater.* **4** 016106

[51] Wang T *et al* 2018 *ACS Appl. Mater. Interfaces* **10** 21061

[52] Thoutam L R, Yue J, Prakash A, Wang T, Elangovan K E and Jalan B 2019 *ACS Appl. Mater. Interfaces* **11** 7666

[53] Yue J, Prakash A, Robbins M C, Koester S J and Jalan B 2018 *ACS Appl. Mater. Interfaces* **10** 21061

[54] Cho A Y and Arthur J R 1975 *Prog. Solid State Chem.* **10** 157–91

[55] Franchi S, Trevisi G, Seravalli L and Frigeri P 2003 *Prog. Crys. Growth Charac. Mater.* **47** 166–95

[56] Brattain W H and Bardeen J 1990 *Phys. Rev. Lett.* **64** 1943–6

[57] Korhonen E, Tuomisto F, Gogova D, Wagner G, Baldini M, Galazka Z, Schewski R and Albrecht M 2015 *Appl. Phys. Lett.* **106** 242103

[58] Vogt P and Bierwagen O 2016 *Appl. Phys. Lett.* **108** 072101

[59] Tsai M Y, Bierwagen O, White M E and Speck J S 2010 *J. Vac. Sci. Technol.* A **28** 354–9

[60] Jalan B, Engel-Herbert R, Wright N J and Stemmer S 2009 *J. Vac. Sci. Technol.* A **27** 461–4

[61] Son J, Moetakef P, Jalan B, Bierwagen O, Wright N J, Engel-Herbert R and Stemmer S 2010 *Nat. Mater.* **9** 482

[62] Oshima T, Okuno T and Fujita S 2007 *Jpn J. Appl. Phys.* **46** 7217–20

[63] Villora E G, Shimamura K, Kitamura K and Aoki K 2006 *Appl. Phys. Lett.* **88** 031105

[64] Oshima Y, Ahmadi E, Kaun S, Wu F and Speck J S 2018 *Semicond. Sci. Technol.* **33** 015013

[65] Kaun S W, Wu F and Speck J S 2015 *J. Vac. Sci. Technol.* A **33** 041508

[66] Han S H, Mauze A, Ahmadi E, Mates T, Oshima Y and Speck J S 2018 *Semicond. Sci. Technol.* **33** 045001

[67] Hao S J *et al* 2019 *J. Appl. Phys.* **125** 105701

[68] Ahmadi E, Koksaldi O S, Kaun S W, Oshima Y, Short D B, Mishra U K and Speck J S 2017 *Appl. Phys. Express* **10** 041102

[69] Okumura H, Kita M, Sasaki K, Kuramata A, Higashiwaki M and Speck J S 2014 *Appl. Phys. Express* **7** 095501

[70] Balykov L and Voigt A 2006 *Surf. Sci.* **600** 3436–45

[71] Calleja E *et al* 1999 *J. Cryst. Growth* **201** 296

[72] Bierwagen O, White M E, Tsai M Y and Speck J S 2009 *Appl. Phys. Lett.* **95** 262105

[73] Vogt P and Bierwagen O 2015 *Appl. Phys. Lett.* **106** 081910

[74] Priyantha W, Radhakrishnan G, Droopad R and Passlack M 2011 *J. Cryst. Growth* **323** 103

[75] Droopad R, Rajagopalan K, Abrokwah J, Adams L, England N, Uebelhoer D, Fejes P, Zurcher P and Passlack M 2007 *J. Cryst. Growth* **301** 139

[76] Ghose S, Rahman S, Hong L, Rojas-Ramirez J S, Jin H, Park K, Klie R and Droopad R 2017 *J. Appl. Phys.* **122** 095302

[77] Sasaki K, Kuramata A, Masui T, Víllora E G, Shimamura K and Yamakoshi S 2012 *Appl. Phys. Express* **5** 035502

[78] Singh M *et al* 2018 *IEEE Electron Device Lett.* **39** 1572

[79] Wong M H, Sasaki K, Kuramata A, Yamakoshi S and Higashiwaki M 2016 *Jpn J. Appl. Phys.* **55** 1202B9

[80] Zhang Z *et al* 2016 *J. Appl. Phys.* **119** 165704

[81] Sasikumar A *et al* 2015 *IEEE Int. Reliability Physics Symp. (Monterey, CA)* IEEE

[82] Farzana E, Ahmadi E, Speck J S, Arehart A R and Ringel S 2018 *J. Appl. Phys.* **123** 161410

[83] Varley J B, Peelaers H, Janotti A and Walle C G V D 2011 *J. Phys. Condens. Matter* **23** 334212

[84] Deák P, Ho Q D, Seemann F, Aradi B, Lorke M and Frauenheim T 2017 *Phys. Rev. B* **95** 075208

[85] Ingebrigtsen M E, Varley J B, Kuznetsov A Y, Svensson B G, Alfieri G, Mihaila A, Badstübner U and Vines L 2018 *Appl. Phys. Lett.* **112** 042104

[86] Zhang Z, Farzana E, Arehart A R and Ringel S A 2016 *Appl. Phys. Lett.* **108** 052105

[87] Neal A T *et al* 2018 *Appl. Phys. Lett.* **113** 062101

[88] Joishi C, Xia Z, McGlone J, Zhang Y, Arehart A R, Ringel S, Lodha S and Rajan S 2018 *Appl. Phys. Lett.* **113** 123501

[89] Pratiyush A S, Krishnamoorthy S, Solanke S V, Xia Z, Muralidharan R, Rajan S and Nath D N 2017 *Appl. Phys. Lett.* **110** 221107

[90] Higashiwaki M, Sasaki K, Kamimura T, Wong M H, Krishnamurthy D, Kuramata A, Masui T and Yamakoshi S 2013 *Appl. Phys. Lett.* **103** 123511

[91] Krishnamoorthy S *et al* 2017 *Appl. Phys. Lett.* **111** 023502

[92] Tang C, Sun J, Lin N, Jia Z, Mu W, Tao X and Zhao X 2016 *RSC Adv.* **6** 78322

[93] Wong M H, Goto K, Kuramata A, Yamakoshi S, Murakami H, Kumagai Y and Higashiwaki M 2017 *75th Device Research Conf. (South Bend, IN)* IEEE 7999413

[94] He H, Li W, Xing H Z and Liang E J 2012 *Adv. Mater. Res.* **535–537** 36

[95] Wong M H, Sasaki K, Kuramata A, Yamakoshi S and Higashiwaki M 2015 *Appl. Phys. Lett.* **106** 032105

[96] Moser N *et al* 2017 *IEEE Electron Device Lett.* **38** 775

[97] Son N T *et al* 2016 *J. Appl. Phys.* **120** 235703

[98] Neal A T, Mou S, Lopez R, Li J V, Thomson D B, Chabak K D and Jessen G H 2017 *Sci. Rep.* **7** 13218

[99] Oishi T, Harada K, Koga Y and Kasu M 2016 *Jpn J. Appl. Phys.* **55** 030305

[100] Orita M, Ohta H and Hirano M 2000 *Appl. Phys. Lett.* **77** 4166

[101] Krishnamoorthy S, Xia Z, Bajaj S, Brenner M and Rajan S 2017 *Appl. Phys. Express* **10** 051102

[102] Kalarickal N K, Xia Z, McGlone J, Krishnamoorthy S, Moore W, Brenner M, Arehart A R, Ringel S A and Rajan S 2019 *Appl. Phys. Lett.* **115** 152106

[103] Edwards D D, Folkins P E and Mason T O 1997 *J. Am. Ceram. Soc.* **80** 253

[104] Jaromin A L and Edwards D D 2005 *J. Am. Ceram. Soc.* **88** 2573

[105] Olivier J and Poirier R 1981 *Surf. Sci.* **105** 347

[106] Irmscher K, Naumann M, Pietsch M, Galazka Z, Uecker R, Schulz T, Schewski R, Albrecht M and Fornari R 2014 *Phys. Status Solidi* **211** 54

[107] Ishizawa N, Miyata T, Minato I, Marumo F and Iwai S 1980 *Acta Crystallogr.* **B36** 228

[108] Marezio M 1966 *Acta Crystallogr.* **20** 723

[109] Bhuiyan A F M U, Feng Z, Johnson J M, Chen Z, Huang H L, Hwang J and Zhao H 2019 *Appl. Phys. Lett.* **115** 120602

[110] Oshima T and Fujita S 2008 *Phys. Status Solidi* **5** 3113–5

[111] Hill V G, Roy R and Osborn E F 1952 *J. Am. Ceram. Soc.* **35** 135

[112] Ahmadi E, Koksaldi O S, Zheng X, Mates T, Oshima Y, Mishra U K and Speck J S 2017 *Appl. Phys. Express* **10** 071101

[113] Okumura H, Kato Y, Oshima T and Palacios T 2019 *Jpn J. Appl. Phys.* **58** SBBD12

[114] Zhang Y *et al* 2018 *Appl. Phys. Lett.* **112** 173502

[115] Sarker J, Zhang Y, Zhu M, Rajan S, Hwang J and Mazumder B 2019 *Microsc. Microanal* **25** 2508

[116] Vogt P, Mauze A, Wu F, Bonef B and Speck J S 2018 *Appl. Phys. Exp* **11** 115503

[117] Vogt P, Brandt O, Riechert H, Lähnemann J and Bierwagen O 2017 *Phys. Rev. Lett.* **119** 196001

[118] Ghosh K and Singisetti U 2017 *J. Appl. Phys.* **122** 035702

[119] Ma N, Tanen N, Verma A, Guo Z, Luo T, Xing H and Jena D 2016 *Appl. Phys. Lett.* **109** 212101

[120] Zhang Y, Xia Z, Mcglone J, Sun W, Joishi C, Arehart A R, Ringle S and Rajan S 2019 *IEEE Trans. Electron Device* **66** 1574–8

[121] Joishi C, Zhang Y, Xia Z, Sun W, Arehart A R, Ringel S, Lodha S and Rajan S 2019 *IEEE Electron Device Lett.* **40** 1241–4

IOP Publishing

Wide Bandgap Semiconductor-Based Electronics

Fan Ren and Stephen J Pearton

Chapter 5

Defects and carrier lifetimes in Ga_2O_3

E B Yakimov and A Y Polyakov

In this chapter we briefly analyse the types of deep electron and hole traps in crystals and films of β-Ga_2O_3 doped with various impurities or subjected to irradiation with high energy particles. The defining features of this material are the very high bandgap close to 5 eV, the low symmetry of its crystalline structure compared to more established semiconductors (monoclinic versus cubic or hexagonal), deep in energy, the very flat valence band, the strong contribution of the ionic component in bonding, and prominent electron–phonon interaction. These features result in even simple structural defects such as vacancies having several nonequivalent configurations, in virtually all centers having multiple charge states within the bandgap, and in holes being converted into self-trapped polaronic states in the absence of suitable shallow acceptor dopants. The absence of band-edge luminescence and sharp excitonic transitions in photoluminescence, and the very wide luminescence bands related to defects, render the standard and very mature methods of defect characterization based on the analysis of the fine structure of photoluminescence (PL) or cathodoluminescence (CL) spectra not particularly informative. The suspicion that holes cannot actively participate in conductivity and photoconductivity processes in gallium oxide puts a big question mark over the standard methods of characterization of nonequilibrium charge carriers kinetics based on the well advanced methods of electron beam induced current (EBIC) or microcathodoluminescence (MCL). The very deep levels of the majority of defects and impurities in gallium oxide predicted by theory make the determination of deep trap spectra very challenging using standard deep level transient spectroscopy with electrical or optical injection (DLTS or ODLTS). In what follows we show that these complications can be, at least in part, overcome. Experimental studies show that, although holes in β-Ga_2O_3 do form self-trapped polaronic states, their binding energy is most likely lower than predicted by theory which renders the holes capable of contributing to conductivity at temperatures above ~120 K, with the diffusion length of nonequilibrium holes being quite high, on the order of 0.4–0.5 µm in good quality material. It is found that the host of deep spectrum characterization techniques previously developed for the characterization of wide bandgap III-nitrides and comprising capacitance and current DLTS, ODLTS, deep levels optical spectroscopy (DLOS), high-temperature/low-frequency admittance spectroscopy, photocapacitance spectroscopy, and capacitance–voltage C–V profiling under monochromatic illumination (LCV) measurements, provides effective instruments for the quantitative assessment of deep trap spectra over the entire bandgap of gallium oxide. Although suitable shallow acceptor dopants are indeed absent in Ga_2O_3, deep acceptors, notably Fe, Mg, and N can be practically used for the formation of high-resistivity buffers or for device isolation in Ga_2O_3-based devices. In several instances, the positive identification of the nature of defects giving rise to deep traps in gallium oxide has been achieved and for many other traps their attribution to native defects or their complexes with defects or impurities has been established based on the results of radiation effect studies. These radiation effect studies have also made it possible to develop some ideas regarding the possible identities of defects that are efficient recombination centers in gallium oxide.

5.1 Introduction

Recent years have seen a growing interest in studies of the properties of Ga_2O_3 and related compounds, manifesting in an avalanche-like increase of the number of published research papers and reviews (some of the recent reviews are listed in [1–4], the number of original papers is so large that it would be counterproductive to try to compile a full list). The reason is that this material system is viewed as an attractive alternative to the more mature III-nitrides and SiC material systems in the realm of high-power/high-temperature/high-frequency devices and solar-blind photodetectors. The members of the Ga_2O_3 family that are most in the public eye are the thermodynamically stable β-polytypes of Ga_2O_3 and $(Al_xGa_{1-x})_2O_3$ $(In_xGa_{1-x})_2O_3$ ternary solutions. The main selling points of these emerging materials are the very high bandgap of β-Ga_2O_3 which is close to 4.8 eV and can be increased or decreased by the addition of, respectively, Al or In. Estimates suggest that the electrical breakdown field of β-Ga_2O_3 can reach ~8 MV cm^{-1}, considerably higher than for GaN or SiC, while preserving the very high saturation velocity of electrons close to 2×10^7 cm s^{-1}. An additional bonus compared to GaN or SiC is that Ga_2O_3 can be prepared in the form of high crystalline quality bulk crystals using all standard liquid growth techniques—Czochralski, edge-defined film-fed growth (EFG), and different versions of gradient-freeze techniques [1–4]—which, it is hoped [4], will be conducive to the commercial fabrication of relatively inexpensive substrates for homo- and heteroepitaxy. Epitaxial films and heterostructures of very good crystalline quality can be produced using various standard growth approaches—halide vapor phase epitaxy (HVPE), different versions of metal–organic vapor phase epitaxy (MOVPE), molecular beam epitaxy (MBE), laser beam deposition (LPD), and even magnetron sputtering (see the recent reviews with references to multiple original papers in [1–4]). The huge potential of the Ga_2O_3-based system has already started to materialize: large diameter n-type and semi-insulating β-Ga_2O_3 substrates and high-quality homo-epitaxial structures fabricated using a combination of EFG and HVPE are commercially available from several vendors, most notably the Tamura Corporation/Novel Crystals Technologies Inc. (Japan) [5]. AlGaO/GaO hetero-structures suitable for the fabrication of heterojunction field effect high electron mobility transistors (HEMTs) are commercially available from Tamura Corporation/Novel Crystals Technologies Inc. (Japan) [5], with the Ga_2O_3 films grown using HVPE and the $(Al_xGa_{1-x})_2O_3$ barriers fabricated using MBE. Such structures have been successfully fabricated by a number of research groups (see e.g. [6–8]). Prototype high-power devices with a breakdown voltage and ON-resistance comparable to advanced GaN or SiC-based devices have been reported, as have Schottky diode controlled field effect transistors (FETs), metal–insulator–semi-conductor FETs (MISFETs), AlGaO/GaO HEMTs with very promising character-istics, and solar-blind photodetectors with very high photosensitivity in the UV range and a high rejection ratio of sensitivity in visible light compared to UV have been described by several research groups (a compilation of achieved results with multiple references can be found in [1, 2]). Given the types of devices that are being developed, one naturally needs detailed information on the dopants and defects suitable for controllably producing n-type and p-type conductivity and semi-insulating crystals and films, on the impact of these defects and impurities on the lifetimes of nonequilibrium charge carriers, on trapping in FETs and HEMTs, on the thermal stability of grown materials, and the impact of irradiation or of operation under high driving currents at high temperatures. Much of this informa-tion is currently already available due to the diligent work of many research groups involved in Ga_2O_3 studies. It should be noted that the work on β-Ga_2O_3 system has many specific features. This is perhaps the first semiconductor material of possible importance for practical applications in electronics that crystallizes in a low-symmetry monoclinic system. The result is the existence of three inequivalent oxygen vacancies, two inequivalent gallium vacancies, and multiple combinations of inequivalent antisite defects, interstitial defects, to say nothing of thee various possible split configurations and variously arranged defect complexes. The very wide bandgap of gallium oxide is conducive to the amphoteric behavior of most defects and impurities and to inherent problems with effectively obtaining both types of conductivity. The ionic nature of the material is favorable to the formation of polaronic states and creates problems with correctly accounting for the electronic and ionic screening [9–11]. The width of the bandgap also results in problems with mapping the defect state distribution throughout the bandgap using standard well established spectroscopic tools, such as deep level transient spectroscopy (DLTS) [12]. Luckily, in this theoretical and experimental work one can draw heavily on the vast experience accumulated when studying other wide bandgap semiconductors and transparent oxides. The theoretical analysis developed for dealing with these systems and based on the density functional theory (DFT) formalism (see e.g. [13]) has been widely used to analyse the band structure, crystalline structure, phonon spectra, defect formation energies, charge transfer levels in the gap, and migration energies of defects in β-Ga_2O_3 (see e.g. [14–19]). The deep level optical spectroscopy (DLOS) [12] combined with steady-state photocapacitance (SSPC) and with capacitance–voltage $(C-V)$ measurements under monochromatic illumination [12] developed and widely used for analysis of defect states in III-nitrides have been successfully applied to defect analysis in β-Ga$_2$O$_3$ (see e.g. [20–26]). Together with DLTS with electrical and optical (ODLTS) injection [12, 21, 23–30], photoinduced current transient spectroscopy (PICTS) [25, 29–31], current DLTS (CDLTS) [25, 30, 31], and admittance spectroscopy [25, 30, 31], these spectroscopic tools allow us to discover much about the defect level spectra in the conducting and semi-insulating crystals and films of β-Ga$_2$O$_3$ and β-Ga$_2$O$_3$ FETs [32] or HEMTs [33]. Hopefully, systematic efforts in that direction will allow us to identify the nature of the major defects contributing to charge trapping and carrier recombination in β-Ga$_2$O$_3$ crystals, films, and heterostructures, albeit at present this work has just begun. These studies, when put in the context of other defect studies in transparent conducting and insulating oxides, certainly have an independent scientific worth and fully justify the efforts undertaken, irrespective of whether or not the widespread implementation of β-Ga$_2$O$_3$-based devices in high-power electronics will happen. An additional interest in β-Ga$_2$O$_3$ studies is inspired by the fact that it is relatively easy to detach thin quasi-two-dimensional nanolayers from the crystals and films of β-Ga$_2$O$_3$ that can be transferred easily to other substrates (see e.g. [32, 34, 36]) and whose properties present much interest in view of certain limitations inherent in the Ga oxide material system, such as the lack of viable p-type conductivity and the low thermal conductivity [1–4]. In what follows we present our view of the current situation of defect studies in Ga$_2$O$_3$. In the first part of this chapter we briefly summarize the results of the theoretical and experimental studies of defects and impurities in β-Ga$_2$O$_3$, and the impact of impurities and defects on conductivity, compensation, and charge trapping in Ga$_2$O$_3$-based structures and devices. Then, in the second section, we try to tie up these results with our current knowledge and understanding of nonequilibrium charge carrier recombination processes in this material.

5.1.1 Summary of the results of theoretical and experimental defect and impurity studies in β-Ga$_2$O$_3$

Several density functional theory (DFT) calculations of the electronic properties of β-Ga$_2$O$_3$ and of the properties of the defects in this material have been undertaken [9–11, 13–19]. Qualitatively, all the results converge to give the same picture. The formalism allows us to correctly predict the band structure of β-Ga$_2$O$_3$, with the indirect bandgap being slightly lower than the direct bandgap and both being close to 4.8 eV. The matrix elements for the indirect electronic transition are, however, considerably lower than for the direct transition, so the material can be considered effectively as a direct bandgap [9–11]. The theory also correctly predicts the experimentally observed lattice parameters and the observed lattice vibration mode frequencies. The distinctive feature of the reported results is that the valence band is quite flat, with a high hole effective mass [10, 11, 13]. Because of the strong ionic component of bonding and the strong electron–phonon interaction in β-Ga$_2$O$_3$, the holes when they are formed show a propensity for turning into highly localized polaronic states [10, 11, 13]. The analysis of the localization of such polaronic states performed in [10, 11, 13] suggests that they are positioned close to the oxygen atoms in the lattice. There are three inequivalent positions of these atoms in the β-Ga$_2$O$_3$ lattice, the so-called O1, O2, and O3 sites [10, 11, 13] of which only the O1 and O2 sites are expected to bind the polarons that have slightly different binding energies in the respective positions (see figure 5.1 taken from [10]). It can be seen that the binding energies of polarons are slightly different and close to 0.8 eV (calculations performed in [34] give a slightly lower value of about 0.53 eV). When electrons are excited from the valence band by above-bandgap light or some other external excitation they form a tightly bound self-trapped exciton with the polaronic state. The configuration-coordinate diagram of respective transition proposed in [34] is depicted in figure 5.2. The energy barrier for the electron transfer to the bound exciton branch is low, close to 0.1 eV according to [34] and can be easily overcome by tunneling. Then the recombination transition occurs with the photon energy of 3.1 eV according to [34] or at two slightly different energies of 3.6 and 3.2 eV according to [10], which explains the total failure to observe the band-edge luminescence in β-Ga$_2$O$_3$ (see e.g. [35]).

These theoretical results have important implications. As shown below there are no shallow acceptors suitable for p-type doping in β-Ga$_2$O$_3$, but from the results of the cited theoretical calculations it follows that even if holes could be created in Ga$_2$O$_3$, e.g. by illumination or by forming a heterojunction with some p-type material, they should not actively participate in current transport. The polaronic states have been invoked by Deak et al [10] to explain a hole-trap-like feature for photon energies near 4.4 eV in DLOS spectra [12, 20, 21]. Such features should not

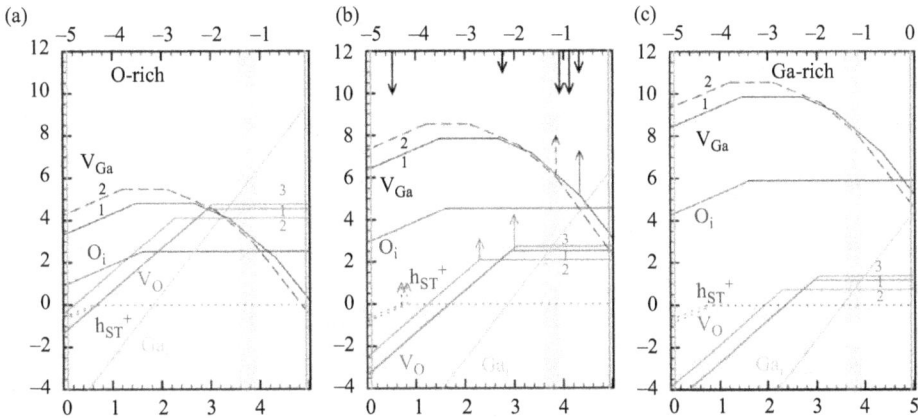

Figure 5.1. Defect formation energies (eV) as a function of the Fermi level position between the valence band maximum (VBM) and the conduction band minimum (CBM) for (a) extremely oxygen-rich, (b) intermediate, and (c) extreme oxygen-poor growth conditions. The Fermi level position is with respect to the VBM. The formation energies for Ga vacancies in two Ga substitutional positions (V$_{Ga1}$ and V$_{Ga2}$), oxygen vacancies for the three inequivalent oxygen atom positions, oxygen interstitials (O$_i$), interstitial Ga (Ga$_i$), and two self-trapped hole polaronic states hST are shown. Reproduced with permission from [10]. Copyright 2017 the American Physical Society.

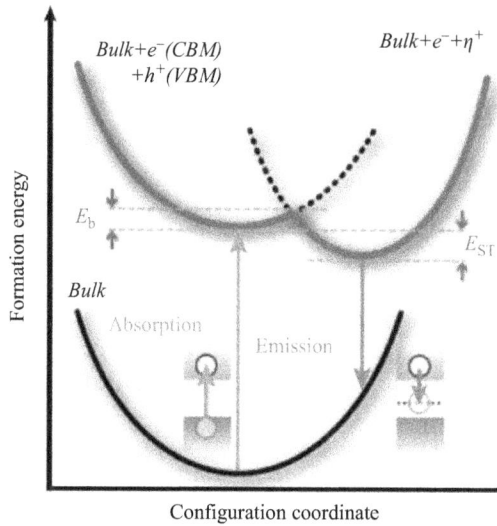

Figure 5.2. Schematic configuration-coordinate diagrams showing energy as a function of the lattice distortion in the formation of a self-trapped hole (STH). The top curve illustrates the barrier E_b between a free hole (h+) and the STH (η+); here, the latter is assumed to be lower in energy by E_{ST}. The optical absorption process creates a free hole at the VBM and an electron at the CBM. Emission results when the electron recombines with the hole, which has become localized due to self-trapping. Reproduced with permission from [34]. Copyright 2017 the American Physical Society.

in principle be observed in β-Ga$_2$O$_3$ where the most shallow acceptor levels available are located above E_v + 1 eV. Also, in [36], the high photosensitivity of β-Ga$_2$O$_3$ Schottky diodes to UV light was ascribed to the appearance under illumination of the immobile built-in positive polaronic charge lowering the Schottky barrier height and thus increasing the current under illumination compared to dark current.

Because of the importance of the matter for the assessment of the prospects of β-Ga$_2$O$_3$-related devices, we performed direct measurements of the binding energy of the polaronic states by measuring ODLTS spectra at low temperatures under above-bandgap light excitation pulses (figure 5.3). In these spectra we detected a prominent hole-trap-like signal near 110 K (figure 5.4) that consisted of two peaks that could be related to the two polaronic states predicted by Deak *et al* [10]. Measurements of the activation energy for this double peak feature yielded the value of 0.2–0.3 eV for both peaks [26]. Because of the doublet structure of the peak and of the very low measured activation energy of the process this feature must belong to the transition of the bound polarons into the true valence band holes that can move in the electric field of the space charge region of the Schottky diode and contribute to the transient capacitance signal in ODLTS (polarons here act similar to hole traps [26]). Recall, however, that the actually measured activation energies are very much lower than predicted by theory and the holes become mobile at quite low temperatures only slightly exceeding ~120 K [26]. The holes should then definitely be mobile at room temperatures. In section 5.2 we present additional experimental evidence for this being the case. These considerations put the prospects of β-Ga$_2$O$_3$

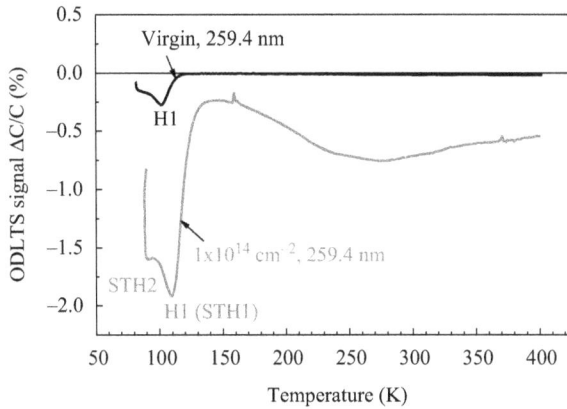

Figure 5.3. ODLTS spectra taken with 259.4 nm wavelength UV LED excitation for a HVPE sample before irradiation (black line) and after irradiation with 20 MeV protons with a fluence of 10^{14} cm^{-2} (red line) at a reverse bias of −1 V and with a 5 s long excitation light pulse. The time windows were 250 ms/2500 ms.

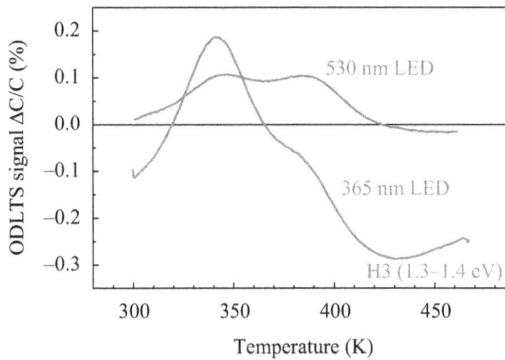

Figure 5.4. High-temperature ODLTS spectra taken for a sample irradiated with 5×10^{13} cm^{-2} 20 MeV protons with 365 nm LED excitation (red line) and 530 nm LED excitation (olive line). Measurements were taken with time windows of 125 ms/1250 ms, a bias of −1 V, a 5 s long light pulse. The H3 hole-trap-like feature was detected with 365 nm excitation, but not with 530 nm excitation.

devices in an entirely different light because there do exist wide bandgap p-type materials that could conceivably form p–n heterojunctions with n-type β-Ga$_2$O$_3$ films (see the discussion in [1]).

One can also mention that in the same paper mentioned above [26] we also observed a hole-trap-like peak corresponding to a process with an activation energy of 1.3–1.4 eV that could be detected with excitation with light whose photon energy was sufficient for transferring electrons from the filled acceptor state near $E_v + 1.4$ eV, but not for more longer-wavelength excitation. Again, the observation of such a feature is difficult to reconcile with the holes in β-Ga$_2$O$_3$ being immobile near room temperature. The problem is, of course, that there exist very few viable ways to determine independently the mobility of such nonequilibrium holes. In [37] the authors independently measured the diffusion length of holes from electron beam induced current and the carrier lifetime from the decay time of the time resolved cathodoluminescence spectra. The latter turned out to be close to 215 ps, which, in conjunction with the measured hole diffusion length of 330 nm, results in a surprisingly high hole mobility of about 200 cm^2 V^{-1} s^{-1}. This, of course, needs more study, but it should be mentioned that the lifetimes calculated from optically probing conduction electrons excited by a fast intrinsic light pulse [38] also resulted in quite fast values of some hundreds of picoseconds.

Returning now to the possibility of intentional p-type doping, it has to be concluded that, in theory, no viable impurities or defects giving rise to shallow acceptor states could be found [15, 16]. Figure 5.5, taken from [16], illustrates the theoretical predictions.

It can be seen that all considered substitutional acceptor impurities give rise to quite deep (0/−) acceptor transition levels totally precluding their possible use in the capacity of shallow acceptor dopants. Moreover, all considered impurities also possess donor (0/+) levels in the lower half of the bandgap that would dominate conductivity for crystals with low-lying Fermi level (the energies of transfer levels slightly differ depending on whether the atom substituting for Ga is located in the tetrahedral Ga1 site or octahedral Ga2 site, while for atoms substituting for oxygen

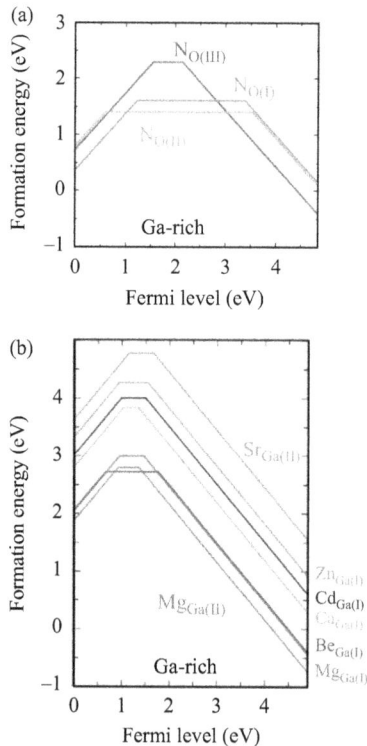

Figure 5.5. Calculated dependences of the formation energy of potential acceptor dopants substituting (a) oxygen on the three possible oxygen sites (the data are for nitrogen) and (b) Ga on the two possible Ga sites. Reproduced from [16] with the permission of AIP Publishing.

(N) there are three inequivalent sites corresponding to the three possible oxygen sites and slightly differing in energy). The situation is complicated by the fact that these deep acceptors can form complexes with hydrogen or other donor impurities [16], and that atoms can be incorporated in the 'wrong' sublattice, occupy simple or split interstitial sites where they perform as donors or acceptors, and can be compensated by donor type native defects. Only when the incorporated deep acceptor impurity absolutely dominates in concentration can the position of the charge transfer levels be unambiguously determined from the Fermi level pinning position for the deep acceptor doped crystal. It should be noted that the absolute values of charge transition levels differ considerably for different calculations and need to be verified by actual measurements. The case of Mg has been experimentally studied in some detail [17, 29, 39].

These papers illustrate the difficulties in constructing the level pattern of Mg in Mg compensated β-Ga$_2$O$_3$. Neal *et al* [39] report the activation energy of resistivity in Mg-doped β-Ga$_2$O$_3$ crystals to be close to 1.1 eV and assume that this energy corresponds to the position of the (0/−) acceptor level of Mg with respect to the valence band (no Hall effect measurements could be reliably performed in such high-resistivity crystals). In our paper [29] we report the activation energy of conductivity to be close to 1.3 eV and conclude from the dependence of the peak magnitude in PICTS spectra on the polarity of the voltage applied to the Schottky diode that the material is high-resistivity n-type. This fact is ascribed to the strong interference of deep donor traps near E_c − 1.35 eV, possibly related to native defects and the possibly lower Mg concentration compared to that in [39]. The peak in PICTS spectra specific to Mg doping was placed in [29] close to E_v + 1.05 eV. The formation of Mg complexes with hydrogen was clearly demonstrated using local vibrational mode (LVM) spectroscopy in [17] and points to additional possible complications.

The case of nitrogen was considered theoretically in [16] in all its complexity, resulting from the existence of different nitrogen donor and acceptor species and the compensation of nitrogen by native defects. It was found that the Fermi level under oxygen-rich growth conditions should be stabilized between E_v + (3–3.5) eV caused by the interaction of nitrogen acceptor species and deep oxygen donors. We are not aware of any detailed studies of the N level positions in β-Ga$_2$O$_3$ films or crystals, but it could be pointed out that nitrogen implantation has been demonstrated to be able to produce thermally stable highly compensated regions in n-type β-Ga$_2$O$_3$ films [40].

It should be noted that the electronic properties of Ga vacancies do not differ strongly from those of standard acceptor dopants occupying the Ga sites (see figure 5.1) and Ga vacancies can be considered as important compensating species in n-type β-Ga$_2$O$_3$ films and crystals. Moreover, it has been demonstrated [41] that, in the absence of shallow donor contamination and under oxygen-rich growth conditions suppressing the formation of compensating oxygen vacancies, undoped β-Ga$_2$O$_3$ films clearly show p-type conductivity at high temperatures and the position of the dominant acceptor species near E_v + 1.1 eV has been ascribed to the singly ionized Ga vacancy acceptors $V_{Ga}{}^{1-}$. Gallium vacancies in a very high concentration on the order of 10^{18} cm^{-3} have indeed been detected in nominally undoped and n-type doped irradiated β-Ga$_2$O$_3$ using positron annihilation [42] and these high values of compensating vacancies illustrate the difficulties in obtaining good electron conductivity of the material.

From the above it is clear that β-Ga$_2$O$_3$ is expected to exist as either a semi-insulator when compensated by acceptor impurities or native defects or as an n-type semiconductor when suitable n-type dopants or native defects prevail. It was for a long time thought that the natural n-type conductivity of undoped β-Ga$_2$O$_3$ comes from the contribution of oxygen vacancy donors. However, theoretical calculations show such defects to be (a) deep donors with ionization energies above 1 eV and (b) to have a high formation energy in n-type material [9]. Rather, as in III-nitrides, the shallow donors are provided by group IV impurities substituting Ga or by group VI impurities substituting oxygen.

Theoretical analysis based on DFT has been undertaken for potential group IV shallow donors in several papers [9, 13, 19]. Through obvious considerations these dopants can occupy oxygen sites where they should behave as doubly charged acceptors or gallium sites where they are expected to be doubly charged donors. Calculations clearly show that occupation of the oxygen sites is not energetically favorable. As for substituting for Ga there are two types of Ga sites, the tetrahedrally coordinated Ga1 site and octahedrally coordinated Ga2 site. As shown by Lany [19], Si and C prefer the Ga1 site irrespective of whether the center is a standard donor or a DX-like acceptor. Ge prefers the Ga1 site when positively charged and Ga2 site when negatively charged. Sn prefers the octahedral site in both charge states. For Si, the charge transfer level e(+/−) from straightforward donor state to the DX-like state lies deep in the conduction band. Ge and Sn possess DX-like negative states at, respectively, E_c − 0.29 eV and E_c − 0.19 eV [9, 13, 19]. Moreover, for Sn on the Ga1 site the DX state is quite deep, close to 0.4 eV and in principle could be observed under very nonequilibrium growth conditions or, for example, after irradiation [19]. Carbon forms a deep DX-like center near E_c − 0.8 eV [19]. Interestingly, all centers in the true donor state show a propensity for forming complexes with Ga vacancy acceptors, thus decreasing the effective donor concentration, particularly under oxygen-rich growth conditions [19] (this can also be a factor when the sample is irradiated with high energy particles). Experimental studies do not confirm the classical DX-like behavior for any of the shallow donors [23, 39], although there are some indications that part of the Si donors can, to a certain degree, perform similarly to DX defects [43]. Also relatively shallow traps with levels near the predicted DX-like acceptor states have been detected in MBE-grown β-Ga$_2$O$_3$ films doped with Ge [20] and deep traps near E_c − 0.28 eV were detected in Si doped β-Ga$_2$O$_3$ films irradiated with α-particles [27]. One wonders if it may be that the normal donor–DX-center transition for donors in β-Ga$_2$O$_3$ is facilitated by higher densities of defects.

Hydrogen on the oxygen site and interstitial hydrogen are expected to behave as shallow donors with the e(+/−) charge transfer level inside the conduction band [9]. However, it has been shown in a recent paper [44], where the higher formation energy states disrupting the Ga bonds rather than interacting with O lone pairs bonds were considered, that the H$_i$ acceptor in such a state can fall in the vicinity of

$E_c - 0.5$ eV, close in energy to the position of the so-called charge neutrality level CNL near $E_c - 0.6$ eV in the material with a high density of defects [44]. The latter demarcates the surface states that are negatively and positively charged and suggests the Fermi level pinning position in, e.g., heavily irradiated material in reasonable agreement with recent experimental results on heavily proton irradiated n-type β-Ga_2O_3 [11].

Experimental studies of the hydrogen states and hydrogen-related vibration modes for hydrogen introduced by high-temperature annealing confirms the hydrogen behavior as a shallow donor, as predicted in [9] (see [45]). On the other hand, when hydrogen is introduced from hydrogen plasma under conditions bound to create considerable surface damage, there are clear indications that hydrogen can be introduced as the H acceptor and form neutral complexes with shallow donors strongly passivating the near-surface hydrogenated region [30].

The electronic and structural properties of some other potentially important defects have also been studied theoretically. It has been found that, in addition to straightforward Ga vacancies, complexes of two adjacent Ga vacancies with an almost interstitial Ga atom can be formed in three inequivalent configurations differing by the site of adjacent oxygen atoms left with dangling bonds [11, 13]. These Ga divacancy–interstitial complexes are denoted as V_{Gai}^a, V_{Gai}^b, or V_{Gai}^c following [11]. Their formation energy is quite low in n-type material under oxygen-rich growth condition. These are amphoteric centers with the charge transfer levels corresponding to triply charged, doubly charged, and singly charged acceptors, and singly charged deep donors. Figure 5.6, taken from [11], shows the dependence of the formation energies and the positions of the charge transfer levels as a function of the Fermi level position for oxygen-rich growth conditions. Naturally, the formation energies become very high under Ga-rich growth conditions when the dominant defects are oxygen vacancies. The properties of triply charged Ga interstitial donors, Ga_i, and Ga_O antisite triply charged donors on three different oxygen sites have also

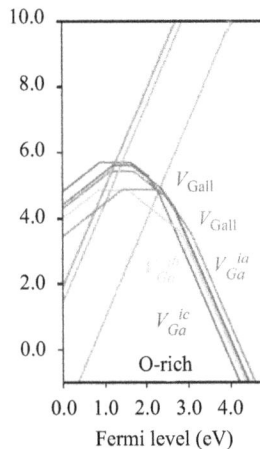

Figure 5.6. Formation energy diagram of gallium vacancy configurations and the most favorable interstitial gallium shown for Ga- O-rich conditions. Reprinted from [11] with permission from AIP Publishing.

been considered [11]. The formation energies were found to be high and such defects are not likely to dominate the behavior of as-grown crystals, but can play a role in strongly nonequilibrium conditions, e.g. under irradiation [11]. Other important defects with a low formation energy in n-type material grown under oxygen-rich conditions are the oxygen interstitials O_i that are deep acceptors with a transition level from the doubly negatively charged to singly negatively charged state (1−/2−) near E_c − 1.2 eV and split oxygen interstitials O_{si} that are deep donors with the level (1+/0) near E_c − 3.3 eV [11].

All primary defects can form complexes with other native defects or with impurity atoms, particularly when the centers comprising the complex are oppositely charged. A good example is provided by the formation of complexes between oxygen vacancies and Ga divacancies with Ga interstitial V_{Gai} and between such divacancies and one, two, or three hydrogen interstitials. Figure 5.7, taken from [11], displays the results of calculations for complexes with three different oxygen vacancies (left panel) and for three types of 'split' Ga vacancies V_{Gai} complexed with two

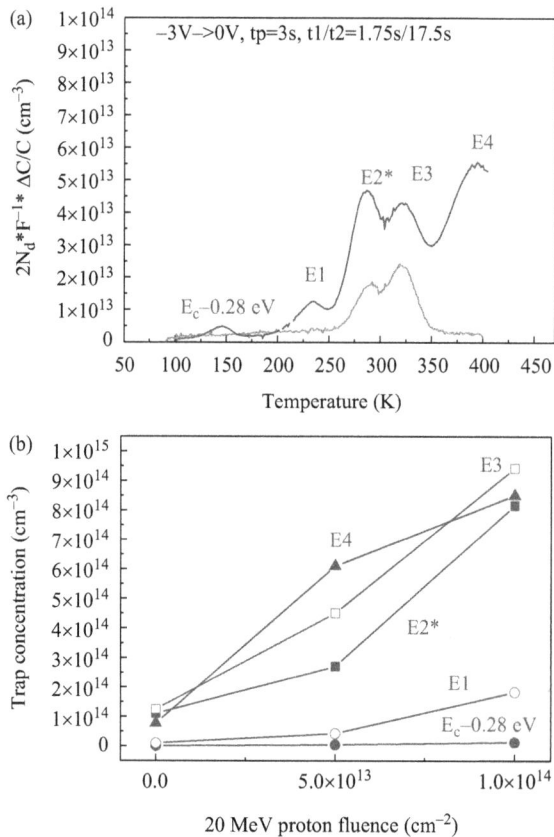

Figure 5.7. (a) DLTS spectra measured after irradiation of HVPE films with 20 MeV protons with a fluence of 10^{14} cm^{-2}. (b) Changes of defect concentrations as a result of 20 MeV irradiation with different fluences. Reproduced from [11] with the permission of AIP Publishing.

hydrogens that decrease the charge of the complex to either -1 or $+1$ depending on the Fermi level position and considerably shift the position of the ($-/0$) level closer to the valence band compared to uncomplexed defects. The results in figure 5.5 are shown for Ga-rich conditions. The formation energies of the complexes with hydrogen will naturally be lower for oxygen-rich growth.

It is useful, of course, to determine how mobile the defects discussed above are. If the defect in question is highly mobile, there is a good chance that it will not be encountered in the isolated form because it would be easy for such a defect to move around the crystal and to form complexes with other defects. On the other hand, defects with a high barrier for movement can be expected to survive as isolated species. Such calculations have been performed in several papers [11, 14, 16]. These calculations show that Ga and oxygen interstitials have a very low migration energy of ~0.1 eV. Oxygen and Ga vacancies require energies of about 1.7 eV, 'split' Ga vacancies V_{Gai} require energies of around 1–1.4 eV to start hopping from site to site. Such defects can in principle be frozen in and detected in spectroscopic studies, while, when produced by irradiation, can be manipulated by high-temperature anneals [11].

One more impurity of interest is Fe, considered in some detail in [46]. Calculations predict Fe to occupy Ga sites and to form acceptor levels in n-type material with charge transfer levels close to $E_c - 0.6$ eV and donor charge transfer levels near $E_v + 0.5$ eV in p-type. The formation energy of Fe acceptors is quite low, particularly under oxygen-rich growth conditions. Thus Fe is a very suitable dopant for compensating n-type conductivity in β-Ga$_2$O$_3$ and producing semi-insulating substrates and films for use in FETs. Tamura Corporation (Japan) is currently fabricating semi-insulating β-Ga$_2$O$_3$ substrates that are commercially available [5].

The results of these theoretical studies are widely used in attempts to identify the localized states detected in grown or variously processed β-Ga$_2$O$_3$ crystals and films. Such states present in parts of the bandgap about 1.5 eV from the conduction or valence band can be reliably detected using present day DLTS/ODLTS analysis of conducting samples with Schottky diodes or by PICTS analysis of semi-insulating crystals. Deeper centers across the entire bandgap can be probed by C–V profiling under monochromatic illumination, by steady-state photocapacitance (SSPC), and by DLOS [12]. Studies of persistent photocapacitance or persistent photoconductivity (PPC) make it possible to detect the presence of centers with strong electron–phonon interaction causing the appearance of appreciable barriers for the capture of electrons and separating the contribution of such centers to PPC effects from quasi-PPC phenomena related simply to the presence of very deep centers in the space charge region that, once recharged, cannot change their charge state other than by thermally emitting the captured carrier or by wiping off the excess charge by flooding the space charge region with charge carriers of opposite sign by application of forward bias [25–27]. These approaches previously developed and tested for III-nitrides (see e.g. [46]) allow us to carry out accurate book-keeping for all major centers in conducting crystals and, under some favorable circumstances, even in not too highly resistive semi-insulating samples (see e.g. [25, 30, 31]).

From the defects detected using DLTS, ODLTS, DLOS, and LCV a more or less consistent picture has emerged as a result of detailed studies performed by several groups [12–20]. A typical DLTS spectrum detected in β-Ga_2O_3 films grown using HVPE and doped with Si is shown in figure 5.7. The main contributions to the electron trap defect spectra always come from the so-called E2* centers and E3 centers. In some films one additionally observes more shallow E1 centers. When DLTS measurements are performed on bulk EFG or Czhochralski crystals the absolutely dominant ones are the so-called E2 centers with levels very close to the E2* traps but a higher electron capture cross-section [11, 23, 46]. For β-Ga_2O_3 films doped with Ge and grown using MBE, two additional shallow traps, E6 and E7, with a low concentration and energy levels near $E_c - 0.18$ eV and $E_c - 0.21$ eV have been detected in DLTS [20]. The results of proton irradiation [11, 24, 26], α-particle irradiation [27], and neutron irradiation [21] consistently show a prominent increase in densities of the E1, E2, and E3 traps, and the emergence of prominent new deeper traps E4 and E5, and shallow traps E8 [11, 21, 24, 26, 27]. The electron traps whose densities are most strongly affected by irradiation are the E2*, E3, and E4 traps (figure 5.7 shows the variations of the trap densities with 20 MeV fluence [27]). There is a certain spread of trap parameters, as reported in different papers, but the similarity in trap behavior with irradiation observed by different groups allows us to attribute the measured centers to the specified class of traps. ODLTS spectrum measurements were performed in [11, 24, 26, 27] and often revealed the presence of hole traps with thermal ionization energy close to 1.3–1.4 eV, traps H3 in table 5.1. These traps required that optical excitation was performed by photons with energy of at least 3.1 eV. The concentration of the traps varied from film to film and increased with irradiation.

In DLOS spectra the absolutely dominant centers always show the optical ionization energies of 2–2.16 eV and 4.37–4.4 eV [12, 20, 21]. In MBE-grown β-Ga_2O_3 films doped with Ge additional traps with optical ionization energies of 3.25 eV and 1.27 eV were detected (see figure 5.8, taken from [20]). The concentrations of the $E_c - 2$ eV centers were consistently high, in the 10^{16} cm^{-3} range, and increased with irradiation with neutrons. The concentration of the $E_c - 3.25$ eV DLOS centers was also quite high in Ge doped films, on the order of 10^{16} cm^{-3}, while the $E_c - 1.27$ eV DLOS centers in Ge-doped films had a low density of some 10^{14} cm^{-3}. Similar centers were detected in neutron irradiated crystals, with their concentration increasing with fast neutron fluence [21]. All three centers showed a high Franc–Condon shift indicative of the strong electron–phonon interaction and a barrier for capture of electrons [12]. The $E_c - 4.4$ eV DLOS feature was very prominent, not sensitive to irradiation [21] and had a low Franc–Condon shift [12, 20, 21].

In LCV spectrum measurements centers were systematically observed with optical thresholds close to 1.35, 2.3, and 3.1–3.4 eV [24–27]. The photocapacitance produced by illumination at low temperatures was persistent up to temperatures above room temperature (see figure 5.9(a)). The photocapacitance produced by illumination with photons having energies below ~3.1 eV could not be quenched by application of electron injection produced by high forward bias pulses. This persistent photocapacitance was due, therefore, to the presence of centers with a

Table 5.1. Characteristics of deep traps in β-Ga$_2$O$_3$.

Trap label	Energy position with respect to E_c (eV)	Capture cross-section (cm^2)	Franc–Condon shift DFC (eV)	Method, references	Possible origin
E1	0.55–0.62	3×10^{-14}	Not measured?	DLTS [11, 12, 20, 21, 24–27]	Introduced by irradiation, native defect
E2	0.74–0.82	3×10^{-15}	Not measured	DLTS [11, 12, 20, 21, 24–27]	Fe$_{Ga1}$
E2*	0.75–0.79	5×10^{-14}	Possibly capture barrier of ~0.3 eV	DLTS [11, 12, 20, 21, 24–27]	Introduced by irradiation, native defect
E3	0.95–1	6×10^{-13}	Not measured	DLTS [11, 12, 20, 21, 24–27]	Introduced by irradiation, native defect
E4	1.2–1.4	10^{-13}	Not measured	DLTS [11, 12, 20, 21, 24–27]	Introduced by irradiation, native defect
E5	1.35	3×10^{-12}	Not measured	DLTS [26]	Introduced by irradiation, native defect
E6	0.28	6×10^{-18}	Not measured	DLTS [26]	Introduced by irradiation, native defect
E7	0.18	5×10^{-19}	Not measured	DLTS [20]	Ge doping in MBE
E8	0.21	10^{-15}	Not measured	DLTS [20]	Ge doping in MBE
H3	3.4	3×10^{-12}*	Not measured	ODLTS [26]	Increases with irradiation, possibly native defect
E_c − 2 eV DLOS	$E_c - (2-2.16)$	n/a	0.52	DLOS [12, 20, 21]	V$_{Gai}$
E_c − 1.27 eV DLOS	$E_c - 1.27$	n/a	0.48	DLOS [12, 20, 21]	Not known
E_c − 3.25 eV DLOS	$E_c - 3.25$	n/a	0.33	DLOS [20]	Not known

(*Continued*)

Table 5.1. (*Continued*)

Trap label	Energy position with respect to E_c (eV)	Capture cross-section (cm^2)	Franc–Condon shift DFC (eV)	Method, references	Possible origin
E_c – 4.4 eV DLOS	E_c – (4.37–4.4)	n/a	0.06	DLOS [12, 20, 21]	STH
1.35 eV LCV/SSPC	E_c – 1.35	n/a	Capture barrier for electrons	LCV/SSPC [24–27]	Native defect
1.6 eV LCV/SSPC	E_c – 1.6	n/a	Capture barrier for electrons	LCV/SSPC [24–27]	Native defect
2.1–2.3 eV LCV/SSPC	E_c – (2.1–2.3)	n/a	Capture barrier for electrons	LCV/SSPC [24–27]	Native defect, possibly the same as E_c –2 eV DLOS
3.1–3.4 eV LCV/SSPC	E_c – (3.1–3.4)	n/a	Capture barrier for electrons	LCV/SSPC [24–27]	Native defect (possibly the same as H3)

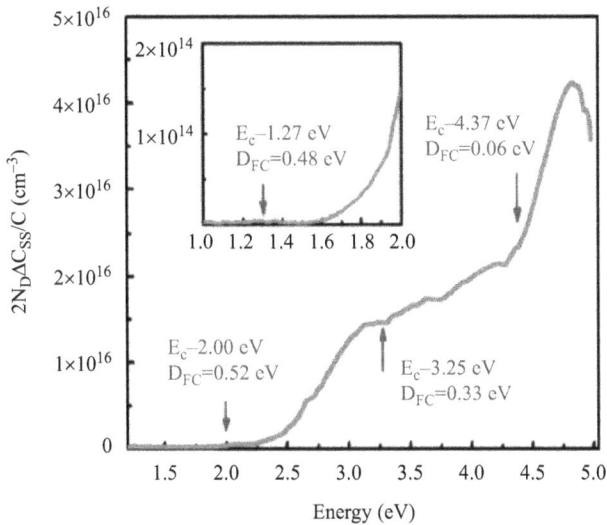

Figure 5.8. DLOS spectra measured for Ge doped samples. Reproduced from [20] with the permission of AIP Publishing.

sizeable barrier for capture of electrons. The barrier height was estimated to be close to 0.5 eV [26]. The partial persistent photocapacitance generated by photons with energies higher than ~3.1–3.4 eV could, on the other hand, be wiped out by forward bias pulses and was owing to the presence of deep acceptor centers about 1.4 eV (optical ionization energy) above the valence band edge not showing a prominent barrier for capture of electrons and giving rise to persistent photocapacitance simply due to the high depth of the center and the absence of electrons in the space charge region [24–27]. All centers detected in LCV measurements increased in concentration upon irradiation with protons or α-particles (figure 5.9(b)) [24, 26, 27]. Table 5.1 summarizes the known characteristics of the traps.

The identity of the observed deep traps has been more or less convincingly established only for the E2 traps and the $E_c - 2$ eV DLOS traps in table 5.1. The E2 traps are now commonly believed to be due to substitutional Fe acceptors on the Ga1 site based on the measured Fermi level pinning position in semi-insulating Fe doped β-Ga_2O_3 substrates [39] in which the presence of Fe was positively confirmed by secondary ion mass spectrometry (SIMS) profiling and by the presence of sharp $^4T_1 \rightarrow {}^6A_1$ intracenter transition in Fe^{3+} near 1.8 eV [47]. Also, a positive correlation between the concentration of Fe by SIMS and the concentration of E2 in HVPE films was observed [23, 28], and the trap signatures in DLTS on conducting samples and in admittance spectra and PICTS measurements in semi-insulating β-Ga_2O_3 crystals were very similar [23, 28, 31, 47]. Because the concentration of the E2 traps is not affected by irradiation they are believed to belong to isolated Fe atoms. In principle, one would expect substitutional Fe to produce doublets in DLTS and quadruplets rather than doublets in the $^4T_1 \rightarrow {}^6A_1$ intracenter transitions near 1.8 eV because of Fe located in the Ga1 and Ga2 sites, but the absence of the second peak

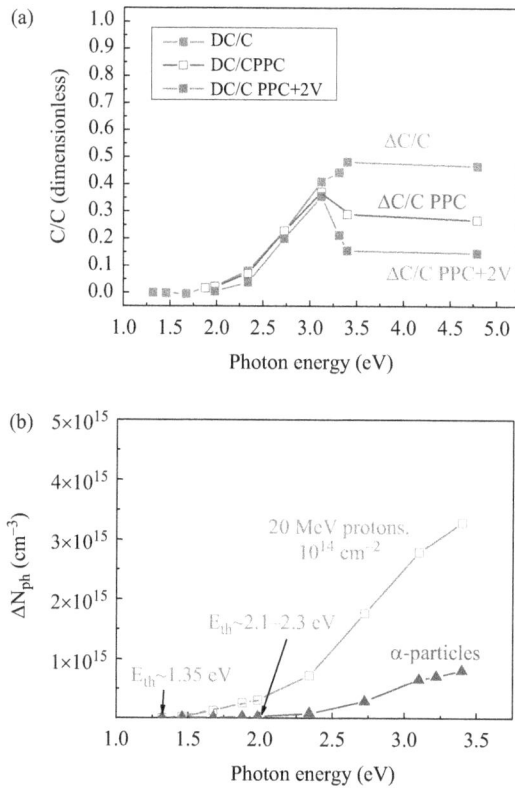

Figure 5.9. (a) Spectra of low temperature photocapacitance ($\Delta C/C$), persistent photocapacitance surviving to room temperature ($\Delta C/C$ PPC), and persistent photocapacitance after the application of 2 V forward bias injection ($\Delta C/C$ PPC+2V) after irradiation with 20 MeV protons. (b) Concentration of centers determined from photocapacitance spectra at room temperature measured after irradiation with 20 MeV protons and with α-particles (before irradiation these samples showed very low concentrations of centers in photocapacitance measurements).

in DLTS and second pair of intracenter transitions in PL/MCL could be due to the predominance of Fe_{Ga1} species that are expected to have much lower formation energy [28]. The energy of the intracenter transition in β-Ga_2O_3 is considerably higher than in GaN and close to that in ZnO [47], which most likely is the reflection of higher ionicity of β-Ga_2O_3 compared to GaN [47, 48]. E2/Fe acceptors are the absolutely dominant deep traps and the major compensating species in undoped bulk n-type β-Ga_2O_3 crystals [31, 47]. They also dominate nonequilibrium charge trapping and current collapse in $(Al_xGa_{1-x})_2O_3/Ga_2O_3$ HEMTs on Ga_2O_3:Fe buffers [33] and in nanobelt back-gated β-Ga_2O_3 on Si/SiO_2 MISFETs [32]. From what is known about the behavior of Fe acceptors in GaN [49], iron acceptors in Ga_2O_3 could also prove to be effective nonradiative recombination centers due to efficient recombination energy dissipation via excited states of Fe^{3+} and Fe^{2+} and that would be interesting to study experimentally and theoretically.

The other centers that have been credibly identified in β-Ga_2O_3 are the 'split' Ga-vacancy–interstitial complexes V_{Gai}. These acceptor defects have been positively identified in scanning transmission electron microscopy (STEM) measurements and their concentration was seen to increase with increased Sn donor doping, as expected from the action mass law. Simultaneously, and in similar concentrations that traced the changes in Sn donor doping, the major DLOS E_c − 2 eV acceptors were detected which allowed these centers to be ascribed to the V_{Gai} defects [22]. Positron annihilation measurements on β-Ga_2O_3 that suggest the presence of high densities of Ga vacancy-like defects [42] could in principle also be related to these V_{Gai} defects.

For all other detected centers, the attributions in various papers are based on comparisons of the positions of the levels in the bandgap with theoretically predicted charge transfer levels. This, however, could be a delicate matter. The positions of predicted levels differ measurably in different papers and the number of possible defects or complexes with charge transfer levels in the energy range of interest is quite high. Moreover, if when accounting for screening in DFT calculations for a limited size cluster of atoms containing a given defect one uses the ionic low-frequency dielectric constant or the electronic dielectric constant, the charge transfer level positions can be very seriously shifted in energy [10, 11]. A good example is given in [10] where it is argued that when such accommodation is made, the level of the V_{Ga1} acceptor can move from E_c − 1.6 eV to E_c − 0.8 eV where it will fall very close to the E2 level of DLTS and this will explain its being the main compensating acceptor. If one did not know from independent experience that the level in question belongs to Fe, one could easily persuade oneself that the center in question is due to V_{Ga1}. Similarly, large corrections are obtained for virtually all charge transition levels [10, 11]. It would seem that, without additional knowledge of the properties of defects, attributions based simply on comparisons of predicted charge transfer levels with experimentally observed level positions can be very misleading. For example, it would help if one knew whether the center is a donor or an acceptor, which can be done by measuring the dependence of the center's activation energy in DLTS on the electric field [50]. For example, our preliminary conclusions regarding the E2* and E3 centers in table 5.1 are that these centers, that dominate the DLTS spectra of HVPE grown material and are the major radiation defects, are deep donors. This probably rules out their attribution to Ga vacancies as proposed in some papers. Independent measurements of the electron capture cross sections at different temperatures from the dependence of the DLTS peak magnitude on the length of the injection pulse [50] shows whether there is a barrier for capture of electrons, which again narrows down the possibilities in defect identification. Such measurements on heavily compensated proton irradiated films suggest that the E2* centers are likely to have a relatively high barrier for the capture of electrons [24], although more detailed and accurate measurements taking into account the Debye tails in electron concentration in the space charge region [51] are necessary. The results of radiation experiments at different temperatures and subsequent annealing are very helpful in narrowing down the pool of defects among which the choice has to be made based on the theoretically calculated patterns of levels. Most of the centers that are listed in table 5.1 show a virtually linear increase with fluences of protons, α-particles, and neutrons [11, 21, 24, 27] suggesting that they are either native point defects per se or their complexes with impurities whose concentration is high and does not limit the introduction rate of defects at high doses. The similarity of behavior for proton, α-particle, and neutron irradiation [11, 21, 24, 27] probably allows the participation of hydrogen in the formation of the E2*, E3, and DLOS $E_c - 2$ eV defects to be ruled out. A comparison of the activation energy of radiation defect annealing at different steps with the theoretically calculated migration energies of defects provides additional guidance for attributions. For example, on such grounds the high-temperature annealing stage with an activation energy of 1.2 eV in proton irradiated β-Ga$_2$O$_3$ allows one to associate the defects responsible for this stage with the hopping of V$_{Gai}$ centers [11]. The Fermi level position that stabilizes after irradiation with high doses when the densities of the radiation defects are much higher than the densities of defects in the pristine state corresponds to the position of the so-called charge neutrality level (CNL) and is approximately determined by the point where the Fermi level dependences of the formation energies of the major donor and acceptor defects cross [11, 16]. In [11] these considerations suggested to the authors that the dominant defects introduced by protons in β-Ga$_2$O$_3$ are the V$_{Ga}$ acceptors and Ga$_i$ and Ga$_O$ donors whose interaction results in the Fermi level pinning near $\sim E_c - 0.6$ eV which should be close to the CNL position. It should be noted, however, that under proton or α-particle irradiation the initial carrier removal rate is much higher than the introduction rates of all individual deep donor and deep acceptor species (combined DLTS, ODLTS, and LCV measurements allow us to obtain these concentrations with reasonable accuracy), while the carrier removal rate is close to the introduction rate of Ga vacancies as obtained from SRIM modeling [27]. This suggested to us that at the initial stages when the densities of radiation defects are much lower than the donor densities an important role is played by shallow donor passivation by forming complexes with Ga vacancies similarly to donor passivation by Ga vacancies during growth as proposed by Lany [19].

As for the centers in the lower half of the bandgap detected in DLTS with optical excitation, by DLOS or by LCV/steady-state photocapacitance (SSPc), their attributions, save for the already discussed DLOS $E_c - 2$ eV centers, is so far poorly understood. It would seem that the LCV/SSPC centers with an optical ionization threshold near 2–2.3 eV [24–27] are similar to the DLOS $E_c - 2$ eV traps. The energies fit reasonably well, both types of centers increase in density with irradiation, and both have a marked barrier for capture of electrons that gives rise to prominent PPC for the 2.3 eV LCV/SSPC centers [12, 20–22, 24–27]. The DLOS $E_c - 4.4$ eV features are very probably due to the self-trapped hole polaronic states simply because such shallow acceptor centers are manifestly lacking in β-Ga$_2$O$_3$. Whether the LCV/SSPC centers with optical thresholds near 3.1–3.4 eV are similar to the H3 traps detected in ODLTS [26, 27] or to the DLOS $E_c - 3.35$ eV [20] traps has yet to be understood. The H3 traps in DLTS were observed with above-bandgap 4.8 eV excitation and with 3.4 eV excitation. Their thermal activation energy in

ODLTS is 1.3–1.4 eV, presumably corresponding to a level near $E_v + (1.3–1.4)$ eV or $E_c - (3.4–3.5)$ eV. This is not too far away from the optical thresholds of the DLOS $E_c - 3.35$ eV [20] or LCV/SSP 3.1–3.4 eV [24–27] traps. The concentration of the H3 and 3.1–3.4 lCV/SSPC traps definitely increases with proton or α-particle irradiation [24–27] suggesting that these could be native defects or their complexes. If one is to be guided by theoretical calculations of the charge transfer level positions, good candidates for the role of such defects could be hydrogenated $(V_{Gai}^{\ b}-2H)^-$ or oxygen split interstitial O_{si}^+ donors [11].

An important question is, of course, which of the centers present in β-Ga$_2$O$_3$ films or introduced by, say, irradiation or as a result of device degradation are efficient Shockley–Hall–Read recombination centers? This matter is discussed in some detail in the second part of this chapter. One of the things that have emerged in the course of our studies is that the diffusion length of nonequilibrium charge carriers L_d significantly decreases after irradiation in tune with the increase of the densities of electron and hole traps [24, 27]. The main radiation defects in the lower half of the bandgap, the DLOS $E_c - 2$ eV/2.3 eV LCV/SSP traps, have a high barrier for the capture of electrons and are not likely candidates for the role of good recombination centers. The introduction rates of the E1, E6, E7, and E8 traps are low, while the E4 and E5 traps have only been observed after irradiation and cannot be dominant recombination centers in pristine films and crystals [11, 24, 27]. Moreover, the E5 traps have only been seen in samples irradiated with α-particles [27]. While the E4 traps were found to undergo structural changes and partial annealing even at relatively low temperatures near 500 K where their DLTS peak is detected [11]. Thus one is left only with the E2* and E3 traps as possible candidates for the role of major recombination centers. If it is confirmed that the E2* centers indeed possess a sizable barrier for the capture of electrons, one is left with only the E3 centers. For these E3 electron traps we found a very reasonable linearity between the $1/L_d^2$ and the E3 as introduced by proton or α-particle irradiation [27]. Further studies are naturally necessary here, but at the moment it appears that these E3 traps are the main candidates for the role of major recombination centers in β-Ga$_2$O$_3$.

It should be noted that nonequilibrium charge carrier lifetimes and diffusion lengths are very important for determining the speed and efficiency of bipolar devices. For mostly unipolar devices dominant among the β-Ga$_2$O$_3$-based rectifiers and FETs the lifetimes do not directly play an important role in device performance, but are a good probe for the overall quality of the material. The exception is, of course, presented by the solar-blind photodetectors. However, in most cases the reported photocurrent build-up and decay times in such devices are inexplicably long, on the order of hundreds milliseconds, and cannot be attributed to the lifetimes governed by common deep traps (see e.g. the review in [1]). One of the explanations offered is that the times are determined by the kinetics of self-trapped holes/polarons [36] as mentioned earlier in this section. We have already discussed the problems with such an explanation. Other possibilities are related to the presence of centers with a high barrier for the capture of electrons. Obviously a lot of additional studies are required here.

5.1.2 Centers active in recombination

5.1.2.1 Photoluminescence and cathodoluminescence

Photoluminescence (PL) and cathodoluminescence (CL) measurements are recognized as a very useful tool for the detection and identification of impurity and intrinsic point defects. The luminescence intensity in β-Ga$_2$O$_3$ is rather high despite the indirect bandgap and numerical investigations carried out using these methods [35, 52–58]. Near-band-edge emission was not usually detected in β-Ga$_2$O$_3$ (the exception may be [59]) and instead a wide emission band centered at about 3–3.3 eV with a large Stokes shift was usually revealed, which was divided into ultraviolet (UV), blue (BL), and green luminescence bands. The green band was detected in the presence of specific impurities such as Be, Ge, Sn, Li, Zr, and Si [52], however, in [60] it was associated with oxygen interstitials. The wide emission band in the UV–BL range was observed in most of the measurements although some dependence on shallow donor concentration was revealed [35, 57]. It seems that the emission band also depends on other defects because in spite of the rather similar spectra observed in different papers, some difference in their position can be revealed. As an example, the spectra measured by us on the samples from different vendors are presented in figure 5.10.

The emission band is mainly featureless and its maximum is practically independent of temperature (figure 5.11), which makes the deconvolution procedure not so reliable. Nevertheless, in some works it was fitted with a few (from two to four) Gaussian peaks, however, the energy position of these bands were close but not exactly the same. In most cases UV and BL bands with energies in the ranges of 3.1–3.6 eV and 2.5–2.8 eV, respectively, are revealed. The similarity of the luminescence spectra measured by different groups can be understood by taking into account the low precision of the deconvolution procedure and a large number of deep acceptor levels near 1 eV from the valence band [16, 60–62].

However, the interpretation of the results obtained is rather ambiguous. The UV band has been shown to be independent of dopants or sample preparation methods and it has been usually attributed to the recombination of free electrons and self-trapped holes or self-trapped excitons, although donor–acceptor pair (DAP) recombination was also considered as a reason for the UV luminescence [57]. The BL band has been usually attributed to DAP recombination involving deep donors and acceptors. As possible donor dopants, Si or Sn, or intrinsic point defects such as oxygen vacancies or Ga interstitial were assumed [52, 53], and as possible acceptors for the BL band Ga vacancies and vacancy complexes [52, 53, 60, 63], interstitial oxygen [60], or deep level impurities were considered. As an indication of oxygen vacancy and oxygen interstitial contribution to BL emission band formation, the effects of post-growth annealing in a nitrogen or oxygen atmosphere [55, 57] or oxygen implantation [64] were considered. However, as discussed above, self-trapping of holes is not so obvious at room temperature. The same can be said about the intrinsic point defect contribution to the emission spectra. Indeed, irradiation is a widely used approach to increase the intrinsic point defect concentration and in β-Ga$_2$O$_3$ irradiated with protons, α-particles, or hydrogen plasma the

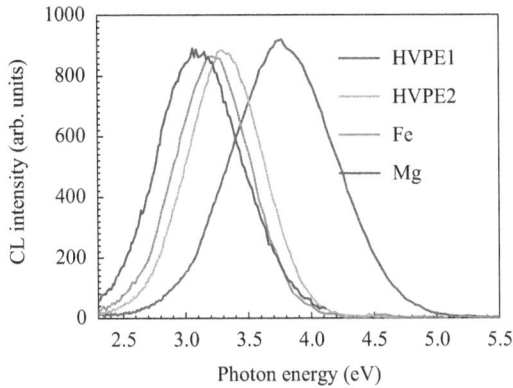

Figure 5.10. CL spectra measured at room temperatures on two different HVPE β-Ga$_2$O$_3$ samples and on β-Ga$_2$O$_3$ doped with Fe or Mg.

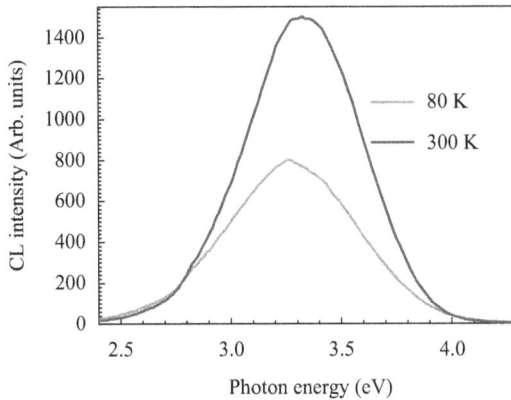

Figure 5.11. CL spectra of HVPE β-Ga$_2$O$_3$ measured at 80 K and 300 K.

deep level concentration was found to increase and the diffusion length to decrease [26, 27, 30, 65]. However, the CL spectrum after such treatments practically does not change (figures 5.12–5.14).

It should be noted that some impurities noticeably affect the UV/BL luminescence band. In HVPE films with a Si concentration of $(2–4) \times 10^{16}$ cm^{-3} the form of the peak practically does not change in the temperature range from 80 to 300 K and its intensity decreases by less than three times (figure 5.11), and in the samples grown using the Czhochralski method doped with Mg both the intensity and form of the spectrum essentially change with temperature (figure 5.15). The spectra measured at room temperature for both samples fitted with Gaussian peaks are shown in figures 5.16 and 5.17, respectively. The spectrum of HVPE β-Ga$_2$O$_3$ can be fitted with three Gaussian peaks $\exp[-(E - E_m)^2/2\sigma^2]$ with energies at maximum E_m of 3.19, 3.29, and 3.73 eV and σ equal to 0.375, 0.27, and 0.13 eV, respectively. The spectrum of Czochralski grown β-Ga$_2$O$_3$ doped with Mg can be fitted with four Gaussian peaks with E_m of 2.55, 2.84, 3.09, and 3.29 eV and σ equal to 0.16, 0.15,

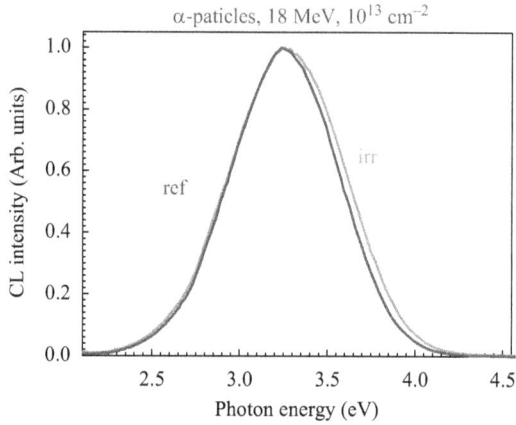

Figure 5.12. CL spectra measured at room temperature on 18 MeV α-particle irradiated and reference β-Ga$_2$O$_3$ samples.

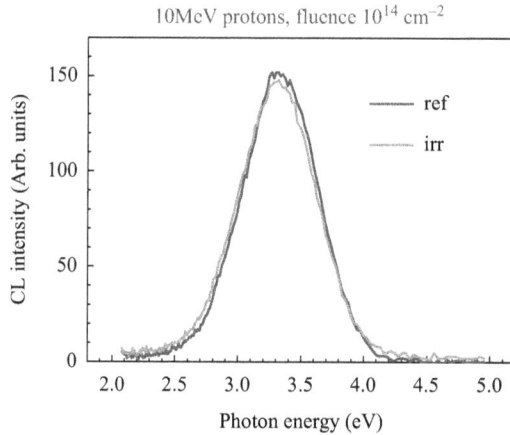

Figure 5.13. CL spectra measured at room temperature on 10 MeV proton irradiated and reference β-Ga$_2$O$_3$ samples.

0.18, and 0.29 eV, respectively. For both samples E_m values are practically independent of temperature while σ increases with temperature. Thus, the pronounced dependence of the CL band maximum on temperature seen in figure 5.15 is determined by different temperature dependences of its components. The temperature dependence of intensity of four emission bands in Mg-doped β-Ga$_2$O$_3$ is shown in figure 5.18. Arrhenius analysis of the intensities yields activation energies of 0.54 and 0.57 eV for the 3.29 eV band and the other three bands, respectively. Thus, it can be assumed that in all transitions one or two donor levels with close energies and different deep acceptor levels are involved. However, as shown above the level with 0.57 eV energy was not detected in such samples.

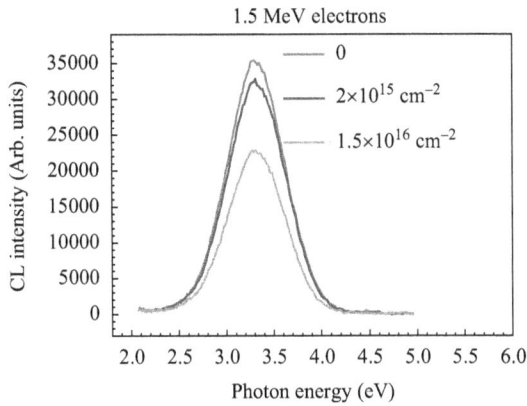

Figure 5.14. CL spectra measured at room temperature on reference and 1.5 MeV electron irradiated β-Ga$_2$O$_3$ samples.

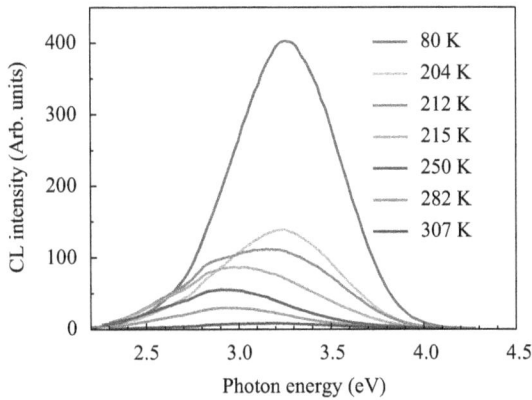

Figure 5.15. CL spectra of Mg-doped β-Ga$_2$O$_3$ measured in the temperature range from 80 K to 307 K.

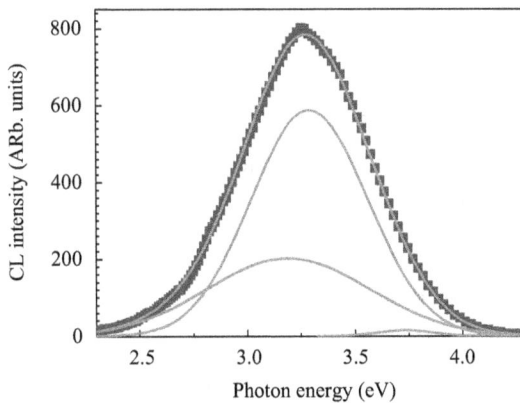

Figure 5.16. Fitting of the CL spectra shown in figure 5.11 (300 K) with three Gaussian peaks with energies of 3.19, 3.29, and 3.73 eV (individual peaks are also shown with lines).

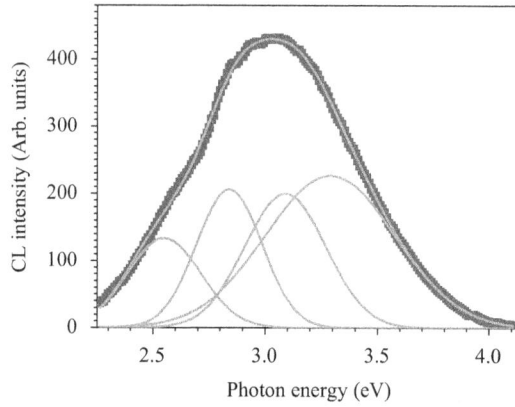

Figure 5.17. CL spectra of Mg-doped β-Ga$_2$O$_3$ measured at 300 K fitted with four Gaussian peaks with energies 2.55, 2.84, 3.09, and 3.29 eV (individual peaks are also shown with lines).

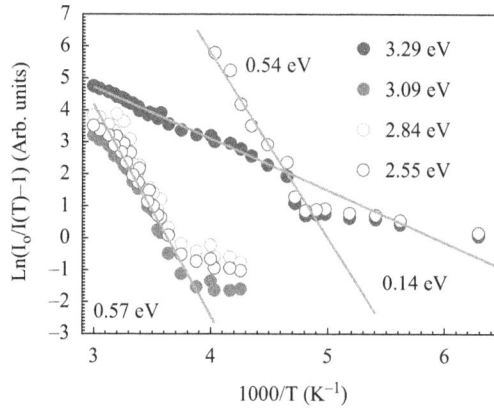

Figure 5.18. Arrhenius plots of four components of the CL band in Mg-doped β-Ga$_2$O$_3$.

The difference between the spectra shown in figures 5.11 and 5.15 can be determined not only by Mg but also by different contents of intrinsic point defects, which can depend on the growth method. Nevertheless, these results show that the careful measurement of temperature dependent CL or PL spectra may open a way for the identification of corresponding defects. The detection of intracenter transitions between the excited states of some impurities should also be mentioned, which also can be used for their identification. Such transitions have already been observed for Eu [66, 67] and Fe [47]. As an example, the spectrum of intracenter transition in β-Ga$_2$O$_3$ doped with Fe measured at room temperature is shown in figure 5.19.

5.1.2.2 Excess carrier lifetime and diffusion length measurements
The excess carrier lifetime τ is the important parameter determining the performance of many semiconductor devices, in particular opto- and photo-electronic devices. In

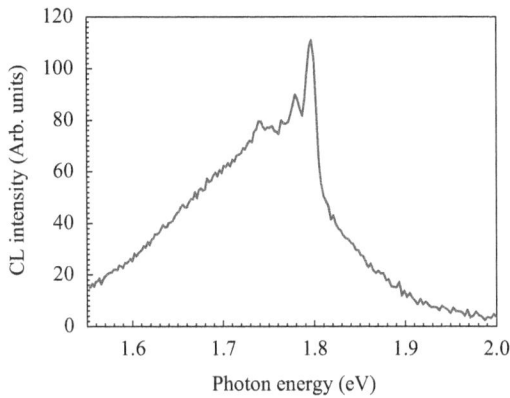

Figure 5.19. Fe-related intracenter transition detected by CL at room temperature.

addition, this parameter is used for the characterization of crystal quality. However, its measurement is not so straightforward and often depends on the method used. In many case the reasons for these discrepancies are fundamental because the lifetime describes a property of a carrier within the semiconductor and therefore in addition to the properties of the semiconductor itself it can be influenced by surfaces, energy barriers, trap distributions, electric fields, interface boundaries, etc. In addition, it should be taken into account that in a common case the recombination lifetime is determined by three main mechanisms: Shockley–Read–Hall recombination, radiative recombination, and Auger recombination. These mechanisms have different dependences on the excess carrier concentration, therefore the measured lifetime can strongly depend on this concentration. As far as the authors know, in β-Ga$_2$O$_3$ the lifetime was estimated mainly by the measurement of photoconductivity or photoluminescence (cathodoluminescence) decay kinetics. The excess carrier concentration dynamics was directly measured only in [38] using a near-IR probe pulse whose transmission through the sample is affected by free-carrier IR absorption. In this work the minority carrier lifetime was estimated to be in the nanosecond range. There has been a lot of discussion concerning the relation between the minority carrier lifetime and the luminescence decay time (see e.g. [68–70]). As shown in [69], the luminescence decay time in general equals the minority carrier lifetime only for a homogeneous and time-independent carrier lifetime. In addition, as shown above, the state-of-the-art β-Ga$_2$O$_3$ crystals contain a large concentration of point defects. The effects of minority carrier traps on the luminescence decay was discussed in [70], where it was shown that trapping/detrapping processes can result in a very slow decay up to the radiative recombination lifetime. Thus, in some papers the PL decay time at room temperature exceeds 10 μs [71]. The luminescence decay can essentially depend on the temperature. Therefore, for definiteness, we limit our discussion to results obtained at room temperature. It was observed that UV band intensity decay is much faster than that of the blue band. Thus, in [53] the UV band intensity was found to decay at room temperature with a time constant shorter than 1 μs while for the BL band it is larger than 1 μs. In [71] the decay of UV/blue emission intensity was found to comprise at least two exponential components. The fastest decay component is 1.5 μs and the other was ten times longer. The decay was found to be practically independent of temperature in the range from 20 K to 300 K. A study of β-Ga$_2$O$_3$ nanoflakes revealed a PL decay time of a few nanoseconds [72]. Probably the fastest luminescence decay of 215 ps for the non-irradiated sample and 138 ps for the 1.5 MeV electron irradiated sample was observed in [37]. Thus, it can be concluded that the lifetime values estimated from the PL or CL intensity decay differ by many orders of magnitude that can be determined by trapping–detrapping effects and/or by rather fast capture of excess carriers by deep traps with subsequent slow donor–acceptor radiative recombination.

It can be expected that the measurements of excess carrier diffusion length L are not so sensitive to the trapping–detrapping processes. The diffusion length is equal to $L = (D\tau)^{0.5}$, where D is in a common case the ambipolar diffusivity, however, at low excitation levels it is equal to the minority carrier lifetime [73, 74]. Thus, the diffusion length is closely related to the excess carrier lifetime. Most measurements of diffusion length in β-Ga$_2$O$_3$ have been carried out using electron beam induced current (EBIC) methods, only in [75] it is estimated from non-steady-state photo-electromotive force measurements. The studied samples included crystals or epilayers grown using different methods and also samples irradiated with high energy particles or electrons and treated with plasma in a dry etching process [24, 27, 30, 37, 65, 76–79]. Depending on the growth method and processing, the diffusion length was found to vary in the range from 50 nm [30] to 600 nm [27], i.e. it is smaller than 1 μm. As discussed in [65, 80, 81], for the measurement of submicron diffusion length values, the method based on fitting the dependence of the current collected in the EBIC mode on beam energy proposed in [82, 83] gives the most reliable diffusion length value. As shown in [84] for a planar structure the collected current can be calculated as

$$I_c = \beta \left\{ \int_0^W h(z)\mathrm{d}z + \int_W^\infty h(z)\psi(z, W)\mathrm{d}z, \right. \tag{5.1}$$

where z is the depth (the distance from the irradiated surface), $h(z)$ is the normalized depth–dose dependence describing the depth dependence of the electron–hole pair generation rate, $\psi(z)$ is the collection probability, which can be calculated as a solution of the homogeneous diffusion equation, W is the depletion region width, and $\beta = I_b E_b \eta / E_i$, where I_b is the beam current, η is the fraction of beam energy absorbed inside the sample, and E_i is the average energy necessary for electron–hole pair creation. For β-Ga$_2$O$_3$ $h(z)$ can be approximated as

$$h(z) = \frac{1.603}{R} \exp\left[-A\left(\frac{z}{R} - 0.22\right)^2\right], \tag{5.2}$$

where $R(\mathrm{nm}) = 7.34 \times E_b \,(\mathrm{keV})^{1.75}$ is the electron range within which 99% of the total absorbed energy is transferred to the target and $A = \begin{cases} 12.86, & z < 0.22R \\ 3.97, & z \geqslant 0.22R \end{cases}$ [65].

Function $h(z)$ calculated for a few beam energies is shown in figure 5.20. The calculated dependences of collected current I_c normalized to the product of beam current and energy on E_b as a function of diffusion length are shown in figure 5.21. It is seen that in the range of diffusion length values typical for β-Ga$_2$O$_3$ the sensitivity of this method to diffusion length is quite good. This method does not need a knowledge of surface recombination velocity and, as shown in [85, 86], in materials with small diffusion length values such fitting also allows us to estimate the local dopant concentration.

The extended defects cannot be a reason for the small values of lifetime and diffusion length in β-Ga$_2$O$_3$ because their density is too small. The observed dependence of diffusion length values on the irradiation or dry etching processes seems to confirm an assumption that some point defects are responsible for the diffusion length in β-Ga$_2$O$_3$. Indeed, it was shown that L decreases more than two times due to Ar or H plasma treatment [30, 77]. High energy electron or proton

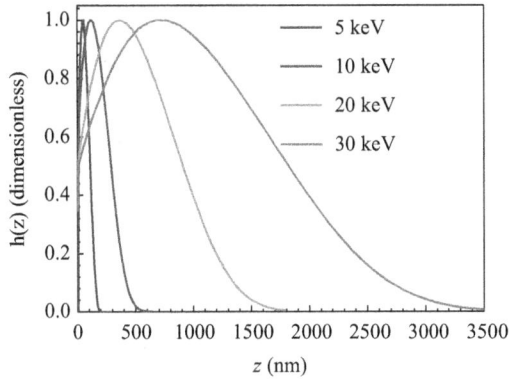

Figure 5.20. Function $h(z)$ calculated for a few beam energy values.

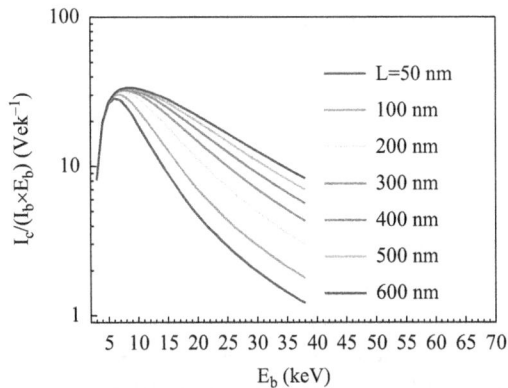

Figure 5.21. Dependences of collected current I_c normalized to the product of beam current I_b and energy E_b calculated for a few values of diffusion length, a donor concentration of 10^{17} cm^{-3}, and a metal thickness of 30 nm.

irradiation also leads to a decrease of L values [27, 37]. Quite good correlation was observed in [27] between the diffusion length decrease due to proton irradiation and an increase of the concentration of E3 defects with energy level $E_c - (0.95–1.05)$ eV. However, the nature of point defects responsible for the low diffusion length in as-grown β-Ga$_2$O$_3$ has not been totally understood until now. Nevertheless, the demonstration of diffusion length dependence on irradiation allows us to use its measurements for monitoring of the damage introduced by processing and for the control of damage annealing.

Another interesting point to discuss is the self-trapping of holes and its effect on the excess carrier transport. It is widely accepted that holes in β-Ga$_2$O$_3$ form localized polarons, where they are localized at lattice distortions. This statement mainly follows from the hybrid functional calculations of hole behavior in the valence band [34, 87]. These calculations predicted that the hole mobility in β-Ga$_2$O$_3$ is very small and can be estimated as 10^{-6} cm^2 V^{-1} s^{-1} at room temperature. Experimental investigations of the stability of self-trapped excitons gave controversial results. Thus, a broad photoluminescence emission band with a large Stokes shift and a form of Urbach tail in absorption spectra were considered in [88] as evidence for the formation of self-trapped excitons. It was concluded that self-trapped excitons are stable in β-Ga$_2$O$_3$ up to room temperature. In [36] the high photo-conductive gain of β-Ga$_2$O$_3$ Schottky photodiodes was explained by lowering of the effective barrier height due to the accumulation of self-trapped holes in the valence band spatially localized near the Schottky contact. However, these experiments cannot be considered as evidence of self-trapped holes because, according to the model proposed in [89], such an explanation needs only deep acceptor traps near the metal–semiconductor interface. In [90], the observed electron paramagnetic resonance (EPR) spectrum was assigned to the self-trapped holes and it was shown that they are stable at low temperatures only and heating to temperatures above 90 K destroys them. As shown above, the binding energy of self-trapped holes seems to be much lower than the predicted one. Therefore, the stability of self-trapped holes and self-trapped excitons consisting of a self-trapped hole and a bound electron at room temperature is now under discussion [24, 91]. Moreover, as discussed in [24, 91], the results of EBIC investigations seem to contradict the assumption of self-trapped immobile holes. Indeed, when an electron beam irradiates the Schottky barrier through the metal, the direction of induced current coincides with the minority carrier flow to the metal but not with the majority flow that should be present if the measured EBIC current is determined by the decrease of Schottky barrier height, as assumed in [36]. In this geometry the measured induced current consists of two components: electrons and holes generated inside the depletion region and holes generated in the quasineutral region outside the depletion region and diffused to it. It is easy to estimate that at zero bias, a beam energy larger than 25 keV, a donor concentration of 10^{17} cm^{-3}, and a diffusion length of 450 nm the current component provided by carriers created inside the depletion region does not exceed 20% of the total measured current. This means that at least 80% of collected current is determined by holes generated outside the depletion region and diffusions to it. Moreover, in the planar-collector geometry of EBIC measurements, when the collected current is totally determined by the mobile holes, the collected current essentially exceeds the beam current even at distances from the depletion region boundary a few times exceeding the size of the generation function. The decay of collected current I_c divided by the beam current I_b as a function of distance from the Schottky barrier measured at 10 keV for two samples with different diffusion length is shown in figure 5.22. Estimations similar to those made in [92] for GaN show that for a beam current of 0.1 nA the excess carrier concentration does not exceed 3×10^{15} cm^{-3} for all beam energies used. Thus, for the conducting samples the low excitation condition can be easily fulfilled. Under such assumptions the ambipolar diffusivity equals $(p + n)/(n/D_h + p/D_e)$ [9, 10], where p and n are electron and hole concentrations and D_e and D_h are their diffusivities, respectively, equal to the minority carrier diffusivity D_h. Thus, the hole flow to the depletion region is mainly determined by their diffusivity and it can be concluded that at room temperature the holes in β-Ga$_2$O$_3$ should be mobile.

Let us try to estimate the hole diffusivity. The hole mobility in β-Ga$_2$O$_3$ was measured experimentally in [41], where a value of 0.2 cm^2 V^{-1} s^{-1} was obtained at 300 K, which corresponds to a diffusivity of 5×10^{-3} cm^2 s^{-1}. Much higher values of about 200 cm^2 V^{-1} s^{-1} for mobility and 5 cm^2 s^{-1} for diffusivity were obtained in [37], where L was measured by the EBIC and τ was obtained from the CL signal decay. This value seems to be unphysically large because it is comparable with the electron mobility measured in this material [93]. The reason for such a high mobility value could be the overestimated value of diffusion length because, as shown in [65, 80, 81], measurements of the diffusion length in the planar-collector geometry can lead to its overestimation, in particular at high beam energies. Nevertheless, as shown above the measured L values are of the order of 100 nm (in some papers they are even larger) thus the assumption that L is about 100 nm seems rather reasonable. For τ a value of about 1 ns was obtained in many works. Moreover, taking into account that some hole traps in β-Ga$_2$O$_3$ have a rather high capture cross-section of about 10^{-13}–10^{-14} cm^2 [29, 38], to provide a τ of about 1 ns the trap concentration

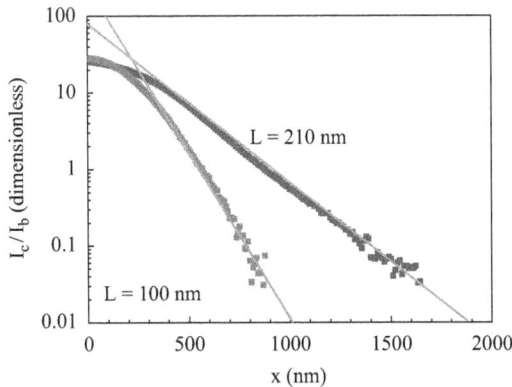

Figure 5.22. $I_c(x)$ dependencies measured on two β-Ga$_2$O$_3$ Schottky diodes with L equal to 100 nm and 210 nm measured in the planar-collector geometry at $E_b = 10$ keV. The corresponding $\exp(-x/L)/x^{0.5}$ dependences are shown by solid lines.

should be about 10^{15}–10^{16} cm^{-3}, which is a reasonable value for the state-of-the-art β-Ga$_2$O$_3$. For such values of L and τ the diffusivity can be estimated as 0.1 cm^2 s^{-1} and the mobility as 4 cm^2 V^{-1} s^{-1}. This value is smaller than that obtained in [37] but a few orders of magnitude larger than the computationally predicted value of 1×10^{-6} cm^2 V^{-1} s^{-1} [34]. Thus, it could be concluded that at least an essential portion of holes in β-Ga$_2$O$_3$ are mobile and their mobility can reach values of 1–10 cm^2 V^{-1} s^{-1}.

5.1.2.3 Extended defects

Extended defect control is an important issue in the development of device technology because they usually produce a negative effect on device performance. In the case of β-Ga$_2$O$_3$, the study of extended defects until now has been superficial. In the as-grown materials the defect structure has been studied using x-ray topography, selective etching, and transmission electron microscopy [94–100]. Three main types of extended defects were revealed: dislocations, nanopipes, and twin lamellae. Nevertheless, it should be stressed that the density of as-grown extended defects in state-of-the-art β-Ga$_2$O$_3$ is extremely low compared to, for example, GaN. This dislocation density can be as small as 10^3 cm^{-2} and the density of other extended defects can be even lower [101–103]. After chemical etching in H$_3$PO$_4$ solution at a temperature of 140 °C at least six different types of etch pits were observed on the (010) surface [95]. Grown-in dislocations are mainly of the screw type with a Burgers vector parallel to $\langle 100 \rangle$ and nanopipes are rod-like defects elongated along [010] [94, 100].

Not much is known about the mechanical properties of β-Ga$_2$O$_3$. Recently, the mechanical properties of β-Ga$_2$O$_3$ under deformation have been studied using nanoindentation [104, 105] and using the deformation of micro-pillars [106]. In [105] the hardness of the β-Ga$_2$O$_3$ thin film produced by RF sputtering was studied as a function of sputtering temperature. It was shown that when this temperature exceeds 500 °C nanocrystalline films of monoclinic β-Ga$_2$O$_3$ were formed and their hardness increased with sputtering temperature. The estimated hardness was about 20 GPa, which is comparable to the hardness of GaN. The hardness of the thin film of a ternary system based on β-Ga$_2$O$_3$ doped with molybdenum and titanium was studied in [107, 108], respectively. It was found that the hardness increases up to 1.5% Ti and then starts to decrease. In the case of Mo it was observed that the film crystallinity decreased with an increase of Mo content and the XRD peak disappeared at 4% of Mo. However, the hardness was found to increase up to 36 GPa at 10% Mo.

The nanoindentation of single crystal β-Ga$_2$O$_3$ showed that slight plastic deformation occurred at room temperature even at a load of 0.2 mN and transmission electron microscopy (TEM) revealed stacking faults along the (200) lattice planes [104]. As the load was increased plastic deformation was significantly increased, as indicated by the large difference of loading and unloading curves and the appearance of pop-ins on the loading curve. Such behavior is similar to that of GaN, in which numerous investigations also revealed pop-in events at room temperature [109–113]. As transmission electron microscopy measurements show, in GaN these events are mainly determined by dislocation glide in the basal and pyramidal slip planes. Estimations of dislocation mobility showed that the activation energy for a glide does not exceed 1 eV, i.e. it is rather low [114]. In β-Ga$_2$O$_3$, in addition to dislocations in the (101) plane, twinning structures with the (−201) plane as the twin boundary were also formed under nanoindentation at room temperature. At a load of about 10 mN cracks were generated and propagated along the (200) planes. These observations correlate well with the analyses carried out in [98], which showed that the most probable slip planes in β-Ga$_2$O$_3$ are the {201}, {101}, and {310} planes. In contrast, the quantum mechanics simulations carried out in [115] showed that the (001)/⟨010⟩ slip system has the lowest ideal shear strength. The introduction of stacking faults at rather small loads means that the stacking fault energy in β-Ga$_2$O$_3$ is relatively low and the mobility of partial dislocations is higher than the mobility of non-dissociated dislocations, similar to the case of 4H-SiC [116, 117]. The deformation of β-Ga$_2$O$_3$ micro-pillars under compression confirmed the introduction of a high density of stacking faults and gliding of stacking faults on the (200) plane [105]. An increase in axial strain resulted in fracture along the (200) lattice plane and micro-cracks along the (200), (001), and (010) lattice planes. It was concluded that the most probable reason for pop-in events on the load–displacement curve is not dislocation glide but the formation of micro-cracks. These observations have shown that β-Ga$_2$O$_3$ can be ductile even at room temperature. Therefore any mechanical treatment of this material, such as slicing or grinding, usually used to remove the damaged layer after slicing should be done very carefully. Indeed, the study of damage introduced by grinding [118] showed that below thin near-surface layers of amorphous material and nanocrystal stacking faults, twins, and dislocations can be revealed using TEM.

It is widely accepted that dislocations and other extended defects are harmful defects that degrade the electrical properties and device performance. However, not so much is known about the electrical properties of extended defects in β-Ga$_2$O$_3$. It was shown that dislocations enhance the reverse leakage current of Schottky barriers [99, 101]. At the same time no correlation between the leakage current and the void defect density was revealed, which allowed the conclusion that void-type defects do not affect the leakage current [99, 101]. It was observed in [119] that incoherent twin boundaries strongly affect the electrical properties of β-Ga$_2$O$_3$, which was explained under the assumption that the traps present on such a boundary captured electrons, which results in electrostatic barrier formation near them and in an electron mobility decrease. However, there is no information concerning the recombination properties of these defects and related to the energy levels. Not much information is available concerning the recombination properties of other extended defects. EBIC measurements show that at least some types of dislocations enhance the excess carrier recombination rate. This can be seen in figure 5.23, which shows that dislocations produce a dark contrast in the EBIC mode. From this figure, the dislocation EBIC contrast can be estimated as 19 ± 3%. The normalized dislocation recombination strength γ/D calculated from this value is about 2. Here $\gamma = v_{th}\sigma_d N_l$ is the dislocation recombination strength, v_{th} is the minority carrier thermal velocity, σ_d and N_l are the capture cross-section and the linear density of defects along a dislocation, and D is the excess carrier diffusivity [120, 121]. For low excitation conditions ($\Delta n/N_d \ll 1$,

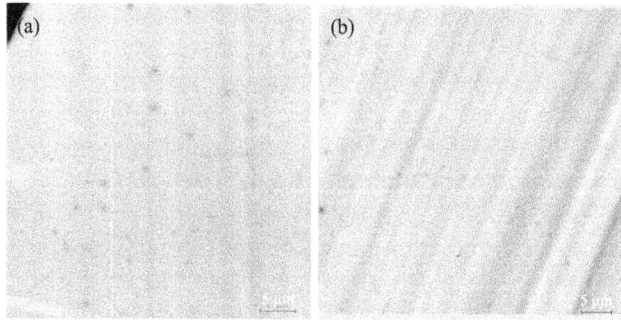

Figure 5.23. EBIC images of β-Ga$_2$O$_3$ with dislocation densities of about (a) 10^7 cm^2 and (b) 10^5 cm^2.

where Δn is the excess carrier concentration and N_d is the donor concentration) D is equal to the minority carrier (hole) diffusivity, which is unknown in β-Ga$_2$O$_3$. Therefore it is difficult to estimate the real dislocation recombination strength and the density of traps along dislocations, but it should be noted that the dislocation contrast is comparable to that in GaN [121]. Nothing is known to date about the specific electrical and optical properties of extended defects, which can allow us to understand the spectra of energy levels corresponding to these defects. Mainly it is determined by a relatively low density of extended defects in the state-of-the-art β-Ga$_2$O$_3$ in comparison to a rather high concentration of impurities and intrinsic point defect related defects.

5.2 Conclusions

The results discussed above suggest that, although holes in β-Ga$_2$O$_3$ do indeed form self-trapped polaronic states, the binding energy of such states is lower than predicted by theory, and free holes in the valence band can contribute to conductivity, charge trapping, and recombination at temperatures higher than some 100 K. Moreover, the diffusion length of nonequilibrium holes in good quality films is quite high, on the order of 0.5 μm, which is good news for the ongoing attempts to combine n-type β-Ga$_2$O$_3$ films in heterojunctions with some wide bandgap p-type materials. This is important in view of the manifest absence of suitable shallow acceptor dopants and defects in β-Ga$_2$O$_3$. Theoretical and experimental studies of the properties of deep acceptor states as compensating agents in the fabrication of high-resistivity substrates and buffer layers and in device isolation in β-Ga$_2$O$_3$-based FETs stress the practically important role of Fe$_{Ga}$ acceptors near $E_c - 0.8$ eV (giving rise to very dominant E2 electron traps in bulk crystals) and Mg$_{Ga}$ acceptors near $E_v + 1.1$ eV. For Fe, a pretty good idea of the position of the Fe^{3+}/Fe^{2+} charge transfer level and some information on the structure of intracenter transitions in Fe^{3+} has been attained. Fe doped semi-insulating substrates are currently commercially available. However, much additional work is necessary in understanding the role of Fe centers in charge trapping and current collapse in the emerging Ga$_2$O$_3$-based FETs. For Mg doping and N doping the actual pattern of acceptor and donor states of the centers has yet to be positively established experimentally and the relative merits of using these dopants for insulation in practical devices compared to Fe doping has yet to be determined. On the one hand, the resistivity of Mg-doped films is expected to be higher than for β-Ga$_2$O$_3$:Fe [29, 31], while N$_O$ acceptors could prove to be a better impurity for conductivity compensation under Ga-rich growth conditions [16]. But whether or not Mg or N doping will find practical applications in the emerging β-Ga$_2$O$_3$ device technology to a large extent depends on their role in current collapse in FETs. For Mg-doped Czochralski grown semi-insulators we argued that their performance in that respect would be inferior to the Fe doped material because the high depth of Mg acceptors would leave high densities of electron traps open for charge capture in the upper half of the bandgap [29, 31].

Among the native-defect-related acceptors the prominent role of Ga vacancy acceptors has been established and in particular the presence of high densities of V$_{Ga}$ divacancy complexes with interstitial Ga has been experimentally established, and the attribution of the prominent centers with optical ionization energy near E_c − 2 eV in DLOS spectra has been verified. The V$_{Ga}$ acceptors are also suspected to be forming complexes with shallow donors thus passivating them at the initial stages of proton and α-particle irradiation [27]. Dominant electron traps with levels near E_c − 0.8 eV (E2*), E_c − 1 eV (E3) have been demonstrated to strongly increase in concentration as a result of irradiation with protons, α-particles, and neutrons and are therefore suspected to be complexes involving native point defects. For the E3 traps a good correlation exists between the changes in their density upon irradiation and the decrease in the diffusion length of nonequilibrium charge carriers, which makes these defects good candidates for the role of major recombination centers in β-Ga$_2$O$_3$. In heavily proton irradiated β-Ga$_2$O$_3$ films and crystals the pinning of the Fermi level is assumed to come from the interaction of Ga vacancy acceptors, Ga interstitials, and Ga antisite donors [11]. Therefore, definitive experimental attribution of the nature of dominant deep traps in β-Ga$_2$O$_3$ is an important scientific and practical task requiring much additional work.

Acknowledgments

It is our pleasure to acknowledge long-time fruitful collaboration in these studies with the groups of Professor Pearton and Professor Ren at the University of Florida, Professor Jiheyon Kim and Professor In-Hwan Lee at Korea University, Dr N Smirnov, Dr I Shchemerov, and Dr A Chernykh at NUST MISiS. The work was supported in part by a grant from the Russian National Science Foundation (Grant No. 19-19-00409)

References

[1] Pearton S J, Yang J, Cary P H, Ren F, Kim J, Tadjer M J and Mastro M A 2018 *Appl. Phys. Rev.* **5** 011301
[2] Pearton S J, Ren F, Tadjer M and Kim J 2018 *J. Appl. Phys.* **124** 220901
[3] Galazka Z 2018 *Semicond. Sci. Technol.* **33** 113001

[4] Mastro M A, Kuramata A, Calkins J, Kim J, Ren F and Pearton S J 2017 *ECS J. Solid State Sci. Technol.* **6** P356–9

[5] Tamura Corporation https://tamuracorp.com/products/

[6] Li J *et al* 2018 *Appl. Phys. Lett.* **113** 041901

[7] Wakabayashi R, Hattori M, Yoshimatsu K, Horiba K, Kumigashira H and Ohtomo A 2018 *Appl. Phys. Lett.* **112** 232103

[8] Zhang Y, Joishi C, Xia Z, Brenner M, Lodha S and Rajan S 2018 *Appl. Phys. Lett.* **112** 233503

[9] Varley J B, Weber J R, Janotti A and Van De Walle C G 2010 *Appl. Phys. Lett.* **97** 142106

[10] Deak P, Ho Q D, Seemann F, Aradi B, Lorke M and Frauenheim T 2017 *Phys. Rev.* B **95** 075208

[11] Ingebrigtsen M E, Kuznetsov A Y, Svensson B G, Alfieri G, Milhaila A, Badstubner U, Perron A, Vines L and Varley J B 2019 *APL Mater.* **7** 022510

[12] Zhang Z, Farzana E, Arehart A R and Ringel S A 2016 *Appl. Phys. Lett.* **108** 052105

[13] Varley J B, Peelaers H, Janotti A and van de Walle C 2011 *J. Phys. Condens. Matter* **23** 334212

[14] Kyrtsos A, Matsubara M and Bellotti E 2017 *Phys. Rev.* B **95** 245202

[15] Kyrtsos A, Matsubara M and Bellotti E 2018 *Appl. Phys. Lett.* **112** 032108

[16] Peelaers H, Lyons J L, Varley J B and Van de Walle C G 2019 *APL Mater.* **7** 022519

[17] Ritter J R, Huso J, Dickens P T, Varley J B, Lynn K G and McCluskey M D 2018 *Appl. Phys. Lett.* **113** 052101

[18] Kananen B E, Halliburton L E, Stevens K T, Foundos G K and Giles N C 2017 *Appl. Phys. Lett.* **110** 202104

[19] Lany S 2018 *APL Mater.* **6** 046103

[20] Farzana E, Ahmadi E, Speck J S, Arehart A R and Ringel S A 2018 *J. Appl. Phys.* **123** 161410

[21] Farzana E, Chaiken M F, Blue T E, Arehart A R and Ringel S A 2019 *APL Mater.* **7** 022502

[22] Johnson J M *et al* 2019 *Appl. Phys. X* **9** 041027

[23] Irmscher K, Galazka Z, Pietsch M, Uecker R and Fornari R 2011 *J. Appl. Phys.* **110** 063720

[24] Polyakov A Y, Smirnov N B, Shchemerov I V, Yakimov E B, Yang J, Ren F, Yang G, Kim J, Kuramata A and Pearton S J 2018 *Appl. Phys. Lett.* **112** 032107

[25] Polyakov A Y, Smirnov N B, Shchemerov I V, Gogova D, Tarelkin S A and Pearton S J 2018 *J. Appl. Phys.* **123** 115702

[26] Polyakov A Y, Smirnov N B, Shchemerov I V, Pearton S J, Ren F, Chernykh A V, Lagov P B and Kulevoy T V 2018 *APL Mater.* **6** 096102

[27] Polyakov A Y *et al* 2018 *Appl. Phys. Lett.* **113** 092102

[28] Ingebrigtsen M E, Varley J B, Kuznetsov A Y, Svensson B G, Alfieri G, Mihaila A, Badstübner U and Vines L 2018 *Appl. Phys. Lett.* **112** 042104

[29] Polyakov A Y, Smirnov N B, Shchemerov I V, Yakimov E B, Pearton S J, Ren F, Chernykh A V, Gogova D and Kochkova A I 2019 *ECS J. Solid State Sci. Technol.* **8** Q3019–23

[30] Polyakov A Y *et al* 2019 *Appl. Phys. Lett.* **115** 032101

[31] Polyakov A Y, Smirnov N B, Shchemerov I V, Pearton S J, Ren F, Chernykh A V and Kochkova A I 2018 *Appl. Phys. Lett.* **113** 142102

[32] Polyakov A Y, Smirnov N B, Shchemerov I V, Chernykh S V, Oh S, Pearton S J, Ren F, Kochkova A and Kim J 2019 *ECS J. Solid State Sci. Technol.* **8** Q3013–8

[33] McGlone J F, Xia Z, Zhang Y, Joishi C, Lodha S, Rajan S, Ringel S A and Arehart A R 2018 *IEEE Electron Device Lett.* **39** 1042–5

[34] Varley J B, Janotti A, Franchini C and Van de Walle C G 2012 *Phys. Rev.* B **85** 081109(R)

[35] Onuma T, Fujioka S, Yamaguchi T, Higashiwaki M, Sasaki K, Masui T and Honda T 2013 *Appl. Phys. Lett.* **103** 041910

[36] Armstrong A M, Crawford M H, Jayawardena A, Ahyi A and Dhar S 2016 *J. Appl. Phys.* **119** 103102

[37] Lee J, Flitsiyan E, Chernyak L, Yang J, Ren F, Pearton S J, Meyler B and Salzman Y J 2018 *Appl. Phys. Lett.* **112** 082104

[38] Koksal O, Tanen N, Jena D, Xing H and Rana F 2018 *Appl. Phys. Lett.* **113** 252102

[39] Neal A T *et al* 2018 *Appl. Phys. Lett.* **113** 062101

[40] Wong M H, Lin C-H, Kuramata A, Yamakoshi S, Murakami H, Kumagai Y and Higashiwaki M 2018 *Appl. Phys. Lett.* **113** 102103

[41] Chikoidze E *et al* 2017 *Mater. Today Phys.* **3** 118–26

[42] Korhonen E, Tuomisto F, Gogova D, Wagner G, Baldini M, Galazka Z, Schewski R and Albrecht M 2015 *Appl. Phys. Lett.* **106** 242103

[43] Iwaya K, Shimizu R, Aida H, Hashizume T and Hitosugi T 2011 *Appl. Phys. Lett.* **98** 142116

[44] Swallow J E N, Varley J B, Jones L A H, Gibbon J T, Piper L F J, Dhanak V R and Veal T D 2019 *APL Mater.* **7** 022528

[45] Qin Y, Stavola M, Fowler W B, Weiser P and Pearton S J 2019 *ECS J. Solid State Sci. Technol.* **8** Q3103–10

[46] Polyakov A Y and Lee I-H 2015 *Mat. Sci Eng.* **94** 1–56

[47] Polyakov A Y, Smirnov N B, Schemerov I V, Chernykh A V, Yakimov E B, Kochkova A I, Tereshchenko A N and Pearton S J 2019 *ECS J. Solid State Sci. Technol.* **8** Q3091–6

[48] Malguth E, Hoffmann A and Phillips M R 2008 *Phys. Status Solidi* B **245** 455–80

[49] Wickramaratne D, Shen J-X, Dreyer C E, Engel M, Marsman M, Kresse G, Marcinkevičius S, Alkauskas A and Van de Walle C G 2016 *Appl. Phys. Lett.* **109** 162107

[50] Li J V and Ferrari G 2018 *Capacitance Spectroscopy of Semiconductors* (Singapore: Pan Stanford) p 437

[51] Pons D 1984 *J. Appl. Phys.* **55** 3644–57

[52] Harwig T and Kellendonk F 1978 *J. Solid State Chem.* **24** 255–63

[53] Binet L and Gourier J 1998 *J. Phys. Chem. Solids* **59** 1241–9

[54] Shimamura K, Villora E G, Ujiie T and Aoki K 2008 *Appl. Phys. Lett.* **92** 201914

[55] Huang C-Y, Horng R-H, Wuu D-S, Tu L-W and Kao H-S 2013 *Appl. Phys. Lett.* **102** 011119

[56] Dong L, Jia R, Xin B and Zhang Y 2016 *J. Vac. Sci. Technol.* A **34** 060602

[57] Onuma T, Nakata Y, Sasaki K, Masui T, Yamaguchi T, Honda T, Kuramata A, Yamakoshi S and Higashiwaki M 2018 *J. Appl. Phys.* **124** 075103

[58] Huynh T T, Lem L L C, Kuramata A, Phillips M R and Ton-That C 2018 *Phys. Rev. Mater.* **2** 105203

[59] Ravadgar P, Horng R-H, Yao S-D, Lee H-Y, Wu B-R, Ou S-L and Tu L-W 2013 *Opt. Express* **21** 24599

[60] Ho Q D, Frauenheim T and Deák P 2018 *Phys. Rev.* B **97** 115163

[61] Sun D, Gao Y, Xue J and Zhao J 2019 *J. Alloys Compd.* **794** 374–84

[62] Lyons J L 2018 *Semicond. Sci. Technol.* **33** 05LT02

[63] Gao H *et al* 2018 *Appl. Phys. Lett.* **112** 242102

[64] Liu C, Berencen Y, Yang J, Wei Y, Wang M, Yuan Y, Xu C, Xie Y, Li X and Zhou S 2018 *Semicond. Sci. Technol.* **33** 095022

[65] Yakimov E B, Polyakov A Y, Smirnov N B, Shchemerov I V, Yang J, Ren F, Yang G, Kim J and Pearton S J 2018 *J. Appl. Phys.* **123** 185704

[66] Peres M, Lorenz K, Alves E, Nogales E, Méndez B, Biquard X, Daudin B, Villora E G and Shimamura K 2017 *J. Phys. D: Appl. Phys.* **50** 325101

[67] Sekiguchi H, Sakai M, Kamada T, Yamane K, Okada H and Wakahara A 2019 *J. Appl. Phys.* **125** 175702

[68] Ahrenkiel R K, Call N, Johnston S W and Metzger W K 2010 *Sol. Energy Mater. Sol. Cells* **94** 2197–204

[69] Maiberg M and Scheer R 2014 *J. Appl. Phys.* **116** 123710

[70] Maiberg M, Hölscher T, Zahedi-Azad S and Scheer R 2015 *J. Appl. Phys.* **118** 105701

[71] Yamaga M, Kishita T, Villora E G and Shimamura K 2016 *Opt. Mater. Express* **6** 3135–44

[72] Pozina G, Forsberg M, Kaliteevski M A and Hemmingsson C 2017 *Sci. Rep.* **7** 421321

[73] Beck M, Streb D, Vitzethum M, Kiesel P, Malzer S, Metzner C and Dohler G H 2001 *Phys. Rev.* B **64** 085307

[74] Bicknell W E 2002 *Infrared Phys. Technol.* **43** 39–50

[75] Bryushinin M A, Sokolov I A, Pisarev R V and Balbashov A M 2015 *Opt. Express* **23** 32736–46

[76] Yang J *et al* 2017 *J. Vac. Sci. Technol.* B **35** 051201

[77] Polyakov A Y *et al* 2019 *APL Mater.* **7** 061102

[78] Modak S, Lee J, Chernyak L, Yang J, Ren F, Pearton S J, Khodorov S and Lubomirsky I 2019 *AIP Adv.* **9** 015127

[79] Modak S, Chernyak L, Khodorov S, Lubomirsky I, Yang J, Ren F and Pearton S J 2019 *ECS J. Solid State Sci. Technol.* **8** Q3050–3

[80] Yakimov E B 2015 *J. Alloys Compd.* **627** 344–51

[81] Yakimov E B 2016 *Jpn J. Appl. Phys.* **55** 05FH04

[82] Wu C J and Wittry D B 1978 *J. Appl. Phys.* **49** 2827–36

[83] Chi J Y and Gatos H C 1979 *J. Appl. Phys.* **50** 3433–40

[84] Donolato C 1985 *Appl. Phys. Lett.* **46** 270–2

[85] Yakimov E B, Vergeles P S, Polyakov A Y, Smirnov N B, Govorkov A V, Lee I-H, Lee C R and Pearton S J 2007 *Appl. Phys. Lett.* **90** 152114

[86] Yakimov E B, Vergeles P S, Polyakov A Y, Smirnov N B, Govorkov A V, Lee I-H, Lee C R and Pearton S J 2008 *Appl. Phys. Lett.* **92** 042118

[87] Gake T, Kumagai Y and Oba F 2019 *Phys. Rev. Mater.* **3** 044603

[88] Yamaoka S and Nakayama M 2016 *Phys. Status Solidi* C **13** 93–6

[89] Katz O, Garber V, Meyler B, Bahir G and Salzman 2001 *Appl. Phys. Lett.* **79** 1417–9

[90] Kananen B E, Giles N C, Halliburton L E, Foundos G K, Chang K B and Stevens K T 2017 *J. Appl. Phys.* **122** 215703

[91] Pearton S J, Yang J, Ren F and Kim J 2019 *Ultra-wide Bandgap Semiconductor Materials* ed M Liao, B Shen and Z Wang (Amsterdam: Elsevier) pp 263–344

[92] Yakimov E B 2018 *J. Surface Investigation* **12** 1000–4

[93] Feng Z, Bhuiyan A F M A U, Karim M R and Zhao H 2019 *Appl. Phys. Lett.* **114** 250601

[94] Nakai K, Nagai T, Noami K and Futagi T 2015 *Jpn J. Appl. Phys.* **54** 051103

[95] Hanada K, Moribayashi T, Koshi K, Sasaki K, Kuramata A, Ueda O and Kasu M 2016 *Jpn J. Appl. Phys.* **55** 1202BG

[96] Hanada K, Moribayashi T, Uematsu T, Masuya S, Koshi K, Sasaki K, Kuramata A, Ueda O and Kasu M 2016 *Jpn J. Appl. Phys.* **55** 030303

[97] Ueda O, Ikenaga N, Koshi K, Iizuka K, Kuramata A, Hanada K, Moribayashi T, Yamakoshi S and Kasu M 2016 *Jpn J. Appl. Phys.* **55** 1202BD

[98] Yamaguchi H, Kuramata A and Masui T 2016 *Superlatt. Microstruct.* **99** 99–103

[99] Kasu M, Oshima T, Hanada K, Moribayashi T, Hashiguchi A, Oishi T, Koshi K, Sasaki K, Kuramata A and Ueda O 2017 *Jpn J. Appl. Phys.* **56** 091101

[100] Masuya S, Sasaki K, Kuramata A, Yamakoshi S, Ueda O and Kasu M 2019 *Jpn J. Appl. Phys.* **58** 055501

[101] Kasu M, Hanada K, Moribayashi T, Hashiguchi A, Oshima T, Oishi T, Koshi K, Sasaki K, Kuramata A and Ueda O 2016 *Jpn J. Appl. Phys.* **55** 1202BB

[102] Kuramata A, Koshi K, Watanabe S, Yamaoka Y, Masui T and Yamakoshi S 2016 *Jpn J. Appl. Phys.* **55** 1202A2

[103] Ohba E, Kobayashi T, Kado M and Hoshikawa K 2016 *Jpn J. Appl. Phys.* **55** 1202BF

[104] Wu Y Q, Gao S and Huang H 2017 *Mater. Sci. Semicond. Process.* **71** 321–5

[105] Battu A K and Ramana C V 2018 *Adv. Eng. Mater.* **20** 1701033

[106] Wu Y Q, Gao S, Kang R K and Huang H 2019 *J. Mater. Sci.* **54** 1958–66

[107] Battu A K, Manandhar S and Ramana C V 2017 *Adv. Mater. Interfaces* **4** 1700378

[108] Battu A K, Manandhar S and Ramana C V 2018 *Mater. Today Nano* **2** 7–14

[109] Kucheyev S O, Bradby J E, Williams J S, Jagadish C, Toth M, Phillips M R and Swain M V 2000 *Appl. Phys. Lett.* **77** 3373–5

[110] Jian S-R 2008 *Appl. Surface Sci.* **254** 6749–53

[111] Huang J, Xu K, Gong X J, Wang J F, Fan Y M, Liu J Q, Zeng X H, Ren G Q, Zhou T F and Yang H 2011 *Appl. Phys. Lett.* **98** 221906

[112] Caldas P G, Silva E M, Prioli R, Huang J Y, Juday R, Fischer A M and Ponce F A 2017 *J. Appl. Phys.* **121** 125105

[113] Vergeles P S, Orlov V I, Polyakov A Y, Yakimov E B, Kim T and Lee I-H 2019 *J. Alloys Compd.* **776** 181–6

[114] Orlov V I, Vergeles P S, Yakimov E B, Li X, Yang J, Lv G and Dong S 2019 *Phys. Status Solidi* A **216** 1900163

[115] An Q and Li G 2017 *Phys. Rev.* B **96** 144113

[116] Yakimov E B, Regula G and Pichaud B 2013 *J. Appl. Phys.* **114** 084903

[117] Orlov V I, Regula G and Yakimov E B 2017 *Acta Material.* **139** 155–62

[118] Gao S, Wu Y, Kang R and Huang H 2018 *Mater. Sci. Semicond. Proces.* **79** 165–70

[119] Fiedler A, Schewski R, Baldini M, Galazka Z, Wagner G, Albrecht M and Irmscher K 2017 *J. Appl. Phys.* **122** 165701

[120] Donolato C 1998 *J. Appl. Phys.* **84** 2656–64

[121] Yakimov E B, Polyakov A Y, Lee I-H and Pearton S J 2018 *J. Appl. Phys.* **123** 161543

IOP Publishing

Wide Bandgap Semiconductor-Based Electronics

Fan Ren and Stephen J Pearton

Chapter 6

Breakdown in Ga$_2$O$_3$ rectifiers—the role of edge termination and impact ionization

Yu-Te Liao, Minghan Xian, Randy Elhassani, Patrick Carey IV, Chaker Fares, Fan Ren, Marko Tadjer and S J Pearton

Ga$_2$O$_3$ rectifiers have shown promising performance, with reverse breakdown voltages >2 kV in small area devices and total forward currents >30 A in large area arrays. The reverse breakdown still generally occurs at the periphery of the Schottky contact, due to electric field crowding in this region. This emphasizes the need for continued development of edge termination methods to reduce the peak fields. We discuss the carrier multiplication mechanisms operative in reverse-biased rectifiers and the various methods employed to mitigate this, including field plates, beveling, and the creation of resistive regions using ion implantation. Under forward bias conditions the rectifiers fail through thermally induced stresses that lead to cracking and epilayer delamination.

6.1 Introduction

Most current interest in Ga$_2$O$_3$ has focused on the β-polytype, although for rectifier applications the α-polytype is also an option, with an even larger bandgap [1–4]. With the exception of its low thermal conductivity, Ga$_2$O$_3$ is well-suited for efficient power switching electronics applications, as well as truly solar blind UV detectors [1–10]. The monoclinic β-Ga$_2$O$_3$ has a bandgap of ~4.8 eV, good chemical and thermal stability, and high optical transparency in both the UV and visible regions [11–18]. Over the past several years, there have been numerous reports of promising performance of rectifiers and transistors, as well as a variety of photodetector structures on thin films, bulk crystals, and nanostructures of β-Ga$_2$O$_3$ [4–7, 14, 16–26]. In a few cases, the devices have been driven to breakdown, and it is clear that this generally occurs at the periphery of the rectifying contact, indicating that edge termination methods are still not optimized. To advance the development of power switching devices the fundamental material science of carrier transport at high electric fields needs to be measured experimentally to obtain the impact ionization coefficients [23, 27–35].

Ga_2O_3 has a very high theoretical breakdown electric field ($\sim$8 MV cm^{-1}), but experimental values are typically less than half this due to the presence of defects in the material and electric field crowding in device structures [10, 26]. As control of crystal growth, device design, and materials processing modules improves, it is expected that the experimental values will move towards the theoretical maximum [35–49].

The wide and ultra-wide bandgap semiconductors are more efficient in power switching applications than Si and this could lead to significant energy savings since over 80% of electronic systems, from communications to industrial manufacturing and e-mobility, require the conversion of primary electricity into another form of electricity [27, 50–56]. The efficiency of energy conversion depends on the ability of power switching transistors to provide low resistance on-state and highly resistive off-state conditions. Ga_2O_3 rectifiers might play a role in hybrid power converters in DC/DC and DC/AC applications [42, 43]. For power switching applications, the operating voltage is limited by the breakdown electric field strength (E_{br}) and the background doping in epitaxial drift layers. The total energy loss is determined by resistive power dissipation during on-state current conduction and the capacitive loss during dynamic switching [50, 51]. In order to obtain a good understanding of the breakdown characteristics of a power device, it is important to know the impact ionization coefficients of electrons and holes as a function of the electric field in the semiconductor [23, 28–35]. There are, as yet, no experimental measurements of these quantities in Ga_2O_3.

6.2 The evolution of rectifier design and performance

The initial reports on Ga_2O_3 rectifiers were based on simple, unterminated vertical geometry designs on bulk substrates, as shown in figure 6.1 [2]. As thick, lightly doped drift regions grown using HVPE became available, this led to much large breakdown voltages. These now incorporate various edge termination methods, including field plates, trenches, beveling, and ion implanted resistive regions [2, 4, 24–26, 53–77]. A summary of rectifier performance is given in table 6.1. The current state-of-the-art is to have a drift layer consisting of 7–20 μm thick lightly Si-doped n-type Ga_2O_3 grown using HVPE on n$^+$ bulk, (−201) Sn-doped (3.6×10^{18} cm^{-3}) single crystal wafers. Diodes have full area back ohmic contacts of Ti/Au (20 nm/80 nm), while the Schottky contacts are often Ni/Au (20 nm/80 nm) on the epitaxial layers. There may be some pre-treatment of the back surface to enhance conductivity and lower the contact resistance. This may be plasma exposure, ozone cleaning, or ion implantation of donor dopants.

6.3 Degradation mechanisms in rectifiers

6.3.1 Reverse bias

Even with field plates or other types of edge termination, rectifiers are observed experimentally to breakdown at the contact periphery [76, 77]. When purposely

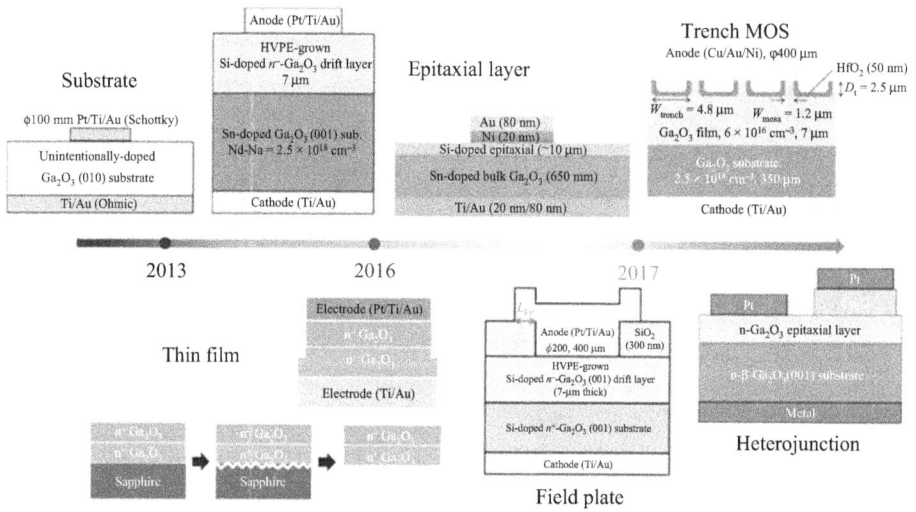

Figure 6.1. Evolution of rectifier structures in Ga_2O_3 in recent years. Reproduced with permission from [2]. Copyright 2018 Springer.

Table 6.1. Summary of the vertical geometry Ga_2O_3 rectifiers on bulk or epilayers reported in the literature.

Reference	Epilayer thickness (μm)	Drift layer doping (cm^{-3})	Edge termination	V_B (V)	R_{ON} ($\Omega\ cm^2$)
Konishi *et al* [4]	10	1.8×10^{16}	Yes—field plate	1076	5.1×10^{-3}
Yang *et al* [25]	10	4.02×10^{15}	No	1600	25×10^{-3}
Yang *et al* [27]	10	2×10^{16}	No	1016	6.7×10^{-3}
Sasaki *et al* [7]	Unintentionally doped substrate	3×10^{16}	No	150	4.3×10^{-3}
Li *et al* [55]	10	2×10^{16}	Trench	1350	15×10^{-3}
Li *et al* [54]	10	2×10^{16}	Trench	2440	25×10^{-3}
He *et al* [67]	Unintentionally doped substrate	2×10^{14}	No	>40	12.5×10^{-3}
Joishi *et al* [61]	2	2.5×10^{17}	Bevel	190	3.9×10^{-3}
Yang *et al* [72]	10	1.33×10^{16}	Yes—field plate	650	1.58×10^{-2}
Yang *et al* [26]	20	2.1×10^{15}	Yes—field plate	2300	0.25
Yang *et al* [71]	7	2×10^{16}	No	466	$0.26–5.9 \times 10^{-4}$
Allen *et al* [69]	8	3.5×10^{16}	Beveled field plate	1100	2×10^{-3}
Lin *et al* [79]	7	10^{16}	N implanted guard ring	1430	4.7×10^{-3}

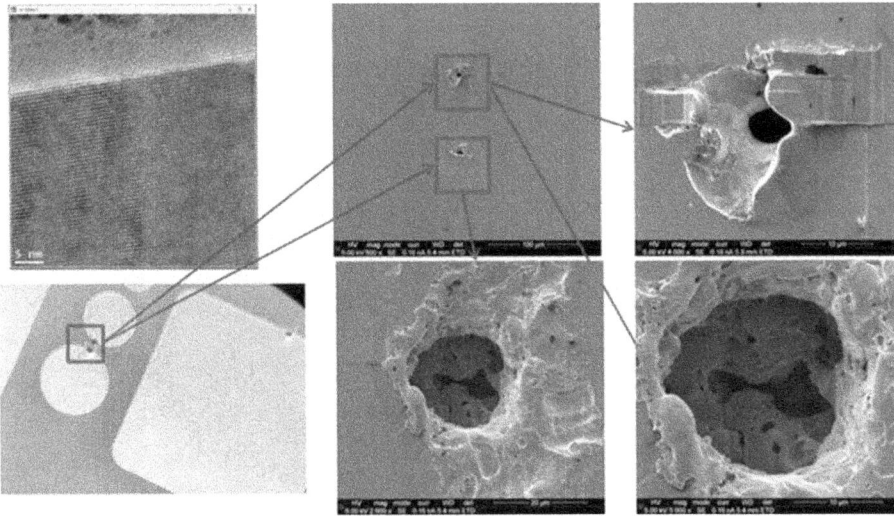

Figure 6.2. TEM lattice image of the rectifier structure (top left), optical image of pits on reverse bias degraded rectifiers (bottom left), and various magnified views of the pits (top and bottom, center and right).

driven to failure at high reverse bias, pits are observed in the high field regions at the edge of the contact. Examples are shown in figure 6.2. The pits appear to result from avalanche failure of the Ga_2O_3 under the high field generated at the edge of the rectifying contact and suggest that further optimization of the edge termination material and geometry is needed to reduce field crowding.

6.3.2 Forward bias

To achieve high reverse blocking voltages in Ga_2O_3 rectifiers, the drift layer must be very lightly doped. This means it is quite resistive and therefore most of the heat generation will occur in this layer by Joule heating, with the highest temperature observed near the Schottky metal–epilayer interface. Devices deliberately tested to failure under forward bias exhibit delamination and cracking of the Ni/Au contact and underlying epitaxial layer due to the low thermal conductivity [76, 77]. This failure mode is different to that under high reverse breakdown conditions, where pits formed by material failure under the high field generated at the edge of the rectifying contact occur. An example is shown in figure 6.3. The failure mechanism has been ascribed to the plastic deformation of the lattice as a result of the device self-heating. Thermal simulations have shown the main mechanism of heat generation in a vertical β-Ga_2O_3 Schottky rectifier [78].

6.3.3 Carrier multiplication mechanisms

Reverse breakdown is caused by avalanche multiplication of charge carriers due to impact ionization between host atoms and high-energy carriers [29–35]. When a high-energy hole or electron in a high electric field collides with an electron in the valence band, it will produce a new electron–hole pair (EHP) [29–34]. This newly

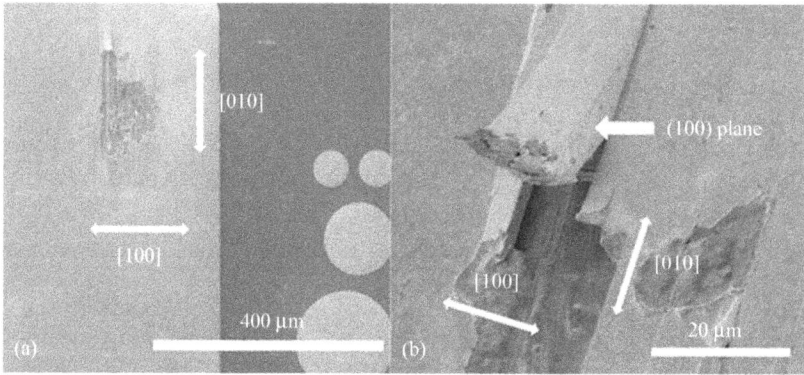

Figure 6.3. SEM image of deliberately induced rupture on a Schottky contact at high forward current condition (a) and with 45° tilting to show delamination of the epilayer (b).

generated EHP will cause other collisions and lead to a multiplication of carriers. Avalanche breakdown is defined to occur when

$$\int_0^{W_D} \alpha_p \exp\left[\int_0^x (\alpha_n - \alpha_p)dx\right]dx > 1$$

$$\alpha_i = \alpha_0 \exp\left(\frac{-b_0}{E}\right),$$

where W_D is the depletion width, α_n and α_p are the ionization rates of electrons and holes, E is the electric field, and b is a constant. The electron ionization rate has been calculated for Ga_2O_3 from a Boltzmann transport equation approach as $\alpha_n = 0.79 \times 10^6 \exp(-2.92 \times 10^7)/E$ (cm^{-1}) [24]. In the typical case of a punch-through junction diode, the breakdown voltage (BV_{PT}) is given by

$$BV_{PT} = E_c W_{PT} - \frac{qN_B W_{PT}^2}{2\varepsilon\varepsilon_0},$$

where E_C is the critical field, N_D is the doping in the drift layer, W_{PT} is the depletion layer thickness at punch-through, and ε is the permittivity. The actual experimental value of the breakdown voltage is far from these theoretical predictions, typically <50%. The presence of defects such as threading dislocations will induce premature breakdown [47–49]. Edge termination is needed to prevent early breakdown and crystal quality should be improved to optimize device performance.

To have a complete understanding of the breakdown characteristics of a power device it is important to know the impact ionization coefficients of electrons and holes as a function of the electric field. Using Chynoweth's equation ($\alpha = ae^{-b/E}$) [31–34], measurements for Ga_2O_3 epitaxial layers grown on bulk Ga_2O_3 substrates should produce values for α and b for the impact ionization coefficient of electrons. It may also be possible to measure the values for holes, since injected holes may transport measurable distances in Ga_2O_3 despite the fact that 'normal' holes form bound polarons.

6.4 Measurement of impact ionization coefficients and their temperature dependence

To obtain a clear understanding of its breakdown characteristics, it is important to know the impact ionization coefficients of Ga_2O_3. It is also important that the breakdown voltage of a device increases with temperature for it to have stable and reliable behavior. This again is determined by the variation of impact ionization coefficients with temperature. The impact ionization coefficients determine the reverse bias characteristics of a device and also influence the forward characteristics. If experimental measurements of the temperature dependence of the reverse bias characteristics show positive temperature coefficients, this would indicate that the breakdown is determined by avalanche breakdown caused by impact ionization. Currently, only a few of the reported rectifiers have shown positive temperature coefficients, with most showing negative values, which correlate with higher concentrations of defects in the substrates and epilayers [4, 57–69]. Electron and, possibly, hole impact ionization rates need to be measured. One approach is to use the pulsed electron beam induced current injection method, in which the rectifier is reverse biased and an electron beam is used to generate carriers to initiate the ionization process. If the impact ionization coefficients for electrons and holes measured at the defective site are higher than those measured at a non-defective site, this would also indicate the breakdown voltage is reduced due to the presence of defects [47–50].

Avalanche multiplication of carriers in reverse-biased diodes was first observed in Si p–n junctions by McKay and McAfee, who demonstrated it was possible to obtain charge multiplication by injecting charge carriers into a p–n junction under high field conditions [29, 30]. They measured the multiplication factor by illuminating a p–n junction. They first measured the $I–V$ characteristics with and without illumination. The multiplication value is obtained by dividing the current with illumination by that of the non-illuminated junction for the same voltage value. They assumed that both electrons and holes multiplied at the same rate, which is now known to be an unjustified assumption. Chynoweth and McKay measured the energy an electron must have to create electron–hole pairs [31] by varying the bias across a p–n junction and determining the potential at the onset of charge multiplication. Breakdown can occur by two mechanisms, each of which requires a critical electric field in the junction transition region [29–35]. The first is the Zener effect, in which energy bands overlap in a heavily doped junction at low voltages, leading to alignment of empty states in the n-side conduction band across the many filled states of the p-side valence band. If there is a small energy barrier separating the bands, tunneling of electrons can occur, contributing to the reverse bias current. As the reverse bias is increased, the barrier between the two bands becomes smaller and the reverse bias current increases. Finally, the device undergoes Zener breakdown at low reverse biases.

The second breakdown mechanism is by avalanche breakdown. For lightly doped junctions, electron tunneling is negligible and the breakdown mechanism involves the impact ionization of host atoms by energetic carriers [33, 34]. As the applied voltage is increased, the electric field in the depletion region increases, accelerating the mobile carriers to higher velocities. At sufficiently high electric fields, these carriers have sufficient energy that their collision with the atoms in the lattice can excite valence band electrons into the conduction band. The process of electron–hole pair generation due to the energy gained from the high field in the semiconductor is called impact ionization. Electron–hole pairs that are generated by the impact ionization undergo acceleration by the existing electric field and they also participate in the creation of further electron–hole pairs. Therefore, impact ionization is a multiplicative process that leads to a cascade of mobile carriers transported through the depletion region leading to a significant current flow through it. Due to the rapid increase in current, the device cannot sustain higher voltages and undergoes avalanche breakdown [33, 34].

One technique to measure the impact ionization coefficients (α) for Ga_2O_3 is based upon the multiplication of carriers generated by a pulsed electron beam in the depletion region of a reverse-biased Schottky diode. The impact ionization coefficients can be obtained from the measured multiplication rates in reversed biased rectifiers. A thin drift region with a low doping density is required to obtain a nearly constant electric field profile within the drift region to correlate the impact ionization coefficients to a particular electric field. Impact ionization coefficients can be extracted from the multiplication data when electron–hole pairs are generated by the excitation beam in the CL/EBIC system in the case of a constant electric field profile in the drift region, from [35]

$$1 - (1/M) = \int \alpha \left[\exp\left(\int -\alpha \, dx \right) \right] dx,$$

where M is the multiplication factor, α is the impact ionization coefficient, and the first integral is over W, the thickness of the drift region. In general, α is a function of electric field (E). However, if the electric field in the depletion region is constant, α becomes independent of x. The equation reduces to $1 - M = -\exp(-\alpha W) + 1$ and the impact ionization coefficient can be obtained using $\alpha = \ln(M)/W$. The experimental set-up is shown schematically in figure 6.4. The multiplication factor (M) is obtained by taking the ratio of the measured pulsed electron beam current to the extrapolated I–V curve from lower reverse bias voltages. Initial numerical simulations show that the extracted impact ionization values depend on the excitation depth. This can be experimentally investigated by varying the electron beam energy. The impact ionization coefficients can be obtained as a function of the electric field (E) by the pulsed electron beam measurement and fitted to Chynoweth's equation $\alpha = ae^{(-b/E)}$ and to Fulop's power law of the form $\alpha = AE^n$, where A is a constant and N is the power index determined from fitting to the experimental data.

Previous work in Si, SiC, GaN, and other materials has determined that the presence of extended defects can enhance the impact ionization coefficients by factors of 2–5×, consistent with the expectation that premature breakdown can be initiated at these sites, depending on their microstructure [35]. The defects may locally enhance the electric field, since there can be states in the gap or strain around

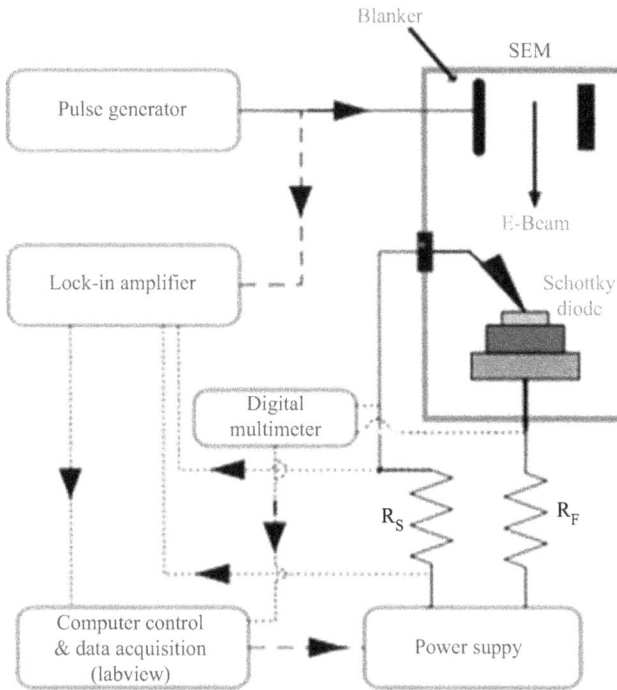

Figure 6.4. Schematic of e-beam injection system for measuring impact ionization coefficients.

the defects that enhance carrier multiplication. EBIC measurements can be used to identify the location of such electrically active defects in Ga_2O_3 Schottky rectifier structures. As the electron beam rasters the surface of the device, the current induced in the semiconductor is collected by the EBIC amplifier. The line scan profile across the diode obtained for increasing reverse biases gives insight into the location of enhanced multiplication due to impact ionization. In Ga_2O_3 rectifier samples with low dislocation density, it is possible to detect individual dislocations as dark spots in EBIC images. Figure 6.5 depicts such an EBIC image taken with a probing beam accelerating voltage of 20 keV and a probing beam current of 0.1 nA.

It is well established that the type of defect has a strong influence on the reverse leakage current in the rectifier. Dislocation defects along the [010] direction were found to act as paths for leakage current, while Si doping did not affect this dislocation-related leakage current [47–49]. In contrast, in the [102] orientation, three types of etch pits were present, consisting of a line-shaped etch pattern originating from a void and extending toward the [010] direction, arrow-shaped pits in the [102] direction and gourd-shaped pits in the [102] direction. Their average densities were estimated to be 5×10^2, 7×10^4, and 9×10^4 cm^{-2}, respectively, but in this orientation there was no correlation between the leakage current in rectifiers and the defects [47–49]. Thus the orientation of the substrate used determines the sensitivity to defect density and this must be accounted for in designing the fabrication masks.

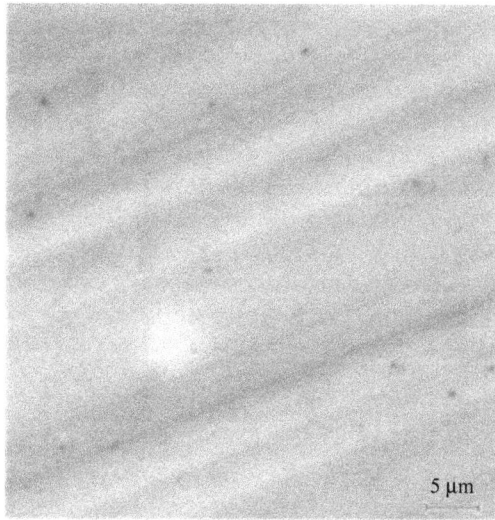

Figure 6.5. EBIC image of a fragment of a β-Ga_2O_3 Schottky barrier. Dislocations can be seen as dark dots.

6.5 Edge termination methods

The reverse breakdown voltages of rectifiers in all wide bandgap semiconductors are still limited by avalanche breakdown at defects and/or surfaces. In SiC rectifiers, there are a variety of edge termination methods employed to smooth out the electric field distribution around the rectifying contact periphery [51, 52], including mesas, junction extension termination, p-type or high resistivity guard rings created by ion implantation, and field plates. The situation is less developed for GaO_3 with a few reports of field-plate termination, some beveling, the use of fin or trench structures, and resistive regions formed by N or Ar implantation [4, 55–77]. In a conventional contact structure, shown in figure 6.6 (top), the edge of the metal contact is a region where electric field crowding occurs. This can be mitigated to some extent by overlapping the contact onto the adjacent dielectric layer, as shown in figure 6.6 (bottom). This is known as a field plate and is the most common edge termination method.

Another approach is to reduce the field gradients by physically beveling the top surface, using controlled etching methods [53, 61, 69]. Obviously, this adds processing complexity and is not widely used. Etching trenches to make fin structures is an effective field reduction method, but has the disadvantage of reducing the active area for current conduction and lowering the total forward current. In other semiconductors used for power electronics, the availability of p-type doping allows for the use of junction termination or guard rings. The lack of p-type doping capability in Ga_2O_3 means that these regions at the edge of the contact must be made with resistive material rather than p-type, as shown in figure 6.7 for guard rings.

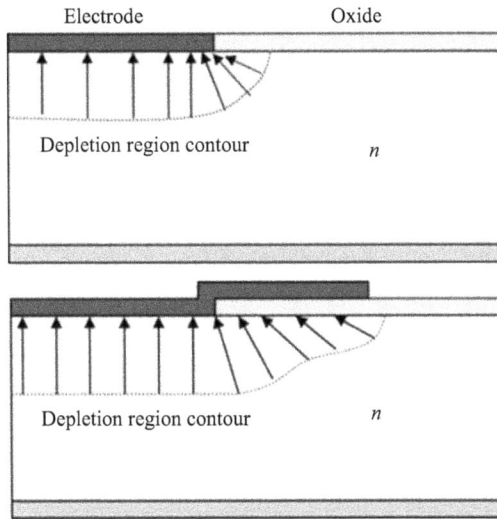

Figure 6.6. Schematic of field crowding at the edge of the anode contact on an unterminated vertical geometry rectifier (top) and the effect of the dielectric overlap in reducing field crowding (bottom).

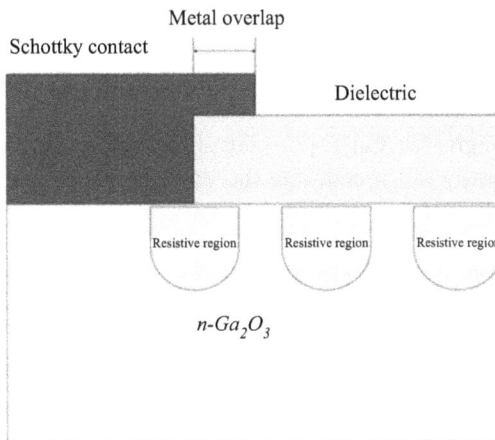

Figure 6.7. Schematic of resistive guard rings and field-plate edge termination.

6.6 The choice of dielectric material for a field plate

The reverse breakdown and leakage current are two characteristics predominantly affected by the field-plate dielectrics. For Ga_2O_3, the primary avenues of improving the reverse breakdown have been through field plating with a single dielectric layer and trenching. The benefit of pursuing a dual stack, such as SiO_2/SiN_x, is the improved reverse breakdown as the SiO_2 absorbs the high electric field and provides for conduction and valence band confinement, while the high-k dielectric lowers the field. Carey *et al* [74] reported a study of three different bi-layer field-plate stacks for vertical geometry Ga_2O_3 rectifiers, namely Al_2O_3/SiN_x, SiO_2/SiN_x, and HfO_2/SiN_x.

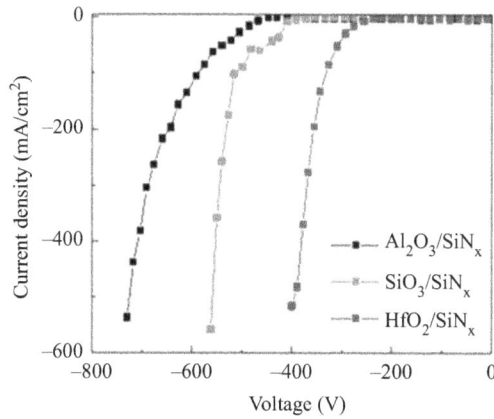

Figure 6.8. Reverse breakdown characteristics for rectifiers with field plates made of bi-layer Al_2O_3/SiN_x, HfO_2/SiN_x, and SiO_2/SiN_x passivation. Reproduced with permission from [74]. Copyright 2019 Electrochemical Society.

Al_2O_3/SiN_x showed superior characteristics, as shown in the reverse breakdown characteristics of figure 6.8, compared to HfO_2/SiN_x and SiO_2/SiN_x. This is a result of the optimized trade-off of bandgap and dielectric constant.

6.7 Summary and conclusions

Edge termination designs for Ga_2O_3 are steadily improving. Failure under reverse bias conditions generally still occurs at the contact edge. Some of the significant milestones reported to date for rectifiers include 2300 V V_B with a 20 μm drift layer (FOM 33.7 MW cm^{-2}), 10 μm drift layer structures with $I_F > 1$ A and reverse breakdown of $V_B = 650$, 2.44 kV breakdown in in small trenches (Cornell), 1.43 kV/ 4.7 mΩ cm^2 (NIICT/Mitsubishi), and finally 33 A arrays under single sweep conditions. Emphasis is needed on thermal management schemes for operation at realistic DC or pulsed conditions [78]. An inductive load test circuit has been used to measure the switching performance of rectifiers with $V_B = 760$–900 V (0.1 cm diameter) and an absolute forward current of 1 A on 8 μm thick drift layers. Switching from 1 A to a reverse off-state voltage of −300 V showed reverse recovery lifetimes, t_{rr}, of 64 ns, similar to SiC. There was no significant temperature dependence of t_{rr} up to 150 °C. There has been terrific progress in rectifier performance in the past two years due to improvements in material quality.

Acknowledgments

The project at UF was sponsored by the Department of the Defense, Defense Threat Reduction Agency, HDTRA1-17-1-011, monitored by Jacob Calkins. Research at NRL was supported by the Office of Naval Research, partially under Award Number N00014-15-1-2392.

References

[1] Mastro M A, Kuramata A, Calkins J, Kim J, Ren F and Pearton S J 2017 *ECS J. Solid State Sci. Technol.* **6** P356

[2] Xue H, He Q, Jian G, Long S, Pang T and Liu M 2018 *Nanoscale Res. Lett.* **13** 290

[3] Guo D *et al* 2014 *Opt. Mater. Express* **4** 1067

[4] Nakagomi S, Momo T, Takahashi S and Kokubun Y 2013 *Appl. Phys. Lett.* **103** 072105

[5] Konishi K, Goto K, Murakami H, Kumagai Y, Kuramata A, Yamakoshi S and Higashiwaki M 2017 *Appl. Phys. Lett.* **110** 103506

[6] Moser N A, McCandless J P, Crespo A, Leedy K D, Green A J, Heller E R, Chabak K D, Peixoto N and Jessen G H 2017 *Appl. Phys. Lett.* **110** 143505

[7] Sasaki K, Higashiwaki M, Kuramata A, Masui T and Yamakoshi S 2013 *Appl. Phys. Express* **6** 086502

[8] Wong M H, Nakata Y, Kuramata A, Yamakoshi S and Higashiwaki M 2017 *Appl. Phys. Express* **10** 041101

[9] Rafique S, Han L and Zhao H 2017 *Phys. Status Solidi* a **214** 1700063

[10] Hu G C, Shan C X, Zhang N, Jiang M M, Wang S P and Shen D Z 2015 *Opt. Express* **23** 13554

[11] Higashiwaki M, Sasaki K, Murakami H, Kumagai Y, Koukitu A, Kuramata A, Masui T and Yamakosh S 2016 *Semicond. Sci. Technol.* **31** 034001

[12] Green A J *et al* 2016 *IEEE Electron Dev. Lett.* **37** 902

[13] Armstrong A M, Crawford M H, Jayawardena A, Ahyi A and Dhar S 2016 *J. Appl. Phys.* **119** 103102

[14] Wong M H, Sasaki K, Kuramata A, Yamakoshi S and Higashiwaki M 2016 *IEEE Electron Dev. Lett.* **37** 212

[15] Sasaki K, Higashiwaki M, Kuramata A, Masui T and Yamakoshi S 2013 *IEEE Electron Dev. Lett.* **34** 493

[16] Wong M H, Sasaki K, Kuramata A, Yamakoshi S and Higashiwaki M 2015 *Appl. Phys. Lett.* **106** 032105

[17] Tadjer M J *et al* 2016 *J. Electr. Mater.* **45** 2031

[18] Rafique S, Han L, Tadjer M J, Freitas J A Jr, Mahadik N A and Zhao H 2016 *Appl. Phys. Lett.* **108** 182105

[19] Tadjer M J, Mahadik N A, Wheeler V D, Glaser E R, Ruppalt L, Koehler A D, Hobart K D, Eddy C R Jr and Kub F J 2016 *ECS J. Solid State Sci. Technol.* **5** 468 2016

[20] Ahn S, Ren F, Kim J, Oh S, Kim J, Mastro M A and Pearton S J 2016 *Appl. Phys. Lett.* **109** 062102

[21] Ahn S, Ren F, Oh S, Jung Y, Mastro M A, Hite J K, Eddy C R Jr, Kim J and Pearton S J 2016 *J. Vac. Sci. Technol.* **B34** 041207

[22] Zeng K, Vaidya A and Singisetti U 2018 *IEEE Electron Dev. Lett.* **39** 1385–8

[23] Son N T *et al* 2016 *J. Appl. Phys.* **120** 235703

[24] Ghosh K and Singisetti U 2014 Calculation of electron impact ionization coefficient in β-Ga_2O_3 *72nd Device Research Conference* (New York: IEEE) 71–2

[25] Yang J, Ahn S, Ren F, Pearton S J, Jang S, Kim J and Kuramata A 2017 *Appl. Phys. Lett.* **110** 192101

[26] Yang J, Ren F, Tadjer M, Pearton S J and Kuramata A 2018 *ECS J. Solid State Sci. Technol.* **7** Q92

[27] Yang J, Ahn S, Ren F, Pearton S J and Kuramata A 2017 *IEEE Electron Dev. Lett.* **38** 906

[28] Cao L, Wang J, Harden G, Ye H, Stillwell R, Hoffman A J and Fay P 2018 *Appl. Phys. Lett.* **112** 262103

[29] McKay K G and McAfee K B 1953 *Phys. Rev.* **91** 1079

[30] McKay K G 1954 *Phys. Rev.* **94** 877

[31] Chynoweth A G and McKay K G 1957 *Phys. Rev.* **108** 29

[32] Chynoweth A G 1958 *Phys. Rev.* **109** 1537

[33] Chynoweth A G and Pearson G L 1958 *J. Appl. Phys.* **29** 1103

[34] Chynoweth A G 1960 *J. Appl. Phys.* **31** 1161

[35] Ozbek A M 2011 Measurement of impact ionization coefficients in GaN *PhD thesis* North Carolina State University, Raleigh, NC

[36] Polyakov A Y, Smirnov N B, Shchemerov I V, Yakimov E B, Yang J, Ren F, Yang G, Kim J, Kuramata A and Pearton S J 2018 *Appl. Phys. Lett.* **112** 032107

[37] Lee J D, Flitsiyan E, Chernyak L, Yang J, Ren F, Pearton S J, Meyler B and Salzman Y J 2018 *Appl. Phys. Lett.* **112** 082104

[38] Yakimov E B, Polyakov A Y, Smirnov N B, Shchemerov I V, Yang J, Ren F, Yang G, Kim J and Pearton S J 2018 *J. Appl. Phys.* **123** 185704

[39] Binet L and Gourier D 1998 *J. Phys. Chem. Solid* **59** 1241

[40] Parisini A and Fornari R 2016 *Semicond. Sci. Technol.* **31** 035023

[41] Wong M H 2016 *Jpn J. Appl. Phys.* **55** 1202B9

[42] Nakai K, Nagai T, Noami K and Futagi T 2015 *Jpn J. Appl. Phys.* **54** 051103

[43] Sasaki K, Wakimoto D, Thieu Q T, Koishikawa Y, Kuramata A, Higashiwaki M and Yamakoshi S 2016 *IEEE Electron Dev. Lett.* **38** 783

[44] Oda M, Kikawa J, Takatsuka A, Tokuda R, Sasaki T, Kaneko K, Fujita S and Hitora T 2015 *Proc. 73rd Annual Device Research Conf. (DRC)* (New York: IEEE) 137–8

[45] Hanada K, Moribayashi T, Uematsu T, Masuya S, Koshi K, Sasaki K, Kuramata A, Ueda O and Kasu M 2016 *Jpn J. Appl. Phys.* **55** 030303

[46] Hanada K, Moribayashi T, Koshi K, Sasaki K, Kuramata A, Ueda O and Kasu M 2016 *Jpn J. Appl. Phys.* **55** 1202BG

[47] Kasu M, Hanada K, Moribayashi T, Hashiguchi A, Oshima T, Oishi T, Koshi K, Sasaki K, Kuramata A and Ueda O 2016 *Jpn J. Appl. Phys.* **55** 1202BB

[48] Oshima T, Hashiguchi A, Moribayashi T, Koshi K, Sasaki K, Kuramata A, Ueda O, Oishi T and Kasu M 2017 *Jpn J. Appl. Phys.* **56** 086501

[49] Ueda O, Ikenaga N, Koshi K, Kuramata A, Moribayashi T, Yamakoshi S, Hanada K M and Kasu M 2016 *Jpn J. Appl. Phys.* **55** 1202BD

[50] Kasu M, Oshima T, Hanada K, Moribayashi T, Hashiguchi A, Oishi T, Koshi K, Sasaki K, Kuramata A and Ueda O 2017 *Jpn J. Appl. Phys.* **56** 091101

[51] Huang A Q 2017 *Proc. IEEE* **105** 2019

[52] She X, Huang A, Lucia O and Ozpineci B 2017 *IEEE Trans. Ind. Electron.* **64** 8193

[53] Gao Y, Li A, Feng Q, Hu Z, Feng Z, Zhang K and Lu X 2019 *Nano Res. Lett.* **14** 8

[54] Li W, Hu Z, Nomoto K, Jinno R, Zhang Z, Tu T Q, Sasaki K, Kuramata A, Jena D and Grace Xing H 2018 2.44 kV Ga_2O_3 vertical trench Schottky barrier diodes with very low reverse leakage current *2018 IEEE Int. Electron Devices Meeting (IEDM)* (New York: IEEE) 8.5.1–4

[55] Li W, Hu Z, Nomoto K, Zhang Z, Hsu J-Y, Thieu Q T, Sasaki K, Kuramata A, Jena D and Xing H G 2018 *Appl. Phys. Lett.* **113** 202101

[56] Sasaki K, Wakimoto D, Thieu Q T, Koishikawa Y, Kuramata A, Higashiwaki M and Yamakoshi S 2017 *IEEE Electron Dev. Lett.* **38** 783

[57] Lin C-H *et al* 2019 Vertical Ga$_2$O$_3$ Schottky barrier diodes with guard ring formed by nitrogen-ion implantation *Compound Semiconductor Week (CSW), Nara, Japan, 2019* pp 1–1

[58] Hu Z *et al* 2018 *IEEE Electron Dev. Lett.* **39** 1564

[59] Yang J, Ren F, Chen Y T, Liao Y T, Chang C W, Lin J, Tadjer M J, Pearton S J and Kuramata A 2019 *IEEE J. Electron Dev. Soc.* **7** 57

[60] Hu Z, Zhou H, Dang K, Cai Feng Y Z, Gao Y, Feng Q, Zhang J and Hao Y 2018 *IEEE J. Electron Dev. Soc.* **6** 815

[61] Joishi C, Rafique S, Xia Z, Han L, Krishnamoorthy S, Zhang Y, Lodha S, Zhao H and Rajan S 2018 *Appl. Phys. Express* **11** 031101

[62] Pomeroy J W *et al* 2019 *IEEE Electron Dev. Lett.* **40** 189

[63] Chatterjee B, Jayawardena A, Heller E, Snyder D W, Dhar S and Choi S 2018 *Rev. Sci. Instrum.* **89** 114903

[64] Zhou H, Maize K, Qiu G, Shakouri A and Ye P D 2017 *Appl. Phys. Lett.* **111** 092102

[65] Cheng Z, Yates L, Shi J, Tadjer M J, Hobart K D and Graham S 2019 *APL Mater.* **7** 031118

[66] Lin C-H, Hatta N, Konishi K, Watanabe S, Kuramata A, Yagi K and Higashiwaki M 2019 *Appl. Phys. Lett.* **114** 032103

[67] He Q *et al* 2018 *IEEE Electron Device Lett.* **39** 556

[68] Wong M H, Morikawa Y, Sasaki K, Kuramata A, Yamakoshi S and Higashiwaki M 2016 *Appl. Phys. Lett.* **109** 193503

[69] Allen N, Xiao M, Yan X, Sasaki K, Tadjer M J, Ma J, Zhang R, Wang H and Zhang Y 2019 *IEEE Electron Dev. Lett.* **40** 1399

[70] Noh J, Alajlouni S, Tadjer M J, Culbertson J C, Bae H, Si M, Zhou H, Bermel P A, Shakouri A and Ye P D 2019 *IEEE J. Electron Dev. Soc.* **7** 914

[71] Yang J *et al* 2019 *Appl. Phys. Lett.* **114** 232106

[72] Yang J, Ren F, Tadjer M, Pearton S J and Kuramata A 2018 *AIP Adv.* **8** 055026

[73] Yang J, Ahn S, Ren F, Pearton S J, Jang S and Kuramata A 2017 *IEEE Electron Device Lett.* **38** 906

[74] Carey P H IV, Yang J, Ren F, Sharma R, Law M and Pearton S J 2019 *ECS J. Solid State Sci. Technol.* **8** Q3221

[75] Chen Y-T, Yang J, Ren F, Chang C-W, Lin J, Pearton S J, Tadjer M J, Kuramata A and Liao Y-T 2019 *ECS J. Solid State Sci. Technol.* **8** Q3229

[76] Yang J *et al* 2019 *ECS J. Solid State Sci. Technol.* **8** Q3028

[77] Yang J, Fares C, Elhassani R, Xian M, Ren F, Pearton S J, Tadjer M and Kuramata A 2019 *ECS J. Solid State Sci. Technol.* **8** Q3159

[78] Sharma R, Patrick E, Law M E, Yang J, Ren F and Pearton S J 2019 *ECS J. Solid State Sci. Technol.* **8** Q3195

[79] Lin C-H *et al* 2019 *IEEE Electron Dev. Lett.* **40** 1487–90

IOP Publishing

Wide Bandgap Semiconductor-Based Electronics

Fan Ren and Stephen J Pearton

Chapter 7

Radiation damage in Ga_2O_3 materials and devices

S J Pearton, Fan Ren, Jihyun Kim, Michael Stavola and Alexander Y Polyakov

The content in this chapter has been adapted from [27] with the permission of the Royal Society of Chemistry.

This chapter provides an overview of radiation damage in β-Ga_2O_3, an ultra-wide bandgap semiconductor that is attracting interest for power electronics and solar-blind ultraviolet detection. Initial studies of proton, electron, neutron, gamma ray, and α-particle damage have been reported for Ga_2O_3. In displacement damage measurements, the carrier removal rates are the highest for α-particles ($\sim 10^3$ cm^{-1}) and the lowest for gamma rays (~ 1 cm^{-1}). Deep electron and hole traps are created, leading to reductions in the minority carrier diffusion length and lifetime. The compensation of shallow donors by gallium vacancies interacting with hydrogen is one of the main carrier loss mechanisms resulting from proton damage. Using rectifiers as a platform for studying radiation effects in β-Ga_2O_3 under conditions relevant to the low earth orbit of satellites shows that the carrier removal rates for proton, electron, and neutron irradiation are comparable to those in GaN of similar doping levels. The main defect created in Ga_2O_3 by proton irradiation is a relaxed V_{Ga}-2H center (a Ga vacancy with two hydrogens attached). Electron paramagnetic resonance (EPR) after irradiation also shows one dominant paramagnetic defect, which has the characteristics of a V_{Ga}-related center. There have been few studies of total ionization dose or single event upsets in Ga_2O_3 and while the former is not expected to be significant because of the lack of gate oxides in most device structures in this material system, single event upsets may be an issue.

7.1 Introduction

There is a strong interest in electronics capable of operation in the harsh environment of space, as well as in terrestrial radiation applications monitoring reactors or stored nuclear waste [1–9]. In the space radiation environment, there are the Van

Allen radiation belts, containing protons with energies up to 500 MeV, as well as electrons up to 10 MeV [1–4]. The fluences of these particles depend on the solar activity [2, 3]. Solar flares produce protons up to 500 MeV and a smaller component of heavy ions up to 10 MeV per nucleon. These are prevalent in the slot region between the main Van Allen belts [1–6]. Finally, there is a near-constant background of galactic cosmic rays consisting of protons and high charge and energy ions, up to 300 MeV per nucleon [1, 4]. Solar flares may expel intense clouds of protons that deliver doses of 0.3–3 Gy/3 days, meaning roughly 50% of the cells of crew members on proposed round trips to Mars would be traversed by at least one galactic cosmic ray with high charge and energy (HZE) [1]. On the old Skylab missions, the typical doses encountered by the crew and equipment were 0.025 Gy for orbits of 250–300 km, at 10 mrad day^{-1} [1–4]. Each pass through the Van Allen belt corresponds to 0.1–0.2 Gy h^{-1} (a passage lasts 10–20 min) [4]. On the Space Shuttle, the dose was 433 mrem/mission average skin dose. To put this in perspective, a CT scan represents 700 mrem per event, a diagnostic x-ray is 100–200 mrem per event, and human natural sources (cosmic and radioactivity) are 80 mrem yr^{-1} [1].

In terms of electronics, there are three main types of radiation damage in semiconductor devices.

1. Total ionization dose (TID) effects are caused by long term exposure to ionizing radiation. These induce changes in the electrical properties of materials that may cause them to operate incorrectly or fail. This is an important effect for insulators (charge build-up), cabling, electronics (surface charge effects), optical elements (lenses and filters), and cryogenics. Since Ga_2O_3 devices are typically not MOS-gate structures [10–13], TID effects are not considered likely to be a significant issue.

2. Displacement damage dose (DDD) effects are due to long term exposure to interactions with non-ionizing energy transfers. These create displacement defects in semiconductors and the introduction of traps deep in the bandgap. Since Ga_2O_3 has a high bond strength, it has a threshold for DDD effects several orders of magnitude larger than conventional semiconductors such as Si and even GaAs [10–19].

3. Single event effects (SEEs) are effects due to a single interaction involving the transit of an ionizing particle through a device [1, 5]. Single event transients (SETs) are a potential reliability issue for devices in space-based RF and power systems [20–23]. SETs occur when incident ions, passing through or close to sensitive volumes in the device, interact with atoms in the semi-conductor, liberating electrons that disturb voltages [1, 5, 7]. These take the form of voltage glitches, whose amplitudes and widths depend on bias conditions and strike location. SETs last on the order of picoseconds to nanoseconds. A recent addition to the probes available for investigating SETs in electronic devices is the pulsed, focused x-ray beam [5, 7]. Similar to ions and to photons in the visible spectrum, x-rays passing through material also produce electron–hole pairs, but the energy-loss mechanisms involved are different. Both result in SETs and an area of investigation is whether those produced by x-rays are similar to those produced by ions or pulsed

laser light. An advantage of x-rays is their ability to penetrate metal over-layers and inject charge underneath the gate, something not possible with visible laser light [5, 7]. SETs are important in digital circuits such as memories or microprocessors, and induce errors, undesired latch-ups, and possible system failure. This will probably be a problem in Ga_2O_3, as has been reported for AlGaN-based high electron mobility transistors (HEMTs) of even high Al content (>70%) [22]. However, to this point, there have been no reported studies of single event effects in Ga_2O_3.

Displacement damage and total ionizing dose are cumulative effects, while single event upsets are stochastic effects, with the latter originating from the protons of solar flares and the radiation belts, as well as the ions from cosmic rays.

β-Ga_2O_3 has a bandgap of 4.8–4.9 eV and high critical electric field (E_c) strength of ~8 MV cm^{-1} [10–19], allowing high temperature and high power device operation. The availability of large, melt-grown substrates at relatively low cost has suggested the possibility of SiC-like performance at Si-like cost [15, 16]. Figure 7.1 shows the crystal structure of the β-polymorph, as well as a large diameter wafer cut from a bulk crystal grown by the edge defined, film-fed growth (EFG) method. The main electronics applications for β-Ga_2O_3 rectifiers and metal–oxide–semiconductor field-effect transistors (MOSFETs) involve power conditioning and switching systems with low power loss during high frequency switching up to the gigahertz regime [24]. Ga_2O_3-based photodetectors are attracting interest as truly solar-blind deep ultraviolet (UV) photodetectors with a cut-off wavelength of <280 nm [9, 14, 16, 18]. These have applications in military systems, air purification, space communications, ozone-layer monitoring, and flame sensing.

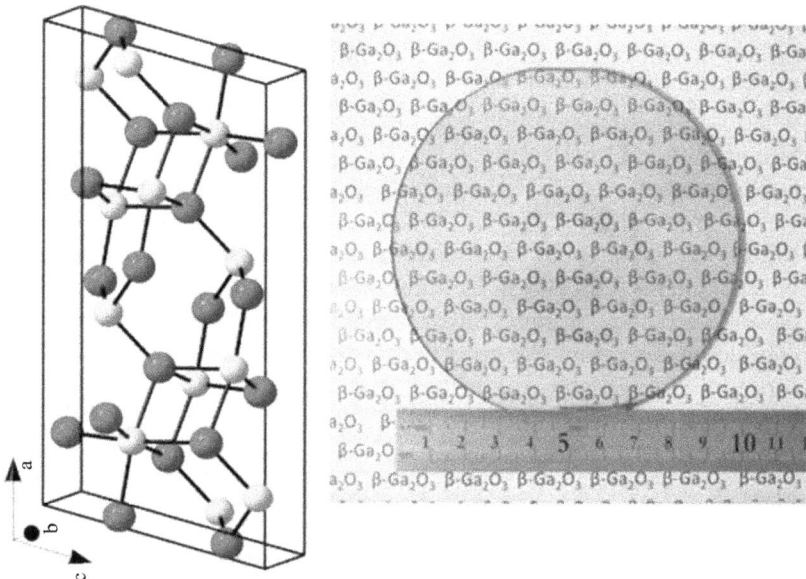

Figure 7.1. (Right) β-Ga_2O_3 crystal structure and (left) large diameter wafer from EFG-grown crystal.

7.2 Basic radiation damage measurement quantities

There are a number of fundamental parameters used in radiation damage studies in semiconductors, which we define below [20–23, 25, 26].

Flux (φ) is the number of particles per unit area and per unit time,

$$\varphi = \text{particles}/(\text{area} \times \text{time}).$$

The time integral of the flux is the fluence (Φ), the number of particles per unit area

$$\Phi = \int \varphi \, dt = \text{particles cm}^{-2}.$$

The dose is the energy deposited by radiation per unit mass, which scales with fluence

$$D = E/M.$$

A particle can transfer (deposit) energy in a semiconductor by ionizing energy loss, which is equivalent to total ionizing dose (TID) or by non-ionizing energy loss (NIEL), which is equivalent to the displacement damage dose (DDD). These are defined as follows:

$$\text{TID} = (\text{energy to ionization})/\text{mass}$$
$$= \text{LET} \times \Phi, \text{ where LET is the linear energy transfer}$$

$$\text{DDD} = (\text{energy to displacements})/\text{mass}$$
$$= \text{NIEL} \times \Phi, \text{ where NIEL is the non-ionizing energy loss.}$$

LET is the energy deposited per unit path length due to ionization, i.e. the Coulomb interaction of the impinging particle with the electrons of the material causing ionization, defined as $\Delta E_{\text{ionization}}/\Delta x \rightarrow \text{LET} = (dE/dx)_{\text{ionization}}$.

NIEL is the energy deposited per length unit due to non-ionizing interaction of the impinging particle with the nuclei of the lattice causing displacement damage, defined as $\Delta E_{\text{displacement}}/\Delta x \rightarrow \text{NIEL} = (dE/dx)_{\text{displacement}}$.

DDD = KERMA/mass = NIEL $\times \Phi$, where KERMA is defined as the kinetic energy imparted by radiation into displacement, i.e. the total kinetic energy released in matter (silicon equivalent). Displacement damage is the result of the transfer of non-ionizing energy to lattice nuclei, causing structural damage to the lattice (defects) by collisions between incoming particles and a lattice nucleus displacing an atom from its original lattice position generating point defects (vacancies and interstitials).

The energy deposited into a block of matter of a certain mass includes the energy to ionization (MeV) = LET (MeV cm^2 mg^{-1}) $\times \Phi$ (cm^{-2}) $\times$ mass (g) $\times 10^3$ and the energy to displacements (MeV) = NIEL (MeV cm^2 mg^{-1}) $\times \Phi$ (cm^{-2}) $\times$ mass (g) $\times 10^3$. Heavy ions lose a small amount of energy per Coulombic collision and the total stopping power dE/dx is given by $dE/dx = (dE/dx)_{\text{ionization}} + (dE/dx)_{\text{nuclear coulombic}} = \text{LET} + \text{NIEL}$. Both of these quantities change along the ion track as the particle slows down.

The displacement energy is manifested as lattice damage to the semiconductors, while the ionizing energy is lost as heat. Incident electrons and positrons may lose a large fraction of their kinetic energy in one collision and are easily scattered to large angles.

In single event upsets, locally, the ionization may be high inside single active/sensitive volumes [1, 6, 7, 22]. A high enough ionization may induce a single event effect. In space applications, electronic devices may receive the direct impacts of HZE cosmic rays during the operational lifetime of a spaceflight. More frequently, energetic neutrons and protons may produce secondary highly ionizing ions. Neutrons are a problem in avionics and at sea level [1, 21–23]. The nature of the device determines the active volumes that may collect charge and the effect on performance when large amounts of charge are deposited.

Ga_2O_3-based UV photoconductors and electronic devices would be subject to fluences of high energy protons, α- particles, and electrons in low earth orbit satellites, as well as neutrons or gamma rays in radiation-hard nuclear or military systems [27]. Each type of radiation produces different types of lattice damage. At high incident energies, the energy of the primary recoils produced by collisions with lattice atoms becomes high enough to give rise to collision cascades and form disordered regions with high defect density in the core. The collision between an incoming ion and a lattice atom displaces the atom from its original lattice position, leading to vacancies, interstitials, and complexes of both, and potentially with impurities in Ga_2O_3 [27–31]. If an incident energetic particle such as a neutron or proton collides with the nucleus of a lattice atom, the primary knock-on atom may be displaced from the lattice if the incident particle has sufficient energy ($E > E_d$), where E_d is the lattice displacement energy [25, 26].

7.2.1 The importance of radiation damage in electronics

Most telecommunications satellites are in geosynchronous orbit, at the outer edge of the second Van Allen radiation belt [3, 4]. Most navigation satellites operate in medium Earth orbit and pass through the heart of that outer radiation belt where they may experience much higher levels of radiation [3, 4]. Most Earth-observation satellites operate in low Earth orbit and may experience higher radiation levels if their orbits traverse the South Atlantic Anomaly or the auroral zones [3, 4]. The variability in the flux of a relativistic electron ($E > 1$ MeV) in the radiation belts is caused by changes in the solar wind through activity on the Sun.

While Ga_2O_3 is not yet used in commercial or government satellites, it is not expected that TID or NIEL will be issues, because the devices do not employ MOS gates and also because of the high bond strength [27]. However, recent experiments on high Al content AlGaN shows that single event upsets are an issue there [22] and one might reasonably expect similar issues with Ga_2O_3, although the low hole mobility may mitigate this somewhat [30–40]. Even in shielded environments, there is a low energy proton spectrum that creates single event upsets, resulting in data corruption and transient disturbance. Within the family of single event upset phenomena, there are classical events that cause a change of state in storage elements, while single event functional interrupts (SEFIs) produce a temporary loss of device functionality and may be recovered through a reset.

It is important to note that the range of high energy protons is much larger than the thickness of typical devices, so even with packaging and any shielding, the ions traverse the entire volume of the device. Figure 7.2 shows the projected range of protons and α-particles in Ga_2O_3 as a function of energy [20]. Typical active layer thicknesses in Ga_2O_3 devices are <20 μm.

7.2.2 Radiation damage in wide bandgap semiconductors

It is well established that Ga_2O_3 is between one and two orders of magnitude more radiation-hard than Si in terms of NIEL with protons. By this, it is meant that for exposure to fluences of the same proton energy, a Ga_2O_3-based device will last 10–100 times longer than a Si device [27]. The wide bandgap and ultra-wide bandgap semiconductors, which include GaN, SiC, AlN, diamond, BN, and Ga_2O_3, all have high bonding energies, which means a lower number of primary defects for a given energy deposition [41]. The atomic displacement energy has been measured in many semiconductors and is empirically determined to be inversely proportional to the volume of the unit cell [1, 8, 25, 26]. β-Ga_2O_3 has been suggested to have similar and perhaps even improved radiation tolerance compared to GaN based on slightly higher displacement energies for Ga and O of 25 eV and 28 eV in β-Ga_2O_3, respectively, compared to 20 eV and 10–20 eV for Ga and N in GaN, respectively [24, 42, 43].

The stability of primary radiation defects in Ga_2O_3 is still being established [44–64]. High energy particle irradiation produces a dominant paramagnetic defect assigned to gallium vacancies (V_{Ga}), with the structure of a hole on an O atom adjacent to a cation vacancy. Korhonen *et al* [32] estimated a V_{Ga} concentration of at least 5×10^{18} cm^{-3} in even as-grown undoped samples. Since theoretical

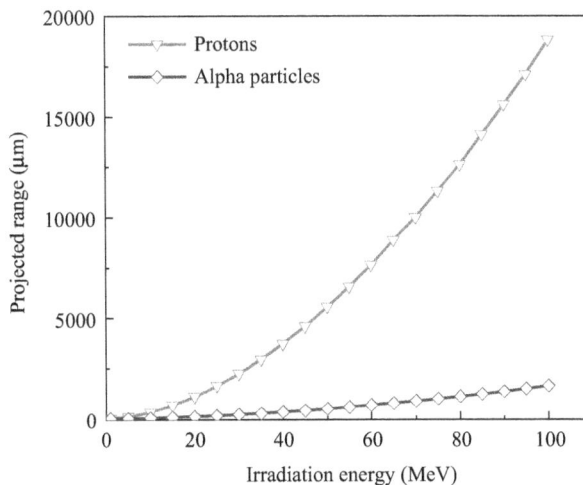

Figure 7.2. Range of protons and α-particles in Ga_2O_3 as a function of energy (1–100 MeV).

calculations predict that these V_{Ga} should be in a negative charge state for n-type samples [28], they will compensate the *n*-type doping. Kananen *et al* [33, 34] showed from electron paramagnetic resonance (EPR) that doubly ionized (V_{Ga}^{2-}) and singly ionized (V_{Ga}^-) acceptors exist at room temperature in CZ Ga_2O_3. They observed singly ionized V_{Ga}^- in neutron irradiated β-Ga_2O_3. The two holes in this acceptor are trapped at individual oxygen ions located on opposite sides of the gallium vacancy. Ga vacancies (V_{Ga}) can exist at tetrahedral (Ga1) and octahedral (Ga2) sites. V_{Ga} forms deep levels at –3 charge state at E_C – 1.62 and 1.83 eV for V_{Ga1} and V_{Ga2}, respectively, and act as deep acceptors [28, 35]. A schematic of the V_{Ga} derived from EPR is shown in figure 7.3.

Oxygen vacancies (O_V) cannot explain the observed unintentional n-type conductivity in Ga_2O_3 since they are deep donors with an ionization energy >1 eV [28, 38]. Theory suggests that unintentional impurities are responsible for n-type conductivity, with H, Si, Ge, Sn, F, and Cl all calculated to be shallow donors [38]. EPR and SIMS identify Si as the primary donor in undoped Ga_2O_3 [35].

7.2.3 Summary of radiation damage studies in Ga_2O_3

7.2.3.1 Proton damage in Ga_2O_3 nanobelt transistors and rectifiers
In back-gated field-effect transistors (FETs) fabricated on exfoliated β-Ga_2O_3 nanobelts, 10 MeV protons caused a decrease of >70% in field-effect mobility [19].

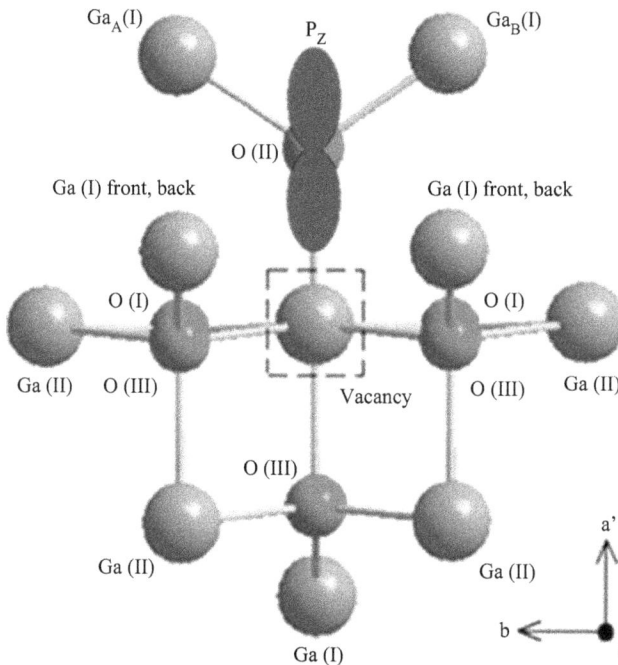

Figure 7.3. Model of the doubly ionized gallium vacancy in β-Ga_2O_3. An unpaired spin (the hole) is localized in a p_z orbital on a threefold oxygen ion, O(II), adjacent to a gallium vacancy (dashed square) at a six-fold Ga (II) site. Reproduced with permission from [33]. Copyright 2017 AIP Publishing.

In addition, there was a positive shift of threshold voltage after proton irradiation at 2 $\times$ 10^{15} cm^{-2}, which corresponds to approximately 10^5 times the intensity of a solar proton event. Doses of $>10^{15}$ cm^{-2} led to significant suppression of drain current, but the damage could be recovered after annealing at 500 °C, similar to that needed for the removal of plasma-induced dry etch damage in Ga_2O_3 [27].

In rectifier structures, Ingebrigtsen *et al* [44] showed that irradiation by 0.6 and 1.9 MeV protons at doses $>2 \times 10^{13}$ cm^{-2} showed complete removal of free charge carriers. Annealing showed a recovery with activation energy of ~1.2 eV. They suggested that the charge carrier removal is a result of Fermi-level pinning due to gallium interstitials (Ga_i), vacancies (V_{Ga}), and antisites (Ga_O), while migration and subsequent passivation of V_{Ga} via hydrogen-derived or V_O defects may be responsible for the recovery [44].

The 10 MeV proton irradiation of vertical geometry Ga_2O_3 rectifiers at a fixed fluence of 10^{14} cm^{-2} produced trap states that reduced the carrier concentration with a carrier removal rate of 235.7 cm^{-1} for protons [27]. Annealing at 300 °C produced a recovery of approximately half of the carriers, while annealing at 450 °C almost restored the reverse breakdown voltage. The minority carrier diffusion length decreased from ~340 nm in the starting material to ~315 nm after proton irradiation [27, 41, 64]. The reverse recovery characteristics showed little change, with values in the range 20–30 ns before and after proton irradiation.

A compilation of trap states in as-grown and proton-irradiated Ga_2O_3 shows these span a large portion of the gap [27, 44, 45, 54, 57, 62, 64–69]. Proton irradiation increases the density of E2* (E_c − 0.75 eV) and E_c − 2.3 eV traps, suggesting these incorporate native defects. Irradiation with 10–20 MeV protons creates deep electron and hole traps, a strong increase in photocapacitance and persistent photocapacitance that partly persists at >300 K [64]. Typical DLTS spectra from samples before and after 10 MeV proton irradiation are shown in figure 7.4. A prominent electron trap near E_c − 1.05 eV is present prior to irradiation. Two minor traps with levels E_c − 0.6 eV and E_c − 0.75 eV are also present. After proton irradiation, the dominant peak in the DLTS spectra is an electron trap with level E_c − 0.75 eV with a prominent shoulder due to the E_c − 1.05 eV electron trap. These are, respectively, the E1, E2*, and E3 electron traps. The concentrations of E2* and E3 increased and a new trap E4 at E_c − 1.2 eV emerged after proton irradiation.

Hole traps in the lower half of the bandgap were investigated using optical injection. Three hole traps, H1 (STH), H2 (ECB), and H3, with activation energies 0.2, 0.4, and 1.3 eV, respectively, were detected [27, 64]. The Arrhenius plots of these traps are shown in figure 7.5. The H1 (STH) peak corresponds to the transition of polaronic states of self-trapped holes (STH) to mobile holes in the valence band. The H2 (ECB) feature was assigned to overcoming of the electron capture barrier (ECB) of centers responsible for persistent photocapacitance at $T < 250$ K. The H3 peak was produced by detrapping of holes from E_v + (1.3–1.4) eV hole traps related to V_{Ga} acceptors. A deep acceptor with an optical ionization threshold near 2.3 eV is probably responsible for high temperature persistent photocapacitance surviving up to temperatures higher than 400 K.

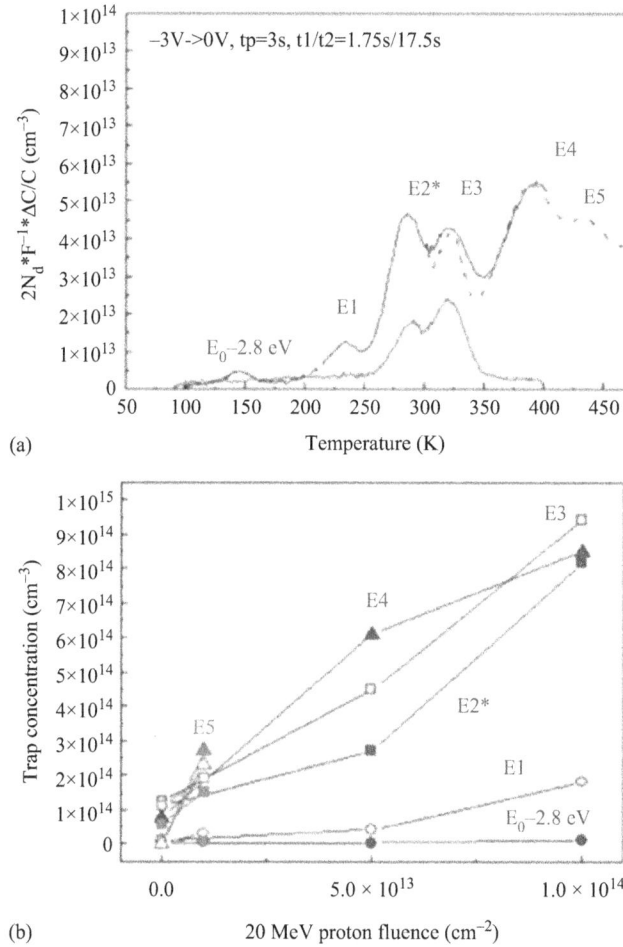

Figure 7.4. (a) DLTS spectra of β-Ga$_2$O$_3$ before (red line) irradiation and after irradiation with 10^{13} cm^{-2} 18 MeV α-particles. (b) Deep trap concentrations as a function of α-particle (red symbols) or 20 MeV proton (blue symbols) fluence. Reproduced with permission from [69]. Copyright 2019 AIP Publishing.

Other studies prior to irradiation of H$^+$ ions show at least four levels, labeled E1–E4, with respective energy levels of 0.56, 0.78, 1.01, and 1.48 eV. The E2^8 level is the most prominent and may be related to iron [44]. Figure 7.6 shows the E2^8 concentration after irradiation is similar to that before irradiation. There was some instability of the concentrations of the different states observed [44].

7.2.3.2 Neutron damage effects

Farzana *et al* [45] showed that neutron irradiation-induced effect Schottky diodes fabricated on EFG UID (010) substrates produced carrier reduction through irradiation. The primary effect was to increase the concentration of a pre-existing state at $E_C - 2.00$ eV and introduce a state at $E_C - 1.29$ eV not observed prior to irradiation. There was a reduction in doping after neutron irradiation, as shown in

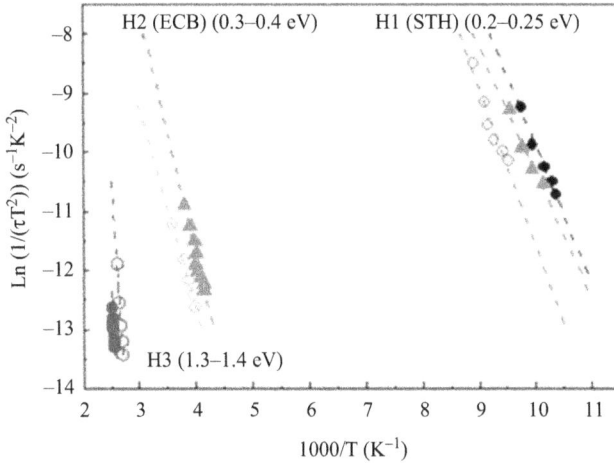

Figure 7.5. Arrhenius plots for three hole-trap-like signals observed in ODLTS with intrinsic and extrinsic excitation. Reproduced from [64].

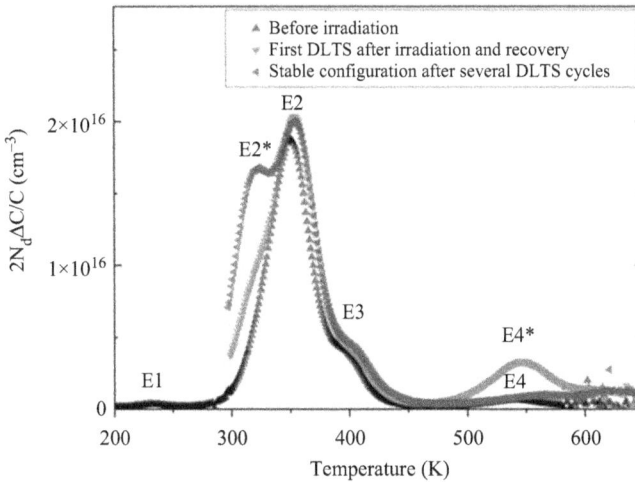

Figure 7.6. DLTS spectra from Ga_2O_3 before and after irradiation with 6×10^{14} cm^2 H^+ ions. Two measurements after irradiation are presented to show the difference between the first measurement and that done after stabilization from several cycles. Reproduced from [44].

figure 7.7(a). Carrier removal rates for neutron irradiation are not straightforward because the neutron spectrum has a broad range of energies, unlike mono-energetic electron or proton beams. Thus, the damage cross-section is a function of neutron energy [46]. Using an analysis of neutron flux and damage cross-section for a particular neutron energy enables the neutron energy spectrum to be approximated by a mono-energetic equivalent fluence convenient for extraction of carrier removal rate. The carrier removal rate for a 1 MeV equivalent neutron fluence was 51 cm^{-1} (figure 7.7(b)). Figure 7.8 shows the neutron-induced $E_C - 1.03$ eV, $E_C - 1.29$ eV, and $E_C - 2.00$ eV concentrations, with trap introduction rates of 1.5, 8, and

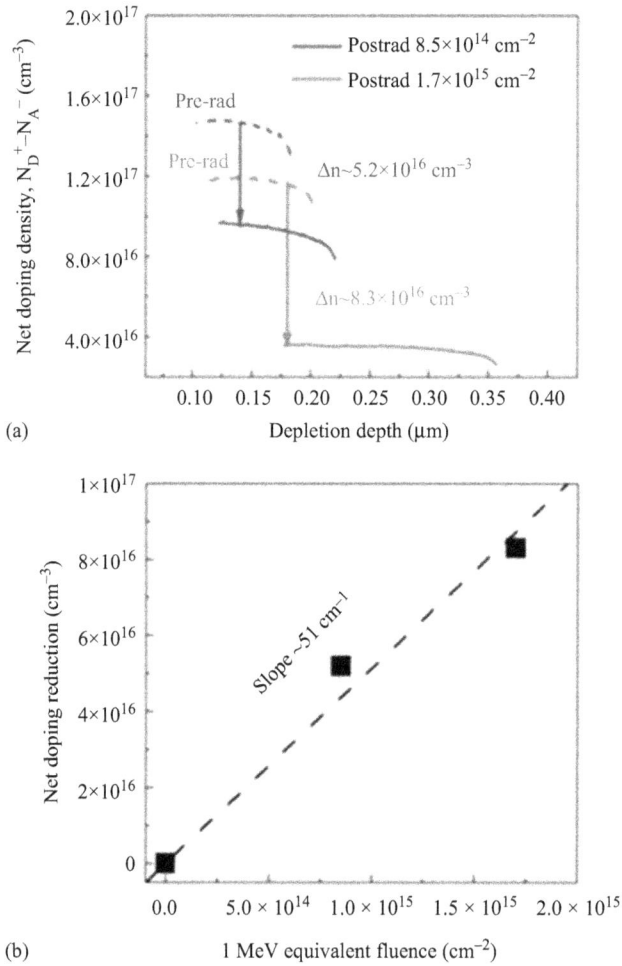

Figure 7.7. (a) Doping concentration prior to and after neutron irradiation, showing reduction of $\sim 5.2 \times 10^{16}$ cm^{-3} and $\sim 8.3 \times 10^{16}$ cm^{-3} obtained for 1 MeV equivalent fluences of 8.5×10^{14} cm^{-2} and 1.7×10^{15} cm^{-2}, respectively. The dashed lines show the pre-irradiation doping profile for the as-grown samples for the respective cases. (b) Net doping reduction versus 1 MeV equivalent fluences, with carrier removal rate ~ 51 cm^{-1}. Reproduced from [45].

9.6 cm^{-1}, respectively [45]. The sum of the individual introduction rates is lower than the total carrier removal rate of 51 cm^{-1}. The displacement cross-section for Ga_2O_3 irradiated by neutrons, given a displacement threshold energy in Ga_2O_3, $E_{Ga}^d = 25$ eV, was $\sigma_{Ga_2O_3}^{disp}$ (1 MeV) $= 92.3$ MeVmb [46, 47].

7.2.3.3 α-particles

Ga_2O_3 vertical geometry rectifiers subjected to 18 MeV α-particle irradiation at fluences of 10^{12}–10^{13} cm^{-2} showed carrier removal rates 406–728 cm^{-1}. These are a factor of 2–3 higher than for 10 MeV protons and two orders of magnitude higher

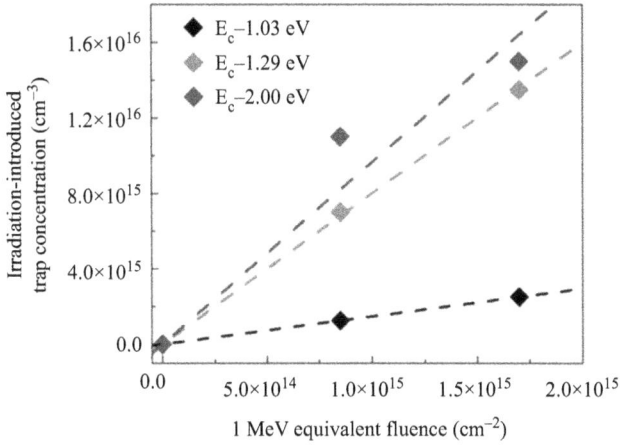

Figure 7.8. Irradiation-induced trap concentration plotted with respect to the 1 MeV equivalent fluences of 8.5×10^{14} cm^{-2} and 1.7×10^{15} cm^{-2} for the $E_C - 1.03$ eV, $E_C - 1.29$ eV, and $E_C - 2.00$ eV states. The slopes provide the introduction rates as 1.5, 8 and 9.6 cm^{-1} for $E_C - 1.03$ eV, $E_C - 1.29$ eV, and $E_C - 2.00$ eV, respectively. Reproduced from [45].

than for 1.5 MeV electron irradiation. The on-state resistance of the rectifiers was more degraded by α-particle irradiation than either the ideality factor or barrier height. The reverse breakdown voltage of the rectifiers increased with α-particle dose as carriers in the drift region were removed by trapping in traps created by the radiation damage, as shown in figure 7.9 [27, 57].

DLTS spectra after irradiation with α-particles showed states at $E_c - 0.28$ eV and $E_c - 1.35$ eV. α-particle irradiation increased the concentration of all traps. The main change was E2* traps whose introduction rate was ~8 cm^{-1} and E3 and E4 traps with an introduction rate of ~20 cm^{-1}. The introduction rates are more than an order of magnitude lower than the carrier removal rates so that, even if all these traps were acceptors, they could not account for the donor decrease after irradiation [27, 64]. The carrier removal rate upon irradiation of β-Ga$_2$O$_3$ with α-particles was close to the introduction rate of acceptor Ga$_v$ determined by SRIM modeling. The most dominant deep electron traps detected in irradiated samples in the upper half of the bandgap were the E2*, E3, and E4 traps with levels from 0.75–1.2 eV from the conduction band edge. The electron removal rates for α-radiation were found to be close to the theoretical production rates of vacancies, whereas the concentrations of major electron and hole traps were much lower, suggesting that the main process responsible for carrier removal is the formation of neutral complexes between vacancies and shallow donors. There was a concurrent decrease in the diffusion length of nonequilibrium charge carriers after irradiation, which correlates with the increase in the density of the main electron traps E2* at $E_c - (0.75-0.78)$ eV, E3 at $E_c - (0.95-1.05)$ eV, and E4 at $E_c - 1.2$ eV. The $E_v + 1.4$ eV acceptors could be related to Ga$_v$ acceptors that have not formed complexes with shallow donors but also the Ga$_v$–acceptor complexes with hydrogen. The levels of electron traps E3 and

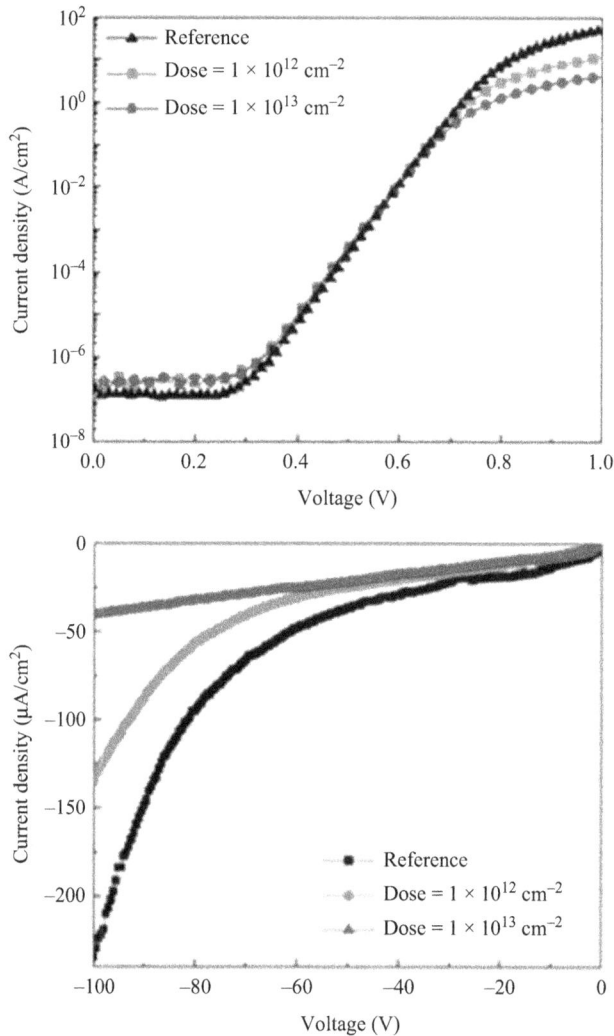

Figure 7.9. Forward (top) and reverse (bottom) current density–voltage characteristics before and after 18 MeV α-particle irradiation with fluences of 10^{12} or 10^{13} cm^{-2}. Reproduced with permission from [57]. Copyright 2018 AIP Publishing.

E4 are close to the two deep oxygen vacancy V_O donors predicted by theory [28, 38] and these could be associated with the lifetime degradation in irradiated β-Ga$_2$O$_3$.

Electron injection on minority carrier transport in β-Ga$_2$O$_3$ Schottky rectifiers with 18 MeV α-particle exposure (fluences of 10^{12}–10^{13} cm^{-2}) was studied from 25 °C to 120 °C [65]. The diffusion length can be significantly increased by electron injection for both reference and irradiated structures, although the rate of its increase is lower after irradiation (figure 7.10). The decrease in activation energy of the electron injection effect on diffusion length for the irradiated sample is attributed to radiation-

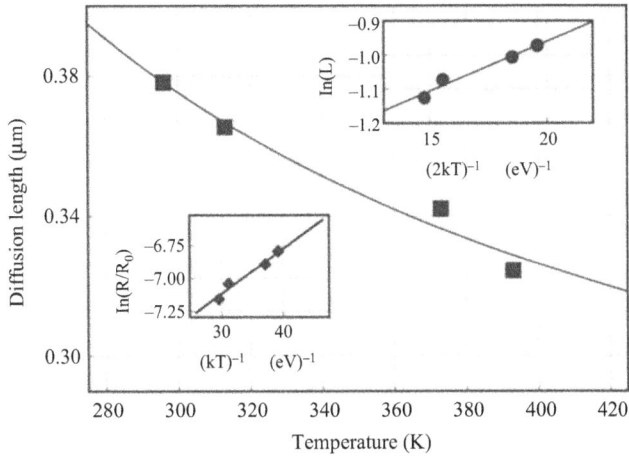

Figure 7.10. Temperature dependence of diffusion length for an α-irradiated Ga_2O_3 rectifier. Inset (top-right): Arrhenius plot, used to calculate the activation energy for diffusion length change with temperature. Inset (bottom): Arrhenius plot used to calculate the activation energy for diffusion length change due to electron injection. Reproduced from [65] copyright 2019 IOP Publishing, reproduced with permission, all rights reserved.

induced generation of additional shallow recombination centers closer to the conduction band edge.

7.2.3.4 Electron irradiation damage

The 1.5 MeV electron irradiation of vertical rectifiers at fluences from 1.79×10^{15} to 1.43×10^{16} cm^{-2} caused a reduction in carrier concentration, with a carrier removal rate of 4.9 cm^{-1}. This compares to a carrier removal rate of $\sim$300 cm^{-1} for 10 MeV protons in the same material. The 2 kT region of the forward current–voltage characteristics increased due to electron-induced damage, with a more than two orders of magnitude increase in on-state resistance at the highest fluence [27, 54]. There was a reduction in reverse current, which scaled with electron fluence. The on–off ratio at -10 V reverse bias voltage was severely degraded by electron irradiation. The changes in device characteristics were accompanied by a decrease in electron diffusion length from 325 to 240 μm at 300 K. Time-resolved cathodoluminescence after 1.5 MeV electron irradiation showed a 210 ps decay lifetime and a reduction in carrier lifetime with increased fluence [27].

7.2.3.5 γ-ray damage

The effects of gamma ray irradiation on wide bandgap semiconductors are much less pronounced than for proton, electron, or neutron irradiation of these same materials. As an example, in SiC, the effective carrier removal rates normalized to non-ionizing energy loss for gamma rays are up to six orders of magnitude less than for α-particles and approximately two orders of magnitude lower than for electrons and neutrons [23, 58]. There was less degradation in metal–oxide semiconductor (MOS) transistors than metal-based gate devices, indicating that the gate oxide is a key determinant of the changes due to gamma ray exposure.

The gamma ray irradiation of lateral depletion mode β-Ga$_2$O$_3$ MOSFETs to doses of 1.6 MGray (SiO$_2$) showed little effect on the output current and threshold voltage. Degradation in the gate oxide was found to limit the overall radiation resistance [66]. β-Ga$_2$O$_3$ Schottky rectifiers consisting of thick (10 μm) epitaxial drift regions on conducting substrates also have a high tolerance to ^{60}Co gamma ray irradiation, with a low carrier removal rate of <1 cm^{-1} for gamma rays. Changes in diode ideality factor, Schottky barrier height, on-resistance, on–off ratio, and reverse recovery time were all minimal for fluences up to 2×10^{16} cm^{-2} (an absorbed dose of 100 kGy (Si)). These results are consistent with gamma-irradiation of Ga$_2$O$_3$ metal–oxide semiconductor field-effect transistors (MOSFETs) where changes were ascribed to damage in the gate dielectric and not to the Ga$_2$O$_3$ itself. The changes induced by gamma ray exposure are much smaller than those caused by proton, electron, α-particle, and neutron damage in the same rectifier structures.

7.2.3.6 x-ray damage
Constant voltage stress of β-Ga$_2$O$_3$ MOS capacitors with Al$_2$O$_3$ gate dielectrics showed increasing electron-trap densities for increasingly positive stress voltages, and hole traps created for irradiation with 10 keV x-ray devices at a dose rate of 31.5 krad (SiO$_2$) min^{-1} under grounded bias conditions [59, 60]. Stress-induced traps were located primarily in the Al$_2$O$_3$ gate dielectric layer and distributed broadly in energy. Oxygen vacancies in the Al$_2$O$_3$ were suggested to be the most likely defects created. The radiation-induced voltage shifts were comparable to or less than those of the MOSFETs exposed to gamma rays discussed above

7.2.4 Dominant defects induced by proton irradiation

There is interest in the properties of hydrogen because it may be a shallow donor in Ga$_2$O$_3$ and therefore contribute to the n-type conductivity, rather than the commonly assumed origin due to native defects such as Ga interstitials or O vacancies, the latter of which are suggested to be deep donors [28, 38–40]. There is experimental support that hydrogen may be a shallow donor from muonium and electron paramagnetic resonance (EPR) measurements [39, 40, 49].

Qin *et al* [67] showed there can be a high concentration of hydrogen in Ga$_2$O$_3$ that is neither electrically or optically active under normal conditions. They suggested that hydrogen shallow donors (possibly the H$_i$ and H$_O$ centers predicted by theory) [28] are candidates for the reservoirs of 'hidden' hydrogen in Ga$_2$O$_3$. Hydrogen can trap at any of the O vacancies in the appropriate charge state and the small bond strength of Ga–H species means that the predicted O–H vibrational frequencies would also lie within the free-electron absorption region of the host and not be visible with IR spectroscopy [52, 53].

EPR spectra from Ga$_2$O$_3$ have been interpreted in different ways. Von Bardeleben *et al* [39, 68] attributed their result to a hole trapped at a Ga(1) vacancy whose structure is identical to the V$_{Ga}$-2H center found in IR absorption, but with the hydrogens absent and a hole trapped at one of the two dangling O sites. Kananen *et al* [33, 34] attributed the same spectrum to a hole trapped at the Ga(2) vacancy. A second spectrum has been assigned to either a hole trapped at a Ga(2) vacancy, or a self-trapped hole [39]. Proton irradiation introduces two main paramagnetic defects which are stable at 300 K. These were not compatible with undistorted V_{Ga} on a tetrahedral or an octahedral site (EPR1 center) or a self-trapped hole center (EPR2 center) [40]. A tetrahedral vacancy has a complex $V_{Ga}(tetra)–Ga_i–V_{Ga}(tetra)$ configuration that agrees well with experiments for the EPR1 center [38]. Figure 7.11 shows one model consistent with the EPR2 center [39]. The uncertainty emphasizes that the identification of the main defects introduced by irradiation is also still open to interpretation of the origin of the EPR spectra.

Data from an IR study by Weiser *et al* [52] and Qin *et al* [67] on samples implanted with hydrogen or deuterium showed that the main defects involve H trapped at a Ga vacancy, the primary member involving a specific two-H configuration. The dominant hydrogen or deuterium absorption lines appear at 3437 and 2545 cm^{-1}, respectively, and were consistent with a model where two H atoms are bonded to a Ga vacancy. When the samples are implanted with hydrogen, additional absorption peaks are observed. As they are annealed, these defects become converted into the 3437 and 2545 cm^{-1} lines at 400 °C. These lines are stable up to 700 °C, where they are then converted into other new lines [67].

The $V_{Ga}(1)$-2H complex is the dominant O–H center in Ga_2O_3 hydrogenated by annealing in H_2 at elevated temperature ($T > 800$ °C) or by proton implantation [52, 53]. This center is thermally stable to 900 °C. The spectra shown in figures 7.12(a) and (b) for samples containing both H and D reveal that the 3427.1 (2546.4) cm^{-1} center contains two identical, weakly coupled, H (D) atoms. These originate from a structure with two identical H (D) atoms bound to a relaxed Ga(1) vacancy (figure 7.12(c)) [67]. For samples quenched from 900 °C, the $V_{Ga}(1)$-2H complex is absent (figure 7.13, spectrum (i)), but additional O–H (O–D) vibrational lines were observed (figure 7.13) [67]. A two-step thermal process, involving annealing in H_2 ambient at >800 °C or greater, followed by cooling to 25 °C and then a second annealing at 400 °C produced the $V_{Ga}(1)$-2H complex (figure 7.13, spectrum (ii)). The interpretation is that there are sources of 'hidden' hydrogen that can be converted into the dominant $V_{Ga}(1)$-2H center by thermal annealing at 400 °C.

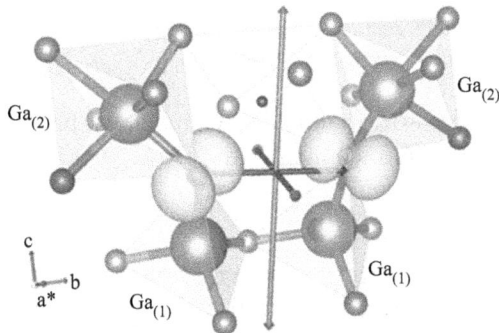

Figure 7.11. V_{Ga}^{2-} (octa) metastable local structure and spin density (yellow). Reproduced from [39].

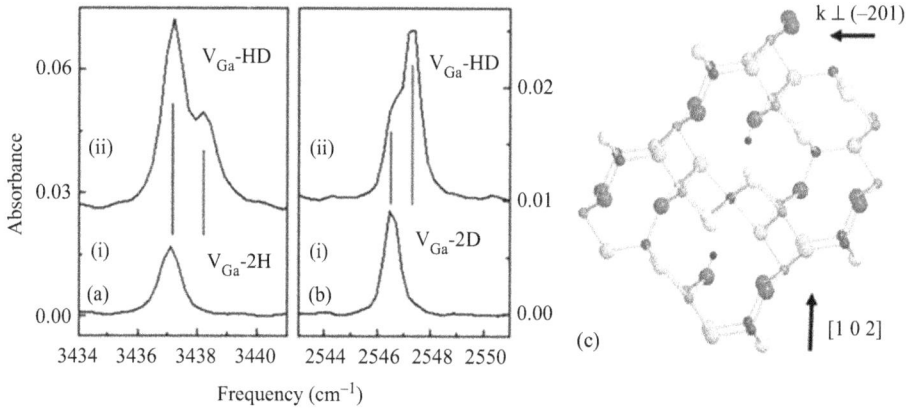

Figure 7.12. Polarized IR absorption spectra (77 K, resolution 0.5 cm^{-1}) for Ga$_2$O$_3$ annealed in H$_2$ and D$_2$ ambients. In panel (a), spectrum (i) is the O–H stretching mode for a sample annealed in H$_2$. In panel (b), spectrum (i) is the O–D stretching mode for a sample annealed in D$_2$. The spectra labeled (ii) in both panels are for a sample annealed in a mixture of H$_2$ and D$_2$. (c) Structure of the V$_{Ga}$(1)-2H complex that gives rise to the 3437.0 cm^{-1} IR line. Reproduced with permission from [67]. Copyright IOP Publishing, all rights reserved.

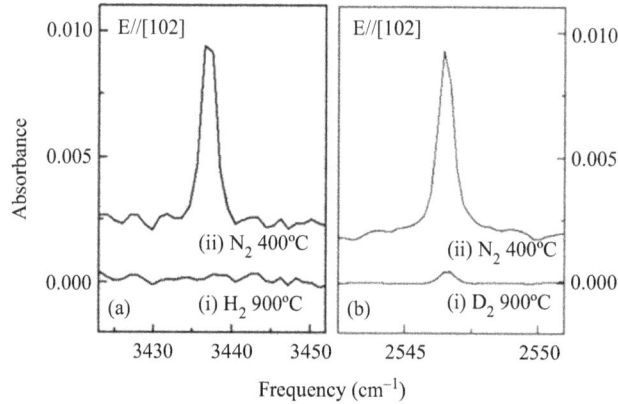

Figure 7.13. (a) IR absorbance spectra for Ga$_2$O$_3$ after two-step annealing treatment to introduce hydrogen. Spectrum (i) was measured for a sample annealed initially in an H$_2$ ambient for 1 h at 900 °C. Spectrum (ii) is for the same sample after a subsequent annealing at 400 °C in flowing N$_2$. (b) shows spectra similar to those shown in (a) except with an initial D$_2$ annealing ambient instead of H$_2$. Reproduced with permission from [67]. Copyright IOP Publishing, all rights reserved.

Possible assignments for this form of hydrogen include interstitial hydrogen shallow donors, other OH species, and H$_2$ molecules [67].

7.3 Conclusions

The initial data on proton, electron, x-ray, gamma, and neutron irradiation of β-Ga$_2$O$_3$ show fairly comparable radiation resistance to GaN under similar conditions. Figure 7.14 shows a compilation of carrier removal rates in Ga$_2$O$_3$ for different types and energy of radiation. The carrier removal rates in irradiated

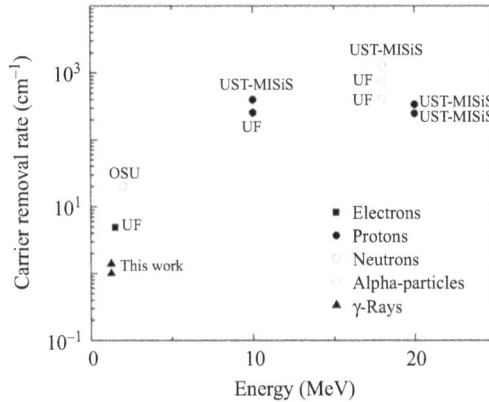

Figure 7.14. Carrier removal rate in Ga_2O_3 as a function of energy for different types of radiation. Data is either from University of Florida (UF) [27, 54, 57, 66], National University of Science and Technology (UST-MISiS) [41, 64, 69], or Ohio State University (OSU) [45].

rectifier structures range from ~5 cm^{-1} for 1.5 MeV electrons to 730 cm^{-1} for 18 MeV α-particles. Thermal annealing at ~500 °C brings a significant recovery towards the initial, un-irradiated characteristics. The dominant defect formed in Ga_2O_3 by annealing in an H_2 ambient or by the implantation of protons is a specific relaxed V_{Ga}-2H structure for the 3437 cm^{-1} line that dominates the infrared absorption spectra. There is significant scope for additional work to determine carrier removal rates at additional energies for each radiation type, to look for dose-rate effects, SEU responses, the role of interfaces in MOS structures, and the annealing stages of the induced defects. In the latter case, it is of importance to know whether *in situ* thermal annealing or forward biased minority carrier injection annealing are effective in Ga_2O_3 devices, since these could be used as simple refresh cycles for radiation damaged devices.

Acknowledgments

The work at UF is partially supported by HDTRA1-17-1-0011 (Jacob Calkins, monitor) and NSF DMR-1856662 (Tania Paskova). The project or effort depicted is sponsored by the Department of the Defense, Defense Threat Reduction Agency. The work at Korea University was supported by the Space Core Technology Development Program (2017M1A3A3A02015033) and the Technology Development Program to Solve Climate Changes (2017M1A2A2087351) through the National Research Foundation of Korea funded by the Ministry of Science, ICT and Future Planning of Korea. The work at NUST MISiS was supported in part by the Ministry of Education and Science of the Russian Federation in the framework of Increase Competitiveness Program of NUST (MISiS) (K2-2014-055). The work at LU is supported by NSF DMR-1901563.

References

[1] Petersen E 2011 *Single Event Effects in Aerospace* (New York: Wiley)

[2] Witulski A F, Ball D R, Galloway K F, Javanainen A, Lauenstein J-M, Sternberg A L and Schrimpf R D 2018 *IEEE Trans. Nucl. Sci.* **65** 1951

[3] Casey M C, Lauenstein J-M, Weachock R J, Wilcox E P, Hua L M, Campola M J, Topper A D, Ladbury R L and LaBel K A 2018 *IEEE Trans. Nucl. Sci.* **65** 269

[4] O'Bryan M V *et al* 2017 Compendium of current single event effects results from NASA Goddard Space Flight Center and NASA Electronic Parts and Packaging Program *IEEE Radiation Effects Data Workshop (REDW)* (New York: IEEE) pp 1–9

[5] Khachatrian A *et al* 2017 *IEEE Trans. Nucl. Sci.* **64** 97

[6] Ionascut-Nedelcescu A, Carlone C, Houdayer A, von Bardeleben H J, Cantin J-L and Raymond S 2002 *IEEE Trans. Nucl. Sci.* **49** 2733

[7] Khachatrian A, Roche N J-H, Ruppalt L B, Champlain J G, Buchner S, Koehler A D, Anderson T J, Hobart K D, Warner J H and McMorrow D 2018 *IEEE Trans. Nucl. Sci.* **65** 369

[8] Pearton S J, Ren F, Patrick E, Law M E and Polyakov A Y 2016 *ECS J. Solid State Sci. Technol.* **5** Q35

[9] Griffin P 2019 *IEEE Trans. Nucl. Sci.* **66** 327

[10] Wenckstern H V 2017 *Adv. Electron. Mater.* **3** 1600350

[11] Mastro M A, Kuramata A, Calkins J, Kim J, Ren F and Pearton S J 2017 *ECS J. Solid State Sci. Technol.* **6** P356

[12] Pearton S J, Yang J, Cary IV P H, Ren F, Kim J, Tadjer M J and Mastro M A 2018 *Appl. Phys. Rev.* **5** 011301

[13] Higashiwaki M and Jessen G H 2018 *Appl. Phys. Lett.* **112** 060401

[14] Chen X, Ren F, Gu S G and Ye J 2019 *Photonics Res.* **7** 381

[15] Reese S B, Remo T, Green J and Zakutayev A 2019 *Joule* **3** 903

[16] Tadjer M J 2018 *ECS Interface* **27** 49–52

[17] Kim J, Oh S, Mastro M A and Kim J 2016 *Phys. Chem. Chem. Phys.* **18** 15760

[18] Kim J, Mastro M A, Tadjer M J and Kim J 2017 *ACS Appl. Mater. Interfaces* **9** 21322

[19] Yang G, Jang S, Ren F, Pearton S J and Kim J 2017 *ACS Appl. Mater. Interfaces* **9** 40471

[20] Ziegler J F, Ziegler M D and Biersack J P 2010 *Nucl. Instrum. Methods Phys. Res.* B **268** 1818

[21] Srour J R and Palko J W 2006 *IEEE Trans. Nucl. Sci.* **53** 3610

[22] Martinez M J, King M P, Baca A G, Allerman A A, Armstrong A A, Klein B A, Douglas E A, Kaplar R J and Swanson S E 2019 *IEEE Trans. Nucl. Sci.* **66** 344

[23] Dodds N A *et al* 2015 *IEEE Trans. Nuclear Sci.* **62** 2440

[24] Pearton S J, Ren F, Tadjer M and Kim J 2018 *J. Appl. Phys.* **124** 222901

[25] Wendler E, Treiber E, Baldauf J, Wolf S, Kuramata A and Ronning C 2015 *Proc. 11th Int. Conf. on the Interaction of Radiation with Solids* (New York: IEEE) p 93

[26] Look D C, Reynolds D C, Hemsky J W, Sizelove J R, Jones R L and Molnar R J 1997 *Phys. Rev. Lett.* **79** 2273

[27] Kim J, Pearton S J, Fares C, Yang J, Ren F, Kim S and Polyakov A Y 2019 *J. Mater. Chem.* C **7** 10

[28] Varley J B, Weber J R, Janotti A and Van de Walle C G 2010 *Appl. Phys. Lett.* **97** 142106

[29] Polyakov A Y, Pearton S J, Frenzer P, Ren F, Liu L and Kim J 2013 *J. Mater. Chem.* C **1** 877

[30] Pearton S J, Deist R, Ren F, Liu L, Polyakov A Y and Kim J 2013 *J. Vac. Sci. Technol.* A **31** 050801

[31] Pearton S J, Ren F, Law M E and Polyakov A Y 2016 *ECS J. Solid State Sci Technol.* **5** Q35

[32] Korhonen E, Tuomisto F, Gogova D, Wagner G, Baldini M, Galazka Z, Schewski R and Albrecht M 2016 *Appl. Phys. Lett.* **106** 242103

[33] Kananen B E, Halliburton L E, Stevens K T, Foundos G K, Chang K B and Giles N C 2017 *Appl. Phys. Lett.* **110** 202104

[34] Kananen B E, Giles N C, Halliburton L E, Foundos G K, Chang K B and Stevens K T 2017 *J. Appl. Phys.* **122** 215703

[35] Son N T *et al* 2016 *J. Appl. Phys.* **120** 235703

[36] Furthmüller J and Bechstedt F 2016 *Phys. Rev.* B **93** 115204

[37] Zacherle T, Schmidt P C and Martin M 2013 *Phys. Rev.* B **87** 235206

[38] Varley J B, Peelaers H, Janotti A and Van de Walle C G 2011 *J. Phys.: Condens. Matter,* **23** 334212

[39] von Bardeleben H J, Zhou S, Gerstmann U, Skachkov D, Lambrecht W R L, Duy Ho Q and Deák P 2019 *APL Mater.* **7** 022521

[40] Deak P, Duy Ho Q, Seemann F, Aradi B, Lorke M and Frauenheim T 2017 *Phys. Rev.* B **95** 075208

[41] Polyakov A Y, Smirnov N B, Shchemerov I V, Gogova D, Tarelkin S A and Pearton S J 2018 *J. Appl. Phys.* **123** 115702

[42] Wendler E, Treiber E, Baldauf J, Wolf S and Ronnig C 2016 *Nucl. Instrum. Meth Phys. Res.* B **379** 85

[43] Dong L, Jia R, Li C, Xin B and Zhang Y 2017 *J. Alloys Compd* **712** 379

[44] Ingebrigtsen M E, Kuznetsov A Y, Svensson B G, Alfieri G, Mihaila A, Badstübner U, Perron A, Vines L and Varley J B 2019 *APL Mater.* **7** 022510

[45] Farzana E, Chaiken M F, Blue T E, Arehart A R and Ringel S A 2019 *APL Mater.* **7** 022502

[46] Chaiken M F and Blue T E 2018 *IEEE Trans. Nucl. Sci.* **65** 1147

[47] Szalkai D, Galazka Z, Irmscher K, Tüttő P, Klix A and Gehre D 2017 *IEEE Trans. Nucl. Sci.* **64** 1574

[48] Janotti A and Van de Walle C G 2007 *Nat. Mater* **6** 44

[49] King P D C and Veal T D 2011 *J. Phys. Condens. Matter* **23** 334214

[50] Ahn S, Ren F, Patrick E, Law M E, Pearton S J and Kuramata A 2016 *Appl. Phys. Lett.* **109** 242108

[51] Ahn S, Ren F, Patrick E, Law M E and Pearton S G 2017 *ECS J. Solid State Sci. Technol.* **6** Q3026

[52] Weiser P, Stavola M, Fowler W B, Qin Y and Pearton S J 2018 *Appl. Phys. Lett.* **112** 232104

[53] Stavola M, Fowler W B, Qin Y, Weiser P and Pearton S J 2018 *Ga₂O₃, Technology, Devices and Applications* ed S Pearton, F Ren and M Mastro (Amsterdam: Elsevier) ch 9 p 191

[54] Yang J, Ren F, Pearton S J, Yang G, Kim J and Kuramata A 2017 *J. Vac. Sci. Technol.* B **35** 031208

[55] Lee J, Flitsiyan E, Chernyak L, Ahn S, Ren F, Yun L, Pearton S J, Kim J, Meyler B and Salzman J 2017 *ECS J. Solid State Sci. Technol.* **6** Q3049

[56] Lee J D, Flitsiyan E, Chernyak L, Yang J, Ren F, Pearton S J, Meyler B and Salzman Y J 2018 *Appl. Phys. Lett.* **112** 082104

[57] Yang J, Fares C, Guan Y, Ren F, Pearton S J, Bae J, Kim J and Kuramata A 2018 *Vac. Sci. Technol.* B **36** 031205

[58] Onoda S, Iwamoto N, Ono S, Katakami S, Arai M, Kawano K and Ohshima T 2009 *IEEE Trans. Nucl. Sci.* **56** 3218

[59] Wong M H, Takeyama A, Makino T, Ohshima T, Sasaki K, Kuramata A, Yamakoshi S and Higashiwaki M 2018 *Appl. Phys. Lett.* **112** 023503 2018

[60] Bhuiyan M A, Zhou H, Jiang R, Zhang E X, Fleetwood D M, Ye P D and Ma T-P 2018 *IEEE Electron Dev. Lett.* **39** 1022

[61] Irmscher K, Galazka Z, Pietsch M, Uecker R and Fornari R 2011 *J. Appl. Phys,* **110** 063720

[62] Farzana E, Ahmadi E, Speck J S, Arehart A R and Ringel S A 2018 *J. Appl. Phys.* **123** 161410

[63] McGlone J F, Xia Z, Zhang Y, Joishi C, Lodha S, Rajan S, Ringel S A and Arehart A R 2018 *IEEE Electron. Dev. Lett.* **39** 1042

[64] Polyakov A Y, Smirnov N B, Shchemerov I V, Pearton S J, Ren F, Chernykh A V, Lagov P B and Kulevoy T V 2019 *APL Mater.* **6** 096102

[65] Modak S, Chernyak L, Khodorov S, Lubomirsky I, Yang J, Ren F and Pearton S J 2019 *ECS J. Solid State Sci. Technol.* **8** Q3050

[66] Yang J C, Koller G J, Fares C, Ren F, Pearton S J, Bae J and Kim J 2019 *ECS J. Solid State Sci. Technol.* **8** Q3041

[67] Qin Y, Stavola M, Fowler W B, Weiser W B and Pearton S J 2019 *ECS J. Solid State Sci. Technol.* **8** Q3103

[68] Skachkov D, Lambrecht W R L, von Bardeleben H J, Gerstmann U G, Duy Ho Q and Deák P 2019 *J. Appl. Phys.* **125** 185701

[69] Polyakov A Y *et al* 2018 *Appl. Phys. Lett.* **113** 092102

IOP Publishing

Wide Bandgap Semiconductor-Based Electronics

Fan Ren and Stephen J Pearton

Chapter 8

Optical properties of Ga_2O_3 nanostructures

Manuel Alonso-Orts, Emilio Nogales and Bianchi Méndez

8.1 Introduction

Following the research activities on gallium oxide, insight into the optical properties of this ultra-wide bandgap (UWBG) semiconductor is of great interest for a wide range of applications. In this chapter, we will focus on the optical properties of undoped and doped Ga_2O_3 single crystals, mainly in the form of micro- and nanomaterials, covering issues related to the absorption, transmission, generation, and confinement of light in Ga_2O_3-based nanostructures. Its wide bandgap ($E_g = 4.9$ eV) results in a wider transparency range and allows interaction with photons from the ultraviolet (UV) to the near-infrared (NIR). As a direct application of this huge bandgap, Ga_2O_3 single crystals, epitaxial films, and nanostructures represent one of the most probable solutions for solar-blind UV photodetectors to date [1]. This and other optoelectronic applications of Ga_2O_3 are covered in chapters 1 and 6. However, keeping in mind the diverse optical behaviour of this material when subject to different dopings, there is still room for applications that can be extended beyond UV optoelectronics. In addition, the light wavelength in optics partially overlaps the nano- and microscale. Hence the design of architectures covering this scale would lead to applications in optical devices such as optical fibres, photonic crystals, waveguides, and optical microcavities, among others.

The structure of this chapter is as follows. We will first summarize the intrinsic properties of Ga_2O_3, such as the electronic band structure, electron–phonon coupling, and the refractive index, which make it an appealing host to study light–matter interaction. Then we will discuss the optical behaviour of undoped Ga_2O_3, which is influenced by the native defect structure that these oxides usually exhibit.

The next section will be devoted to Ga_2O_3 doped with several impurities relevant to optical properties. So far, most research activity has been devoted to Sn, Si, and Ge doping with the aim of controlling the electrical conductivity. However, some links to optical properties can be found due to the impurity–defect interaction, which could consequently modify the electronic energy levels in the bandgap. Regarding nanostructures, impurity incorporation in nanowires or nanobelts could also affect the morphology of the final product, which could have implications for the optical properties as well. In this section, we will also review the possibility of Ga_2O_3 doping with optically active ions, such as transition metal ions, rare-earth ions, and alkaline ions.

The last section is devoted to optical confinement in Ga_2O_3 microstructures. Its refractive index allows waveguiding of the luminescence through the material under suitable geometries, such as microwires. In addition, the possibilities of building optical microcavities will be presented. Finally, we will summarize and provide an outlook on the prospective work related to Ga_2O_3 nanostructures regarding optical applications.

8.2 Optical parameters of Ga_2O_3

The optical properties observed in crystals reflect the many different ways in which light can interact with matter. The simple picture of incident light incoming into a crystal along with its reflection, propagation, and transmission serves in a first classification of optical phenomena. In addition, the propagation of light through a crystal can provoke additional optical phenomena, such as refraction, absorption, luminescence, and scattering, in which strong changes in the output can occur with respect to the incoming light.

8.2.1 Optical processes in semiconductors and insulators

The optical phenomena can be quantified by some parameters that describe the optical properties of the crystal at the macroscopic level. For example, the coefficient of reflection, R, is defined as the ratio of the reflected power to the power incident on the surface. This parameter is linked to the coefficient of transmission, T, which is defined as the ratio of the transmitted power to the incident power. In the case of no absorption or scattering, the law of conservation of energy states that $R + T = 1$. On the other hand, the propagation of light through a transparent material is commonly described by the refractive index, $n = c/v$, where c and v stand for the light velocity in free space and in the medium, respectively. The refractive index depends on the frequency of the light. This is known as dispersion. However, in many transparent media, the dispersion effects are small in the visible spectral region and a single value is often taken as the refractive index.

The absorption of light is quantified by the coefficient of absorption α that gives us a measure of the fraction of absorbed power when propagating along a unit of length of the crystal. It occurs when the frequency of the light is resonant with the transition frequencies of the atoms or ions in the medium. Let us consider z the direction of propagation, then the intensity (optical power per unit area) at position z is $I(z)$ is related to the incident intensity, I_0, by Beer's law: $I(z) = I_0 \exp(-\alpha z)$. Scattering is another phenomenon that could occur during light propagation and refers to the change of the direction of propagation and to the eventual change of frequency of the propagating light. This is caused by variations of the refractive index on a length scale much smaller than the wavelength of the light. For example, the presence of impurities or defects in the crystal can act as scattering centres and redirect light in other directions. The attenuation of the light caused by scattering can be described in a way similar to absorption according to $I(z) = I_0 \exp(-N\sigma_s z)$, where N is the number of scattering centres per unit volume and σ_s is the scattering cross-section of the particular scattering centre.

Luminescence refers generally to the spontaneous emission of light by excited atoms in a material. One of the ways to obtain excited atoms is precisely through the absorption of light of suitable wavelength. Once the atom is excited, it usually takes some time to be relaxed via radiative (spontaneous luminescence) or non-radiative processes in the crystal. Luminescence can occur while exciting light is propagating through the crystal, as we will see later. In principle, it is emitted in all directions and can display colours or luminescence bands different to that of the incoming light. Spontaneous emission is a quantum process and hence it cannot be described easily by a 'coefficient of luminescence' as in the previously described optical phenomena. Einstein formulated the quantum theory of radiative absorption and emission, in which besides light absorption and spontaneous emission, the phenomenon of stimulated emission was first postulated. However, before the advent of quantum mechanics, George Stokes discovered that luminescence is usually downshifted in frequency relative to the absorption, a phenomenon known as Stokes shift. The magnitude of this shift is determined by the energy levels of the atoms in the medium and most commonly through interaction with phonons. In the case of crystalline solids, the electronic band structure and the vibronic bands play a role in the luminescence processes, as will be shown in the subsequent sections on Ga_2O_3 crystals.

Finally, as is well known from optics, absorption and refraction can be incorporated in just one parameter called the complex refractive index, $\tilde{n}$, where the imaginary part, κ, called the extinction coefficient, takes into account the light absorption through the relationship $\alpha = 4\pi\kappa/\lambda$, with λ being the free space wavelength of the light. This is derived by considering the propagation of plane electromagnetic waves through a medium of refractive index: $\tilde{n} = n + i\kappa$. In many cases, it is easier to calculate the electrical permittivity ($\varepsilon = \varepsilon_1 + i\varepsilon_2$) instead of the refractive index. Again, the imaginary part of the dielectric function is related to absorption or losses. Under specific conditions, stimulated emission can result in a positive optical gain, i.e. negative absorption, $g = -\alpha > 0$.

To describe optical phenomena a semi-classical physics approach is necessary, in which light is considered a classical electromagnetic wave whereas the crystal is modelled according to quantum mechanics that pictures the electronic band structure. Otherwise, the absorption or luminescence process cannot be fully explained in the classical model of solids, for example. On the other hand, light is often also described as a beam of photons. Since the light–matter interaction in a crystalline solid involves photons, electrons, and ion coupling effects, we will first summarize the crystalline and electronic band structures of Ga_2O_3 and then we will move on to the peculiarities of nanomaterials where surface and size effects must also be considered.

8.2.2 Gallium oxide as an optical material

In order to assess the optical properties of Ga_2O_3 nanostructures, we first have to consider some basic properties of the material, such as the crystal symmetries of Ga_2O_3, the electronic and the vibrational band structure, and the effects of dopants and other defects. All of these properties would be reflected in light propagation, absorption, and emission experiments. In addition, optical effects supported by phonons and electron–phonon coupling effects between the electronic levels of single atoms/defects with the vibrational modes would provoke broadening in the absorption and emission spectra, as well as the Stokes shift.

Among the five crystallographic phases reported for Ga_2O_3, the monoclinic phase, β-Ga_2O_3, is the most stable and well-known one [2, 3]. Recently, however, there has also been interest in other phases, such as the corundum, hexagonal α-, defective spinel, cubic γ-, and orthorhombic ε-phases, but their study is not yet as well developed as that of β-Ga_2O_3. For example, very recent works report the coexistence of both β- and ε-Ga_2O_3 thin films on a sapphire substrate depending on the parameters during MBE or CVD growth [4, 5]. The conventional unit cell of monoclinic β-Ga_2O_3 consists of eight gallium atoms surrounded by oxygen atoms where four Ga atoms exhibit an octahedral coordination and the other four a tetrahedral coordination. Figure 8.1 shows the crystalline structure of β-Ga_2O_3, where the two kinds of Ga coordination are highlighted. In this structure, the b-axis in real space coincides with the C_{2h} symmetry axis, which confers a preferred direction and hence anisotropy to the crystal structure. This translates into orientation-dependent physical properties, including the optical properties. In particular, electronic and thermal conductivity, as well as optical absorption, show anisotropic behaviour, with the b-axis a distinctive direction [6, 7].

An excellent summary of the basic properties of Ga_2O_3 is provided in the recent review [8]. Concerning the electronic state structure, one key parameter is the magnitude of the energy bandgap E_g that sets the absorption band edge. Regarding

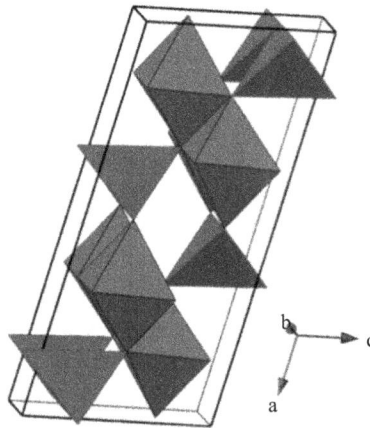

Figure 8.1. The crystalline structure of monoclinic Ga_2O_3. Green and blue label $Ga1$ (octahedral) and Ga_2 (tetrahedral) positions in the lattice, respectively.

vibrational modes, the characteristic energies are in the infrared (IR) range [9]. Ga_2O_3 has the widest energy bandgap, about 4.7–4.9 eV, among the family of semiconducting oxides, which allows optical transparency in a very wide range of the optical spectrum, from the UV to the IR spectral region. On the other hand, the equivalent wavelength of this energy bandgap is about 250 nm, overlapping the upper limit of the nanoscale. This means that additional features can be explored in the light interaction with Ga_2O_3 nanostructures as a consequence of this size–wavelength matching.

The electronic band structure of undoped single crystals of Ga_2O_3 has been studied theoretically and experimentally. Theoretical approaches used first principles calculations based on density functional theory (DFT) and the experimental work was carried out using high-resolution angle-resolved photoelectron spectroscopy in a synchrotron facility [10]. First principles calculations of the electronic band structure usually refer to perfect bulk crystals and recent functional models provide very accurate results in agreement with experiments. Along with the band structure, optical related parameters, such as the tensor of the dielectric function, have also been calculated by these methods [11]. On the other hand, the experiments to characterize the optical properties of Ga_2O_3 by several groups have mostly been carried out in thin films [12–14].

Light absorption and refraction can be assessed by the determination of the refractive index of the material, which is linked to the transmission and reflection coefficients in an insulator. Experimental measurements of the transmittance and reflectance curves of amorphous Ga_2O_3 thin films enabled us to calculate the refractive index [14]. Alternatively, generalized spectral ellipsometry also allows determining the dielectric function experimentally. These techniques allow the fitting of the real part of the refractive index dispersion to the Cauchy model, $n = n_\infty + B/\lambda^2 + C/\lambda^4$ shown in table 8.1, for the spectral range between approximately 290–900 nm (1.35–4.25 eV). The model fitting provides a high-dielectric constant ε_∞ of 3.57 (corresponding to $n_\infty = 1.89$) for the optical transparency range.

Thin films, however, are not able to show the optical anisotropy derived from the monoclinic lattice of Ga_2O_3. To evaluate the full dielectric tensor, single crystals as samples and polarized light beams are required. Ueda *et al* measured some electrical and optical properties in Ga_2O_3 single crystals [15], finding a clear anisotropy of the values of electrical conductivity, mobility, and absorption edge along the b- and c-axes. In particular, figure 8.2(a) shows the optical transmission spectra of a single crystal of 159 µm thickness for incident light polarized in several orientations. The

Table 8.1. Parameters of the Cauchy formula for the refractive index in the transparent spectral range.

	n_∞	B (nm^2)	C (nm^4)
Electron beam deposited films [13]	1.891	1.1×10^4	4.8×10^8
Sputtered films [13]	1.883	1.1×10^4	3.59×10^8
Amorphous films [14]	1.835	3.22×10^4	4.94×10^8

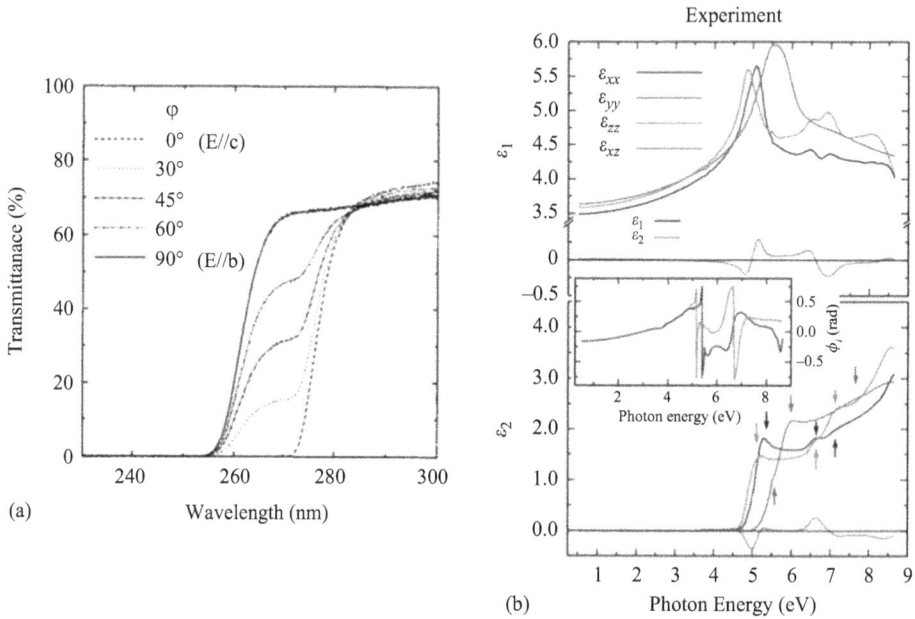

Figure 8.2. (a) Optical transmission spectra of single crystal Ga_2O_3 under incident light polarized in several directions. Reproduced with permission from [15]. Copyright 1997 AIP Publishing. (b) Dielectric tensor components (ε_{xx}, ε_{yy}, ε_{zz}, and ε_{xz}) measured using spectroscopy ellipsometry. Inset: rotation angle of the dielectric axes around the y-axis for the real (black) and imaginary (red) part. The y-direction corresponds to the symmetry axis. Reproduced with permission from [16]. Copyright 2015 AIP Publishing.

fundamental absorption edge was found to be 253 nm (4.9 eV) and 270 nm (4.59 eV) for **E**//b and **E**//c, respectively, demonstrating the optically biaxial anisotropic behaviour of the monoclinic phase. More recently, spectral ellipsometry measurements have been used to determine the full dielectric tensor, including off-diagonal components in the spectral range of 0.5–8.5 eV [16]. Figure 8.2(b) displays the experimental determined values for the tensor components for both the real and the imaginary parts of the dielectric tensor, showing the anisotropic effects. From these results it should also be noted that the bandgap value differs slightly from the b to c direction in the lattice.

It is well known that the presence of defects and impurities introduces electronic levels in the bandgap of semiconductors that affect the electro-optical properties. Oxygen vacancies are mainstream native defects in Ga_2O_3 and in other semi-conducting oxides, such as ZnO, SnO_2, TiO_2, In_2O_3, etc. They are usually considered to be responsible for the n-type conductivity in most of them and are also attributed the origin of visible luminescence bands. A lot of work has also been carried out and is currently undergoing on the theoretical and experimental study of structural point defects in Ga_2O_3, such as oxygen or gallium vacancies, gallium interstitials, or unintentionally doped foreign atoms in the lattice on substitutional or interstitial sites [17, 18]. The presence of defects has consequences on the electronic

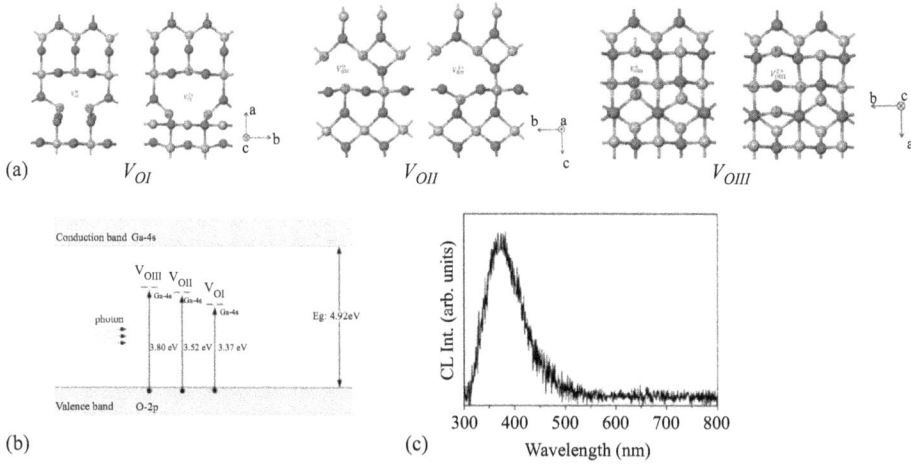

(a) V_{OI} V_{OII} V_{OIII}

(b) (c)

Figure 8.3. (a) Atomic structure relaxations of oxygen vacancies: V_{OI}, V_{OII}, and V_{OIII}. (b) A possible absorption process related to oxygen vacancies. Reproduced with permission from [18]. Copyright 2017 Springer. (c) CL spectrum of undoped Ga_2O_3 microcrystals. Reproduced with permission from [19]. Copyright 2011 Elsevier.

band structure, since they locally break the symmetries and usually lead to electronic levels within the bandgap.

Figure 8.3 shows the lattice relaxations due to different oxygen vacancies in Ga_2O_3. The consequences of these defects on the optical properties have been studied and a model for the light absorption in Ga_2O_3 has been suggested (figure 8.3(b)) [18]. The luminescence of undoped Ga_2O_3 has been studied in the past in films and single crystals and some models have been proposed [20, 21]. Complementary EPR measurements were performed to ascertain the particular nature of the point defects responsible for the UV–blue luminescence in undoped Ga_2O_3 [20]. Figure 8.3(c) shows the room temperature cathodoluminescence spectrum of a Ga_2O_3 nanowire obtained by a thermal evaporation method, which shows a composed band centred at about 360 nm (3.4 eV) related to intrinsic defects involving oxygen vacancies [19].

8.2.3 Ga₂O₃ nano- and microstructures

Most of the work devoted to Ga_2O_3 has been focused on thin films or bulk materials. However, research on Ga_2O_3-based nanomaterials seems mandatory in the nano-technology era. This approach adds additional variables in engineering physical properties. Here, not only doping and defects, but also shape, morphology, and dimensionality can influence the material performance and obtain synergetic effects.

In the case of nanowires, the growth method affects the orientation of the nanowire and other features. For example, Ga_2O_3 nanowires along the $\langle 001 \rangle$ direction have been obtained by the arc-discharge method [22], while $\langle 110 \rangle$ oriented nanowires are produced by PECVD [23]. The anisotropy of the lattice allows the achieving nanowires oriented along $\langle 40{-}1 \rangle$ by using a vertical radio-frequency furnace [24]. The different orientations would provide specific physical properties along the main axis and specific surface properties due to the anisotropy of the crystalline structure. Interestingly, size and orientation-dependent effects have been reported in Raman and infrared spectrum measurements of Ga_2O_3 nanobelts and nanorods of sizes in the nanoscale [23–26].

Quantum confinement effects in many semiconductors allow a size tuning of the optical properties due to the energy blue shift of the near band-edge luminescence as the crystal size decreases. Theoretical simulations have shown that freestanding Ga_2O_3 nanolayers do not show quantum confinement, but they do when embedded in Al_2O_3 barriers [27]. However, there are other phenomena related to nanoscale dimensions that can be exploited. The photoluminescence band position was tuned from UV to blue in γ-Ga_2O_3 nanoparticles with sizes between 3.3 and 6.0 nm, respectively [28], which was related to a reduced donor–acceptor separation. One of the main advantages of nanomaterials is that the surface and interface properties play a key role in determining several device features. In particular, the huge surface/volume ratio makes them of great interest for chemical sensing, catalysis, and batteries since the chemical processes key to these devices take place at the surface [29].

Photonics applications, such as nano-waveguiding or micro-optical resonators in the visible–UV range, also take advantage of one-dimensional nano- and micro-materials [19, 30–32] as a result of the refractive index difference between the material and its surroundings, as will be exemplified below.

8.3 Luminescence of doped Ga_2O_3

The near band edge luminescence of Ga_2O_3 should be in the UV, around 4.9 eV and below, but it is difficult to detect. However, native defects such as oxygen vacancies, gallium vacancies, and interstitials, and self-trapped holes introduce electronic levels in the bandgap that provide radiative recombination paths. These produce UV luminescence bands around 3.4–3.6 eV and 3.2 eV and blue bands around 2.8–3.0 eV both in the bulk and in nanostructures, as shown, for example, in figure 8.3(c) [19, 21]. However, the doping of semiconductors with suitable impurities can alter the optical properties of the oxides, which allows tailoring of their luminescence behaviour. A well-known example is sapphire (undoped Al_2O_3, with a transparency range from 200 nm to 6 μm) and ruby (Cr doped Al_2O_3) [33], in which the addition of chromium produces two strong absorption bands in the yellow/green and in the blue region, which gives ruby its characteristic red colour. In addition, the presence of Cr^{3+} ions in ruby results in the well-known red, sharp R lines, which were used to obtain the first laser [34]. In fact, some of the first reports on gallium oxide were related to its optical properties and particularly to its luminescence when doped with Cr^{3+} [35, 36]. In addition, thanks to the ultra-wide bandgap of this oxide, adding the appropriate dopant creates luminescence bands in a very wide range, from the UV—above 280 nm—to the IR. Figure 8.4 shows some of the luminescence lines and bands, as well as the dopant used to create each of them in nanocrystalline β-Ga_2O_3 [37].

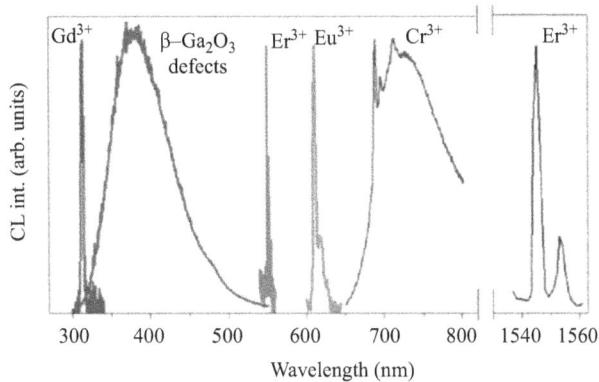

Figure 8.4. Overlapped room temperature CL spectra of Ga_2O_3 nanostructures doped with different optically active ions. The figure illustrates the wide range of wavelength emissions covered, from the UV to the IR, depending on the particular impurity. The native defect band is also plotted for the sake of completeness. Reproduced with permission from [37]. Copyright 2012 SPIE.

On the other hand, unintentional impurities could remain after the growth process, such as Si, adding electronic levels to the system. In many works, the correlation between the n-type character of unintentionally doped (UID) Ga_2O_3 and the UV–blue luminescence bands has been studied [21]. The most accepted model considers that blue luminescence is related to a donor–acceptor process involving deep donors, probably oxygen vacancies. Its intensity is reduced and/or even suppressed when the electrical conductivity increases. On the other hand, the UV luminescence is mainly attributed to the recombination of free electrons and self-trapped holes or self-trapped excitons [38].

In addition, in the case of nanomaterials incorporation of dopants during growth, i.e. *in situ* doping, may influence the morphology and additional features can be studied. Let us consider in this section the influence of some impurities in the optical properties of Ga_2O_3 nano- and microstructures. In some cases, both optical and morphology features have been altered.

8.3.1 Transition metal ion doping (Cr, Ni, Mn, and Zn)

One of the more widely studied families of optically active elements in nanocrystalline gallium oxide is the transition metals family. These ions allow a wide tunability range of the luminescence wavelength by adequately selecting the element. Miyata *et al* showed this in thin film electroluminescent displays, which were doped with Cr, Co, Mn, or Sn, obtaining red, green, and blue broad bands at room temperature through electroluminescence [39].

Chromium ion, Cr^{3+}, is an efficient, red–IR emitter in several hosts. In addition, the behaviour of its energy levels when included in octahedral-symmetry substitutional sites has been studied intensely, mostly because of the applications in laser materials, such as ruby [34]. In that material, with a strong crystal field strength, Cr^{3+} luminescence yields the well-known sharp R lines due to $^2E–^4A_2$ intra-ionic transitions at room and low temperatures. In the case of β-Ga_2O_3, this ion replaces gallium cation sites with the octahedral sites [36] and its energy levels and their resulting luminescence emission properties are well described by the Tanabe–Sugano diagram [40]. As the crystal field strength is intermediate in this case, only the sharp R lines are observed at low temperatures, while a broad, phonon-assisted band due to $^4T_2$–$^4A_2$ intra-ionic transitions are also observed at room temperature superposed on the R lines. This fact suggests possible applications in tunable lasers, as reported in a study on bulk β-Ga_2O_3:Cr [41].

More recently, chromium doped monoclinic gallium oxide nanowires were obtained via a vapour–solid growth mechanism and their optical characteristics were analysed in detail by Nogales $et\ al$ [42]. In particular, the complete configurational coordinate diagram for Cr^{3+} ions in this host was obtained in that work by combining photoluminescence (PL) and photoluminescence excitation (PLE) measurements between 10 K and room temperature. From those results, time decays of 2.8 ms and 24 μs for the ^2E–$^4A_2$ and $^4T_2$–$^4A_2$ transitions, respectively, an energy difference between the minima of the ^2E and $^4T_2$ levels of $\Delta E = 60$ meV, and a Huang–Rhys factor $S \approx 5$ were obtained. The intense room temperature luminescence band from β-Ga_2O_3:Cr nano- and microwires was subsequently used to functionalize them for photonic applications, such as waveguides and resonant optical cavities [30–32, 43], as will be shown in section 8.4.

On the other hand, a very long afterglow, around 4 h, has been reported for the room temperature, broad luminescence band of β-Ga_2O_3:Cr nanowire assemblies obtained through a hydrothermal process [44] after irradiation with UV light. The long afterglow was assigned to the thermal release of electrons trapped in defect levels surrounding the emitting ions. These luminescence properties of chromium doped, monoclinic gallium oxide nanoparticles have been used for biomedical purposes [45, 46].

Nickel ions have also been studied in several reports as dopants for gallium oxide nanostructures, in this case nanoparticles, as they present very interesting properties for potential use in fields such as optical communications, broadband amplifier, tunable lasers, biological diagnosis, or magnetic information. Ye $et\ al$ synthesized Ni^{2+} doped gallium oxide nanoparticles via a hydrothermal route [47]. Concentrations of the dopant were 1, 3, or 5 mol % and the nanoparticle sizes were in the few hundreds of nanometres range. A strong near-infrared luminescence band, centred around 1400–1450 nm, was reported and assigned to the characteristic Ni^{2+} $^3T_{2g}(^3F)$–$^3A_{2g}(^3F)$ intra-ionic transition. The absorption bands were the ones expected for this ion when substituting gallium in octahedral sites of monoclinic gallium oxide. A slight redshift and reduction of the lifetime were shown when increasing Ni concentration, which was assigned to a decrease of the crystal field strength.

On the other hand, Zhou $et\ al$ showed that transparent glass-ceramics containing β-Ga_2O_3:Ni^{2+} nanocrystals present a strong infrared luminescence band centred around 1200 nm with a relatively long RT luminescence lifetime [48]. Furthermore, they showed that this system also results in IR broadband optical amplification, as shown in figure 8.5(a) as a function of pump power and in figure 8.5(b) as a function

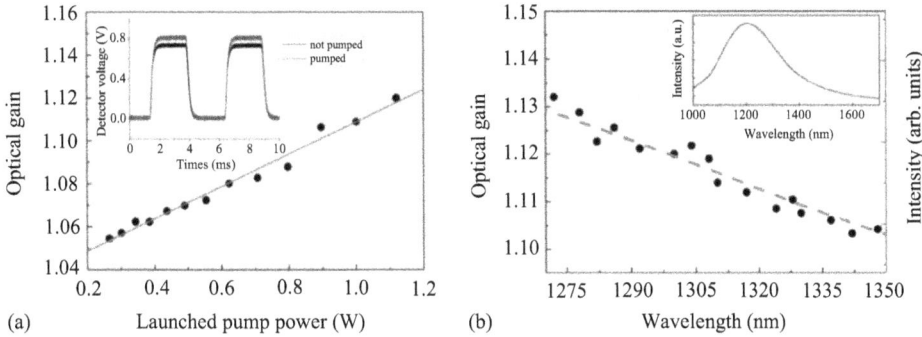

Figure 8.5. Optical gain of transparent glass ceramics containing β-Ga$_2$O$_3$:Ni^{2+} nanocrystals as a function of (a) pump power at 1300 nm and (b) seed beam wavelength at an excitation power of 1.12 W. The inset shows fluorescence spectrum excitation at 980 nm. Reproduced with permission from [49]. Copyright 2007 Optical Society.

of seed beam wavelength [49]. Optical gain was assigned to the nickel doped nanocrystals.

Manganese doping has also attracted attention due to its luminescent and magnetic properties. Song *et al* analysed the physical properties of gallium oxide nanowires doped with this element [50] *in situ* during a vapour-phase growth method. These nanostructures presented diameters of several tens of nanometres. A broad green PL emission band due to the $^4T_1$–$^6A_1$ intra-ionic transition in Mn^{2+}($^3d_5$) was observed, which was broadened and red-shifted from 536 to 559 nm when the temperature rose from 10 to 300 K. Manganese doped nanowires obtained by a similar method were also studied [51]. A difference was observed between microrods with diameters around 1 μm and nanowires with lateral dimensions around 100 nm. The former showed a CL broad, blue-green band, while the latter emitted an orange band centred around 590 nm. This can be compared to the above-mentioned 559 nm band obtained by Song *et al* and the shift can be related to the fact that one was obtained using CL and the other using PL, techniques that use quite different excitation conditions.

Finally, zinc has been introduced into gallium oxide nano- and microstructures as a possible candidate for p-type doping and its influence on the luminescence properties of the material has been analysed. In particular, CL, PL, PLE, and time-resolved PL were used to characterize the samples. It was observed that the relative intensities between the UV and the blue luminescence bands strongly differ between Zn doped and undoped samples, resulting in a shift to lower energies of the emission in both CL and PL on the Zn doped micro- and nanostructures with respect to the undoped reference [52]. Wang *et al* reported similar results in CL for Zn doped Ga$_2$O$_3$ thin films; again, the 2.7 eV (blue) band has an increase in relative intensity with respect to the higher energy bands as the Zn content increases. In addition, a green band at 2.4 eV appears for high Zn content (5% or over) [53].

8.3.2 Rare-earth ion doping (Er, Eu, Gd, Tb, Dy, and Nd)

Another commonly studied family of optically active dopants in monoclinic gallium oxide nanostructures is the rare-earth elements. Rare-earth ions produce characteristic narrow emission lines from the UV to the near-infrared when hosted in crystals. However, there are some issues in their application due to the thermal quenching effects of their emission at room temperature. Wide bandgap materials minimize this effect [54] and therefore Ga_2O_3 is a very interesting host for these ions for device applications such as emissive displays, as very low thermal quenching at room temperature is expected. Miyata *et al* showed that thin film electroluminescent displays doped with Pr, Ce, Ho, Er, Dy, Eu, Sm, and Nd showed characteristic rare-earth-related sharp lines spanning from blue to red at room temperature [39]. Gallium oxide nanocrystals have been doped with Er, Eu, Gd, Tb, Dy, and Nd.

Europium has been doped into monoclinic gallium oxide (β-Ga_2O_3:Eu) nanostructures, obtaining intense red emission lines mainly due to $^5D_0$–^7F_J transitions by *in situ* doping during the synthesis process. These techniques include the sol–gel [55, 56] hydrothermal method followed by calcination resulting in nanoparticles [57] and hollow nanospheres [58], the combustion method [59], thermal evaporation using the liquid–solid growth mechanism of nanowires [60], solvothermal synthesis [61], the liquid-phase precursor (LPP) method to dope nanoparticles with Eu concentrations 1–5 mol % [62], and electrospinning to obtain nanofibres with europium concentrations ranging from 0.5 to 5 mol % [63]. In these last two cases, the maximum rare-earth-related luminescence intensity was obtained for Eu^{3+} 3 mol % concentrations. On the other hand, doping with Eu^{3+} has been also obtained after the fabrication of nanowires, either by thermal diffusion [60] or ion implantation followed by thermal annealing to recover the crystal quality of the nanowires and optically activate the ions [64, 65].

γ-Ga_2O_3 nanocrystals have also been doped with europium. Either colloidal synthesis [66, 67] or thermal decomposition reaction [68] procedures were used, obtaining the above-mentioned characteristic sharp, red Eu^{3+} luminescence lines. Furthermore, Layek *et al* achieved Eu^{2+} doping by reduction of the Eu^{3+} doped nanocrystals [67], resulting in a dramatic change of the luminescence properties, from the sharp red lines of Eu^{3+} to a broader violet band due to $Eu^{2+}4f_6^5d_1 \rightarrow {}^4f_7$ intra-ionic transition, as shown in figure 8.6. Co-doping with the two ions by colloidal synthesis was also achieved and the electroluminescence of the nanoparticles resulted in white light emission [69]. This could have ready applications in solid-state lighting.

Terbium has been doped into both β-Ga_2O_3 [61, 70, 71] and γ-Ga_2O_3 [66, 68] nanocrystals, resulting in sharp green luminescence lines due to $^5D_4$–^7F_J intra-ionic transitions. In two of the cases, Eu and Tb co-doping was achieved [61, 68].

Emission lines have been obtained from Er^{3+} doped β-Ga_2O_3 nanocrystals obtained by solution combustion synthesis [72] and nano- and microwires obtained by the vapour–solid growth method [30, 60]. The most intense lines when doping with these ions are in the green $^4S_{3/2}$–$^4I_{15/2}$ transition or infrared $^4I_{13/2}$–$^4I_{15/2}$ transition regions. In addition, Dy^{3+} related blue luminescence from nanoparticles [73],

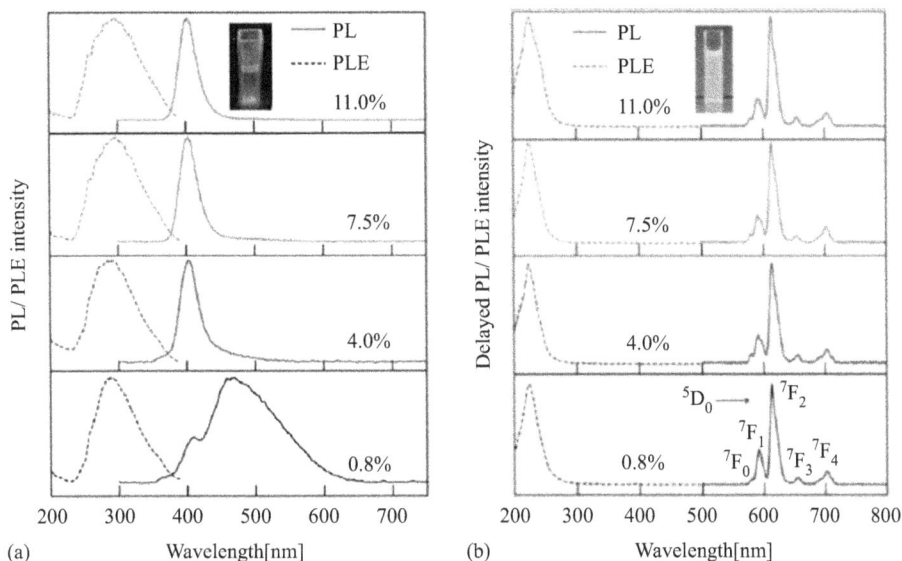

Figure 8.6. PL (solid lines) and PLE (dashed lines) from γ-Ga$_2$O$_3$ nanocrystals with different concentrations of (a) Eu^{2+} and (b) Eu^{3+}. Reproduced with permission from [67]. Copyright 2015 American Chemical Society.

Nd^{3+} infrared from a nanocrystalline film [74], and Gd^{3+} ultraviolet from *nanowires* [64, 65] have also been obtained.

8.3.3 Sn and Si doping

Doping gallium oxide with Si or Sn allows control of its n-type conductivity over several orders of magnitude, as shown in bulk materials [75, 76]. Therefore, research on the influence of these dopants on the properties of nanostructured gallium oxide is of great interest. We will first focus on the resulting luminescence and morphological changes induced by the presence of Sn during the growth of nanowires using the vapour–liquid–solid (VLS) growth mechanism, with Au as a catalyst. Maximenko *et al* showed that nanowires presenting a strong Sn concentration gradient resulted in severe differences on the luminescence, shifting from blue to green due to an increase in Sn concentration along the wire axis to a few Sn at % [77]. At a certain point of the nanowires, the composition shifted to Ga doped SnO$_2$ and a red luminescence band was observed. In other work, Sn doping enhanced the formation of branched gallium oxide nanowires by VLS where Sn itself was the catalyser [43]. A green band was observed in these nanowires, attributed to the presence of Sn dopant. Cr and Sn codoping was subsequently achieved in these nanowires, resulting in the above-mentioned Cr^{3+} related red band, while the rest of the UV and visible bands were completely quenched [43]. In the case of highly Sn doped β-Ga$_2$O$_3$/SnO$_2$ nanowires grown by VLS, red emission was obtained when the doping concentration was higher than 10 at. % [78].

On the other hand, β-Ga$_2$O$_3$:Si nanowires were grown by the VLS growth mechanism where silicon oxide droplets acted as the catalyst [79]. Si doping induced

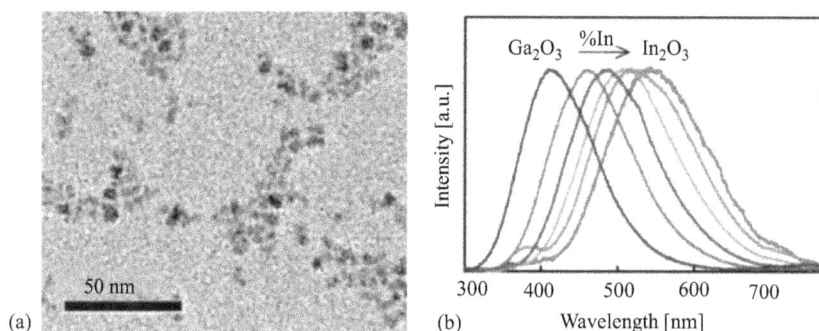

Figure 8.7. (a) A TEM image of gallium indium oxide nanocrystals with 5 at. % In. (b) PL spectra from alloyed nanocrystals with different In contents. Reproduced with permission from [83]. Copyright 2011 American Chemical Society.

a redshift of the CL spectrum due to a relative increase of the blue band with respect to the UV one.

8.3.4 Al and In doping and alloying—ternary oxides

Semiconductor alloys are key to tailoring their optoelectronic properties. Al and In are the natural elements to span a very wide range of gallium oxide based optoelectronic properties. Actually, there is a huge interest in developing $(Al_xGa_{1-x})_2O_3/Ga_2O_3$ heterostructures for electronic devices in order to spatially confine carriers in two dimensions.

The vapour–solid (VS) growth mechanism has been used to grow elongated nanostructures of the monoclinic phase. Zig-zag shaped nanowires were obtained in one case, with a broad blue emission band [80] while nanobelt growth was promoted in another case [81]. In the latter work, segregation of In to the edge of the planar structures was observed. A clear blue/UV intensity ratio dependence on the In content was observed up to 6 In at. %; the blue band was enhanced with respect to the UV when increasing In content. A comparison between In doped and Al doped nanowires showed that the blue–UV band is shifted 0.2 eV in In doped and 0.4 eV in Al doped β-Ga_2O_3, in comparison to undoped samples [82].

On the other hand, alloyed $Ga_{2-x}In_xO_3$ colloidal nanoparticles have been achieved, resulting in a broad luminescence band that was tuned from violet to orange/red, as shown in figure 8.7 [83]. In that work, the unalloyed nanocrystals with diameters of a few nanometres were γ-Ga_2O_3.

8.3.5 Other dopants

Song *et al* showed that nitrogen doping of monoclinic gallium oxide—obtained by annealing in an ammonia atmosphere nanowires previously grown by thermal evaporation—results in an intense, broad, red emission band, as shown in figure 8.8 [84], which was ascribed to the recombination of electrons trapped in oxygen-vacancy-related donor states and holes trapped in N-dopant-related acceptor states. Subsequent work showed similar results in N doped nanostructures [85, 86].

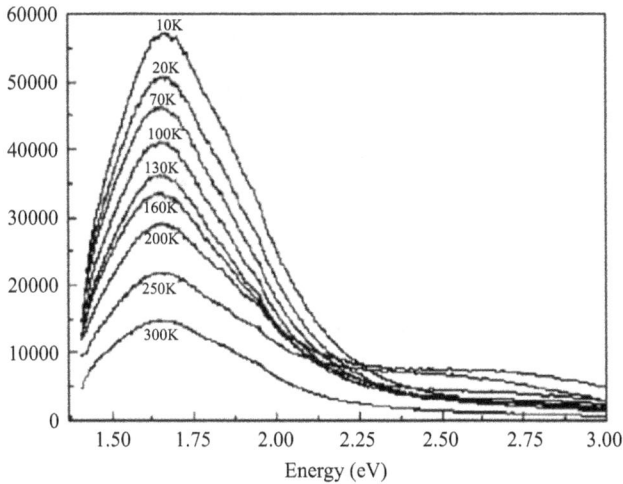

Figure 8.8. Temperature evolution of the PL spectra for 2.60 at. % N doped gallium oxide nanowires. Reproduced with permission from [84]. Copyright 2009 American Physical Society.

Finally, the Li^+ ion has also been studied as a dopant in β-Ga_2O_3 nanowires, obtained by the VS growth mechanism, due to their potential applications in optoelectronics or energy conversion [87]. The work showed that, similar to the cases of Sn or In, this dopant influences both the morphology—inducing growth of micropyramids in the direction perpendicular to the nanowire axis—and the luminescence properties of elongated nanostructures obtained using thermal evaporation methods. Indeed, an intense, sharp emission peak centred on 717 nm was obtained, which has been related to the presence of Li dopants within the oxide [87].

8.4 Optical confinement in Ga_2O_3 microstructures

Light amplification by stimulated emission of radiation (LASER) devices have revolutionized our world and are found in many fields, from fibre-optic communications to laser surgery. In the last decade, the push for miniaturization has provided new amplification devices such as vertical-cavity-surface-emitting lasers and photonic crystal lasers, some of which are already used commercially. In order to continue in this path, one of the main areas of research is the semiconductor nanowire, which can simultaneously act as an optical gain medium and an optical cavity and can be electrically integrated in an optoelectronic nanodevice [88].

Current nanowire based nanolasers are being extensively developed. So far, II–VI (ZnO, MgO, CdS, etc) and III–V (InP, GaN, etc) semiconductors have been proved suitable for these applications due to their excellent optical properties, offering lasing from the near UV to the near IR depending on the material chosen [89, 90]. Nanoscale tunability, which is needed for many photonic devices such as light emitting devices or optical switches, has been achieved on some of these nanolasers by controllable stoichiometry. Reversibly wavelength-tunable nanowire lasers have also been recently obtained [91].

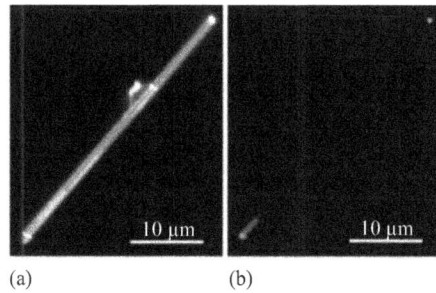

(a)   (b)

Figure 8.9. (a) Optical image of a Cr doped β-Ga$_2$O$_3$ microwire. (b) Micro-PL image of the wire in (a) illuminated with $\lambda = 436$ nm at its lower end, showing its intense red–IR luminescence. An intense red spot is observed on its other end (top right of the image). Reproduced with permission from [30]. Copyright 2007 AIP Publishing.

Research on other emergent semiconductors, as for UWBG Ga$_2$O$_3$, should provide new possibilities for these devices such as the possibility of tuning the emission range from the broad UV–blue band of the undoped material to the red–NIR band of Cr doped gallium oxide.

Generally, a semiconductor nanowire cavity will only operate in single mode when its diameter is around the wavelength of light in the material or smaller. However, in the subwavelength range, diffraction effects must be kept in mind. When the optical field extends outside the nanowire, the amount of reflection and feedback is limited and hence the lasing threshold of the nanowire increases [88].

8.4.1 Waveguiding in Ga$_2$O$_3$

Total internal reflection of light in a micro- or nanowire produces waveguiding as long as the wire has small surface roughness. This light can be generated by the wire due to photoluminescence (PL) or can be injected into the wire from another source. Gallium oxide's high refractive index (1.9–2.0 in the visible region) makes this material suitable for light guiding. This property was studied experimentally in Ga$_2$O$_3$ micro- and nanowires doped with Cr, Er [30], and Eu [60]. As shown in sections 3.3 and 3.4, Cr doped Ga$_2$O$_3$ emits in the 650–850 nm (red–NIR) range due to the well-known sharp R lines superposed on the broad $^4T_2$–$^4A_2$ band, while Eu doped and Er doped Ga$_2$O$_3$ have emissions peaked at 610 nm (orange) and on the 540–570 nm (green) range, respectively, due to their intra-ionic transitions. In Cr doped Ga$_2$O$_3$, the luminescence is efficiently guided towards the other end of the nanowire, as shown in figure 8.9. In addition, undoped, Er doped, and Cr doped microwires are able to transmit incident visible light with wavelengths different to the absorption peaks of each material system with a direction nearly parallel to the wire's longitudinal axis [30].

8.4.2 Fabry–Pérot resonant cavities

Light emitted and guided by micro- and nanowires can be partly reflected at their ends if these are flat enough to act as mirrors due to the refractive index change from the nanowire end to the air. Consequently, the whole microwire can sustain Fabry–Pérot resonances. These occur for every light wavelength (λ) matching with an integer number (N) of half-waves over the wire length (L), i.e. $L = N\lambda/2n$, where n is the refractive index of the wire. López et al experimentally demonstrated Fabry–Pérot resonances in Ga_2O_3 micro- and nanowires [31]. Cr doped wires showed a weak modulation that depended on the length of the wire. For example, in figure 8.10 (a) and (b), wire A is longer than wire B and so the adjacent maxima have a lower separation than wire B. The dependence between the adjacent maxima separation ($\Delta\lambda$) and the wire length is quantified by the well-known Fabry–Pérot mode spacing equation:

$$\Delta\lambda = \frac{\lambda^2}{2Ln_g},\qquad(8.1)$$

where n_g is the group refractive index. This dependence was demonstrated for wires of lengths ranging from 40 to 270 μm [31, 60], and by plotting $\Delta\lambda$ versus the inverse of L (figure 8.10(c)), the group refractive index of Cr doped Ga_2O_3 was measured, giving rise to a value of 2.0 ± 0.1.

8.4.3 Distributed Bragg reflector based microcavities

The main drawback of a nanowire Fabry–Pérot cavity is the relatively low reflection at the wire end. Assuming normal incidence and a refractive index of 1.9, the Ga_2O_3–air boundary reflectivity is only 10%. This low reflectivity can be widely increased by the fabrication of the so-called distributed Bragg reflectors (DBR) at the wire ends. A focused ion beam (FIB) can fabricate these periodic structures with dimensions of about the wavelength of light in order to generate stop bands, i.e. wavelength gaps in which light cannot propagate through the DBR, which can be tunable across the whole visible spectrum depending on the DBR parameters, in

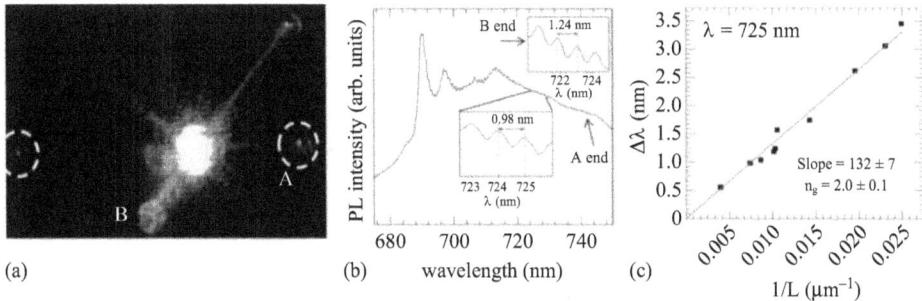

Figure 8.10. (a) Optical image of two Cr doped β-Ga_2O_3 crossing wires under laser excitation (orange spot) at the crossing point. The luminescence is guided towards both ends of both wires. (b) PL spectra from point A shown in (a). The insets show the separation between adjacent maxima around 725 nm for both points A and B. (c) The spacing between adjacent modes versus $1/L$ for $\lambda = 725$ nm, showing a linear dependence, as expected for Fabry–Pérot resonances. n_g is obtained from the slope of the linear fitted curve. Reproduced with permission from [31]. Copyright 2012 AIP Publishing.

particular the periodicity of the DBR. This can be understood by applying the Bragg condition

$$\lambda_B = \frac{2n_{eff}\Lambda}{m},\tag{8.2}$$

where λ_B is the Bragg wavelength, which lies in the midpoint of the stop band, Λ is the periodicity of the DBR, n_{eff} is its effective refractive index, and m is an integer. The wavelength reflectivity tunability of the DBRs in nanowires was demonstrated experimentally and by numerical simulations based on finite element methods by Fu *et al* [92] for GaN nanowires.

Following a similar procedure, several DBR cavities have been fabricated by FIB on Cr doped Ga_2O_3 microwires and the resonant modes produced have been studied by means of experimental analysis, and analytical and finite difference time domain (FDTD) simulations [32, 93]. The tunability of these microcavities has been proven, as the DBR parameters and the length of the microcavities determines the wavelength position and wavelength separation of the resonances, respectively.

Figure 8.11(a) shows the SEM images of the FIB milled microcavities. The aforementioned weak Fabry–Pérot modulations observed in the unpatterned microwires are strongly enhanced in the cavities, as shown by the micro-PL measurements displayed in figure 8.11(b). The reflectivity of a cavity can be obtained experimentally by using the expression that links it to the finesse (F):

$$F = \frac{\pi\sqrt{R}}{1 - R}.\tag{8.3}$$

The finesse is obtained by calculating the ratio between the difference in wavelengths of two adjacent peaks and the full width at half maximum (FWHM) of the peak. By using equation (8.3), the reflectivity of each resonance is obtained. Experimental reflectivities up to 70%–78% were obtained on the DBR Ga_2O_3 cavities, which are good values for optical confinement [32, 93]. The cavities have side dimensions in the range of 510–1150 nm and lengths in the range of 4.9–15.2 μm. Keep in mind that, in order to avoid diffraction losses, the theoretical minimum dielectric nanocavity lateral dimension for red light ($\lambda \sim$700 nm) in Ga_2O_3 ($n \sim 2$) is around 350 nm.

The PL spectra also show other interesting properties of these modes. Polarized PL measurements reveal different positions and relative intensities of the resonances when the polarizer is placed parallel to or perpendicular to the longitudinal axis of the microwire, as shown in figure 8.11(c). In addition, another set of secondary peaks with different separation between adjacent maxima can also be detected for a given polarization set-up (the green dotted lines in figure 8.11(c)).

To explain these experimental results, analytical models were used. As the microwires are almost rectangular, the different field distributions inside a dielectric rectangular waveguide can be used. It is well known that the field of a rectangular waveguide can be treated as the sum of transverse electric (TE) fields and transverse magnetic (TM) fields. However, unlike in a metallic waveguide, modes in a rectangular semiconductor microwire, which acts as a dielectric strip waveguide,

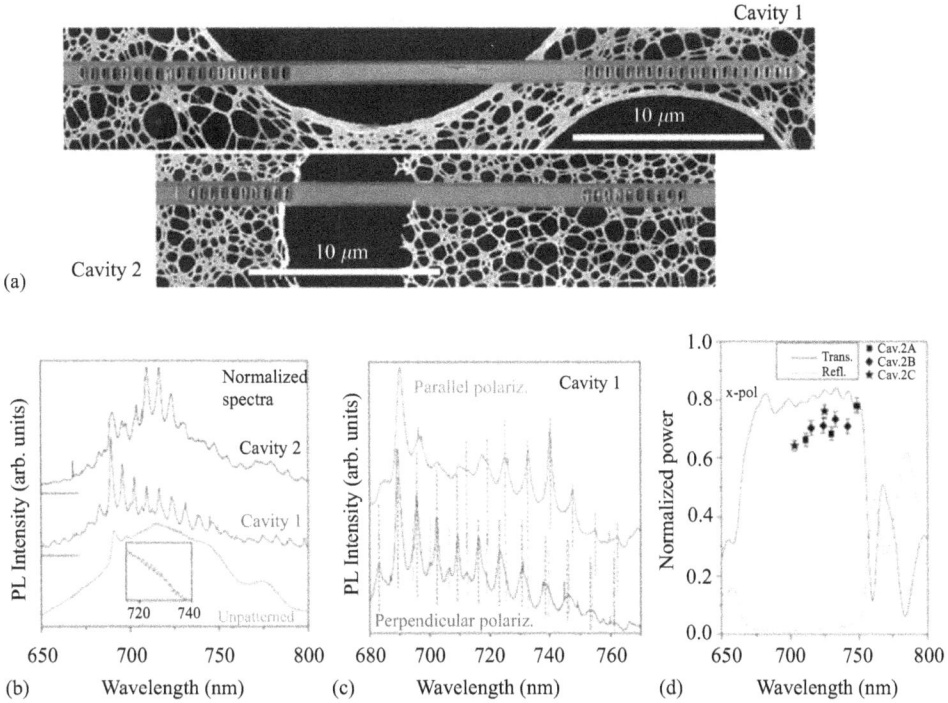

Figure 8.11. (a) SEM image of the fabricated microcavities in [32]. Cavity 1 has a length of 15.2 μm and 20 holes on each side, while cavity 2 has a length of 15 μm and ten holes on each side. (b) Micro-PL spectra from unpatterned wire and cavities 1 and 2 with a perpendicular polarization set-up. (c) Micro-PL spectra from cavity 1 in perpendicular and parallel polarization set-ups. Dashed blue, green, and red lines indicate the E_{11}^{x}, E_{12}^{x}, and E_{11}^{y} identified modes, respectively. (d) FDTD-simulated reflected power spectrum for perpendicular polarization set-up for a given microwire with a given DBR compared with individual reflectivities obtained from the experimental resonances of three cavities of that same DBR-fabricated microwire. Panels (a), (b), and (c) reproduced with permission from [32]. Copyright 2018 American Physical Society. Panel (d) reproduced with permission from [93]. Copyright 2019 SPIE.

do not have exact analytical solutions. The authors used the approximate analytical Marcatili model, and the computational Goell model when the Marcatili model is no longer valid, in order to identify the experimental resonant modes and obtain their wavevector components.

For a perpendicular polarization configuration, the dominant mode is the fundamental E_{11}^{x} mode (the blue dashed lines in figure 8.11(c)) while the dominant mode for parallel polarization is the fundamental E_{11}^{y} mode (the red dashed lines in figure 8.11(c)). Less intense resonances are due to their higher modes. The requirement for longitudinal resonance in a Fabry–Pérot cavity, $k_z l = n\pi (l \in Z)$, is fulfilled for all the identified resonances. In all the cavities, the effective lengths obtained using this method are notably higher than the physical lengths of the cavities. This is explained by the well-known effective penetration depth of the confined wave into the DBRs.

Finally, FDTD simulations were performed in order to design future DBR-based cavities on microwires of different sizes and shapes, which could guide light in different wavelength ranges. Figure 8.11(d) displays the transmittance and the reflectivity of the designed DBRs in the red–NIR range (solid and dashed lines, respectively) as well as the experimental results (dots), showing good agreement between them. This procedure paves the way for future optimized micro- and nanocavity designs in Ga_2O_3 and other transparent semiconductor materials.

In summary, waveguiding has been proven for Ga_2O_3 micro- and nanowires with different dopants, namely Cr, Er, and Eu. A small surface roughness and large enough lateral dimensions are required in order to achieve optical confinement in these structures. Cr doped Ga_2O_3 has been proven so far to guide light effectively and produce Fabry–Pérot resonances. Tunable optical DBR-based microcavities with high reflectivity have been fabricated by FIB and their resonances have been analysed experimentally, analytically, and via simulations. This and other fabrication and characterization procedures should be implemented in micro- and nanowires with other dopants, as well as undoped Ga_2O_3, allowing emission and confinement in the NIR, visible, and UV range. The morphological and optical properties of these structures should be further explored and optimized in order to achieve optical gain in Ga_2O_3 based microcavities. In addition, waveguiding and optical confinement in more complex structures, such as core–shell or coupled nanowire structures, would result in additional optical properties.

8.5 Summary, outlook, and prospective work

This chapter reviews the main aspects regarding the optical properties of gallium oxide nanostructures. This oxide is experiencing a re-birth since the achievement of high quality substrates and their potential in high-power electronics. These applications mainly encompass its electronic properties and optical properties have received less attention. In addition, the recent development of Ga_2O_3 nanomaterials adds another parameter to play with and spans the field of applications combining optics and electronics. This oxide is transparent up to the UV and the luminescence properties of Ga_2O_3 thin films and bulk crystals have been studied some years ago. Models for luminescence mechanisms have been proposed, albeit interactions between structural defects and dopants are still unclear. The synthesis of Ga_2O_3 in materials and structures with dimensions in the micro- and nanoscale has allowed us to obtain a deeper understanding of the optical properties of this oxide.

An extensive analysis of the dopants that have been introduced into gallium oxide nanostructures for luminescence purposes shows that the two most studied families so far are the rare earths—in particular Eu and Tb—and transition metals—in particular Cr, Ni, and Mn—showing strong luminescence emissions in a very wide energy range. Sn and Si, two n-type dopants, have also been explored in order to modify the luminescence properties, as well as isoelectronic In and Al, which result in some cases in alloys, allowing tunability of the emission bands. It should be mentioned that much work is currently ongoing on the synthesis and characterization of Al–Ga–O alloys as electronic materials to achieve electron confinement in two dimensions. Undoubtedly, it will also be of interest from the optical point of view. Finally, other few works have been devoted to nitrogen or Li, which result in red and IR luminescence bands or peaks, respectively.

Regarding optical confinement, gallium oxide micro- and nanowires doped with some elements such as Eu have shown waveguiding properties, and Cr doped gallium oxide tunable optical microcavities with reflectivities up to 78% have been fabricated using FIB and have been analysed experimentally, analytically, and using simulations. The study of Ga_2O_3 as a gain material is still in its early stages and, so far, no optical resonances on the green, blue, and UV parts of the spectrum have been reported.

Keeping in mind the high transparency range of the material, its diverse optical behaviour with respect to different dopants and, in the micro- and nanoscale, the variety of morphologies and heterostructures that have been reported, new designs based on Ga_2O_3 with improved optical properties should pave the way for their application in novel nanoscale optoelectronic and photonic devices.

References

[1] Chen X, Ren F, Gu S and Ye J 2019 Review of gallium-oxide-based solar-blind ultraviolet photodetectors *Photonics Res.* **7** 4

[2] Roy R, Hill V G and Osborn E F 1952 Polymorphism of Ga_2O_3 and the system Ga_2O_3–H_2O *J. Am. Chem. Soc.* **74** 719

[3] Geller S 1960 Crystal structure of β-Ga_2O_3 *J. Chem. Phys.* **33** 676

[4] Kracht M *et al* 2017 Tin-assisted synthesis of ε-Ga_2O_3 by molecular beam epitaxy *Phys. Rev. Appl.* **8** 054002

[5] Zhuo Y, Chen Z, Tu W, Ma X, Pei Y and Wang G 2017 β-Ga_2O_3 versus ε-Ga_2O_3: control of the crystal phase composition of gallium oxide thin film prepared by metal–organic chemical vapor deposition *Appl. Surf. Sci.* **420** 802–7

[6] Ricci F, Boschi F, Baraldi A, Filippetti A, Higashiwaki M, Kuramata A, Fiorentini V and Fornari R 2016 Theoretical and experimental investigation of optical absorption anisotropy in β-Ga_2O_3 *J. Phys. Cond. Matter.* **28** 224005

[7] Slomski M, Blumenschein N, Paskov P P, Muth J F and Paskova T 2017 Anisotropic thermal conductivity of β-Ga_2O_3 at elevated temperatures: effect of Sn and Fe dopants *J. Appl. Phys.* **121** 235104

[8] Pearton S J, Yang J, Cary P H IV, Ren F, Kim J, Tadjer M J and Mastro M A 2018 A review of Ga_2O_3 materials, processing and devices *Appl. Phys. Rev.* **5** 011301

[9] Schubert M *et al* 2016 Anisotropy, phonon modes, and free charge carrier parameters in monoclinic β-gallium oxide single crystals *Phys. Rev.* **B 93** 125209

[10] Mohamed M, Janowitz C, Unger I, Manzke R, Galazka Z, Uecker R, Fornari R, Weber J R, Varley J B and Van de Walle C G 2010 The electronic structure of β-Ga_2O_3 *Appl. Phys. Lett.* **97** 211903

[11] He H, Orlando R, Blanco M, Pandey R, Amzallag E, Baraille I and Rerat M 2006 First-principles study of the structural, electronic, and optical properties of Ga_2O_3 in its monoclinic and hexagonal phases *Phys. Rev.* **B 74** 195123

[12] Passlack M, Hunt N E J, Schubert E F, Zydzik G J, Hong M, Mannaerts J P, Opila R L and Fischer R J 1994 Dielectric properties of electron-beam deposited Ga$_2$O$_3$ films *Appl. Phys. Lett.* **64** 2715

[13] Rebien M, Henrion W, Hong M, Mannaerts J P and Fleischer M 2002 Optical properties of gallium oxide thin films *Appl. Phys. Lett.* **81** 250

[14] Al-Kuhaili M F, Durrani S M A and Khawaja E E 2003 Optical properties of gallium oxide films deposited by electron-beam evaporation *Appl. Phys. Lett.* **83** 4533

[15] Ueda N, Hosono H, Waseda R and Kawazoe H 1997 Anisotropy of electrical and optical properties in β-Ga$_2$O$_3$ single crystals *Appl. Phys. Lett.* **71** 933

[16] Sturm C, Furthmüller J, Bechstedt F, Schmidt-Grund R and Grundmann M 2015 Dielectric tensor of monoclinic Ga$_2$O$_3$ single crystals in the spectral range 0.5–8.5 eV *APL Mater.* **3** 106106

[17] Varley J B, Weber J R, Janotti A and Van de Walle C G 2010 Oxygen vacancies and donor impurities in β-Ga$_2$O$_3$ *Appl. Phys. Lett.* **97** 142106

[18] Dong L, Jia R, Xin B, Peng B and Zhang Y 2017 Effects of oxygen vacancies on the structural and optical properties of β-Ga$_2$O$_3$ *Sci. Rep.* **7** 40160

[19] Nogales E, Méndez B and Piqueras J 2011 Assessment of waveguiding properties of gallium oxide nanostructures by angle resolved cathodoluminescence in a scanning electron microscope *Ultramicroscopy* **111** 1037–42

[20] Binet L and Gourer D 1998 Origin of the blue luminescence of β-Ga$_2$O$_3$ *J. Phys. Chem. Solids* **59** 1241–9

[21] Shimamura K, Víllora E G, Ujiie T and Aoki K 2008 Excitation and photoluminescence of pure and Si-doped beta-Ga$_2$O$_3$ single crystals *Appl. Phys. Lett.* **92** 201914

[22] Choi Y C, Kim W S, Park Y S, Lee S M, Bae D J, Lee Y H, Park G-S, Choi W B, Lee N S and Kim J M 2000 Catalytic growth of β-Ga$_2$O$_3$ nanowires by arc discharge *Adv. Mater.* **12** 746

[23] Rao R, Rao A M, Xu B, Dong J, Sharma S and Sunkara M K 2005 Blueshifted Raman scattering and its correlation with the [110] growth direction in gallium oxide nanowires *J. Appl. Phys.* **98** 094312

[24] Gao Y H, Bando Y, Sato T, Zhang Y F and Gao X Q 2002 Synthesis, Raman scattering and defects of β-Ga$_2$O$_3$ nanorods *Appl. Phys. Lett.* **81** 2267

[25] Dai L, Chen X L, Zhang X N, Jin A Z, Zhou T, Hu B Q and Zhang Z 2002 Growth and optical characterization of Ga$_2$O$_3$ nanobelts and nanosheets *J. Appl. Phys.* **92** 1062

[26] Geng B Y, Liu X W, Wei X W, Wang S W and Zhang L D 2005 Low-temperature growth of β-Ga$_2$O$_3$ nanobelts through a simple thermochemical route and their phonon spectra properties *Appl. Phys. Lett.* **87** 113101

[27] Peelaers H and Van de Walle C G 2017 Lack of quantum confinement in Ga$_2$O$_3$ nanolayers *Phys. Rev.* B **96** 081409(R)

[28] Wang T, Farvid S S, Abulikemu M and Radovanovic P V 2010 Size-tunable phosphorescence in colloidal metastable γ-Ga$_2$O$_3$ nanocrystals *J. Am. Chem. Soc.* **132** 9250

[29] Lin H J, Baltrus J P, Gao H, Ding Y, Nam C Y, Ohodnicki P and Gao P X 2016 Perovskite nanoparticle-sensitized Ga$_2$O$_3$ nanorod arrays for CO detection at high temperature *ACS Appl. Mater. Interfaces* **8** 8880–7

[30] Nogales E, Garcia J A, Mendez B and Piqueras J 2007 Doped gallium oxide nanowires with waveguiding behavior *Appl. Phys. Lett.* **91** 133108

[31] López I, Nogales E, Méndez B and Piqueras J 2012 Resonant cavity modes in gallium oxide microwires *Appl. Phys. Lett.* **100** 261910

[32] Alonso-Orts M, Nogales E, San Juan J M, Nó M L, Piqueras J and Méndez B 2018 Modal analysis of β-Ga$_2$O$_3$:Cr widely tunable luminescent optical microcavities *Phys. Rev. Appl.* **9** 064004

[33] McCarthy D E 1967 Transmittance of optical materials from 0.17 μ to 0.30 μ *Appl. Optics* **6** 1896–8

[34] Maiman T H 1960 Stimulated emission of radiation in ruby *Nature* **187** 493–4

[35] Peter M and Schawlow A L 1960 Optical and paramagnetic resonance spectra of Cr^{+++} in Ga$_2$O$_3$ *Bull. Am. Phys. Soc.* **5** 158

[36] Tippins H H 1965 Optical and microwave properties of trivalent chromium in β-Ga$_2$O$_3$ *Phys. Rev.* **137** A865

[37] Nogales E, López I, Méndez B, Piqueras J, Lorenz K, Alves E and García J A 2012 Doped gallium oxide nanowires for photonics *Proc. SPIE* **8263** 82630B

[38] Onuma T, Nakata Y, Sasaki K, Masui T, Yamaguchi T, Honda T, Kuramata A, Yamakoshi S and Higashiwaki M 2018 Modeling and interpretation of UV and blue luminescence intensity in β-Ga$_2$O$_3$ by silicon and nitrogen doping *J. Appl. Phys.* **124** 075103

[39] Miyata T, Nakatani T and Minami T 2000 Gallium oxide as host material for multicolor emitting phosphors *J. Lumin.* **87–89** 1183–5

[40] Henderson B and Imbusch G F 1989 *Optical Spectroscopy of Inorganic Solids* (Oxford: Clarendon)

[41] Vivien D, Viana B, Revcolevschi A, Barrie J D, Dunn B, Nelson P and Stafsudd O M 1987 Optical properties of β-Ga$_2$O$_3$:Cr^{3+} single crystals for tunable laser applications *J. Lumin.* **39** 29–33

[42] Nogales E, García J A, Méndez B and Piqueras J 2007 Red luminescence of Cr in β-Ga$_2$O$_3$ nanowires *J. Appl. Phys.* **101** 033517

[43] López I, Nogales E, Méndez B, Piqueras J, Peche A, Ramírez-Castellanos J and González-Calbet J M 2013 Influence of Sn and Cr doping on morphology and luminescence of thermally grown Ga$_2$O$_3$ nanowires *J. Phys. Chem.* C **117** 3036–45

[44] Lu Y Y, Liu F, Gu Z and Pan Z 2011 Long-lasting near-infrared persistent luminescence from β-Ga$_2$O$_3$:Cr^{3+} nanowire assemblies *J. Lumin.* **131** 2784–7

[45] Wang X S, Situ J Q, Ying X Y, Chen H, Pan H F, Jin Y and Du Y Z 2015 β-Ga$_2$O$_3$:Cr^{3+} nanoparticle: a new platform with near infrared photoluminescence for drug targeting delivery and bio-imaging simultaneously *Acta Biomater.* **22** 164–72

[46] Wang X S 2015 Multi-functional mesoporous β-Ga$_2$O$_3$:Cr^{3+} nanorod with long lasting near infrared luminescence for *in vivo* imaging and drug delivery *RSC Adv.* **17** 12886–9

[47] Ye S, Zhang Y, He H, Qiu J and Dong G 2015 Simultaneous broadband near-infrared emission and magnetic properties of single phase Ni^{2+}-doped β-Ga$_2$O$_3$ nanocrystals via mediated phase-controlled synthesis *J. Mater. Chem.* C **3** 2886

[48] Zhou S, Feng G, Wu B, Jiang N, Xu S and Qiu J 2007 Intense infrared luminescence in transparent glass-ceramics containing β-Ga$_2$O$_3$:Ni^{2+} nanocrystals *J. Phys. Chem.* C **111** 7335–8

[49] Zhou S F, Dong H F, Feng G F, Wu B T, Zeng H P and Qiu J R 2007 Broadband optical amplification in silicate glass-ceramic containing β-Ga$_2$O$_3$:Ni^{2+} nanocrystals *Opt. Express* **15** 5477

[50] Song Y P, Wang P W, Xu X Y, Wang Z, Li G H and Yu D P 2006 Magnetism and photoluminescence in manganese–gallium oxide nanowires with monoclinic and spinel structures *Physica* E **31** 67–71

[51] Gonzalo A, Nogales E, Lorenz K, Víllora E G, Shimamura K, Piqueras J and Méndez B 2017 Raman and cathodoluminescence analysis of transition metal ion implanted Ga_2O_3 nanowires *J. Lumin.* **191** 56–60

[52] López I, Alonso-Orts M, Nogales E and Méndez B 2018 Structural and luminescence properties of Ga_2O_3:Zn micro and nanostructures *Phys. Status Solidi* a **215** 1800217

[53] Wang X H, Zhang F B, Saito K, Tanaka T, Nishio M and Guo Q X 2014 Electrical properties and emission mechanisms of Zn-doped β-Ga_2O_3 films *J. Phys. Chem. Solids* **75** 1201–4

[54] Favennec P N, Haridon H L, Salvi M, Muotonnet D and Le Guillo Y 1989 Luminescence of erbium implanted in various semiconductors *Electron. Lett.* **25** 718–9

[55] Kim J S, Kim H E, Park H L and Kim G C 2004 Luminescence intensity and color purity enhancement in nanostructured β-Ga_2O_3:Eu^{3+} phosphors *Solid State Commun.* **132** 459–63

[56] Kim J S, Kim H E, Kwon A W, Park H L and Kim G C 2007 Effect of initial pH on nanophosphor β-Ga_2O_3:Eu^{3+} prepared through sol–gel process *J. Lumin.* **122** 710–3

[57] Liu G, Duan X, Li H and Liang D 2008 Preparation and photoluminescence properties of Eu-doped Ga_2O_3 nanorods *Mater. Chem. Phys.* **110** 206–11

[58] Kang B K, Mang S R, Lim H D, Song M K, Song Y H, Go D H, Jung M K, Senthil K and Yoon D H 2014 Synthesis, morphology and optical properties of pure and Eu^{3+} doped β-Ga_2O_3 hollow nanostructures by hydrothermal method *Mater. Chem. Phys.* **147** 178–83

[59] Zhu H, Li R, Luo W and Chen X 2011 Eu^{3+}-doped β-Ga_2O_3 nanophosphors: annealing effect, electronic structure and optical spectroscopy *Phys. Chem. Chem. Phys.* **13** 4411–9

[60] Nogales E, Méndez B, Piqueras J and García J A 2009 Europium doped gallium oxide nanostructures for room temperature luminescent photonic devices *Nanotechnology* **20** 115201

[61] Sinha G and Patra A 2009 Generation of green, red and white light from rare-earth doped Ga_2O_3 nanoparticles *Chem. Phys. Lett.* **473** 151–4

[62] Kim M O, Yang B and Yoon D 2013 Structural and optical characterization of Eu^{3+} doped β-Ga_2O_3 nanoparticles using a liquid-phase precursor method *J. Nanosci. Nanotech.* **13** 5556–60

[63] Zhao J, Zhang W, Xie E, Ma Z, Zhao A and Liu Z 2011 Structure and photoluminescence of β-Ga_2O_3: Eu^{3+} nanofibers prepared by electrospinning *Appl. Surf. Sci.* **257** 4968–72

[64] Nogales E, Hidalgo P, Lorenz K, Méndez B, Piqueras J and Alves E 2011 Cathodoluminescence of rare earth implanted Ga_2O_3 and GeO_2 nanostructures *Nanotechnology* **22** 285706

[65] López I, Lorenz K, Nogales E, Méndez B, Piqueras J, Alves E and García J A 2014 Study of the relationship between crystal structure and luminescence in rare-earth-implanted Ga_2O_3 nanowires during annealing treatments *J. Mater. Sci.* **49** 1279–85

[66] Wang T, Layek A, Hosein I D, Chirmanov V and Radovanovic P V 2014 Correlation between native defects and dopants in colloidal lanthanide-doped Ga_2O_3 nanocrystals: a path to enhance functionality and control optical properties *J. Mater. Chem.* C **2** 3212

[67] Layek A, Yildirim B, Ghodsi V, Hutfluss L N, Hedge M, Wang T and Radovanovic P V 2015 Dual europium luminescence centers in colloidal Ga_2O_3 nanocrystals: controlled *in situ* reduction of Eu (III) and stabilization of Eu (II) *Chem. Mater.* **27** 6030–7

[68] Wawrzynczyk D, Nyk M and Samoc M 2015 Synthesis and optical characterization of lanthanide-doped colloidal Ga_2O_3 nanoparticles *Chem. Phys.* **456** 73–8

[69] Yu C, Cao M, Yan D, Lou S, Xia C, Xuan T, Xie R J and Li H 2018 Synthesis of Eu^{2+}/Eu^{3+} Co-doped gallium oxide nanocrystals as a full colour converter for white light emitting diodes *J. Colloid Interface Sci.* **530** 52–7

[70] Kang B K, Mang S R and Yoon D H 2013 Synthesis and characteristics of pure β-Ga_2O_3 and Tb^{3+} doped β-Ga_2O_3 hollow nanostructures *Mat. Lett.* **111** 67–70

[71] Zhao J, Zhang W, Xie E, Liu Z, Feng J and Liu Z 2011 Photoluminescence properties of β-Ga_2O_3:Tb^{3+} nanofibers prepared by electrospinning *Mat. Sci. Eng.* B **176** 932–6

[72] Biljan T, Gajović A and Meić Z 2008 Visible and NIR luminescence of nanocrystalline β-Ga_2O_3:Er^{3+} prepared by solution combustion synthesis *J. Lumin.* **128** 377–82

[73] Shen W Y, Pang M L, Lin J and Fang J 2005 Host-sensitized luminescence of Dy^{3+} in nanocrystalline β-Ga_2O_3 prepared by a Pechini-type sol–gel process *J. Electrochem. Soc.* **152** H25–8

[74] Podhorodecki A, Banski M, Misiewicz J, Lecerf C, Marie P, Cardin J and Portier X 2010 Influence of neodymium concentration on excitation and emission properties of Nd doped gallium oxide nanocrystalline films *J. Appl. Phys.* **108** 063535

[75] Víllora E G, Shimamura K, Yoshikawa Y, Ujiie T and Aoki K 2008 Electrical conductivity and carrier concentration control in β-Ga_2O_3 by Si doping *Appl. Phys. Lett.* **92** 202120

[76] Baldini M, Albrecht M, Fiedler A, Irmscher K, Klimm D, Schewski R and Wagner G 2016 Semiconducting Sn-doped β-Ga_2O_3 homoepitaxial layers grown by metal organic vapour-phase epitaxy *J. Mater. Sci.* **51** 3650–6

[77] Maximenko S I, Mazeina L, Picard Y N, Freitas J A Jr, Bermúdez V M and Prokes S M 2009 Cathodoluminescence studies of the inhomogeneities in Sn-doped Ga_2O_3 nanowires *Nano Lett.* **9** 3245–51

[78] Zervos M, Othonos A, Gianetta V, Travlos A and Nassiopoulou A G 2015 Sn doped β-Ga_2O_3 and β-Ga_2S_3 nanowires with red emission for solar energy spectral shifting *J. Appl. Phys.* **118** 194302

[79] Díaz J, López I, Nogales E, Méndez B and Piqueras J 2011 Synthesis and characterization of silicon-doped gallium oxide nanowires for optoelectronic UV applications *J. Nanopart. Res.* **13** 1833–9

[80] Su Y, Gao M, Meng X, Chen Y, Zhou Q, Li L and Feng Y 2009 Synthesis of In-doped Ga_2O_3 zigzag-shaped nanowires and optical properties *J. Phys. Chem. Solids* **70** 1062–5

[81] López I, Utrilla A D, Nogales E, Méndez B, Piqueras J, Peche A, Ramírez-Castellanos J and González-Calbet J M 2012 In-doped gallium oxide micro- and nanostructures: morphology, structure, and luminescence properties *J. Phys. Chem.* C **116** 3935–43

[82] Nogales E, Sánchez B, Méndez B and Piqueras J 2009 Cathodoluminescence study of isoelectronic doping of gallium oxide nanowires *Superlatt. Microst.* **45** 156–60

[83] Farvid S S, Wang T and Radovanovic P V 2011 Colloidal gallium indium oxide nanocrystals: a multifunctional light-emitting phosphor broadly tunable by alloy composition *J. Am. Chem. Soc.* **133** 6711–9

[84] Song Y P, Zhang H Z, Lin C, Zhu Y W, Li G H, Yang F H and Yu D P 2004 Luminescence emission originating from nitrogen doping of β-Ga_2O_3 nanowires *Phys. Rev.* B **69** 075304

[85] Chang L W, Yeh J W, Li C F, Huang M W and Shih H C 2009 Modulation of luminescence emission spectra of N-doped β-Ga_2O_3 nanowires by thermal evaporation *Thin Solid Films* **518** 1434–8

[86] Peng Y, Yu N, Xiang Y, Liu J, Cao L and Huang S 2018 One-step hydrothemal synthesis of nitrogen doped β-Ga_2O_3 nanostructure and its optical properties *J. Nanosci. Nanotechnol.* **18** 5654–9

[87] López I, Alonso-Orts M, Nogales E, Méndez B and Piqueras J 2016 Influence of Li doping on the morphology and luminescence of Ga_2O_3 microrods grown by a vapor–solid method *Semicond. Sci. Technol.* **31** 115003

[88] Zimmler M A, Capasso F, Müller S and Ronning C 2010 Optically pumped nanowire lasers: invited review *Semicond. Sci. Technol.* **25** 024001

[89] Arbiol J and Chong Q 2015 *Semiconductor Nanowires: Materials, Synthesis, Characterization and Applications* (Cambridge: Woodhead Publishing)

[90] Eaton S W, Fu A, Wong A B, Ning C Z and Yang P 2016 Semiconductor nanowire lasers *Nat. Rev. Mater.* **1** 16028

[91] Zhuge M H *et al* 2019 Fiber-integrated reversibly wavelength-tunable nanowire laser based on nanocavity mode coupling *ACS nano* **13** 9965–72

[92] Fu A, Gao H, Petrov P and Yang P 2015 Widely tunable distributed Bragg reflectors integrated into nanowire waveguides *Nano Lett.* **15** 6909–13

[93] Alonso-Orts M, Nogales E, San Juan J M, Nó M L and Méndez B 2019 Exciting and confining light in Cr doped gallium oxide *Proc. SPIE* **10919** 109191s

Chapter 9

Band alignment of various dielectrics on Ga_2O_3, $(Al_xGa_{1-x})_2O_3$, and $(In_xGa_{1-x})_2O_3$

C Fares, F Ren, Max Kneiß, Holger von Wenckstern, Marius Grundmann and S J Pearton

Understanding the band alignment of a dielectric to a semiconductor is one of the most important requirements to design and optimize metal–oxide–semiconductor field-effect transistor and metal–insulator–semiconductor field-effect transistor devices. For these heterostructure designs, wide bandgap oxides can be used as gate dielectrics and narrow energy bandgap conductive oxides can be utilized to reduce contact resistance between the ohmic contacts and semiconductor surface. In this review, we report on the determination of the band alignment of various insulating and semiconducting dielectrics onto Ga_2O_3, $(Al_xGa_{1-x})_2O_3$, and $(In_xGa_{1-x})_2O_3$. We also discuss factors that influence the band alignment of these heterostructures including the deposition method, film impurities, and the effects of post-deposition annealing. Thus far, the band alignment results for Ga_2O_3 provide a good starting point for the optimization and future commercialization of Ga_2O_3 based devices.

9.1 Introduction

The employment of wide bandgap materials such as β-Ga_2O_3 is paramount within power switching technologies in advanced military applications, electric power transmission, general industrial machinery, and renewable energy systems [1–9]. Over the past few decades, several wide bandgap semiconductors such as GaN (3.4 eV) and SiC (3.3 eV) have been studied and introduced commercially. GaN and SiC have started replacing Si power devices in some applications since wide bandgap materials can yield lower energy losses than Si devices and higher breakdown voltages [10]. One major drawback for SiC and GaN based technologies is the high cost of the growth process which corresponds to a high substrate cost. In comparison to these commercially available materials, β-Ga_2O_3 is an emerging wide bandgap semiconductor with the potential for low-cost growth and substrates [11, 12]. In addition to the potential for low-cost substrates, Ga_2O_3's ultra-wide bandgap of ~4.8 eV, theoretical breakdown field of ~8 MV cm^{-1}, and ability to operate at high temperatures greater than 250 °C hold promise for power devices that are more efficient and have higher breakdown voltages than current commercially available devices [7, 13–17]. Figure 9.1 compares Ga_2O_3's material properties with other candidate semiconductors for power electronics. As shown in the radar plot, Ga_2O_3's most notable weakness is its relatively low thermal conductivity which will need to be addressed in an effective manner.

Despite Ga_2O_3's recent entry as a power-electronic candidate, there have already been impressive demonstrations of vertical and lateral geometry Ga_2O_3 rectifiers, with total forward currents greater than 30 A, breakdown voltages above 2 kV, and switching of currents of up to 0.7 A [15, 18–23]. These devices reported within the literature have primarily been composed of pure Ga_2O_3. In addition to these devices, alloying Ga_2O_3 with Al or In during the growth process can be used to tune its bandgap to be smaller or larger, which is highlighted in figure 9.2.

9.1.1 $(Al_xGa_{1-x})_2O_3$

The β-$(Al_xGa_{1-x})_2O_3/Ga_2O_3$ heterostructure has shown impressive and improved performance for heterostructure field-effect transistors (HFETs) [3, 13, 24–30]. The upper limits for the solubility of Al in β-Ga_2O_3 has been commonly reported to be in the range 67%–78% [3, 29, 31–33] when grown using a number of methods, including molecular beam epitaxy and pulsed laser deposition [25, 28, 33–39]. Additionally, several groups have reported methods to calculate the strain in pseudomorphic $(Al_xGa_{1-x})_2O_3$ heterostructures on bulk β-Ga_2O_3 substrates [37, 38]. Recently, there have been reports of high-quality $(Al_xGa_{1-x})_2O_3/Ga_2O_3$ HFETs yielding enhanced electron mobility due to the formation of a two-dimensional electron gas (2DEG)

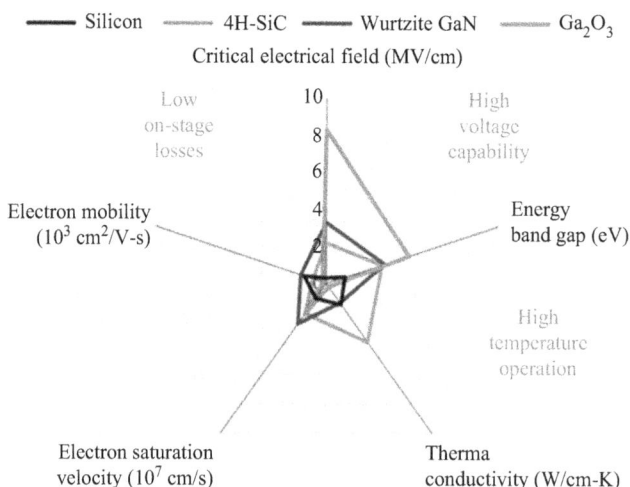

Figure 9.1. Ga_2O_3's material properties compared to those of other commercially available semiconductors.

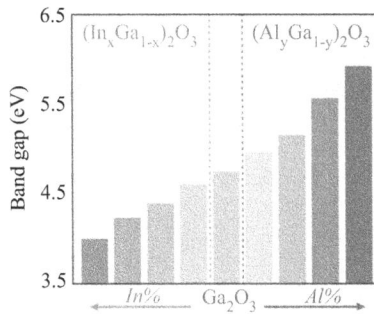

Figure 9.2. Bandgap tunability of Ga_2O_3 by incorporating In or Al to form ternary alloys.

[25–30, 40]. For example, Zhang *et al* [41] utilized Hall measurements to verify the formation of a 2DEG in a modulation-doped β-$(Al_xGa_{1-x})_2O_3/Ga_2O_3$ structure with a channel mobility of 143 cm^2 V^{-1} s^{-1} at 300 K and 1520 cm^2 V^{-1} s^{-1} at 50 K. If such devices continue to be optimized, they may prove useful for RF power device applications.

9.1.2 $(In_xGa_{1-x})_2O_3$

In addition to $(Al_xGa_{1-x})_2O_3$, the incorporation of In_2O_3 into Ga_2O_3 to form $(In_xGa_{1-x})_2O_3$ allows for a wide range of realizable bandgaps for this material system. Due to this tunability, $(In_xGa_{1-x})_2O_3$ is emerging as a candidate in photodetectors, transparent electrodes on optoelectronic devices, gas sensors, and heterostructure transistors [3, 42–46]. By incorporating $(In_xGa_{1-x})_2O_3$ along with the previously mentioned $(Al_xGa_{1-x})_2O_3$ it is expected that a high sheet carrier density 2DEG can be induced by the polarization differences at the ε-$(In,Ga)_2O_3/\varepsilon$-$(Al,Ga)_2O_3$ heterointerface [47]. In order for such $(In_xGa_{1-x})_2O_3$ based structures to be realized, reports have focused on the behavior of charge accumulation layers, miscibility gaps, and native defect behavior as monoclinic Ga_2O_3 is alloyed with cubic In_2O_3 [3, 42–45, 47–62]. Thus far, several groups have investigated the crystal structure and phase stability of $(In_xGa_{1-x})_2O_3$, which can show four-, five-, or six-fold coordinated cation sites [3, 42]. To grow $(In_xGa_{1-x})_2O_3$, numerous synthesis methods have been reported, including pulsed laser deposition (PLD), molecular beam epitaxy, sol–gel processing, metal–organic chemical vapor deposition, and sputtering, each yielding different phase stabilities [3, 39, 42–46, 49, 53, 62–70].

The addition to these alloyed derivatives to the Ga_2O_3 family allows for the possibility of heterostructure based device architectures of which an example is shown in figure 9.3. The choice of dielectric used in these heterostructure based devices (MOSFETs, FETs, MISFETS, etc) and in field plates for rectifiers is crucial for optimizing device performance [71]. Additionally, the effect of the dielectric deposition technique and the thermal budget of the dielectric/semiconductor heterostructure must be understood for Ga_2O_3 based technologies to enter the commercial sector [72–74].

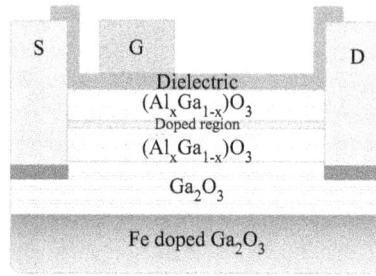

Figure 9.3. A typical $(Al_xGa_{1-x})_2O_3/Ga_2O_3$ HFET where gate insulator selection is crucial.

In this chapter, we summarize the band alignment of various dielectrics onto Ga_2O_3, $(Al_xGa_{1-x})_2O_3$, and $(In_xGa_{1-x})_2O_3$. We will also address the effect of the deposition method on these band offsets and the effect of post-deposition annealing.

9.2 Band alignment principles

When a dielectric is selected and then deposited onto the semiconductor of interest, their bandgaps will align in one of three ways. The three commonly accepted types of band alignment are type I (straddled gap), type II (staggered gap), and type III (broken gap) [71, 75]. The band alignment type that the two materials form will determine the ability or restriction for carrier transport across the interface as well as the electrostatic potential within the heterostructure. For example, in device designs that utilize a gate dielectric, one of the most crucial parameters for the dielectric/ semiconductor system of interest is that the dielectric acts as a barrier to both holes and electrons to prevent leakage current [71, 76–78]. Additional requirements for gate dielectrics are that the deposited dielectric must not react with the semiconductor surface, and it must be thermodynamically stable and provide a high-quality interface with a low trap density to ensure good carrier mobility [79–84].

The electronic structure of an oxide–semiconductor heterostructure and how each material aligns with one another can be represented visually using an energy band diagram. The valence band (E_v) is the highest occupied electron state and is commonly referred to as the valence band maximum (VBM). The conduction band (E_c) is the onset of the upper band of allowed electron states. If an electron within the conduction band of the semiconductor were to move toward the dielectric, the probability that it will be blocked increases as the barrier height increases. This barrier height is determined by the difference between the semiconductor and dielectric conduction band maximum, known as the conduction band offset (CBO), and is the energy barrier to electron transport across the interface. Likewise, the difference between the semiconductor and the dielectric valence band maximum is known as the valence band offset (VBO) and is the energy barrier to hole transport across the interface. The knowledge and determination of these two offsets are some of the primary goals of band alignment studies and are crucial in optimizing Ga_2O_3 based devices.

9.3 Measuring band offset

In order to determine the band alignment between two materials in a heterostructure, various methods including internal photoemission spectroscopy [85, 86], external photoemission spectroscopy [87, 88], and x-ray photoelectron spectroscopy (XPS) have been used. Out of these methods, the XPS based technique initially developed by Kraut *et al* [89] has proven to be an accurate and effective method for many material systems. In addition to band alignment information, the interfacial properties between the two heterostructure materials can be studied during the XPS process. To measure the band alignment using this method, three samples are needed. First, precise core level and valence band edge data must be measured from bulk samples of each of the two materials in the heterostructure. Second, the same core level locations measured in the bulk samples must be measured within a heterostructure of these two materials. The shift of the core level binding energy locations within the heterostructure as compared to the initial bulk binding energies can be used to determine the valence band offset via the following equation:

$$\Delta E_V = \left(E_{core}^1 - E_{VBM}^1 \right) - \left(E_{core}^2 - E_{VBM}^2 \right) - \left(E_{core}^1 - E_{core}^2 \right). \tag{9.1}$$

One crucial requirement for the heterostructure sample is that the top dielectric layer be sufficiently thin so that the XPS can detect not only core levels from the top material but also core levels from the bottom material of the heterostructure. In our experience, 1–3 nm of dielectric seems to be effective at accomplishing this task [73, 83]. Once the valence band offset has been determined, the conduction band offset can be calculated using the measured ΔE_V along with the measured bandgap of each material as shown in the following equation (methods to determine the bandgap of each material will be discussed in the following section):

$$\Delta E_C = E_g^{\text{Dielectric}} - E_g^{\text{Ga}_2\text{O}_3} - \Delta E_V. \tag{9.2}$$

Figure 9.4 shows an example of how these core level measurements in the bulk and heterostructure samples can be used to determine the band offset.

9.4 Bandgap determination

9.4.1 Onset of inelastic losses using XPS

In order to determine how the conduction and valence bands of a dielectric and semiconductor structure align, a precluding task is to determine the respective bandgap of each material. The technique utilized to measure a material's bandgap depends on the type of material to be studied and tool availability/practicality. In the case of Ga_2O_3 band alignment studies, most reports have measured the band offsets using XPS. Since XPS is such a common technique for these types of studies, it makes sense that the bandgaps of Ga_2O_3, $(Al_xGa_{1-x})_2O_3$, and $(In_xGa_{1-x})_2O_3$ have also been measured using XPS, often within the same session that the band alignment is measured.

Determining a material's bandgap using XPS can be done by examining the onset of inelastic losses in a core level atomic spectrum [90, 91]. These inelastic collisions

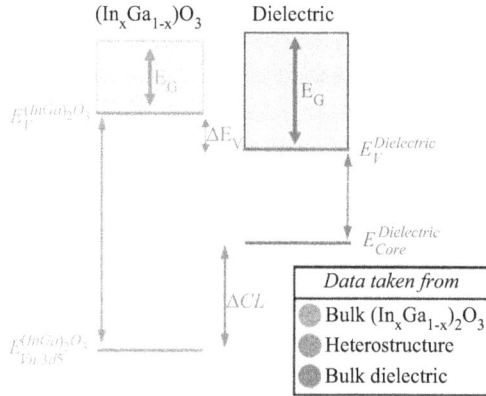

Figure 9.4. Example of how XPS measurements from a bulk semiconductor sample, bulk dielectric sample, and a heterostructure sample of the two can be used to determine the band alignment.

occur during the photoexcitation and photoemission of electrons from the sample. Within the three-step photoemission model, an electron that has been photoemitted during the XPS process undergoes three transitions after photoexcitation before reaching the tool's electron analyzer [92]. In the first transition, the electron is excited by an x-ray. Next, the electron moves to the surface of the material and, in the final step, escapes into the vacuum. During this process, single particle excitations can occur from band-to-band transitions (the initial ejected electron excites another electron from the valence band to the conduction band) and result in an inelastic loss process with a single particle energy loss that can be described using the following equation:

$$\Delta E_{sp} = E_f - E_i, \tag{9.3}$$

where E_i is the initial state of the excited electron and E_f is the final state [93]. For semiconducting materials such as Ga_2O_3, this energy loss feature is equivalent to the energy that an electron requires to be excited from the valence band into the conduction band. Therefore, the fundamental lower limit of the inelastic loss feature is equal to the bandgap energy.

Figure 9.5 shows an example of how this technique is used to determine the bandgaps of (a) $(Al_xGa_{1-x})_2O_3$ and (b) $(In_xGa_{1-x})_2O_3$ at various compositions of interest [72, 94]. In figure 9.5, the separation between the core level peak energy and the onset of inelastic (plasmon) losses in each O1s photoemission spectra was measured. In principle, any core level peak will exhibit the same inelastic losses and the chosen peak must be sufficiently large in order to obtain an acceptable signal-to-noise ratio for the loss feature [95].

9.4.2 Reflection electron energy loss spectroscopy

Another method that can be used for bandgap determination is reflection electron energy loss spectroscopy (REELS). The basic principles of this technique are similar to the previously mentioned XPS based technique. As electrons are reflected off a

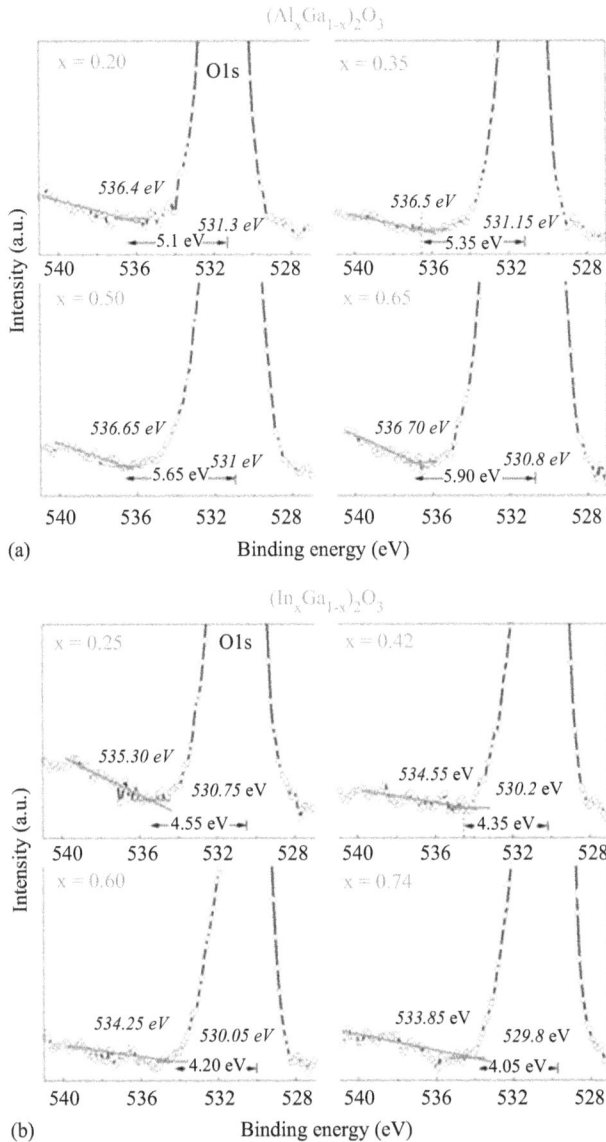

Figure 9.5. Bandgap of (a) $(Al_xGa_{1-x})_2O_3$ and (b) $(In_xGa_{1-x})_2O_3$ determined using the onset of the plasmon loss feature in a O1s photoemission spectrum. The intensities are in arbitrary units (a.u.).

material's surface, some of the electron's energy is lost at certain energy bands related to the excitation of the material's electrons. If enough electrons lose energy from exciting other substrate electrons from the valence band to the conduction band, the difference between a linear fit of the leading plasmon peak and the zero energy with the background can be used to determine the material's bandgap. Figure 9.6 shows the bandgap of two common dielectrics, SiO_2 and $HfSiO_2$, measured using REELS. With regard to the bandgap of Ga_2O_3, RHEELS data

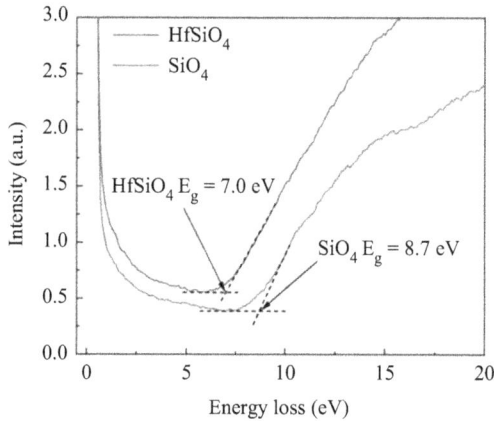

Figure 9.6. Bandgap of SiO_2 and $HfSiO_4$ determined by reflection electron energy loss spectra. The intensities are in arbitrary units (a.u.).

and optical transmittance have shown a bandgap value of 4.6 eV [2, 82, 96]. Other values up to 4.8 eV have been reported using both direct and indirect gap measurements [97, 98].

9.4.3 Ultraviolet–visible spectroscopy

The final method we will discuss to determine the bandgap of a material is ultraviolet–visible spectroscopy (UV–vis). UV–vis is based on the absorption of light (usually over a wavelength range of 200–500 nm) within the material of interest. The wavelength data can be plotted against the UV–vis measured absorption and used in conjunction with a Tauc plot with the relation shown in the following equation:

$$(\alpha h\upsilon)^{1/n} = B(h\upsilon - E_g),\tag{9.4}$$

where E_g is the bandgap, $h\upsilon$ is the known photon energy, α is the absorption coefficient, B is a constant, and n is selected based on whether the material is direct or indirect bandgap and if it is an allowed or forbidden transition. Another way to utilize this technique is to directly relate the wavelength to the bandgap using $E = hc/\lambda$ whenever absorption reaches the background level.

9.5 Choice of dielectric

Figure 9.7 shows the various dielectrics that could be used within Ga_2O_3 based devices. Ideally, a high dielectric constant material that also has a wide bandgap to provide sufficient offsets for Ga_2O_3 is desirable. However, as seen in figure 9.7, the bandgap of a material and its dielectric constant are usually inversely proportional. Due to Ga_2O_3's wide bandgap of 4.6 eV that can be tuned to over 6.0 eV with the alloying of aluminum during the growth process, the choice of dielectric is very limited. Two of the most commonly utilized dielectrics on Ga_2O_3 to date have been

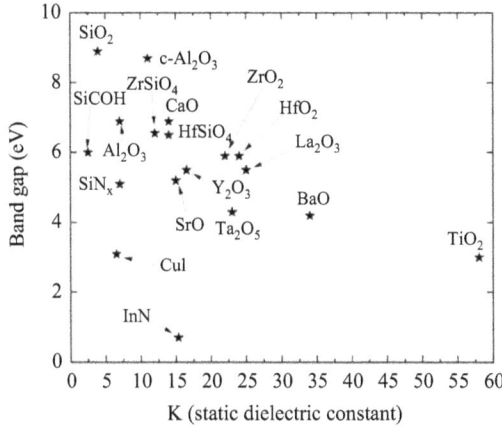

Figure 9.7. Dielectric candidates as a function of their bandgap and dielectric constant.

SiO$_2$ and Al$_2$O$_3$ [8, 30, 83, 96, 99–108]. SiO$_2$ has a large bandgap of 8.7 eV that can provide large band offsets, blocking hole and electron transfer. Although Al$_2$O$_3$ (6.9 eV) has a smaller bandgap than SiO$_2$, its dielectric constant is higher, making it advantageous in various applications. In the case of Ga$_2$O$_3$ based MOSFETs or similar device structures, higher dielectric constant (κ) materials can lower the effect of interface defects, lower the device's power consumption, and increase the capacitance density of the gate oxide. The relation of dielectric constant to capacitance (C) is highlighted in equation (9.5) and the relation of device current is correlated to capacitance (C) in equation (9.6):

$$C \approx \frac{\kappa \varepsilon_0}{t} \tag{9.5}$$

$$I \approx \frac{W}{2L} \mu C \left(V_g - V_t \right)^2, \tag{9.6}$$

where t is the thickness of the oxide, κ is the relative permittivity of the insulator, ε_0 is the permittivity of free space, W is the transistor width, μ is the carrier mobility, V_g is the gate voltage, L is the channel length, and V_t is the threshold voltage.

Several of the dielectric films with a higher dielectric constant (κ) are deposited in polycrystalline form. Polycrystalline films allow for impurity diffusion due to grain boundaries and a larger degree of leakage current which is undesirable for device performance. There are fewer choices of useful high-κ films with band offsets larger than 1 eV for both the conduction and valence bands. In the following sections, we will discuss the band alignment of various dielectrics onto Ga$_2$O$_3$, (Al$_x$Ga$_{1-x}$), and (In$_x$Ga$_{1-x}$)$_2$O$_3$ and explore the effects of the deposition method and annealing on some of these structures.

9.6 Reported band offsets

In this section, we will summarize the work involving band alignment studies on Ga_2O_3, $(Al_xGa_{1-x})_2O_3$, and $(In_xGa_{1-x})_2O_3$. Within the literature reports we will discuss, there is typically some variation for band offsets even for identical material systems. This is due to the fact that there are other factors at play, including interfacial disorder, strain and stress of the materials, metallic contamination, hydrogen/carbon contamination, and surface termination [72–74]. Despite knowing the variables that can contribute to deviations between reports, the sparse literature discussing the band alignment and interfaces of Ga_2O_3 and its alloyed derivatives makes exact determination of the differences between the results of different papers difficult at this point in time. Therefore, it is imperative for the reader to be cautious when reading reports on the band alignment of these material systems until further research can reinforce the initial findings and Ga_2O_3 crystal growth and alloying are optimized.

9.6.1 Gate dielectrics on Ga_2O_3, $(Al_xGa_{1-x})_2O_3$, and $(In_xGa_{1-x})_2O_3$

For gate dielectrics on Ga_2O_3 based structures, a material with a high-dielectric constant and a bandgap large enough to offset Ga_2O_3's valance and conduction band by at least 1 eV is desirable. The choices of dielectric become even more restricted when searching for gate dielectric candidates on $(Al_xGa_{1-x})_2O_3$, due to its larger bandgap that can exceed 6.0 eV depending on the aluminum concentration. We will now review Al_2O_3, $LaAl_2O_3$, SiO_2, HfO_2, and $HfSiO_4$ to determine which dielectrics could provide the gate with sufficient energy barriers.

9.6.2 Al_2O_3

Starting with one of the most used gate dielectrics for Ga_2O_3, several groups have studied the band offset of Al_2O_3 on Ga_2O_3. The variance between these reports is approximately 0.4–0.6 eV in conduction band offsets for well-controlled deposition methods. Kamimura *et al* [99] deposited Al_2O_3 onto Ga_2O_3 using plasma enhanced atomic layer deposition (PEALD) and measured a conduction and valence band offset of 1.5 and 0.7 eV for the heterostructure, respectively. Hattori *et al* [109] found that pulsed laser deposition (PLD) deposited γ-Al_2O_3 yielded conduction band and valence band offsets of 1.9 and 0.5 eV, respectively. One factor that can lead to variation in band alignment is the choice of deposition method, a comparison of ALD deposited dielectrics and sputtered dielectrics allows us to identify and observe some of these potential mechanisms. Carey *et al* studied the band alignment differences between sputtered and ALD deposited Al_2O_3 on Ga_2O_3, the bandgaps of all materials were measured using REELS and the offsets for the heterostructure were measured using XPS [96]. It was found that the ALD deposited Al_2O_3 had a type I alignment with a valence band offset of 0.07 ± 0.20 eV and the conduction band offset is then 2.23 ± 0.6 eV, whereas the sputter deposited film has a definite type II alignment with a valence band offset of -0.86 eV and a conduction band

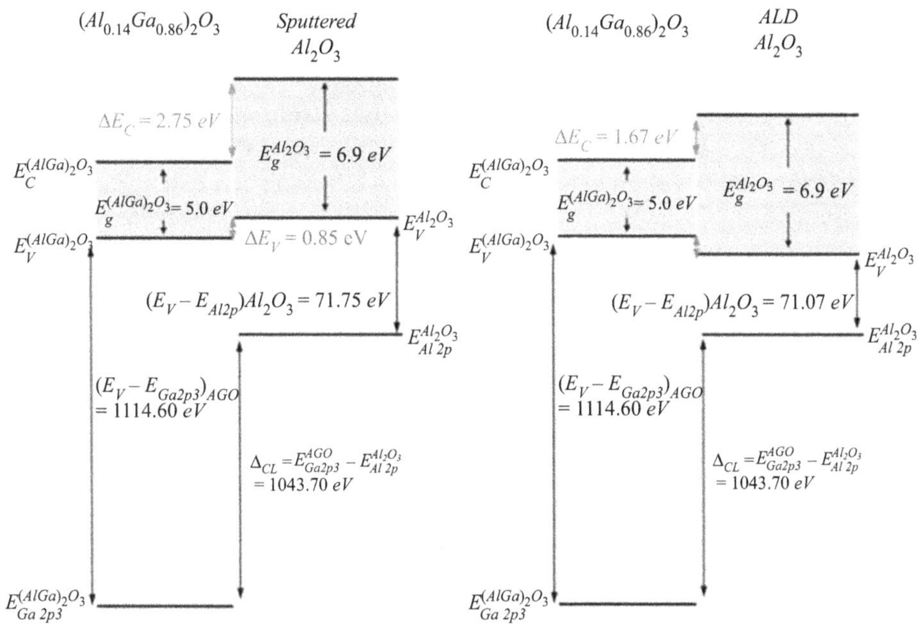

Figure 9.8. Band diagrams for the $Al_2O_3/(Al_{0.14}Ga_{0.86})_2O_3$ heterostructure in which the Al_2O_3 was deposited by ALD or sputtering and the $(Al_{0.14}Ga_{0.86})_2O_3$ was grown using MBE.

offset of 3.16 eV. Further work is required to produce consistent band alignment results for this material system.

For Al_2O_3 on $(Al_xGa_{1-x})_2O_3$, similar band alignment properties are reported. One study compared the band alignment of ALD deposited Al_2O_3 and sputtered Al_2O_3 on $(Al_{0.14}Ga_{0.86})_2O_3$ grown by molecular beam epitaxy. Additional details on the growth and deposition parameters can be found in a previous report [102]. The bandgap of the β-$(Al_{0.14}Ga_{0.86})_2O_3$ was determined to be 5.0 ± 0.3 eV from the onset of the plasmon loss feature in the O1s photoemission spectrum. After measuring the VBMs and core levels for the bulk $(Al_{0.14}Ga_{0.86})_2O_3$, bulk ALD or sputtered Al_2O_3, and $Al_2O_3/(Al_{0.14}Ga_{0.86})_2O_3$ heterostructures, the band offsets were determined and are shown in figure 9.8. For the sputtered films, the alignment is type II, a staggered gap with a valence band offset of −0.85 eV and conduction band offset of 2.75 eV. For comparison, the ALD deposited Al_2O_3 has type I alignment with a valence band offset of 0.23 ± 0.04 eV and a conduction band offset of 1.67 ± 0.30 eV. The conduction band offset would provide good electron confinement, whereas the valence band offset is too small for effective hole confinement. The small valence band offset is less of an issue because there are currently no effective p-type dopants to be used within Ga_2O_3. The difference in band alignment in this case and the previous case on pure Ga_2O_3 is most likely due to interfacial defects from sputter induced damage or cation effects since the high-resolution survey scans of all materials showed no traces of metallic contamination. These results prove that a heterojunction's band alignment is strongly dependent on processing conditions which have not yet been optimized for Ga_2O_3 or $(Al_xGa_{1-x})_2O_3$.

Another report investigated the band alignment of ALD Al_2O_3 on $(Al_xGa_{1-x})_2O_3$ grown by PLD. In this study, several aluminum concentrations were studied ranging from $x = 0.2$–0.65. Detailed growth parameters and deposition parameters for this heterostructure can be found in a previous report [110]. The bandgap of each studied composition of $(Al_xGa_{1-x})_2O_3$ was obtained from XPS energy loss measurements of the O1S peak and is shown in figure 9.5. REELS was used to determine the bandgap of the ALD deposited Al_2O_3, yielding a value of 6.9 eV. XPS survey scans were taken to ensure all samples were free from impurities and high-resolution XPS scans were used to measure the core level deltas in the Al_2O_3/β-$(Al_xGa_{1-x})_2O_3$ hetero-structures. These measured deltas along with the initial core levels and valence band maxima are shown in table 9.1.

The valence band offsets for Al_2O_3 were 0.23 ± 0.05 eV for $(Al_{0.2}Ga_{0.8})_2O_3$, 0.28 ± 0.05 eV for $(Al_{0.35}Ga_{0.65})_2O_3$, 0.33 ± 0.06 eV for $(Al_{0.5}Ga_{0.5})_2O$, and 0.33 ± 0.06 eV for $(Al_{0.65}Ga_{0.35})_2O_3$. Using the experimentally measured bandgap of Al_2O_3, the conduction band offsets for this system are 1.57, 1.27, 0.92, and 0.67 eV for $x = 0.2$, 0.35, 0.5, and 0.65, respectively. These offsets yield a type I alignment, shown in figure 9.9, with marginal electron confinement at high Al concentrations in $Al_2O_3/(Al_xGa_{1-x})_2O_3$.

The final heterostructure system we will discuss for Al_2O_3 is the band alignment on $(In_xGa_{1-x})_2O_3$, also referred to here as IGO. The IGO was grown using PLD in the same manner as the previously mentioned graded $(Al_xGa_{1-x})_2O_3$. Additional growth and fabrication details can be found in a previous report [111]. The indium concentration varied between 0.16 and 0.86 and showed a slight S-shaped dependence along the gradient direction, in agreement with calculations [39]. Along lines perpendicular to the gradient direction, the In concentration was constant. Energy-dispersive x-ray spectroscopy (EDX) was used for the spatially resolved chemical analysis. XPS survey scans of all samples showed that only lattice constituents were present. We measured the bandgaps of the four $(In_xGa_{1-x})_2O_3$ compositions, shown previously in figure 9.5. The respective bandgaps were 4.55 eV for $(In_{0.25}Ga_{0.75})_2O_3$,

Table 9.1. Summary of the measured reference and heterostructure peaks for Al_2O_3 on $(Al_xGa_{1-x})_2O_3$ (eV).

Aluminum concentration	Reference $(Al_xGa_{1-x})_2O_3$ Core level peak (Ga 2p$_{3/2}$)	VBM	Core–VBM	Reference Al_2O_3 Core level peak (Al 2p)	VBM	Core–VBM	Thin Al_2O_3 on $(Al_xGa_{1-x})_2O_3$ Δ core level (Ga 2p$_{3/2}$–Al 2p)	Valence band offset
$(Al_{0.20}Ga_{0.80})_2O_3$	1118.50	3.6	1114.90	74.32	3.25	71.07	1043.60	0.23
$(Al_{0.35}Ga_{0.65})_2O_3$	1118.35	3.3	1115.05	—	—	—	1043.70	0.28
$(Al_{0.50}Ga_{0.50})_2O_3$	1118.10	2.9	1115.20	—	—	—	1043.80	0.33
$(Al_{0.65}Ga_{0.35})_2O_3$	1118.00	2.6	1115.40	—	—	—	1044.00	0.33

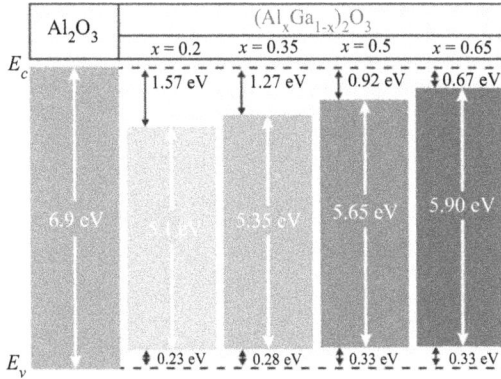

Figure 9.9. Band alignment diagram for the $Al_2O_3/(Al_xGa_{1-x})_2O_3$ heterostructure in which the Al_2O_3 was deposited by ALD and the AGO was grown using pulsed laser deposition.

4.35 eV for $(In_{0.42}Ga_{0.58})_2O_3$, 4.20 eV for $(In_{0.60}Ga_{0.40})_2O_3$, and 4.05 eV for $(In_{0.74}Ga_{0.26})_2O_3$. These are in general agreement with the reported values for PLD films in a similar composition range (3, 42). Table 9.2 shows the core level deltas and valence band maxima (VBMs) for the Al_2O_3 and the $(In_xGa_{1-x})_2O_3$ from linear fitting of the leading edge of the valence band. These values were used to determine the band alignment of the samples directly after deposition.

After measuring the band alignment of the as-deposited Al_2O_3/IGO samples. Sections from the samples were annealed at 600 °C under N_2 ambient for 30 s in a rapid thermal annealing system, and then the band alignment was remeasured. This temperature was chosen since it is at the high end of ohmic contact alloying conditions, as well as tuning of the resistance in implant isolation regions for inter-device isolation. Therefore, it represents a realistic test of interface stability during the thermal budget encountered during device processing.

Figure 9.10 shows the measured shift in valence band offsets for the annealed $Al_2O_3/(In_xGa_{1-x})_2O_3$ heterostructures compared to their as-deposited values. Across the full composition range, the annealing process resulted in a large decrease in the valence band offset. At higher indium concentrations, the valence band shift was shown to be larger. The cause of band alignment is probably due to interfacial chemistry changes. TEM images before and after annealing are shown in figure 9.11. After annealing, no significant crystal or morphology changes are shown in the In-rich end, yet it has the largest shift in band alignment. The Ga-rich samples reveal a more pronounced morphology change within the $(In_xGa_{1-x})_2O_3$ after annealing but show less change in valence band offset than the In-rich samples. A potential cause of this post-annealing crystallinity change is that In_2O_3 is thermodynamically less stable than Ga_2O_{3+}, with Gibbs energies of formation of -198.6 and -238.6 kcal mol^{-1} for In_2O_3 and Ga_2O_3, respectively [112, 113]. Additionally, the diatomic bond strength of In–O bonds is weaker compared to Ga–O bonds, causing them to break more easily. Knowing this, the band alignment shift in the In-rich sample could be larger due to the relative instability of indium compared to gallium within the IGO. In summary, despite crystallinity changes within the gallium-rich samples, the

Table 9.2. Summary of the measured reference and heterostructure peaks for Al_2O_3 on $(In_xGa_{1-x})_2O_3$ (eV).

Indium concentration	Reference $(In_xGa_{1-x})_2O_3$			Reference Al_2O_3			Thin Al_2O_3 on $(In_xGa_{1-x})_2O_3$	
	Core level peak (In $3d_{5/2}$)	VBM	Core–VBM	Core level peak (Al 2p)	VBM	Core–VBM	Δ core level (In $3d_{5/2}$–Al 2p)	Valence band offset
$(In_{0.25}Ga_{0.75})_2O_3$	444.65	2.50 ± 0.15	442.15	74.32	3.25	71.07	370.20	0.88
$(In_{0.42}Ga_{0.58})_2O_3$	444.40	2.25 ± 0.15	442.15	—	—	—	370.10	0.98
$(In_{0.60}Ga_{0.40})_2O_3$	444.35	2.25 ± 0.15	442.10	—	—	—	369.90	1.13
$(In_{0.74}Ga_{0.26})_2O_3$	444.20	2.10 ± 0.15	442.10	—	—	—	369.80	1.23

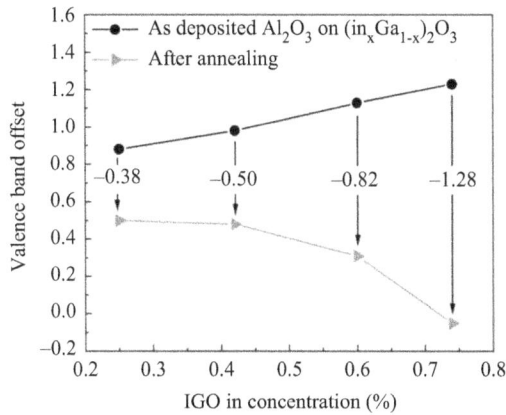

Figure 9.10. Valence band offsets for as-deposited and annealed $Al_2O_3/(In_xGa_{1-x})_2O_3$ heterostructures as a function of indium concentration.

Figure 9.11. Cross-section TEM images of (a) the substrate-epi region of Ga-rich sample prior to annealing, (b) the same region after annealing, (c) In-rich sample after annealing, and (d) a magnified image of this same region.

interfacial chemistry dominates the band alignments as determined by the surface-sensitive XPS.

Figure 9.12 shows the band alignment of the (a) as-deposited $Al_2O_3/(In_xGa_{1-x})_2O_3$ heterostructures compared to their (b) annealed values. For the as-deposited band alignment, the Al_2O_3 yields sufficient offsets for the conduction and valence bands, allowing for good carrier confinement for all compositions of $(In_xGa_{1-x})_2O_3$. After annealing, the valence band offsets were reduced for all compositions studied, with the high-indium concentration samples showing the largest degree of change. The band alignment is still type I for $x = 0.25$–0.6 after annealing and shifts to type II for the $x = 0.74$ sample.

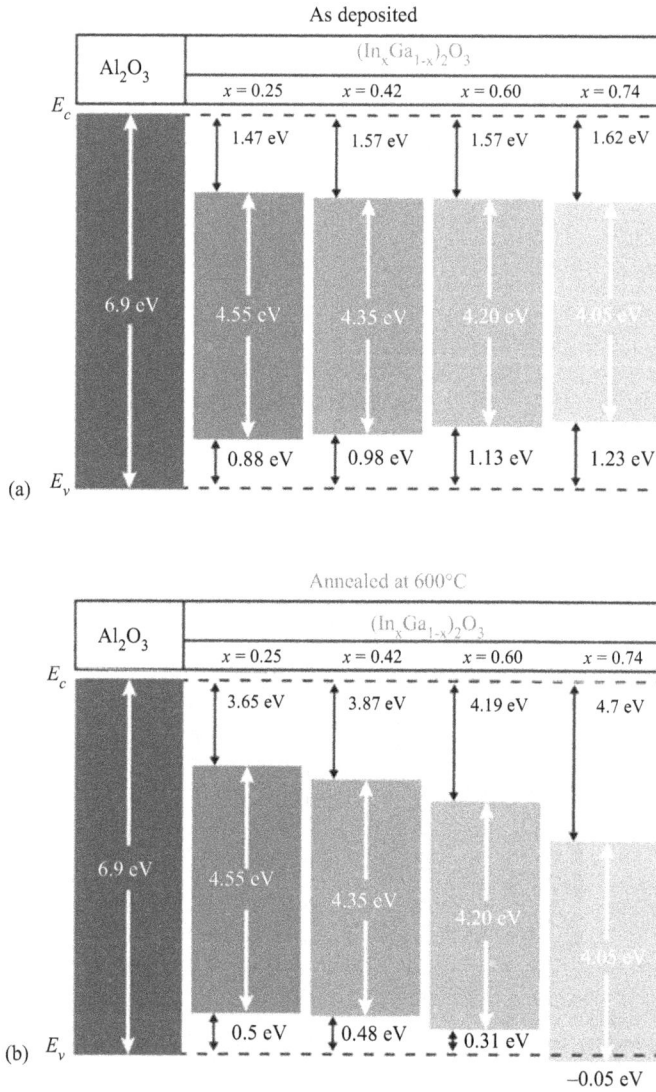

Figure 9.12. Band diagrams for the (a) as-deposited and (b) annealed $Al_2O_3/(In_xGa_{1-x})_2O_3$ heterostructures as a function of indium concentration.

9.6.3 SiO$_2$ and HfSiO$_4$

Like Al_2O_3, SiO_2 has drawn significant interest as a gate dielectric candidate for Ga_2O_3 based devices. Konishi *et al* examined plasma enhanced chemical vapor deposition (PECVD) deposited SiO_2 and found valence band and conduction band offsets of 1.0 and 3.1 eV [103]. Jia *et al* also studied the SiO_2/Ga_2O_3 heterostructure using XPS where the SiO_2 was deposited via PEALD at 300 °C. He found a valence band and conduction band offset of 0.43 and 3.63 eV, respectively [114, 115]. Just like Al_2O_3, we see a large dependence on the deposition method on the final offset.

Another benefit of SiO_2 is that it has been shown to be a thermally stable dielectric on Ga_2O_3 up to 1000 °C [57, 116]. In sharp contrast, the Al_2O_3–Ga_2O_3 phase system does not possess the same thermal stability as SiO_2 [3, 117].

In addition to studying the band offsets of SiO_2, another viable option is $HfSiO_4$ which has a significantly higher dielectric constant and would therefore allow for the utilization of thicker films. To study this, we deposited $HfSiO_4$ by alternating layers of HfO_2 and SiO_2 using ALD and an ICP source of 300 W. The same deposition conditions, substrate, and dielectric thickness were used during our SiO_2/Ga_2O_3 study. The bandgap of SiO_2 was found to be 8.7 ± 0.4 eV and $HfSiO_4$ had a bandgap of approximately 7 ± 0.3 eV. Therefore, the difference in the bandgap between Ga_2O_3 and $HfSiO_4$ is 2.4 eV and between Ga_2O_3 and SiO_2 it is 4.1 eV. Both dielectrics have met the requirement of a 2 eV difference in bandgap to potentially provide a suitable barrier to both hole and electron transport depending on their alignment.

For the ALD deposited SiO_2, the valence offset is calculated to be 1.23 ± 0.2 eV, and the corresponding conduction band offset is 2.87 ± 0.4 eV [103, 114]. This offset agrees with Konishi *et al*'s reported value for PECVD SiO_2 within experimental error [103]. However, the result is not consistent to Jia *et al*, who also used plasma enhanced ALD and reported a valence band offset of 0.43 eV. For $HfSiO_4$, a valence band offset of 0.02 ± 0.003 eV and a conduction band offset of 2.38 ± 0.50 eV were measured.

There have also been reports on the band alignment of SiO_2 on $(Al_xGa_{1-x})_2O_3$. Feng *et al* [105] investigated the band alignment for ALD SiO_2 on $(Al_xGa_{1-x})_2O_3$ with $x = 0$–0.49 and found the valence band offsets to be from 1.5 to 0.8 eV for this composition range. Feng also reported band offsets for atomic layer deposited HfO_2 and SiO_2 on $(Al_xGa_{1-x})_2O_3$ with $x = 0$–0.53. Although these studies are from the same research group, differences of up to 0.5 eV in the final band alignment were found for SiO_2 at similar Al contents in $(Al_xGa_{1-x})_2O_3$.

Another study investigated the band alignment and thermal stability of ALD SiO_2 on $(Al_xGa_{1-x})_2O_3$ with $x = 0.2$–0.65 grown by PLD. Detailed growth parameters and deposition parameters for this heterostructure can be found in a previous report [110]. After the heterostructures were fabricated and initial band alignment measurements were taken, the heterostructures were annealed to simulate potential device processing steps. Figures 9.13(a) and (b) show the band diagrams for the SiO_2/$(Al_xGa_{1-x})_2O_3$ heterostructures as deposited and after annealing at 600 °C for 5 min in N_2 ambient, respectively. For the as-deposited SiO_2, the entire studied AGO composition range can provide excellent confinement of electrons due to adequate offsets in both the valence and conduction bands. After annealing, the offsets in the conduction band increase, therefore still providing electron confinement. However, at aluminum concentrations above 50% hole confinement would be marginal after annealing due to smaller valence band offsets. Despite the smaller valence band offsets, the SiO_2/$(Al_xGa_{1-x})_2O_3$ band alignment is type I before and after annealing across the entire composition range studied.

The final heterostructure system we will discuss for SiO_2 is the band alignment onto $(In_xGa_{1-x})_2O_3$. After 'as-deposited' band alignment measurements were taken,

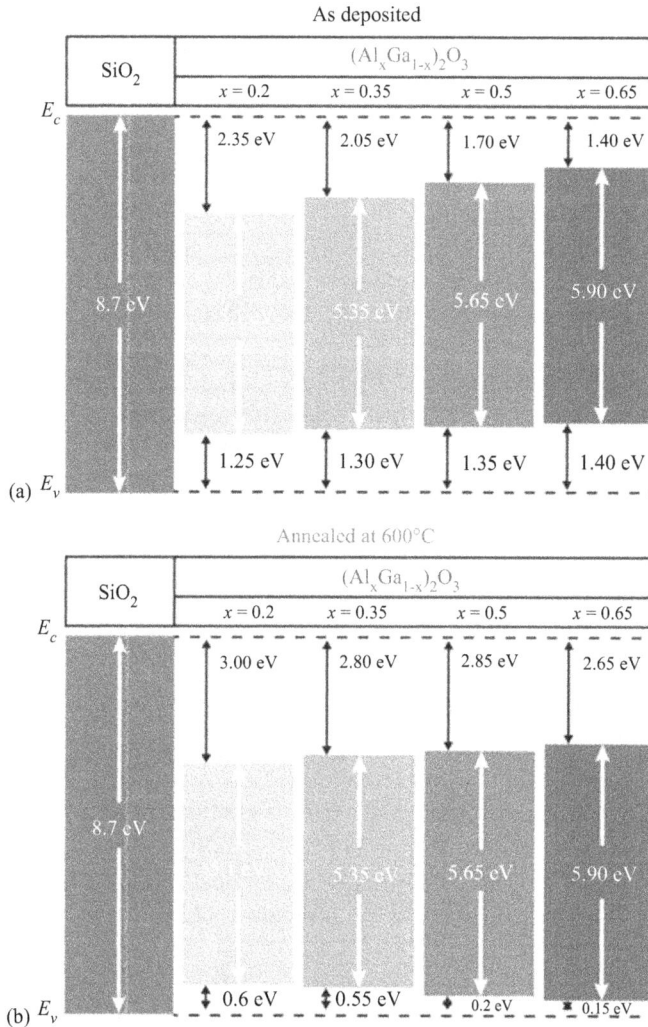

Figure 9.13. Band diagrams for the (a) as-deposited and (b) annealed $SiO_2/(Al_xGa_{1-x})_2O_3$ heterostructures as a function of aluminum concentration.

a rapid thermal annealing system was utilized to anneal the $SiO_2/(In_xGa_{1-x})_2O_3$ heterostructures at 450 °C and 600 °C under N_2 ambient for 5 min. The band alignments of the heterostructures were measured after the initial fabrication and after each annealing cycle. The annealing temperatures were selected to simulate potential IGO based device fabrication steps. The two separate temperature annealings were performed to examine the thermal stability of the heterostructure band alignment.

High-resolution XPS spectra for the as-deposited $(In_xGa_{1-x})_2O_3$ to SiO_2 core delta regions are shown in figures 9.14 (a) and (b). After these initial measurements were taken, the $SiO_2/(In_xGa_{1-x})_2O_3$ heterostructures along with bulk ALD

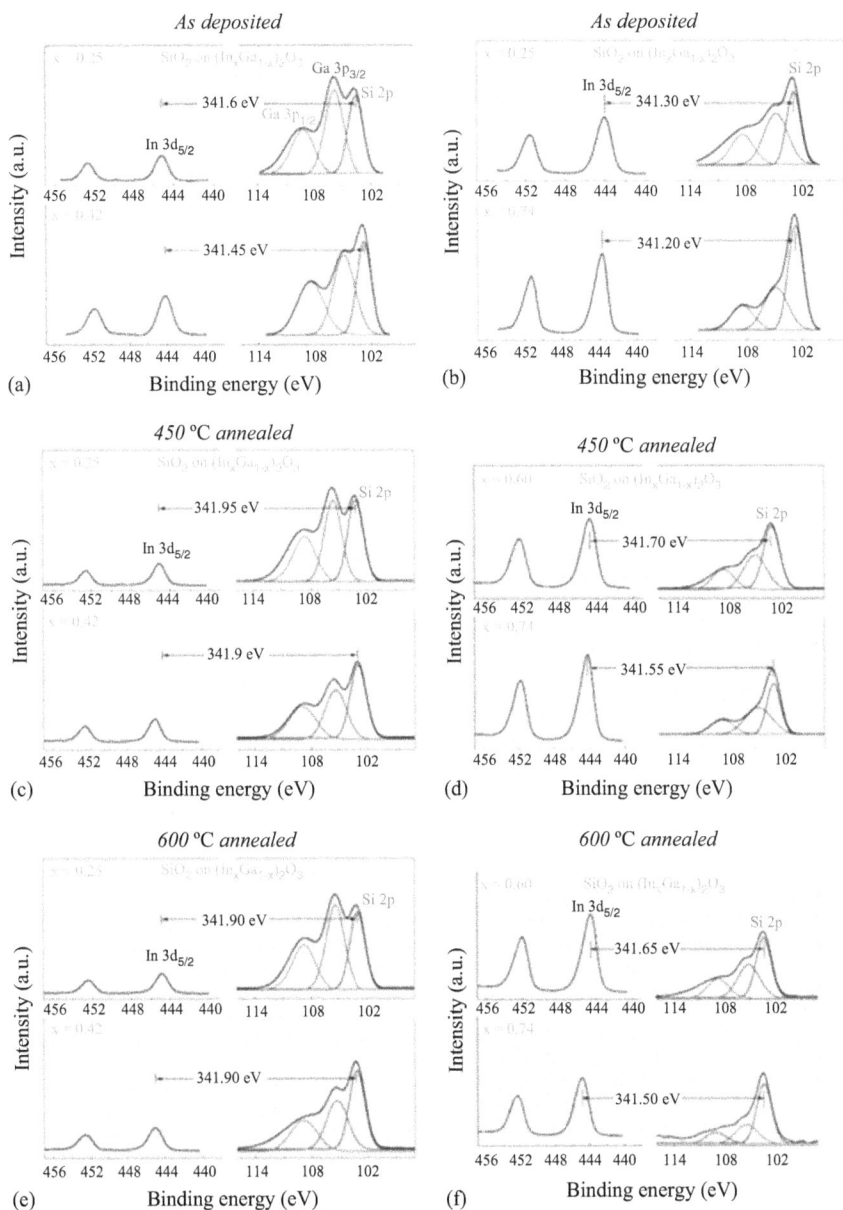

Figure 9.14. High-resolution XPS spectra for (a) and (b) $(In_xGa_{1-x})_2O_3$ to SiO_2 core delta regions as deposited, (c) and (d) after annealing at 450 °C for 5 min in N_2 ambient, and (e) and (f) after annealing at 600 °C for 5 min in N_2 ambient. The intensity is in arbitrary units (a.u.).

deposited SiO_2 and bulk $(In_xGa_{1-x})_2O_3$ were annealed at 450 °C for 5 min in N_2 ambient. High-resolution XPS measurements were taken again after 450 °C annealing and are shown in figures 9.14(c) and (d). A second annealing step at 600 °C was performed on these same samples and post-annealing XPS data are

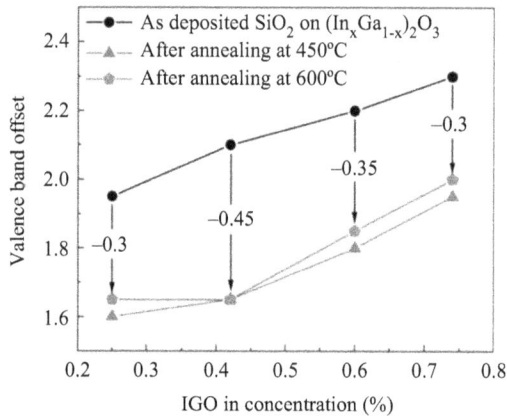

Figure 9.15. Valence band offsets for the as-deposited and annealed $SiO_2/(In_xGa_{1-x})_2O_3$ heterostructures as a function of indium concentration.

shown in figures 9.14(e) and (f). The $SiO_2/(In_xGa_{1-x})_2O_3$ valence band offset shift after both annealing steps is shown in figure 9.15. After the first annealing at 450 °C, the band alignment shift was between 0.3 and 0.45 eV for all of the studied compositions. After the final annealing at 600 °C, the band alignment for all of the compositions investigated remained essentially the same from the band values obtained after the first annealing at 450 °C. As a function of indium composition in $(In_xGa_{1-x})_2O_3$, the band alignment shifts in this study are relatively constant and are likely caused by interfacial chemistry changes between the $(In_xGa_{1-x})_2O_3$ and SiO_2. These chemical changes are attributed to compositional change at the interface and dipole formation leading to changes in valence band offset, as commonly reported in other systems [118–122]. The $SiO_2/(In_xGa_{1-x})_2O_3$ band diagrams before annealing and after 600 °C annealing for 5 min in N_2 ambient are shown in figures 9.16 (a) and (b). In the as-deposited $SiO_2/(In_xGa_{1-x})_2O_3$ heterostructures, the alignment is type I with the SiO_2 providing large offsets in both the valence band and the conduction band. After 600 °C annealing, the band alignment shifts marginally. Despite the band alignment shift, the confinement is still type I and greater than 1 eV for all studied compositions. These offsets allow for good carrier confinement at all compositions of $(In_xGa_{1-x})_2O_3$ and reinforce the acceptable thermal stability of SiO_2 as a potential dielectric for this material system.

In conclusion, the band alignment of SiO_2 on Ga_2O_3 based structures has been measured by multiple groups and is shown to sufficiently block both hole and electron transport. However, significant differences in the final band alignment can be influenced by the dielectric deposition conditions. A deeper understanding of the interface and quality of these thin films will provide insight into what parameters of the deposition could be causing these differences. $HfSiO_4$ provides decent electron confinement, but only marginal hole confinement. Like lanthanum aluminate, $HfSiO_4$ may see use as a passivation layer to protect the Ga_2O_3 from environmental factors that may negatively impact device performance or in multi-oxide gate stacks to increase the capacitance. A summary of the band alignment of gate dielectric

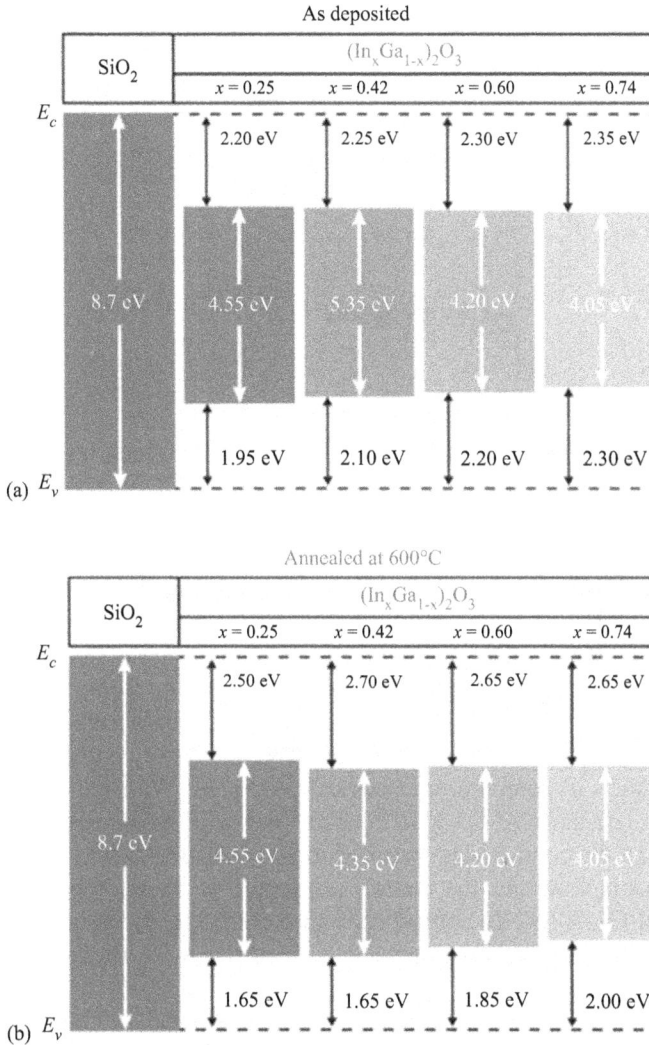

Figure 9.16. Band diagrams for the $SiO_2/(In_xGa_{1-x})_2O_3$ heterostructure (a) as deposited and (b) after annealing at 600 °C for 5 min in N_2 ambient.

candidates on pure Ga_2O_3 is shown in figure 9.17. Future work should be done to study additional gate dielectrics on Ga_2O_3 alloyed with aluminum or indium.

9.6.4 Indium tin oxide and aluminum zinc oxide

When working with a new material such as Ga_2O_3, one of the difficulties is forming low-resistance ohmic contacts. Narrow gap materials such as indium tin oxide (ITO) and aluminum zinc oxide (AZO) have the potential to alleviate that issue by reducing the barrier of electron or hole transport between the semiconductor and

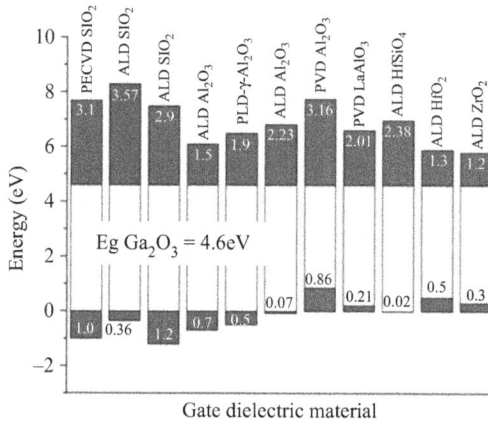

Figure 9.17. Summary of gate dielectric/Ga_2O_3 band alignment studies.

contact metal. Additionally, both materials can withstand the high processing temperatures which are found in numerous semiconductor processes [123–129].

For ITO on Ga_2O_3, the band offset was determined to be nested type I alignment with a valence and conduction band offset of -0.78 ± 0.21 eV and -0.32 ± 0.1 eV, respectively. The nested offset of this heterostructure suggests that ITO may provide an intermediate step in the conduction band between the fermi level of the Ga_2O_3 and a metal contact. To verify this theory, Carey *et al* fabricated transmission line measurement (TLM) structures consisting of 20/80 nm Ti/Au contacts and 10/20/80 nm ITO/Ti/Au contacts to compare the electrical properties. With the ITO contact, a low specific contact resistance of 6.3×10^{-5} Ω cm^2 and a transfer resistance of 0.6 Ω mm were found when the contact was annealed to 600 °C [130]. AZO interlayers have also proven effective in lowering the contact resistance of Ti/Au contacts on n-type, Si implanted β-Ga_2O_3. A specific contact resistance of 2.82×10^{-5} Ω cm^2 was reported for annealing at 400 °C, while Ti/Au contacts without the AZO interlayer did not lead to ohmic behavior [131, 132].

A recent study has explored the band alignment of ITO and AZO onto $(Al_xGa_{1-x})_2O_3$, in which ohmic contact formation will be even more difficult than on Ga_2O_3 due to its larger bandgap. In this study, 1.5 nm AZO or ITO was deposited by RF magnetron sputtering on the MBE-grown $(Al_{0.14}Ga_{0.86})_2O_3/Ga_2O_3$ structures and thicker layers (150 nm) were deposited onto quartz substrates to be used as bulk reference samples. Additional growth and fabrication details can be found in a previous report [80]. The measured bandgaps for the AZO and ITO were 3.2 ± 0.30 eV and 3.5 ± 0.30 eV, respectively, from REELS data. The bandgap of the β-$(Al_{0.14}Ga_{0.86})_2O_3$ was determined to be 5.0 ± 0.3 eV from XPS O1s based electron energy loss measurements. After XPS measurements were taken on all samples, the band alignment for both heterostructure systems was calculated.

Figure 9.18 illustrates the determined band alignment of the ITO/β-$(Al_{0.14}Ga_{0.86})_2O_3$ and AZO/β-$(Al_{0.14}Ga_{0.86})_2O_3$ heterostructures. Both are type I nested systems. For the ITO/β-$(Al_{0.14}Ga_{0.86})_2O_3$ heterostructure, the values are -1.18 ± 0.20 eV for the valence

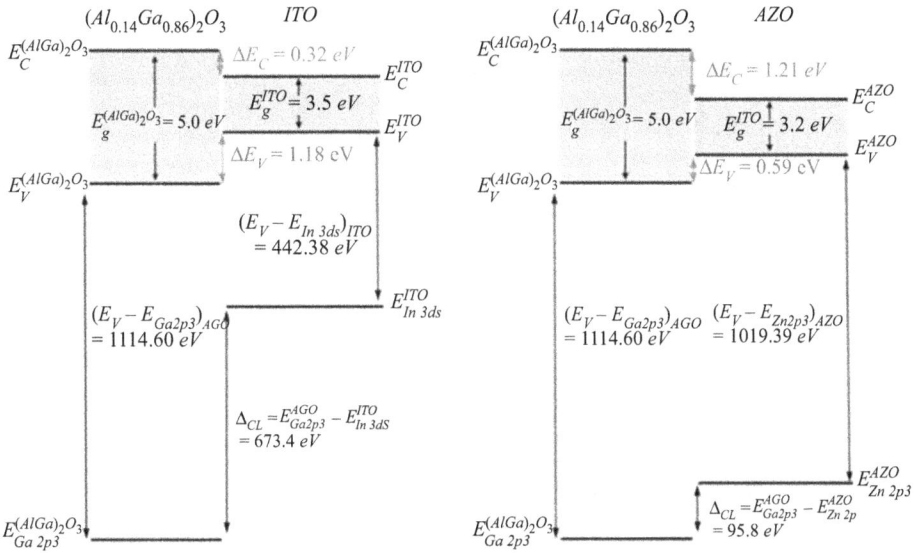

Figure 9.18. Band diagrams for ITO (left) and AZO (right) sputtered onto $(Al_{0.14}Ga_{0.86})_2O_3$.

band offset and -0.32 ± 0.05 eV for the conduction band offset. For the AZO/β-$(Al_{0.14}Ga_{0.86})_2O_3$ structure, the valence band offset is -0.59 ± 0.10 eV and the conduction band offset is -1.21 ± 0.25 eV. These were obtained using the differences in bandgaps and the directly measured valence band offset, i.e. $\Delta E_C = E_g^{AZO\ or\ ITO} - E_g^{AlGaO} - \Delta E_V$. The band offsets are negative for both oxides on β-$(Al_{0.14}Ga_{0.86})_2O_3$, ensuring good electron and hole transport and hence the ability to reduce contact resistance when used as an interlayer in metal stacks on this wide bandgap material.

9.6.5 InN

Another low bandgap material that can be used to make low-resistance ohmic contacts on Ga_2O_3 is InN. InN is a strong candidate due to its ability to be grown highly n$^+$ which could allow its function as a regrown barrier for an ohmic contact on Ga_2O_3 [133, 134]. We have explored InN deposition on β-Ga_2O_3 and measured the band offsets of this heterojunction. The band offsets of InN grown on β-Ga_2O_3 are used to report an experimental band diagram of this heterojunction. Information on the deposition and growth parameters can be found in a previous report [135].

XPS survey scans of a Ga_2O_3 control substrate and two InN samples are shown in figure 9.19. Signature peaks with high intensity from the expected elements (Ga, O, In, N, and C) were present. The O1s peak was severely suppressed in the InN scans but not completely absent, indicating that a small amount of oxygen is present on the InN surface even with storing the samples under a vacuum. For the 1.5 nm InN film on Ga_2O_3, a small amount of oxidation was expected due to the brief exposure of the InN films to atmosphere. Any potential oxygen content within the InN, particularly at the interface, could notably affect the measured band offsets.

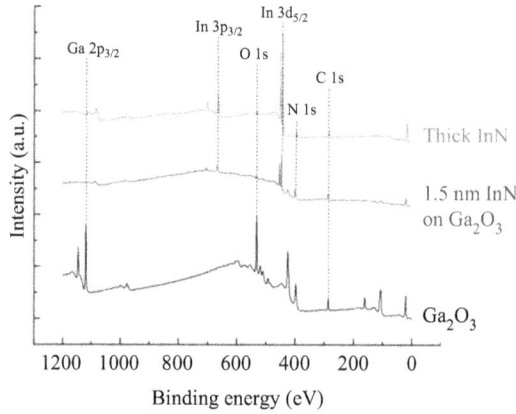

Figure 9.19. XPS survey scans of thick InN, 1.5 nm InN on Ga_2O_3, and a reference Ga_2O_3 sample, offset for readability. The intensity is in arbitrary units (a.u.).

Therefore, InN growth optimization and further study are needed if this material is to be used as a contact reducer for Ga_2O_3.

After measuring the VBMs, reference sample core levels, and core level deltas in the heterostructure, the valence band offset ΔE_V was calculated to be -0.55 ± 0.11 eV. This measured valence band offset for InN is similar to the theoretical valence band offset for In_2O_3/Ga_2O_3, as recently reported by Swallow *et al* [136]. Although interfacial oxygen was likely present in our films, XPS depth-resolved profiling did not indicate a significant amount of oxygen within the films. Additionally, since our growth substrate (Ga_2O_3) contains oxygen, interfacial oxygen was impossible to quantify. If the effect of interfacial oxygen is eliminated, we would reasonably expect that the VBO for InN to Ga_2O_3 in future reports to increase by up to about 1 eV. To calculate the conduction band offset in the InN–Ga_2O_3 heterostructure, prior to knowledge of the bandgap of the two films is required. The bandgap of Ga_2O_3 was measured to be 4.6 eV using REELS and the bandgap of InN was assumed to be 0.7 eV based on values reported in the literature by our group and others [133, 137]. Therefore, the conduction band offset ΔE_C was determined to be -3.35 ± 0.11 eV. Using the calculated values for ΔE_C and ΔE_V, a type I band diagram was determined. This value for the conduction band offset is the largest reported for semiconductor heterojunctions on β-Ga_2O_3 to date and is shown in figure 9.20.

9.6.6 CuI

The final material system we will discuss for Ga_2O_3 and $(Al_xGa_{1-x})_2O_3$ is their band alignment with CuI. Interest in this system is due to the lack of practical p-type conduction in Ga_2O_3 based structures [138]. This asymmetry in p-type versus n-type doping is common for many wide bandgap materials [139–141]. A lack of p-type doping significantly limits the type of devices that can be fabricated. In addition to device architectures, certain edge termination options are simply unrealizable without p-type doping.

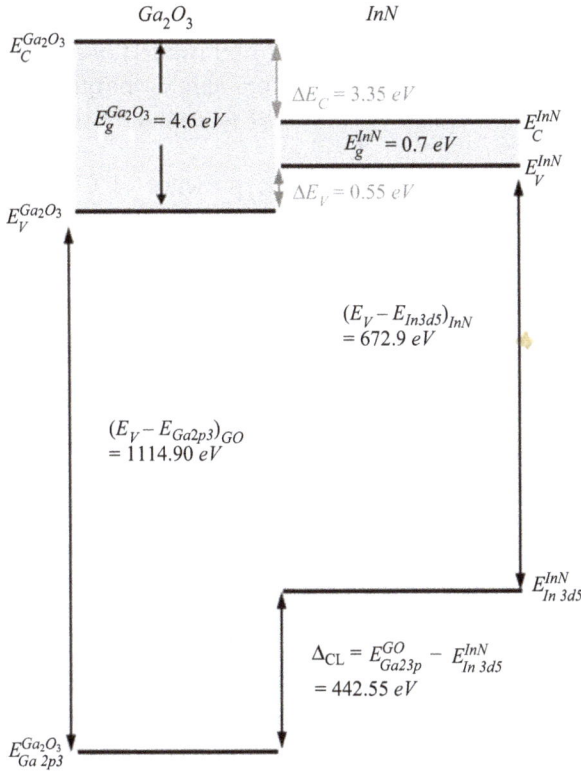

Figure 9.20. Band diagram for the InN–Ga$_2$O$_3$ heterostructure.

To counter this restriction, one option is to utilize heterojunctions of Ga$_2$O$_3$ with p-type semiconductors such as NiO, SiC, CuI, Cu$_2$O, or diamond [139–145]. Watahiki *et al* [145] showed 1.49 kV pn diodes with a specific on-resistance of 8.2 mΩ cm^2 by utilizing sputtered p-Cu$_2$O. Several other reports have shown heterostructures of Ga$_2$O$_3$ with p-Cu$_2$O [82, 145]. Another option is p-type CuI [139–145]. CuI is a wide-gap semiconductor that has been used to form hetero-junctions with n-type oxides and has also been used as a transparent electrode to improve hole collection in organic solar cells. Koehler *et al* studied p-CuI/n-Ga$_2$O$_3$ rectifying heterojunctions and showed a room temperature on–off ratio of >10^4 with an ideality factor of $n = 1.18$. To further understand CuI–Ga$_2$O$_3$ heterostructures, we report on CuI's band alignment with Ga$_2$O$_3$.

The CuI films were fabricated using two steps [139–145]. The first step was to deposit Cu using RF magnetron sputtering on Ga$_2$O$_3$ and (Al$_x$Ga$_{1-x}$)$_2$O$_3$ using a three-inch diameter target of pure copper. After copper deposition, the samples were placed in a petri dish cover and mounted with a Teflon holder. 99.999% iodine particles were set into the dish and heated to 120 °C for 5 min using a hotplate. After iodination, the samples were placed directly into the XPS system to obtain the chemical state of the thick CuI reference, reference Ga$_2$O$_3$, reference (Al$_x$Ga$_{1-x}$)$_2$O$_3$, and the heterostructures for both CuI/(Al$_x$Ga$_{1-x}$)$_2$O$_3$ and CuI/Ga$_2$O$_3$. Additional fabrication details can be found in [82]. The bandgap of CuI was assumed to be 3.1 eV based on the previously published literature [139–145]. Deviation in the actual CuI bandgap to the reported values could marginally affect the band alignment result. The measured values used to calculate the band alignment are shown in table 9.3.

XPS high-resolution scans confirmed the presence of CuI, with Cu in the +1 oxidation state (Cu 2p) and iodine in the −1 oxidation state (I 3d). The Cu 3p and iodine 4d revealed that CuI was stoichiometric throughout the top 20 nm of the film. This shows that during the iodization process, the iodine vapor reacts with more than just the Cu surface and creates a uniform CuI film.

Using the measured data, figure 9.21 shows the determined band alignments of the CuI/β-Ga_2O_3 and CuI/β-$(Al_{0.14}Ga_{0.86})_2O_3$ heterostructures. The CuI/β-Ga_2O_3 alignment is a type I nested system, while the CuI/β-$(Al_{0.14}Ga_{0.86})_2O_3$ alignment is a type II system. For the CuI/β-Ga_2O_3 structure, the valence band offset is -0.25 ± 0.04 eV and the conduction band offset is 1.25 ± 0.25 eV. For the CuI/β-$(Al_{0.14}Ga_{0.86})_2O_3$ system, the values are 0.05 ± 0.10 eV for the valence band offset and 1.85 ± 0.35 eV for the conduction band offset. These were obtained using the differences in bandgaps and the directly measured valence band offset. Based on these offset values, the band alignment for CuI on Ga_2O_3 is favorable for hole transport across the heterointerface. Additionally, the low processing temperature of the CuI is attractive due to the minimal formation of interface states. To further confirm CuI's potential with Ga_2O_3, electrical measurements are needed to show minority carrier injection at low biases in heterojunction samples. Figure 9.22 shows a summary of low gap dielectrics on Ga_2O_3.

9.7 Conclusion

In this chapter, we have discussed the band alignment of various dielectrics on Ga_2O_3, $(Al_xGa_{1-x})_2O_3$, and $(In_xGa_{1-x})_2O_3$. Due to the wide bandgaps of Ga_2O_3 and $(Al_xGa_{1-x})_2O_3$, the number of potential gate dielectrics that can provide a sufficient barrier to hole and electron transport is limited. Of the dielectrics with bandgaps able to meet this requirement, SiO_2 and Al_2O_3 appear to be the most promising options for gate dielectrics, with Al_2O_3 being the most commonly used gate dielectric for all β-Ga_2O_3 based power electronics, thus far. For gate dielectrics on $(Al_xGa_{1-x})_2O_3$ and $(In_xGa_{1-x})_2O_3$, SiO_2 and Al_2O_3 show the same promising band alignment properties. For all the dielectric and semiconductor structures examined, the band alignment is a strong function of the deposition method and conditions, with ALD being the most reproducible deposition technique. Post-deposition annealing was shown to have a large effect on the final band alignment of a dielectric/semiconductor system, indicating that the thermal budget of a chosen heterostructure must be carefully considered when designing devices. The band offsets of both AZO and ITO are negative on β-$(Al_{0.14}Ga_{0.86})_2O_3$, ensuring good electron and hole transport and hence the ability to reduce contact resistance when used as an interlayer in metal stacks on this wide bandgap material. There is still significant work needed in order to optimize and eventually commercialize Ga_2O_3

Table 9.3. Summary of measured core levels used to calculate the band alignment of CuI on Ga_2O_3 and $(Al_{0.14}Ga_{0.86})_2O_3$ (eV).

| Substrate | Reference | | | | Reference CuI | | | | Thin CuI on Ga_2O_3 or $(Al_{0.14}Ga_{0.86})_2O_3$ | |
	Core level	VBM	Core level peak	Core–VBM	Core level	VBM	Core level peak	Core–VBM	Δ core level Ga $2p_{3/2}$–Cu $2p_3$	Valence band offset
Ga_2O_3	Ga $2p_{3/2}$	3.20	1118.1	1114.9	Cu $2p_3$	0.95	932.8	931.85	182.8	0.25
$(Al_{0.14}Ga_{0.86})_2O_3$	Ga $2p_{3/2}$	3.00	1117.6	1114.6	Cu $2p_3$	0.95	932.8	931.85	182.8	−0.05

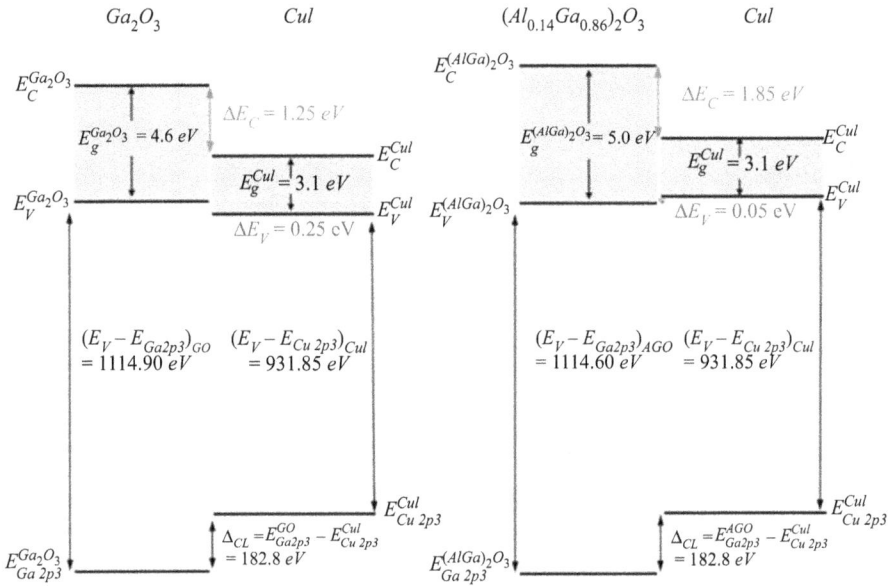

Figure 9.21. Band diagrams for CuI on Ga_2O_3 (left) and CuI on $(Al_{0.14}Ga_{0.86})_2O_3$ (right).

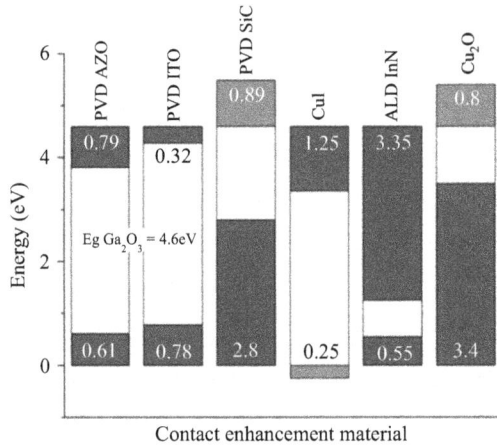

Figure 9.22. Summary of band alignment studies focused on contact enhancement dielectrics on Ga_2O_3.

based devices; however, the progress discussed in this chapter is a promising start to this endeavor.

References

[1] Higashiwaki M and Jessen G H 2018 *Appl. Phys. Lett.* **112** 060401

[2] Mastro M A, Kuramata A, Calkins J, Kim J, Ren F and Pearton S J 2017 *ECS J. Solid State Sci. Technol.* **6** P356–9

[3] von Wenckstern H 2017 *Adv. Electron. Mater.* **3** 1600350

[4] Pearton S J, Yang J, Cary P H, Ren F, Kim J, Tadjer M J, Mastro M A and Cary P H 2018 *Appl. Phys. Rev.* **51** 011301

[5] Rafique S, Han L, Neal A T, Mou S, Tadjer M J, French R H and Zhao H 2016 *Appl. Phys. Lett.* **109** 132103

[6] Zeng K and Singisetti U 2017 *Appl. Phys. Lett.* **111** 122108

[7] Tadjer M J *et al* 2016 *J. Electron. Mater.* **45** 2031–7

[8] Bae J, Kim H W, Kang I H, Yang G and Kim J 2018 *Appl. Phys. Lett.* **112** 122102

[9] Okur S 2017 *J. Vacuum Technol.* **5** 31–5

[10] Millan J, Godignon P, Perpina X, Perez-Tomas A and Rebollo J 2014 *IEEE Trans. Power Electron.* **29** 2155–63

[11] Reese S B, Remo T, Green J and Zakutayev A 2019 *Joule* **3** 903–7

[12] Tadjer M J 2018 *Electrochem. Soc. Interface* **27** 49–52

[13] Galazka Z *et al* 2017 *ECS J. Solid State Sci. Technol.* **6** Q3007–11

[14] Higashiwaki M *et al* 2016 *Appl. Phys. Lett.* **108** 133503

[15] Hu Z *et al* 2018 *Appl. Phys. Lett.* **113** 122103

[16] Green A J *et al* 2017 *IEEE Electron Device Lett.* **38** 790–3

[17] Wong M H, Sasaki K, Kuramata A, Yamakoshi S and Higashiwaki M 2016 *IEEE Electron Device Lett.* **37** 212–5

[18] Lin C-H *et al* 2019 *IEEE Electron. Device Lett.* **40** 1487–90

[19] Yang J, Ren F, Tadjer M, Pearton S J and Kuramata A 2018 *ECS J. Solid State Sci. Technol.* **7** Q92–6

[20] Yang J, Ren F, Pearton S J and Kuramata A 2018 *IEEE Trans. Electron Devices* **65** 2790–6

[21] Yang J, Ren F and Chen Y 2019 *IEEE J. Electron Devices Soc.* **7** 57–61

[22] Yang J, Ren F, Tadjer M, Pearton S J and Kuramata A 2018 *AIP Adv.* **8** 055026

[23] Yang J *et al* 2019 *Appl. Phys. Lett.* **114** 232106

[24] Pearton S J, Yang J, Cary P H, Ren F, Kim J, Tadjer M J and Mastro M A 2018 *Appl. Phys. Rev.* **5** 011301

[25] Zhang Y *et al* 2018 *Appl. Phys. Lett.* **112** 173502

[26] Krishnamoorthy S *et al* 2017 *Appl. Phys. Lett.* **111** 023502

[27] Ahmadi E, Koksaldi O S, Zheng X, Mates T, Oshima Y, Mishra U K and Speck J S 2017 *Appl. Phys. Express* **10** 071101

[28] Oshima T, Kato Y, Kawano N, Kuramata A, Yamakoshi S, Fujita S, Oishi T and Kasu M 2017 *Appl. Phys. Express* **10** 035701

[29] Oshima T, Okuno T, Arai N, Kobayashi Y and Fujita S 2009 *Jpn J. Appl. Phys.* **48** 070202

[30] Wakabayashi R, Hattori M, Yoshimatsu K, Horiba K, Kumigashira H and Ohtomo A 2018 *Appl. Phys. Lett.* **112** 232103

[31] Hill V G, Roy R and Osborn E F 1952 *J. Am. Ceram. Soc.* **35** 135–42

[32] Jaromin A L and Edwards D D 2005 *J. Am. Ceram. Soc.* **88** 2573–7

[33] Kaun S W, Wu F and Speck J S 2015 *J. Vac. Sci. Technol.* A **33** 041508

[34] Schmidt-Grund R, Kranert C, von Wenckstern H, Zviagin V, Lorenz M and Grundmann M 2015 *J. Appl. Phys.* **117** 165307

[35] Kranert C, Jenderka M, Lenzner J, Lorenz M, von Wenckstern H, Schmidt-Grund R and Grundmann M 2015 *J. Appl. Phys.* **117** 125703

[36] Oshima Y, Ahmadi E, Badescu S C, Wu F and Speck J S 2016 *Appl. Phys. Express* **9** 061102

[37] Grundmann M 2017 *Phys. Status Solidi* **254** 1700134

[38] Grundmann M 2018 *J. Appl. Phys.* **124** 185302

[39] von Wenckstern H, Zhang Z, Schmidt F, Lenzner J, Hochmuth H and Grundmann M 2013 *Cryst. Eng. Comm.* **15** 10020

[40] Zhang F, Saito K, Tanaka T, Nishio M, Arita M and Guo Q 2014 *Appl. Phys. Lett.* **105** 162107

[41] Zhang Y, Xia Z, Mcglone J, Sun W, Joishi C, Arehart A R, Ringel S A and Rajan S 2019 *IEEE Trans. Electron Devices* **66** 1574–8

[42] von Wenckstern H 2018 *in Gallium Oxide* ed S Pearton, F Ren and M Mastro (Oxford: Elsevier) pp 119–48

[43] Zhang F, Li H, Arita M and Guo Q 2017 *Opt. Mater. Express* **7** 3769

[44] Zhang Z, Von Wenckstern H, Lenzner J, Lorenz M and Grundmann M 2016 *Appl. Phys. Lett.* **108** 243505

[45] Kokubun Y, Abe T and Nakagomi S 2010 *Phys. Status Solidi* **207** 1741–5

[46] Zhang F, Saito K, Tanaka T, Nishio M and Guo Q 2014 *Solid State Commun.* **186** 28–31

[47] Maccioni M B and Fiorentini V 2016 *Appl. Phys. Express* **9** 041102

[48] Janowitz C *et al* 2011 *New J. Phys.* **13** 085014

[49] Schmidt-Grund R, Kranert C, Böntgen T, Von Wenckstern H, Krauß H and Grundmann M 2014 *J. Appl. Phys.* **116** 053510

[50] Edwards D D, Folkins P E and Mason T O 1997 *J. Am. Ceram. Soc.* **80** 253–7

[51] Wang V, Xiao W, Ma D M, Liu R J and Yang C M 2014 *J. Appl. Phys.* **115** 043708

[52] Peelaers H, Steiauf D, Varley J B, Janotti A and Van De Walle C G 2015 *Phys. Rev.* B **92** 235201

[53] Wang X, Chen Z, Saito K, Tanaka T, Nishio M and Guo Q 2017 *J. Alloys Compd.* **690** 287–92

[54] King P D C, Veal T D, Payne D J, Bourlange A, Egdell R G and McConville C F 2008 *Phys. Rev. Lett.* **101** 116808

[55] Yang F, Ma J, Luan C and Kong L 2009 *Appl. Surf. Sci.* **255** 4401–4

[56] Baldini M, Albrecht M, Gogova D, Schewski R and Wagner G 2015 *Semicond. Sci. Technol.* **30**

[57] Vogt P and Bierwagen O 2016 *APL Mater.* **4** 086112

[58] Oshima T and Fujita S 2008 *Phys. Status Solidi* **5** 3113–5

[59] Fuchs F and Bechstedt F 2008 *Phys. Rev.* B **77**

[60] Kranert C, Lenzner J, Jenderka M, Lorenz M, Von Wenckstern H, Schmidt-Grund R and Grundmann M 2014 *J. Appl. Phys.* **116** 155107

[61] Patzke G and Binnewies M 2000 *Solid State Sci.* **2** 689–99

[62] Phillips J M *et al* 1994 *Appl. Phys. Lett.* **65** 115–7

[63] Wang A, Edleman N L, Babcock J R, Marks T J, Lane M A, Brazis P R and Kannewurf C R 2002 *J. Mater. Res.* **17** 3155–62

[64] Grundmann M, Frenzel H, Lajn A, Lorenz M, Schein F and von Wenckstern H 2010 *Phys. Status Solidi* **207** 1437–49

[65] Lin W T, Ho C Y, Wang Y M, Wu K H and Chou W Y 2012 *J. Phys. Chem. Solids* **73** 948–52

[66] Zhang F, Saito K, Tanaka T, Nishio M and Guo Q 2014 *J. Alloys Compd.* **614** 173–6

[67] Kneiß M, Storm P, Benndorf G, Grundmann M and Von Wenckstern H 2018 *ACS Comb. Sci.* **20** 643–52

[68] von Wenckstern H, Splith D, Werner A, Müller S, Lorenz M and Grundmann M 2015 *ACS Comb. Sci.* **17** 710–5

[69] Von Wenckstern H, Splith D, Purfürst M, Zhang Z, Kranert C, Müller S, Lorenz M and Grundmann M 2015 *Semicond. Sci. Technol.* **30** 024005

[70] Minami T, Takeda Y, Kakumu T, Takata S and Fukuda I 1997 *J. Vac. Sci. Technol.* A **15** 958–62

[71] Robertson J 2000 *J. Vac. Sci. Technol.* B **18** 1785

[72] Fares C, Ren F, Hays D C, Gila B P and Pearton S J 2019 *ECS J. Solid State Sci. Technol.* **8** Q3001–6

[73] Hays D C, Gila B P, Pearton S J and Ren F 2017 *Appl. Phys. Rev.* **4** 021301

[74] Hays D C, Gila B P, Pearton S J, Trucco A, Thorpe R and Ren F 2017 *J. Vac. Sci. Technol.* **35** 011206

[75] Robertson J 2002 *MRS Bull.* **27** 217–21

[76] Higashiwaki M, Sasaki K, Murakami H, Kumagai Y, Koukitu A, Kuramata A, Masui T and Yamakoshi S 2016 *Appl. Phys. Lett.* **108** 133503

[77] Tsao J Y *et al* 2018 *Adv. Electron. Mater.* **4** 1600501

[78] Oh S, Kim J, Ren F, Pearton S J and Kim J 2016 *J. Mater. Chem.* C **4** 9245–50

[79] Polyakov A Y *et al* 2018 *Appl. Phys. Lett.* **113** 092102

[80] Fares C, Ren F, Lambers E, Hays D C, Gila B P and Pearton S J 2019 *Semicond. Sci. Technol.* **34** 025006

[81] Fares C, Ren F, Lambers E, Hays D C, Gila B P and Pearton S J 2018 *ECS J. Solid State Sci. Technol.* **7** P519–23

[82] Fares C, Ren F, Hays D C, Gila B P, Tadjer M, Hobart K D and Pearton S J 2018 *Appl. Phys. Lett.* **113** 182101

[83] Fares C, Ren F, Lambers E, Hays D C, Gila B P and Pearton S J 2018 *J. Vac. Sci. Technol.* B **36** 061207

[84] Carey P H, Ren F, Hays D C, Gila B P, Pearton S J, Jang S and Kuramata A 2017 *J. Vac. Sci. Technol.* **35** 041201

[85] Puthenkovilakam R, Carter E A and Chang J P 2004 *Phys. Rev.* B **69** 155329

[86] Adamchuk V K and Afanas'ev V V 1992 *Prog. Surf. Sci.* **41** 111–211

[87] Werner W S M 2001 *Surf. Interface Anal.* **31** 141–76

[88] Waldrop J R, Kowalczyk S P, Grant R W, Kraut E A and Miller D L 1981 *J. Vac. Sci. Technol.* **19** 573–5

[89] Kraut E A, Grant R W, Waldrop J R and Kowalczyk S P 1980 *Phys. Rev. Lett.* **44** 1620–3

[90] Nichols M T, Li W, Pei D, Antonelli G A, Lin Q, Banna S, Nishi Y and Shohet J L 2014 *J. Appl. Phys.* **115** 094105

[91] Penn D R 1977 *Phys. Rev. Lett.* **38** 1429–32

[92] Hüfner S 2003 *Photoelectron Spectroscopy: Principles and Applications* 3rd edn (Berlin: Springer) p 662

[93] Nozières P and Pines D 1959 *Phys. Rev.* **113** 1254–67

[94] Fares C, Kneiss M, von Wenckstern H, Grundmann M, Tadjer M J, Ren F, Hays D, Gila B P and Pearton S J 2019 *ECS Trans.* **92** 79–88

[95] Bell F G and Ley L 1988 *Phys. Rev.* B **37** 8383–93

[96] Carey P H, Ren F, Hays D C, Gila B P, Pearton S J, Jang S and Kuramata A 2017 *Vacuum* **142** 52–7

[97] Peelaers H and Van de Walle C G 2015 *Phys. Status Solidi* **252** 828–32

[98] Varley J B, Weber J R, Janotti A and Van De Walle C G 2010 *Appl. Phys. Lett.* **97** 142106

[99] Kamimura T, Sasaki K, Hoi Wong M, Krishnamurthy D, Kuramata A, Masui T, Yamakoshi S and Higashiwaki M 2014 *Appl. Phys. Lett.* **104** 192104

[100] Peelaers H, Varley J B, Speck J S and Van de Walle C G 2018 *Appl. Phys. Lett.* **112** 242101

[101] Dong H *et al* 2018 *AIP Adv.* **8** 065215

[102] Fares C, Ren F, Lambers E, Hays D C, Gila B P and Pearton S J 2019 *J. Electron. Mater.* **48** 1568–73

[103] Konishi K, Kamimura T, Wong M H, Sasaki K, Kuramata A, Yamakoshi S and Higashiwaki M 2016 *Phys. Status Solidi* **253** 623–5

[104] Zhou H, Maize K, Qiu G, Shakouri A and Ye P D 2017 *Appl. Phys. Lett.* **111** 092102

[105] Feng Z, Feng Q, Zhang J, Li X, Li F, Huang L, Chen H-Y, Lu H-L and Hao Y 2018 *Appl. Surf. Sci.* **434** 440–4

[106] Feng Z *et al* 2018 *J. Alloys Compd.* **745** 292–8

[107] Hung T-H, Sasaki K, Kuramata A, Nath D N, Sung Park P, Polchinski C and Rajan S 2014 *Appl. Phys. Lett.* **104** 162106

[108] Chabak K D *et al* 2016 *Appl. Phys. Lett.* **109** 213501

[109] Hattori M *et al* 2016 in *Jpn J. Appl. Phys.* **55** 025002

[110] Fares C, Kneiß M, von Wenckstern H, Tadjer M, Ren F, Lambers E, Grundmann M and Pearton S J 2019 *ECS J. Solid State Sci. Technol.* **8** P351–6

[111] Fares C, Kneiß M, von Wenckstern H, Grundmann M, Tadjer M, Ren F, Lambers E and Pearton S J 2019 *APL Mater.* **7** 071115

[112] Feng X, Li Z, Mi W and Ma J 2016 *Vacuum* **124** 101–7

[113] Martinez E, Grampeix H, Desplats O, Herrera-Gomez A, Ceballos-Sanchez O, Guerrero J, Yckache K and Martin F 2012 *Chem. Phys. Lett.* **539–540** 139–43

[114] Jia Y, Zeng K, Wallace J S, Gardella J A and Singisetti U 2015 *Appl. Phys. Lett.* **106** 102107

[115] Zeng K, Jia Y and Singisetti U 2016 *IEEE Electron Device Lett.* **37** 906–9

[116] Kita K, Suzuki E and Mao Q 2019 *ECS Trans.* **92** 59–63

[117] Pearton S J, Ren F, Tadjer M J and Kim J 2018 *J. Appl. Phys.* **124** 220901

[118] Yadav M K, Mondal A, Das S, Sharma S K and Bag A 2019 *J. Alloys Compd.* **819** 153052

[119] Chiam S Y, Chim W K, Ren Y, Pi C, Pan J S, Huan A C H, Wang S J and Zhang J 2008 *J. Appl. Phys.* **104** 063714

[120] Nguyen N V, Xu M, Kirillov O A, Ye P D, Wang C, Cheung K and Suehle J S 2010 *Appl. Phys. Lett.* **96** 052107

[121] Sun L *et al* 2017 *Nanoscale Res. Lett.* **12** 102

[122] Fares C, Islam Z, Haque A, Kneiß M, Von Wenckstern H, Grundmann M, Tadjer M, Ren F and Pearton S J 2019 *ECS J. Solid State Sci. Technol.* **8** P751–6

[123] Liu H, Avrutin V, Izyumskaya N, Özgr Ü and Morkoç H 2010 *Superlattices Microstruct.* **48** 458–84

[124] Ginley D S and Perkins J D 2011 *Handbook of Transparent Conductors* (New York: Springer) pp 1–25

[125] Minami T and Miyata T 2008 *Thin Solid Films* **517** 1474–7

[126] Suzuki A, Matsushita T, Aoki T, Yoneyama Y and Okuda M 2001 *Jpn J. Appl. Phys.* **40** L401

[127] Loureiro J *et al* 2014 *J. Mater. Chem.* A **2** 6649–55

[128] Dasgupta N P, Neubert S, Lee W, Trejo O, Lee J R and Prinz F B 2010 *Chem. Mater.* **22** 4769–75

[129] Minami T 2005 *Semicond. Sci. Technol.* **20** S35

[130] Carey P H, Yang J, Ren F, Hays D C, Pearton S J, Kuramata A and Kravchenko I I 2017 *J. Vac. Sci. Technol.* **35** 061201

[131] Carey P H, Yang J, Ren F, Hays D C, Pearton S J, Jang S, Kuramata A and Kravchenko I I 2017 *AIP Adv.* **7** 095313

[132] Carey P H, Ren F, Hays D C, Gila B P, Pearton S J, Jang S and Kuramata A 2017 *Vacuum* **141** 103–8

[133] Nepal N, Mahadik N A, Nyakiti L O, Qadri S B, Mehl M J, Hite J K and Eddy C R 2013 *Cryst. Growth Des.* **13** 1485–90

[134] Nepal N, Anderson V R, Hite J K and Eddy C R 2015 *Thin Solid Films* **589** 47–51

[135] Fares C, Tadjer M J, Woodward J, Nepal N, Mastro M A, Eddy C R, Ren F and Pearton S J 2019 *ECS J. Solid State Sci. Technol.* **8** Q3154–8

[136] Swallow J E N, Varley J B, Jones L A H, Gibbon J T, Piper L F J, Dhanak V R and Veal T D 2019 *APL Mater.* **7** 022528

[137] Wu J, Walukiewicz W, Yu K M, Ager J W, Haller E E, Lu H, Schaff W J, Saito Y and Nanishi Y 2002 *Appl. Phys. Lett.* **80** 3967–9

[138] Chikoidze E *et al* 2017 *Mater. Today Phys.* **3** 118–26

[139] Grundmann M, Schein F-L, Lorenz M, Böntgen T, Lenzner J and von Wenckstern H 2013 *Phys. Status Solidi A* **210** 1671

[140] Grundmann M, Klüpfel F, Karsthof R, Schlupp P, Schein F L, Splith D, Yang C, Bitter S and Von Wenckstern H 2016 *J. Phys. D: Appl. Phys.* **49** 213001

[141] Pishtshev A and Karazhanov S Z 2017 *J. Chem. Phys.* **146** 064706

[142] Liu A, Zhu H, Park W-T, Kang S-J, Xu Y, Kim M-G and Noh Y-Y 2018 *Adv. Mater.* **30** 1802379

[143] Schein F L, Von Wenckstern H and Grundmann M 2013 *Appl. Phys. Lett.* **102** 092109

[144] Yang C, Kneiß M, Schein F L, Lorenz M and Grundmann M 2016 *Sci. Rep.* **6** 21937

[145] Watahiki T, Yuda Y, Furukawa A, Yamamuka M, Takiguchi Y and Miyajima S 2017 *Appl. Phys. Lett.* **111** 222104

Part II

Gallium Nitride/Aluminum Nitride

IOP Publishing

Wide Bandgap Semiconductor-Based Electronics

Fan Ren and Stephen J Pearton

Chapter 10

The effect of growth parameters on the residual carbon concentration in GaN high electron mobility transistors: theory, modeling, and experiments

Indraneel Sanyal and Jen-Inn Chyi

10.1 Introduction

10.1.1 Opportunities and challenges for GaN based devices

III-nitride wide bandgap semiconductors such as GaN and its alloys have emerged as one of the mainstream solutions for high power switching devices and radio frequency amplifiers, among several other applications [1]. Judging from Johnson's figure of merit (JFOM), which is a measure of the ultimate device high frequency capability of the material, and Baliga's figure of merit (BFOM), which is a measure of the power device capability of the material, GaN based transistors offer great potential in both above-mentioned sectors because of its high breakdown field and high electron velocity. Figure 10.1 summarizes various current transistor technologies for RF applications in terms of their breakdown voltage and current gain cut-off frequency. As can be seen in the figure, GaN field-effect transistors exhibit JFOMs as high as 7 THz V with the promise of delivering above 10 THz V, while Si, GaAs, and InP-based transistors, whose performance is typically limited to 1 THz V and below [1, 2]. Further developments in material growth and device processing technologies could take GaN devices to their full potential as manifested the intrinsic properties of the material.

To date, however, the device reliability of GaN HEMTs is still a concern and the subject of extensive research and development [3]. Defects are inherent in III-nitride heterostructures, partly due to the lattice mismatch between the substrate and epilayers and partly due to the use of different growth techniques to grow the epilayers, such as metal–organic chemical vapor deposition (MOCVD), molecular beam epitaxy (MBE), and hydride vapor phase epitaxy (HVPE). Dislocations and

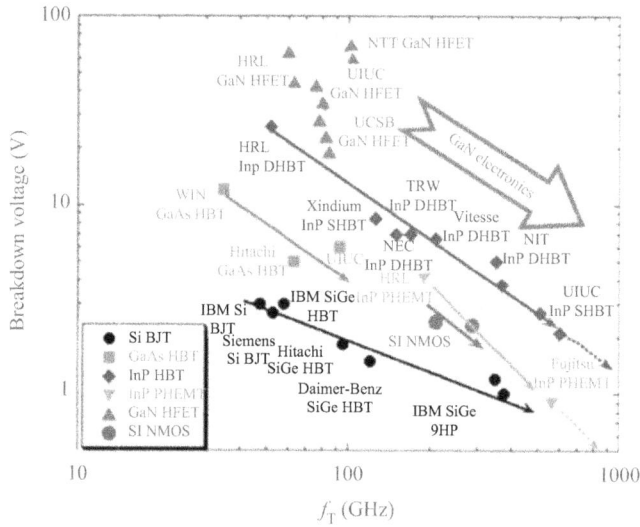

Figure 10.1. Breakdown voltage versus current gain cut-off frequency performance (JFOM) of different transistor technologies. Reproduced with permission from [1]. Copyright 2016 Elsevier.

point defects, such as vacancies, antisites, complexes, interstitials, and impurities of carbon, oxygen, silicon, and so forth in the epilayers, have substantial effects on device performance [4]. As a result, current collapse, the kink effect, and high gate leakage current are often observed [5]. While there is still an ongoing debate on the physical origin of those experimentally observed device degradations [6], this chapter presents an in-depth discussion on residual carbon incorporation in the epitaxial growth of GaN using MOCVD and its influence on the transport properties of GaN epilayers.

10.1.2 Carbon impurity and related defects in GaN

Carbon incorporation in MOCVD grown GaN is inevitable due to the use of metal–organic precursors. The role of carbon related defects in the electrical and optical properties of GaN based HEMTs is still under extensive investigation. So far, intentional doping of carbon in GaN has been used to achieve semi-insulating GaN [7], as carbon substituting N in GaN (C_N) acts as a deep acceptor with an energy level of 0.9 eV above the valence band maximum. It is also involved in the widely observed yellow luminescence [8]. If carbon occupies the cation site in GaN (C_{Ga}) it acts as a donor [6]. Carbon can also incorporate as an interstitial defect with high formation energy. Carbon at the N site is likely to be stable from the atomic size point of view. However, the chemical potential of Ga and N and the associated Fermi level are more crucial and therefore determine the site preference density functional theory (DFT) as shown in figure 10.2 [6]. As can be seen in the figure, the defect formation energy of C_N is higher in the N-rich condition than the Ga-rich one.

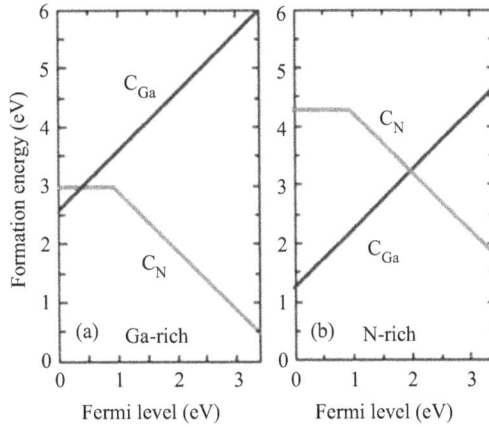

Figure 10.2. Carbon formation energy in Ga- and N-rich growth conditions as a function of Fermi level energy. Reproduced with permission from [8]. Copyright 2010 American Physical Society.

Moreover, the formation energy of C_N decreases monotonically with increasing Fermi level energy. The scenario is the opposite for C_{Ga}, where the formation energy is higher in the Ga-rich condition due to a sufficient availability of Ga atoms. C_{Ga}, i.e. C at the Ga site, acts as a donor by giving an electron in the conduction band (CB) as a substitutional carbon.

A Ga atom is much bigger than a C atom, which leads to local lattice relaxation of about 26% around the defect. Due to this reason, C_{Ga} is also a DX-like center and only stable for Fermi level energy above the CB minimum. On the other hand, C at the N site may have a $\varepsilon(0/-)$ charge state, i.e. neutral if unoccupied (C_N^0) and negatively charged if occupied (C_N^-). Moreover, the size difference between N and C is small, which results in only about 2% lattice relaxation of the Ga–N bond length compared to 26% relaxation in case of C at the Ga site. C_N is also responsible for the yellow luminescence observed in carbon doped GaN with optical emission at 2.14 eV. The optical absorption energy is usually found by analysing the configuration-coordinate diagram. Figure 10.3 shows such a configuration-coordinate diagram for C_N defects in GaN. The distance between a C atom and a Ga atom along the c-axis is represented by the coordinate. The optical absorption by the initial state of C_N^- starts at 2.6 eV and reaches 2.95 eV. Then the transition from C_N^0 to C_N^- by optical emission is likely to occur at 2.14 eV, which is the so-called yellow luminescence observed in GaN with a considerable amount of C impurity [8]. In the next section, we explore the incorporation of residual carbon into GaN due to various growth conditions and provide an in-depth theoretical explanation of the underlying physics of defect incorporation.

10.2 Correlation between carbon concentration and growth conditions

10.2.1 The effects of MOCVD growth parameters

The incorporation of carbon into a GaN layer grown by MOCVD mostly depends on the growth conditions, such as the V/III ratio, reactor pressure, growth

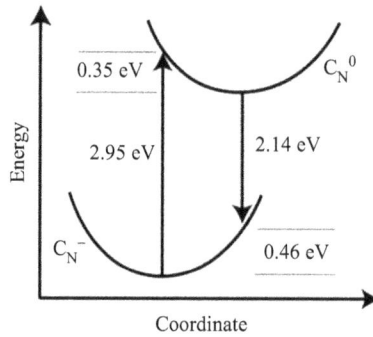

Figure 10.3. Configuration-coordinate diagram for the CN impurity in GaN showing optical absorption and emission. Reproduced with permission from [8]. Copyright 2010 American Physical Society.

Table 10.1. The trend of carbon incorporation with GaN growth parameters.

Growth parameter	Carbon concentration
NH$_3$ partial pressure ⬆	⬇
H$_2$/N$_2$ ratio ⬆	⬆
TMG versus TEG	TEG ⬇
V/III ⬆	⬇
Reactor pressure ⬆	⬇
Temperature ⬆	⬇

temperature, the type of carrier gas, and Ga precursors. Table 10.1 summarizes the experimentally observed trend of carbon incorporation for each of the growth parameters. The change in carbon concentration with the change in NH$_3$ partial pressure, V/III ratio, reactor pressure, growth rate, and temperature have been studied extensively. Several reports suggest that carbon incorporation decreases with an increase in NH$_3$ partial pressure, V/III ratio, reactor pressure, and temperature [9–11], whereas an increased growth rate also increases carbon incorporation [11]. Additionally, the choice of carrier gas and metal precursor also considerably influence carbon incorporation.

The relationship between C incorporation and NH$_3$ flow rate is shown in figure 10.4. It can be seen as that, as the NH$_3$ flow rate increases from 1 to 2 slm, the C concentration decreases from more than 5×10^{17} cm^{-3} to below 2×10^{17} cm^{-3} without significantly changing the growth rate, particularly at low to medium

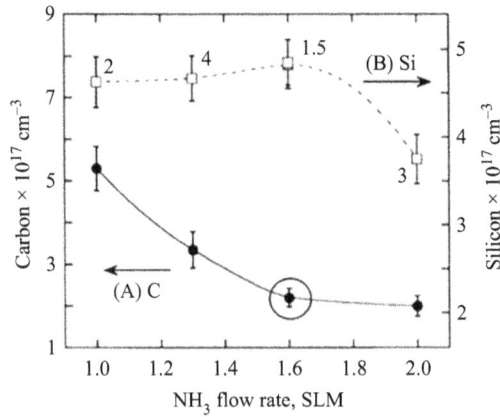

Figure 10.4. Carbon concentration as a function of NH₃ flow rate. Reproduced with permission from [11]. Copyright 2002 Elsevier.

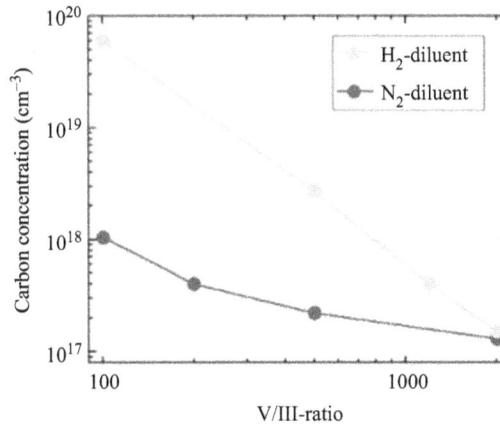

Figure 10.5. Carbon concentration as a function of V/III ratio in nitrogen and hydrogen ambient. Reproduced with permission from [9]. Copyright 2016 American Physical Society.

NH₃ flow. Moreover, any change in NH₃ flow also changes the V/III ratio, as MOCVD GaN is usually grown with a high V/III ratio. Therefore, it is expected that a higher V/III ratio should also lower the carbon contamination, which is evidenced in figure 10.5 [9]. Furthermore, as can be seen in the figure, the C concentration could be very different in H_2 and N_2 ambient, particularly at a lower V/III ratio. The GaN grown in N_2 ambient results in an almost two orders of magnitude lower C concentration at a low V/III ratio. However, at a high V/III ratio, such as 2000 or higher, the C incorporation is similar in GaN layers grown in N_2 and H_2 ambient. Nonetheless, nitrogen is still preferred over hydrogen as a carrier gas, in particular when low carbon is desirable as a very high V/III ratio may deteriorate the surface morphology and/or material quality, as observed by the authors.

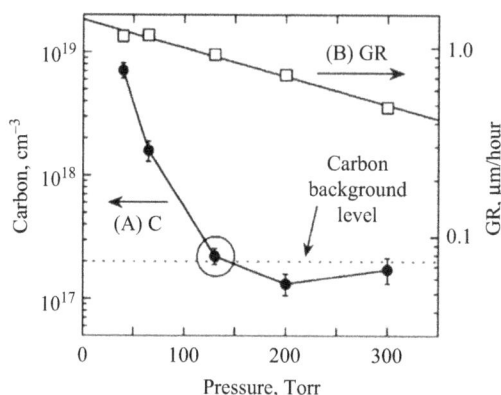

Figure 10.6. Carbon concentration as a function of reactor pressure. Reproduced with permission from [11]. Copyright 2002 Elsevier.

The reactor pressure and growth temperature are also critical for controlling residual C incorporation in the GaN layer. The reactor pressure and growth temperature show similar trends where the C concentration decreases with increasing pressure or temperature. For example, figure 10.6 indicates a reduction in C concentration by a factor of 50 when the reactor pressure increases from 40 to 300 Torr [11].

Similarly, as shown in figure 10.7, C concentration decreases by a factor of 25 as the growth temperature increases from 955 °C to 1065 °C. Apart from the growth parameters discussed above, the choice of Ga precursor can also significantly influence the C incorporation in the as-grown GaN layer.

Figure 10.8 shows the secondary ion mass spectroscopy (SIMS) depth profiles of residual carbon impurities in GaN samples where the thin GaN layer is grown either with TMG (structure A) or TEG (structure B), both in N_2 ambient. However, as indicated in the panel on the right, the bulk GaN buffer is grown with TMG in both of the structures. Aligning the profiles and growth sequence of these samples, a decrease in carbon concentration from $\sim 4 \times 10^{17}$ cm^{-3} to $\sim 2 \times 10^{17}$ cm^{-3} is observed at the end of the thick GaN buffer after changing the carrier gas from H_2 to N_2. This further confirms the trend shown in figure 10.5. After this approximately 100 nm GaN transition layer, a further reduction in carbon concentration is achieved by the use of a TEG precursor, as shown by the red line in figure 10.8. The TEG-grown GaN channel in structure B has an order of magnitude lower background carbon than that of the TMG-grown GaN channel in structure A [12].

The advantage of reducing the C impurity in the channel using a N_2 carrier gas and TEG precursor is evident experimentally from the transport properties of these two heterostructures. By changing the TMG to the TEG precursor for the channel of structure B, the electron mobility increases from 1740 cm^2 V^{-1} s^{-1} to 1820 cm^2 V^{-1} s^{-1}. Moreover, the corresponding two-dimensional electron gas (2DEG) concentration also increases from 1.15×10^{13} to 1.26×10^{13} cm^{-2} in structure B. This could be due to the

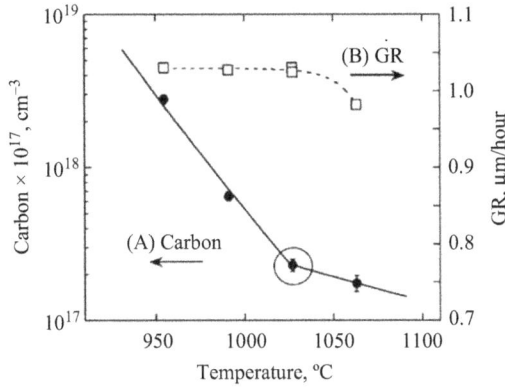

Figure 10.7. The effect of growth temperature on residual carbon incorporation. Reproduced with permission from [11]. Copyright 2002 Elsevier.

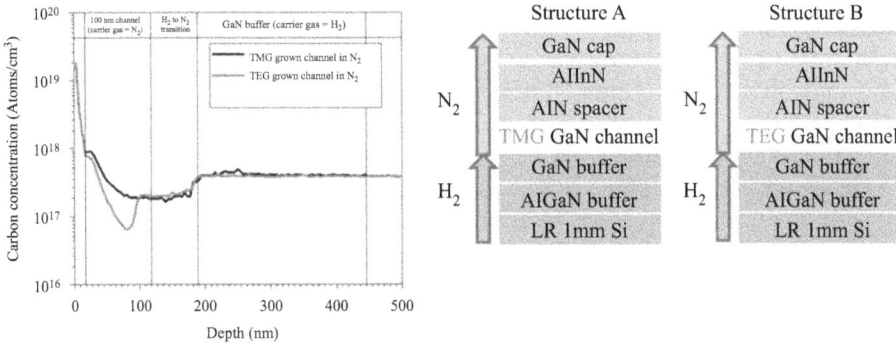

Figure 10.8. Left: Carbon incorporation in the TMG- and TEG-grown GaN channel in N_2 ambient. Right: Schematics of the HEMT structures.

reduction in acceptor-like traps in the channel. As a result, the sheet resistance also decreases from 312 to 271 $\Omega/\square$ in structure B.

The same effect of the precursor on the transport properties is also observed in AlGaN/GaN HEMTs. Shown in figure 10.9 are two AlGaN/GaN HEMTs of identical structure except that the GaN channel and AlGaN barrier are grown with TMG in H_2 for structure C and TEG in H_2 for structure D. In this case, temperature-dependent Hall measurements are conducted to better observe the effect of residual carbon on the transport properties. The electron mobility is mainly dominated by optical and acoustic phonon scattering at room temperature, while the effects of defect and/or impurity scattering are more prominently observed at low temperatures. Reducing the Hall measurement temperature down to 10 K, as shown in figure 10.9 (right), electron mobility increases from 7900 cm^2 V^{-1} s^{-1} in structure C to 9360 cm^2 V^{-1} s^{-1} in structure B at 10 K, and the corresponding 2DEG concentration also increases from 1.16×10^{13} to 1.28×10^{13} cm^{-2} [13]. This is consistent with that observed for structures A and B discussed above. Moreover, it is noted that the 2DEG concentrations in the samples are nearly constant with

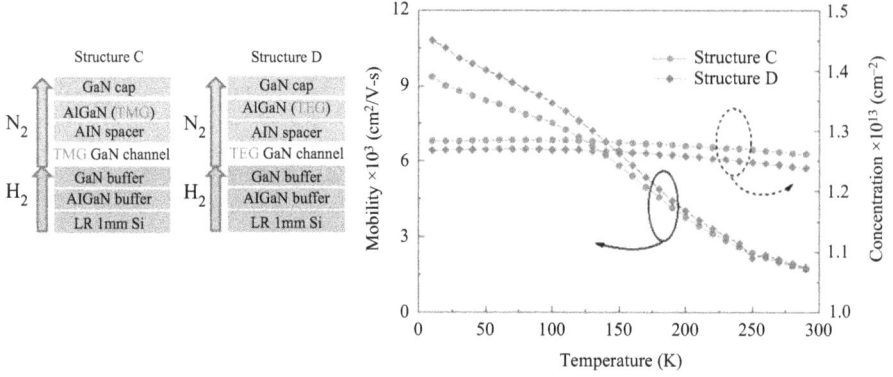

Figure 10.9. Schematics of AlGaN/GaN heterostructures with the GaN channel grown with TMG in H$_2$ (structure C) and TEG in H$_2$ (structure D). The temperature-dependent transport properties of structures C and D (right).

temperature, indicating that the 2DEG is induced by the polarization charge of the heterostructure. The next section discusses the physical mechanism of C incorporation in detail with the help of DFT and a thermodynamic growth model.

10.3 Theory and modeling of carbon incorporation

The correlations between C incorporation and the MOCVD growth parameters as discussed in the previous section have been experimentally reported and studied over the years. However, it is not until recently that several researchers have started investigating their theoretical aspects in order to understand the C incorporation mechanism in epitaxial GaN [14–19]. This section discusses the theory and modeling of C incorporation in GaN.

10.3.1 The surface reconstruction of GaN

The first step towards DFT modeling of C incorporation in GaN is to understand the surface reconstruction under different growth conditions. Knowledge of the surface reconstruction is needed to realize the adsorption–desorption characteristics of the precursors in the gas phase on the GaN growth surface. The surface with the lowest Gibb's energy is most likely to appear under equilibrium. The energy of formation of a reconstructed surface can be expressed following the work by Kusaba *et al* [15] as

$$G = E_{\text{adsorption}} - n_{\text{Ga}}\mu_{\text{Ga}}^{\text{Gas}} - \frac{1}{2}n_{\text{N}}\mu_{\text{N}_2}^{\text{Gas}} - \frac{1}{2}n_{\text{H}}\mu_{\text{H}_2}^{\text{Gas}}, \qquad (10.1)$$

where n and μ represent the number of respective adatoms on the reconstructed GaN surface and their chemical potentials, respectively. $E_{\text{adsorption}}$, the total adsorption energy of the ideal and reconstructed GaN surface, can be expressed as follows:

$$E_{\text{adsorption}} = E_{\text{surface}}^{\text{reconstructed}} - E_{\text{surface}}^{\text{ideal}} - n_{\text{Ga}}E_{\text{Ga}}^{\text{Gas}} - \frac{1}{2}n_{\text{N}}E_{\text{N}_2}^{\text{Gas}} - \frac{1}{2}n_{\text{H}}E_{\text{H}_2}^{\text{Gas}}. \qquad (10.2)$$

Here $E_{\text{surface}}^{\text{reconstructed}}$ and $E_{\text{surface}}^{\text{ideal}}$ are the total energies of the reconstructed and ideal GaN surfaces, respectively, and $E_{\text{Ga}}^{\text{Gas}}$, $E_{\text{N}_2}^{\text{Gas}}$, and $E_{\text{H}_2}^{\text{Gas}}$ are the total energies of the respective molecules in the gas phase. It is worth noting that the difference between the total adsorption energy ($E_{\text{adsorption}}$) and the chemical potential energy ($\mu_{\text{Ga},\text{H}_2,\text{N}_2}^{\text{Gas}}$) defines the adsorption–desorption behavior of the adatoms on the GaN surface. For example, adsorption (desorption) of adatoms on (from) the growth surface occurs if $E_{\text{adsorption}}$ is less (more) than the cumulative sum of the total chemical potential energy, resulting in negative (positive) Gibb's energy.

Now, there are two types of energy terms we need to calculate in order to solve equation (10.1). One is the adsorption energy term ($E_{\text{adsorption}}$) and the other is the chemical potential energy terms of the respective gas molecules. The adsorption energy is calculated by calculating the ground state energy of any particular electronic configuration using the DFT. An in-depth discussion of DFT modeling of the III-nitride heterostructure is beyond the scope of this chapter. However, interested readers may find [16] useful in this regard [20]. On the other hand, the chemical potential energy, i.e. the free energy of ideal gas per particle, is a strong function of epitaxial growth conditions and the adsorption–desorption behavior of the adatoms on the epilayer surface. Therefore, it is important to consider the specific growth conditions to obtain a correct model of the adatom behavior on the surface. Kangawa et al proposed a thermodynamic model to calculate the chemical potential by taking into account the growth parameters as follows [21]:

$$\mu = k_{\text{B}}T\ln\left(\frac{gk_{\text{B}}T}{p}\xi_{\text{trans}}\xi_{\text{rot}}\xi_{\text{vib}}\right). \tag{10.3}$$

ξ_{trans}, ξ_{rot}, and ξ_{vib} are known as the translational, rotational, and vibrational partition functions, respectively. A detailed discussion of the partition functions can be found elsewhere [15, 21]. We can now consider various possibilities of surface reconstruction based on equations (10.1) and (10.3).

Kusaba et al considered various possible reconstructed GaN surfaces, as shown in figure 10.10, to determine the most stable configuration [15]. The ideal 2×2 GaN surface is shown in figure 10.10(a), which is a single bilayer of Ga and N atoms, whereas an extra Ga atom, as shown in figure 10.10(b), is on the top, bonded with

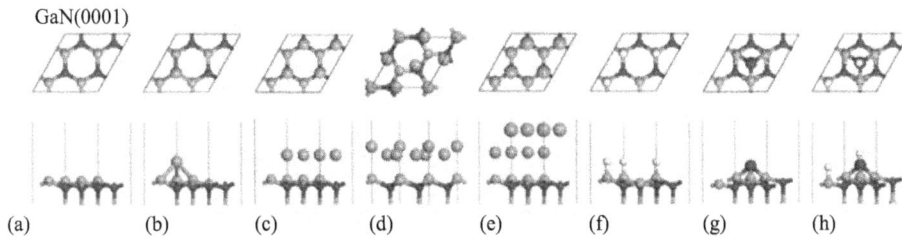

Figure 10.10. Different possible GaN surface reconstructions during growth: (a) ideal, (b) Ga adatom, (c) Ga adlayer, (d) pseudo-(1×1) surface, (e) Ga bilayer, (f) 3Ga-H, (g) N adatoms, and (h) N_{ad}-H+Ga-H. Reproduced from [15]. Copyright IOP Publishing, reprinted with permission, all rights reserved.

three other surface Ga atoms. Figure 10.10(c) presents the case where a complete Ga adlayer is present on top of each N atom. Another possibility is considered in figure 10.10(d), where, a pseudo-1 × 1 surface reconstruction is illustrated. There is also a possibility that the surface may have two adlayers, i.e. a Ga bilayer as shown in figure 10.10(e), similar to figure 10.10(c). It is also possible to have various other surface reconstructions, such as those shown in figure 10.10(f)–(h). For example, a more complex scenario, where a hydrogen terminated surface with three hydrogen atoms is bonded with three Ga surface atoms, presented as 3Ga-H in figure 10.10(f). The surface with a single N atom in the middle, commonly referred to as the H3 site, is shown in figure 10.10(g). Furthermore, an N adatom bonded with a H atom at the same H3 site together with an H adatom bonded with a surface Ga atom is considered in figure 10.10(h).

As discussed above, among all these different surface configurations, the surface with the lowest Gibb's energy is more likely to form a stable reconstructed surface under a certain growth condition. A surface phase diagram, as shown in figure 10.11, can be helpful in this regard [15]. It captures the change in surface reconstruction with a large variation in growth conditions. Usually, GaN is grown under N_2 or H_2 ambient. Figure 10.11 shows a noticeable difference in the reconstructed surface with different carrier gases. As we can see, the GaN grown in N_2 ambient results in an adsorption of Ga adatoms (denoted as Ga_{ad} (T4)) at a low growth temperature (see figure 10.10(b)). As the temperature increases close to 1200 °C, the ideal GaN surface appears by desorbing the Ga adatoms. The transition temperature at which the Ga adatom surface reconstruction changes to the ideal GaN surface reconstruction decreases to 1100 °C as the V–III ratio increases from 1000 to 5000.

In the case of using H_2 carrier gas, three different surface reconstructions are possible at different growth temperatures and V–III ratios. For instance, the same Ga adatom-rich surface can be formed for a V–III ratio below 1000 and a growth temperature up to 1200 °C. A hydrogen terminated GaN surface, denoted as 3Ga-H in figure 10.10(f), is most likely to appear for a V–III ratio above 1000 and a growth temperature less than 1150 °C. However, for very high temperature such as 1200 °C

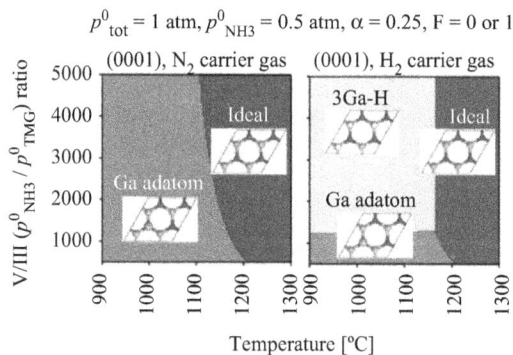

Figure 10.11. Surface phase diagram in the (0001) direction in H_2 and N_2 ambient. Reproduced from [15]. Copyright 2017 IOP Publishing, reprinted with permission, all rights reserved.

or above, the surface approaches its ideal form by desorbing the extra adatoms on the surface at such a high growth temperature. In summary, at a typical GaN growth temperature of 1050 °C with a typical V–III ratio of 1000 or above, it is likely to form a Ga-rich surface (Ga_{ad} (T4)) under N_2 ambient as shown in figure 10.10(b), and hydrogen terminated GaN surface, i.e. 3Ga-H, in H_2 ambient as shown in figure 10.10(f).

10.3.2 The effects of carrier gas

The effects of carrier gases on C incorporation are discussed in the previous section where it is shown that the C incorporation in H_2 ambient could be as much as two orders of magnitude higher than that of N_2 ambient. Therefore, it is crucial to understand their physical origin.

The probability of defect incorporation into a pure crystal depends on its formation energy. Considering carbon substituting N (C_N), TMG as the Ga precursor, NH_3 as the source of active N, and H_2 as the carrier gas, the formation energy of C_N in a pure GaN crystal grown using MOCVD can be expressed as [17]

$$
E^f(C_N) = \left[E^{bulk}_{GaN, C_N} - E^{bulk}_{GaN} \right] + \left[\mu_{NH_3}(T, P) + \frac{1}{2}\mu_{H_2}(T, P) - \mu_{CH_4}(T, P) \right]
$$
$$
+ RT \ln \frac{P_{NH_3}}{P_{CH_4}} + \frac{1}{2} RT \ln \frac{P_{H_2}}{P}, \tag{10.4}
$$

where E^{bulk}_{GaN,C_N} is the energy of a pure GaN crystal containing one carbon defect, E^{bulk}_{GaN} is the energy of a pure GaN crystal, P denotes the standard pressure, and μ represents the chemical potentials of the respective species. The C concentration can then be calculated as follows:

$$
C = C_0 \exp\left(-\frac{E^f(C_N)}{k_B T} \right). \tag{10.5}
$$

Therefore, the higher the formation energy is, the lower the defect incorporation. The last term in (10.4) indicates that higher H_2 partial pressure should increase the defect formation energy. In contrast, as shown in figure 10.5, GaN grown in H_2 contains more carbon than in N_2. Furthermore, there is no correlation between the carrier gas partial pressure and C incorporation. However, the partial pressure of hydrogen is also related to the decomposition of NH_3 into nitrogen and hydrogen. Therefore, an increase of the V–III ratio will also increase the H_2 partial pressure and thus decrease the carbon concentration. This possibility will be discussed in detail in the thermodynamic modeling in the next section. The rest of this section is devoted to understanding the atomic level changes in the crystal when an impurity is incorporated into the crystal in H_2 and N_2 ambient.

In order to predict the C incorporation in a crystal, accurate calculation of the formation energy from equation (10.4) is required. Kempisty *et al* show that the impurity incorporation also depends on the characteristics of the epitaxial surface and the sub-surface layers underneath during the growth [17]. The charged surface states, energy band bending, and electric field at the surface along with the complex nature of the surface reconstruction also plays an important role, as the band energy near the surface is different from the bulk layer. Consequently, the formation energy is also different. Therefore the surface reconstruction, as discussed in the previous section, must be taken into account in order to explain the experimentally observed trend in C incorporation with different carrier gases.

Recalling the different surface reconstructions from section 10.3.1, the Ga adsorbed surface Ga_{ad} (T4) or hydrogen adsorbed surface 3Ga-H appears under N_2 and H_2 ambient, respectively, with a typical GaN growth temperature of 1050 °C and a V–III ratio higher than 1100. The energy band diagram with these two different surface reconstructions of a pure n-type GaN is shown in figure 10.12. As we can see, the valence band (VB) bends upwards due to the high density of surface states within the bandgap in the (0001) direction, where the surface is covered by Ga adatoms, i.e. (Ga_{ad} (T4)) in figure 10.12(a). In the case of a hydrogen covered surface (3Ga-H) in figure 10.12(b), no such high density states are present inside the bandgap, resulting in a flat VB in the same (0001) direction. However, the H adatoms generate surface states below the VB maxima. In both cases, the unsaturated Ga adatoms generate empty surface states above the conduction band maxima [17].

Now if we put a carbon atom on the (0001) surface and move it to the bulk through each atomic layer, we will be able to calculate the energy change (ΔE) of the GaN crystal, as shown in figures 10.13(a) and (b). As can be seen, for the Ga_{ad} (T4)

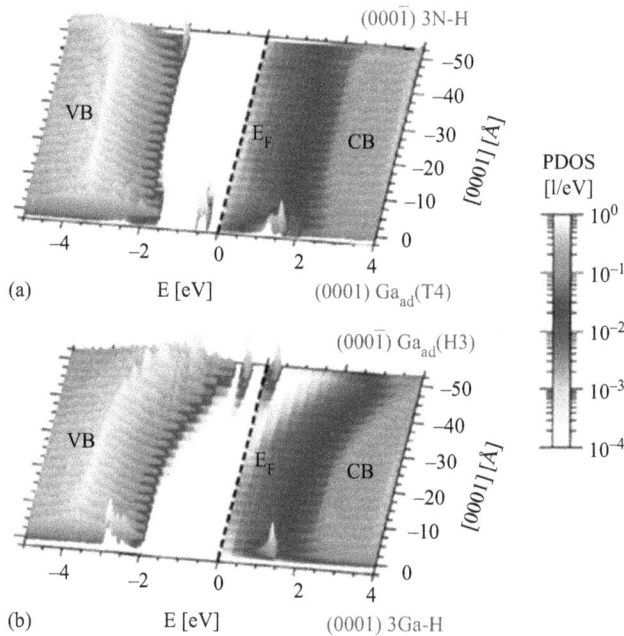

Figure 10.12. Structure of the energy band of pure n-type GaN with different reconstructed surfaces in H_2 and N_2 ambient. Reproduced with permission from [17]. Copyright 2017 American Physical Society.

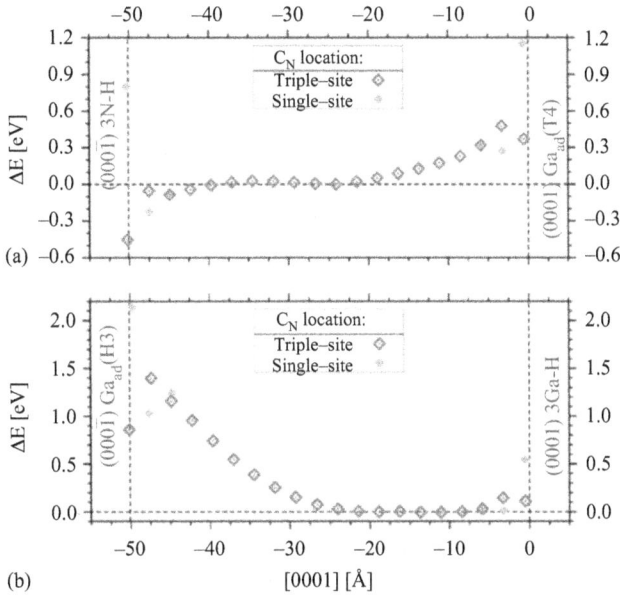

Figure 10.13. Energy change in the GaN layer with substitutional C defects as a function of the carbon position, which is shifted from the surface in the (0001) direction towards the bulk. Reproduced with permission from [17]. Copyright 2017 American Physical Society.

surface, i.e. when the GaN is grown in N_2 ambient, the energy is higher at the surface in the (0001) direction and reduces gradually as the carbon atom moves into the bulk (figure 10.13(a)). In other words, a potential barrier exists in the surface and sub-surface region, which results in lower C incorporation [17].

In contrast, the energy barrier of the 3Ga-H surface is almost the same as the bulk and much lower than the Ga_{ad} (T4) surface. Here the energy band is flat, as shown in figure 10.13(b). Thus the dopant energy at the surface is almost the same as in the bulk. The defect formation energy is also low. This explains that the C_N concentration in GaN is lower under N_2 ambient due to a higher formation energy barrier and a higher concentration of C_N is expected in GaN grown in H_2 ambient due to the comparatively lower energy barrier [17].

This model describes the physical mechanism of C incorporation in GaN under H_2 and N_2 ambient, which is in line with the experimental observations discussed in the previous section. However, it does not explain the characteristics of C incorporation in H_2 and N_2 ambient with different V–III ratios, as shown in figure 10.5. The next section considers the thermodynamic model of C impurities in GaN under different growth conditions.

10.3.3 The thermodynamic model of impurity incorporation

In general, the defect formation energy determines the incorporation of an impurity in a crystal and thereby its density. In the case of acceptor-like substitutional carbon defects at nitrogen sites in GaN, the formation energy may be described by [22]

$$E^{\mathrm{f}}\!\left(\mathrm{C}_{\mathrm{N}}^{q}\right) = E_{\mathrm{tot}}\!\left(\mathrm{C}_{\mathrm{N}}^{q}\right) - E_{\mathrm{tot}}(\mathrm{GaN}) - \mu_{\mathrm{C}} + \mu_{\mathrm{N}} + q(E_{\mathrm{F}} + E_{\mathrm{V}} + \Delta v), \qquad (10.6)$$

where $E^{\mathrm{f}}(\mathrm{C}_{\mathrm{N}}^{q})$ is the formation energy of carbon impurities in charged state q, $E_{\mathrm{tot}}(\mathrm{C}_{\mathrm{N}}^{q})$ is the total energy of a GaN crystal containing a single substitutional carbon at the N site, $E_{\mathrm{tot}}(\mathrm{GaN})$ is the total energy of the same GaN crystal without any defects, μ_{C} and μ_{N} are the chemical potentials of carbon and nitrogen, respectively, E_{F} is the Fermi level with respect to the valence band maximum (E_{V}), and Δv is a correction parameter. In the case of MOCVD growth of GaN, the residual carbon concentration is a strong function of the growth temperature, V–III ratio, choice of carrier gas, and metal precursors used. The chemical potentials, μ_{C} and μ_{N}, reflect the dependence of carbon incorporation on the growth conditions, such as temperature, V–III ratio, and carrier gases, and can be determined by the thermodynamic growth models. Therefore, an in-depth understanding of the relationship between the change in the growth parameters and the corresponding change in the chemical potentials is needed in order to explain the dependence of carbon incorporation on the carrier gases as evidenced in figure 10.5.

The change in C_{N} formation energy due to the change in growth parameters with respect to a reference growth condition can be obtained from equation (10.6) [14], where

$$\Delta E^{\mathrm{f}}\!\left(\mathrm{C}_{\mathrm{N}}^{q}\right) = \Delta\mu_{\mathrm{N}} - \Delta\mu_{\mathrm{C}}. \qquad (10.7)$$

The corresponding residual carbon concentration can be given by equation (10.5) as discussed above and which is rewritten below for convenience

$$\mathrm{C}_{\mathrm{N}} = (\mathrm{C}_{\mathrm{N}})_{\mathrm{ref}} \exp\!\left(-\frac{\Delta E^{\mathrm{f}}\!\left(\mathrm{C}_{\mathrm{N}}^{q}\right)}{k_{\mathrm{B}}T}\right). \qquad (10.5)$$

Furthermore, during GaN growth, the Ga and N chemical potentials are related to each other such that $\Delta\mu_{\mathrm{Ga}} + \Delta\mu_{\mathrm{N}} = 0$ or $\Delta\mu_{\mathrm{N}} = -\Delta\mu_{\mathrm{Ga}}$.

Therefore,

$$\Delta E^{\mathrm{f}}\!\left(\mathrm{C}_{\mathrm{N}}^{q}\right) = -\Delta\mu_{\mathrm{Ga}} - \Delta\mu_{\mathrm{C}}. \qquad (10.8)$$

In any case, finding μ_{N}, μ_{Ga}, or μ_{C} from the thermodynamic growth model is not straightforward. However, they can be estimated by a combination of experiments at different growth conditions and their relationship with the thermodynamic model. The chemical potential physically represents the corresponding change in Gibb's free energy at a given growth condition and can be expressed as [14]

$$\mu_{\mathrm{III}} = \Delta G_{\mathrm{IIIN}} - \Delta G_{\mathrm{III}},$$

where ΔG represents the change in Gibb's free energy of the respective elements. In this case, we are interested in μ_{Ga}, which can be expressed as

$$\mu_{\mathrm{Ga}} = \Delta G_{\mathrm{GaN}} - \Delta G_{\mathrm{Ga}}. \qquad (10.9)$$

Furthermore, a good estimation of the change in chemical potential of Ga (μ_{Ga}) may be obtained from the actual growth parameters by the following equation [14]:

$$\Delta\mu_{Ga} = \left[k_B T \log\left(\frac{P_{eq}^{Ga}}{P_{vapor}^{Ga}} \right) \right]_{X2} - \left[k_B T \log\left(\frac{P_{eq}^{Ga}}{P_{vapor}^{Ga}} \right) \right]_{X1}, \tag{10.10}$$

where X1 and X2 refer to two different growth conditions, and P_{par}^{Ga} and P_{vapor}^{Ga} are the equilibrium partial pressure of Ga on the growth surface and equilibrium vapor pressure of Ga at a given temperature, respectively. P_{par}^{Ga} and P_{vapor}^{Ga} are related through a fourth-order polynomial and can be estimated from the MOCVD growth parameters. The source of nitrogen during GaN growth is NH_3, which decomposes into H_2 and N_2 at the growth temperature. An understanding of the NH_3 decomposition mechanism is also needed to understand the relationship of carbon incorporation with the carrier gases. The following set of equations describes the NH_3 decomposition at the GaN growth temperature [23]:

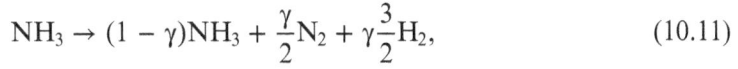

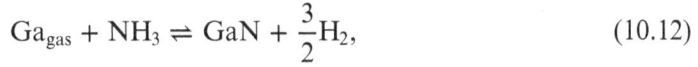

$$NH_3 \rightarrow (1-\gamma)NH_3 + \frac{\gamma}{2}N_2 + \gamma\frac{3}{2}H_2, \tag{10.11}$$

$$Ga_{gas} + NH_3 \rightleftharpoons GaN + \frac{3}{2}H_2, \tag{10.12}$$

$$K = \frac{P_{H_2}^{3/2} a_{GaN}}{P_{eq}^{Ga} P_{NH_3}}, \tag{10.13}$$

where K is called the growth constant, P_{H_2}, P_{Ga}, and P_{NH_3} are the H_2, Ga, and NH_3 partial pressure, respectively, and a_{GaN} is the activity the of GaN.

As evidenced by the SIMS measurements in figure 10.5, the GaN layer grown with the TMG precursor and H_2 carrier gas shows a higher carbon concentration than the GaN layer grown in N_2 ambient. This is true for a lower V–III ratio, typically below 1000. However, the trend could be the opposite in the case of a very high V–III ratio, e.g. 2000 or higher. To explain the relation between C incorporation and a V–III ratio under H_2 and N_2 ambient, let us define three different growth regimes. Regime 1 refers to GaN growth in H_2 ambient with a V–III ratio less than 2000. Regime 2 refers to GaN growth in N_2 ambient with the same V–III ratio less than 2000, and regime 3 refers to the GaN growth in H_2 or N_2 ambient with a V–III ratio higher than 2000. Figure 10.14 describes the C incorporation in all three regimes. The ammonia decomposition is poor in H_2 ambient, therefore, according to equation (10.13), an increase in NH_3 partial pressure, i.e. the ammonia flow and in turn the V–III ratio, decreases the Ga partial pressure to maintain the growth constant (figure 10.14, middle panel). According to equation (10.10), we see a decrease in the Ga chemical potential, as shown in figure 10.14 (right panel). Hence, the formation energy of C_N is much lower with a low V–III ratio and increases sharply with increasing V–III ratio in H_2 ambient, thereby resulting in lower carbon incorporation.

In contrast, NH_3 decomposes much faster in N_2 ambient. Therefore, higher NH_3 partial pressure also increases H_2 partial pressure due to rapid decomposition of

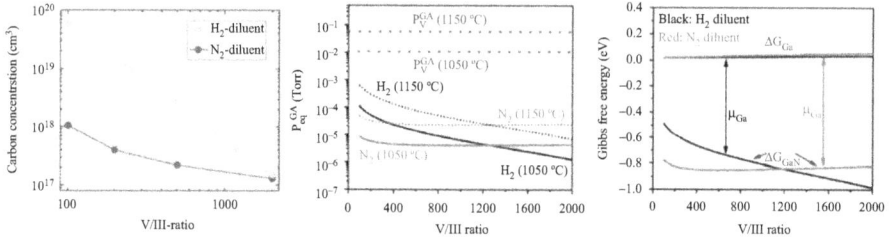

Figure 10.14. Left: C concentration as a function of the V–III ratio in H_2 and N_2 ambient. Middle: The change in Ga partial pressure in N_2 and H_2 ambient as a function of the V–III ratio as obtained from equation (10.10). Right: The corresponding change in Gibbs' free energy as calculated from equation (10.9). Reproduced with permission from [14]. Copyright 2017 American Physical Society.

NH_3, as depicted by equation (10.11). This balances the reaction constant K in equation (10.13). Thus P_{eq}^{Ga} remains unchanged at a higher V–III ratio (figure 10.14, middle) panel. At a low V–III ratio, the increase in P_{H_2} is not significant after NH_3 decomposition due to lower NH_3 flow, resulting in an increase in Ga partial pressure. As a result, the formation energy increases with decreasing V–III ratio but remains unchanged at a high V–III ratio. Therefore, at a low V–III ratio, typically below 1000, the formation energy of substitutional carbon at N sites is higher in N_2 ambient than that of H_2 ambient while keeping all other growth parameters identical. Furthermore, there is a crossover of P_{eq}^{Ga} at higher V–III ratios when H_2 carrier gas is used, which may also result in low C incorporation in H_2 ambient with a very high V–III ratio.

10.3.4 The effects of the Ga precursor

Carbon incorporation is inherent in MOCVD GaN due to the use of metal–organic precursors. The most common Ga precursors for growing GaN are trimethylgallium (TMG) and triethylgallium (TEG). Both of them fully decompose during GaN growth, releasing either methyl or ethyl radicals in the case of TMG or TEG, respectively. The way these radicals interact with the GaN surface depends on their decomposition mechanisms during growth. Therefore, a detailed understanding of TMG and TEG decompositions is needed to interpret their roles in C incorporation into epitaxial GaN.

10.3.4.1 TMG decomposition mechanism
The TMG decomposition mechanism on a GaN surface as reported by An *et al* is schematically shown in figure 10.15 [24]. At the beginning, TMG is physically absorbed on an N-rich Ga surface during the growth, as illustrated in figure 10.15(a). This requires -5.51 kcal mol^{-1} as the heat of reaction or adsorption energy. In this process, the distance between the Ga atom in TMG and the N atom on the surface is 0.2272 nm, which is longer than the Ga–N bond length in a GaN crystal, indicating a weaker interaction between the nitrogen and the gallium empty orbitals. A hydrogen atom either from the surface (in the case of a 3Ga-H surface) or

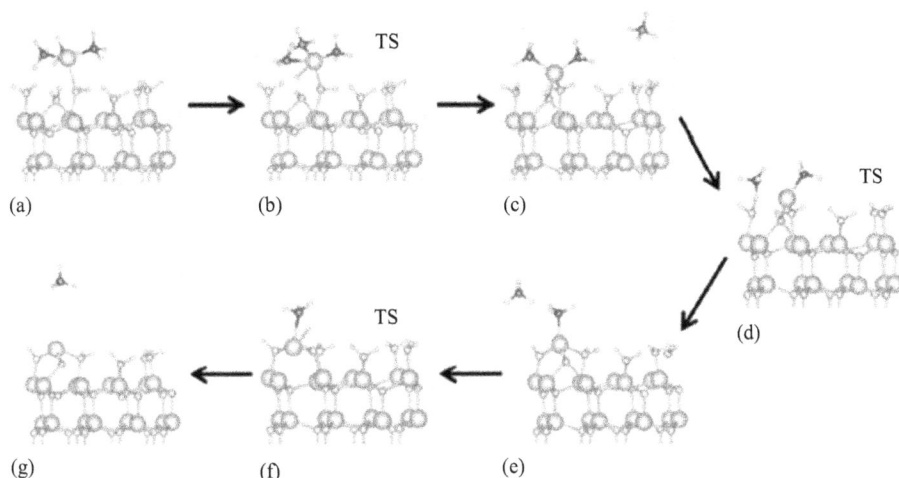

Figure 10.15. TMG decomposition mechanism on the N-rich Ga surface in the (0001) growth direction. Reproduced with permission from [24]. Copyright 2015 American Chemical Society.

from the NH_3 decomposition reacts with one CH_3 radical and forms the very stable CH_4. CH_4 is then released from the surface, requiring -23.49 kcal mol^{-1} as the heat of reaction and 29.5 kcal mol^{-1} as the reaction barrier. After the chemical adsorption of TMG on the surface, the transition from TMG to dimethylgallium (DMG) occurs by releasing the CH_4 from the surface, as shown in figures 10.15(b) and (c). Meanwhile, the Ga–N bond reduces to 0.2058 nm and the distance between Ga and C increases from 0.2035 to 0.2297 nm. In the next step, another CH_3 from the DMG forms CH_4 by reacting with one more H atom followed by the formation of monomethylgallium (MMG). This process requires a higher reaction barrier of 42.54 and -2.3 kcal mol^{-1} as the heat of reaction. The MMG is likely to form on a T4 site, where MMG is bonded with the three N atoms on the surface, as indicated in figures 10.15(c) and (d). The T4 site refers to the Ga layer growth site and hence no diffusion of the Ga adatoms is needed for the growth. The remaining CH_3 in the MMG requires one more H atom to form CH_4 and leaves a bare Ga, as shown in figure 10.15(g). However, as shown in figure 10.15(f), the surface H atom is far from the CH_3 in the MMG. Therefore, the last CH_3 will most likely attach to a gas phase H atom. The reaction barrier for this reaction is 29.70 kcal mol^{-1}, which is less than the reaction barrier required for the transition from DMG to MMG. The energy barrier for the transition from DMG to MMG, as discussed above, is 42.54 kcal mol^{-1}, which is the highest energy required in one transition during the process of TMG decomposition. The only consequences is that the TMG decomposition does not favor low temperature growth. However, at a typical growth temperature of above 1000 °C, the TMG is fully decomposed into bare Ga and results in crystalline GaN [24].

10.3.4.2 TEG decomposition mechanism

There are two probable decomposition mechanisms of TEG during GaN growth. They are (i) direct decomposition, such as TMG, and (ii) beta-hydride elimination. Beta-hydride elimination is favored over direct decomposition. There are three steps involved in beta-hydride elimination. In the first step, one C_2H_5 decomposes into $C_2H_4 + H$. C_2H_4 is then dissociated with a reaction barrier of 30 kcal mol^{-1} and a heat of reaction of 26.10 kcal mol^{-1}. In this transition, TEG ($Ga(C_2H_5)_3$) is transformed into $Ga(C_2H_5)_2H$. We note that the reaction energy in this step is very similar to that of CH_4 dissociation from TMG. The second C_2H_4 dissociation from $Ga(C_2H_5)_2H$ needs even a higher reaction barrier of 55.12 and 25.85 kcal mol^{-1} as the heat of reaction. This is followed by the dissociation of the last C_2H_4 radical from the $Ga(C_2H_5)H_2$. This reaction needs to overcome a high-energy barrier of 80.03 kcal mol^{-1} and a heat of reaction of 26.69 kcal mol^{-1}. Such a high energy for dissociation of C_2H_4 from $Ga(C_2H_5)H_2$ does not suggest the use of a TEG precursor for low temperature GaN growth over its TMG counterpart.

In the case of direct decomposition, TEG decomposes into diethylgallium (DEG) to monomethylgallium (MEG) by dissociating C_2H_5 in each step, similar to the direct decomposition of TMG by releasing CH_3. C_2H_5 then reacts with one H atom from the gas phase and releases C_2H_6 from the surface. The reaction barrier of the first C_2H_5 dissociation from the TEG is 27.3 kcal mol^{-1}. This energy barrier is comparable to both beta-hydride elimination with a reaction barrier of 30 kcal mol^{-1} and direct decomposition of TMG with a reaction barrier of 29.50 kcal mol^{-1}. Therefore, there is a possibility that both beta-hydride elimination and direct decomposition occur during the growth process [24].

At this point, it is clear that TMG decomposes into Ga and CH_4 during growth, whereas TEG decomposes into Ga and either C_2H_4 or C_2H_6 based on the decomposition mechanism. In any case, the CH_4, C_2H_4, or C_2H_6 is the source of residual carbon in the as-grown GaN layer.

10.3.4.3 Carbon incorporation by TMG and TEG

In order to explain the experimental observations of C incorporation on TMG- and TEG-grown GaN, we recall equation (10.2):

$$\Delta E^f\left(C_N^q\right) = \Delta\mu_N - \Delta\mu_C,$$

where $\Delta\mu_C$ can be evaluated from the methyl or ethyl radicals in the following ways:

$CH_4 \rightleftharpoons C + 2H_2$, $C_2H_4 \rightleftharpoons 2C + 2H_2$, and $C_2H_6 \rightleftharpoons 2C + 3H_2$. The chemical potential of carbon μ_C is proportional to $-z\mu_{H_2} + \mu_{C_2H_6/CH_4}$, where z is a constant to maintain the generality of the expression [14]. Therefore, a lower chemical potential of ethyl or methyl radicals will result in a higher C_N formation energy and thereby lower C incorporation. Figure 10.16 shows the chemical potentials of methane and ethane as a function of their mole fraction [25]. As can be seen from the figure, ethane (C_2H_6) has a lower chemical potential than methane (CH_4), therefore leading to a lower residual carbon in the as-grown GaN layer.

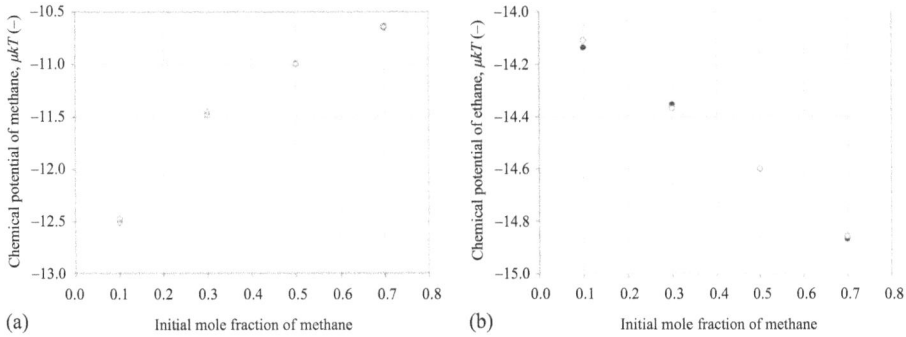

Figure 10.16. Chemical potential of (a) methane and (b) ethane as a function of their initial mole fraction. Reproduced with permission from [25]. Copyright 2017 Taylor and Francis.

10.4 Conclusions

Residual carbon in GaN shows profound effects on the transport properties of HEMTs, such as electron mobility and 2DEG concentration due to its deep acceptor-like characteristics in the material point of view. Both theoretical and experimental results show that carbon related defects in MOCVD grown GaN originate from the metal–organic sources and therefore are highly dependent on the choice of metal precursors and growth parameters, such as the V–III ratio, reactor pressure, and growth temperature. It is clearly observed that carbon incorporation decreases with increasing NH_3 partial pressure, V–III ratio, reactor pressure, and growth temperature. Additionally, the choice of carrier gas and metal precursor also considerably influences the C incorporation into GaN. The C impurity concentration of a GaN layer grown in nitrogen ambient with a low V–III ratio can be reduced up to two orders of magnitude compared to the layer grown in hydrogen ambient. Furthermore, TEG-grown GaN can have a significantly lower carbon background than TMG-grown GaN. The interplay between the residual carbon background and the growth parameters can be understood at the atomic level with the help of density functional theory and thermodynamic models. The theoretical models suggest that the change in chemical potentials due to the change in growth parameters significantly influences the defect formation energy, which in turn controls the carbon incorporation rate. Moreover, the surface reconstruction in different growth conditions also influences the defect formation energy. Although residual C incorporation in MOCVD grown GaN is inevitable, a deep understanding of C related defects and their incorporation mechanism should help in reducing the defect concentration below a certain level such that their adverse effects on device performance may be ignored.

References

[1] Palacios T, Mishra U K and Sujan G K 2016 GaN-based transistors for high-frequency applications *Reference Module in Materials Science and Materials Engineering* (Amsterdam: Elsevier)

[2] Rosker M J, Albrecht J D, Cohen E, Hodiak J and Chang T 2010 DARPA's GaN technology thrust *2010 IEEE MTT-S Int. Microwave Symp.* (New York: IEEE) pp 1214–7

[3] Acurio E *et al* 2018 Reliability improvements in AlGaN/GaN Schottky barrier diodes with a gated edge termination *IEEE Trans. Electron Devices* **65** 1765–70

[4] Diallo I C and Demchenko D O 2016 Native point defects in GaN: a hybrid-functional study *Phys. Rev. Appl.* **6** 064002

[5] Jia Y, Xu Y, Lu K, Wen Z, Huang A-D and Guo Y-X 2018 Characterization of buffer-related current collapse by buffer potential simulation in AlGaN/GaN HEMTs *IEEE Trans. Electron Devices* **65** 3169–75

[6] Bisi D *et al* 2013 Deep-level characterization in GaN HEMTs—part I: advantages and limitations of drain current transient measurements *IEEE Trans. Electron Devices* **60** 3166–75

[7] Koller C, Pobegen G, Ostermaier C, Huber M and Pogany D 2017 The interplay of blocking properties with charge and potential redistribution in thin carbon-doped GaN on n-doped GaN layers *Appl. Phys. Lett.* **111** 032106

[8] Lyons J L, Janotti A and Van de Walle C G 2010 Carbon impurities and the yellow luminescence in GaN *Appl. Phys. Lett.* **97** 152108

[9] Kaess F *et al* 2016 Correlation between mobility collapse and carbon impurities in Si-doped GaN grown by low pressure metal–organic chemical vapor deposition *J. Appl. Phys.* **120** 105701

[10] Kusaba A, Li G, Kempisty P, Von Spakovsky M R and Kangawa Y 2019 CH_4 adsorption probability on GaN (0001) and (000−1) during metal–organic vapor phase epitaxy and its relationship to carbon contamination in the films *Materials (Basel)* **12** 972

[11] Koleske D, Wickenden A, Henry R and Twigg M 2002 Influence of MOVPE growth conditions on carbon and silicon concentrations in GaN *J. Crystal Growth* **242** 55–69

[12] Chyi I S 2018 Improving the performance of AlInN/GaN and AlInGaN/GaN HEMTs by using a triethylgallium-grown channel layer and barrier *2018 Int. Symp. on Growth of III-Nitride (ISGN-7)* (New York: IEEE)

[13] Chen Y, Sanyal I and Chyi J 2019 Enhanced electrical properties of AlInN/AlN/GaN heterostructure using $Al_xGa_{1-x}N/Al_yGa_{1-y}N$ superlattice *2019 Compound Semiconductor Week (CSW)* (New York: IEEE) pp 1–2

[14] Reddy P *et al* 2017 Point defect reduction in MOCVD (Al)GaN by chemical potential control and a comprehensive model of C incorporation in GaN *J. Appl. Phys.* **122** 245702

[15] Kusaba A *et al* 2017 Thermodynamic analysis of (0001) and GaN metal−organic vapor phase epitaxy *Jpn J. Appl. Phys.* **56** 070304

[16] Sekiguchi K *et al* 2017 First-principles and thermodynamic analysis of trimethylgallium (TMG) decomposition during MOVPE growth of GaN *J. Crystal Growth* **468** 950–3

[17] Kempisty P *et al* 2017 DFT modeling of carbon incorporation in GaN(0001) and GaN (0001⁻) metalorganic vapor phase epitaxy *Appl. Phys. Lett.* **111** 141602

[18] Stegmüller A, Rosenow P and Tonner R 2014 A quantum chemical study on gas phase decomposition pathways of triethylgallane (TEG, $Ga(C_2H_5)_3$) and tert-butylphosphine (TBP, $PH_2(t-C_4H_9)$) under MOVPE conditions *Phys. Chem. Chem. Phys.* **16** 17018–29

[19] Danielsson Ö, Li X, Ojamäe L, Janzén E, Pedersen H and Forsberg U 2016 A model for carbon incorporation from trimethyl gallium in chemical vapor deposition of gallium nitride *J. Mater. Chem.* C **4** 863–71

[20] Freysoldt C *et al* 2014 First-principles calculations for point defects in solids *Rev. Modern Phys.* **86** 253

[21] Kangawa Y, Ito T, Taguchi A, Shiraishi K and Ohachi T 2001 A new theoretical approach to adsorption–desorption behavior of Ga on GaAs surfaces *Surf. Sci.* **493** 178–81

[22] Van de Walle C G and Neugebauer J 2004 First-principles calculations for defects and impurities: applications to III-nitrides *J. Appl. Phys.* **95** 3851–79

[23] Mita S, Collazo R, Rice A, Dalmau R F and Sitar Z 2008 Influence of gallium super-saturation on the properties of GaN grown by metalorganic chemical vapor deposition *J. Appl. Phys.* **104** 013521

[24] An Q, Jaramillo-Botero A, Liu W-G and Goddard W A III 2015 Reaction pathways of GaN (0001) growth from trimethylgallium and ammonia versus triethylgallium and hydrazine using first principle calculations *J. Phys. Chem.* C **119** 4095–103

[25] Tan S J, Do D and Nicholson D 2017 A new kinetic Monte Carlo scheme with Gibbs ensemble to determine vapour–liquid equilibria *Mol. Simul.* **43** 76–85

IOP Publishing

Wide Bandgap Semiconductor-Based Electronics

Fan Ren and Stephen J Pearton

Chapter 11

High Al-content AlGaN-based HEMTs

Albert G Baca, B A Klein, A M Armstrong, A A Allerman, E A Douglas and R J Kaplar

The status and outlook for AlGaN-channel high electron mobility transistors (HEMTs) are reviewed as a new and important research topic. Figure-of-merit (FOM) analysis shows encouraging comparisons relative to today's state-of-the-art GaN devices for high Al content and even more so at elevated temperatures. The critical electric field (E_C) fuels the AlGaN HEMT FOM for high Al composition and electron mobility acts as a drag. The average gate–drain electric field at breakdown is so far substantially better in AlGaN-channel devices compared to GaN. Although numerically inferior to GaN, the electron mobility is less temperature sensitive for AlGaN compared to GaN. The challenges for AlGaN are associated with current density constraints arising from both mobility (dominated by ternary alloy scattering) and the difficulty of making reasonable ohmic contacts to high Al content materials, but also due to thermal conductivity. Nevertheless, considerable progress has been made recently. Excellent I_{ON}–I_{OFF} current ratios have been reported for Schottky-gated structures, in some cases exceeding 10^{11}. The high Al-content transistors also outperform GaN in one important aspect at high temperature by maintaining I_{ON}–I_{OFF} ratios of $\sim 10^6$ at 500 °C. Specific contact resistivity (ρ_c) approaching $\rho_c \sim 2 \times 10^{-6}\ \Omega\ cm^2$ to AlGaN devices with 70% Al content in the channel have been reported. Improvements in contact resistance, channel length, and tailoring the threshold voltage have enabled a considerable increase in the current density, which has now reached 0.6 A mm^{-1}. The depletion-mode radio frequency (RF) performance in high Al-content transistors is showing f_T and f_{max} in the tens of GHz range for submicron gates, while enhancement-mode devices have been reported and are aimed towards power switching applications. An initial e-mode reliability study is supportive of the F stability in these HEMTs and initial radiation results have also been reported.

doi:10.1088/978-0-7503-2516-5ch11

11.1 Introduction

A high electron mobility transistor [1] operates on the basis of separating the ionized donor impurities from the mobile electrons in a semiconductor device design so that electron transport in a two-dimensional electron gas (2DEG) is unencumbered by ionized impurity scattering. Mobility is therefore decoupled from conductivity in a field effect transistor (FET). Many applications have been envisioned for HEMTs, but their enduring use came about for low noise RF amplifiers, initially with GaAs and InGaAs channels [2, 3], enabled by the impurity-free conduction in the 2DEG. The introduction of GaAs-based power amplifiers also occurred in the 1980s, but HEMT-based power amplifiers would come into prominence with the wide bandgap (WBG) AlGaN/GaN transistors [4, 5].

Originally lacking a native substrate, much of the early research on AlGaN/GaN involved growth and fabrication on sapphire substrates. Heteroepitaxial GaN growth on substrates without lattice constants matched to that of hexagonal GaN have resulted in high densities of threading dislocations, commonly exceeding 10^9 cm^{-2}. These threading dislocations were viewed with great concern and clouded the prospects for gaining sufficient reliability in GaN devices. With hindsight, neither the performance nor reliability turned out to be significantly affected by dislocations. Innovations involving SiN passivation [6, 7], field plates [8, 9], and migration to SiC, Si, or free-standing GaN substrates resulted in impressive RF power density demonstrations [8–11] arising from a 10× larger critical electric field (E_C) and a thermal conductivity improvement over sapphire substrates. Research into reliability topics and other issues associated with technological viability concerns ensued. Topics such as transient current conductivity degradation in response to large electric fields pulses [12] were vigorously pursued for many years prior to commercial introduction. Meanwhile, multiple RF applications, including military radars and cell phone base stations, began adopting GaN-based solutions [13, 14].

Power switching applications, in contrast, developed more slowly. These applications required normally-off (enhancement-mode) transistors in contrast to RF devices. In addition, the desire for Si substrates in this cost-sensitive market resulted in less commonality than initially expected for RF and power GaN technologies. A divergence in the learning curve resulted for many research topics. GaN on Si development brought a renewed study of the effect of the substrate on buffer quality, while high breakdown required thick buffers. Voltage derating due to transient on-state conductivity also depended uniquely on the substrate and buffer technology. All of these issues have been mitigated to a great extent and GaN power switching devices are now becoming commercially available. However, the technology is complex and will require specialized GaN-based converter design skills [13]. GaN power switching technology will also have particular advantages in fast switching, but may be voltage limited to less than 600 V.

Although the WBG semiconductors have now achieved commercial success in these electronic applications, research ideas to further their development continue to be vigorously pursued. One example is polarization doping for improved power gain linearity [15]. A second example is the implementation of multiple, stacked quantum well channels for HEMTs that may greatly increase current density and RF performance while also lessening current collapse [16]. As these innovations can also apply to any HEMT device, research is also underway in search of better materials. In particular, research is underway on the ultra-wide bandgap (UWBG) semiconductors [17], those with a bandgap greater than that of GaN, 3.4 eV. These possibilities include diamond [18], Ga_2O_3 [19, 20], BN, and the high Al-content AlGaN alloys [21–23] with bandgaps approaching 6.2 eV. This chapter reviews the nascent research into AlGaN HEMTs aimed at power and RF electronics. Aspects of this work have been previously reviewed in [21].

The high Al-content AlGaN HEMT field is an evolving and exciting research area. Section 11.2 makes the case for theoretical performance advantages based on FOM analysis. Section 11.3 describes the challenges and status of making low resistivity ohmic contacts to UWBG semiconductors. Section 11.4 describes AlGaN-channel HEMT research with particular attention on the historical development. Section 11.5 describes breakdown voltage comparisons in high Al-content transistors. Section 11.6 describes other HEMT topics, including pulsed I–V characterization, reliability, high-temperature operation, and radiation environments.

11.2 Figures-of-merit suggest performance advantages for AlGaN-channel HEMTs

11.2.1 Power switching figures-of-merit

Researchers often face the difficult question of how to compare a new material with that of an incumbent technology. FOMs, if chosen judiciously, can be useful for these types of comparisons. No single FOM is ideal for this type of analysis because all have shortcomings. For power switching applications, the Baliga FOM (BFOM), also referred to as the unipolar FOM, is often used:

$$\text{BFOM} = \frac{V_B^2}{R_{sp,on}} = \frac{\varepsilon \mu E_C^3}{4}, \tag{11.1}$$

where V_B is the breakdown voltage, $R_{sp,on}$ is the specific on-resistivity (the on-resistance of a transistor multiplied by its area), ε is the permittivity, μ is the electron mobility, and E_C is the critical electric field for breakdown in a semiconductor. The BFOM equation applies to vertical devices such as p–i–n diodes. A similar FOM that applies to lateral devices is known as the lateral FOM (LFOM). Derivation of the BFOM and that of its lateral equivalent are given in [24]. The rationale for using this FOM is that it accounts for conduction losses in a power switching circuit in a way that accounts for the breakdown voltage capability. The similar FOM for lateral devices is given by

$$\text{LFOM} = \frac{V_B^2}{R_{sp,on}} = q \mu n_s E_C^2, \tag{11.2}$$

where n_s is the sheet carrier concentration in the 2DEG. Both FOMs represent conduction loss associated with the transistor design for a given breakdown voltage and, therefore, they can be experimentally compared by measuring V_B and $R_{sp,on}$. Without knowing that they represent V^2/R it is not obvious that $\varepsilon\mu E_C^3$ has the same units as $q\mu n_s E_C^2$. However, since they do and are directly comparable, the potentials of widely different semiconductor materials are compared after calculating the FOMs with the appropriate constants. In the power switching field, it is common to compare Si, SiC, GaN, AlN, and $Al_x Ga_{1-x}N$ alloys even though Si and SiC are vertical devices and the III-N materials can be either lateral or vertical.

Some comments regarding the use of the conduction loss FOMs are in order since comparisons using them are often used and generally valuable. Nevertheless, it should be understood that any comparisons are based on best case scenarios that represent the full potential of a material and such an outcome is not guaranteed, or even likely in many cases. Such comparisons are made to guide the research into emerging materials that may one day challenge the incumbent technology. Unfortunately, no guide other than intuition helps to estimate how closely a given material will approach its theoretical best. Some of the shortcomings of using the the BFOM and LFOM are that they do not account for the independent benefits of high frequency switching capability nor do they account for switching losses, nor any differences in thermal conductivity. In fact, thermal conductivity in $Al_x Ga_{1-x}N$ alloys is approximately 7× smaller than in GaN [25], necessitating HEMT designs that limit the thickness of AlGaN layers and buffers.

The FOMs of equations (11.1) and (11.2) are based on four material parameters, two of which (ε and n_s) vary little with Al content. The first key parameter is the mobility. Many researchers will unfavorably view a mobility that does not exceed $300 \text{ cm}^2 \text{ V}^{-1} \text{ s}^{-1}$ and for that reason will not consider the ternary $Al_x Ga_{1-x}N$ alloys as suitable candidates for power switching applications. This viewpoint is shortsighted, as will be seen, in particular at elevated temperatures. Figure 11.1 illustrates a mobility plot in the shape of a bathtub, with $Al_x Ga_{1-x}N$ mobility less than $300 \text{ cm}^2 \text{ V}^{-1} \text{ s}^{-1}$ for $0.15 < x < 0.9$. Figure 11.1 also shows that polar optical phonon scattering is the limiting factor in the total room temperature mobility for GaN and is also an important contributor to mobility for AlN. In contrast, alloy scattering limits the room temperature mobility for the ternary $Al_x Ga_{1-x}N$ alloys. In spite of its low room temperature mobility, the comparisons improve for the $Al_x Ga_{1-x}N$ alloys at elevated temperatures. Alloy scattering is relatively insensitive to temperature, while polar optical phonon scattering is strongly temperature-dependent. This would suggest that $Al_x Ga_{1-x}N$ channel HEMTs could be less temperature sensitive than GaN-channel HEMTs.

The second key material parameter is the critical electric field, E_C. Device operation at a high enough voltage will result in avalanche breakdown, the condition where the avalanche multiplication coefficient M exceeds unity and will rapidly increase with the electric field. In [21] the derivation of E_C based on the evaluation of the expression for M in a uniformly doped one-dimensional junction is reviewed. As alluded to in the introduction, E_C is also strongly dependent on the bandgap E_G. Most researchers cite the work of Hudgins *et al* [26], which asserted that $E_C \sim E_G^{2.0}$ for indirect-gap

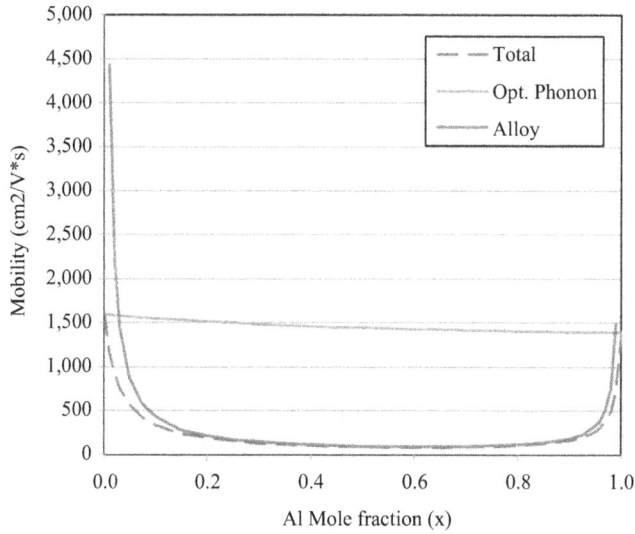

Figure 11.1. Calculated low-field mobility in the 2DEG of an AlN/AlGaN heterostructure, plotted as a function of Al fraction at 300 K. Reproduced with permission from [24]. Copyright 2017 Electrochemical Society.

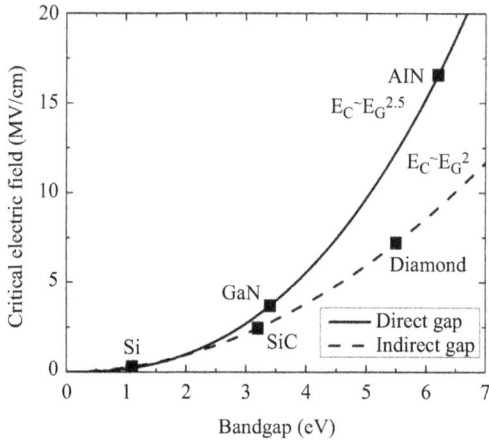

Figure 11.2. Predicted critical electric field as a function of bandgap based on the analysis of Hudgins *et al* [26], plotted as a function of the electric field. Reproduced from [21] with permission of AIP Publishing.

semiconductors and $E_C \sim E_G^{2.5}$ for direct-gap semiconductors. This analysis simplifies the picture somewhat, as E_C also depends on doping and temperature, which were not accounted for in Hudgins *et al*'s work [26]. Further, while the specific values of the exponents as well as the distinct exponents for direct- and indirect-gap materials are topics of active research, there is little doubt that critical electric field E_C is a *strong function* of the bandgap E_G. This relation is plotted in figure 11.2 for Si and several WBG and UWBG semiconductors.

E_C appears as the third power in the BFOM and as the square in the LFOM. While this would seem to suggest that the BFOM would be higher for materials that can have both vertical and lateral devices, e.g. the $Al_xGa_{1-x}N$ alloys, n_s can be quite large in a lateral device. Under the assumption $n_s \sim 1 \times 10^{13}$ cm^{-2}, figure 11.3 shows that the room temperature LFOM is larger than the BFOM for GaN and that both the lateral and vertical device FOMs are larger than the SiC BFOM. The trend is similar for other practical choices of n_s. However, the room temperature FOMs for $Al_xGa_{1-x}N$ alloys are smaller than GaN's for $x < 0.5$ for the BFOM and for $x < 0.8$ for the LFOM. Figure 11.3 also shows that the BFOM is favored over the LFOM at $x > 0.5$.

The FOM analysis favors the $Al_xGa_{1-x}N$ alloys at all compositions at 500 °C, as illustrated in figure 11.4. Both the lateral and vertical device FOMs are greater than those for GaN. In addition, both the BFOM and LFOM are better that the 25 °C BFOM for SiC. The true potential for HEMTs based on the $Al_xGa_{1-x}N$ alloys lies in the high Al-content compositions of $x > 0.8$ at room temperature or at all x for 500 °C.

11.2.2 RF figures-of-merit

Not surprisingly, an FOM based on conduction loss is not the best choice for an RF device. In that case, the most widely used FOM for RF performance is the Johnson figure-of-merit (JFOM) [27]:

$$\text{JFOM} = V_B f_T = \frac{E_C v_{\text{sat}}}{2\pi}, \tag{11.3}$$

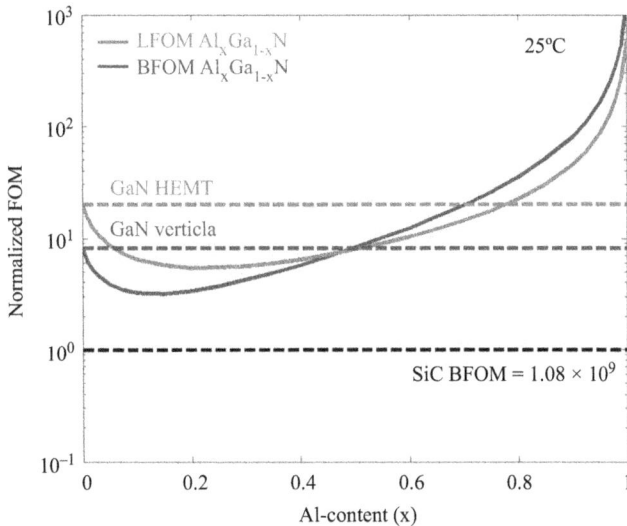

Figure 11.3. 25 °C LFOM and BFOM versus Al content for AlGaN-channel HEMTs, normalized to the 25 °C SiC BFOM value.

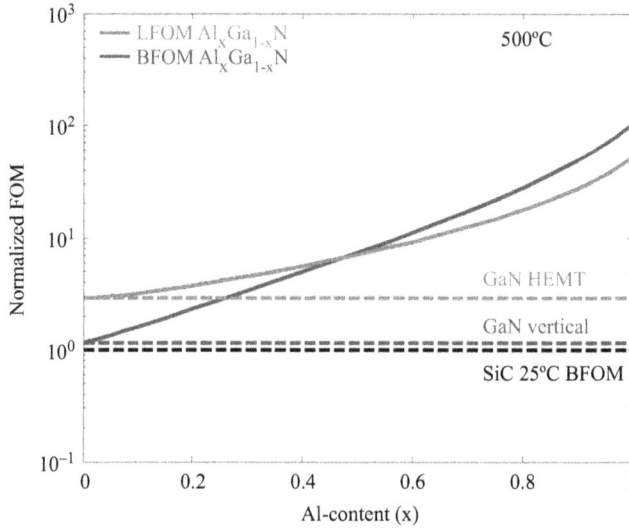

Figure 11.4. 500 °C LFOM and BFOM versus Al content for AlGaN-channel HEMTs, normalized to the 25 °C SiC BFOM value.

where f_T is the unity-gain cut-off frequency, the frequency where the current gain is reduced to one, v_{sat} is the saturation velocity, and the other symbols are defined previously. The JFOM was derived in [21] and the main assumption is that of a uniform electric field profile. This is a similar assumption to that used deriving the LFOM of equation (11.2), and with similar caveats. Also assumed is that a sufficiently high electric field exists such that the transit time is determined by the saturation electron velocity, which is the case for typical RF device geometry. The JFOM is therefore dependent on E_C and v_{sat}. The increased E_C for AlGaN compared to GaN increases the JFOM in a linear relation, rather than the square or the cube in the FOM for power switching devices.

The carrier velocity has been studied using Monte Carlo calculations [28], which are illustrated in figure 11.5. This plot shows the anticipated electron saturation velocity in AlGaN alloys. A recent paper has reported $v_s = 3.8 \times 10^6$ cm s^{-1} for $x = 0.7$ [29], somewhat lower than expected from figure 11.5. However, the decrease in v_{sat} with alloying is far weaker than it is for the low-field mobility (compare figure 11.1 and figure 11.5.), resulting in an increase in the JFOM for AlGaN relative to GaN over the majority of the Al composition range, and particularly so at a high Al percentage. A calculation of the JFOM as a function of Al% is shown in figure 11.6.

Breakdown and high frequency metrics (high JFOM) enable a high RF power density in transistors. An estimate of the power density as articulated in [21] for AlGaN relative to a similar GaN device may be made as follows. The maximum power density (W mm^{-1}) is estimated as $P_{max} = J_{max}V_{max}/8$, where J_{max} is the maximum current density (amps per unit gate width) and V_{max} is the maximum voltage, which depends on the breakdown voltage. From figure 11.2, it is seen that

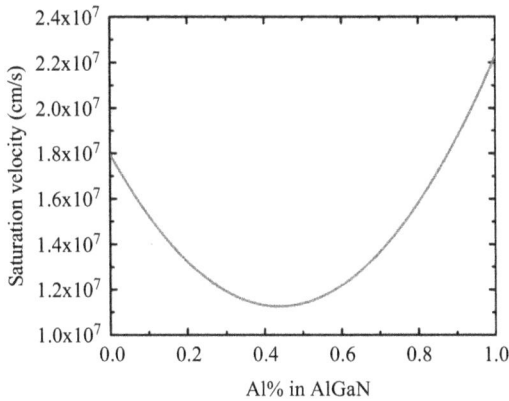

Figure 11.5. Expected electron saturation velocity in AlGaN derived from Monte Carlo calculations. Reproduced with permission from [24]. Copyright 2017 Electrochemical Society.

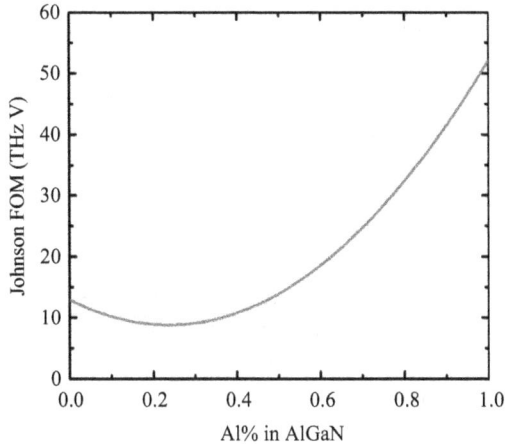

Figure 11.6. Calculation of JFOM as a function of Al% for AlGaN. Reproduced with permission from [24]. Copyright 2017 Electrochemical Society.

E_C for AlGaN can be up to four times that of GaN, implying that V_B will be up to four times higher than GaN. Further, the sheet carrier density for an Al-rich heterostructure can be twice that of an $Al_{0.3}Ga_{0.7}N/GaN$, a typical composition for GaN-channel RF HEMTs, due to the stronger piezoelectric properties of AlGaN [30]. Assuming parity in electron saturation velocity (figure 11.5), a doubling of the current density J_{max} is possible. The 4× increase in V_{max} times the 2× increase in J_{max} yields an expected 8× increase in RF power density for high Al-content AlGaN relative to a GaN-channel transistor.

11.3 Ohmic contacts for high Al-content AlGaN

One of the major challenges for high Al-content AlGaN involves ohmic contacts. High resistance of the source and drain electrodes and possibly an offset voltage if the contact is Schottky-like, lead to reduced current and thus limited performance. Increasing the Al content in AlGaN transistors lowers the electron affinity and causes large Schottky barriers to form at the metal–semiconductor interface [31]. High Al composition, therefore, makes producing effective ohmic contacts increasingly difficult. Although the topic of this chapter is HEMTs, ohmic contact research on high Al-content light-emitting diodes (LEDs) and on other transistor types is included because of the likelihood that complex process integration is ultimately involved in a good solution to the problem. Approaches to ohmic contacts reported in the literature include planar, Si-implanted, recess etched, recess etched plus graded regrowth, and polarization-doped contacts, each of which are elaborated upon in the following paragraphs.

The earliest reported AlGaN-channel HEMTs used $Al_yGa_{1-y}N/Al_xGa_{1-x}N$ heterostructures (described as y/x HEMTs or heterostructures) and also utilized a Si implant in the ohmic contact regions prior to metallization [32]. The implant was chosen to reduce the width of the Schottky barrier the electrons must traverse to cross between the HEMT barrier and metal electrodes. Contacts to several Al-channel compositions of 0%, 16%, and 38% were made, achieving, respectively, contact resistivity (ρ_c) of 6.9×10^{-7}, 9.68×10^{-6}, and 5.28×10^{-3} Ω cm^2 [32–35]. The room temperature implant energy was 50 keV with a dose of 1×10^{15} cm^{-2}. A post-implant annealing for 5 min at 1150 °C–1200 °C activated the implanted Si. A Ti/Al metal stack was deposited, lifted off, and thermally annealed.

Si ion implant doping of AlN was also used to make a conductive channel in an AlN metal–semiconductor field effect transistor (MESFET) [36]. A post-implant annealing at 1230 °C was followed by a Ti/Al/Ni/Au metallization for the source and drain electrodes and then another annealing for ohmic contact alloying for 30 s at 800 °C. Despite the complex processing, the room temperature drain current densities were limited to about 60 μA mm^{-1} by the source and drain contacts.

Planar ohmic contacts arise from a metal stack that is deposited onto a planar AlGaN barrier. Two types of metallurgical reactions are commonly cited as contributing to the contact formation mechanism [37]: (i) metal spiking through the barrier and into the channel, involving an interfacial reaction between the semiconductor and the metal, or (ii) the formation of TiN from Ti-containing metal stacks, which causes N vacancies in the semiconductor and the formation of a tunnel junction.

Other planar ohmic contact options to high Al-content AlGaN heterostructures (unenhanced by Si implantation) have also been reported [38–40]. A detailed study with a Zr (15 nm)/Al (60 nm)/Mo (35 nm)/Au (50 nm) metal stack and a 950 °C, 30 s annealing produced ρ_c of 4.8×10^{-2} Ω cm^2 [40] for an 86/51 heterostructure, and 2.6×10^{-4} for a 55/30 heterostructure [38]. The rapid thermal annealing (RTA) temperatures ranged from 850 °C to 1000 °C and the thicknesses of the Al and Zr were varied between 60–180 nm and 5–25 nm, respectively. This study concluded that Zr is fundamentally different than a Ti based contact, as transmission electron micrograph (TEM) images showed that Zr was able to spike more deeply into the AlGaN-channel layer.

Planar V/Al/V/Au contacts for LEDs were studied using molecular beam epitaxy (MBE) grown Si-doped AlGaN films with Al composition ranging from 0% to 100% [41]. Contacts made from structures with Al compositions between 0% to 70% all had ρ_c of around $1-3 \times 10^{-6}$ Ω cm^2; nevertheless, those with the higher Al compositions of 96% and 100% had much higher ρ_c of 0.35 and 1.01 Ω cm^2, respectively. Similarly, a planar V/Al/Ni/Au stack contacting a Si-doped Al$_{0.62}$Ga$_{0.38}$N bulk channel structure, 2 μm thick, was reported, achieving a ρ_c of 1.13×10^{-6} Ω cm^2 [42]. A third type of LED structure, one utilizing digital alloys with an average composition of $x = 0.66$, an average Si-doping of 2×10^{19} cm^{-3}, and using a Ti/Al/Ti/Au metal stack, showed ρ_c slightly higher than those of the uniform composition LEDs, with $\rho_c \sim 5.6 \times 10^{-5}$ Ω cm^2 [43].

A comparative study of five planar ohmic metal stacks commonly used in GaN- or AlGaN-channel HEMTs was made with 45/30 HEMTs [37]. The metal stacks were: (i) Ti/Al/Ni/Au, (ii) Ti/Al/Mo/Au, (iii) Zr/Al/Mo/Au, (iv) Nb/Ti/Al/Mo/Au, and (v) V/Al/V/Au. Annealing under ambient nitrogen was performed with temperatures that varied from 850 °C to 950 °C for 30 s. Ti/Al/Ni/Au and Ti/Al/Mo/Au produced the lowest ρ_c when annealed at 950 °C, with respective values of 3×10^{-6} and 2.5×10^{-5} Ω cm^2. HEMTs utilizing Ti/Al/Ni/Au were fabricated and tested from −50 °C to 200 °C. They showed ohmic behavior throughout the temperature range with decreasing ρ_c and increasing sheet resistance as the temperature was increased.

Linear ohmic contacts to a 70% → 85% graded polarization-doped FET (PolFET) structure have been reported [44]. A planar contact stack with Zr/Al/Mo/Au was annealed successively at 700 °C and then 1000 °C for 30 s and resulted in a ρ_c of 1.1×10^{-3} Ω cm^2. This structure is more amenable to ohmic contact formation because the PolFET has a polarization-doped conductive channel that extends to the AlGaN surface. In contrast, the standard HEMT has the barrier sandwiched between the metal contacts and channel layer and does not conduct mobile charge near the surface.

Another approach to improve ohmic contacts is to decouple the contact region epitaxy from the main heterostructure. Such an approach enables high Al-content devices in the transport regions and buffer, while having more conductivity in the contact areas based on GaN or low Al-content AlGaN through selective epitaxy.

Low-resistance ohmic contacts to a polarization graded AlGaN FET (0 → 50% Al) by regrowth of reverse graded and highly doped n-type AlGaN (50 → 0% Al) by MBE were realized [45]. An SiO$_2$ mask was used to selectively regrow the reverse graded contacts in the source and drain regions, leaving the access and gate regions untouched. A ρ_c of 8.28×10^{-8} Ω cm^2 was measured and no annealing was required to achieve this result. This work illustrates the use of bandgap engineering to remove abrupt conduction band offsets, while maintaining the desirable attributes of the original heterostructure. Such concepts were then ported to AlGaN structures with higher Al content.

Ohmic contacts with ρ_c of 1.9×10^{-6} Ω cm^2 on an Al$_{0.75}$Ga$_{0.25}$N MISFET structure were also reported [31] using the concept described in [45]. Unannealed Ti/Al/Ni/Au metal stacks were made into a heavily doped n$^+$ reverse graded (75→0%)

contact layer, grown directly after the MESFET channel in a single growth process encompassing the buffer, the MESFET channel, and the reverse graded contact layer. The MESFET gate was constructed after the graded layer was dry etched in the gate and access regions of the device. Low ρ_c was achieved, but maximum current densities were limited to 60 mA mm^{-1}, which the authors attributed to a low channel mobility of 16 cm^2 V^{-1} s^{-1}. A later paper by the same group for a similar structure and approach, but using a more complex process integration with both metal–organic chemical vapor deposition (MOCVD) and MBE, showed improved current densities to ~350 mA mm^{-1}, but with a higher ρ_c of 3.5 × 10^{-4} Ω cm^2 [46].

An etch and regrowth approach for contacts for an 85/66 HEMT structure was used by removing the barrier layer in the areas of the ohmic contacts and re-growing heavily doped (n$^+$) GaN in its place [47]. The Ti/Al/Ni/Au contacts were annealed at 850 °C to achieve a ρ_c of 5 × 10^{-3} Ω cm^2. This technique was compared to a recess etch plus contact scheme and a planar contact scheme, all with the same metallization. In the two cases without the etch and regrowth process, highly nonlinear contacts were observed. A similar etch and regrowth technique was also reported for a 100/85 HEMT [48], using a combination of a contact region dry etch, followed by an n$^+$ GaN regrowth, and Ti/Al/Ni/Au contacts annealed at 850 °C. The resulting ρ_c was measured at 8.6 Ω cm^2.

Figure 11.7 plots the ρ_c results for the different ohmic contact schemes. HEMT contact resistivity (red diamonds) are fitted with a line to illustrate how HEMTs exhibit a logarithmically increasing ρ_c with linear increase in Al content, demonstrating the difficulty of working with high Al-content materials. The metal–insulator–semiconductor FETs (MISFETs), LED test structures, and polarization-doped graded FETs have contact resistances well beneath the fit line,

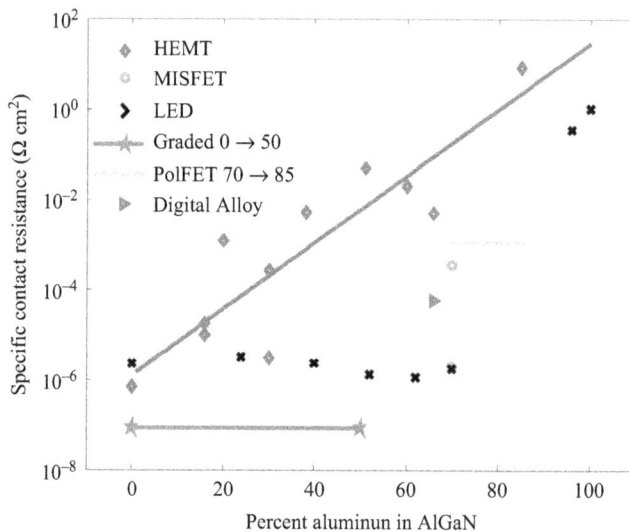

Figure 11.7. Specific contact resistance for HEMTs, MISFETs, LED test structures, and graded PolFETs. Reproduced from [21] with permission of AIP Publishing.

suggesting that standard planar HEMTs have a disadvantage with ohmic contacts. Table 11.1 summarizes the ohmic contact work of various researchers.

11.4 AlGaN-channel HEMTs

11.4.1 Early work in AlGaN-channel HEMTs

Nanjo *et al* initiated the research into AlGaN-channel HEMTs and cited the need for improved power conversion and RF devices to motivate it [32]. They recognized that the advantages of a large AlN bandgap coupled with a comparable saturation velocity would result in better JFOM and BFOM for AlN compared to GaN. However, the ohmic contact challenges with high Al-content AlGaN no doubt influenced their decision to first study HEMTs with lower Al composition. Their 40/20 HEMT had high enough Al to require a modified approach to ohmic contacts. The application of a GaN-based Si implantation process selective to the source and drain regions [49] would realize sufficiently good ohmic contacts to AlGaN. Their HEMT with Ti/Al-based ohmic contacts and the selective area implants realized a much greater drain current (I_D) against the unimplanted HEMT. Such results enabled reasonable current density for demonstration of the first AlGaN-channel transistor. I_D of 120 mA mm^{-1} with gate voltage $V_G = 2$ V, gate length $L_G = 1$ μm, and source–drain length $L_{SD} = 4$ μm were achieved. This initial result was modest, but so too was the I_D for the first AlGaN/GaN HEMT. It achieved $I_D = 50$ mA mm^{-1} in an initial report of a long-channel device ($L_G = 4$ μm, $L_{SD} = 10$ μm) [4] while later devices would exceed 1 A mm^{-1}.

Supporting evidence would be needed to establish the premise of voltage enhancement in AlGaN transistors. Detailed comparisons of breakdown voltages in many types of devices and heterostructure compositions are presented in section 11.5, but the first evidence of compelling breakdown enhancements came early [33]. The breakdown voltage was evaluated for Al composition from $x = 0$ (GaN) to $x = 0.38$, as seen in figure 11.8 [33]. Increasing breakdown with higher Al content was realized, but a trade-off between breakdown voltage and current density also existed. The average breakdown field was 1.6 MV cm^{-1} and exceeded the GaN control by about 8× [33]. The HEMTs of [33] did not utilize field plates, which for GaN HEMTs can enable an average breakdown field near 1 MV cm^{-1}. Improvements in breakdown from the use of optimized field plates can be expected for both GaN- and AlGaN-channel HEMTs.

HEMT characteristics with an increased Al content (86/51) were also reported. A maximum I_D of 25 mA mm^{-1} was realized for $V_G = +4$ V [40]. The I–V curves also show a slight offset voltage that was not seen in earlier HEMTs with $x \leqslant 0.38$. As articulated in section 11.3, the offset voltage occurs due to Schottky-like ohmic contacts, with $\rho_c = 4.8 \times 10^{-2}$ Ω cm^2 using Zr/Al/Mo/Au metal stacks [40]. When the HEMTs were compared against AlGaN/GaN controls over the 25 °C–300 °C temperature range, I_D was found to decline less than in the GaN-channel control, as illustrated in figure 11.9 [40]. Temperature associated mobility degradation, larger for GaN than AlGaN, was given as the causal reason for the greater temperature I_D decline of the GaN-channel HEMT. The temperature insensitive dependence of the

Table 11.1. A summary of ohmic contact research results on AlGaN-channel transistors and LEDs.

Al% y/x or x or (x→y)	Ohmic metal	Contact resistance (Ω cm²)	Mobility (cm² V⁻¹ s⁻¹)	Contact method	Reference
39/16	Ti/Al	9.68×10^{-6}	645	Si ion implant, RTA unspecified	[35]
40/20	Ti/Al	1.2×10^{-3}	—	Si ion implant, post Si RTA, RTA	[32]
53/38	Ti/Al	5.28×10^{-3}	—	Si ion implant, post Si RTA, RTA	[33]
39/16	Ti/Al	1.79×10^{-5}	—	Si ion implant, post Si RTA, RTA	[33]
18/0	Ti/Al	6.96×10^{-7}	—	Si ion implant, post Si RTA, RTA	[33]
100/60	Zr/Al/Mo/Au	1.9×10^{-2}	—	RTA 950 °C	[39]
55/30	Zr/Al/Mo/Au	2.6×10^{-4}	—	RTA 950 °C	[38]
51/23	Zr/Al/Mo/Au	$\sim 10^{-4}$	—	RTA 950 °C	[38]
86/51	Zr/Al/Mo/Au	$\sim 9 \times 10^{-3}$	—	RTA 950 °C	[38]
86/51	Zr/Al/Mo/Au	4.8×10^{-2}	—	RTA 950 °C, 30 s	[40]
0→50 graded PolFET	Ti/Au	8.28×10^{-8}	690	Reverse graded 50→0 n$^+$ MBE regrown contacts, no RTA	[45]
0, 24, 40, 52, 70, 96, 100 LED structures	V/Al/V/Au	2.2×10^{-6}, 3.2×10^{-6}, 2.4×10^{-6}, 1.3×10^{-6}, 1.8×10^{-6}, 0.35, 1.01	—	Annealing optimized for each composition	[41]
100/85	Ti/Al/Ni/Au	8.6 Ω cm² (1900 Ω mm)	250	Contact region dry etch, n$^+$ GaN regrowth, RTA 850 °C	[48]
85/66	Ti/Al/Ni/Au	5×10^{-3}	~160	Dry etch recess, n+ GaN regrowth	[47]
75 (MISFET)	Ti/Al/Ni/Au	1.9×10^{-6}	16	Reverse graded 75→0 n^{++} contacts, no RTA, etch away between source and drain	[31]

(Continued)

Table 11.1. (*Continued*)

Al% y/x or x or (x→y)	Ohmic metal	Contact resistance (Ω cm²)	Mobility (cm² V⁻¹ s⁻¹)	Contact method	Reference
100 (MESFET)	Ti/Al/Ni/Au	—	130	Si implant, implant RTA, 1230 °C, metal RTA, 800 °C	[36]
62 (LED)	V/Al/Ni/Au	1.13×10^{-6}	49	RTA, 900 °C, 100–200 m	[42]
66 (dig. alloy)	Ti/Al/Ti/Au	5.6×10^{-5}	—	Annealing, 700 °C, 120 m	[43]
45/30	Ti/Al/Ni/Au	3×10^{-6}	—	RTA, 900 °C	[37]
	Ti/Al/Mo/Au	2.5×10^{-5}		RTA, 900 °C	
	Zr/Al/Mo/Au			RTA, 850 °C–950 °C	
	Nb/Ti/Al/Mo/ Au			RTA, 850 °C–950 °C	
	V/Al/V/Au			RTA, 850 °C–950 °C	
70→85 graded PolFET	Zr/Al/Mo/Au	1.1×10^{-3}	210	Successive RTAs: 700 °C and 1000 °C	[44]
70 (MISFET)	Ti/Al/Ni/Au	3.5×10^{-4}	90	Reverse graded 70→0 n^{++} cap, no RTA, etch cap between source and drain	[46]

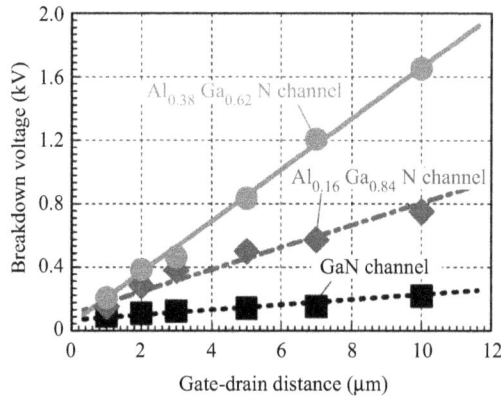

Figure 11.8. Dependence of breakdown voltage on gate–drain separation for AlGaN and GaN transistors. Reproduced with permission from [33]. Copyright 2008 AIP Publishing.

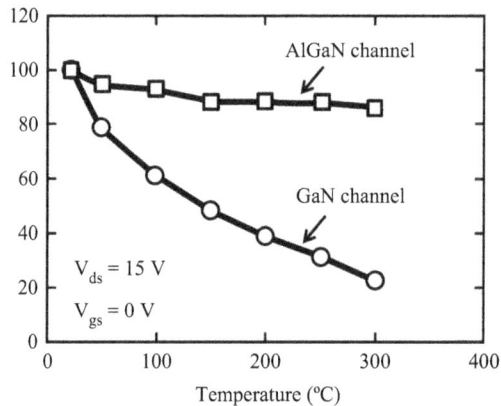

Figure 11.9. Temperature dependence of the drain saturation current for an 86/51 HEMT. The current is normalized at room temperature. Reproduced with permission from [40]. Copyright 2010 IOP Publishing, all rights reserved.

AlGaN-channel HEMT may confer an advantage for high-temperature device operation and is discussed in more detail in section 11.6.3.

This early AlGaN HEMT research highlighted breakdown voltage enhancement, examined the trade-off with current density, and uncovered the reduced I_{MAX} temperature dependence. Other research with low- and mid-range of Al composition includes choosing the Al composition so as to detect the desired wavelength of UV emission for transistor-based photodetection [50]. This work featured improved ohmic contacts [37] and resulted in a long-channel 45/30 HEMT ($L_G = 2\ \mu m$) with a 50 nm barrier, a conventional field plate, SiN passivation, and $L_{SD} = 10\ \mu m$. The current density was $70\ mA\ mm^{-1}$ in a long-channel device at $V_G = +2$ V and over $120\ mA\ mm^{-1}$ at $V_G = +4$ V, a bias condition which takes the HEMT beyond the onset of gate leakage [50].

11.4.2 Recent work in high Al-content HEMTs

For many years Nanjo *et al* were practically the only ones researching AlGaN-channel HEMTs and one may wonder whether the low AlGaN mobility made this research difficult to promote. As recent work has focused on the upper range of Al composition, more groups began to participate in AlGaN HEMT research. Aside from the benefits inherent in the LFOM, high Al content is compatible with AlN, which may be the long-term best substrate for high Al-content transistors [39, 40], despite its immaturity. Computational analysis also suggested that HEMTs with AlN barriers and AlGaN channels could effectively achieve more positive V_{TH} while still maintaining the low power switching losses obtained in GaN-based HEMTs [51]. Despite the attractiveness of AlN, the majority of epitaxial growth efforts have been conducted on sapphire substrates [32–35, 37, 38, 44, 46–48, 50, 52–62] using AlN templates due to the cost of AlN substrates.

Among the highest channel bandgaps reported in transistors are HEMTs with 100/85 heterostructures with a channel bandgap of 5.8 eV [48], PolFETs graded up to AlN composition [62], and AlN-channel MESFETs with E_G of 6.2 eV [36]. The 100/85 HEMT achieved 1.7 mA mm^{-1} current density at $V_G = +3$ V. The AlN MESFETs [36] only reached 60 μA mm^{-1} at 250 °C and $V_G = +4$ V. The current density limitation for both transistors was due to high contact resistivity. A PolFET with unintentionally doped Al$_x$Ga$_{1-x}$N graded over Al compositions $0.6 \leqslant x \leqslant 1.0$ over 75 nm achieved $I_D = 188$ mA mm^{-1} at $V_G = 10$ V and $V_D = 18$ V, as illustrated in figure 11.10. This current density, quite high for an FET with surface AlN, was achieved despite the Schottky-like ohmic contact apparent in the offset voltage of figure 11.10, with a ρ_c of ~0.01 Ω cm^2. The average mobility for this PolFET was estimated to be $\mu = 320$ cm^2 V^{-1} s^{-1} [62]. Although not a HEMT, the PolFET is likewise capable of high mobility because polarization doping occurs in the absence of impurity doping.

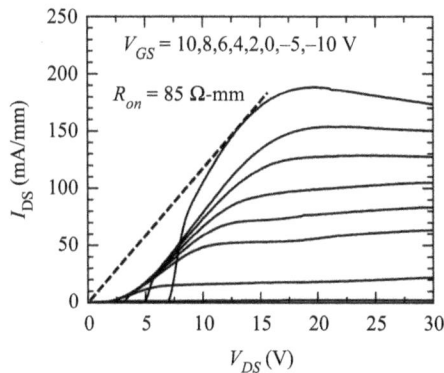

Figure 11.10. DC output characteristics of 60–100 PolFET. The drain current was negative for $V_{GS} > 4$ V at low V_{DS} due to the forward gate current. Reproduced with permission from [62]. Copyright 2019 AIP Publishing.

A 100/60 HEMT with long-channel dimensions of $L_G = 6$ μm and $L_{SD} = 14$ μm and a 20 nm barrier achieved 40 mA mm^{-1} current density at $V_G = +2$ V, also limited by contact resistivity and an offset voltage [39]. The challenge for Al composition approaching AlN is two-fold. First, Si-doping of $Al_xGa_{1-x}N$ by epitaxial growth or ion implantation becomes problematic as x approaches 1 due to an increase in resistivity from Si self-compensation and an increase of the donor ionization energy for $x \geqslant 0.8$ [63, 64], whereby Si transitions from a shallow donor to the DX center. Second, the fabrication of low resistivity ohmic contacts is increasingly challenging with more Al incorporation, as reviewed in section 11.3.

These obstacles notwithstanding, there is a presumption that contact resistivity will improve over time and that the complex fabrication approaches described in section 11.3, e.g. etch and regrowth processes, will realize success at even higher Al composition. In addition, HEMTs with slightly lower Al composition have achieved a measure of success even with conventional planar ohmic contacts. Long-channel ($L_G = 2$ μm) 85/70 HEMTs with a 25 nm barrier were reported with Ni/Au Schottky gates with a field plate, SiN passivation, Ti/Al/Ni/Au ohmic contacts, and a 10 μm source–drain channel length [53]. They realized a current density of 46 mA mm^{-1} at $V_G = +4$ V, albeit with an offset voltage still present in the I_D–V_D plot. The long-channel 85/70 HEMTs achieved an impressive I_{ON}–I_{OFF} ratio of 8×10^9 at room temperature, measurement limited up to 50 °C, and illustrated in figure 11.11 at temperatures from −50 °C to +200 °C [53]. The excellent I_{ON}–I_{OFF} ratios are typical of many types of Schottky-gated Al-rich transistors, as even AlGaN/GaN transistors tend to require insulated gates or p$^+$ gates to achieve such ratios. Even with 80 nm gates using an 85/70 heterostructure with a 30 nm barrier layer, a similarly high I_{ON}–I_{OFF} ratio of 10^9 was reported. This transistor displayed a considerably softer pinch-off due to short-channel effects and could benefit from a back barrier design [57].

The very large I_{ON}–I_{OFF} ratio likely exceeds measurement limits until such a temperature that intrinsic leakage rises to pA levels. In 85/70 HEMTs the leakage

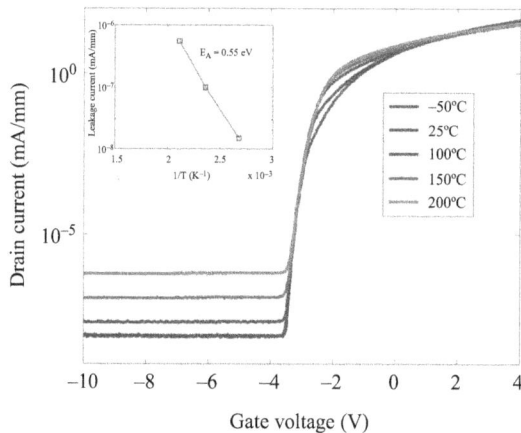

Figure 11.11. Transfer characteristics over temperature for an 85/70 HEMT with $V_D = 10$ V. Reproduced with permission from [50], copyright 2020, The Electrochemical Society.

current increases with temperature with an activation energy $E_A = 0.55$ eV, as seen in the inset of figure 11.11. The leakage current follows the Frenkel–Poole equation for conduction by electrons traversing the source to drain as they are captured and emitted from traps. The Frenkel–Poole equation accounts for emission from a barrier whose height is modulated by the applied electric field arising from the drain voltage. The existence of the Frenkel–Poole mechanism in high Al-content HEMTs is readily evident because of the extremely low gate leakage current, in contrast to Schottky-gated AlGaN/GaN HEMTs. The latter do not reach comparably low levels of I_D leakage current nor similarly high I_{ON}–I_{OFF} ratios, probably because this drain leakage current is often dominated by gate-mediated tunneling mechanisms [65] that do not seem to be present in high Al-content transistors.

Recent reports include HEMTs with substantially greater current densities. A long-channel ($L_G = 1.8$ μm) 85/65 HEMT with a 27 nm barrier layer and $L_{SD} = 5.5$ μm achieved 250 mA mm^{-1} at $V_D = +4$ V with a Si-doped barrier, despite being overdoped with a soft pinch-off [54]. Shrinking the channel dimension has been another approach for increased current density. An 85/70 HEMT with a 30 nm barrier, 80 nm gates, $L_{DS} = 2.6$ μm, and planar contacts with an offset voltage achieved 160 mA mm^{-1} of open channel current, $V_G = 10$ V [57]. Although this open channel current incurs considerable forward biased gate leakage, it may be appropriate for an RF loadline (section 11.4.5) whose gate leakage is considerably smaller with $V_D \sim$ 5–6 V. A 75/60 HEMT with regrown, reverse graded contacts achieved 460 mA mm^{-1} with a 30 nm barrier, 130 nm gates, and $L_{DS} = 1.2$ μm [58]. The regrown, reverse graded contacts made a significant difference, even though they did not match the best low contact resistance of MISFETs with similar reverse graded contacts [31].

11.4.3 Enhancement-mode HEMTs

Power switching applications require normally-off transistors for fail-safe operation. Examples of these have recently been reported for high Al-content transistors [55, 56]. Two different approaches to realize e-mode operation are both similar to methods previously described in AlGaN/GaN transistors [66, 67]. The former used a p$^+$ region to deplete 2DEG charge under the gate [66] and the latter introduced F ions into the interface between the gate metal and the barrier layer, likewise to deplete 2DEG charge under the gate [67]. Both methods selectively deplete charge under the gate while leaving intact the conductivity of the HEMT access region.

The heterostructure layers and device cross-section in an e-mode, p$^+$ gated 45/30 HEMT are illustrated in figure 11.12 and p$^+$ gates have not yet been realized at higher Al content. The HEMT reached 98 mA mm^{-1} for $L_G = 1.5$ μm and $L_{SD} = 10$ μm, at $V_G = 8$ V and $V_D = 10$ V [55]. The transfer characteristic of this long-channel HEMT is shown in figure 11.13. The V_{TH} is slightly more negative than that of a true e-mode HEMT, possibly due to insufficient Mg-doping of the p$^+$ layer. The gate–source diode current of this HEMT only reaches 1 nA mm^{-1} at $V_G = +8$ V and $V_D = +10$ V compared to 50 nA mm^{-1} at $V_G = +3$ V, for GaN-channel HEMTs with p$^+$ gates [68]. Such low leakage allows for a larger forward

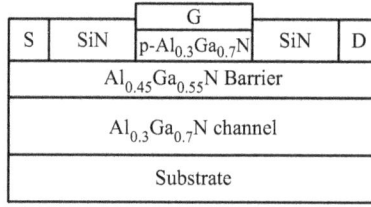

Figure 11.12. Cross-section schematic of AlGaN-channel HEMT with a p-AlGaN gate. The Schottky metallization is Ni/Au, while the source and drain are Ti/Al/Ni/Au. Reproduced with permission from [55]. Copyright 2019 AIP Publishing.

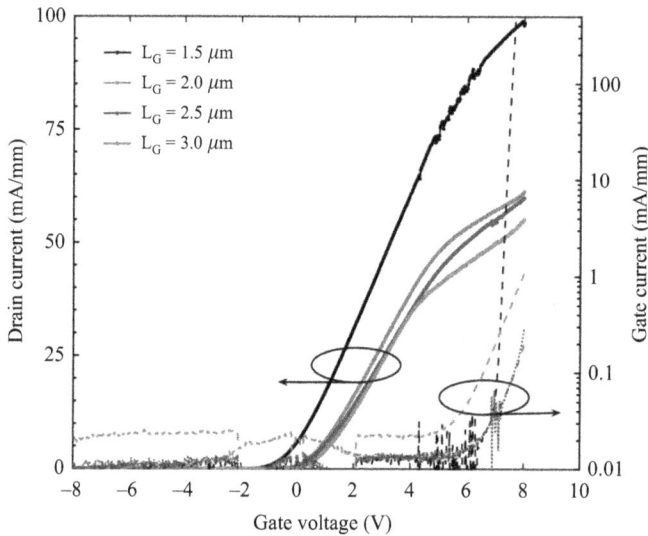

Figure 11.13. Transfer characteristics for a 45/30 HEMT with a p^+ gate at $V_D = 10$ V with varying L_G. Maximum I_D decreases with increasing gate length. In addition, gate leakage current up to $V_G = 8$ V is exceptionally low. Reproduced with permission from [55]. Copyright 2019 AIP Publishing.

voltage swing for a 45/30 e-mode transistor, compared to GaN, leveraging its larger bandgap and allowing for good power switching capability with a more positive V_{TH}.

An F treatment process similar to that in [67] was applied to a 85/70 heterostructure [56]. A reactive ion etch (RIE) with CHF_3 and SF_6 gases was used to etch an opening in SiN to place the gate metal between the source and drain. This standard fabrication procedure allows for incorporating field plates in depletion-mode (normally-on) HEMTs. In different regions of the wafer designated for e-mode HEMTs, the etch was continued for an extra period of time after clearance of the SiN in order to incorporate F ions in the near surface region of the barrier layer and the amount of F ions incorporated is proportional to the time. Thus, the gate region was *both* recess etched and also embedded with a controllable amount of fixed negative charge in the region of the metal–semiconductor interface. A +5.6 V V_{TH} shift between the d- and e-mode was realized with this procedure, as seen in the

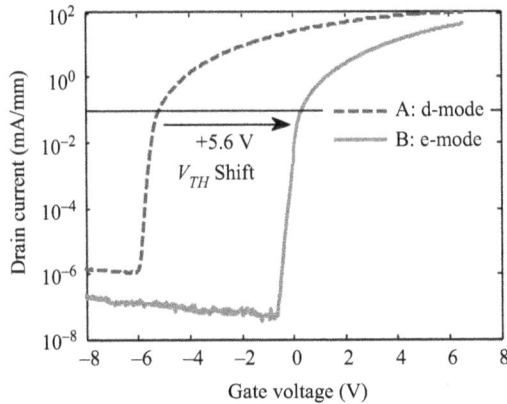

Figure 11.14. Transfer characteristics for 85/70 HEMTs with (red solid line) and without F treatment (blue dashed line), indicating the V_{TH} shift for F treatments. Reproduced with permission from [56]. Copyright 2019 AIP Publishing.

transfer characteristic of figure 11.14. An analysis that included an analytical determination of the recess depth concluded that the recess etch contributed approximately 60% to ΔV_{TH}. The analysis also indicated that the effect of the F ions at the gate–AlGaN interface contributed the rest of ΔV_{TH} [56]. Examination of figure 11.14 shows that the subthreshold slope and I_D leakage current remained comparable to or lower than the d-mode control HEMT after the F treatment. However, the onset of gate leakage in the HEMT occurs at approximately $V_G = +3$ V and is comparable in magnitude to that in the d-mode HEMT. Although an improvement over the similar type of GaN-channel HEMTs, the high Al-content e-mode HEMT with the F treatment has less voltage swing prior to the onset of gate leakage current than 45/30 e-mode HEMT with a p$^+$ gate.

11.4.4 Toward high current density in AlGaN-channel HEMTs

In order to realize the promise inherent in the large bandgaps of UWBG semiconductors, higher transistor breakdown voltage is desired without sacrificing too much in drain current. The breakdown voltage properties for high Al-content transistors are evaluated in section 11.5 and the current density results are summarized in this section. Ideally the high Al-content HEMTs would approach the current density of AlGaN/GaN HEMTs, which can reach 1–2 A mm^{-1}. Even with lower mobility and non-ideal contacts, the AlGaN-channel HEMTs are making progress in reaching higher current density. The DC performance results have been tabulated for AlGaN-channel HEMTs in table 11.2 and the main contributors to I_{MAX} are identified. Low contact resistivity is certainly needed. Examination of table 11.2 shows that $x \leqslant 0.3$ already allows good contact resistivity and progress has been made in certain high Al-content transistors as well. High I_{MAX} requires aggressive dimensional scaling, which will minimize the effect of mobility with extremely small source–gate dimensions. Electron velocity in the gate and gate–drain regions can be increased with high electric fields determined by

Table 11.2. A summary of DC performance characteristics of AlGaN-channel HEMTs along with a comparison to a high current MESFET and MOS-HEMT.

Al content y/x	Barrier thickness (nm)	I_{MAX} (mA/mm)	V_D (@I) (V)	V_G (@I) (V)	V_{TH} (V)	L_G (μm)	L_{SD} (μm)	ρ_c (Ω cm²)	n_s ($\times 10^{12}$ cm^{-2})	μ (cm² V^{-1} s^{-1})	R_{sheet} (Ω/□)	$I_{ON}-I_{OFF}$	Reference
40/20	20	130	5	2	−3	1	4	1.2×10^{-3}	6.9	—	—	—	[32]
39/16	—	155	—	2	—	1	4	1.8×10^{-5}	3.3	—	—	—	[33]
53/38	—	114	6	2	−3	1	4	5.3×10^{-3}	5.4	—	—	—	[33]
39/16	16	150	5	2	−1	1	4	9.7×10^{-6}	2.2	645	—	—	[35]
86/51	25	25	15	4	−5.5	9	22	4.8×10^{-2}	—	—	2900	—	[40]
40/15	25	340	7.5	2	−4	1	4	4.5×10^{-5}	7.9	460	1720	—	[34]
100/60	20	38	7.5	2	−4	6	14	1.9×10^{-2}	—	—	3800	—	[39]
100/85	48	1.7	10	2	−4.9	2	14	8.5	6	250	4200	10^7	[48]
45/30	50	70	10	2	−1.3	2	10	—	—	—	3500	8×10^8	[50]
85/70	25	46	10	4	−2.2	2	10	3.4×10^{-2}	9	—	2200	8×10^9	[53]
85/65	27	250	25	4	−14	1.8	5.5	7.0×10^{-2}	11	284	1800	—	[54]
85/70	30	160	10	10	−2	0.08	2.6	$\sim10^{-2}$	7.2	390	2200	10^9	[57]
75/60	30	460	12	2	−7	0.13	1.2	$\sim10^{-4}$	8.5	175	8000	20	[58]
45/30	50	100	8	8	−0.5	1.5	10	1.4×10^{-6}	—	—	2800	10^9	[55]
85/70	31	35	7	6.6	0.0	3	10	8.4×10^{-2}	14	240	1900	$>10^8$	[56]
70	MESFET	440	15	0	−20	.25	1.3	1.3×10^{-4}	28	56	6.4	10^7	[59]
65/40	MOS-HEMT	600	17	6	−10	2	6	1.4×10^{-5}	9	430	1.9	5×10^7	[52]

dimensions and biasing to the extent allowable by reliability and design/manufacturing considerations. However, the small electric fields of the source access region result in high source resistance that researchers seek to minimize by small dimensions. Finally, mobility limitations are minimized in applications that allow a more negative V_{TH}, possibly improving the performance of RF transistors. With good ohmic contacts, the I_D–V_D slope is resistance limited by mobility in the source access region. So apart from minimizing the source access region length, a larger knee voltage allows for more current density. Since $V_{KNEE} \sim (V_G - VTH)$, a more negative V_{TH} is helpful for attaining the high current density that benefits RF devices and a more positive V_G is helpful for both power switching devices and RF devices. Thus, the largest current densities tabulated in table 11.2 have been achieved for HEMTs with two or more of these factors, i.e. $V_{TH} \ll -7$ V or $L_{SD} < 2\ \mu m$ or $\rho_c \ll 1 \times 10^{-4}\ \Omega\ cm^2$ or $x \leqslant 0.15$ Al content.

High Al-content MESFET devices are among the leaders in current density. MESFETs are not used in applications where HEMTs of the same channel material are available because lower mobility associated with ionized impurity scattering lowers the current density for a comparable V_{TH}. Nevertheless, they are providing a vehicle for perfecting complex process integration schemes for ohmic contacts and are illustrating the possibilities for high current density. High Al-content MESFETs with a reverse graded ohmic contact have achieved a high current density of 440 mA mm^{-1} for $V_G = 0$ V [59] with a 40 nm thick $Al_{0.7}Ga_{0.3}N$ MESFET channel, doped n-type with $Si = 7 \times 10^{18}\ cm^{-3}$. The ohmic contact layer was reverse graded from 0.7 to 0.3. Contact resistivity of $\rho_c \sim 1.3 \times 10^{-4}\ \Omega\ cm^2$ contributed to the realization of this current density. It is reasonable to expect that process integration using a combined MOCVD/MBE process used in the MESFET work will be applied to HEMTs and lead to further improvements.

In light of excellent I_{ON}–I_{OFF} ratios with Schottky gates, gate insulators may not be needed to reduce gate current in high Al-content HEMTs. However, HEMTs with gate insulators have also attained excellent research results aimed at high current density. Silicon dioxide (SiO_2) was used as a gate insulator in an $Al_{0.65}Ga_{0.35}N/Al_{0.40}Ga_{0.60}N$ heterostructure-based transistor [52]. The SiO_2 was deposited by plasma-enhanced chemical vapor deposition (PECVD). A high forward V_{GS} of ~6 V was achieved, resulting in a high saturated drain current density around 600 mA mm^{-1} for $V_{TH} \sim -10$ V. The reverse biased gate current of the MOS-gated HEMT was roughly four orders of magnitude lower than that of a control HEMT.

11.4.5 RF performance of high Al-content HEMTs

Good RF performance is predicated on the JFOM, as described in section 11.2, and it favors high values for E_C and v_{sat}. Additionally, other factors that optimize current density are important, such as those discussed in section 11.4.4. Only a couple of reports on RF performance exist for high Al-content AlGaN-channel HEMTs. These results are summarized in table 11.3 and so far exceed those of high Al-content MESFETs and PolFETs [59].

Table 11.3. Reported RF properties of high Al-content transistors and some other useful secondary properties.

Al% y/x or $x\rightarrow y$	Transistor type	f_T (GHz)	f_{max} (GHz)	I_{MAX} (mA mm^{-1})	V_D (@I) (V)	V_G (@I) (V)	L_G (μm)	L_{SD} (μm)	ρ_c (Ω cm^2)	$I_{ON}-I_{OFF}$	Freq. (GHz)	P_{max} (W mm^{-1})	PAE (%)	Reference
85/70	HEMT	28.4	18.5	160	10	10	0.08	2.6	~0.01	10^9	3	.38	11	[57]
75/60	HEMT	40	58	460	12	2	0.13	1.2	~10^{-4}	20	10	1.8	22	[58]

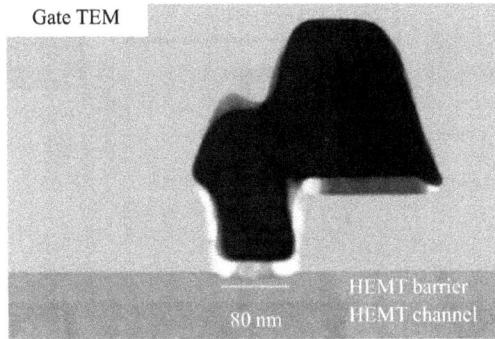

Figure 11.15. A transmission electron micrograph cross-section of an 80 nm gate for an 85/70 RF HEMT. Reproduced with permission from [57]. Copyright 2019 IEEE.

Small signal RF performance was reported for a HEMT with 80 nm gates and planar ohmic contacts [57]. An 85/70 HEMT with a 30 nm thick barrier, 80 nm gates, and $L_{DS} = 2.6$ μm, as illustrated in the TEM image of figure 11.15, was evaluated for RF performance. f_T of 28.4 GHz and f_{max} of 18.5 GHz was reported. A field plate structure into SiN was filled with metal with an 80 nm base, a 100 nm extension to the source, and a 200 nm extension to the drain, rather than the typical mushroom gate. This unconventional structure may have been a factor in realizing $f_T > f_{max}$ when the opposite is expected. Large signal RF performance was characterized with a 3 GHz power sweep using on-wafer load pull techniques. When biased at $V_D = 20$ V and $V_G = 3.75$ V, a 2×50 μm^2 HEMT delivered 0.38 W mm^{-1} with 11% power-added-efficiency (PAE). Short-channel effects limited the drain biasing (and power output) to values well below the intrinsic breakdown. High contact resistivity, estimated as ~0.01 Ω cm^2, limited both the small and large signal performance.

The other HEMT RF results achieved even better performance in a 75/60 HEMT which operated at a relatively high current density of 460 mA mm^{-1}. Compact dimensions along with reverse graded ohmic contacts were key aspects of the work [58]. The fabrication of this HEMT, whose cross-section is illustrated in figure 11.16, involved an MOCVD HEMT structure growth and an etch and regrowth procedure for ohmic contacts. The MOCVD growth used methods described in [52, 54]. The source and drain regions were then etched down into the AlGaN-channel layer using a Cl-based plasma. Next an MBE regrowth was performed starting with 100 nm of n$^+$ Al$_{0.6}$Ga$_{0.4}$N followed by 50 nm of heavily doped reverse-composition graded AlGaN ($x = 0.6 \rightarrow 0$). The Si-doping needs to be high enough (sheet charge ~ 1.1×10^{14} cm^{-2}) to overcome the positive polarization charge from reverse (high-to-low) composition grading. This procedure resulted in a relatively good contact resistivity of $\rho_c \sim 1 \times 10^{-4}$ Ω cm^2. Small and large signal performance were characterized for the compact HEMT with $L_G = 130$ nm, $L_{SG} = 300$ nm, and L_{SD} of 1.2 μm. The small signal measurement showed a current/power gain cut-off frequency of 40 GHz/58 GHz, respectively [58]. A 10 GHz continuous wave signal load pull measurement was performed and a maximum output power of 1.8 W mm^{-1} with 22% PAE and a power gain of 3.8 GHz was realized. The

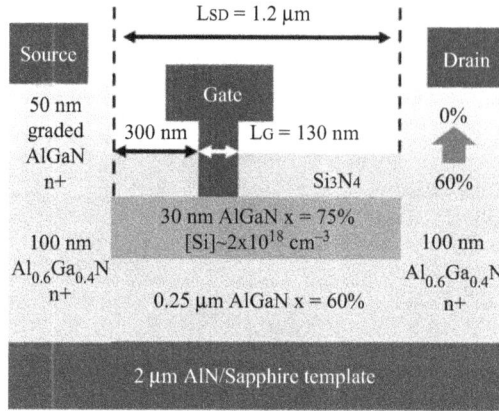

Figure 11.16. Schematic cross-section of a 75/60 HEMT grown with MOCVD followed by etch and regrowth of the ohmic contact using MBE. Reproduced with permission from [58]. Copyright 2019 IOP Publishing, all rights reserved.

HEMT was operated in class A with a bias condition of $V_D = 23$ V and $V_G = -3$ V. Still, self-heating due to the large access resistance at both the source and load sides of the HEMT as well as the relatively low breakdown voltage were the main limiting factors on RF performance. Several factors, including a more compact geometry, decent HEMT mobility (175 cm^2 V^{-1} s^{-1}), decent ρ_s, and a greater voltage swing for higher current (associated with a more negative V_{TH}), contributed to the improved RF performance [58].

A simulation study was used to estimate what RF performance might be achieved without the usual dimensional or contact resistivity constraints of high Al-content HEMTs [69]. A 100/70 HEMT structure with $L_G = 30$ nm was simulated. It incorporates a 78% back barrier to address short-channel effects and implements reduced lateral dimensions of $L_{SG} = 100$ nm and $L_{DG} = 500$ nm to state-of-the-art values that are not common in III–V technology. By minimizing the effect of mobility-dominated source resistance in this manner, the intrinsic potential of the technology was explored. Assuming a negligible contact resistance, simulations indicated that 18 W mm^{-1} with 55% PAE could be achieved at 30 GHz as illustrated in figure 11.17. These results support the notion that RF performance for high Al-content HEMTs can exceed that for AlGaN/GaN.

11.5 Breakdown properties of high Al-content transistors

A higher critical electric field needs to translate into higher breakdown voltage in order to realize the potential in the large bandgaps of UWBG semiconductors. Ideally this is accomplished without sacrificing too much drain current in a future state where the fundamental and practical limitations imposed by mobility and by high contact resistivity have been overcome. In this section, we report the progress towards achieving a higher breakdown field. In terms of breakdown characteristics, HEMTs may not be much different from MESFETs and PolFETs and we have included these types of devices in the compilation of results in table 11.4.

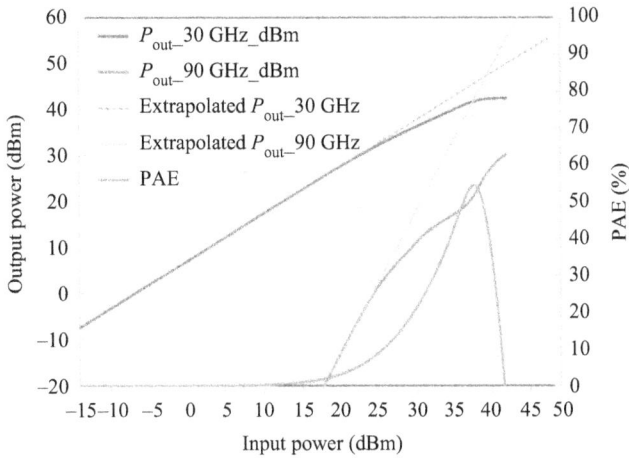

Figure 11.17. Simulated power performance for a 100/70 HEMT with a $x = 0.78$ back barrier (not intended to reflect a finalized design, but rather an estimate of the expected OPI3). Reproduced with permission from [69]. Copyright 2019 IOP Publishing, all rights reserved.

Table 11.4. Reported values for the average breakdown field for high Al-content transistors.

Al content, x or y/x or $x \rightarrow y$	Device type	Breakdown voltage (V)	L_{G} (μm)	L_{GD} (μm)	Breakdown field (MV cm^{-1})	Reference
53/38	HEMT	1650	1	10	1.65	[33]
53/38	HEMT	463	1	3	1.54	[33]
39/16	HEMT	381	1	3	1.27	[33]
39/16	HEMT	>350	1	2	1.75	[35]
86/51	HEMT	1800	9	15	1.2	[40]
40/15	HEMT	1700	1	10	1.7	[34]
100/85	HEMT	810	2	10	0.81	[48]
100	MESFET	2370	2	25	0.95	[36]
70	MOSFET, Al_2O_3	620	1	1.7	3.65	[46]
75	MOSFET, Al_2O_3	224	0.7	1.1	2.04	[31]
85/65	HEMT	770	1.8	9	0.86	[54]
85/70	HEMT	510	2	2	2.56	[53]
85/70	HEMT	495	2	2.0	2.48	[57]
75/60	HEMT	45	0.13	0.77	0.58	[58]
70→85	PolFET	554	2	3.2	1.73	[44]
60→100	PolFET	620	2	2.7	2.30	[62]

The first report on a 40/20 HEMT in 2008 [32] did not include breakdown characteristics. A subsequent publication compared breakdown voltage in 39/16 and 40/20 HEMTs for which the 39/16 HEMT contained an improved high-temperature AlN buffer and a higher breakdown voltage [35]. The improved result was realized by reducing the trapped charge in the buffer, which enabled suppression of off-state leakage current. A comparison of the breakdown voltages of both 39/16 and 53/38 HEMTs for variable gate-to-drain spacings against a GaN HEMT control did show breakdown voltage enhancement, as seen in figure 11.8 of section 11.4.1 [33]. The control, a GaN-channel HEMT, only realized a breakdown voltage 1/8 of the AlGaN channel for similar L_{DS} and without field plate structures for either HEMT. Once field plates were incorporated in a subsequent publication, a 40/15 HEMT reached 1700 V breakdown, as seen in figure 11.18, for $L_{GD} = 10$ μm (1.7 MV cm^{-1}) in HEMTs that reached 340 mA mm^{-1}. The V_B of the GaN control saturated at 400 V at $L_{GD} = 5$ μm (0.8 MV cm^{-1}) [34], but with a higher current density of 620 mA mm^{-1} [34].

The breakdown voltages for AlGaN-channel HEMTs are tabulated in table 11.4. The reported breakdown field is the average electric field at breakdown (breakdown voltage divided by L_{GD}). In lateral devices such as HEMTs, the average electric field may not be a good indicator of breakdown, since peaks in the electric field that are observed at the gate edge can greatly affect the average. Therefore, the average breakdown field always understates E_C, but it still can be a useful practical guide. Studies examining factors affecting the peak electric field in high Al-content transistors are yet to be reported. In table 11.4, most reports indicate breakdown fields >1 MV cm^{-1}, a typical good value for AlGaN/GaN. Four other reports of >2 MV cm^{-1} are also seen in table 11.4, with the best at 3.65 MV cm^{-1}. All of these values fall short of the $E_C \sim$ 17 MV cm^{-1} estimate for AlN from the Hudgins $E_C \sim E_G^{2.5}$ relation [26], but all HEMTs are expected to fall short of this estimate, not just Al-rich ones. The reason

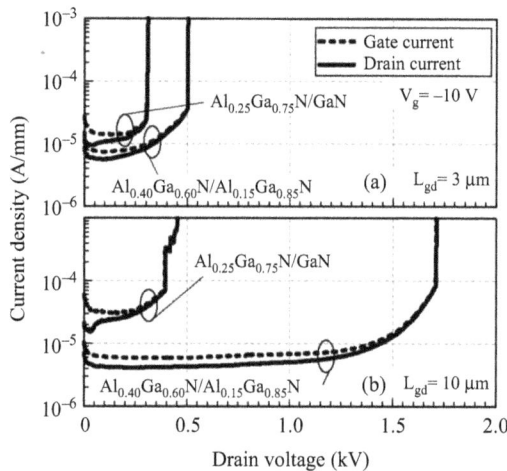

Figure 11.18. Off-state I_D–V_D and I_G–V_D curves in 40/16 and 25/0 (GaN-channel) HEMTs showing gate-dominant leakage at breakdown. Reproduced with permission from [34]. Copyright 2013 IEEE.

is the same: peak electric fields are considerably higher than the average. Another problem for lateral devices is that they will not repeatedly enter and exit an avalanche condition without catastrophic damage as do vertical devices such as p–n junction diodes [70].

We next contrast off-state I–V curves for Al-rich and GaN-based transistors with a couple of examples. Schottky-gated AlGaN/GaN HEMTs can incur ever increasing gate leakage current rather than avalanche as the current responsible for breakdown. These currents are a possible cause of catastrophic damage. MISFETs or metal–oxide–semiconductor FETs (MOSFETs) might reduce the gate leakage, but at a possible price of dealing with reliability issues associated with trapped charge in gate insulators. Two examples of gate-dominant drain current are shown in figures 11.19 and 11.20. Figure 11.19 shows gate-dominant leakage for low

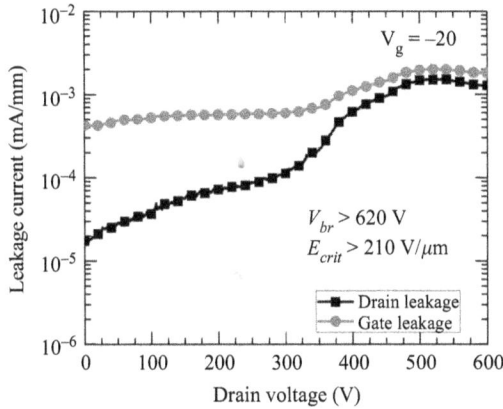

Figure 11.19. Off-state I_D–V_D and I_G–V_D curves in a 60–100 PolFET showing gate-dominant leakage at breakdown. Reproduced with permission from [62]. Copyright 2019 AIP Publishing.

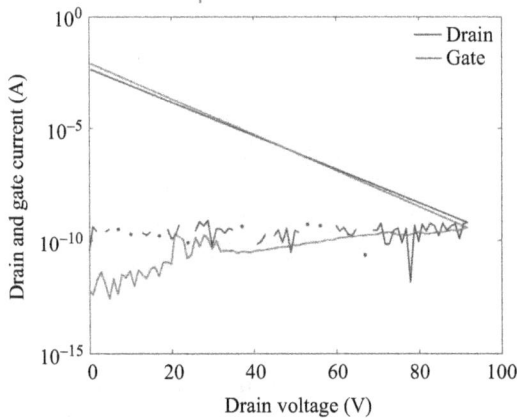

Figure 11.20. Off-state I_D–V_D and I_G–V_D curves in a 85/70 HEMT. Reproduced with permission from [57]. Copyright 2019 IEEE.

x, e.g. GaN and $x = 0.16$. The sharp rise in breakdown current arises from gate leakage reaching critical values. High Al-content PolFETs can also evidence gate-dominant leakage at breakdown, as in figure 11.19. In spite of this leakage, they still have breakdown fields comparable to other high Al-content HEMTs. High Al-content HEMTs can also show source-dominant leakage at breakdown, as in figure 11.20. Since high Al-content transistors are less prone to gate leakage than Schottky-gated AlGaN/GaN HEMTs, breakdown is influenced by the breakdown strength of the AlGaN, as one would desire. Still, breakdown mechanisms are not well characterized and further study is warranted.

Table 11.4 indicates the potential for high breakdown in high Al-content transistors, even as considerably varying results are apparent. Several factors need addressing in order to sustain this progress. Flashover between metal electrodes at HEMT breakdown has been observed even when devices are immersed in fluorine [54]. Therefore, researchers may have not yet learned how to properly treat the non-UWBG portions of their devices which must also sustain peak electric fields that are much higher than those in other technologies. In fact, typical GaN-based passivation dielectrics such as SiN have lower E_C than the UWBG semiconductors. This illustrates that research into more robust dielectrics needs to be a part of future research efforts. Buffer leakage may also be contributing to breakdown, in particular for short channel gates [58]. Methods to address short channel effects can also be borrowed from the experience of GaN and other technologies.

11.6 Other nascent AlGaN HEMT research

As an emerging technology, the reliability and trapping properties of high Al-content HEMTs are of interest in order to better evaluate their promise. Even niche applications sometimes provide the impetus for promoting research of a technology that ends up advancing the technology in ways that make a compelling case for broader adoption. This section reviews the initial research results for a variety of topics: pulsed characterization of HEMTs, initial reliability experiments, extremely high-temperature operation, and radiation properties.

11.6.1 Pulsed I–V

When degradation studies are undertaken, one must distinguish between permanent and repeatable changes. The repeatable ones may arise from trapping and detrapping associated with capture and emission processes. These types of repeatable changes may affect the performance metrics of the high Al-content HEMTs. They must also be interpreted correctly to properly analyse a reliability study.

A common effect called 'current collapse' was exhaustively studied during the development of AlGaN/GaN transistors. It is associated with electron trapping after application of a bias such as a high drain voltage and it refers to a temporary reduction in drain current that results immediately after removal of the bias, ostensibly due to electron trapping at the surface, in the barrier, or in the buffer of a HEMT [12]. Drain current recovery will depend on the emission time constants of the electron or hole traps when biased in the on-state at low drain and gate voltages. 'Temporary' can be a long enough time that affects the switching or amplifying current available in an application. The recovery time depends on a number of factors, including the energy barrier for electron emission.

These types of studies were initiated in 65/40 HEMTs with an undoped $Al_{0.4}Ga_{0.6}N$ channel (500 nm) and a Si-doped $Al_{0.65}Ga_{0.35}N$ barrier (25 nm) [71]. Both were grown on a 3 μm AlN buffer on a sapphire substrate. The HEMT was passivated with SiN and a Ni/Au gate without a field plate was reported. HEMTs were pulsed with quiescent conditions ($V_{GS,Q}$, $V_{DS,Q}$) of (-15 V, 20 V) with pulses varying from 500 ns to 1 ms and a duty cycle of 100. Figure 11.21(b) shows that the SiN passivation eliminates current collapse compared to the unpassivated HEMT of figure 11.21(a). Further examination of the pulse width shows that as the pulse width increases from 100 ns to 1000 μs, the degree of current collapse is reduced [71]. The data were analysed with gated transmission line model methods [72], where the time-

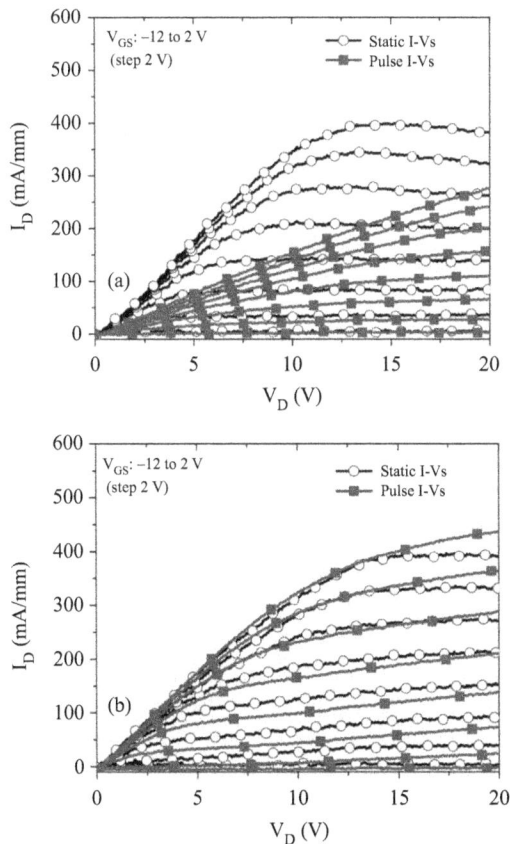

Figure 11.21. Static and pulsed I–V characteristics of (a) a HFET without passivation and (b) a Si_3N_4 passivated HFET. Pulse duration T_{ON} = 500 ns, pulse period T = 500 μs. Reproduced with permission from [71]. Copyright 2019 IOP Publishing, all rights reserved.

dependent effects of channel resistance were separated into contributions from the gate and the access regions in both the linear and the saturation regions. HEMTs with equivalent access region geometries and varying gate lengths were analysed for the slope and intercepts of $R_{CH}(L_G)$ to decouple contributions from the gate and gate–drain access resistance. Access region traps were mainly responsible for the current collapse effects, similar to prior work in AlGaN/GaN HEMTs.

11.6.2 Reliability

Generally, full reliability studies are not warranted in the early development of a technology. However, understanding the degradation processes of transistors can influence technology choices. By incorporating reliability metrics in addition to performance metrics as part of the development process, important answers to questions regarding the technological viability against incumbent technologies are provided. Such is the case for fluorine treated HEMTs for e-mode transistors. RIE plasma fluorine treatment was reported for the fabrication of enhancement-mode transistors in AlGaN/GaN HEMTs [67] and a similar process was successful in accomplishing the same purpose in Al-rich HEMTs [56], as described in section 4.3. For AlGaN/GaN, the mechanism of V_{TH} shift was attributed to F incorporation in the near surface region of AlGaN, presumably from F^- ion implantation from the F plasma of the RIE and a similar mechanism was found to be important in high Al-content transistors. With its high electron affinity, the fluorine incorporating into the near surface region of the AlGaN barrier manifests as a negative ionic charge at the gate–semiconductor interface. Charge balance requires 2DEG depletion of electrons which provide a more positive V_{TH} than HEMTs with identical epilayers and no F treatment. Among the many reports of F treated AlGaN/GaN HEMTs, only one paper on the stability of F treatment was reported. Stability was found for 288 h of thermal stress at 350 °C and for high electric fields at room temperature [73]. No evidence was reported showing stability when *both* bias and high-temperature stress were applied. The lack of reliability reports or commercial devices so long after the initial report raises doubt of the viability of the F treatment approach for AlGaN/GaN. Questions of the attractiveness of this approach for high Al-content transistors are therefore worth considering. A case can be made for high Al-content AlGaN in light of the considerably higher cohesive energy per bond for AlN compared to GaN (2.88 versus 2.24 eV/bond) [23] and a degradation study can quantify whether the expected greater stability is meaningful.

The F stability in high Al-content e-mode HEMTs whose V_{TH} shift is partly due to a F treatment of the near surface region of the HEMT's gate prior to gate metal deposition has been examined [74]. Reversed biased HEMTs showed a small (~0.5 V) V_{TH} instability in both e-mode HEMTs and d-mode controls at 190 °C and $V_D = 50$ V, then relative stability for over 3 h, as seen in figure 11.22. Forward or neutral bias HEMTs showed even greater stability. The early time instability affected d- and e-mode HEMTs similarly, indicating an instability factor that was common to both types of HEMTs. This was taken as evidence of the bias-temperature stability of the F

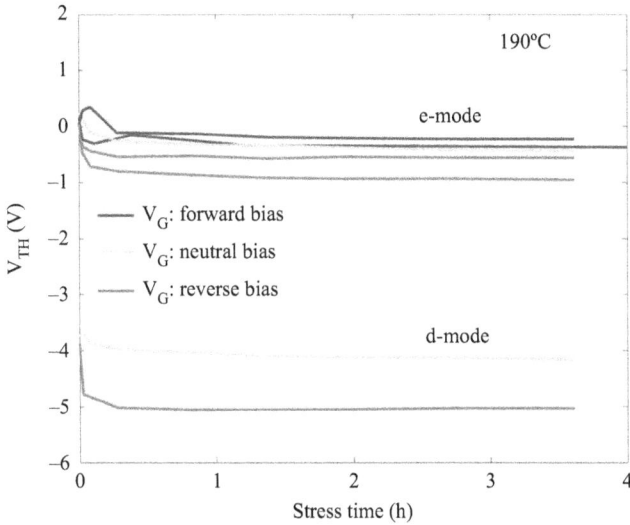

Figure 11.22. V_{TH} stability measured during stress interruptions for e-mode and d-mode HEMTs stressed at 190 °C under forward, neutral, and reverse biased gates, all with $V_D = 50$ V. Reproduced with permission from [74]. Copyright 2019 IEEE.

treatment for $Al_{0.85}Ga_{0.15}N$ barriers with near surface fluorine. This type of claim for fluorine treated AlGaN/GaN so far lacks similar evidence.

11.6.3 Extreme temperature operation

Potential uses of high-temperature capable devices include hydrothermal and oil drilling systems, aircraft electromechanical assemblies, and space applications [75]. Many reports on high-temperature operation of WBG transistors can be found, including a number that reach or exceed 500 °C in GaN [76–79] and SiC [80–82]. Si MOSFET technologies should not be used above 300 °C, a temperature where the intrinsic carrier concentration exceeds $n_i \sim 1 \times 10^{14}$ cm^3. Temperatures above 300 °C may be considered the domain of the WBG and UWBG semiconductors. Using the n_i metric the WBG semiconductors might operate up to 1200 °C–1400 °C, as seen in figure 11.23. If true, the UWBG semiconductors would seem to be unnecessary for high-temperature electronics. However, other factors will set the upper temperature of operation. These include mundane factors such as the temperature compatibility of other necessary materials, but also the basic physical mechanisms of transistor operation, such as midgap trap dominated leakage current (e.g. Frenkel–Poole emission), which will impose lower practical temperature limits on Si and WBG semiconductors. In fact, most semiconductors do not operate close to the limits based on n_i. Si complementary metal–oxide–semiconductor (CMOS) technology products do not operate above 250 °C and are usually rated much lower. In addition to the n_i limits, the temperature dependence of parameters such as electron mobility, saturation velocity, thermal conductivity, etc, must be part of the analysis.

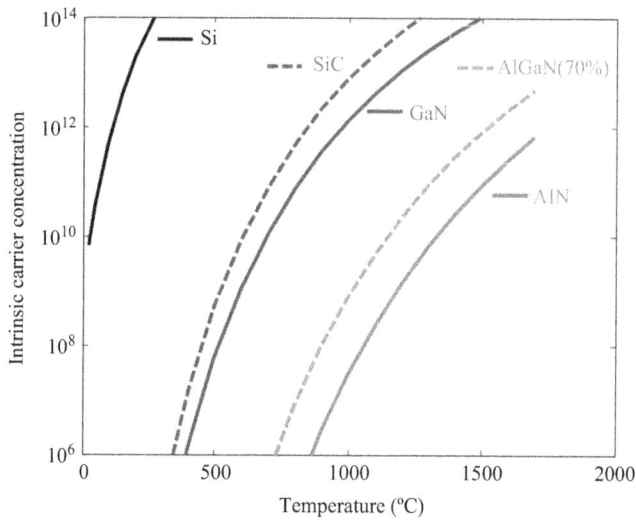

Figure 11.23. Intrinsic carrier concentration versus temperature for Si, SiC, GaN, $Al_{0.7}Ga_{0.3}N$, and AlN.

High Al-content AlGaN might become more promising for high-temperature electronics than SiC devices and materials. The HEMT barrier becomes more insulator-like due to its higher bandgap, thereby improving the high-temperature junction operation [83]. Calculations also show that the transport properties of AlGaN-channel HEMTs decline modestly compared to their binary counterparts in high-temperature environments [24]. Although materials such as AlGaN that are dominated by alloy scattering have lower room temperature mobilities, alloy scattering has a lower temperature dependence, thus damping the decline in transport properties with temperature. Conversely, for the optical polar phonon scattering that dominates the binaries, including GaN and AlN, transport is more strongly dependent on temperature.

The published literature gives insight into two critical trends that limit device performance: (i) increased temperature is associated with increased off-state drain leakage, whether source- or gate-dominated, and (ii) increased temperature is also associated with decreased drain saturation current. The decrease in mobility over temperature [24] can contribute greatly to the reduction in saturation drain current. The forward biased gate current leakage is due to thermionic emission in Schottky gates and p–n junctions, while reverse bias leakage is more complex. Temperature stable drain current, source-dominant subthreshold current, and low forward and reverse gate leakage over a wide temperature range are all highly desirable traits for a suitable high-temperature transistor.

A comparison of the drain current density of an 86/51 device and a GaN-channel HEMT over temperatures between 25 °C–300 °C was reported [40]. The 86/51 HEMT experienced only a 15% reduction in drain current, whereas the GaN-channel had an 80% reduction. A stronger mobility degradation for GaN was found to be responsible for this difference. Lower composition (51/24) HEMTs, were compared to a GaN-channel HEMT over temperature [84]. Again, greater drain current stability over temperature was observed, evidenced by only a 40% reduction for the $Al_{0.24}Ga_{0.76}N$ channel compared to a 60% reduction of drain current for the GaN channel. The temperature variable characterization of a 100/60 HEMT from 25 °C to 300 °C showed virtually no degradation in the drain current, which was stable at about 40 mA mm^{-1} [39]. Compared to a GaN-channel HEMT, which show an 80% reduction in drain current, these transistors were exceptionally stable over temperature.

Temperature-based characteristics from -50 °C to 200 °C were reported for 45/30 and 85/70 HEMTs [50, 53]. Both exhibited high I_{ON}–I_{OFF} current ratios over the entire temperature range, with the 45/30 remaining invariant at $I_{ON}/I_{OFF} \sim 10^8$, and the 85/70 I_{ON}–I_{OFF} ratio degrading slightly from 8×10^9 to 5×10^7 (25 °C and 200 °C, respectively). The high Al content in the 85/70 device was associated with a smaller change of its room temperature saturation current compared to the 45/30 device, with only a 17% reduction compared to 63%.

A 65/40 metal–oxide–semiconductor HEMT (MOS-HEMT) made with a PECVD-deposited SiO_2 gate oxide was studied over temperature [52]. The device's saturation current dropped only 25% at 250 °C from an initial room temperature density of 400 mA mm^{-1}. Over the 25 °C–250 °C temperature range, the I_{ON}–I_{OFF} ratio dropped by three orders of magnitude, from 5×10^7 to 5×10^4, which was due to the rise in off-state leakage current over temperature. The 25 °C ratio was approximately four orders of magnitude better than standard HFET devices.

The maximum temperature at which high Al-content devices have operated were observed for an 85/70 HEMT operated over the 25 °C to 500 °C temperature range [83, 85]. A modest 58% reduction in saturation current density was observed over this temperature range. The I_{ON}–I_{OFF} ratio of $>2 \times 10^{11}$ at room temperature decreased to 3×10^6 at 500 °C. The 500 °C I_{ON}–I_{OFF} ratio is more than 1000× better than GaN, as seen in figure 11.24. The changed slope of the I_{ON}–I_{OFF} ratio at 350 °C

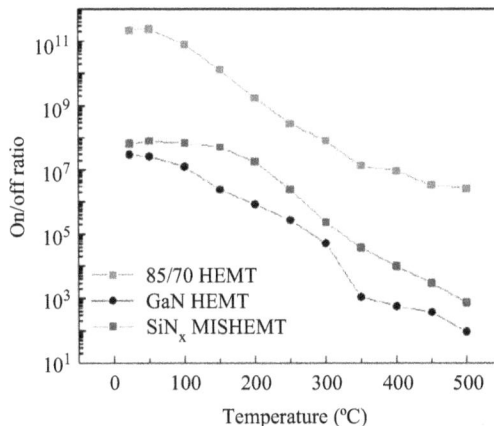

Figure 11.24. I_{ON}–I_{OFF} ratio for an 85/70 HEMT compared to a GaN HEMT ($Al_{0.25}Ga_{0.75}N$) and a SiN_x gate insulator MISHEMT. Reproduced with permission from [83]. Copyright 2019 IEEE.

Table 11.5. Changes to high-temperature I_{ON}–I_{OFF} ratio and saturation drain currents for AlGaN-channel HEMTs.

(Al%) y/x	Temperature range (°C)	I_{ON}/I_{OFF} change	Change in saturation current	Reference
25/0	25 to 300	Not reported	61% reduction	[84]
51/24	25 to 300	Not reported	37% reduction	[84]
45/30	−50 to 200 (25 to 200)	$10^8 \rightarrow 10^8$	63% reduction (45% reduction from 25 °C)	[50]
65/40	25 to 250	$5 \times 10^7 \rightarrow 5 \times 10^4$	~25% reduction	[52]
85/70	−50 to 200	$8 \times 10^9 \rightarrow 5 \times 10^7$	17% reduction	[53]
86/51	25 to 300	Not reported	20% reduction	[40]
100/60	25 to 300	Not reported	4% reduction	[39]
85/70	25 to 500	$> 2 \times 10^{11} \rightarrow 3 \times 10^6$	58% reduction	[83]

and higher would seem to indicate that 85/70 AlGaN HEMTs maintain an I_{ON}–I_{OFF} ratio above 10^5 when extrapolating to 500 °C–800 °C. A similar extrapolation for GaN indicates that its outlook becomes untenable in this difficult temperature range. The GaN extrapolation is consistent with the results of Daumiller *et al*, who found that GaN heterostructure material degrades above 600 °C [77].

Even though these reports from several different research groups are only a starting point, they illustrate the potential advantages of AlGaN channel relative to GaN-channel devices at high temperatures. Much remains to be learned about compatible, temperature-tolerant materials, as well as other factors. Nevertheless, the excellent off-state properties and modest declines in mobility over a large temperature range are encouraging continued research into AlGaN transistors. A summary of high-temperature results of high Al-content transistors is given in table 11.5.

11.6.4 Radiation performance

Electronics in space-based applications will encounter ionizing particles such as high energy protons, alpha particles, and electrons. Concerns in these types of environments include the radiation-induced formation of point defects in the crystal and interfaces, phase changes such as amorphization, and elemental mixing at heterojunction interfaces during ionized dose exposure. All these effects have the potential to degrade device performance [86]. Simplifying a complicated picture somewhat, protons can induce distributions of individual point defects through displacement damage and heavy ion strikes can result in cascades of point defects, whose extent depends on the ion energy. Single event upset effects from both protons and heavy ions can also occur at the device level due to induced transients and bit flips.

Researchers often seek to harden transistors against extreme radiation environments by leveraging intrinsic material properties. Due to the higher bond strength in wide bandgap semiconductors, the energy required for an incoming charged particle to cause atom displacement (e.g. bulk or interface damage) is higher than materials with lower bond strength [87, 88]. Empirically, the energy required for atomic displacement is inversely proportional to the volume of the unit cell [87, 88]. The best prior examples of intrinsic radiation hardness are the wide bandgap materials, such as SiC, GaN, and Ga_2O_3, which have exhibited exceptional robustness in extreme environmental conditions [89–93]. With AlN having an almost 8% smaller unit cell than GaN, it is expected that increasing the Al content in AlGaN, approaching that of AlN, will result in improved radiation tolerance over GaN devices. This section reviews the initial experiments evaluating high Al-content HEMTs.

The effect of proton radiation (2.5 MeV) was investigated for 45/30 HEMT devices with a gate length of 2 μm [89]. A total ionizing dose of 131 Mrad was estimated from stopping range of ions in matter (SRIM) simulations with a maximum fluence of 1×10^{14} p cm^{-2} at 2.5 MeV. A comparison of pre-irradiation and post-irradiation device characteristics in figure 11.25 shows an increase in leakage current, a decrease in threshold voltage, and otherwise stable electrical properties such as a subthreshold slope up to a fluence of 1×10^{13} p cm^{-2}.

Positive charge trapping should result in a V_{TH} shift and increased leakage current. Bulk donor-like defect states that are ionized due to irradiation damage in the buffer layer could also contribute to a similar device response. Further increases in proton fluence will result in further V_{TH} increase, and I_{MAX} decrease. Bulk defects in the AlGaN layer could result in compensation of the positive traps even at lower fluences. Although future studies will need to identify and verify defect and trap formation, 45/30 HEMTs exhibit a high damage threshold, which is a consistent

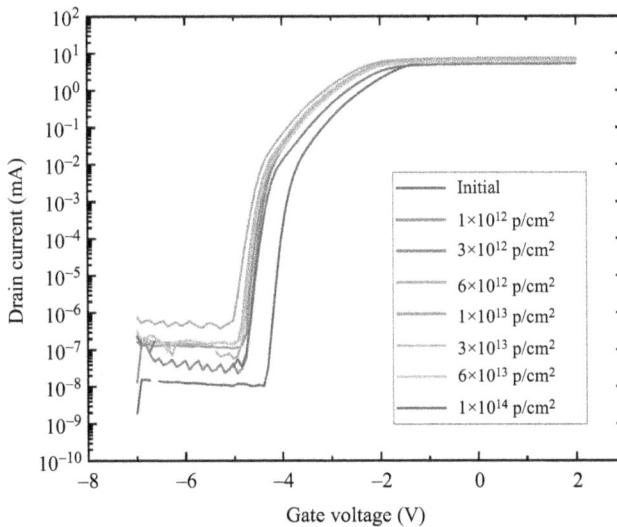

Figure 11.25. Plot of drain current versus gate voltage showing the example response to 2.5 MeV proton irradiation. The total ionizing dose was estimated to be 131 Mrad ($Al_{0.45}Ga_{0.55}N$) at the maximum fluence of 1 $\times 10^{14}$ p cm^{-2}. Reproduced with permission from [89]. Copyright 2019 IEEE.

observation in WBG devices such as GaN HEMTs. Increasing the Al content further is expected to extend these trends.

Another radiation effect known as single event burnout (SEB) arises when devices are hit by a single energetic particle. SEB can result in transients, data corruption, or high current which can cause catastrophic device failure. The voltage at which catastrophic failure occurs due to a single energetic particle can be a function of the ion and the energy. Therefore, SEB as a function of linear energy transfer (LET) was investigated in an exploratory study [89]. Device gate-to-drain spacings of 4, 8, and 10 μm were evaluated for both 45/30 (type I) and 85/70 HEMTs (type II). The study included ions such as F, Si, Ti, Fe, Ge, Br, and Au. Irradiation and *in situ* testing was carried out the Tandem Van de Graff facility at Brookhaven National Laboratory. All devices were packaged in lidless ceramic dual inline packages to allow for direct exposure, utilizing zero insertion force socketed test boards, and placed into a vacuum chamber. V_G was set below V_{TH} (−6 V) to pinch off the channel, with a fluence of 2.5×10^7 ions cm^{-1} (0.25 ions μm^{-1}), resulting in more than 3×10^4 ions on average striking the active area in each device. The expected normal environment breakdown voltage for the two heterostructures was ~200 V and ~500 V for 45/30 and 85/70 HEMTs, respectively. As the LET (MeV-cm^2 mg^{-1}) increased for both heterostructures, the SEB voltage (V_{SEB}) dropped, as seen in figure 11.26. Ions as heavy as Br only reached ~25% of the expected breakdown voltage for Al$_{0.3}$Ga$_{0.7}$N channel HEMTs. The response observed for the Al$_{0.3}$Ga$_{.7}$N channel with Br

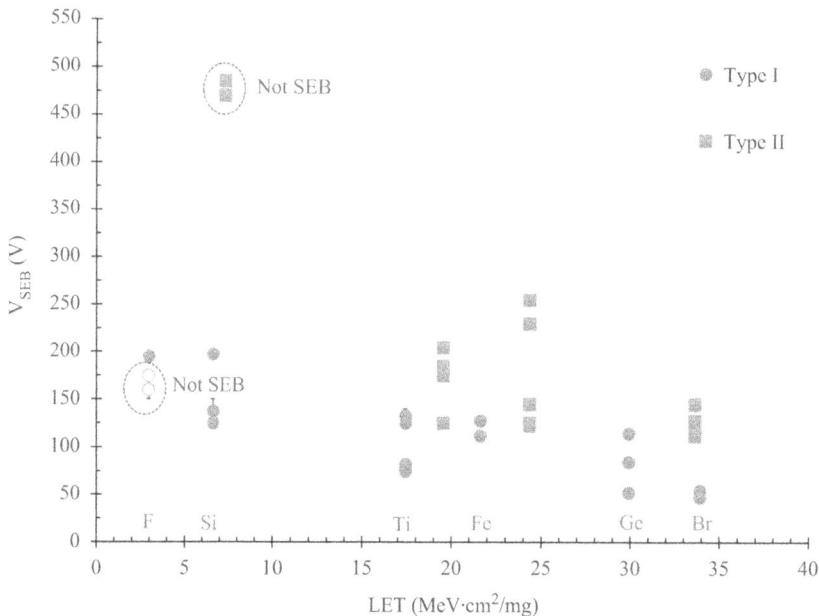

Figure 11.26. Plot of the minimum SEB voltage for type I (45/30) and type II (85/70) HEMTs ($L_{GD} = 5 \ \mu$m). The data from the experiments are described in [89]. Two devices of each type failed for non-SEB breakdown and are labeled and circled in red. Reproduced with permission from [21]. Copyright 2020 AIP Publishing.

(LET = 33.9 MeV-cm^2 mg^{-1}) resulted in V_{SEB} ~50 V. For the Al$_{0.7}$Ga$_{.3}$N channel, a correspondingly similar LET for Ge (33.6 MeV-cm^2 mg^{-1}), resulted in a substantially higher V_{SEB} ~125 V. With V_{SEB} for both cases exhibiting approximately 25% of the expected breakdown voltage of an un-irradiated device, the similar response highlights the considerable increase in operating voltage that can be attained with a higher Al content.

11.7 Summary

High Al-content AlGaN offers great potential as the material of choice for the next generation of HEMTs for power switching and radio frequency applications based on FOM analysis. The key parameter enabling this potential is the critical electric field for avalanche breakdown, which may reach 17 MV cm^{-1} for AlN. Although both the LFOM and the JFOM are favorable for room temperature Al$_x$Ga$_{1-x}$N HEMTs at high enough x, it will take time to realize this promise. At elevated temperatures, the FOM advantage becomes much more pronounced, in particular for power devices, because alloy scattering in AlGaN (and hence its mobility) is somewhat insensitive to temperature, whereas GaN mobility drops considerably due to the prominence of phonon scattering. This suggests that early AlGaN-rich transistor applications may be at high temperatures. Several key metrics give an indication of the status of current high Al-content HEMTs. Current densities exceeding 600 mA mm^{-1} have been achieved and further improvements can be expected with improvements in ohmic contacts and greater emphasis on dimensional scaling. Average breakdown fields have reached 3.6 MV cm^{-1} and excellent I_{ON}–I_{OFF} ratios have been reported, in some cases exceeding 10^{11}. f_T and f_{max} have been demonstrated in the tens of gigahertz range for devices with short gate lengths (<150 nm) and performance is expected to increase further as ohmic contact technology improves. These metrics have been achieved largely for Schottky-gated HEMTs, in contrast to traditional GaN-channel HEMTs, which use insulating gates to achieve high I_{ON}–I_{OFF} ratios and reduce forward gate current. Saturation current density is considerably less sensitive to temperature for high Al-content HEMTs than for GaN-channel HEMTs and superior I_{ON}–I_{OFF} metrics are also maintained over temperature. High Al-content AlGaN HEMT research has grown in recent years, with many new and noteworthy results. Much research is still needed towards furthering the technological development in topics that include the effect of thermal conductivity on current density, characterizing high voltage performance at elevated temperatures, and characterizing trapping and detrapping characteristics.

Acknowledgement

Sandia National Laboratories is a multi-program laboratory managed and operated by National Technology & Engineering Solutions of Sandia, LLC (NTESS), a wholly owned subsidiary of Honeywell Corporation, for the US Department of Energy's National Nuclear Security Administration under contract DE-NA0003525. The views expressed in the article do not necessarily represent the views of the US Department of Energy or the United States Government.

References

[1] Mimura T 2002 *IEEE Trans. Micro. Theory Tech* **50** 780

[2] Chao P C, Palmateer S C, Smith P M, Mishra U K, Duh K H G and Hwang J C M 1985 *IEEE Electron. Device. Lett.* **6** 531

[3] Duh K H G, Chao P C, Ho P, Kao M Y, Smith P M, Ballingall J M and Jabra A A 1989 *Proc. IEEE MTT-S Int. Microwave Symp. (Long Beach, CA, USA, 13–15 June)* p 805

[4] Khan M A, Bhattarai A, Kuznia J N and Olson D T 1993 *Appl. Phys. Lett.* **63** 1214

[5] Khan M A, Chen Q, Shur M S, Dermott B T, Higgins J A, Burm J, Schaff W and Eastman L F 1996 *Electron. Lett.* **32** 357

[6] Green B M, Chu K K, Chumbes E M, Smart J A, Shealy J R and Eastman L F 2000 *IEEE Electron. Dev. Lett.* **21** 268

[7] Edwards A P, Mittereder J A, Binari S C, Katzer D S, Storm D F and Roussos J A 2005 *IEEE Electron. Dev. Lett.* **26** 225

[8] Ando Y, Okamoto Y, Miyamoto H, Nakayama T, Inoue T and Kuzuhara M 2003 *IEEE Electron. Dev. Lett.* **24** 289

[9] Wu Y-F, Saxler A, Moore M, Smith R P, Sheppard S, Chavarkar P M, Wisleder T, Mishra U K and Parikh P 2004 *IEEE Electron. Dev. Lett.* **25** 117

[10] Johnson J W, Piner E L, Vescan A, Therrien R, Rajagopal P, Roberts J C, Brown J D, Singhal S and Linthicum K J 2004 *IEEE Electron. Dev. Lett.* **25** 459

[11] Chu K K *et al* 2004 *IEEE Electron. Dev. Lett.* **25** 596

[12] Binari S C, Klein P B and Kazior T E 2002 *Proc. IEEE* **90** 1048

[13] Jones E A, Wang F and Costinett D 2016 *IEEE J. Emer. Sel. Top. Power Electron.* **4** 707

[14] Mishra U K, Shen L, Kazior T E and Wu Y-F 2008 *Proc. IEEE* **96** 287

[15] Bajaj S, Yang Z, Akyol F, Park P S, Zhang Y, Price A L, Krishnamoorthy S, Meyer D J and Rajan S 2017 *IEEE Trans. Electron. Dev.* **64** 3114

[16] Shinohara K, King K, Carter A D, Regan E J, Aria A, Bergman J, Urteaga M and Brar B 2018 *IEEE Electron. Dev. Lett.* **39** 18

[17] Tsao J Y *et al* 2018 *Adv. Electron. Mater.* **4** 1600501

[18] Kohn E and Denisenko A 2007 *Thin Solid Films* **515** 4333

[19] Higashiwaki M, Sasaki K, Kuramata A, Masui T and Yamakoshi S 2014 *Phys. Status Solidi A* **211** 21

[20] Pearton S J, Yang J, Cary P H, Ren F, Kim J, Tadjer M J and Mastro M A 2018 *Appl. Phys. Rev.* **5** 011301

[21] Baca A G, Armstrong A M, Klein B A, Douglas E A, Allerman A A and Kaplar R J 2020 *J. Vac. Sci. Technol.* A **38** 020803

[22] Kaplar R J, Allerman A A, Armstrong A M, Crawford M G, Dickerson J R, Fischer A J, Baca A G and Douglas E A 2017 *ECS J. Sol. State Science Technol.* **6** Q3061

[23] Pearton S J, Douglas E A, Shul R J and Ren F 2020 *J. Vac. Sci. Technol.* A **38** 020802

[24] Coltrin M E, Baca A G and Kaplar R J 2017 *ECS J. Sol. State Science Technol.* **6** S3114

[25] Daly B C, Maris H J, Nurmikko A V, Kuball M and Han J 2002 *J. Appl. Phys.* **92** 3820

[26] Hudgins J L, Simin G S and Khan M A 2003 *IEEE Trans. Power Electronics* **18** 907

[27] Johnson E 1965 *RCA Rev.* **26** 163

[28] Bellotti E and Bertazzi F 2012 *J. Appl. Phys.* **111** 103711

[29] Klein B A *et al* 2019 *J. Electronics Mat.* **48** 5581

[30] Bernardini F, Fiorentini V and Vanderbilt D 1997 *Phys. Rev.* B **56** R 10024

[31] Bajaj S, Akyol F, Krishnamoorthy S, Zhang Y and Rajan S 2016 *Appl. Phys. Lett.* **109** 133508

[32] Nanjo T, Takeuchi M, Suita M, Abe Y, Oishi T, Tokuda Y and Aoyagi Y 2008 *Appl. Phys. Express* **1** 011101

[33] Nanjo T, Takeuchi M, Suita M, Oishi T, Abe Y, Tokuda Y and Aoyagi Y 2008 *Appl. Phys. Lett.* **92** 263502

[34] Nanjo T, Imai A, Suzuki Y, Abe Y, Oishi T, Suita M, Yagyu E and Tokuda Y 2013 *IEEE Trans. Electron. Dev.* **60** 1046

[35] Nanjo T, Takeuchi M, Imai A, Suita M, Oishi T, Abe Y, Yagyu E, Kurata T, Tokuda Y and Aoyagi Y 2009 *Electron. Lett.* **45** 1346

[36] Okumura H, Suihkonen S, Lemettinen J, Uedono A, Zhang Y, Piedra D and Palacios T 2018 *Jpn J. Appl. Phys.* **57** 04FR11

[37] Klein B A *et al* 2017 *ECS J. Sol. State Sci. Technol.* **6** S3067

[38] Yafune N, Hashimoto S, Akita K, Yamamoto Y and Kuzuhara 2011 *Jpn J. Appl. Phys.* **50** 100202

[39] Yafune N, Hashimoto S, Akita K, Yamamoto Y, Tokuda H and Kuzuhara M 2014 *Electron. Lett.* **50** 211

[40] Tokuda H, Hatano M, Yafune N, Hashimoto S, Akita K, Yamamoto Y and Kuzuhara M 2010 *Appl. Phys. Exp.* **3** 121003

[41] France R, Xu T, Chen P, Chandrasekaran R and Moustakas T D 2007 *Appl. Phys. Lett.* **90** 062115

[42] Mori K, Takeda K, Kusafuka T, Iwaya M, Takeuchi T, Kamiyama S, Akasaki I and Amano H 2016 *Jpn J. Appl. Phys.* **55** 05FL03

[43] Yun J, Choi K, Mathur K, Kuryatkov V, Borisov B, Kipshidze G, Nikishin S and Temkin H 2006 *IEEE Electron. Dev. Lett.* **27** 22

[44] Armstrong A M *et al* 2018 *Jpn J. Appl. Phys.* **57** 074103

[45] Park P S, Krishnamoorthy S, Bajaj S, Nath D N and Rajan S 2015 Recess-free nonalloyed ohmic contacts on graded AlGaN heterojunction FETs *IEEE Electron. Dev. Lett.* **36** 226

[46] Bajaj S *et al* 2018 *Electron. Dev. Lett.* **39** 256

[47] Douglas E A, Reza S, Sanchez C, Koleske D, Allerman A, Armstrong A M, Kaplar R J and Baca A G 2017 *Phys. Status Solidi* A **214** 1600842

[48] Baca A G, Armstrong A M, Allerman A A, Douglas E A, Sanchez C A, King M P, Coltrin M E, Fortune T R and Kaplar R J 2016 *Appl. Phys. Lett.* **109** 033509

[49] Suita M, Nanjo T, Oishi T, Abe Y and Tokuda Y 2006 *Phys. Stat. Sol.* C **3** 2364

[50] Baca A G, Armstrong A M, Allerman A A, Klein B A, Douglas E A, Sanchez C A and Fortune T R 2017 *ECS J. Sol. State Sci. Technol.* **6** S3010–S3013

[51] Hahn H, Reuters B, Kalisch H and Vescan A 2013 *Semi. Sci. Technol.* **28** 074017

[52] Hu X, Hwang S, Hussain K, Floyd R, Mollah S, Asif F, Simin G and Khan A 2018 *IEEE Electron. Dev. Lett.* **39** 1568

[53] Baca A G, Klein B A, Allerman A A, Armstrong A M, Douglas E A, Stephenson C A, Fortune T R and Kaplar R J 2017 *ECS J. Sol. State Sci. Technol.* **6** Q161

[54] Muhtadi S, Hwang S M, Coleman A, Asif F, Simin G, Chandrashekhar M B S and Khan A 2017 *IEEE Electron. Dev. Lett.* **38** 914

[55] Douglas E A, Klein B, Allerman A A, Baca A G, Fortune T and Armstrong A M 2019 *J. Vac. Sci. Technol.* B **37** 021208

[56] Klein B A, Douglas E A, Armstrong A M, Allerman A A, Abate V M, Fortune T R and Baca A G 2019 *Appl. Phys. Lett.* **114** 112104

[57] Baca A G, Klein B A, Wendt J R, Lepkowski S M, Nordquist C D, Armstrong A M, Allerman A A, Douglas E A and Kaplar R J 2019 *IEEE Electron. Dev. Lett.* **40** 17

[58] Xue H, Lee C H, Hussain K, Razzak T, Abdullah M, Xia Z, Sohel S H, Khan A, Rajan S and Lu W 2019 *Appl. Phys. Express* **12** 066502

[59] Hu X *et al* 2018 *76th Device Res. Conf.*
Razzak T *et al* 2018 *Electron. Lett.* **40** 1245

[60] Muhtadi S, Hwang S, Coleman A, Asif G, Lunev A, Chandrashekhar M V S and Khan A 2017 *Appl. Phys. Lett.* **110** 171104

[61] Muhtadi S, Hwang S, Coleman A, Asif F, Lunev A, Chandraashekhar M V S and Khan A 2017 *Appl. Phys. Lett.* **110** 193501

[62] Armstrong A M *et al* 2019 *Appl. Phys. Lett.* **114** 052103

[63] Trinh X T, Nilsson D, Ivanov I G, Janzén E, Kakanakova-Georgieva A and Son N T 2014 *Appl. Phys. Lett.* **105** 162106

[64] Mehnke F, Trinh X T, Pingel H, Wernicke T, Janzen E, Son N T and Kneissl M 2016 *J. Appl. Phys.* **120** 145702

[65] Miller E J, Yu E T, Waltereit P and Speck J S 2004 *Appl. Phys. Lett.* **84** 537

[66] Hu X, Simin G, Yang J, Khan M A, Gaska R and Shur M S 2000 *Electronics Lett.* **36** 753

[67] Cai Y, Zhou Y, Lau K M and Chen K J 2006 *IEEE Trans. Electron. Dev.* **53** 2207

[68] Rossetto I, Meneghini M, Hilt O, Bahat-Treidel E, De Santi C, Dalcanale S, Wuerfl J, Zanoni E and Meneghesso G 2016 *IEEE Trans. Electron. Dev.* **63** 2334

[69] Reza S, Klein B A, Baca A G, Armstrong A M, Douglas E A and Kaplar R J 2019 *Jpn J. Appl. Phys.* **58** SCCD04

[70] Carrano J C *et al* 2000 *Appl. Phys. Lett.* **76** 924

[71] Mollah S, Gaevski M, Hussain K, Mamun A, Floyd R, Hu. X, Chandrashekhar M V S, Simin G and Khan A 2019 *Appl. Phys. Express* **12** 074001

[72] Simin G, Koudymov A, Tarakji A, Hu X, Yang J, Khan M A, Shur M S and Gaska R 2001 *Appl. Phys. Lett.* **79** 2651

[73] Yi C, Wang R, Huang W, Tang W C-W, Lau K M and Chen K J 2007 *Proc. Int. Electron Device Meeting* p 389

[74] Baca A G, Klein B A, Armstrong A M, Allerman A A, Douglas E A, Fortune T R and Kaplar R J 2019 Stability in fluorine-treated Al-rich high electron mobility transistors with 85% Al-barrier composition *2019 IEEE Int. Rel. Phys. Symp. (IRPS) (Monterey, CA, USA)* pp 1–4

[75] Son K, Liao A, Lung G, Gallegos M, Hatake T, Harris R D, Scheick L Z and Smythe W D 2010 *Nanosci. Nanotechnol. Lett.* **2** 89

[76] Hou M, Jain S R, So H, Heuser T A, Xu X, Suria A J and Senesky D G 2017 *J. Appl. Phys.* **122** 195102

[77] Daumiller I, Kirchner C, Ebeling M K K J and Kohn E 1999 *IEEE Electron. Dev. Lett.* **20** 448

[78] Maier D *et al* 2010 *IEEE Trans. Dev. Mat. Reliability* **10** 427

[79] Arulkumaran S, Egawa T, Ishikawa H and Jimbo T 2002 *Appl. Phys. Lett.* **80** 2186

[80] Spry D, Neudeck P, Chen L, Chang C, Lukco D and Beheim G 2016 *SPIE Defense + Security 2016* (Baltimore, MD: SPIE)

[81] Neudeck P G, Garverick S L, Spry D J, Chen L-Y, Beheim G M, Krasowski M J and Mehregany M 2009 *Phys. Status Solidi* A **206** 2329

[82] Neudeck P G, Spry D J, Chen L, Prokop N F and Krasowski M J 2017 *IEEE Electron. Dev. Lett.* **38** 1082

[83] Carey P H IV, Ren F, Baca A G, Klein B A, Allerman A A, Armstrong A M, Douglas E A, Kaplar R J, Kotula P G and Pearton S J 2019 *J. Electron. Dev. Soc.* **7** 444

[84] Hatano M *et al* 2010 *Proc. CS MANTECH (Portland, OR, USA)* p 101

[85] Carey P H IV, Ren F, Baca A G, Klein B A, Allerman A A, Armstrong A M, Douglas E A, Kaplar R J and Pearton S J 2020 *J. Vacuum Sci. Technol.* B **38** 033202

[86] Nord J, Nordlund K, Keinonen J and Albe K 2003 *Nucl. Instr. Meth. Phys. Res.* B **202** 93

[87] Ionacscut-Nedelcescu A, Carlone C, Houdayer A, von Bardeleben H J, Cantin J-L and Raymond S 2002 *IEEE Trans. Nuclear Sci.* **49** 2733

[88] Kim J, Pearton S J, Fares C, Yang J, Ren F, Kim S and Polyakov A Y 2019 *J. Mat. Chem.* C **7** 10

[89] Martinez M J, King M P, Baca A G, Allerman A A, Armstrong A A, Klein B A, Douglas E A, Kaplar R J and Swanson S 2019 Radiation response of AlGaN-channel HEMTs *IEEE Trans. Nuclear Sci.* **66** 344

[90] Sellin P J and Vaitkus J 2006 *Nucl. Instr. Meth. Phys. Res.* A **557** 479

[91] Polyakov A Y *et al* 2012 *J. Vac. Sci. Technol.* B **30** 6

[92] Pearton S J, Hwang Y-S and Ren F 2015 *J. Mater.* **67** 1601

[93] Weaver B D, Anderson T J, Koehler A D, Greenlee J D, Hite J K, Shahin D I, Kub F J and Hobart K D 2016 *ECS J. Sol. State Sci. Technol.* **5** Q208

IOP Publishing

Wide Bandgap Semiconductor-Based Electronics

Fan Ren and Stephen J Pearton

Chapter 12

Understanding interfaces for homoepitaxial GaN growth

Jennifer Hite and Michael A Mastro

The recent availability of commercial bulk GaN substrates has enabled vertical GaN devices to become a reality through homoepitaxial growth. However, additional materials challenges need to be understood and surmounted to fully develop the technology. The first step to this is forming an understanding of the influences of the interface on growth and device performance. This chapter will examine the effects of surface preparation and initiation on growth morphology, impurity incorporation, and interface composition, as well as how these issues affect device performance.

12.1 Introduction

GaN-based devices have become ubiquitous in daily life with the commercialization and widespread adoption of light emitting diodes (LEDs) [1–3]. In addition, they have been highly useful as electrical devices in RF and microwave applications in the form of high electron mobility transistors (HEMTs) [4, 5]. As such, one may ask why research still continues on this successful material. The answer is that, although successful, the electrical devices are still not meeting the theoretical limits of the material. The best GaN-based HEMT device technology is still only attaining around a third of the theoretical estimated breakdown field (>3 MV cm^{-1}) [6].

The main culprits are two-fold: heteroepitaxial growth and lateral device geometry. Current commercial devices are grown on non-native substrates, mainly Si, SiC, and sapphire. None of these are lattice-matched or thermally matched with GaN and require varying degrees of buffer layers and engineered strain in order to produce epitaxial layers. Si substrates have the additional issue of diffusion of Ga into the substrate without appropriate buffer layers, while sapphire substrates are highly insulating, restricting their use to only lateral conducting devices. The two most direct consequences of the lattice and thermal mismatch are defectivity in the

doi:10.1088/978-0-7503-2516-5ch12

material, mainly in the form of dislocations, and major limitations on the thickness of the GaN layers. The critical thickness for cracking of a GaN film is generally around 5 μm, but varies a bit with substrate material and buffer layers [7]. It generally trends lower (~2 μm) when grown on Si substrates, even with multiple strain engineered buffer layers [8]. Dislocations are scattering sites in lateral devices for both electrons and phonons, which reduce mobility and decrease thermal conductivity [9–11]. A schematic of the most common lateral device, the high electron mobility transistor (HEMT), is shown in figure 12.1(a). In this geometry, the conduction path of lateral devices is very close to the surface, allowing surface states to impair the operation of the substrate. Additionally, extremely high fields are formed at the gate edge, requiring extensive thermal engineering efforts, which have been the subject of multiple large programs over the past decade [12, 13].

To alleviate this, and allow devices to reach the full potential of the material, a vertical device geometry grown on a thermally and lattice-matched substrate with low dislocations is needed. Figure 12.1(b) shows a very simple schematic for a vertical diode. The advantage to this geometry is that the conduction path flows vertically through both the epilayer and substrate, meaning the characteristics are determined by the intrinsic bulk properties of the nitrides. Additionally, this geometry allows for a distributed field, meaning less thermal management and, with the use of native, lattice-matched substrates, the ability to grow thick epitaxial layers. This is important, as thick layers with low impurity concentration are required to reach high breakdown fields, as shown in the plot of avalanche and punch-through breakdown of GaN Schottky diodes in figure 12.2 [14]. Unlike SiC, Si, GaAs, and other semiconductor workhorses, a native bulk substrate for GaN has been historically unavailable. Breakthroughs in hydride vapor phase epitaxy and ammonothermal growth techniques have enabled the recent development and commercialization of bulk GaN substrates [15–19]. However, the availability of native substrates does not automatically enable high quality homoepitaxial layers. Taking heteroepitaxial growth recipes and directly applying them to native substrates without forming an understanding of the GaN substrate interface surface and characteristics of the initial stages of growth does not necessarily lead to successful growth. The effects of the structural and chemical characteristics of the

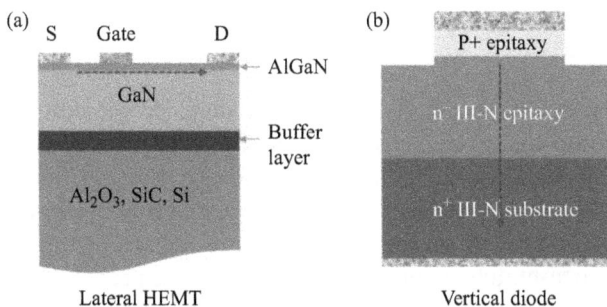

Figure 12.1. (a) Basic schematic for a GaN-based lateral HEMT for comparison with (b) a vertical p–n diode. The red dotted line denotes the electron conduction path through the device.

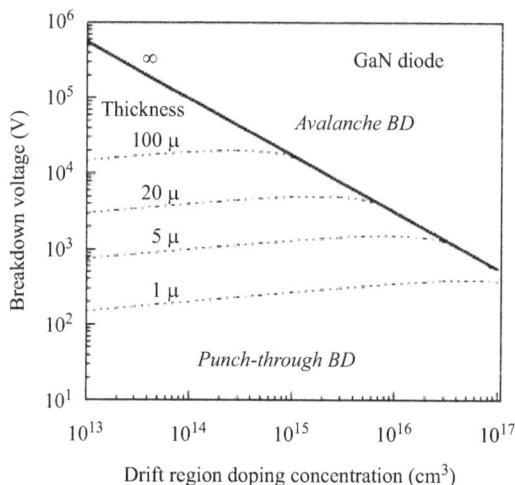

Figure 12.2. Calculated breakdown voltage as a function of doping concentration and thickness of the drift region in GaN n⁻n⁺ diodes. Reproduced with permission from [14]. Copyright 2000 Elsevier.

substrate–epitaxial interface on growth and device performance will be covered herein for GaN oriented in the +c (0001) direction by metal–organic chemical vapor deposition (MOCVD).

12.2 Surface interface structure

In epitaxial growth, the structure of the starting substrate surface is extremely important. Although the surface can be changed slightly during the ramp to growth temperature and during initiation of the growth, many of the initial surface features are set by the substrate manufacturer's product specifications. These characteristics include wafer offcut, bow, and surface roughness. In addition to these standard specifications, there are additional challenges in the uniformity of impurities within the substrates that are starting be noticed [20]. The technology to produce bulk GaN substrates is not fully mature and control over these characteristics is developing at the same time as the manufacturers are scaling their technology to larger substrate diameters.

12.2.1 Offcut

The basic mechanism of single-crystal growth stems from atoms attaching in an orderly manner onto the crystal surface at an edge [21, 22]. For epitaxial growth, a regular surface of steps is the preferred surface to start this. In heteroepitaxial growth, these steps are seeded by the spiral growth around dislocations (figure 12.3(a)). For growth on a native substrate, where the starting substrate is crystalline and lattice-matched, an alternative method to provide a stepped surface is possible. A periodic stepped surface can be produced by cutting and polishing the original crystal substrate at an angle to the basal plane (figure 12.3(b)). Such substrates are referred to as offcut or mis-cut. The benefit is that by using a stepped

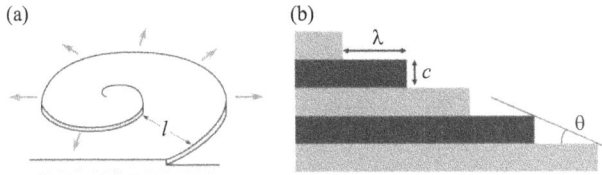

Figure 12.3. (a) Schematic of steps originating spirally from a dislocation, which is the source of growth steps in heteroepitaxy. Reproduced with permission from [23]. Copyright 2016 Elsevier. (b) Illustration of an offcut crystal, where the step width, λ, is set by the geometric relationship between step height, c (lattice constant), and offcut angle, θ.

surface, the growth conditions can be chosen such that the growth of the steps outpaces the growth at dislocations, producing a smooth film with each atomic layer forming as a 2D sheet, where the terrace width is determined by the offcut angle.

The offcut angle has a direct impact on the morphology of epitaxial films. On vicinal substrates (0° offcut), screw dislocations will still dominate as step producers, resulting in a rough surface decorated by hillocks [24]. With HVPE GaN substrates containing $\sim10^6$ dislocations/cm^2, this results in a very rough morphology. With increasing offcut (toward m), the growth morphology smoothens, resulting in regularly stepped surfaces at 0.5°–1° offcut angles [25–27]. At offcuts >1°, although the epitaxial layers appear macroscopically smooth under optical microscopy, more detailed characterization shows that the surfaces are stepped, but the steps are bunched [25, 28]. This step-bunching means that the regular bi-layer stacking of the surface is interrupted, with varying step heights. However, at these smaller offcut angles, step meandering is sometimes observed in the epitaxial films. In homoepitaxial GaN growth using HVPE, Fujikura and Konno showed that step meandering at offcuts under a degree could be reduced at low V–III ratios or high growth temperatures (figure 12.4) [29].

The offcut not only affects homoepitaxy morphology, but it can also influence incorporation rates of different species in the films. The most straightforward effect is on InGaN composition. Increasing the offcut reduces the incorporation of indium into the films [25, 30]. The effect on AlGaN composition seems negligible [31, 32], however, it has been reported that increasing the offcut decreases crack generation in AlGaN films on GaN [25]. Additionally, the screw dislocation density in the AlGaN films is independent of the offcut, but edge dislocation density decreases at offcuts greater than 1.4° [31]. For n-type dopants, no real change is seen in silicon or oxygen incorporation with changing offcut [33, 34]. However, magnesium (Mg) incorporation has a few conflicting reports. Several reports show no increase in Mg concentration in the films with changing offcut, however, at the same time, they show an increase in hole concentration and reduced resistivity with increased offcut angle [33, 35]. Suski *et al* report this increase in p-type carrier concentration to be due to decreasing compensating donors in the films [33], while Jiang *et al* attribute it to higher carbon incorporation in the p-GaN, observed in SIMS. Grenko *et al* report an opposite effect on Mg incorporation, showing a reduction in Mg concentration with increasing offcut, using PL, CL, and SIMS to study their films. However, the

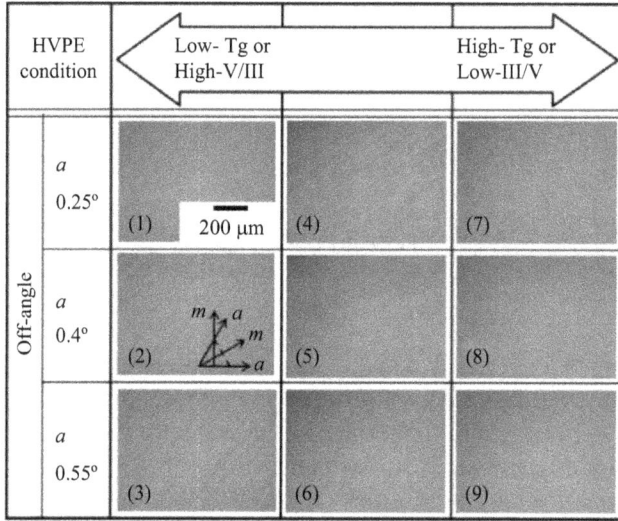

Figure 12.4. Growth parameter dependence of step meandering in HVPE homoepitaxial growth of 20–30 μm on substrates with different offcuts. Reproduced with permission from [29]. Copyright 2018 AIP Publishing.

samples also had an undoped GaN layer grown on top of the p-GaN layers and SIMS showed that Mg migrated to the p-GaN/undoped GaN interface. This effect was more pronounced with increased offcut angles [31]. Revisiting the effect of offcut on carbon incorporation, although Jiang *et al* saw a significant effect of offcut on carbon concentration in p-GaN layers, the effect on n-GaN layers was much smaller, reducing from 7×10^{16} to 5×10^{16} cm^{-3} when decreasing the offcut by $0.4°$ [35]. The C suppression with increased offcut was attributed to reduced step motion velocity. This may mean that the C incorporation is highly dependent on growth parameters, as another report finds the opposite relationship on unintentionally doped films, namely an increase in C with increasing offcut. In this case, the increased carbon is attributed to an increasing number of steps, and a lower energy barrier for C to incorporate at steps than elsewhere on the surface [34].

12.2.2 Wafer bow

Offcut is not the only consideration for surface preparation, in particular for bulk GaN substrates grown by HVPE. A majority of the commercially available GaN substrates use this method to grow thick layers on non-native substrates, sometimes patterned, use dislocation guttering techniques to reduce the defects, then separate the GaN layers from the non-native seed. Since growth occurs on a non-native substrate some lattice or thermal mismatch exists and when growing thick layers (>400 μm) lead the wafers to bow [36, 37]. In order to produce flat substrates, these layers must be polished. However, with the original bow, this can result in variations in step density across the wafer surface, all of which is illustrated in figure 12.5. This, in turn, affects the uniformity of the homoepitaxy grown on such surfaces in roughness and, potentially, impurity incorporation.

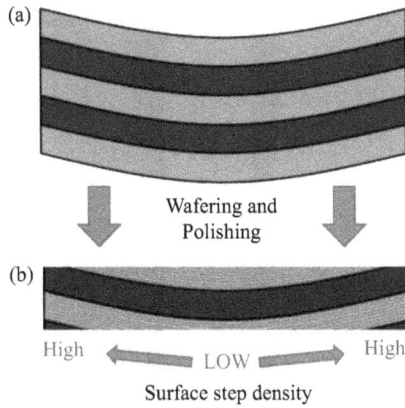

Figure 12.5. Schematic illustrating the effect of wafering and polishing a bowed wafer on step density. The bowed wafer (a) shows curvature over the entire lattice in the case of compressive stress at the wafer surface. A view of it after wafering and polishing is shown in (b), with additional lattice lines drawn to make the change in step density more apparent. This case results in many more steps on the wafer edge than in the center.

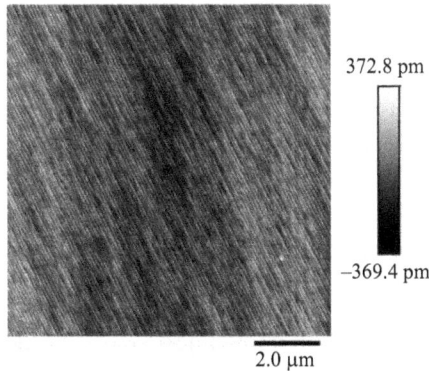

Figure 12.6. A representative example of a small-scale AFM scan ($10 \times 10 \ \mu m^2$) of an incoming substrate.

12.2.3 Surface polish and morphology

Most commercial substrates come with root mean square (RMS) roughness specifications that are on the Ångström scale. AFM scans of these substrates generally show regular, stepped surfaces, as required for step-flow growth. However, each manufacturer uses a different, normally proprietary, recipe, and the results are not always equivalent. On a small scale ($10 \times 10 \ \mu m^2$) the surface appears smooth and regularly stepped, as required by the specifications. An example of this is shown in figure 12.6. However, the macro-scale roughness also needs to be taken into consideration, in particular for large devices. At scan lengths of hundreds of microns, these surface differences between wafers and vendors is quite dramatic, examples of which are shown in figure 12.7 [20]. The best way to characterize the incoming GaN substrates is not just on a small, micro-scale, but also using large-scale, rapid techniques such as optical profilometry, Raman mapping, and PL

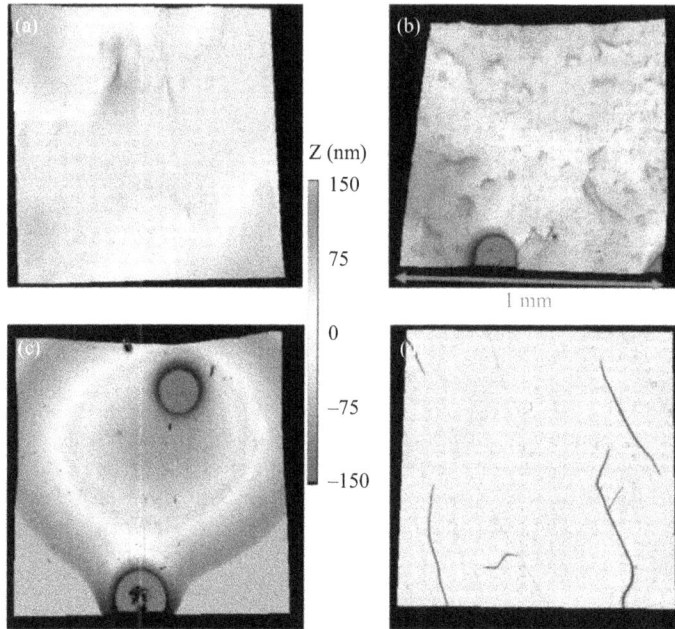

Figure 12.7. Examples of larger scale optical profilometry scans (1×1 mm^2) of several substrates, using the same height scale. The difference in roughness is considerable on these scales, with large features present in (b), but a relatively smooth surface on (a). The substrate in (c) shows large wafer bow and (d) is perfectly flat, despite cracking that occurred during dicing. Reproduced with permission from [20]. Copyright 2018 Elsevier.

mapping in order to see these differences, as well as some startling impurity uniformity issues within the wafers [20]. The reason the macro-scale is also important is that the epitaxy conforms to the surface and increasingly magnifies these features with increasing thickness [34]. For vertical devices, where the homoepitaxial layers will need to be 100 μm thick, this will result in very rough surfaces.

12.3 Chemical interfaces

Homoepitaxial growth using MOCVD has several advantages over molecular beam epitaxy: a higher growth rate, reasonable vacuum levels, and large-scale commercialization. For cleaning the interface, it is particularly advantageous, as ramping the substrates to temperature is done under an ammonia atmosphere with some hydrogen and/or nitrogen as the carrier gases. Not only does this keep the GaN substrates from decomposing, but it also serves to remove C and O from those surfaces without additional *ex situ* steps. Until recently, most MOCVD growers believed that was the end of the story: the MOCVD homoepitaxial interface is generally clean of C and O, as seen in the bottom two plots of figure 12.8. However, with homoepitaxial growth and regrowth of p–n junctions, a small snag has come to the attention of the community. This is in the form of a spike of Si at the growth (or regrowth) interface, as illustrated in figure 12.8 [31, 34, 38, 39]. This ubiquitous spike

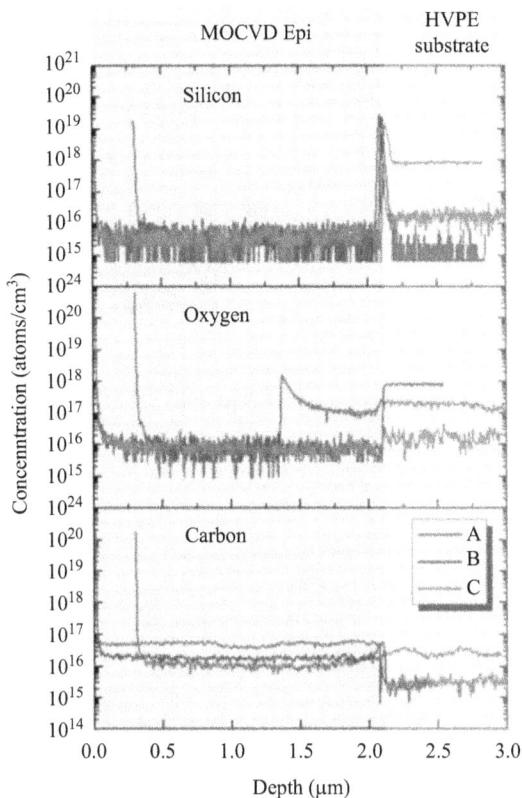

Figure 12.8. SIMS data showing MOCVD regrowth on three different bulk GaN HVPE substrates in a single run. Although the interface between the substrate and MOCVD homoepitaxy is rather clean, a spike in Si at that interface reaches levels of 10^{19} atoms/cm^3 before falling back down to unintentionally doped levels. Reproduced with permission from [34]. Copyright 2018 Elsevier.

reaches very high levels in growth on bulk substrates, but is also seen on GaN templates on sapphire. The source of the Si is still under consideration. Several sources have been debated, including impurities left behind from the polishing process, coatings on the wafer carrier, Si from inside the reactor, excess Si left during decomposition of the substrate during ramping conditions, and something mysterious coming from the atmosphere. Experiments from Hite *et al* and Pickrell *et al* suggest that it is an atmospheric effect [38, 39]. These have shown the only time a major spike is observed is when the interface is exposed to atmosphere—not during ramp, not during growth interrupts held at high temperature, and not when kept under a vacuum. This is shown in figure 12.9 [38]. The Si spike observed in samples that are exposed to atmosphere also have a dependence on the Si concentration in the original substrate. Figure 12.10 shows this, where GaN templates grown on sapphire with higher levels of n-type doping do have a larger spike at the interface than the unintentionally doped films when exposed to atmosphere prior to regrowth. However, the largest change in Si incorporation is for the unintentionally doped sample.

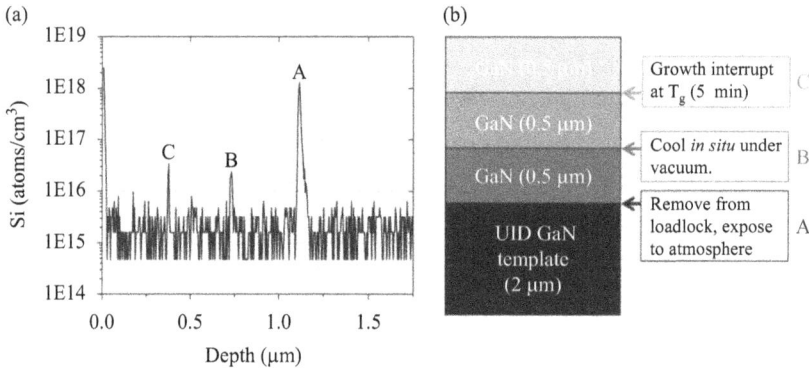

Figure 12.9. (a) SIMS data and (b) structure schematic illustrating Si incorporation at the regrowth interface of a GaN surface (A) exposed to atmosphere, (B) cooled and regrown under vacuum, and (C) 5 min growth interrupt held at growth temperature. Reproduced with permission from [38]. Copyright 2018 IOP Publishing.

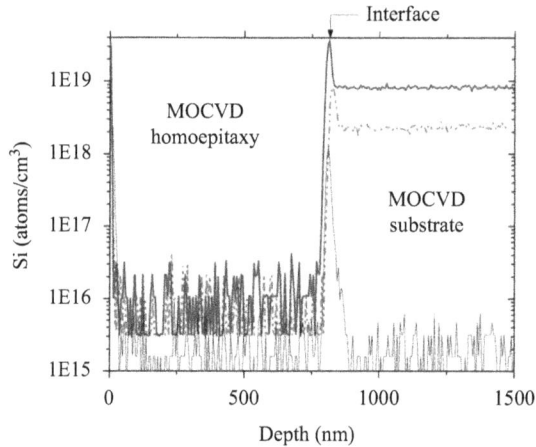

Figure 12.10. SIMS data showing the homoepitaxial growth interface on three substrates with different doping levels: undoped (gray), 2×10^{18} cm^{-3} (dotted red), and 8×10^{18} cm^{-3} (thick blue). The largest spike, in relation to the levels in the substrate, is found in the undoped sample.

12.4 Effects on device performance

In the previous sections the effects of the surface and interface were discussed with regard to growth morphology, species incorporation, and interfacial layers. All of these have effects on the performance of devices fabricated from such homoepitaxial films.

12.4.1 Surface morphology effects

The impact of surface morphology is a first order issue. Immediately, it can be observed that changes in offcut and step density from wafer bow and polish cause uniformity issues across wafer surfaces. The electrical effect, as observed in Schottky diodes, is found in areas of large-scale roughness at the wafer surface. For devices

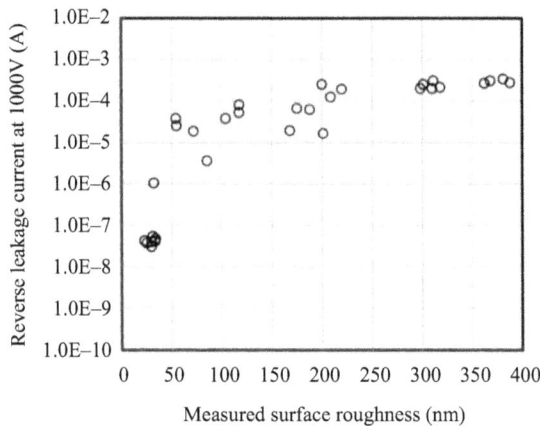

Figure 12.11. Plot of leakage current versus surface roughness. Reproduced with permission from [41]. Copyright 2015 Elsevier.

fabricated in regions where the RMS roughness over a 200×200 um is over 50 nm, an increased leakage current and decreased breakdown field is observed [40]. Smoother regions of the same wafer showed higher breakdown fields, with simple structures lacking field management reaching critical breakdown fields of 1.4 MV cm^{-1}. In p–n diodes on homoepitaxial growth, a non-optimal offcut produces rough surfaces, which also show high leakage and fail high-temperature reverse-bias (HTRB) reliability stress tests [41]. This direct impact of roughness on reverse leakage current is plotted in figure 12.11. This report predicted an even more stringent cut-off for surface roughness, 25 nm, measured on long length scales. As such, the results indicate that surface roughness can be employed as a metric to screen epitaxial layers, where substantial variation of offcut angle across the substrate acts as a potential limitation to epitaxial layer uniformity and repeatability.

12.4.2 Chemical interface effects

For vertical devices, an n+ substrate is normally required; therefore the conduction caused by a Si spike at the initial growth of a drift layer on the bulk substrate is not an important issue. However, this is an issue for lateral devices on semi-insulating bulk substrates as well as any p-type regrowth required in more advanced vertical devices. It has been shown that a lateral HEMT on a regrown buffer has increased off-state leakage due to the presence of Si at the interface when compared to a non-regrown structure [38]. In regrown p-GaN layers forming a p–i–n diode, the Si interfacial layer causes the device to act like a Zener tunnel diode, with temperature independent leakage current [39].

12.5 Conclusion

Understanding the growth surface preparation and interface is extremely important in developing homoepitaxial growth of GaN for vertical devices where thick films, regrowth, and high purity layers are required. Preparing the surface morphology is the first step. For smooth growth, regular steps controlled across the surface are provided by a slight offcut of the native substrate crystal. If the offcut is too low, hillocks form. If it is too high, step-bunching can occur. For GaN, optimal offcuts have been reported between 0.5° and 1°, but this can vary with growth conditions. A rough morphology can occur due to variations in offcut due to substrate bow and defects in wafer polishing. This roughness has a direct impact on vertical diodes, reducing the breakdown voltage and increasing leakage current. Additionally, surface cleaning and growth initiation must be taken into consideration, in particular for regrowth on low-doped layers, as a Si interfacial layer appears at all regrowth interfaces after exposure to atmosphere. This still needs to be understood and solved. This layer can also cause increased leakage current in p–n diodes. However, even with these considerations, high voltage vertical devices are already being demonstrated, which are coming closer to the theoretical breakdown field of GaN than lateral devices.

Acknowledgements

Work at the US Naval Research Laboratory is supported under the Office of Naval Research.

References

[1] Nakamura S 2015 Nobel lecture: background story of the invention of efficient blue InGaN light emitting diodes *Rev. Mod. Phys.* **87** 1139–51

[2] Pimputkar S, Speck J S, DenBaars S P and Nakamura S 2009 Prospects for LED lighting *Nat. Photonics* **3** 180–2

[3] Li G, Wang W, Yang W, Lin Y, Wang H, Lin Z and Zhou S 2016 GaN-based light-emitting diodes on various substrates: a critical review *Rep. Prog. Phys.* **79** 056501

[4] Ishida T 2011 GaN HEMT technologies for space and radio applications *Microwave J.* **54** 56–66

[5] Mishra U K, Parikh P and Wu Y F 2002 AlGaN/GaN HEMTs—an overview of device operation and applications *Proc. IEEE* **90** 1022–31

[6] Asubur J, Tokuda H and Kuzuhara M 2016 Pushing the GaN HEMT towards its theoretical limit *Compound Semiconductors* November 21 (accessed 12 Jan 2020) https://compoundsemiconductor.net/article/100602-pushing-the-gan-hemt-towards-its-theoretical-limit.html

[7] Etzkorn E V and Clarke D R 2004 Cracking of GaN films *Int. J. High Speed Electron. Syst.* **14** 63–81

[8] Feltin E, Beaumont B, Laugt M, deMierry P, Vennegues P, Lahreche H, Leroux M and Gibart P 2001 Stress control in GaN grown on silicon (111) by metal–organic vapor phase epitaxy *Appl. Phys. Lett.* **79** 3230–2

[9] Speck J S and Rosner S J 1999 The role of threading dislocations in the physical properties of GaN and its alloys *Physica* B **273–274** 24–32

[10] Hite J K, Gaddipati P, Meyer D J, Mastro M A and Eddy C R Jr 2014 Correlation of threading screw dislocation density to GaN 2DEG mobility *Electron. Lett.* **50** 1722–4

[11] Termentzidis K, Isaiev M, Salnikova A, Belabbas I, Lacrois D and Kioseoglou J 2018 Impact of screw and edge dislocations on the thermal conductivity of individual nanowires and bulk GaN: a molecular dynamics study *Phys. Chem. Chem. Phys.* **20** 5159–72

[12] Cho J, Li Z, Asheghi M and Goodson K E 2015 Near-junction thermal management: thermal conduction in gallium nitride composite substrates *Annual Review of Heat Transfer* vol 18 ed C L Tien (Danbury, CT: Begell House) ch 2, pp 7–45

[13] Oh S K, Lundh J S, Shervin S, Chatterjee B, Lee D K, Choi S, Kwak J S and Ryou J H 2019 Thermal management and characterization of high-power wide-bandgap semiconductor electronic and photonic devices in automotive applications *J. Electron. Packaging* **141** 0208011

[14] Pearton S J, Ren F, Zhang A P and Lee K P 2000 Fabrication and performance of GaN electronic devices *Mater. Sci. Eng.* R **30** 55–212

[15] Amano H 2013 Progress and prospect of the growth of wide-band-gap group III nitrides: development of the growth method for single-crystal bulk GaN *Jpn J. Appl. Phys.* **52** 050001

[16] Liu L and Edgar J H 2002 Substrates for gallium nitride epitaxy *Mater. Sci. Eng.* R **37** 61–127

[17] Dwilinski R, Doradzinski R, Garczynski J, Sierzputowski L P, Puchalski A, Kanbara Y, Yagi K, Minakuchi H and Hayashi H Bulk ammonothermal GaN *J. Cryst. Growth* **311** 3015–8

[18] Hashimoto R, Letts E R, Key D and Jordan B 2019 Two inch GaN substrates fabricated by the near equilibrium ammonothermal (NEAT) method *Jpn J. Appl. Phys.* **58** SC1005

[19] Bockowski M, Iwinska M, Amilusik M, Fijalkowski M, Lucznik B and Sochacki T 2016 Challenges and future perspectives in HVPE-GaN growth on ammonothermal GaN seeds *Semicond Sci. Tech.* **31** 093002

[20] Gallagher J C, Anderson T J, Luna L E, Koehler A D, Hite J K, Mahadik N A, Hobart K D and Kub F J 2019 Long range, non-destructive characterization of GaN substrates for power devices *J. Cryst. Growth* **506** 178–84

[21] Frank F C 1949 The influence of dislocations on crystal growth *Disc. Faraday Soc.* **5** 48–54

[22] Burton W K, Cabrera N and Frank F C 1951 The growth of crystals and the equilibrium structure of their surfaces *Phil. Trans. Royal Soc. London* A **243** 299–358

[23] Uwaha M 2016 Introduction to the BCF theory *Prog. Cryst. Growth Char. Mater.* **62** 58–68

[24] Oehler F, Zhu T, Rhode S, Kappers M J, Humphreys C J and Oliver R A 2013 Surface morphology of homoepitaxial c-plane GaN: hillocks and ridges *J. Cryst. Growth* **383** 12–8

[25] Sarzynski M, Leszczynski M, Krysko M, Domagala J Z, Czernecki R and Suski T 2012 Influence of GaN substrate off-cut on properties of InGaN and AlGaN layers *Cryst. Res. Tech.* **47** 321–8

[26] Kizilyalli I C, Edwards A P, Aktas O, Prunty T and Bour D 2015 Vertical power p–n diodes based on bulk GaN *IEEE Trans. Electron. Dev.* **62** 414–22

[27] Huang L, Liu F, Zhu J, Kamaladasa R, Preble E A, Paskova T, Evans K, Porter L, Picard Y N and Davis R F 2012 Microstructure of epitaxial GaN films grown on chemomechanically polished GaN(0001) substrates *J. Cryst. Growth* **347** 88–94

[28] Xu X, Vaudo R P, Flynn J, Dion J and Brandes G R 2005 MOVPE homoepitaxial growth on vicinal GaN(0001) substrates *Phys. Status Sol.* A **202** 727–31

[29] Fujikura H and Konno T 2018 Roughening of GaN homoepitaxial surfaces due to step meandering and bunching instabilities and their suppression in hydride vapor phase epitaxy *Appl. Phys. Lett.* **113** 152101

[30] Zhou K, Ren H, Liu J, Ikeda M, Ma Y, Gao S, Tang C and Yang H 2016 Surface morphology and optical properties of InGaN/GaN multiple quantum wells grown on freestanding GaN (0001) substrates *Superlatt. Microstruct.* **100** 968–72

[31] Grenko J A, Reynolds C L, Barlage D W, Johnson M A L, Lappi S E, Ebert C W, Preble E A, Paskova T and Evans K R 2010 Physical properties of AlGaN/GaN heterostructures grown on vicinal substrates *J. Electron. Mater.* **39** 504–16

[32] Leach J H, Biswas N, Paskova T, Preble E A, Evans K R, Wu M, Ni X, Li X, Ozgur U and Morkoc H 2011 Effect of substrate offcut on AlGaN/GaN HFET *Proc. SPIE* **7939** 79390E

[33] Suski T *et al* 2010 Hole carrier concentration and photoluminescence in magnesium doped InGaN and GaN grown on sapphire and GaN misoriented substrates *J. Appl. Phys.* **108** 023516

[34] Hite J K, Anderson T J, Luna L E, Gallagher J C, Mastro M A, Freitas J A and Eddy C R 2018 Influence of HVPE substrates on homoepitaxy of GaN grown by MOCVD *J. Cryst. Growth* **498** 352–6

[35] Jiang L *et al* 2019 Influence of substrate misorientation on carbon impurity incorporation and electrical properties of p-GaN grown by metal–organic chemical vapor deposition *Appl. Phys. Exp.* **12** 055503

[36] Paskova T, Backer L, Bottcher T and Hommel D 2007 Effect of sapphire-substrate thickness on the curvature of thick GaN films grown by hydride vapor phase epitaxy *J. Appl. Phys.* **102** 123507

[37] Foronda H M, Romanov A E, Young E C, Robertson C A, Beltz G E and Speck J S 2016 Curvature and bow of bulk GaN substrates *J. Appl. Phys.* **120** 035104

[38] Hite J K, Mastro M A, Luna L E and Anderson T J 2018 Understanding interfaces in homoepitaxial GaN growth *ECS Trans.* **86** 15–9

[39] Pickrell G W, Armstrong A M, Allerman A A, Crawford M H, Cross K C, Glaser C E and Abate V M 2019 Regrown vertical GaN p–n diodes with low reverse leakage current *J. Electron. Mater.* **48** 3311–6

[40] Hite J K, Anderson T J, Mastro M A, Luna L E, Gallagher J C, Myers-Ward R L, Hobart K D and Eddy C R 2017 Effect of surface morphology on diode performance in vertical GaN Schottky diodes *ECS J. Sol. State Sci. Tech.* **6** S3103–5

[41] Kizilyalli I C, Bui-Quang P, Disney D, Bhatia H and Aktas O 2015 Reliability studies of vertical GaN devices based on bulk GaN substrates *Microelectron. Reliab.* **55** 1654–61

Chapter 13

Gas sensors based on wide bandgap semiconductors

Kwang Hyeon Baik and Soohwan Jang

Wide bandgap semiconductors are one of the most suitable materials for applications in diverse gas sensors due to their excellent reliability, durability, and long lifetime. There are growing demands for highly sensitive and robust wide bandgap semiconductor-based sensors in high temperature and harsh radiation environments, as well as mild conditions, for the detection of ethanol, ammonia, carbon dioxide, and hydrogen gas species. Among various wide bandgap semiconductors, GaN has superior material properties such as chemical and mechanical hardness, excellent carrier transport, radiation robustness, and a large signal-to-noise ratio. Notably, the two-dimensional electron gas channel of the AlGaN/GaN heterostructure is easily affected by AlGaN surface charge change, which is a key component in GaN-based gas sensors. In this work, various sensing properties of gas sensors based on the AlGaN/GaN heterostructure are investigated. The effects of the gas species and the concentration, temperature, and humidity on the sensing characteristics are discussed. In addition, an encapsulation method to solve humidity problems in AlGaN/GaN-based gas sensors is suggested.

13.1 Introduction

Wide bandgap (WBG) semiconductor-based gas sensors have gained in importance for the applications of indoor and outdoor environmental monitoring, medical analyte sensing, food safety, and security. They have superior advantages in durability, reliability, lifetime, and robustness over conventional silicon-based gas sensors. Typical WBG materials include ZnO, Ga_2O_3, and GaN. Intensive research has been devoted to the WBG based gas sensors over the last decade. The number of papers published each year on ZnO, Ga_2O_3, and GaN gas sensors is shown in figure 13.1.

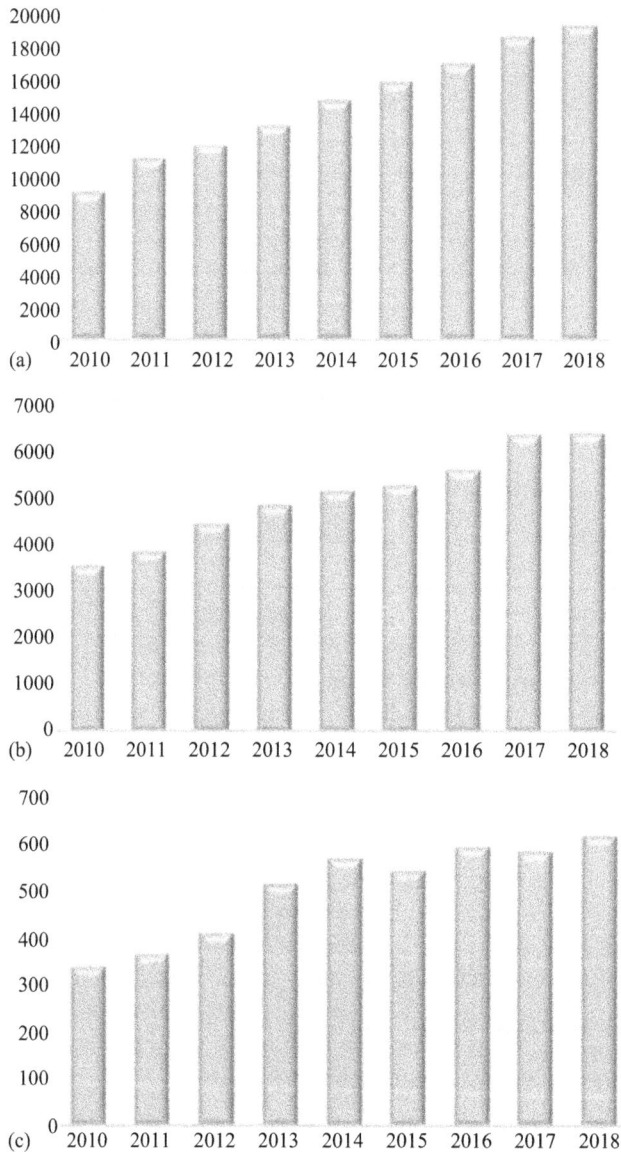

Figure 13.1. The number of published papers on (a) ZnO, (b) Ga_2O_3, and (c) GaN-based gas sensor per year since 2010.

ZnO is a II–V compound semiconductor with a direct bandgap energy of 3.37 eV and a large exciton biding energy of 60 meV [1–15]. One of the main advantages of ZnO is the facile synthesis of ZnO nanorods via hydrothermal growth, which is a low-cost, non-toxic, low temperature, and scalable process. In hydrothermal growth, the growth rate of the c-plane is faster than that of the m-plane, thus ZnO nanorods along the c-axis are generally formed on seed particles. The ZnO nanorods with a high surface-to-volume ratio are beneficial for improving gas sensing properties.

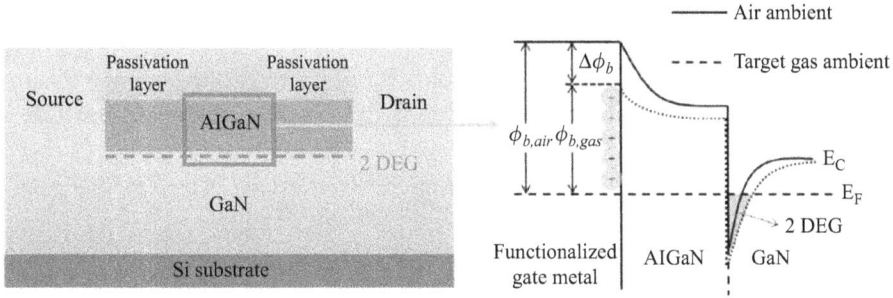

Figure 13.2. Schematic diagram of an AlGaN/GaN HEMT-based gas sensor with its energy band diagram.

However, ZnO is vulnerable to harsh chemical environments. It is easily etched under even weak acid conditions, which causes reliability issues for a gas sensor using this material.

Gallium oxide (Ga_2O_3) has excellent thermal and chemical stability. β-phase Ga_2O_3 has a very wide energy bandgap of 4.9 eV and chemical reactivity to a wide variety of gases. Many types of Ga_2O_3-based gas sensors for high temperature applications to monitor exhaust gases of automobiles, flue gases of incinerators, pollutant gases of refinery plants, and explosive gases from military applications have been reported [16–25]. Nevertheless, its low thermal conductivity and the difficulty in doping and growing heterostructures limit the practical application of Ga_2O_3 in semiconductor-based gas sensors [26–28].

Gallium nitride (GaN) is one of the most suitable material systems for semiconductor-type gas sensors. Due to its superior material properties, including a wide bandgap of 3.4 eV, physical and chemical robustness, excellent carrier transport, and radiation hardness, GaN-based gas sensors exhibit reliable and stable operation with a high signal-to-noise ratio at high temperatures and in harsh radiative environments [29–39]. One of the biggest merits of a GaN-based semiconductor is the availability of the AlGaN/GaN heterostructure. In this structure, a two-dimensional electron gas (2DEG) channel forms at the interface between the AlGaN and GaN with an electron mobility of more than 1600 cm^2 V^{-1} s^{-1} due to spontaneous polarization and piezoelectric effects [35, 36, 39]. The electron conductivity of the 2DEG is very sensitive to charge changes on the top AlGaN surface. This unique property leads to high current changes of the sensor which are created by a catalytic reaction of target gas on the active layer and amplified through the high electron mobility transistor (HEMT) [35, 40]. Typically, the field effect transistors with higher drain currents exhibit better sensitivity toward gas detection. With 30% Al concentration in the AlGaN layer, electron densities in the channel are found to be 5–10 times higher compared to those of GaAs or InP HEMTs [41, 42].

There are two types of AlGaN/GaN heterostructure-based sensors. One is an AlGaN/GaN HEMT-based sensor and the other is the Schottky diode sensor on an AlGaN/GaN heterostructure. Figure 13.2 shows a schematic diagram of the HEMT sensor with the energy band diagram. The gate contact is composed of catalytic materials on which the specific reaction occurs only with the target gas [29–34, 38–44].

The ions produced by the surface reaction diffuse into the interface and accumulate between the gate material and AlGaN layer. The drain current flowing from the source to drain through the 2DEG channel is very sensitive to the change of surface charge on the AlGaN surface. The charge change at the interface of the gate and AlGaN enhances or depletes the 2DEG channel depending on the type of reaction (oxidation or reduction). Hence, by monitoring the drain current change, the target gas is detected by the AlGaN/GaN HEMT-based gas sensor.

The basic structure of the Schottky diode sensor based on the AlGaN/GaN heterostructure is shown in figure 13.3. In this device, the Schottky contact consists of catalytic metal or oxide which promotes only the reaction of the target gas molecules. The products of the reaction and charged ions diffuse into and accumulate at the AlGaN interface. The ion accumulation induces the increase or the reduction of the Schottky barrier height depending on the charge change at the surface of the AlGaN. As a result, we can detect a target gas by monitoring the current change of the AlGaN/GaN-based Schottky diode.

In this chapter, AlGaN/GaN heterostructure-based sensors to detect various gas species including ethanol, ammonia, carbon dioxide, and hydrogen are reported. A silver (Ag) catalyst on the AlGaN/GaN HEMT was used for the detection of ethanol. To detect ammonia and carbon dioxide gases, ZnO nanorods were employed as the sensing material. The current signal of the sensor decreased upon ammonia exposure, but the increase of the current was observed under carbon dioxide ambient. In addition, the solution for signal distortion of the gas sensor by the humidity effect will be suggested in the case of the AlGaN/GaN HEMT-based hydrogen sensor with platinum as the catalytic sensing material.

13.2 An AlGaN/GaN HEMT-based ethanol sensor

Ethanol is one of the most commonly used ingredients in alcoholic beverages, chemicals, cosmetics, and medical products as well as being a renewable biofuel. In recent decades, there have been increasing demands for highly sensitive and stable ethanol sensors to detect ethanol gas in chemical, food, and biomedical production [45–47]. A variety of ethanol gas sensors have been developed, mostly based on metal–oxide semiconductors such as SnO_2, CuO, Fe_2O_3, TiO_2, WO_3, ZnO, and NiO. The resistive-type solid-state gas sensors are attractive because of their fast

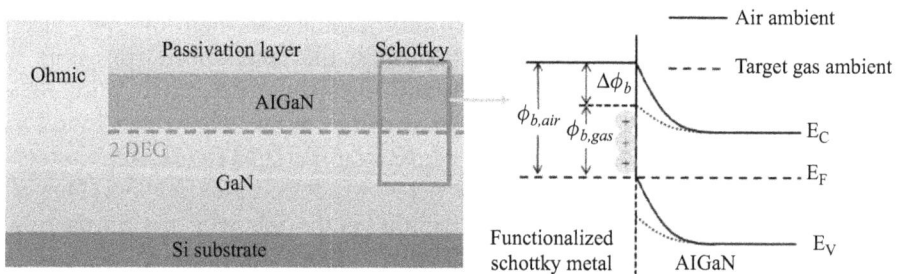

Figure 13.3. Schematic diagram of the AlGaN/GaN heterostructure-based Schottky diode gas sensor with its energy band diagram.

and high sensitivity, low power consumption, and high compatibility with mobile electronic devices [48–50]. Interestingly, recent studies on ethanol gas sensors have shown promise for enhancing the sensitivity and selectivity using nanostructures, noble metal addition, catalyst doping, and composite core–shell structures with ZnO/SnO_2 and Ag/SnO_2 [51–53]. In this work, diode sensors with silver (Ag) Schottky contacts based on AlGaN/GaN HEMTs were fabricated to investigate the response to ethanol gas with varying concentration and operating temperature.

The AlGaN/GaN HEMT heterostructures have been grown on c-plane sapphire substrates using a metal–organic chemical vapor deposition (MOCVD) system with trimethyl-aluminum, trimethyl-gallium, and ammonia as aluminum, gallium, and nitrogen precursors, respectively. The epitaxial layers consisted of a 2 μm thick undoped GaN layer, an undoped 35 nm thick $Al_{0.3}GaN$ layer, and a 2.5 nm thick GaN capping layer. The sheet resistance was measured to be 330 Ω/square from the Hall-effect measurement. The sheet carrier density of 2DEG was $1.06 \times 10^{13}/cm^2$ with a carrier mobility value of 1900 cm^2 V^{-1} s^{-1}.

The fabrication of the AlGaN/GaN HEMT device began with the Ohmic contact formation with a Ti/Al/Ni/Au multi-layer, which was deposited using an e-beam evaporator and formed through a lift-off process. The Ohmic contact metallization was annealed at 850 °C for 1 min in nitrogen flow. Si_3N_4 with a thickness of 200 nm was deposited using plasma enhanced chemical vapor deposition for device isolation. The Si_3N_4 window for active regions was opened by soaking in buffered oxide etchant. The Schottky contact of AlGaN/GaN HEMT was e-beam evaporated with 10 nm thick Ag film. The metal contact pads for wire bonding were deposited with Ti/Au evaporation. Electrical current–voltage ($I–V$) characteristics of the diodes were recorded using an Agilent 4155C semiconductor parameter analyzer before and after the exposure of ethanol gas balanced with nitrogen in a gas test chamber.

Figure 13.4(a) shows the $I–V$ characteristics of the Ag Schottky diode sensors based on the AlGaN/GaN heterostructure before and after exposure to 5.87% ethanol in N_2 at 250 °C. The inset figure presents the corresponding $I–V$ curve in log scale. The current decreases after ethanol exposure, mainly due to an increase in the

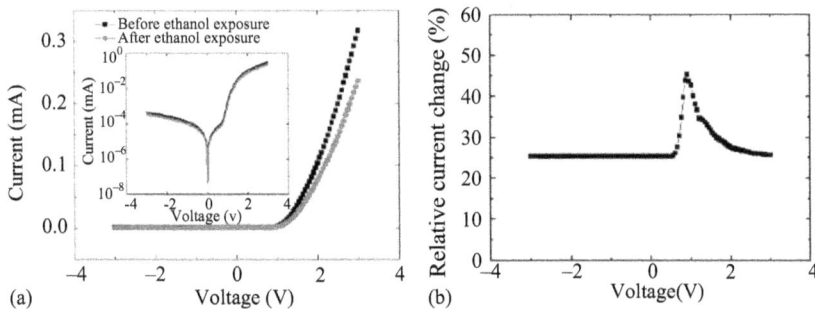

Figure 13.4. (a) $I–V$ characteristics from the Ag-AlGaN/GaN diode in dry N_2 or during exposure to 5.87% ethanol in N_2 at 250 °C. The inset is the semi-log plot of $I–V$. (b) Relative current change as a function of bias voltage. Reproduced with permission from [45], copyright IOP Publishing, all rights reserved.

effective Schottky barrier height. Figure 13.4(b) shows the relative current change in percentage as a function of bias voltage. A maximum gas response of 45.4% was observed at the forward bias of 0.9 V. The $I-V$ characteristics of a Schottky diode can be expressed as follows based on the thermal emission model:

$$J = A^* T^2 e^{\left(\frac{-e\varnothing_b}{kT}\right)} e^{\left(\frac{eV}{nkT}\right)},$$

where V, k, and T are the applied voltage, Boltzmann's constant, and temperature, A^* is the effective Richardson's constant, and n is the diode ideality factor. The Schottky barrier height and the ideality factor can be extracted by fitting the measured $I-V$ data before and after ethanol gas exposure in nitrogen ambient.

Figure 13.5 shows the barrier height change as a function of time when the diode was exposed to 5.87% ethanol and switching back to nitrogen at the temperature of 250 °C. The barrier height increased from 0.603 to 0.656 eV ($\Delta\Phi_B = 53$ meV), resulting in total current decrease. The increase of the Schottky barrier height is in sharp contrast to the case of hydrogen sensing using Pt Schottky diodes on the AlGaN/GaN heterostructure.

The ethanol sensing mechanism can be explained as follows. The ethanol molecules were adsorbed on the Ag surface of the diode sensor. The decomposition reactions of silver oxide (Ag_2O) formed in the device fabrication steps result in chemisorbed oxygen ions and electron release on the Ag surface. The molecular adsorption of ethanol vapors can be facilitated by adsorbed oxygen ions as shown in figure 13.6. The decomposition reaction of Ag_2O and resultant oxygen ions are more likely to occur at high temperatures above 200 °C. Thus, the Ag film on the AlGaN surface can enhance the catalytic oxidation of ethanol vapor. The ethanol molecules can adsorb, react with oxygen ions, and produce C_2H_4O, H_2O, and electron release. The negative charging in Ag film results in the increase of Schottky barrier height, inducing the current decrease at both forward and reverse bias. The overall reaction is expressed as follows: $C_2H_5OH + O^- \rightarrow C_2H_4O + H_2O + e^-$.

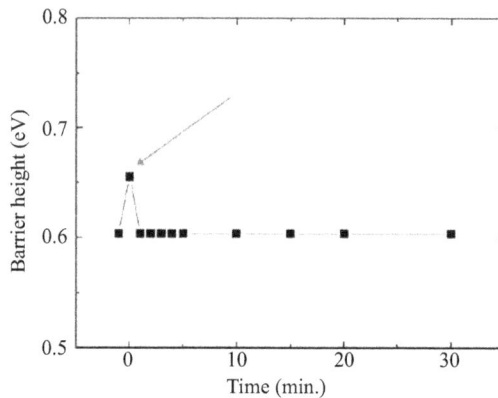

Figure 13.5. Change in effective barrier height as a function of time upon the injection of ethanol gas into the test chamber, followed by a return to dry N_2. Reproduced with permission from [45], copyright IOP Publishing, all rights reserved.

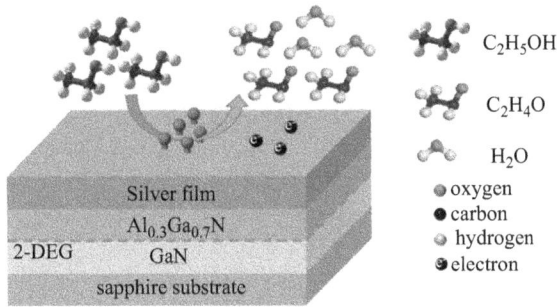

Figure 13.6. Schematic illustration of the oxidation of ethanol on Ag film in the gate region of the diode. Reproduced with permission from [45], copyright IOP Publishing, all rights reserved.

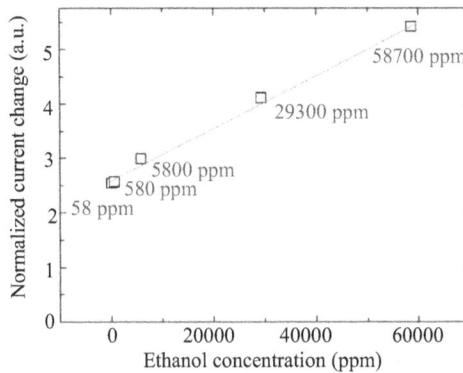

Figure 13.7. Normalized current change in Ag-AlGaN/GaN diodes as a function of ethanol concentration at 250 °C. Reproduced with permission from [45], copyright IOP Publishing, all rights reserved.

Figure 13.7 shows that the normalized current changes as a function ethanol concentration at 250 °C. The current response increases linearly in the range of 58 to 58 700 ppm. It is notable that there was no significant change in diode currents upon exposure to ethanol below 250 °C. At the temperature of 300 °C, the current showed a slighter decrease than that at 250 °C, possibly due to increased Ag reactions. The AlGaN/GaN diodes with the Ag Schottky contact showed comparable and good sensing performance in terms of the operating temperature and sensitivity. The nitride-based diodes are advantageous because they show highly reliable sensing performance at high temperatures and can be easily integrated with microelectronic devices that are suitable for remote monitoring systems.

13.3 AlGaN/GaN HEMT-based ammonia sensor

Ammonia (NH_3) is a colorless and hygroscopic gas with a pungent odor. According to the US Occupational Safety and Administration, 15 min exposure under 35 ppm or 8 h under 25 ppm may put general laborers at risk of fatal health conditions. Patients who have inhaled ammonia gas suffer from burning of the respiratory tract, bronchiolar and alveolar edema, and airway destruction. The use of ammonia as a reactant or product includes uses in fertilizers, explosives, textiles, foods, dyes, refrigerants, and chemical additives. In these applications, ammonia detection is required for the safety of workers and the quality control of the product. Recently, there has been growing interest in the ammonia slip of diesel engines with increasingly strict emissions standard. Nitrogen oxide (NO_x) emitted from the exhaust of diesel engines is one of the main pollutant gases resulting in acid rain and the formation of ground-level ozone [29, 30, 54]. NO_x emission can be controlled by selective catalyst reduction (SCR) systems to meet the increasingly stringent new emissions regulations. The SCR treatment reduces NO_x emissions through a chemical reaction in which NO_x reacts with ammonia to produce nitrogen and water on a vanadium catalyst [1, 29, 31]. It is important for gas sensors to detect a low concentration of NH_3 at high temperatures with selectivity, stability, reproducibility, and short response time to maintain the optimal reaction conditions and to prevent an ammonia slip in the SCR feedback loop. Therefore, the challenge is to develop thermally stable, low-cost, and compact ammonia gas sensors, in particular with selectivity over various concentrations of oxygen in the exhaust line [29, 30].

The MOCVD growth of AlGaN/GaN HEMT structures and the fabrication process of the Schottky diode sensor were described above. The Schottky contact area was patterned by conventional photolithography and a buffered oxide etchant for the subsequent ZnO nanorod growth on the AlGaN surface. ZnO nanorod growth on the Schottky region begins with preparation of the nano-crystal seeds. The ZnO nano-crystal seed solution was mixed by adding 30 mM NaOH in methanol to a 10 mM zinc acetate dihydrate solution at 60 °C for a 2 h period. The ZnO seed solution was spin-coated onto the active region of the diode and the sample was heated at 300 °C on a hot plate for 30 min in air. The ZnO crystalline seed coated diodes were then immersed in an aqueous mixture of 20 mM zinc nitrate hexahydrate and 20 mM hexamethylenetetramine, and put in the oven at 95 °C for 3 h for the ZnO nanorod growth [2, 3]. After the nanorod growth, the device was removed from the solution, thoroughly rinsed with de-ionized water, and dried with nitrogen gas. Finally, interconnection and pad contacts were formed by lift-off of *e*-beam deposited Ti/Au (20/100 nm). The devices were exposed to ammonia of which the concentrations are 0.1–2 ppm at temperatures from 25 °C to 300 °C, and the diode current–voltage characteristics were measured for the ambient with different concentrations of ammonia at different temperatures.

Figure 13.8 shows the current change responses of the ZnO nanorod functionalized diode to 0.1–2 ppm exposure at 25 °C and 300 °C with an applied bias voltage of 5 V. The diode current decreases upon ammonia exposure, and the current change is larger at higher temperature for the same concentration of ammonia. The ZnO nanorod diode with large surface area can detect 0.1 ppm ammonia at both 25 °C and 300 °C. Under the ammonia ambient, NH_3 molecules react with negatively charged oxygen ions on the ZnO nanorod surface, which induces relatively more negative (less positive) charge build-up on the AlGaN surface [29, 30]. Hence, the Schottky barrier height at the interface increases resulting in the current decrease. This is opposite to the current change when detecting reducing gases such as

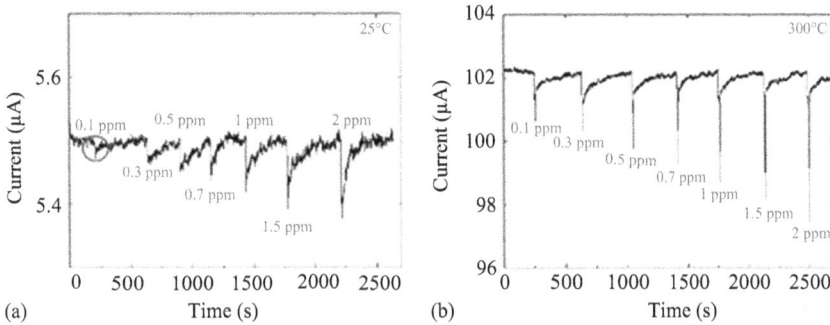

Figure 13.8. Current response of the AlGaN/GaN heterostructure diode with a ZnO nanorod Schottky contact to 0.1–2 ppm NH_3 exposure at (a) 25 °C and (b) 300 °C. The applied bias voltage was 5 V.

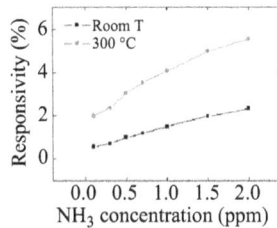

Figure 13.9. Responsivity of the AlGaN/GaN-based ZnO nanorod diode for 0.1–2 ppm ammonia exposure at 25 °C and 300 °C with an applied bias of 5 V.

hydrogen [1, 16, 30, 32–36, 40, 55–59]. Figure 13.9 presents the responsivity of the nanorod diode sensor as a function of ammonia concentration at 25 °C and 300 °C. The responsivity of the sensors is the difference in current under dry air ambient relative to the various concentrations of ammonia. The linear responsivity was found in the measurement range of 0.1–2 ppm for both 25 °C and 300 °C.

In addition, ZnO nanorod functionalization was employed in the gate region of AlGaN/GaN HEMT. The device fabrication process was similar to that for the AlGaN/GaN heterostructure-based ZnO nanorod Schottky diode as described earlier. The mesa, Ohmic contact, metal pad for probing, and hydrothermal growth of ZnO nanorods on the gate region were made at each step [3].

Figure 13.10 shows the drain current–voltage characteristics of the HEMT sensor with ZnO nanorods on the gate region for dry air and 2 ppm ammonia exposures at 25 °C, 100 °C, 200 °C, and 300 °C, and the time–current response to 2 ppm ammonia at 25 °C. As the temperature increases, the drain current decreases due to carrier scattering in the 2DEG channel. As seen in figure 13.10, the drain current reduced upon 2 ppm ammonia exposure in contrast to the case of reducing gas. When the device with ZnO nanorods is exposed to an ammonia environment, negatively charged oxygen ions adsorbed on the ZnO nanorod surface react with NH_3 gas molecules, inducing the reduction of positive charge which was balanced with negative oxygen ions on the AlGaN surface. This enhancement of relatively negative charge on the AlGaN interface reduces the electron carrier concentration in

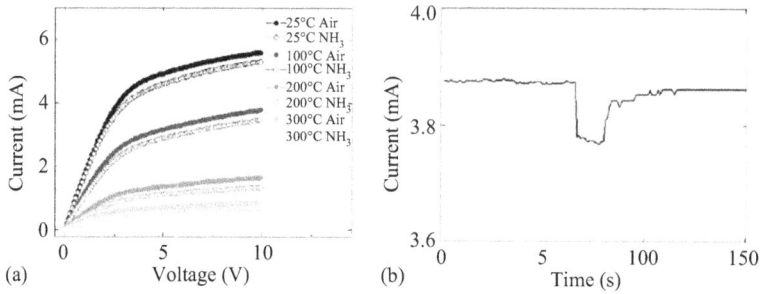

Figure 13.10. (a) Current–voltage characteristics of an AlGaN/GaN HEMT sensor under dry air and 2 ppm NH$_3$ ambient at 25 °C, 100 °C, 200 °C, and 300 °C. (b) Current response of the device to 2 ppm NH$_3$ exposure at 25 °C.

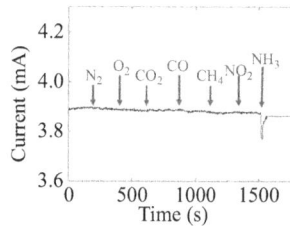

Figure 13.11. Response of the sensor to exposures of O$_2$ (100%), CO$_2$ (10%), CO (0.1%), CH$_4$ (4%), NO$_2$ (0.05%), and NH$_3$ (2 ppm) at 25 °C.

the 2DEG channel, resulting in the drain current drop of the sensor after ammonia exposure [30].

As clearly seen in figure 13.11, the ZnO nanorod sensor is completely selective for 2 ppm NH$_3$ over O$_2$ (100%), CO$_2$ (10%), CO (0.1%), CH$_4$ (4%), and NO$_2$ (0.05%) under the same detection conditions as used for the NH$_3$. The concentrations of the other gases were chosen in the range of US health exposure limits set by the National Institute for Occupational Safety and Health [30, 34, 60]. The ZnO nanorods on the gate region play a catalytic role only for the decomposition reaction of ammonia. The sensor was able to detect 100 ppb ammonia at both 25 °C and 300 °C, which is a similar detection capability to the AlGaN/GaN-based ZnO Schottky diode sensor.

13.4 The AlGaN/GaN HEMT-based carbon dioxide sensor

With an increase of fossil fuel usage, the atmospheric concentration of carbon dioxide has been increased to 400 ppm, while it was 280 ppm before the industrial revolution [61]. Carbon dioxide is a colorless and odorless gas with a higher density than normal air. It is easily accumulated on the ground in an enclosed construction environment. Concentrations of 7% to 10% may cause acute symptoms in an exposed casualty, such as unconsciousness, even in the presence of a high concentration of oxygen. It is necessary to detect promptly the specific amounts of carbon dioxide in the ambient for workplace safety and the health of workers. As well as the health of individuals, carbon dioxide is also used in the chemical, biomedical, energy, and food industries as a reactant, intermediate, and final product. Conventionally, a non-dispersive infrared absorption (NDIR) sensor is used to detect carbon dioxide, but it requires large, complicated equipment with high power consumption [62]. The AlGaN/GaN-based carbon dioxide sensor has many advantages, including low-cost, high sensitive, good reliability, compact size, and low power consumption compared to the conventional NDIR [37].

The MOCVD growth of the AlGaN/GaN HEMT structure and the fabrication process of the Schottky diode sensor were described previously. The gate areas for ZnO nanorod growth were opened using buffered oxide etchant. ZnO nanorods were selectively grown on the region using conventional photoresist patterning and simple hydrothermal methods with a growth solution of zinc nitrate hexahydrate and methenamine. The current–voltage characteristics of the AlGaN/GaN-based carbon dioxide sensor were measured in the gas test chamber with various concentration, temperature, and flow rate conditions.

Figure 13.12 shows the drain current–voltage characteristics of the fabricated AlGaN/GaN HEMT-based ZnO nanorod sensor for air and 10% carbon dioxide exposures. The device structure is very similar to the previously mentioned AlGaN/GaN heterostructure-based ammonia sensor with ZnO nanorods on the gate region. However, the drain current increases dramatically under CO_2 ambient in contrast to the case of oxidizing NH_3 gas. Carbon dioxide molecules decompose into carbon monoxide and oxygen on the ZnO surface. In this reaction, negatively charged oxygen ions accumulate on the ZnO nanorods, inducing a positive charge increase at the AlGaN interface. Therefore, the 2DEG channel between AlGaN and GaN is enhanced and the drain current of the device increases with exposed CO_2 molecules.

As the measurement temperature reached 150 °C, the ZnO nanorod sensor started to show a current response to 10% carbon dioxide. The drain current change as a function of time for carbon dioxide concentrations of 500, 1000, 2000, 5000, 10 000, 30 000, 50 000, 70 000, and 100 000 ppm at 150 °C and 300 °C is shown in figure 13.13. As the exposed concentration of carbon dioxide increases, the drain current change increases. The lowest detection limit of the ZnO nanorod sensor was 500 ppm for both 150 °C and 300 °C, and the current response at the higher temperature is more distinctive through providing more energy to overcome the activation energy of the reaction at higher temperature.

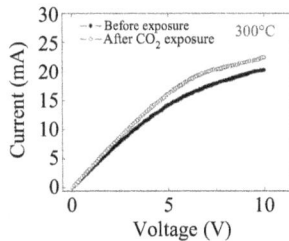

Figure 13.12. The drain current–voltage characteristics of the AlGaN/GaN-based ZnO nanorod FET sensor under air and 10% CO_2 ambient at 300 °C.

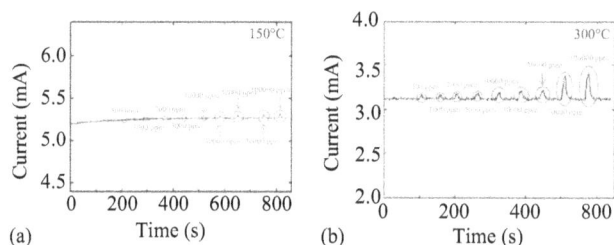

Figure 13.13. The current response of the device to 500, 1000, 2000, 5000, 10 000, 30 000, 50 000, 70 000, and 100 000 ppm CO_2 at (a) 150 °C and (b) 300 °C.

13.5 The AlGaN/GaN HEMT-based hydrogen sensor with a water blocking layer

There is also growing interest in highly sensitive hydrogen sensors in high humidity, in particular their use in fuel cell vehicles and hydrogen breath tests in medical diagnosis. One of the crucial concerns for the implementation of AlGaN/GaN-based hydrogen sensors is the degradation of sensitivity in a humid ambient. A decrease in hydrogen sensitivity is clearly observed, which is directly proportional to the relative humidity. In a humid ambient, water molecules adsorb on the surface and thus reduce the adsorption sites available for hydrogen molecules. The effects of relative humidity on hydrogen sensing can be mitigated by the use of active filters, membranes, or encapsulation layers which can block the access of water molecules to the surface. In previous studies, polymer-based composites have shown promise in reducing the influence of relative humidity in hydrogen–humid air mixtures [30, 32, 34, 53]. Polymer encapsulation layers include poly-methyl methacrylate (PMMA), polytetrafluoroethylene (PTFE), polydimethylglutarimide (PMGI), and polyimide films. In this section, we have studied the hydrogen sensing and detection in humid ambient using PMGI and PMMA on AlGaN/GaN Schottky diodes.

PMGI (SF-13, Microchem) was diluted with cyclopentanone for resist thinning, spin-coated at 2000 rpm for 40 s, and then baked at 180 °C for 3 min in air ambient. A PMGI layer with a thickness of 123 nm was formed on the fabricated diode. The contact window was opened for probe tip contact. For spin-coating of PMMA on the device surface, PMMA (molecular weight; 996 000) was dissolved in anisole with a concentration of 40 mg ml^{-1}, and the solution was mixed for 12 h. 150 nm thick PMMA was spin-coated on the fabricated device at 4000 rpm for 30 s. The schematic diagram of Pt Schottky diodes on AlGaN/GaN HEMT with a PMMA or PMGI encapsulation layer is shown in figure 13.14.

Figure 13.15(a) shows the *I–V* characteristics of a Pt Schottky diode on a AlGaN/GaN heterostructure before and after exposure to 500 ppm H_2 in N_2 for 20 s in a dry ambient. The forward current increased with response to hydrogen gas exposure, which is induced by the reduction of Schottky barrier height. It is known that hydrogen molecules absorb and dissociate on the Pt surface, diffuse to the AlGaN layer, and the dipole formation thus results in the lowering of the barrier height. Figure 13.15(b) shows the *I–V* curves of the Pt Schottky diode when exposed to

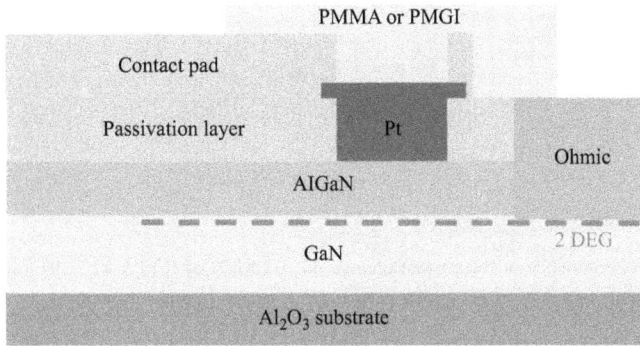

Figure 13.14. Schematic device structure of Pt Schottky diodes on AlGaN/GaN HEMT with encapsulation layers.

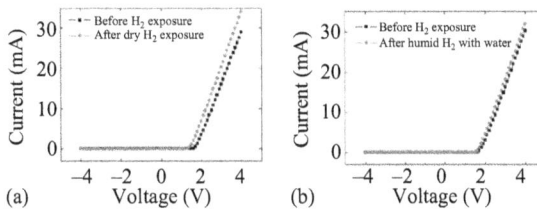

Figure 13.15. Forward and reverse I–V characteristics of unencapsulated Pt Schottky diodes at 25 °C before and after exposure to (a) dry 500 ppm H_2 in N_2 or to (b) humid 500 ppm H_2.

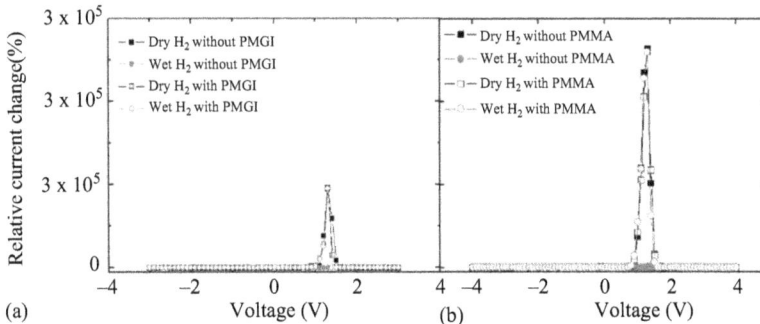

Figure 13.16. Relative current change in percentage of (a) PMGI and (b) PMMA encapsulated Pt Schottky diodes on AlGaN/GaN as a function of bias voltage for dry and humid H_2 exposure.

500 ppm H_2 for 20 s in 100% relative humidity (RH). Clearly, the diode sensors did not respond to hydrogen in humid H_2 ambient. This humidity effect is consistent with the previous reports that the response and signal of Pt- or Pd-based gas sensors decreases drastically in the presence of water vapor [32–34]. It is known that the chemisorption of water vapor is promoted on the Pt surface, thus leading to decreased surface active sites for hydrogen chemisorption.

Figures 13.16(a) and (b) show the relative current change in percentage of the diodes with and without (a) PMGI and (b) PMMA encapsulation layers as a

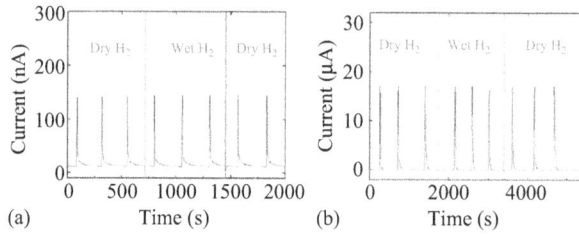

Figure 13.17. Time dependence of the current change of (a) PMGI or (b) PMMA encapsulated diodes with cycling of the ambient for 5 s from N_2 to 500 ppm H_2 in a dry or humid ambient.

function of bias voltage under 500 ppm H_2 exposure in dry conditions and 100% RH. One can see a drastic decrease of H_2 sensitivity in the presence of water vapor. The maximum value of relative current change percentage dropped by about a factor of 8 at 100% RH level. It clearly indicates that water vapors adsorb on the catalytically active sites and present hydrogen molecules from adsorbing and dissociating on the Pt surface [32, 33, 54]. In sharp contrast, the hydrogen response of PMGI and PMMA encapsulated diode sensors showed almost the same detection sensitivity in the range of $\sim 10^{5\%}$ in the humid ambient as the value in dry H_2 conditions. The encapsulation with a PMGI or PMMA layer can effectively resolve the water cross-sensitivity problems by allowing hydrogen molecules to pass through the encapsulation layers while selectively blocking the penetration of water molecules.

Figures 13.17(a) and (b) show the time-dependent responses of the forward current for (a) PMGI and (b) PMMA encapsulated diodes in both dry and humid 500 ppm H_2 at room temperature. The H_2 cycling response in dry and humid ambient for both the diode sensors was recorded, when exposed under 500 ppm H_2 for 5 s and purged with pure N_2 when switching off hydrogen. Both PMGI and PMMA coated diode sensors showed an almost identical current response in an alternating ambient of dry and humid 500 ppm H_2. It clearly proves that PMMA and PMGI layers can significantly reduce the humidity effect because of the effective filtration of water molecules.

In order to investigate the thermal stability and working temperature of the encapsulation layers, the hydrogen response of the PMGI coated diodes was tested with increasing temperature up to 300 °C. Figure 13.18 shows the change of maximum relative current as a function of temperature for diodes with a PMGI layer upon exposure to 500 ppm hydrogen in both the dry and humid ambient. The PMGI coated diodes exhibited almost the same maximum relative current in percentage in the range of 25 °C and 300 °C. The hydrogen response peaked at 200 °C and decreased at higher temperatures. While the diode sensitivity is likely to be limited by the efficiency of hydrogen dissociation, the hydrogen-induced dipole layers are not stable enough to have an effect on the barrier height of the Pt catalytic layer. It can be concluded that the encapsulation layers with PMMA and PMGI as coatings or membranes can be stable moisture barriers under 200 °C under a humid

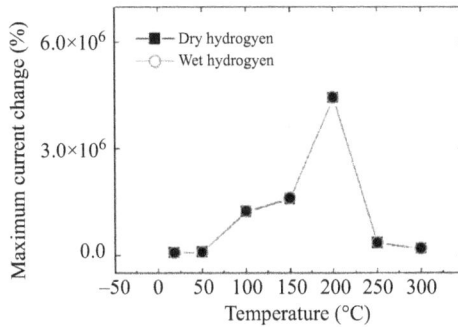

Figure 13.18. The variation of maximum current change in percentage of PMGI encapsulated diodes as a function of temperature upon exposure to 500 ppm H_2 in both dry and humid ambient.

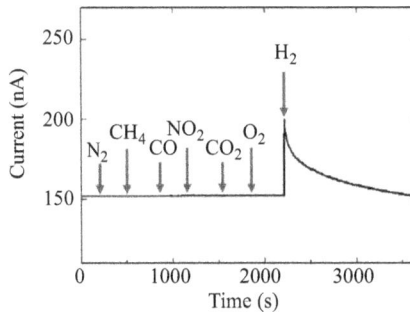

Figure 13.19. The time-dependent current change of PMGI encapsulated Schottky diodes upon the sequential exposure to gas species, including N_2 (100%), CH_4 (4%), CO (0.1%), NO_2 (0.05%), CO_2 (10%), and O_2 (100%) at 25 °C.

ambient, which help block water vapor diffusion on active adsorption sites on the Pt surface.

Figure 13.19 shows the time-dependent current change of PMGI encapsulated Schottky diodes when exposed to various gas species as follows: N_2 (100%), CH_4 (4%), CO (0.1%), NO_2 (0.05%), CO_2 (10%), and O_2 (100%), and 500 ppm H_2. The diode sensors with the PMGI layer did not show any response to other gas species. We believe that the hydrogen-specific response results from the hydrogen permeation into the encapsulation layer. H_2 molecules (0.298 nm) are known to have one of the smallest kinetic diameters in comparison to other gas species, including CH_4 (0.380 nm), CO (0.376 nm), NO_2 (0.340 nm), CO_2 (0.330 nm), and O_2 (0.346 nm). Note that the PMMA encapsulated diode sensors also showed the same hydrogen selectivity toward other gases.

13.6 Conclusion

AlGaN/GaN HEMTs with 2DEG induced by piezoelectric and spontaneous polarization between the AlGaN and GaN layers show excellent sensitivity to changes in surface charges created by detecting materials, which is amplified through the unique AlGaN/GaN HEMT structure. By functionalizing the gate region of

FET sensors or the Schottky contact area of diode sensors, specific target gas molecules are detected through surface charge transfer reaction.

A silver Schottky contact was used to sense ethanol gas on the AlGaN/GaN-based Schottky diode. The device showed the maximum sensitivity to ethanol at 250 °C, and presented a stable and recoverable current response to ethanol vapor resulting from a Schottky barrier height increase. The barrier height increase under ethanol ambient at 250 °C was 0.053 eV. This result indicates the potential possibility of an AlGaN/GaN-based Schottky diode with an active silver layer to monitor food fermentation and spoilage, and alcohol levels in many industrial applications with low power consumption in compact size.

ZnO nanorod functionalized AlGaN/GaN Schottky diodes and FET sensors detected low ammonia concentrations of 0.1–2 ppm at temperatures in the range 25 °C–300 °C. The diode and FET currents decreased upon ammonia exposure due to oxygen ion consumption on the ZnO nanorod surface. The absolute current change and responsivity for ammonia exposure increased with measurement temperature. The simple fabrication and excellent thermal stability of the ZnO/AlGaN/GaN diode and FET sensors indicates that these have potential for applications such as automobile exhaust sensing.

An AlGaN/GaN HEMT-based carbon dioxide sensor with ZnO nanorods on the gate region was fabricated. The device showed the distinctive sensitivity to 500 ppm CO_2 at 300 °C and fast reliable repeatability to cyclic exposures of various concentrations of carbon dioxide. Interestingly, the sensor with ZnO nanorods exhibited a current increase upon carbon dioxide exposure in contrast to the case of ammonia. The released oxygen ion from the decomposition reaction of CO_2 on the ZnO nanorod surface enhances the 2DEG channel conductivity, resulting in a drain current increase. These opposite characteristics would be beneficial for multi-gas sensors with different functionalization catalyst arrays.

One of the issues in semiconductor-based gas sensors is humidity. When the device is exposed to a humid ambient, water molecules block the active sites of the functionalizing catalyst and distort the sensor signals. PMMA or PMGI encapsulation enables a Pt-AlGaN/GaN hydrogen sensor to eliminate the deleterious effect of moisture in the sensing ambient on the detection of hydrogen, maintaining the same sensitivity level as the device without a moisture barrier. The PMGI coated sensors showed excellent selectivity over other common gases such as N_2, CH_4, CO, NO_2, CO_2, and O_2 at 25 °C. By using thermally stable PMMA and PMGI layers at high temperature, the devices showed robust sensing characteristics up to 100 °C and 300 °C, respectively. Since the these polymer layers are commonly available and easily applied to the semiconductor surface using simple spin-on processing, they represent a good choice as a moisture barrier layer on the hydrogen sensor applications.

Acknowledgments

This research was supported by the Basic Science Research Program through the National Research Foundation of Korea (NRF) funded by the Ministry of

Education (2018R1D1A1A09083988, 2017R1D1A3B03035420), and Nano·Material Technology Development Program through the National Research Foundation of Korea (NRF) funded by the Ministry of Science, ICT and Future Planning (2015M3A7B7045185).

References

[1] Jang S, Jung S and Baik K H 2019 *ECS J. Solid State Sci. Technol.* **8** Q85

[2] Jang S, Park J, Kim S and Baik K H 2016 *J. Nanosci. Nanotechnol.* **16** 11599

[3] Kim J, Baik K H and Jang S 2016 *Curr. Appl Phys.* **16** 221

[4] Jang S, Son P, Kim J, Lee S and Baik K H 2015 *Opt. Mater. Express* **5** 231271

[5] Son P, Baik K H and Jang S 2015 *ECS Solid State Lett.* **4** Q1

[6] Baik K H, Kim H and Jang S 2014 *Thin Solid Films* **569** 1

[7] Baik K H, Kim H, Kim J, Jung S and Jang S 2013 *Appl. Phys. Lett.* **103** 091107

[8] Wright J S *et al* 2007 *App. Surface Sci.* **253** 3766

[9] Jang S, Chen J J, Ren F, Yang H, Han S, Norton D P and Pearton S J 2006 *J. Vac. Sci. Technol.* B **24** 690

[10] Wang H, Kang B S, Chen J, Anderson T, Jang S, Ren F, Kim H S, Li Y J, Norton D P and Pearton S J 2006 *Appl. Phys. Lett.* **88** 102107

[11] Lopatiuk-Tirpak W V, Schoenfeld L, Chernyaka F X, Xiu J L, Liu S, Jang F, Ren S J, Pearton and Osinsky A 2006 *Appl. Phys. Lett.* **88** 202110

[12] Chen J J, Jang S, Anderson T J, Ren F, Li Y, Kim H, Gila B P, Norton D P and Pearton S J 2006 *Appl. Phys. Lett.* **88** 122107

[13] Chen J J, Jang S, Ren F, Li Y, Kim H, Norton D P, Pearton S J, Osinsky A, Chu S N G and Weaver J F 2006 *J. Electron. Mater.* **35** 516

[14] Ip K, Thaler G T, Yang H, Han S Y, Lia Y, Norton D P, Pearton S J, Jang S and Ren F 2006 *J. Cryst. Growth* **287** 149

[15] Jang S, Chen J, Kang B S and Ren F 2005 *Appl. Phys. Lett.* **87** 222113

[16] Jang S, Jung S, Kim J, Ren F, Pearton S J and Baik K H 2018 *ECS J. Solid State Sci. Technol.* **7** Q3180

[17] Yang G, Jang S, Ren F, Pearton S J and Kim J 2017 *ACS Appl. Mater. Interfaces* **9** 40471

[18] Carey P H, Yang J, Ren F, Hays D C, Pearton S J, Jang S, Kuramata A and Kravchenko I I 2017 *AIP Adv.* **7** 095313

[19] Jang S, Jung S, Beers K, Yang J, Ren F, Kuramata A, Pearton S J and Baik K H 2018 *J. Alloy. Compd.* **731** 118

[20] Carey P H IV, Ren F, Hays D C, Gila B P, Pearton S J, Jang S and Kuramata A 2017 *Appl. Surf. Sci.* **422** 179

[21] Carey P H IV, Ren F, Hays D C, Gila B P, Pearton S J, Jang S and Kuramata A 2017 *Jpn J. Appl. Phys.* **56** 071101

[22] Carey P H, Ren F, Hays D C, Gila B P, Pearton S J, Jang S and Kuramata A 2017 *J. Vac. Sci. Technol.* **35** 041201

[23] Carey P H, Ren F, Hays D C, Gila B P, Pearton S J, Jang S and Kuramata A 2017 *Vacuum* **142** 52

[24] Carey P H, Ren F, Hays D C, Gila B P, Pearton S J, Jang S and Kuramata A 2017 *Vacuum* **141** 103

[25] Yang J, Ahn S, Ren F, Pearton S J, Jang S, Kim J and Kuramata A 2017 *Appl. Phys. Lett.* **110** 192101

[26] Stepanov S I, Nikolaev V I, Bougrov V E and Romanov A E 2016 *Rev. Adv. Mater. Sci.* **44** 63

[27] Víllora E G, Shimamura K, Yoshikawa Y, Ujiie T and Aoki K 2008 *Appl. Phys. Lett.* **92** 202120

[28] Galazka Z *et al* 2014 *J. Cryst. Growth* **404** 184

[29] Jung S, Baik K H, Ren F, Pearton S J and Jang S 2017 *J. Vac. Sci. Technol.* **35** 042201

[30] Jung S, Baik K H, Ren F, Pearton S J and Jang S 2018 *ECS J. Solid State Sci. Technol.* **7** Q3020

[31] Halfaya Y, Bishop C, Soltani A, Sundaram S, Aubry V, Voss P L, Salvestrini J P and Ougazzaden A 2016 *Sensors* **16** 273

[32] Jung S, Baik K H, Ren F, Pearton S J and Jang S 2018 *ECS J. Solid State Sci. Technol.* **7** Q3009

[33] Jung S, Baik K H, Ren F, Pearton S J and Jang S 2017 *IEEE Sens. J.* **17** 5817

[34] Jung S, Baik K H, Ren F, Pearton S J and Jang S 2017 *IEEE Electron Device Lett.* **38** 657

[35] Kim H and Jang S 2013 *Curr. Appl Phys.* **13** 1746

[36] Kim H, Lim W, Lee J, Pearton S J, Ren F and Jang S 2012 *Sens. Actuator B* **164** 64

[37] Jung S, Baik K H and Jang S 2017 *ECS Trans.* **77** 121

[38] Astbury G R and Hawksworth S J 2007 *Int. J. Hydrog. Energy* **32** 2178

[39] Pearton S J, Ren F, Wang Y L, Chu B H, Chen K H, Chang C Y, Lim W, Lin J and Norton D P 2010 *Prog. Mater. Sci.* **55** 1

[40] Jang S, Jung S and Baik K H 2018 *Thin Solid Films* **660** 646

[41] Luo B *et al* 2002 *Appl. Phys. Lett.* **80** 1661

[42] Morkoç H, Cingolani R and Gil B 1999 *Solid-State Electron.* **43** 1909

[43] Eickhoff M *et al* 2003 *Phys. Status Solidi C* **6** 1908

[44] Hsu C S, Chen H I, Chang C F, Chen T Y, Huang C C, Chou P C and Liu W C 2012 *Sens. Actuator B* **165** 19

[45] Jung S, Baik K H, Ren F, Pearton S J and Jang S 2017 *J. Electrochem. Soc.* **164** B417

[46] Kolmakov A, Zhang Y, Cheng G and Moskovits M 2003 *J. Adv. Mater.* **15** 997

[47] Timmer B, Olthuis W and Van Den Berg A 2005 *Sens. Actuator B* **107** 666

[48] Aswal D K and Gupta S K 2007 *Science and Technology of Chemiresistor Gas Sensors* (Hauppauge, NY: Nova)

[49] Das S and Jayaraman V 2014 *Prog. Mater. Sci.* **66** 112

[50] Zhang J, Qin Z, Zeng D and Xie C 2017 *Phys. Chem. Chem. Phys.* **19** 6313

[51] Pourfayaz F, Mortazavi Y, Khodadadi A and Ajami S 2008 *Sens. Actuator B* **130** 625

[52] Wu R, Lin D, Yu M, Chen M H and Lai H 2013 *Sens. Actuator B* **178** 185

[53] Tharsika T, Thanihaichelvan M, Haseeb A S M A and Akbar S A 2019 *Front. Mater* **6** 122

[54] Guo P and Pan H 2006 *Sens. Actuator B* **114** 762

[55] Jang S, Lee S and Baik K H 2017 *Jpn J. Appl. Phys.* **56** 051001

[56] Baik K H, Kim J and Jang S 2017 *Sens. Actuator B* **238** 462

[57] Jang S, Kim J and Baik K H 2016 *J. Electrochem. Soc.* **163** B456

[58] Jang S, Son P, Kim J, Lee S and Baik K H 2016 *Sens. Actuator B* **222** 43

[59] Baik K H, Kim H, Lee S, Lim E, Pearton S J, Ren F and Jang S 2014 *Appl. Phys. Lett.* **104** 072103

[60] Hong J, Lee S, Seo J, Pyo S, Kim J and Lee T 2015 *ACS Appl. Mater. Interfaces* **7** 3554

[61] Monastersky R 2013 *Nature* **497** 13

[62] Vincent T A and Gardner J W 2016 *Sens. Actuator B* **236** 954

IOP Publishing

Wide Bandgap Semiconductor-Based Electronics

Fan Ren and Stephen J Pearton

Chapter 14

Modeling of AlGaN/GaN pH sensors

Madeline Esposito, Erin Patrick and Mark E Law

14.1 Introduction

While GaN-based devices have gained much attention for high power and high frequency electronic applications, they are also attractive as sensors. Given their strong sensitivity due to their large electron mobility, HEMT sensors have a fast response time and are currently the cheapest technology available for sensing [1]. The main design difference between the GaN-based high electron mobility transistor (HEMT) for sensing applications versus its common electronic applications is an un-metallized and 'open' gate contact that allows ions to interact with the semi-conductor surface. The sensing capability originates from measuring changes in the drain current resulting from ions interacting at the open gate of this device. Given that the surface of GaN is inert to etching and its material properties include nontoxicity and a large energy bandgap, it is an ideal candidate for sensing in harsh environments, ranging from *in vivo* applications to radiation-hard sensors.

The need for inexpensive and reliable diagnostic testing is apparent, with approximately 160 000 cases of prostate cancer [2] and 180 000 cases of breast cancer each year in the United States alone. Human exhaled breath condensate (EBC) has been shown to be a reliable detector of human pH [3–8], where human pH is a strong indicator of the metabolic state and can indicate various forms of cancer [9–11], kidney injury [12], and glucose issues such as diabetes [9]. Human pH can range from 5 to 8, where a typical healthy human pH is between 7 and 8 [13]. Measuring pH can also be used to detect various drugs and medications which is valuable in measuring medication tolerance and reliability [1]. Accurate measure-ment of human pH through EBC can effectively replace uncomfortable, expensive, and time-consuming blood and urine tests.

The need for computational optimization emanates from the lack of a general study of AlGaN/GaN HEMT device limitations in the experimental literature. Simulation facilitates exploration of the device's full range and identifies device parameters that result in maximal performance. Testing these aspects of the

doi:10.1088/978-0-7503-2516-5ch14

GaN-based HEMT chemical sensors through simulation reduces the testing time and cost, furthering the future development of commercializing these devices in the biosensor market. This work builds upon a simple fundamental model of biological and chemical sensing of GaN-based HEMTs and incorporates specific adsorption with the theory of an electrical double layer (EDL).

14.2 Background

14.2.1 Experimental review

Interest in GaN-based devices evolved due to their unique ability to serve in applications of radiation hardness [14]. Their investigation began with gas sensors, as GaN and SiC were known to be sensitive to a number of gases [14]. There has been much activity in studying GaN-based transistors for gas sensing [15–25] where its success led to the idea of sensing liquid chemicals in electrolyte rather than gaseous chemicals [14]. Thus far, 21 experimental studies have been conducted on these pH sensing GaN-based HEMTs with little variation of device and material parameters. These experiments collectively provide very little comprehensive understanding for optimal device operating conditions. Because this field is novel, there is much to investigate and comparatively limited literature to review.

In 2003 Steinhoff *et al* pioneered the application of GaN-based transistors for pH sensing. GaN-based sensing [23] is an attractive alternative to Si-based ion-sensitive field effect transistors (ISFETs) [24, 25] since ISFET's oxide layer, SiO_2, contributes to a low sensor response and poor sensor stability. Prior to Steinhoff *et al*, some work on III-nitride-based chemical sensors was performed, however, the ion sensitivity of GaN surfaces in aqueous solutions was not systematically investigated [26, 27]. Steinhoff *et al* studied the electrical response of deposited and thermally grown oxide on GaN surfaces to variations of hydrogen ion, H^+, concentrations in electrolyte solutions [23]. This study measured the chemical response of the gate surface to changes in the electrolyte composition through adjusting an Ag/AgCl reference electrode to compensate for the ion-induced changes in the channel or drain current at a constant applied drain bias. Since this work, there have been proof-of-concept studies [1, 9, 28–46] with a GaN-based device for pH sensing while only testing a few parameters at a time and all with varied parameters (mobility, gate dimensions, AlGaN thickness/molarity, varied GaN oxide, etc), making it nearly impossible to collectively compare and observe trends for device optimization.

Sensitivity is the standard metric to compare the response for these chemical sensors. Steinhoff *et al* were the first to report linear behavior with pH for this device with sensitivities (57.3 mV pH^{-1} for GaN:Si/GaN:Mg and 56.0 mV pH^{-1} for GaN/ AlGaN/GaN HEMT) close to the Nernstian response to H^+ ions [23]. These sensitivities were reflected in a drain current decrease with increasing pH. The Nernst equation is the theoretical value of the potential change with pH and is expressed as follows [47], where k is Boltzmann's constant, T is temperature, and q is electron charge,

$$\Delta\psi_0 = -2.303\frac{kT}{q}\Delta\text{pH}. \tag{14.1}$$

This equation may be regarded as a modified version of the Nernst equation relating change in potential, φ_0, to a change in pH, ΔpH [47], where at room temperature, 25 °C, the equation is simplified to

$$\Delta\psi_0 = -0.0589\Delta\text{pH} = -58.9\ \frac{\text{mV}}{\text{pH}}. \tag{14.2}$$

However, transistor output is not commonly a unit of potential, but rather one of current. Sensor response is measured through a change in drain, or channel, current per change in pH (mA mm pH^{-1} or mA pH^{-1}). It is a well-accepted notion that improving the sensitivity is the main objective to increasing the accuracy and effectiveness of the sensor.

The Yates *et al* site-binding model for the EDL at the oxide/electrolyte interface is widely accepted as the basis for the chemical interactions at the open gate of the sensor [47, 48]. In the site-binding model, there are a set number of surface sites available for chemical interactions as the hydrogen ions within the electrolyte react with the oxygen in the oxide. GaO is commonly present as the native oxide of the open gate. These reactions have forward and reverse reaction rates, the values of which are not understood as well as the value for the number of surface sites available.

While there are a handful of groups who perform experimental work, only two other groups have completed simulation work with one paper each. This study [49] expands upon simulation work from Bayer *et al* [50] and presents a one-dimensional, more extensive model. Throughout this simulation work, we will only be addressing the steady-state response. However, it is clear from the literature that the transient response will need to be studied with simulation work in the future once a steady-state simulation is successful.

14.2.2 Simulation review

As stated previously, only two other groups have published work on simulating the AlGaN/GaN HEMT in sensing pH. First, Bayer *et al* executed a one-dimensional simulation that linked a change in adsorbed charge on the interface to a change in electron density of the conducting channel, resulting in a change in the drain current [50]. In their work, transport of the ions in the electrolyte was not included, however, a simplified model of the charge in the electrolyte was accounted for through the Poisson–Boltzmann equation. To execute one-dimensional effects, it was assumed that the charge distribution was laterally homogeneous. While this work was novel, it has limitations to be expanded past one dimension to fully represent the pH sensing mechanism of this device. This group also incorporated a reference electrode in their work. The Bayer *et al* simulation study tested values of material parameters such as the number of available surface sites, N_S, and dissociation chemical reaction rate constants, K_1 and K_2, to increase sensitivity. They started with N_S as 8×10^{14} cm^{-2}

to mimic Al_2O_3 as recent experiments indicate the thermal oxide on top of the GaN, Ga_xO_y, behaves similarly to Al_2O_3 [23, 51]. This work also tested the N-face polarity of the GaN material and the molar fraction of the AlGaN layer predicting that the N-face polarity of the GaN material will significantly improve the sensing features as well as a reduction in the Al content within the AlGaN layer. Lastly, Bayer *et al* ended the study with an assertion that the area of the open gate exposed to the electrolyte should be as large as possible relative to the total surface of the device to produce an effective sensor, yet they did not publish evidence to support this reasonable claim. While the simulation study of Bayer *et al* excels in its results, the structure modeled did not include the EDL.

The next and more recent simulation study was presented this past year, 2018, by Anvari *et al* [52]. Anvari *et al* implemented a triple layer model (TLM) developed by Sverjensky *et al* [53, 54] which integrated the electrostatics of the water dipole to the site-binding model [47, 48]. This approach has answered questions of the water dipole effect on the reaction where the selectivity of the surface to pH is explained with the TLM equilibrium rates which varied in previous experimental work [23, 55]. Further, Anvari *et al* conclude that the GaN-based sensor is not as sensitive to ionic type and strength and needs to be coupled with a membrane on its surface to add this functionality [56]. In this work, predictions regarding the relationship between sensitivity and thickness of the surface oxide layer were made where the thin layer of oxide at the surface accounts for the surface chemistry and reaction. The thickness of the AlGaN layer was shown to have a large effect on the 2DEG charge, which is due to the polarization charge, where a decrease in the thickness decreases the charge.

Both Bayer *et al* and Anvari *et al* lack a comparison to the reports of current-based sensitivity, instead of the potential change with pH. The current-based sensitivity is favorable to report as sensors more readily report current. The simulation work of both Bayer *et al* and Anvari *et al* were completed with finite-difference methods in one dimension. This approach has made it difficult to expand beyond one dimension to examine the geometry dependence of the HEMT sensor. Our past work presented two-dimensional finite-element models, however, both lack the inclusion of the EDL and complex electrochemical kinetics to accurately model the surface reaction [57, 58]. The finite-element method with the EDL simulation work we propose can be expanded beyond one dimension for further modeling and the development of the GaN-based HEMT chemical sensor.

To implement the EDL within our past work, a brief review of modeling the EDL with specific adsorption was necessary. Orazem's dissertation work presented an EDL model of a semiconductor–electrolyte interface for a solar cell [59–62] which was an extension of Grahame's model [63]. This work was computed with a finite-difference model and focused on the semiconductor modeling for generation and recombination of the solar cell. What is most interesting is how the EDL was modeled as a space charge region, along with the semiconductor. This work also included charge-transfer reactions to allow current to flow from the semiconductor to the solution, which is not studied in this work. However, further simulations with the EDL theory and a semiconductor response are nonexistent. The work found has been focused on metal–electrolyte systems, and effectively modeling electrodes in electrolytes. Guldbrand *et al* presented a Monte Carlo study of the force between two charged surfaces [64]. Boda *et al* present another Monte Carlo simulation study proving that the Gouy–Chapman theory is not sufficient and the intermolecular potentials, in particular for water, need to be added in future models [65]. In another study, for the application of a biosensor measuring protein, Henderson and Boda further support their previous work which predicted that the Gouy–Chapman–Stern (GCS) model is not enough to accurately model ion selectivity and adsorption of ions [66]. Scaramuzza *et al* studied electrode and electrolyte interfaces in COMSOL, MATLAB, and HSPICE, yet all simulation models lacked specific adsorption [67]. The work presented by Wu *et al* computed numerical simulations to examine the faradaic and charging current influence on impedance spectroscopy and the impedance response of a rotating disk electrode, however, these models did not include specific adsorption as well [68, 69].

14.3 Simulation methodology: an open-gate high electron mobility transistor as pH sensor

The Poisson equation and the continuity of charge carriers are the device equations used to solve charge transport within the semiconductor and electrolyte. Electrons and holes comprise the charge carriers within the semiconductor materials whereas ions solely encompass the charge carriers in the electrolyte. The device simulator solves for solution variables originating from two partial differential equations: Poisson's equation for electrostatics (14.3) and charge continuity (14.4). The electrostatic potential, Ψ, is continuous throughout the device structure. The total charge in a material is σ_{total}, c_i refers to the concentration of a charged species i, and N_i is the flux density of the species modeled with drift and diffusion related expressions corresponding to each material:

$$\nabla^2 \Psi = -\frac{\sigma_{\text{total}}}{\epsilon} \tag{14.3}$$

$$\frac{\partial c_i}{\partial t} = -\nabla N_i. \tag{14.4}$$

14.3.1 Device structure

Figure 14.1(a) illustrates a two-dimensional rendering of the structure tested through the simulation framework. Figure 14.1(b) displays a cross-sectional view with labeled polarization charges. The device structure is comprised of a 3 µm electrolyte, a 2 Å outer Helmholtz region (OHR), a 2 Å inner Helmholtz region (IHR), a 15 nm AlGaN layer, a 1.8 µm GaN buffer layer, and a 1.3 µm substrate layer of AlN and SiC. All the layers are of realistic dimensions, with the exception of the electrolyte layer which is scaled large enough to simulate transport characteristics.

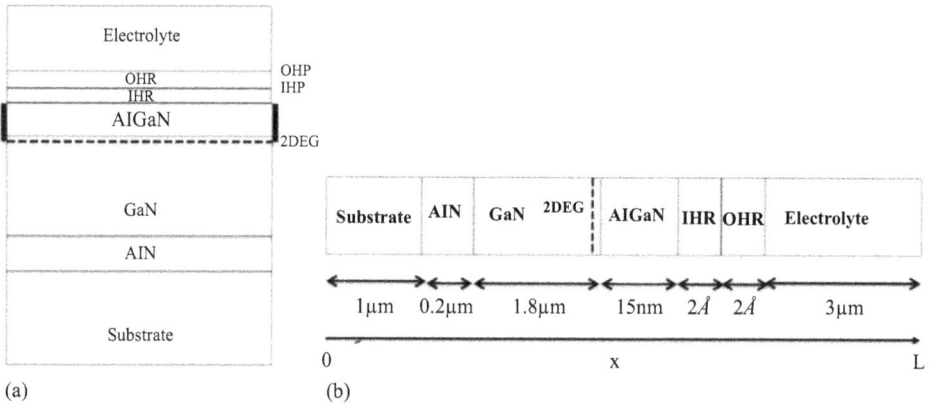

Figure 14.1. Two-dimensional rendering of the GaN-based HEMT structure simulated in FLOODS. (a) Two-dimensional design of the GaN-based HEMT structure. The un-metallized gate area includes the electrolyte and double layer regions. (b) Vertical cross-section displaying the thickness of each layer for the GaN-based HEMT.

14.3.2 Two-dimensional electron gas

The Ga-face polarization model was used in which the AlGaN undergoes tensile strain to result in an induced piezoelectric positive polarization charge, σ_{pol_2}, at the AlGaN–GaN interface [70, 71]. The bottom of the GaN buffer layer has a positive charge, σ_{pol_1}, with a value of 2.2×10^{13} cm^{-2}. The spontaneous polarization charge cancels at the AlGaN–GaN interface and the remaining piezoelectric polarization of the AlGaN produces a net positive charge, σ_{pol_2}, of 1.06×10^{13} cm^{-2}. The net negative polarization charge, which is the sum of the spontaneous and piezoelectric polarization charges, is implemented as σ_{pol_3}, valued at 3.26×10^{13} cm^{-2}, located at the top of the AlGaN. These polarization charge values used in the device structure for simulation testing were acquired from the Ambacher group [70].

14.3.3 Electrolyte

The Poisson equation within the electrolyte is defined in equation (14.5) with permittivity, ε, (vacuum permittivity, 8.85×10^{-14} F cm^{-1} times relative permittivity of electrolyte, 78), the elementary charge q, 1.6×10^{-19} C, and the ion concentrations sodium [Na$^+$], chloride [Cl$^-$], hydrogen [H$^+$], and arbitrary anion to react with hydrogen [A$^-$]:

$$\epsilon \nabla^2 \Psi = q([\mathrm{Na^+}] - [\mathrm{Cl^-}] + [\mathrm{H^+}] - [\mathrm{A^-}]). \tag{14.5}$$

Ion continuity equations for the electrolyte are depicted in equations (14.6)–(14.9) which determine changes in the ion concentration relative to time due to drift and diffusion where the first term is drift and the second term is diffusion. Therefore $\mu_{\mathrm{Na^+}}$, $\mu_{\mathrm{Cl^-}}$, $\mu_{\mathrm{H^+}}$, and $\mu_{\mathrm{A^-}}$ are the mobilities and $D_{\mathrm{Na^+}}$, $D_{\mathrm{Cl^-}}$, $D_{\mathrm{H^+}}$, and $D_{\mathrm{A^-}}$ are the diffusion coefficients of the ions [Na$^+$], [Cl$^-$], [H$^+$], and [A$^-$], respectively. To calculate diffusivity from mobility, Einstein's relation is applied:

$$\frac{\partial[\text{Na}^+]}{\partial t} = \nabla(q\mu_{\text{Na}^+}[\text{Na}^+]\nabla\Psi + (D_{\text{Na}^+}\nabla[\text{Na}^+])) \tag{14.6}$$

$$\frac{\partial[\text{Cl}^-]}{\partial t} = \nabla(q\mu_{\text{Cl}^-}[\text{Cl}^-]\nabla\Psi + (D_{\text{Cl}^-}\nabla[\text{Cl}^-])) \tag{14.7}$$

$$\frac{\partial[\text{H}^+]}{\partial t} = \nabla(q\mu_{\text{H}^+}[\text{H}^+]\nabla\Psi + (D_{\text{H}^+}\nabla[\text{H}^+])) \tag{14.8}$$

$$\frac{\partial[\text{A}^-]}{\partial t} = \nabla(q\mu_{\text{A}^-}[\text{A}^-]\nabla\Psi + (D_{\text{A}^-}\nabla[\text{A}^-])). \tag{14.9}$$

The supporting electrolyte is NaCl with a concentration of 100 mM. HA characterizes an acid in the electrolyte where H^+ is a hydrogen cation and A^- is a corresponding anion. In this simulation testing, the hydrogen concentration ranges from 10^{-2} mM to 10^{-9} mM. The pH is determined through the negative log of the hydrogen concentration. This simulation ensured electroneutrality in the electrolyte by setting a condition where the concentration of A^- always equals H^+. The flux of species due to convection is assumed to be negligible and is not modeled in this simulation. To correctly compute the drift and diffusion current balance, the Scharfetter–Gummel method is used to discretize the charge flow [72].

14.3.4 Semiconductor

The Poisson equation in the AlGaN, GaN, and substrate semiconductor layers of the device are expressed as

$$\epsilon\nabla^2\Psi = q(n - p + N_\text{D} - N_\text{A}). \tag{14.10}$$

The variables n and p are the electron and hole carrier concentrations, and N_D and N_A are the donor and acceptor impurity concentrations. The GaN layer was assigned a donor doping of 10^{16} cm^{-2} to model n-type doping whereas smaller donor concentration values of 10^{12} cm^{-2} were assigned to model unintentionally doped regions. No acceptor doping was simulated in this study.

The continuity equations for electrons and holes in the semiconductor material are dependent on quasi-Fermi energy levels, ϕ_fn and ϕ_fp, as follows:

$$\frac{\partial n}{\partial t} = -nq\mu_\text{n}\nabla^2\phi_\text{fn} \tag{14.11}$$

$$\frac{\partial p}{\partial t} = -pq\mu_\text{p}\nabla^2\phi_\text{fp}. \tag{14.12}$$

This computation captures drift–diffusion physics and is most compatible for simulating heterostructure based devices. The quasi-Fermi energy levels are continuous throughout the semiconductor and are analogous to electrochemical potentials.

The Maxwell–Boltzmann distributions for electron and hole concentrations with respect to the quasi-Fermi levels and the conduction, E_c, and valence, E_v, energy bands are expressed as follows:

$$n = N_c e^{\frac{-(E_c - \phi_{fn})}{V_t}} \tag{14.13}$$

$$p = N_v e^{\frac{-(E_v - \phi_{fn})}{V_t}}. \tag{14.14}$$

N_c and N_v are the density of states for electrons and holes, respectively, and V_t is the thermal voltage at room temperature (0.0259 eV). In this work, the temperature was held constant at room temperature. In this simulation, the electrostatic potential, Ψ, defines the vacuum energy level, which is related to the conduction band through electron affinity: $E_c = \Psi - \chi$. This method can be correlated to multiple semiconductor materials with different energy band gaps to be given a similar reference energy level. A low-field mobility model was used after the Farahmand group [73]. Since this device is based on a wide-bandgap material, neither recombination nor generation of charge carriers occurs in the simulation.

14.3.5 Inner and outer Helmholtz regions

Given that charge is assumed not to be retained within the IHR and OHR, the corresponding Laplacian of the electrostatic potential is equivalent to zero. The Laplace equation for these material layers is expressed as

$$\epsilon \nabla^2 \Psi = 0. \tag{14.15}$$

It is important to note that these regions exist only for the purpose of simulating the EDL effect in finite-element modeling (FEM). In EDL theory, these regions are not physical. This is due to ions only existing at the planes, the inner Helmholtz plane (IHP) and outer Helmholtz plane (OHP). Between the planes is the atomic spacing where an ion cannot physically fit between the two planes of charge. To simulate this in FLOODS, an FEM tool, the IHR and OHR had to be created, however we designed the grid spacing to exist only at these planes so an ion could exist only at the boundaries of the regions. Effectively the ions simulated in these regions could exist only at the IHP and OHP to accurately simulate the EDL.

14.3.6 Interface regions

The strain and spontaneous polarization charges within the semiconductor discussed previously are added as a sheet of charge (of a constant value) to the Poisson equation at their designated material interfaces. The interface of most interest and complexity is the electrolyte–AlGaN since it is the area for the sensing mechanism. The GaN-based HEMT pH sensing capability originates from the hydrogen ions available at the un-metallized gate surface altering conductive channel current. The site-binding model for oxides [47, 48] was used to simulate the amphoteric nature of the electrolyte and semiconductor interface:

$$[S - OH_2^+] \rightleftharpoons [S - OH] + [H^+] \tag{14.16}$$

$$[S - OH] \rightleftharpoons [S - O^-] + [H^+]. \tag{14.17}$$

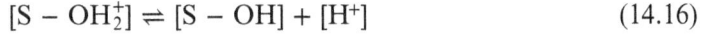

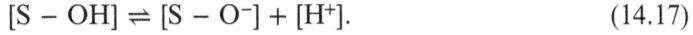

S denotes an oxide molecular site with OH, a bonded hydroxyl group, resulting in creating positively charged, SOH_2^+, and negatively charged, SO^-, surface sites, respectively, from neutral sites, SOH. While the AlGaN is not an oxide, it is clearly observed that a thin native oxide will form at the surface of the AlGaN layer. This is where the reaction will occur and it is simulated as such.

These chemical reactions are executed in FLOODS via the following rate equations:

$$K_1[S - O^-][H^+] - K_2[S - OH] + \frac{\partial}{\partial t}[S - O^-] = 0 \tag{14.18}$$

$$K_3[S - OH][H^+] - K_4[S - OH_2^+] + \frac{\partial}{\partial t}[S - OH_2^+] = 0 \tag{14.19}$$

$$[S - OH] = N_S - [S - O^-] - [S - OH_2^+]. \tag{14.20}$$

The differential equations for reactions (14.16) and (14.17) to be computed with adsorption, K_1 and K_3, and dissociation, K_2 and K_4, rate constants are expressed in equations (14.18) and (14.19). The reaction rate constants were tuned as fitting parameters to match the existing literature and achieve a maximum concentration of positive surface sites at low pH and a maximum concentration of negative surface sites at high pH. The boundary condition for the number of available surface sites is expressed in equation (14.20). N_S is the number of available sites at the surface, which was set to a maximum of 1×10^{15} cm^{-2}. Bayer et al validate this assumption with work from Bergveld et al [50, 51]. Therefore, all the possible sites, being positive, negative, or neutrally charged, must be equal to the number of available bonding sites, N_S.

Both the IHR and OHR are still treated as insulators but now there is transport of H$^+$ ions from the electrolyte through the OHR to adsorb on the IHP. Since the OHR is treated as an insulator, it has no ion concentration defined within it. Therefore, in our simulations we allowed H$^+$ ions to transport from the electrolyte into the OHR to be adsorbed on the IHP.

It is well accepted that reactions are dependent on potential and this is expressed as the reaction rates being exponentially dependent on potential [74, 75]. A previous work studied the effects of the equilibrium rate having an exponential dependence on potential in the steady-state [49]. The exponential dependence was implemented on the forward reaction rate and is expressed as

$$K_f = K_{fo*} \exp[\frac{(1 - \beta)F}{RT}](\Psi_{\text{Bulk electrolyte}} - \Psi_{\text{IHP}}). \tag{14.21}$$

It is critical to note that the potential dependence on IHP is with respect to the potential at the bulk of the electrolyte. In our FLOODS simulations the potential in the bulk electrolyte was set to the Fermi energy level of the semiconductor to keep the Fermi level flat for a neutral pH of 7. When the pH is shifted, the Fermi level is shifted with it, changing the boundary condition of the potential for the bulk electrolyte, as discussed previously. Faraday's constant is represented as F. Temperature is written as T and R is Boltzmann's constant. The symmetry factor, β, was assumed to be 0.5 and this approximation is well accepted for this work [59, 75].

14.3.7 Boundary conditions

The electrostatic potential and quasi-Fermi level solution variables are defined to specific values using Dirichlet boundary conditions at two points of the device: the bottom of the substrate and the top of the electrolyte. The substrate is given a reference potential of zero. However, the electrolyte contact is fixed at 3.7 eV to match the HEMT Fermi level. Yet when the pH of the electrolyte changes, the Nernst value of 59 mV is added to or subtracted from the electrolyte. The electrolyte potential is pinned to the pH potential value with respect to the HEMT equilibrium Fermi energy level. The solution variables for ion concentrations are fixed to their bulk value at the boundary of the electrolyte and the thickness of the electrolyte is more than ten Debye lengths.

14.4 Results: a pH GaN-based HEMT sensor with EDL and specific adsorption finite-element modeling

The HEMT sensor was simulated in FLOODS and modeled via a 1D structure to study the effect of equilibrium rates, AlGaN surface charge ($\sigma_{AlGaN/IHR}$), and number of available surface sites (N_S, used in the adsorption chemical reaction). Also included in this chapter are the preliminary results of drain bias dependence. A comparison of simulated and experimental current-based sensitivity is discussed in the following section.

Three major parameters were tested in this 1D simulation, namely equilibrium rates, AlGaN surface charge, and number of available surface sites, to achieve a sensor response. Only variations in the number of available surface sites resulted in a sensor response. Once a sensor response was achieved, the effects of the remaining two parameters were tested to understand their relationship with the sensor response. The key figure of merit discussed in this work to quantify the sensor response is the concentration of charge in the channel, or 2DEG channel charge in HEMT devices, since it is directly proportional to current. Current in a channel for a linear region of operation can be expressed as

$$I = \frac{Q\mu V_{DS}W}{L}. \tag{14.22}$$

The 2DEG channel charge concentration is noted as Q, the mobility is noted as μ, L is the channel gate length, W is the channel gate width, and V_{DS} is the applied drain bias. For the remainder of this chapter, only the concentration of 2DEG channel charge, Q, is examined as all other values are assumed constant for this 1D simulation. The number of surface sites available, N_S, is largely dependent on the material at the top of the semiconductor, which is the interface with the electrolyte. This top layer is typically a native oxide. N_S is the physical displacement of how many adsorption sites can take place per area on the oxide surface. Since a bond length is fixed, there is a physical limitation to the number of surface sites available. Since FLOODS could not simulate above an N_S value of 9×10^{14} cm^{-2} due to convergence issues, below is an investigation of this physical limitation of 10^{15} cm^{-2}:

$$N_S = \frac{10^{15}}{\text{cm}^2} * \frac{1\text{cm}^2}{10^{16}\text{Å}^2} = \frac{1}{10\text{Å}^2}. \tag{14.23}$$

10^{15} surface sites in an area of 1 cm^2 is equal to one surface site for an area of 10 Å^2. The lattice constant for GaN is 3.186 Å and 5.186 Å depending on the growth and plane of the AlGaN/GaN stack [76], and the area could be 10.15 Å^2 or 26.89 Å^2. If we consider that the hydrogen bond length of a water molecule is 2 Å, this physical limitation for the number of surface sites is reasonable. Therefore, the value calculated in equation (14.23) is a maximum value of the number of available sites, predicting the most available surface sites physically possible for this oxide.

14.4.1 Understanding the 2DEG as a sensor response

Since a maximum number of available adsorbed sites was previously determined and a minimum value was needed to achieve a sensor response, further investigation into this range of available surface sites was explored to determine a relationship between the number of surface sites and the sensor response. In a small range where N_S is varied between 10^{11} and 10^{13} cm^{-2}, the response of the channel charge, diffusion charge, and interface charge were studied. An expected result was observed, where the interface charge increased to compensate for the new increasing N_S value. A slight change in the channel charge behavior is observed yet the total change in channel charge concentration with pH, or sensitivity, does not increase in magnitude to increase the sensor response. This phenomenon is due to the diffusion charge still compensating for the interface charge. Higher values of N_S were evaluated to determine the effect on channel charge, diffusion charge, and interface charge. These results are shown in figure 14.2. With the largest value of N_S there is a significant shift from negative to positive in the total diffuse charge resulting in increased change in the channel charge with pH, or in other words, an increased sensitivity. Further investigation into the potential profiles are reviewed to explain this result.

The potential as a function of depth, or potential profile, for a low value of N_S, 10^{11} cm^{-2}, is displayed in figure 14.3 where the full system (device and electrolyte) depth is in part (a) and part (b) is zoomed in for a potential profile of the EDL structure (at the electrolyte–semiconductor interface). In part (a), -3 μm is the bulk of the electrolyte solution and $+3$ μm is the backside of the SiC. The potential of the bulk electrolyte is controlled by the pH where an ideal Nernst behavior (a 59 mV decrease in potential with increasing pH) was controlled in the simulation boundary conditions. In this case, where N_S is low, there is a small change in potential at the electrolyte–semiconductor surface. The same potential profile was also studied for the case where N_S is large, 9×10^{14} cm^{-2}, and is shown in figure 14.4. The zoomed in

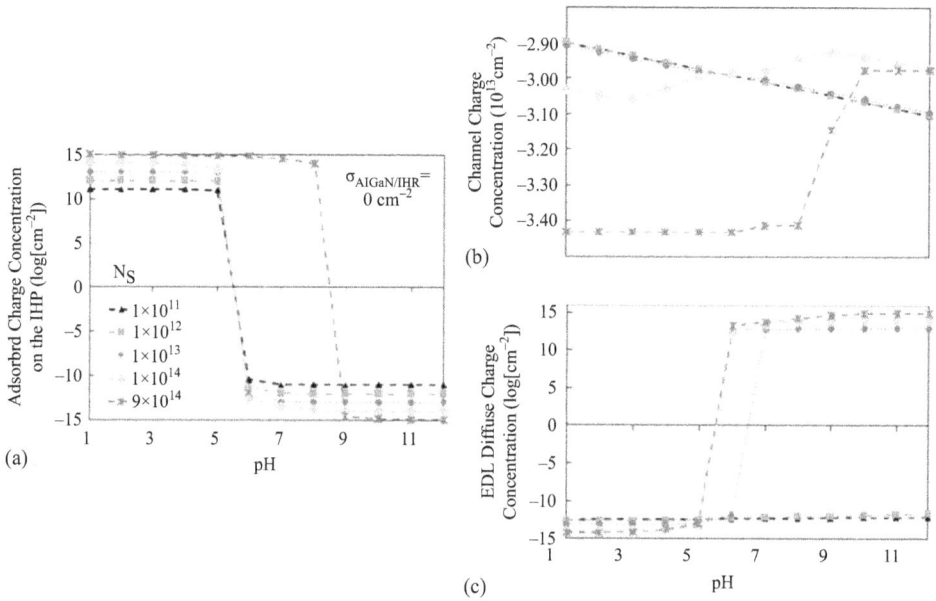

Figure 14.2. HEMT sensor response to extreme values for the number of available surface sites, N_S. (a) The adsorbed charge concentration at the IHP. (b) The electron concentration, 2DEG channel charge, and concentration in the conducting channel. (c) The total charge concentration in the diffuse part of the EDL.

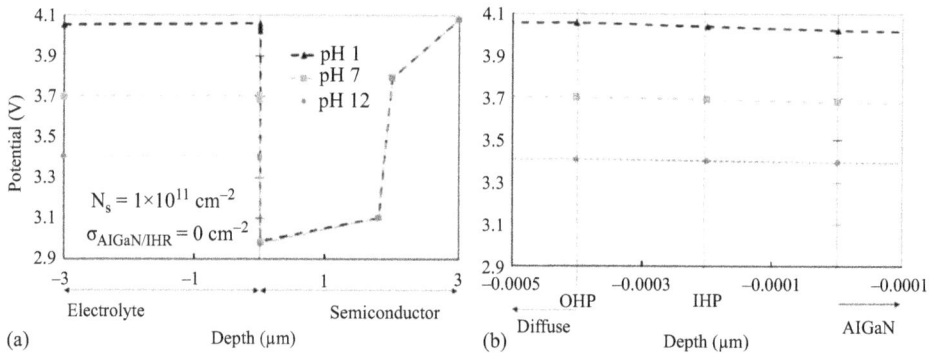

Figure 14.3. HEMT sensor potential profile for a low value of N_S. (a) The potential profile for the whole system. At -3 μm is the bulk of the electrolyte solution and at $+3$ μm is the end of the semiconductor SiC substrate. (b) The potential profile zoomed in to examine the potential profile in the EDL. At -0.0005 μm is part of the diffuse region of the EDL and at $+0.0001$ μm is the AlGaN surface of the semiconductor.

profile (in part (b)) shows that the potential is influenced by the concentration of charge adsorbed at the IHP and provides clarity for figure 14.2. The large jump in channel charge is explained through the large potential jump between low and high pH values.

To further explain the potential profile in figure 14.4, the concentration of charges in each material was examined for N_S equal to $9 \times 10^{14}\,\text{cm}^{-2}$. A graphical conclusion for the sum of charge in each area is shown in figure 14.5. Part (a) is a rendering of

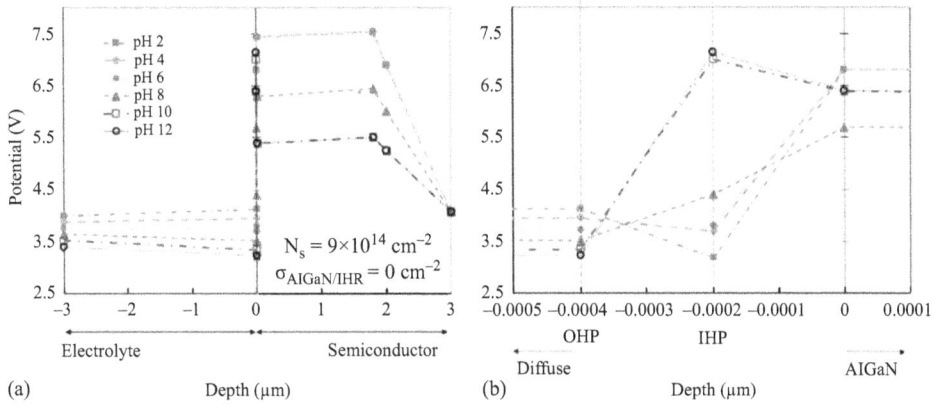

Figure 14.4. HEMT sensor potential profile for an extremely high value of N_S. (a) The potential profile for the whole system. At -3 μm is the bulk of the electrolyte solution and at $+3$ μm is the end of the semiconductor SiC substrate. (b) The potential profile zoomed in to examine the potential profile in the EDL. At -0.0005 μm is part of the diffuse region of the EDL and at $+0.0001$ μm is the AlGaN surface of the semiconductor.

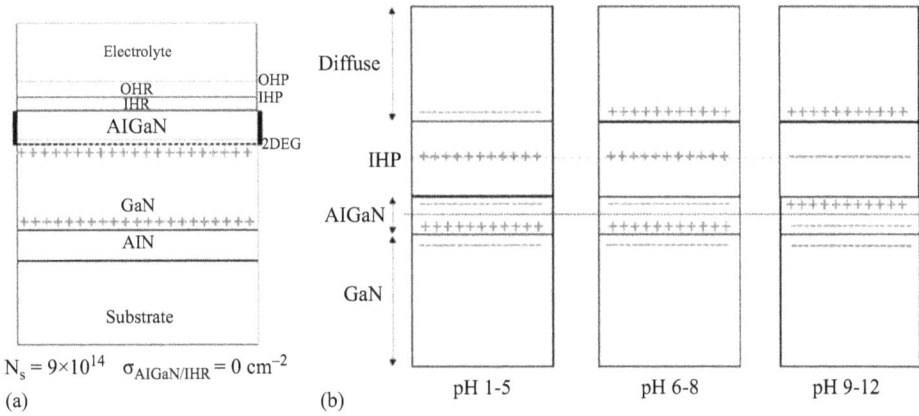

Figure 14.5. Inverted AlGaN charge in HEMT. (a) HEMT sensor structure with strain polarization charges and (b) zoomed in structure for three major pH ranges with positive and negative charge concentrations accounted for in each range.

the HEMT sensor structure simulated with the corresponding AlGaN–GaN HEMT polarization charges. Part (b) describes the balance of charge in each material for a given pH range. As shown in figure 14.5, for the pH range of 1 to 5, a large positive charge is adsorbed at the IHP, with a large negative diffuse charge and large negative channel charge. A separation of positive and negative charge in the AlGaN region is observed. To maintain electroneutrality, the AlGaN region becomes two regions of charge, where the top is negative to balance the adsorbed charge and the bottom is positive to balance the 2DEG charge. In the next pH range, 6 to 8, the adsorbed charge is less in magnitude to switch from positive to negative (SOH_2^+ to SO^-) and the diffuse charge becomes net positive. In the last pH range, 9 to 12, the adsorbed charge is a net negative charge and the AlGaN separation of charge is reversed, where previously it was negative then positive charge and now it is positive and then negative charge (from top down). However, this negative charge in the bottom of the AlGaN, closest to the AlGaN–GaN interface, for this pH range 9–12 is negligible since it is much smaller in magnitude and results in an overall net positive charge in the AlGaN material. In summary the adsorbed charge is fixed where its polarity depends on the pH value. The system has a set amount of free charges to compensate for the polarity and concentration of net adsorbed charge. The regions of free charge are the diffuse layer and AlGaN layer, where both are competing and compensating charge based on coulombic effects. This explains why the 2DEG is not responding to the large adsorbed charge—the AlGaN layer is effectively inverted and masking the 2DEG from the adsorbed charge. The minimum and maximum values of N_S for the sensor response are presented in figure 14.2. From this analysis, N_S largely dominates the sensor response of all the parameters explored in this simulation study.

14.4.2 Equilibrium reaction rate

Of great interest in sensor response is the effect of equilibrium rate dependence on the potential compared to an equilibrium rate as a constant value. The range of N_S was further explored within the bounds previously discussed and the equilibrium rate with an exponential dependence on potential, presented in figure 14.6, is compared with equilibrium rates as a constant value, presented in figure 14.7. The magnitude of adsorbed charge concentration plays a significant role in the channel response and is key to tuning the optimum sensitivity of the device. As N_S increases, the adsorbed charge increases, where the 2DEG charge increases as well (except for the inverted charge case discussed previously with N_S equal to 9×10^{14} cm^{-2}). The N_S was set to a reasonable value, 3×10^{14} cm^{-2}, to achieve a sensor response (or a change in the 2DEG charge) and the AlGaN surface charge was set to 0 cm^{-2} (zero charge), where the two methods of equilibrium rates were compared. These results

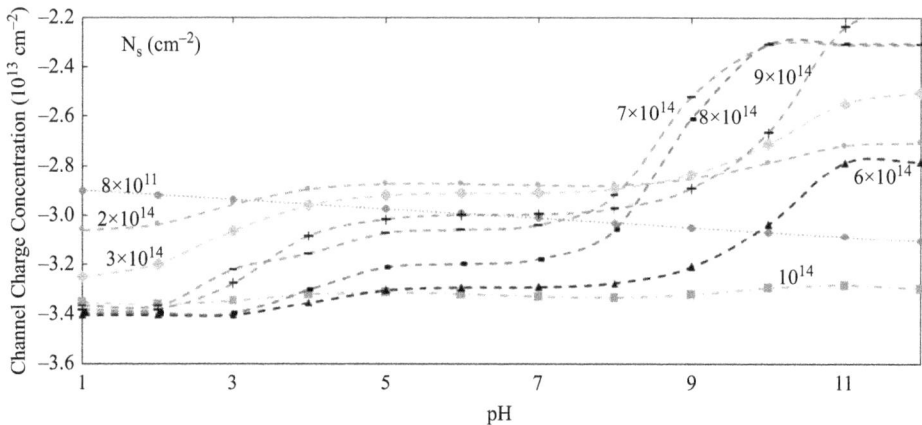

Figure 14.6. Channel charge concentration with varied available adsorbed charge, N_S, for an equilibrium rate with an exponential dependence on potential. The AlGaN surface charge was kept at zero.

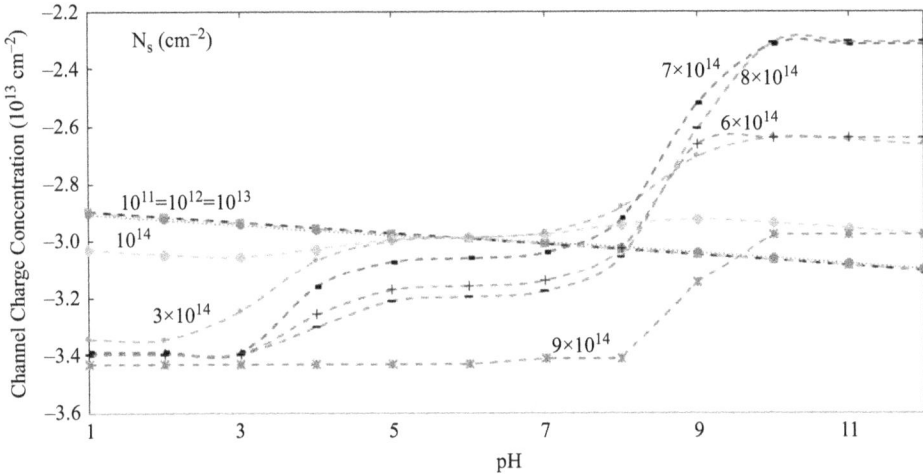

Figure 14.7. Channel charge concentration with varied available adsorbed charge, N_S, for an equilibrium rate expressed as a constant. The AlGaN surface charge was kept at zero.

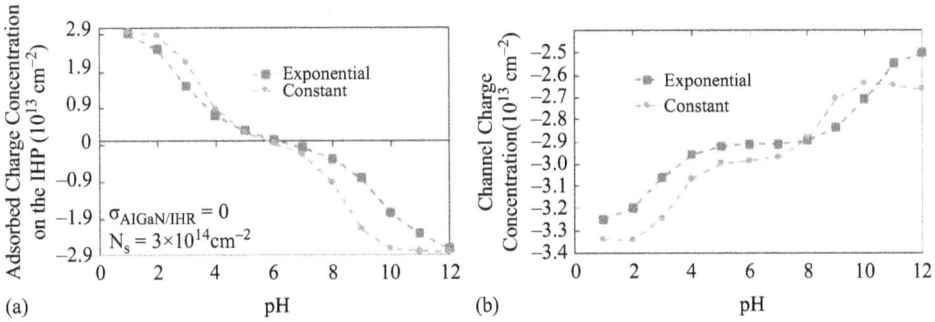

Figure 14.8. Two methods of implementing equilibrium rates: constant versus exponential dependence on potential. (a) Adsorbed charge concentration on the IHP. (b) Channel charge concentration with respect to pH.

are shown in figure 14.8 where the exponential dependence 'smooths' the adsorbed charge curve approaching linearity (in part (a)); yet the 2DEG response (part (b)) does not follow this trend. The exponential dependence on potential in the equilibrium rates pushes the response towards being linear, however, the channel charge concentration response with pH does not become linear. Using an equilibrium rate that is exponentially dependent on potential is more realistic than using a constant value [59]. Hence forth the remainder of the results will examine cases only with the equilibrium rate as an exponential dependence on potential.

For this method of modeling, the equilibrium rates as an exponential dependence on potential, the potential in the bulk electrolyte, OHP, IHP, and AlGaN surface are presented in figure 14.9. It is interesting to observe the potential of the bulk electrolyte mirroring the OHP where the IHP potential mirrors the AlGaN surface

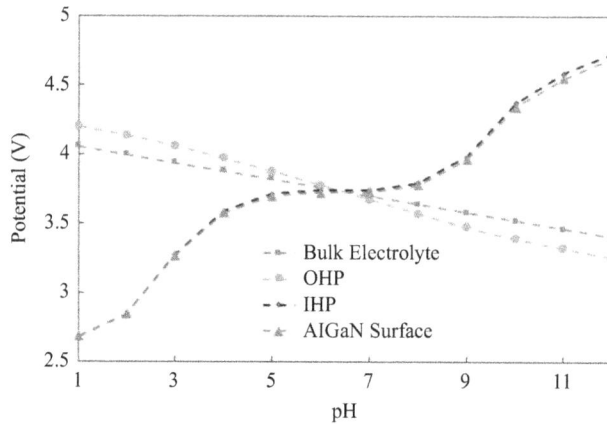

Figure 14.9. Potential at the edge of the electrolyte (or bulk electrolyte), OHP, IHP, and surface of the AlGaN with respect to pH.

potential. This behavior indicates adsorbed charge on the IHP has a direct relationship with the surface of the AlGaN layer. This is expected since the distance between the two interfaces is very small at only 2 Å.

14.4.3 Passivation charge

The results presented previously have a surface charge concentration at the top of the AlGaN layer set to 0 cm^{-2}. However, passivating the surface with different charges unintentionally in an experimental environment is inevitable and it is important to understand the effect of different surface charges on the sensor response. The AlGaN surface charge concentration was explored with a reasonable N_S value (3×10^{14} cm^{-2}). The results are shown in figures 14.10 and 14.11. The value of the AlGaN surface charge concentration shifts the magnitude of 2DEG channel charge concentration. As the AlGaN surface charge concentration increases, the 2DEG charge concentration reduces (shifts up towards zero). Recall that the HEMT total polarization charge below the AlGaN layer equals $+3.26 \times 10^{13}$ cm^{-2}, therefore an AlGaN surface charge of -3.26×10^{13} cm^{-2} would make the HEMT device electrically neutral. Interestingly, a large amount of charge, $+2.26 \times 10^{13}$ cm^{-2}, decreased the sensitivity or the change in 2DEG charge concentration with pH. This was due to a substantial amount of charge in the AlGaN layer which is again masking the 2DEG from responding and compensating for the charge at the surface and prevents a large response from the 2DEG channel. The change in passivation charge does not largely affect the adsorbed charge depicted in figure 14.11(a). In figure 14.11(b), for the extreme pH values (both high and low pHs), the total interface charge concentration is largely dominated by one of the adsorbed charge species; at low pH, the total interface charge is dominated by positive adsorbed charge (negative charge for high pH).

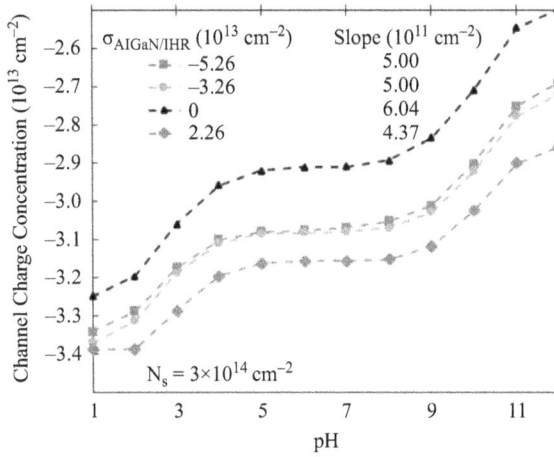

Figure 14.10. Channel charge concentration for a high N_S value (3×10^{14} cm^{-2}) at varied surface charge concentration on the top of the AlGaN layer.

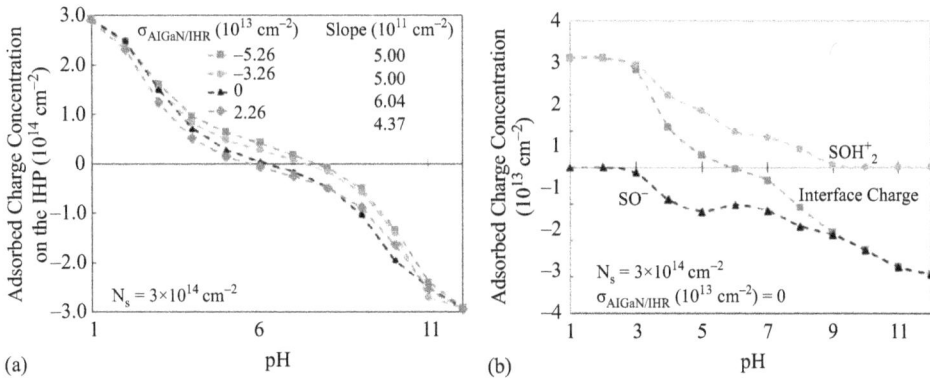

Figure 14.11. Adsorbed charge concentration for varied AlGaN surface charge concentrations. (a) Adsorbed charge concentration at the IHP with a reasonable N_S value (3×10^{14} cm^{-2}) for varied AlGaN surface charge concentrations. (b) Concentrations for specific adsorbed charges with respect to pH for a given AlGaN surface charge concentration set to 0.

14.4.4 Linear 2DEG sensor response

In the work presented thus far, a linear response for the channel charge with pH is not observed. For a model to be linear, the slope must be constant. With a reasonable N_S value (i.e. N_S equal to 3×10^{14} cm^{-2}), the transition region of adsorbed charge from positive to negative, roughly around the pH range of 5 to 8, breaks the constant slope (refer back to figures 14.13–14.15). This is due to a large shift in the magnitude of adsorbed charge. To make the response more linear for a large N_S value, the transition region needs to be minimized so that this small amount of charge occurs only at one pH instead of three or more. It would be desirable to have a high N_S value where the 2DEG channel charge concentration can have a larger range, however, this sacrifices a linear response.

Another approach to achieving a linear 2DEG response is to cut off the sharp slopes observed at extreme pH values. A narrow window of figure 14.8 is presented in figure 14.12 where a smaller pH range, from 4 to 9, is examined to observe a near-linear 2DEG response. From this narrowed pH range, a linear line is fitted to these points, which is a method reported by experimental results from Huang *et al* and Dong *et al* [33, 38]. It is worth noting the experimental work from Huang *et al* and Dong *et al* mirror the same non-linear trend ('S' curve) observed in the simulation presented here.

The concentration of charge for each different region throughout the device/system is examined for each N_S value and is shown in figure 14.13 for a high pH value of 11 and a middle pH value of 7. For a large pH value of 11 at higher values of N_S, the AlGaN charge at the bottom (which is closest to the 2DEG), increased in

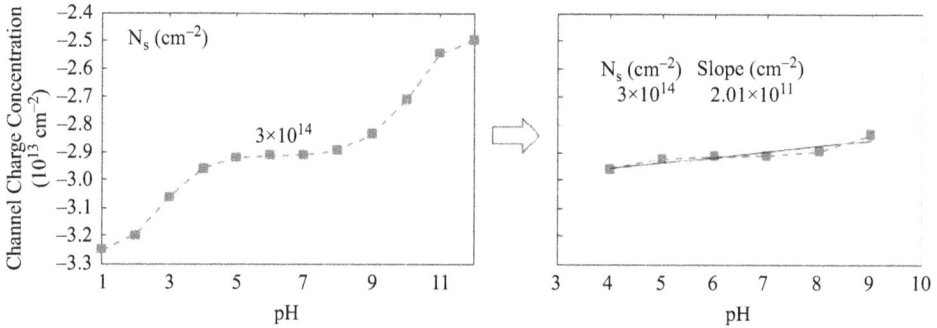

Figure 14.12. Channel charge concentration versus pH with N_S value 3×10^{14} cm^{-2} for an AlGaN surface charge of zero. The pH range is narrowed to achieve a near-linear response to compare to reported experimental activities.

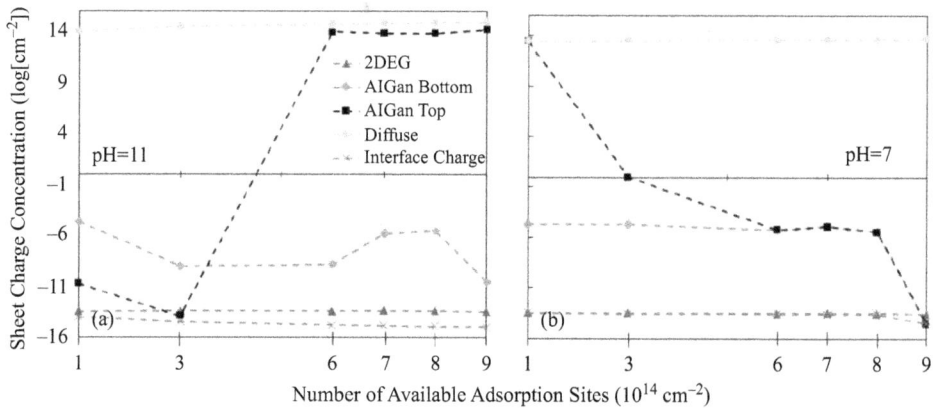

Figure 14.13. The concentration of charges throughout the device with respect to N_S values for a high pH value of 11 and a middle range pH value of 7.

orders of magnitude. This explains the large shift in the 2DEG charge for varied N_S at a large pH value in figures 14.13 and 14.14. At a pH of 7, the charge in the AlGaN layer is always negative with N_S and is assisting the 2DEG charge in compensating for adsorbed charge. This is due to the 2DEG charge being effectively capped out or being the largest it can be. In order to maintain electroneutrality, the AlGaN layer compensates for adsorbed charge. The potential profile was examined for an optimum N_S value of 3×10^{14} cm^{-2} and is presented in figure 14.14. The potential for pH 6 is roughly flat through the EDL and to the surface of the semiconductor, indicating a true equilibrium potential value. The potential of the remaining pHs mirror each other but experience cross-over regions as expected with the switch from positive to negative values of adsorbed charge. Due to the observed non-linear response, a consistent sensitivity (or slope) of potential change with pH could not be achieved.

14.4.5 Drain bias

Finally, an applied bias was placed on the back of the AlGaN layer to mimic the effect of a 2D source–drain bias. The range of applied bias tested was −0.5 V to +0.5 V. The 0 to +0.5 V bias range is reasonable given the experimental results [22, 23, 28, 30, 32–35, 37–39, 77]. There were convergence issues for a reasonable N_S value leaving only a small N_S of 10^{11} cm^{-2} to be tested. The sensor response did not change for this case, as expected since the N_S was so small, yet the two methods of equilibrium rate as constant versus exponential dependence on potential were compared. With constant equilibrium rates, little changed and the result was uninteresting. With exponential dependence on potential in the equilibrium rate, the response changed drastically with applied drain bias. These results are shown in figure 14.15. Three pH values were compared (a low value of 1, a middle value of 7, and a high value of 12) where the true adsorption behavior of these pHs were only

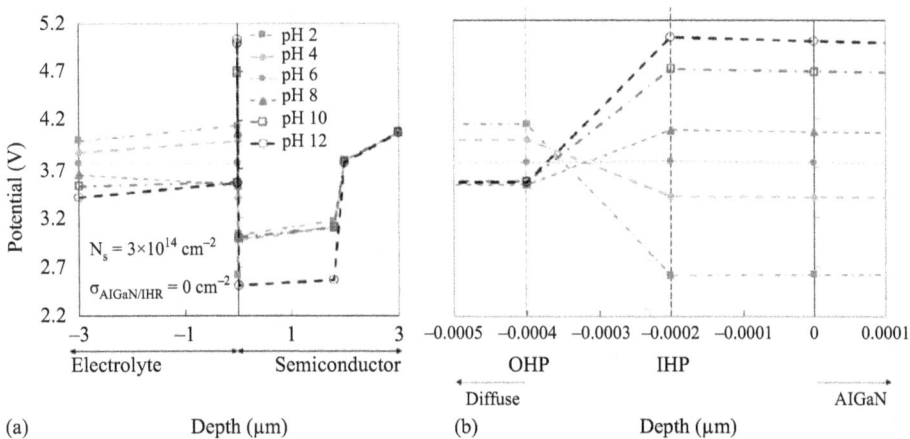

Figure 14.14. Potential as a function of depth for an N_S value of 3×10^{14} cm^{-2}. The x-axis is the lateral direction of the simulation system. (a) shows the entire system while (b) zooms in on the solid–electrolyte interface.

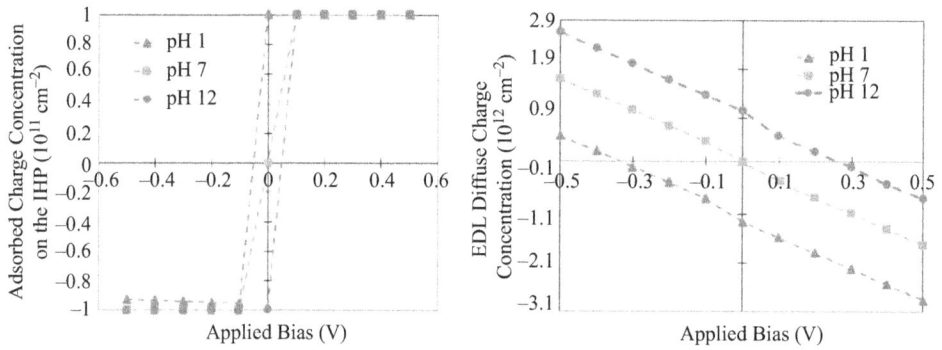

Figure 14.15. Applied bias with equilibrium rate potential exponential dependence. (a) Adsorbed charge concentration on the IHP as a function of applied bias on the AlGaN backside with the equilibrium rate as an exponential dependence on potential for varied pH values.

observed at 0 V applied bias. Once a bias of + or −0.1 V was applied, the reaction was forced to an extreme behavior of the adsorption model. This is reasonable if we recall pH 1 has a high potential value and pH 12 has a low potential (recall Nernst behavior), where at a low pH (high potential) the adsorbed charge is largely positive and negative for high pH. If we compare the −0.5 V extreme to the pH 12, it is rational that the adsorption model would be an extreme value of negatively adsorbed charge. The same is true for a +0.5 V bias to compare to pH 1 with a large positive adsorbed charge. On either side of zero bias, the smallest applied bias of 0.1 V is large enough to swing the adsorption model to its extreme and fully adsorbed case, since the equilibrium rate is exponentially dependent on potential.

Since only a small N_S value could be simulated, the sensor response was not achieved and the channel charge did not change with this drastic charge in the adsorption model. However, it is expected with larger and more reasonable N_S values, where the sensor response occurs, this effect would drastically change the sensor response. In a previous study without the EDL but in 2D, the interface charge was found to vary spatially. At a given pH and applied drain bias, the interface charge would be positive on one side of the channel and have a gradual shift to negative on the other side of the channel [57].

This is expected to be a result of the applied drain bias having an effect on the adsorption model. However, this study did not have an exponential dependence on potential in the equilibrium rate. Given that this drain bias dependence is spatial, if it were to be combined with an exponential dependence on potential in the equilibrium rate, the interface charge would vary significantly. Based on this first approximation, the source to drain bias for a 2D model with exponential equilibrium rate dependence on bias would play a significant role in the channel charge. It is expected that, instead of the gradual shift from positive to negative observed in the previous work [57], a dramatic shift from positive to negative interface charge would occur and this could lead to significant reliability issues in the sensor response. Overall the sensor response was observed to be largely dependent on N_S (compared to the variables studied here: AlGaN surface charge and equilibrium rate).

14.5 Comparing simulation work with experimental results

While most experimental works report the potential sensitivity as mV pH^{-1}, to compare with Steinhoff *et al* [23] and to compare to the maximum Nernst theoretical value (-59 mV pH^{-1}), a few papers report sensitivity as a change in the channel current. Reporting the sensitivity as a change in current is more useful than a change in potential since the sensor can more readily measure a change in current as it outputs a current value. Tables 14.1 and 14.2 provide a synopsis of these current measured sensitivities and how our simulation results compare to them.

The simulated sensitivities are expressed by equation (14.24) and calculated from the simulated 2DEG charge concentration with the equilibrium rate as an exponential dependence on potential where a maximum sensitivity was achieved with the AlGaN surface charge concentration set to 0 cm^{-2} and the N_S set to 3×10^{14} cm^{-2}. The 2DEG charge, or channel charge, concentration is noted as Q, the mobility of the 2DEG is noted as μ, L_{cm} is the channel gate length in centimeters, W_{cm} is the channel gate width in centimeters, and V_{DS} is the applied drain bias. The 2DEG

Table 14.1. Summary of experimental work.[a]

Author	Oxide	pH range	Gate dimensions	Mobility (cm^2 V^{-1} s^{-1})	Drain bias (V)
Kang *et al* [9, 39]	Native oxide and Sc_2O_3	3–10	2×150 μm^2	980	0.25
Abidin *et al* [37]	Native oxide	1.7–11.9	40×490 μm^2	1860	0.5
Brazzini *et al* [35]	Native oxide	4–10	1.2×0.5 mm^2	400	0.2
Dong *et al* [38][a]	Native oxide	4–8	13×250 μm^2 175×250 μm^2 175×50 μm^2	1730	1.5
Huang *et al* [33]	Native oxide	4–10	15×15 μm^2	750–1150	0.25

[a] The exposed gate length was 10 μm for all source to drain dimensions listed.

Table 14.2. Summary of experimental sensitivity with simulated FLOODS sensitivity.[a]

Author	Experimental sensitivity	FLOODS sensitivity
Kang *et al* [9, 39]	Native oxide: 70 μA Sc_2O_3: 37 μA pH^{-1}	0.588 mA pH^{-1}
Abidin *et al* [37]	1.9 mA pH^{-1} or 3.88 mA mm^{-1} pH^{-1}	1.825 mA pH^{-1} or 3.725 mA mm^{-1} pH^{-1}
Brazzini *et al* [35]	2.83 μA pH^{-1}	1.074 μA pH^{-1}
Dong *et al* [38][a]	1.35 mA pH^{-1} 21.29 μA pH^{-1} 7.08 μA pH^{-1}	1.599 mA pH^{-1} 23.75 μA pH^{-1} 4.75 μA pH^{-1}
Huang *et al* [33]	0.923 mA mm^{-1} pH^{-1}	0.617 mA mm^{-1} pH^{-1}

[a] A FLOODS pH range of 4 to 9 with equilibrium rates with a exponential dependence on potential was used for the sensitivity calculation.

charge concentration was taken from these 1D simulations and the remaining parameters were matched with its respective experiment to compare sensitivity values. The current is measured as amperes and can be written, for a linear region of operation, as

$$I(A) = \frac{Q\mu V_{DS} W_{cm}}{L_{cm}}.$$

(14.24)

Overall FLOODS simulation sensitivities were within reasonable percent error of experimental work except for the work reported by Kang *et al* [9, 39]. Since the surface charge of the AlGaN layer accompanied by the presence and thickness of an oxide layer is not discussed in detail for most of the experimental works, including that reported by Kang *et al*, it is likely that these factors are the root of any discrepancies. Without further information from these studies, an exact justification or investigation cannot be made. It is worth noting that the work by Abidin *et al* [37] reported sensitivities at a large drain bias where it is assumed that this sensor was operating in a saturation region. The charge in the channel is largely inverted and pinched-off where the current is reduced by a factor of 2. Our simulated results were halved to reflect this change from the linear to saturation region of the transistor operation. It is also worth noting that the work by Dong *et al* [38] showed a varied gate length (the distance between the drain and source contacts) but the exposed gate length remained the same. In this case, the drain bias is different for the region of the exposed surface when the distance between the drain and source contacts is increased. We approximated a reasonable drain bias voltage based on their figures to compare with our simulated sensitivity.

14.6 Future work

The HEMT sensor with the EDL and specific adsorption models could thus far only be simulated in one dimension in FLOODS due to convergence issues associated with tuning and solving a solution variable with an exponential dependence on potential. To expand this model to 2D will require mathematical correction in the exponential function to improve convergence. 2D models should be explored further to combine the electrochemical reaction exponential dependence on potential to test the influence of the spatial source to drain current on the adsorption model. The drain to source potential variation is expected to contribute voltage to the exponential dependence on potential of the equilibrium rates. The reaction, and therefore the HEMT sensor response, could vary exponentially along the direction of the gate length, or conducting channel. This could significantly reduce the window of sensor operation since the exponential swing occurs rapidly.

Another idea for further work is to explore different oxides with varied thicknesses and quality. Since the thickness of oxides can vary in the fabrication process (native oxide versus thermally grown versus deposited oxide), a large oxide thickness is expected to cause a decrease in the sensitivity as the electrostatic force between the 2DEG charge and the adsorbed charge is decreased. A study of the limitations for oxide thicknesses of this nature should be explored. The quality of oxides is known to differ with varied fabrication methods and it is expected that the quality of oxides can also have an effect on sensitivity [9, 39]. If a poor oxide is used (one with less oxygen available for hydrogen to bond with) it can hinder the adsorption reaction at the surface and can cause a degradation in the 2DEG response. Kang *et al* compares two oxides, a high-quality oxide deposited versus a native oxide [9, 39]. If the deposited oxide is of higher quality and leads to higher stability, the thickness of the oxide is also much larger for deposited oxide than a native oxide and the magnitude of the sensitivity is lower. Hence a simulation study of these two mechanisms to evaluate the limitations between the two can lead to the production of an oxide for optimum sensitivity.

In addition, combining knowledge of the oxide effects on the sensor response with varied electrolyte solutions of varied molecule size and dipole moment can lead to a better understanding of the sensor response. There is sufficient experimental work studying the effect of different supporting electrolytes and the selectivity of the GaN to anions [28–31, 36]. In FLOODS simulations, the charges in the electrolyte are treated as electrons and holes, where the size does not play a role in the diffusion. It is expected that the size and orientation of the molecules will affect the reaction rate of the adsorbed charge and this would vary with different supporting electrolytes. The TLM implemented by Anvari *et al* [52] analyzes the selectivity of the surface to pH, the ionic strength, and ionic types, and are all explained by the equilibrium rates since the TLM takes water molecule effects into account with charge to be adsorbed on the oxide surface. This method seems more accurate to model the reaction at the oxide surface and is complex. If it could be implemented in FLOODS and compared with the simpler EDL with specific adsorption used in this project, it could be better understood if the TLM complex model is needed or if our EDL model will suffice. Based on experimental results, the EDL model appears more accurate, however, the TLM discusses dipole moment and molecule size to make the reaction more complex and understanding this effect on the sensor response can be useful. Furthermore, this model should be combined with a study of the various oxides (grown versus deposited, thickness, quality, etc) to better understand how to increase the sensor response.

Finally, this work studied only the equilibrium rate since only steady-state cases are explored, which is a progressive first step in simulating this sensor. Transient simulations should be built to understand the balance between the forward and reverse reaction rates. These transient simulations can be integrated with an understanding of various oxides. In Kang *et al* [39] a native oxide was compared with deposited scandium oxide, where the resolution was compared and other experimental work also present sensor current as a function of time [1, 9, 29, 32, 35, 38, 77]. Simulating these effects with transient analysis can lead to a better understanding of the 2DEG response to optimize the GaN-based HEMT pH sensor, ultimately building a more complete model of this sensor.

References

[1] Chu B H *et al* 2010 Wireless detection system for glucose and pH sensing in exhaled breath condensate using AlGaN/GaN high electron mobility transistors *IEEE Sens. J.* **10** 64–70

[2] Kramer B S, Brown M L, Prorok P C and Gohagan J K 1993 Prostate cancer screening: what we know and what we need to know *Ann. Intern. Med.* **119** 914–23

[3] Kullmann T *et al* 2007 Exhaled breath condensate pH standardised for CO_2 partial pressure *Eur. Respir. J.* **29** 496–501

[4] Machado R F *et al* 2005 Detection of lung cancer by sensor array analyses of exhaled breath *Am. J. Respir. Crit. Care Med.* **171** 1286–91

[5] Horváth I, Hunt J and Barnes P J 2005 Exhaled breath condensate: methodological recommendations and unresolved questions *Eur. Respir. J.* **26** 523–48

[6] Vaughan J *et al* 2003 Exhaled breath condensate pH is a robust and reproducible assay of airway acidity *Eur. Respir. J.* **22** 889–94

[7] Kullmann T, Barta I, Antus B, Valyon M and Horvath I 2008 Environmental temperature and relative humidity influence exhaled breath condensate pH *Eur. Respir. J.* **31** 474–5

[8] Namjou K, Roller C and McCann P 2006 The breathmeter—a new laser device to analyze your health *IEEE Circuits Devices Mag.* **22** 22–8

[9] Kang B S *et al* 2008 Role of gate oxide in AlGaN/GaN high-electron-mobility transistor pH sensors *J. Electron. Mater.* **37** 550–3

[10] Chen K H *et al* 2008 c-erbB-2 sensing using AlGaN/GaN high electron mobility transistors for breast cancer detection *Appl. Phys. Lett.* **92** 192103

[11] Kang B S *et al* 2007 Prostate specific antigen detection using AlGaN/GaN high electron mobility transistors *Appl. Phys. Lett.* **91** 112106

[12] Wang H T *et al* 2007 Electrical detection of kidney injury molecule-1 with AlGaN/GaN high electron mobility transistors *Appl. Phys. Lett.* **91** 222101

[13] Hunt J F *et al* 2000 Endogenous airway acidification: implications for asthma pathophysiology *Am. J. Respir. Crit. Care Med.* **161** 694–9

[14] Pearton S J *et al* 2004 GaN-based diodes and transistors for chemical, gas, biological and pressure sensing *J. Phys. Condens. Matter* **16** R961–94

[15] Schalwig J, Muller G, Ambacher O and Stutzmann M 2001 Group-III-nitride based gas sensing devices *Phys. Status Solidi* A **185** 39–45

[16] Eickhoff M *et al* 2003 Electronics and sensors based on pyroelectric AlGaN/GaN heterostructures—Part B: Sensor applications *Phys. Status Solidi* C **0** 1908–18

[17] Eickhoff M, Neuberger R, Steinhoff G, Ambacher O, Muller G and Stutzmann M 2001 Wetting behaviour of GaN surfaces with Ga- or N-face polarity *Phys. Status Solidi* B **228** 519–22

[18] Kim J, Gila B P, Chung G Y, Abernathy C R, Pearton S J and Ren F 2003 Hydrogen-sensitive GaN Schottky diodes *Solid-State Electron.* **47** 1069–73

[19] Kim J, Ren F, Gila B P, Abernathy C R and Pearton S J 2003 Reversible barrier height changes in hydrogen-sensitive Pd/GaN and Pt/GaN diodes *Appl. Phys. Lett.* **82** 739–41

[20] Kim J, Gila B P, Abernathy C R, Chung G Y, Ren F and Pearton S J 2003 Comparison of Pt/GaN and Pt/4H-SiC gas sensors *Solid-State Electron.* **47** 1487–90

[21] Ambacher O *et al* 2003 Electronics and sensors based on pyroelectric AlGaN/GaN heterostructures *Phys. Status Solidi* C **0** 1878–907

[22] Kang B S *et al* 2003 Effect of external strain on the conductivity of AlGaN/GaN high-electron-mobility transistors *Appl. Phys. Lett.* **83** 4845–7

[23] Steinhoff G, Hermann M, Schaff W J, Eastman L F, Stutzmann M and Eickhoff M 2003 pH response of GaN surfaces and its application for pH-sensitive field-effect transistors *Appl. Phys. Lett.* **83** 177–9

[24] Bergveld P 1970 Development of an ion-sensitive solid-state device for neurophysiological measurements *IEEE Trans. Biomed. Eng.* **BME-17** 70–1

[25] Bergveld P 1972 Development, operation, and application of the ion-sensitive field-effect transistor as a tool for electrophysiology *IEEE Trans. Biomed. Eng.* **BME-19** 342–51

[26] Schalwig J, Muller G, Eickhoff M, Ambacher O and Stutzmann M 2002 Gas sensitive GaN/AlGaN-heterostructures *Sens. Actuators* B **87** 425–30

[27] Stutzmann M *et al* 2002 GaN-based heterostructures for sensor applications *Diam. Relat. Mater.* **11** 886–91

[28] Mehandru R *et al* 2004 AlGaN/GaN HEMT based liquid sensors *Solid-State Electron.* **48** 351–3

[29] Neuberger R, Muller G, Ambacher O and Stutzmann M 2001 High-electron-mobility AlGaN/GaN transistors (HEMTs) for fluid monitoring applications *Phys. Status Solidi* A **185** 85–9

[30] Alifragis Y, Volosirakis A, Chaniotakis N A, Konstantinidis G, Iliopoulos E and Georgakilas A 2007 AlGaN/GaN high electron mobility transistor sensor sensitive to ammonium ions *Phys. Status Solidi* A **204** 2059–63

[31] Chaniotakis N A, Alifragis Y, Georgakilas A and Konstantinidis G 2005 GaN-based anion selective sensor: probing the origin of the induced electrochemical potential *Appl. Phys. Lett.* **86** 164103

[32] Kokawa T, Sato T, Hasegawa H and Hashizume T 2006 Liquid-phase sensors using open-gate AlGaN/GaN high electron mobility transistor structure *J. Vac. Sci. Technol.* B **24** 1972

[33] Huang H-S, Lin C-W and Chiu H-C 2008 High sensitivity pH sensor using $Al_xGa_{1-x}N$/GaN HEMT heterostructure design *IEEE International Conference on Electron Devices and Solid-State Circuits* p 4

[34] Niigata K, Narano K, Maeda Y and Ao J-P 2014 Temperature dependence of sensing characteristics of a pH sensor fabricated on AlGaN/GaN heterostructure *Jpn J. Appl. Phys.* **53** 11RD01

[35] Brazzini T, Bengoechea-Encabo A, Sánchez-García M A and Calle F 2013 Investigation of AlInN barrier ISFET structures with GaN capping for pH detection *Sens. Actuators* B **176** 704–7

[36] Podolska A *et al* 2010 Ion versus pH sensitivity of ungated AlGaN/GaN heterostructure-based devices *Appl. Phys. Lett.* **97** 012108

[37] Abidin M S Z, Hashim A M, Sharifabad M E, Rahman S F A and Sadoh T 2011 Open-gated pH sensor fabricated on an undoped-AlGaN/GaN HEMT structure *Sensors* **11** 3067–77

[38] Dong Y *et al* 2018 AlGaN/GaN heterostructure pH sensor with multi-sensing segments *Sens. Actuators* B **260** 134–9

[39] Kang B S *et al* 2007 pH sensor using AlGaN/GaN high electron mobility transistors with Sc_2O_3 in the gate region *Appl. Phys. Lett.* **91** 012110

[40] Varghese A, Periasamy C, Bhargava L, Dolmanan S B and Tripathy S 2019 Linear and circular AlGaN/AlN/GaN MOS-HEMT-based pH sensor on Si substrate: a comparative analysis *IEEE Sens. Lett.* **3** 1–4

[41] Pyo J-Y *et al* 2018 AlGaN/GaN high-electron-mobility transistor pH sensor with extended gate platform *AIP Adv.* **8** 085106

[42] Xing J *et al* 2018 Influence of an integrated quasi-reference electrode on the stability of all-solid-state AlGaN/GaN based pH sensors *J. Appl. Phys.* **124** 034904

[43] Wang L, Li L, Zhang T, Liu X and Ao J-P 2018 Enhanced pH sensitivity of AlGaN/GaN ion-sensitive field effect transistor with Al_2O_3 synthesized by atomic layer deposition *Appl. Surf. Sci.* **427** 1199–202

[44] Parish G *et al* 2019 Role of GaN cap layer for reference electrode free AlGaN/GaN-based pH sensors *Sens. Actuators* B **287** 250–7

[45] Zhang H, Tu J, Yang S, Sheng K and Wang P 2019 Optimization of gate geometry towards high-sensitivity AlGaN/GaN pH sensor *Talanta* **205** 120134

[46] Ding X *et al* 2018 Molecular gated-AlGaN/GaN high electron mobility transistor for pH detection *Analyst* **143** 2784–9

[47] Yates D, Levine S and Healy T 1974 Site-binding model of the electrical double layer at the oxide/water interface *J. Chem. Soc. Faraday Trans.* **1** 1807–18

[48] Levine S and Smith A L 1971 Theory of the differential capacity of the oxide/aqueous electrolyte interface *Discuss. Faraday Soc.* **52** 290

[49] Esposito M G 2018 *Fundamental Modeling and Simulation of AlGaN/GaN High Electron Mobility Transistors pH Sensing Technologies* (Gainesville, FL: University of Florida)

[50] Bayer M, Uhl C and Vogl P 2005 Theoretical study of electrolyte gate AlGaN/GaN field effect transistors *J. Appl. Phys.* **97** 033703

[51] Bergveld P and Sibbald A 1988 *Comprehensive Analytical Chemistry: Analytical and Biomedical Applications of Ion-selective Field-effect Transistors* 23 (Amsterdam: Elsevier)

[52] Anvari R, Spagnoli D, Umana-Membreno G A, Parish G and Nener B 2018 Theoretical study of the influence of surface effects on GaN-based chemical sensors *Appl. Surf. Sci.* **452** 75–86

[53] Sahai N and Sverjensky D A 1997 Evaluation of internally consistent parameters for the triple-layer model by the systematic analysis of oxide surface titration data *Geochim. Cosmochim. Acta.* **61** 2801–26

[54] Sverjensky D A and Fukushi K 2006 Anion adsorption on oxide surfaces: inclusion of the water dipole in modeling the electrostatics of ligand exchange *Environ. Sci. Technol.* **40** 263–71

[55] Das A *et al* 2013 GaN thin film based light addressable potentiometric sensor for pH sensing application *Appl. Phys. Express* **6** 036601

[56] Steinhoff G, Purrucker O, Tanaka M, Stutzmann M and Eickhoff M 2003 $Al_xGa_{1-x}N$ —a new material system for biosensors *Adv. Funct. Mater.* **13** 841–6

[57] Sciullo M, Choudhury M, Patrick E and Law M E 2016 Optimization of GaN-based HEMTs for chemical sensing: a simulation study *ECS Trans.* **75** 259–64

[58] Patrick E, Choudhury M and Law M E 2015 Simulation of the pH sensing capability of an open-gate GaN-based transistor *ECS Trans.* **69** 15–23

[59] Orazem M E 1993 *Mathematical Modeling and Optimization of Liquid-Junction Photovoltaic Cells* (Berkeley: University of California)

[60] Orazem M E and Newman J 1986 Photoelectrochemical devices for solar energy conversion *Modern Aspects of Electrochemistry* (Berlin: Springer) pp 61–112

[61] Orazem M E and Newman J 1984 Mathematical modeling of liquid-junction photovoltaic cells: II. Effect of system parameters on current-potential curves *J. Electrochem. Soc.* **131** 2574

[62] Orazem M E and Newman J 1984 Mathematical modeling of liquid-junction photovoltaic cells *J. Electrochem. Soc.* **131** 2569

[63] Grahame D C 1947 The electrical double layer and the theory of electrocapillarity *Chem. Rev.* **41** 441–501

[64] Guldbrand L, Jönsson B, Wennerström H and Linse P 1984 Electrical double layer forces. A Monte Carlo study *J. Chem. Phys.* **80** 2221–8

[65] Boda D, Chan K-Y and Henderson D 1998 Monte Carlo simulation of an ion–dipole mixture as a model of an electrical double layer *J. Chem. Phys.* **109** 7362–71

[66] Henderson D and Boda D 2009 Insights from theory and simulation on the electrical double layer *Phys. Chem. Chem. Phys.* **11** 3822

[67] Scaramuzza M, Ferrario A, Pasqualotto E and De Toni A 2012 Development of an electrode/electrolyte interface model based on pseudo-distributed elements combining COMSOL, MATLAB and HSPICE *Procedia Chem.* **6** 69–78

[68] Wu S-L, Orazem M E, Tribollet B and Vivier V 2014 The influence of coupled Faradaic and charging currents on impedance spectroscopy *Electrochim. Acta.* **131** 3–12

[69] Wu S-L, Orazem M E, Tribollet B and Vivier V 2015 The impedance response of rotating disk electrodes *J. Electroanal. Chem.* **737** 11–22

[70] Ambacher O *et al* 1999 Two-dimensional electron gases induced by spontaneous and piezoelectric polarization charges in N- and Ga-face AlGaN/GaN heterostructures *J. Appl. Phys.* **85** 3222–33

[71] Mishra U K, Parikh P and Wu Y-F 2002 AlGaN/GaN HEMTs—an overview of device operation and applications *Proc. IEEE* **90** 1022–31

[72] Scharfetter D L and Gummel H K 1969 Large-signal analysis of a silicon Read diode oscillator *IEEE Trans. Electron Devices* **16** 64–77

[73] Farahmand M *et al* 2001 Monte Carlo simulation of electron transport in the III-nitride wurtzite phase materials system: binaries and ternaries *IEEE Trans. Electron Devices* **48** 535–42

[74] Bard A J and Faulkner L R 2001 *Electrochmeical Methods: Fundamentals and Applications* 2nd edn (New York: Wiley)

[75] Newman J and Thomas-Alyea K E 2004 *Electrochemical Systems* 3rd edn (Hoboken, NJ: Wiley)

[76] Levinshtein M E, Rumyantsev S L and Shur M S 2001 *Properties of Advanced Semiconductor Materials: GaN, AlN, InN, BN, SiC, SiGe* (Hoboken, NJ: Wiley)

[77] Fang J-Y, Hsu C-P, Kang Y-W, Fang K C, Ren F and Wang Y-L 2013 Investigation of the current stability of AlGaN/GaN high electron mobility transistors in various liquid/solid interface on the gate area *ECS Trans.* **58** 3–7

IOP Publishing

Wide Bandgap Semiconductor-Based Electronics

Fan Ren and Stephen J Pearton

Chapter 15

The potential and challenges of *in situ* microscopy of electronic devices and materials

Zahabul Islam, Nicholas Glavin and Aman Haque

This chapter briefly overviews the reliability of gallium nitride (GaN)-based high electron mobility transistors (HEMTs) using *in situ* electron microscopy. Due to their superior breakdown voltage, energy bandgap, and electron mobility, AlGaN/GaN HEMTs have drawn tremendous attention in recent years. However, there are still several reliability concerns impeding the realization potential of these devices. This chapter introduces the current trends and reliability issues in GaN HEMT devices in the introduction. Then we briefly describe the material properties and the working principle of these devices. In the subsequent section, we discuss the advanced techniques used to characterize HEMT devices. The focus is shifted to *in situ* transmission electron microscopy (*in situ* TEM) techniques applied on GaN HEMT devices. This includes insight on the preparation of electron-transparent samples, the available commercial holders, the results obtained from *in situ* TEM reliability studies, and their comparison with recently published data. This chapter ends with a discussion of the future directions to fulfill the potential of the *in situ* TEM technique for reliability studies.

15.1 Introduction

High electron mobility transistors (HEMTs) are heterostructure field-effect transistors primarily composed of III–V semiconductor groups. At the heterojunction of a HEMT, carriers are confined by a quantum well due to the band discontinuation at the interface. This acts as a two-dimensional electron gas (2DEG). Gallium arsenide (GaAs), gallium nitride (GaN), and their compounds are two primary categories widely used in HEMTs. The GaAs-based HEMTs were first introduced in 1971 and since then they have drawn tremendous attention in the community due to their high carrier density and mobility [1]. The GaN-based HEMT is the latest addition, introduced by Khan *et al* [2] in 1994. Although the GaAs HEMT has higher electron mobility compared to GaN-based HEMTs, the narrow bandgap of GaAs limits their applications in power electronics. In contrast, GaN has a wide bandgap and a high breakdown voltage, which makes it a better candidate for high-power and high-frequency applications. Figure 15.1 shows a comparison between competing semiconductors [3].

Figure 15.1 indicates that GaN has a one order of magnitude higher breakdown voltage, and both the bandgap and electron saturation velocities are higher compared to the competitor, GaAs. The thermal conductivity of GaN is also higher. These material properties make GaN a potential candidate for next-generation high-power and high-frequency applications. GaN exhibits more radia-tion-tolerance compared to SiC-based devices which makes it an attractive choice for space applications [4–8]. Due to these superior properties compared to other competing semiconductors, the scientific community has seen a surge in GaN-based HEMT applications in wireless communications, radar, high voltage electronics, automotive industries, and space, to name only a few fields.

Over the last decade, both the material quality and device performance of GaN have been improved significantly [9] due to the MOCVD [10] and MBE [11] growth techniques, which facilitate the formation of high-speed conduction channels with good carrier confinement. Even though GaN HEMT devices outperform GaAs-based HEMTs under high-power conditions, they suffer different types of reliability issues, which may lead to device degradation. These are reported for GaN-based HEMTs under different operating conditions [12–18]. For example, thermo-mechanical stress could develop in the device layer due to the electric and thermal fields [19, 20] at high-power conditions. Degradation due to stress can be classified into two categories, namely electrical and physical. Electrical degradation such as current collapse, an increase of leakage current, and nonlinear behavior of trans-conductance could be considered as an outcome of hot electrons, traps [14, 21–24], or inverse piezoelectric effects [25]. However, the exact mechanism of electrical degradation is still debated and may be strongly influenced by details of the operating condition. The literature [21] shows that under DC and RF stressing

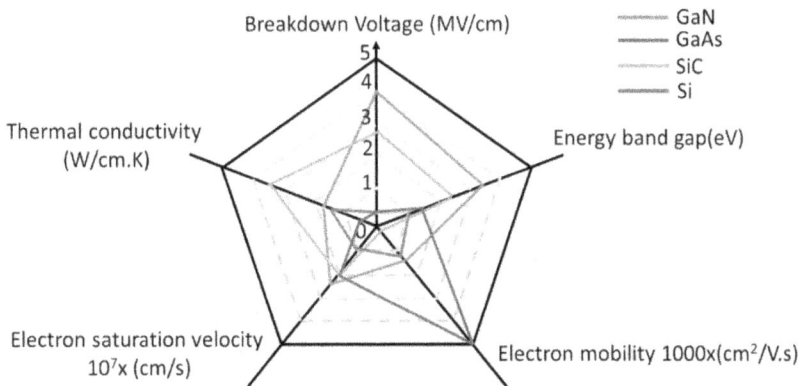

Figure 15.1. Comparison of the properties of competing semiconductors.

conditions, hot electrons may play an important role in the reduction of output current and transconductance, and the increase in channel resistance and knee voltage. The presence of traps and their role in electrical degradation has been also reported [22]. This study reported the formation of traps in the drain–gate region due to the hot electron effects. However, a recent study [25] contradicts this, with traps or hot electrons [26, 27] inducing degradation [20, 21]. Instead of hot electrons or traps, the study shows that above a critical drain–gate voltage crystallographic defects could be generated due to the inverse piezoelectric effects which causes the electrical degradation of the device [28, 29]. Physical, i.e. mechanical, degradation is another type of stress induced degradation which could act as a source of traps and lead to the electrical failure of devices [20, 30]. It is important to note that the devices experience very large residual stress during fabrication processing. Thermomechanical or electromechanical stresses, when added to the residual stress, could trigger the damage phenomena described above.

Although GaN HEMTs exhibit outstanding performance, their long-term reliability issues need to be addressed. The conventional approach is to analyse the electrical characterization data to detect failure signatures. This is followed by post-mortem microscopic visualization using atomic force (AFM), micro-Raman, cathodoluminescence, infrared, thermoreflectance, scanning electron (SEM), and transmission electron (TEM) microscopes. Instead of post-failure analysis, an electron microscope, such as the TEM, could provide useful information on the degradation mechanism of the HEMT during real-time operation. TEM has the capability to provide useful information on structural, chemical, and electrical degradation at high resolution while the device is in operation. Thus, *in situ* TEM will allow researchers to characterize HEMTs both qualitatively and quantitatively. In this chapter, the unique potentials and challenges of *in situ* TEM on the reliability study of GaN-based HEMT devices are described.

15.2 Materials and characterization techniques

15.2.1 The material properties and working principle of AlGaN/GaN HEMTs

GaN and its ternary alloys, such as AlGaN, have three types of structure, namely wurtzite, zinc-blend, and rock-salt. Among these three structures, wurtzite is thermodynamically more stable [31]. Depending on growth methods GaN could have two different types of structure, as shown in figure 15.2. Figure 15.2 shows the GaN structure with both the Ga-face and N-face structure. Interestingly, both the surface properties as well as the polarization properties of these two types of GaN are different [32]. Additionally, the Ga-face structure is more robust in a corrosive environment (acid or base) compared to the N-face structure [32]. It has been reported [33] that growth conditions can directly control the type of atom (Ga or N) faced structure. For example, Ga-face is commonly observed during MOCVD growth and N-face structures during MBE methods [33]. It is well known that a covalent bond exists between Ga and N in the GaN structure; figure 15.2 shows such covalent bonding between Ga and N atoms. Due to the higher electronegativity of N, cationic (Ga^+) and anionic (N^-) characteristics appear in GaN, which acts as a

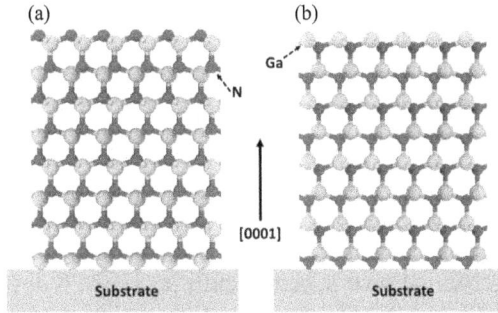

Figure 15.2. Growth direction of GaN showing Ga- and N-face structures.

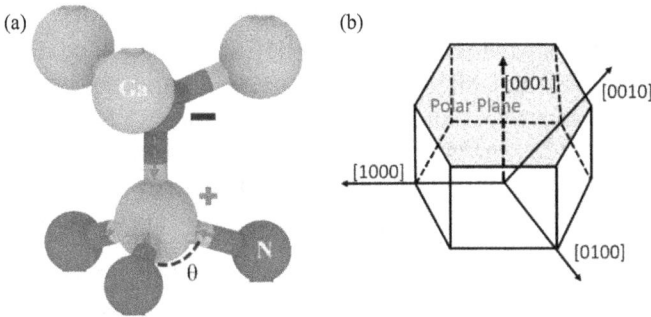

Figure 15.3. (a) Covalent bonding in GaN showing the cation (Ga$^+$) and anion (N$^-$), and (b) the polarization plane along the basal plane, i.e. the [0001] direction in GaN.

source for the microscopic polarization in the GaN structure. This microscopic polarization results in macroscopic polarization in the GaN structure in the absence of inversion symmetry. The basal plane, i.e. [0001], does not have any inversion symmetry and causes spontaneous polarization (P_{SP}) in the GaN structure as shown in figure 15.3(a). Due to this spontaneous polarization along the growth direction, free carriers are generated at the AlGaN/GaN interface. However, other non-polar planes (1$\bar{1}$00 and 11$\bar{2}$0) or semi-polar planes (11$\bar{2}$2) either do not have any polarization or generate a small amount of polarization due to the inversion symmetry. Alongside iconicity, a change in lattice parameter can also develop spontaneous polarization in GaN. For example, the ratio of the equilibrium lattice parameter in GaN is $c/a = 1.633$, whereas c and a indicate the lattice parameter along the [0001] (i.e. the basal plane or growth direction) and [1000] plane (perpendicular to the growth direction, figure 15.3(b)). Deviation from the ideality of the lattice parameter also induces spontaneous polarization in the basal plane [34–41]. Lattice mismatch at the heterostructure interface is a common phenomenon, which causes tension and compression in the different layers of the heterostructure. For example, the lattice parameter of AlGaN is smaller compared to GaN in an AlGaN/GaN heterostructure, which can induce tensile strain in AlGaN and compressive strain in the GaN layer. Due to this additional strain, another type of polarization appears in the heterostructure, which is known as piezoelectric polarization (P_{PE}).

Both spontaneous polarization and piezoelectric polarization are shown in figure 15.4(a)–(d). However, strain in GaN could relax due to its thicker dimension and nullify the piezoelectric polarization in the GaN buffer layer. This scenario has been shown in figure 15.4(e), where net polarization in the AlGaN/GaN hetero-structure is the summation of spontaneous and piezoelectric polarization in the AlGaN layer, and only spontaneous polarization in the GaN layer. This net polarization causes the appearance of an electron gas, also known as a two-dimensional electron gas (2DEG), at the interface of the AlGaN/GaN heterostructure. This 2DEG has been shown schematically in HEMT devices (figure 15.5).

Typically, AlGaN/GaN HEMTs consist of gate, drain, and source terminals as shown in figure 15.5(a). As mentioned earlier, the 2DEG forms at the interface of AlGaN and GaN due to the net polarization. A band diagram of the AlGaN/GaN heterostructure is shown in figure 15.5(b), where E_F and E_C indicate the Fermi level and conduction band at the interface of the HEMT. The bandgap of AlGaN and GaN are different, and this difference in conduction band energy (ΔE_c) causes the bending of the conduction band as shown in figure 15.5(b). Above a critical thickness of AlGaN [42], the conduction band offset forces the energy level to fall below the Fermi level, and thus a triangular-shaped quantum well of electrons (shown by the red dotted circles in figure 15.5(b)) forms at the GaN side of the

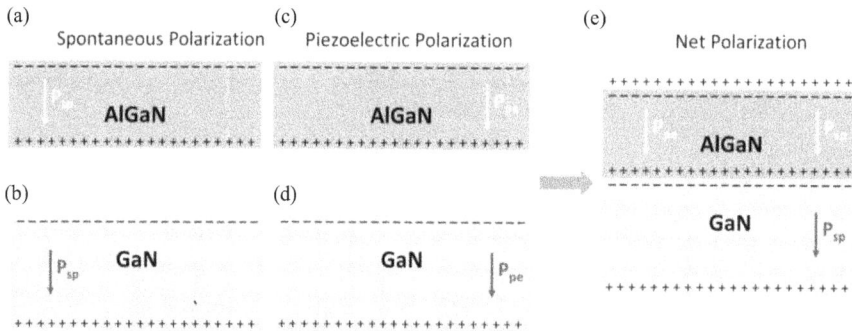

Figure 15.4. Polarization in Ga-face AlGaN/GaN HEMT: (a) spontaneous polarization in AlGaN, (b) spontaneous polarization in GaN, (c) piezoelectric polarization in AlGaN, (d) piezoelectric polarization in GaN, and (e) net polarization in a HEMT.

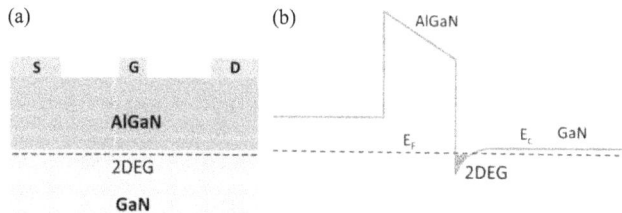

Figure 15.5. (a) A cross-sectional view of GaN HEMT and (b) quantum well-confined 2DEG at the interface of the Ga-face GaN HEMT.

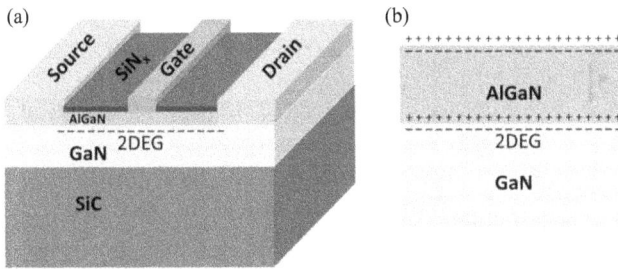

Figure 15.6. Schematic showing an AlGaN/GaN HEMT: (a) 3D view of the HEMT and (b) 2D schematic showing the 2DEG at the interface of AlGaN and GaN.

interface. This electron sheet forms the channel of the HEMT device and is known as two-dimensional electron gas (2DEG). To identify the source of 2DEG several studies have been conducted [25, 43–46]. It can be concluded that alongside superior material properties, both conduction band offset and polarization properties make GaN HEMTs suitable for high-power and high-frequency applications.

A simplified 3D view of the AlGaN/GaN structure and corresponding 2DEG gas have are shown in figure 15.6(a) and 15.6(b), respectively. The GaN HEMT structure is composed of multilayer materials, as shown in figure 15.6(a), which contains the substrate, nucleation layer, GaN buffer layer, spacer, AlGaN layer, cap layer, and passivation layer. The following paragraphs briefly describe the functions, available materials, and properties of each layer:

Substrate. GaN HEMTs are usually fabricated on silicon (Si), silicon carbide (SiC), sapphire (Al$_2$O$_3$), etc. Current technology limits the use of the GaN substrate due to its high synthesis cost [47]. SiC (in light brown in figure 15.6(a)) is the preferred substrate due to its superior thermal conductivity, low lattice mismatch, thermal expansion coefficient, and excellent mechanical properties. However, SiC is expensive compared to Si and sapphire. Thus, silicon on a polycrystalline SiC substrate could be another viable cost-effective solution for high-power and high-frequency applications [48].

Nucleation layer. The nucleation layer is a very thin layer usually put above the substrate layer to mitigate the lattice mismatch between the substrate and GaN buffer layer. GaN, AlN, and AlGaN are the preferred materials. High resistivity of the nucleation layer is desirable to avoid any leakage current through the substrate. The nucleation layer can also reduce threading dislocations in the GaN buffer layer. In practice, low-temperature GaN is used as a nucleation layer to avoid dislocation formation in the GaN buffer layer.

GaN buffer layer. A high-quality GaN buffer layer (in cyan in figure 15.6(a)) with low defect density and high resistivity is desirable for the fabrication of high-quality GaN HEMT devices. A high-quality surface of GaN is also necessary to have a good interface which potentially facilitates electron mobility and carrier confinement at the interface. The thickness of the buffer layer can affect 2DEG at the interface. Thus increased buffer layer thickness can prevent dislocation formation and enhanced 2DEG mobility [15, 49–52].

Spacer layer. A spacer layer is introduced between the AlGaN and GaN layers to improve the 2DEG mobility. Spacer materials are usually made of AlN or non-intentionally doped AlGaN with the same alloy composition that has already been used in the AlGaN. The spacer layer can reduce coulombic scattering at the interface and thus significantly increase the 2DEG mobility.

Cap layer. A cap layer made of GaN is introduced between the gate and AlGaN layer to prevent gate leakage and to enhance the breakdown voltage. GaN, AlGaN, or a multilayer of GaN/AlGaN/GaN could be used as a cap layer.

AlGaN layer. AlGaN is the primary source of 2DEG in the GaN HEMT. Both the thickness and Al concentration control the sheet carrier density. Below a certain thickness (known as the critical thickness) no 2DEG forms at the interface. Again, with the increment of aluminum concentration carrier density also increases. However, Al alloying concentration is limited due to the de-alloying effects which can have detrimental effects on the film's quality.

Passivation layer. The passivation layer is introduced in the GaN HEMT to avoid the neutralization of surface states by trapped electrons. It also prevents the formation of a virtual gate. Dielectric materials such as Si_3N_4, SiO_2, MgO, NiO, SiON, etc are used as a passivation layer. A passivated device has advantages over a non-passivated device. For example, a device with a passivation layer shows higher transconductance, drain current output, output power density, and drain–gate breakdown voltage, to name only a few effects.

15.2.2 Material and device characterization using the *in situ* TEM technique

Electron microscopes are widely used for post-failure analysis of electronic devices [53–63]. Nanoscale materials and devices can be characterized at high resolution using SEM and TEM. Such microscopes can be equipped with energy dispersive spectroscopy (EDS), electron energy loss spectroscopy (EELS), and selected area electron diffraction (SAED) and can be powerful tools for device characterization. However, until recently [64–66], electron microscopes were used only for post-failure analysis. Thus, their true potentials were mostly unexploited as post-failure analysis cannot exactly pinpoint the fundamental mechanisms behind the nucleation and evolution of damage. Thus, the researcher must employ both theory and intuition to connect the gap between the characterization data and microscopy. This critical roadblock could be solved effectively with *in situ* TEM, which can provide both qualitative and quantitative data during the real-time operation of electronic devices. The procedure is challenging for various reasons, such as specimen preparation, specimen holders with biasing, and the dimensional or boundary condition differences between electron-transparent and bulk devices. These challenges have been resolved by state-of-the-art specimen holders (figure 15.7). This chapter therefore focuses more on specimen preparation, experimental set-up for three-point biasing, and scaling of device physics. It is important to note that the latest technologies allow *in situ* TEM experiments with two point electrical biasing, mechanical straining, ion irradiation, thermal heating/cooling, etc, stimuli, but *in situ* TEM studies involving transistors have been non-existent until recently [64, 65, 67].

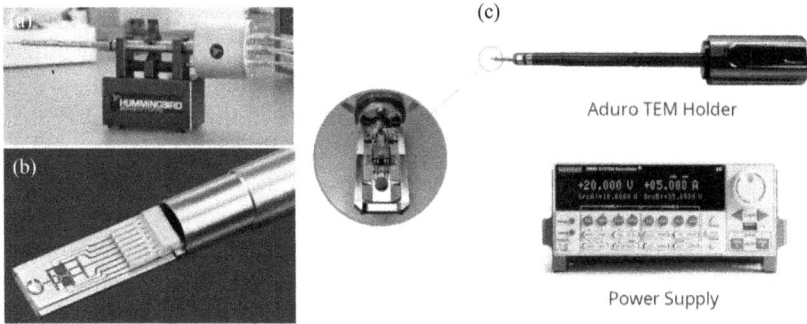

Figure 15.7. *In situ* TEM holder for electrical biasing experiments. (a) and (b) Hummingbird Scientific. (Reproduced with permission from [68]. Copyright 2004 Hummingbird Scientific). (c) Protochip. (Reproduced with permission from [69]. Copyright 2002 Protochip).

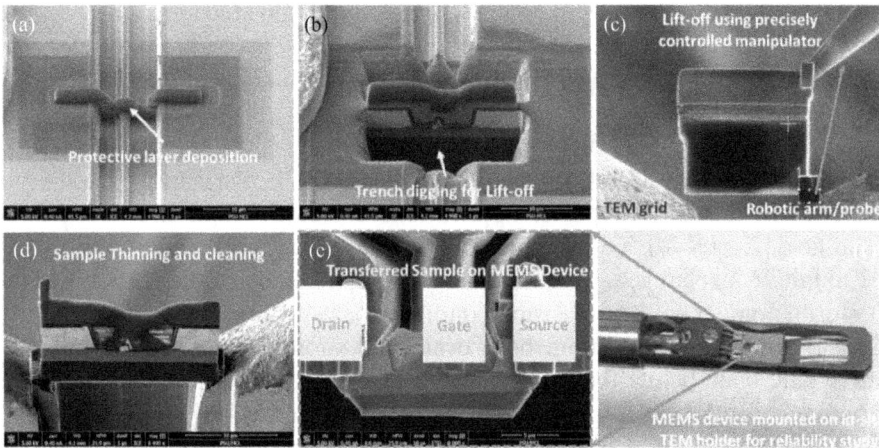

Figure 15.8. Details of the GaN HEMT sample preparation technique using FIB for an *in situ* TEM reliability study.

As mentioned previously, *in situ* TEM experiments are challenging due to the compactness of the TEM chamber, the requirements of electron-transparent samples, and their transfer to the testing chip or specimen holder. To prepare electron-transparent (nominally 100 nm thick) specimens, a coupon from a GaN HEMT device can be lifted out using a focused ion beam (FIB) as shown in figure 15.8. The sample preparation requires three important steps: (i) electron-transparent 100 nm thin sample preparation (figure 15.8(a)–(d)); (ii) sample transfer on a micro-electromechanical system (MEMS) device (figure 15.8(e)) [70, 71]; and (iii) wire bonding of an MEMS device [72] on a TEM chip carrier, and careful placement of a chip on a TEM holder with biasing capability (figure 15.8(f)). The FIB-based coupon lift out involves deposition of a protective layer to define a thicker section in the GaN HEMT device and subsequent trench cutting (figure 15.8(a) and (b)). The coupon is lifted off using a manipulator and placed on a TEM grid (figure 15.8(c)). The thicker coupon requires a series of ion milling steps to be thinned down to electron transparency (figure 15.8(d)). High current such as 21 nA can be used to lift out the sample from the bulk HEMT (figure 15.8(b)), however, thinning down requires much smaller current in the range of 0.79 nA–34 pA (values of currents are based on a 4 mm working distance). The thickness of the sample needs to be tracked at regular intervals during the thinning down process. Additionally, both accelerating voltage and current require adjustment depending on the sample thickness during the milling. After having an electron-transparent sample, the final step requires a low accelerating voltage and current to transfer the electron-transparent sample from the TEM grid to the MEMS device. A low accelerating voltage such as 5 kV can be used without introducing any beam damage and re-deposition on the sample. Once the sample is transferred on the MEMS device, the sample is ready for the loading on the *in situ* TEM holder (figure 15.8(f)) and further electrical characterization using a transmission electron microscope (TEM) equipped with EDS, EELS, SAED, bright field (BF), and dark field (DF) imaging capability.

15.3 AlGaN/GaN HEMT reliability study

15.3.1 Degradation in the GaN HEMT

Although the GaN-based HEMT exhibits outstanding performance, reliability is a major concern during high-power applications due to the high electrical field and thermal field [25, 73–77]. To overcome these issues an in-depth understanding of the physical failure mechanism is required. The reliability study of the GaN HEMT can be performed under different biasing conditions, as shown in figure 15.9.

Case I: Under this condition, both drain and source are grounded and a high negative gate bias (below the threshold voltage) is applied (point I in figure 15.9). The outcome is a lower output current, as shown in figure 15.9 by point I.

Case II: The device is under high current and low field conditions. The gate voltage at this condition is above the threshold voltage, the drain voltage is positive, and the source is grounded (point II in figure 15.9).

Case III: At this biasing condition the device is under high field and high current conditions, as shown by point III in figure 15.9. The drain voltage is sufficiently high and the gate voltage is above the threshold voltage.

Case IV: As shown in figure 15.9, the device is under high field but low current conditions. The drain voltage is sufficiently high, however the gate voltage is below the threshold voltage. Thus, the output current is sufficiently low due to the off-state biasing mode.

In all the above-mentioned cases, the device will suffer electrical degradation. However, due to the low current condition self-heating might be negligible in case I and case IV. Depending on the operating conditions different types of degradation could take place in a GaN HEMT, as shown in figure 15.10. For example, a hot electron could generate traps in the device layer and thermally activated processes such as degradation of the Ohmic contact and lattice defects can appear during high-power operating conditions. The following section briefly describes reliability issues in GaN HEMT at different operating conditions.

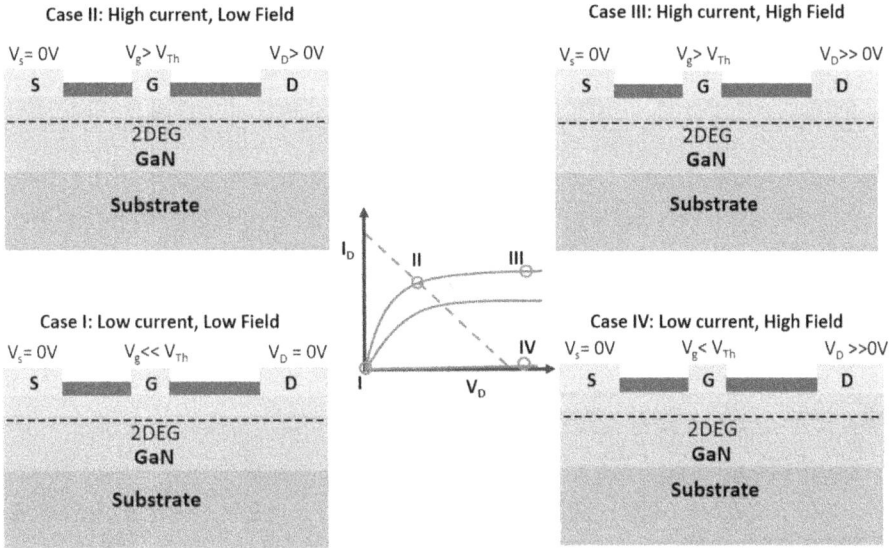

Figure 15.9. Schematic showing different biasing conditions of the GaN HEMT device.

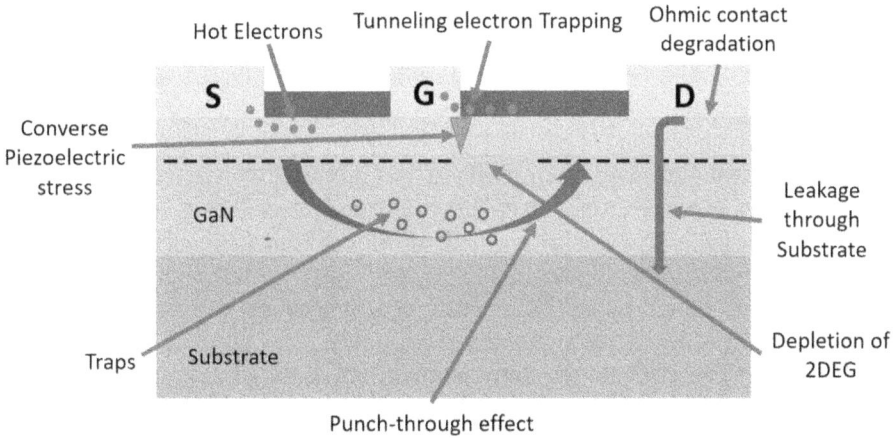

Figure 15.10. Schematic showing reliability issues in GaN HEMT.

15.3.1.1 Electrical stress

GaN HEMTs can be operated under different biasing conditions, such as high power, on-state, and off-state. In the high-power mode, the device is biased at high drain voltage to obtain high drain current, while the gate voltage is kept above the threshold voltage. During on-state biasing the drain current reaches the saturation value and the device runs above the threshold voltage. Additionally, during off-state biasing, the drain voltage is sufficiently high while the drain current is low due to the large negative bias (below threshold voltage) applied at the gate. It is well known that in a GaN HEMT the AlGaN layer is under tensile stress while the GaN layer experiences compressive stress. This stress limits the thickness of the grown AlGaN layer. Again, GaN and its ternary alloys are piezoelectric materials in nature. Thus, when a high field is applied to the device they suffer from converse piezoelectric stress [14, 25, 73, 78]. Under a high biasing condition, these stresses are enough to induce cracks and voids in the device layer, and the device eventually fails.

15.3.1.2 Thermal stress

The thermal field can significantly affect the device threshold voltage shift and output current reduction. Under high DC stressing conditions, the formation of an additional interfacial layer has been reported which might increase the Schottky barrier height [79, 80] at the gate contact. Due to the high temperature increase in the device the inter-diffusion of metallic components can take place, which increases contact resistance [80, 81]. Sublimation of GaN can also take place due to the high thermal field [66].

15.3.1.3 Hot electron effects

The output current during high-power, on-state operation can energize the electrons (also known as hot electrons) enough to induce traps in the channel, barrier, or at the interface. Thus, hot electron induced traps can degrade device performance by collapsing the drain current.

15.3.1.4 Trap effects

Traps are crystal defects that are generated by a high electrical field, converse piezoelectric stress, hot electrons, and tensile strain in the AlGaN layer at the gate edge of the drain–gate side [24, 82, 83]. Both current collapse and threshold voltage shifts are reported outcomes of trap effects in GaN HEMT devices. Traps can be generated due to nitrogen (N) vacancies, damage induced by plasma processing, dislocation at the interface due to residual strain, hot electrons, piezoelectric effects, etc. Traps can create a virtual gate in the device by capturing tunneling electrons at the gate edge of the drain–gate side. The formation of this virtual gate further reduces the output drain current by increasing channel resistance and causes current collapse. A passivation layer can mitigate the detrimental effect of the virtual gate, and recently significant efforts have been made in the design and processing of the passivation layer [84]. Trap induced current collapse can be prevented by introducing a field plate at the gate [85]. A field plate can also increase the breakdown voltage of the device. Current collapse prevention can be achieved by minimizing tunneling electrons from the gate to the channel layer. The field plate can also resist the occupation of the trapped electron at the surface, thus preventing the current collapse.

15.3.2 AlGaN/GaN HEMT characterization techniques

Typical reliability studies of HEMT involve voltage/current stressing of the device followed by microscopy of the failed device to investigate structural and chemical degradation. *In situ* TEM studies of HEMTs have appeared in the literature only recently [65, 70, 71]. In this section, we will discuss the trends and issues in GaN HEMT reliability studies using both post-failure and real-time monitoring.

Electroluminescence. In a HEMT device, an electron–hole pair can recombine under a high electric field due to a hot electron and subsequently photoemission can take place. A detector with high gain can measure this photoemission and can further map the hot electron current distribution in the device. This technique is known as electroluminescence and can be used to investigate dislocation, other types of defects, and degradation of the Schottky interface. A typical electroluminescence study is shown in figure 15.11 [86].

Cathodoluminescence. In a direct bandgap semiconductor, an electron beam can induce photoemission due to the recombination of an electron–hole pair. The local band structure and charge carrier diffusion rate is directly related to the emitted photon energy. Thus, the cathodoluminescence technique can map temperature [88] and identify dislocations and other types of structural defects [89, 90].

Raman spectroscopy. At high-power conditions, the thermal field can degrade device performance. Thus, thermal characterization is important from a device reliability perspective. It is well known that the Raman peak shift is a function of temperature. Thus, a correlation can be developed between the known temperature and Raman shift, which is further used to determine the local temperature at different operating conditions of HEMT. Compared to infrared (IR) measurements, micro-Raman a has resolution smaller than 1 μm [91]. Thus, micro-Raman thermometry could provide useful thermal information of a HEMT with gate and drain dimensions in the micrometer range. Figure 15.12 shows temperature measurements using micro-Raman techniques reproduced from the literature [91].

Thermoreflectance. Reflectance of a material surface is a function of temperature. In a thermoreflectance microscope, the change in reflectance due to the temperature change is measured. The thermoreflectance coefficient which determines the change in reflectance is a basic material property and depends on the material, illumination wavelength, and ambient conditions. Thus, using this property thermal mapping can be performed in a HEMT device [92, 93].

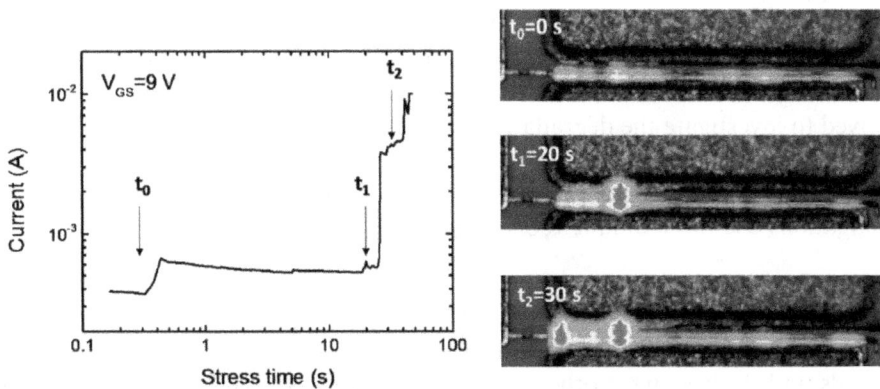

Figure 15.11. Current variation measured during a constant voltage experiment (VGS = 9 V). Right: False color electroluminescence (EL) pattern reporting the distribution of light emitted by the gate junction during stress, at different time intervals. Reproduced with permission from [87] (open access).

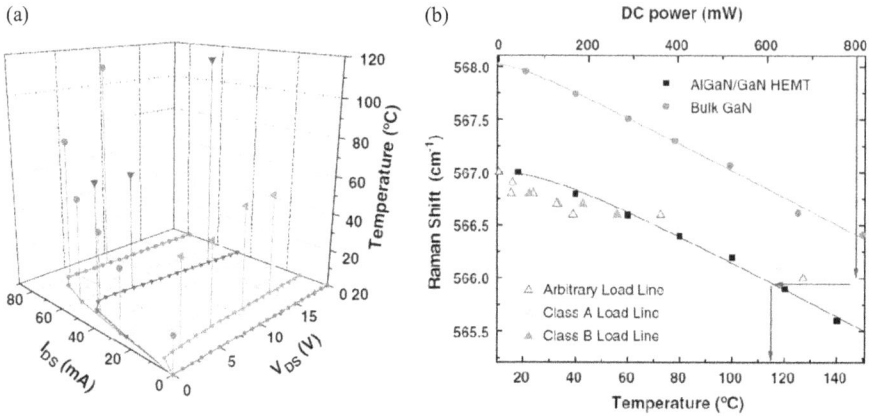

Figure 15.12. (a) 3D plot showing channel temperature and device characteristics. (b) Active heating (self-heating) temperature at various load lines. Raman shift versus temperature plot of the FS–HVPE GaN and the HEMT device layer under passive heating. Top-left: Raman shift as a function of active heating. Bottom-left: Raman shift as a function of passive heating. The device temperature can be obtained by comparing the two top curves. The FS–HVPE results are shown to illustrate the stress introduced by the heterostructure. Reproduced with permission from [91]. Copyright 2006 Elsevier.

Laser scanning electron microscope. The internal structure and interfaces of GaN HEMT can be investigated using a laser scanning electron microscope (LSM). This technique allows the flatness of the metal–semiconductor interface to be measured [94]. Due to the transparency of the SiC substrate, the interface of the HEMT device can be tracked from the backside using the LSM technique.

Electron microscope (EM). Electron microscopes provide atomic resolution images that can be deployed to investigate electronic materials and devices [52, 95, 96]. An electron microscope can be equipped with EDS, EELS, SAED, BF, and DF which make an EM an excellent analytical tool for electronic materials and device characterization. Both scanning and transmission electron microscopes (SEM and TEM) could be used as a quantitative as well as qualitative tools. Figures 15.13(a) and 15.13(b) show SEM images of failed devices. These devices were stressed for different time durations and the corresponding crack formation is shown in figures 15.13(a) and 15.13(b). Additionally, TEM has also been widely deployed to investigate the degradation mechanism of the GaN HEMT as shown in figure 15.13(c). A cross-sectional image of HEMT shows electrochemical oxidation during on-state stressing experiments (figure 15.13(c)). However, most of the investigations use these unique tools only for post-failure analysis.

15.3.3 GaN HEMT reliability study using an *in situ* TEM study

Electron microscopes (EMs) are widely used to study the reliability of electronic materials and devices due to their analytical as well as atomic resolution imaging capability. Most of the reliability studies are performed outside the EM and failed specimens are investigated to figure out the failure mechanism. Thus, the true potential of EMs for device characterization requires further attention. However,

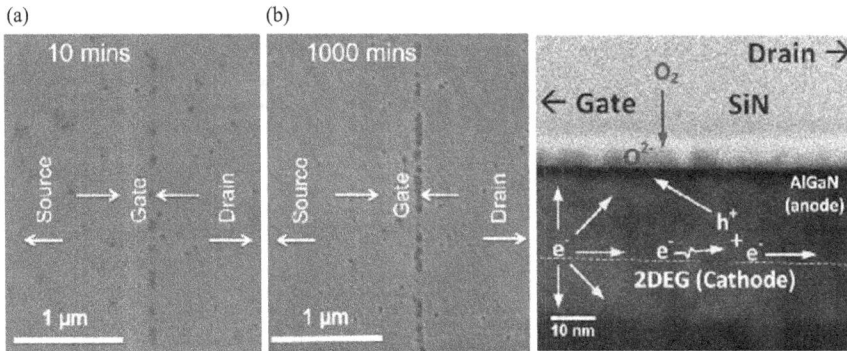

Figure 15.13. SEM images of the pits after stressing at $V_{DGstress}$ = 50 V for (a) 10 min and (b) 1000 min [96], with permission of AIP Publishing, and (c) TEM images of the electrochemical oxidation mechanism of AlGaN under the on-state condition Reproduced with permission from [95]. Copyright 2016 Elsevier.

recently an *in situ* TEM study has been employed to investigate GaN HEMTs at different biasing conditions [65, 70, 71] to capture the degradation mechanism in real time. Determination of the exact failure modes using *ex situ* techniques and post-failure analysis is strenuous, whereas *in situ* microscopy could be a great help to identify the failure modes in GaN HEMT devices due to its real-time visualization and quantitative capability. It is well known that the degradation of GaN HEMT devices is a functions of the electrical, mechanical, and thermal fields. In section 15.3.1 it has been shown that a GaN HEMT can be operated in four different modes. Among them case II and case III suffer from both high electrical and high thermal fields, known as the 'on-state' mode, whereas case I and case IV are known as 'off-state' mode operation and experience only high electrical fields. Recent studies [65, 66, 70] have considered both modes of operation. Again, due to their high-power effectiveness, radiation-tolerant properties, and compact size, GaN HEMTs are an attractive choice for space applications. However, outer-space devices undergo heavy ion irradiation which could cause significant lattice damage in such devices [97]. Thus, it is necessary to investigate irradiation-induced defect generation and their effects on device performance. The present study also discusses the effects of irradiation-induced damage on GaN HEMT performance using an *in situ* TEM study. In the following sections, we will briefly discuss the reliability issues of GaN HEMTs both at the 'on'- and 'off'-state modes using *in situ* TEM techniques.

15.3.3.1 On-state failure mechanism

During on-state mode operation the electrical field and current are high enough to induce Joule heating, thus the thermal field is unavoidable. In this high electrical and thermal field, self-heating and hot electron effects are dominant and cause failure in the device [98, 99]. Figure 15.14 shows such a reliability study of GaN HEMT devices using *in situ* TEM. A BF TEM image of the GaN HEMT sample before on-state reliability testing is shown in figure 15.14. A significant amount of bend contour in the device, as well as the substrate layer, is observed in figure 15.14(a).

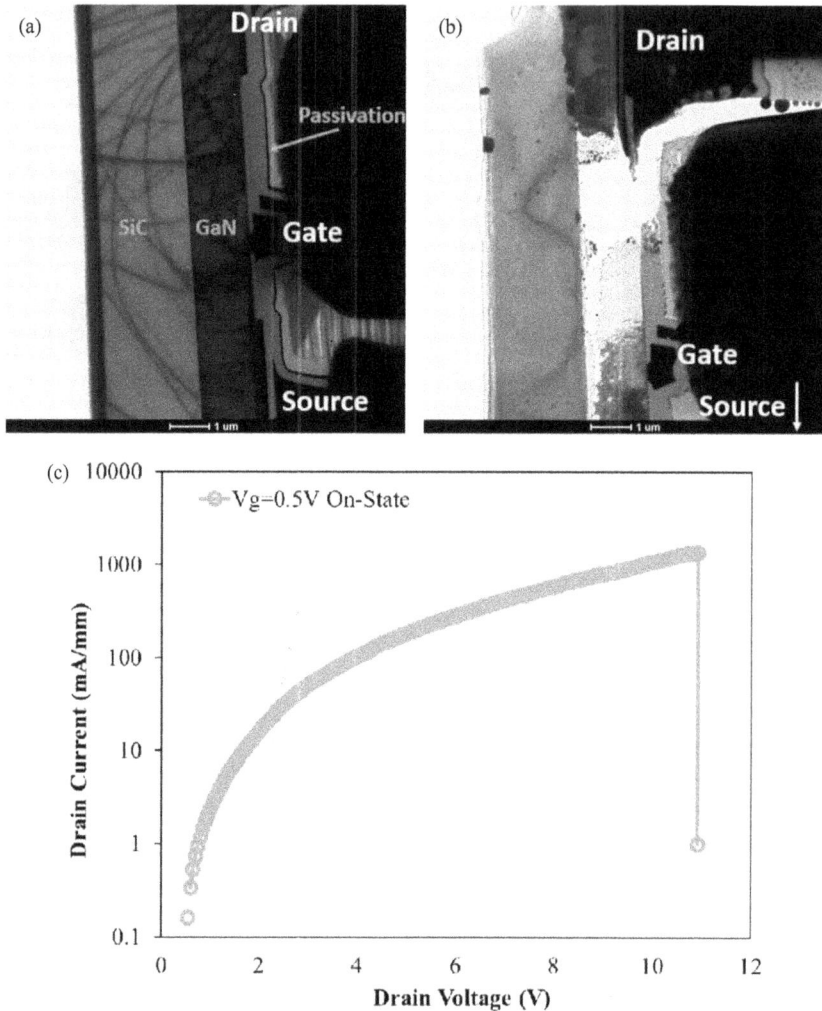

Figure 15.14. *In situ* TEM reliability study of a GaN HEMT: (a) electron-transparent sample before current stressing, (b) device after failure, and (c) output characteristic curve obtained from current stressing.

For a reliability study, electrical connections are made at the gate, drain, and source pad during sample preparation (figure 15.8). FIB deposited carbon is used to accomplish this goal, which further introduces stress in the sample. As the sample is in a plane strain condition, thus it generates a bend contour in the sample due to the residual stress. The threshold voltage of this device is −3.1 V and 1 V gate bias is applied to make sure that the device is at the on-state mode. Figure 15.14(b) shows a failed device after on-state characterization and the corresponding output characteristic (i.e. drain current versus drain voltage) is shown in figure 15.14(c). The device fails at 10.8 V drain bias as shown in figure 15.14(c), and the current density at the failure point is in good agreement with the reported value [100]. As mentioned previously, during on-state biasing the drain current is sufficiently high and Joule heating plays a dominant role in device degradation. This is reflected in the present study as shown in figure 15.14(b). At 10.8 V drain biasing, the current density is high enough to sublimate the buffer layers, as shown in figure 15.14(b). A metal–semiconductor diffusion is also observed near the drain electrode, which indicates that the thermal field is very high during on-state biasing. A similar incidence has been reported in the literature [66] during on-state biasing.

Figure 15.15(a) shows microstructural changes observed due to the thermal degradation in the device layer. A high thermal field during on-state operation is confirmed by the amorphization of the passivation layer (figure 15.15(b)) which requires very high temperature as the recrystallization temperature of SiN_x is in the range of 1200 °C–1400 °C [101]. The diffused ring pattern (figure 15.15(c)) obtained from SAED clearly indicates the amorphization and polycrystalline nature of the buffer layer. This structural change could be partially attributed to the very high thermal field accompanied by high current density.

Figure 15.16(a) presents hot electron induced structural degradation in the buffer layer. Under high electrical biasing conditions, an electron could achieve high kinetic energy (known as a hot electron) which can create traps and lattice defects in the buffer layer. With the progression of time these defects could percolate and induce structural defects. This phenomenon is evident in figure 15.16(a), where we notice interface breaching at the GaN/SiC substrate. Weak bonding at the interface of GaN and SiC makes it vulnerable to hot electron induced defects, thus the interface easily separates during high drain bias. Again, a high thermal field can evaporate GaN and small crystallites of gallium nanoparticles may appear, as shown in figure 15.16(c). Due to the lower binding energy of the N atom compared to the Ga atom [102, 103], the N atom may diffuse out of GaN and facilitate Ga nanoparticle formation.

The true potential of GaN HEMT can be realized by capturing the failure incidence during the operation as shown in figure 15.17. Unlike post-failure analysis, figure 15.17 shows failure initiation and propagation during real-time operation, thus it can pinpoint the exact failure mechanism. Due to the high biasing condition associated with the high thermal field, a metal pool forms near the drain area during

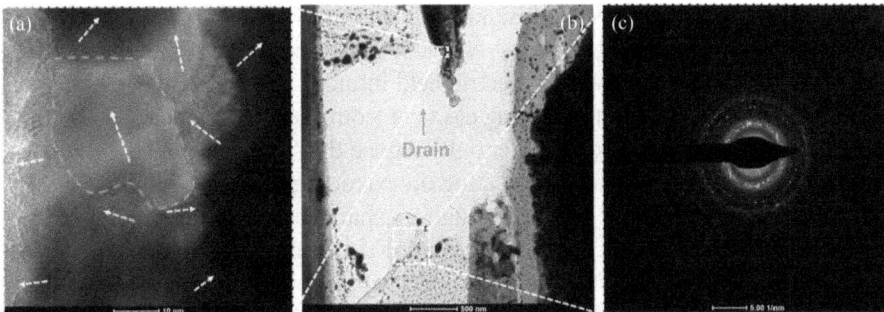

Figure 15.15. (a) Recrystallization of the passivation layer, (b) evaporation of the buffer layer due to a high thermal field, and (c) diffraction pattern indicating the transformation of GaN from the crystalline to amorphous state.

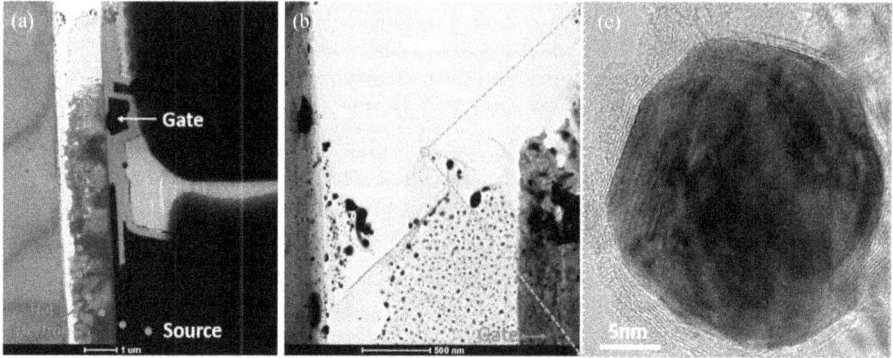

Figure 15.16. (a) Hot electron induced failure at the source side, (b) evaporation of the buffer layer and formation of small crystallites (nanoparticles) due to the high thermal field, and (c) high-resolution image of a spherical crystallite.

the on-state operation at 10.8 V (figure 15.17(a)), and this metal pool quickly diffuses through the GaN buffer layer, as shown in figures 15.17(b) and 15.17(c). Subsequent diffusion and evaporation continue, as shown in figures 15.17(d)–(f). After 12 s from the diffusion initiation, further diffusion starts in the substrate layer, as shown in figure 15.17(g), and propagates very rapidly through the substrate layer (figures 15.17(h) and 15.17(i)). It is evident that a high thermal field originating from a high drain current is responsible for the degradation during on-state operation. During on-state operation, as the gate leakage is minimum no apparent degradation is observed at the gate (figure 15.17). However, a hot electron ejected from the source during high-power operation can breach the GaN–SiC interface as observed in figure 15.17(i) near the source side.

15.3.3.2 Off-state failure mechanism

During off-state operation the gate voltage is kept below the threshold voltage of the device, and the corresponding current is significantly lower compared to an on-state mode operation. Details of the off-state biasing condition have been shown in the previous section (section 15.3.1, figure 15.9) by case I and case IV. Due to the low output current thermal effects are insignificant during off-state operation. However, a high electric field at the gate can create converse piezoelectric stress (figure 15.10), thus the device may fail due to this high electric field induced piezoelectric stress. Alongside converse piezoelectric stress, a tunneling electron from the gate to device layer could be captured at the surface, and could potentially reduce the surface state. This reduction in surface state changes the 2DEG by creating an extending depletion region. Due to this extending depletion region, a punch-through mechanism may dominate device degradation in off-state conditions. A hot electron could be another viable source to induce degradation during off-state biasing. A hot electron acts as a source of traps that can percolate over time and eventually the device fails. Recently [70] an *in situ* TEM reliability study has captured these failure modes during real-time operation. Such an *in situ* TEM experiment has been depicted in figure 15.18.

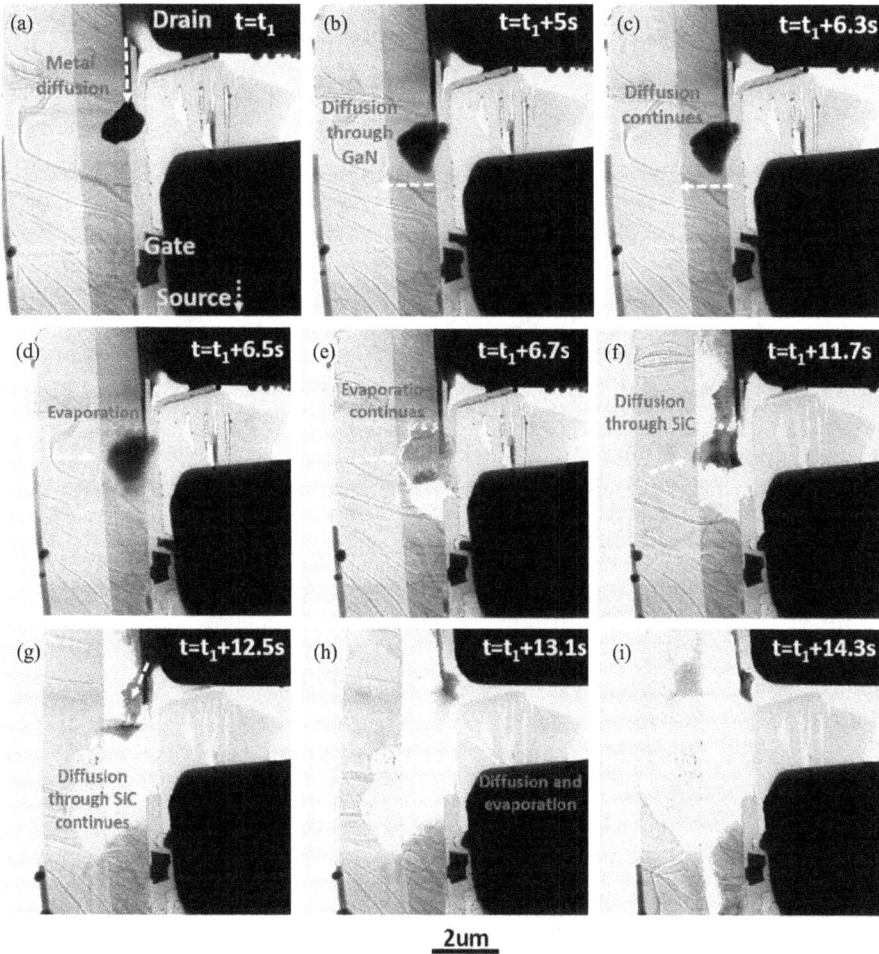

Figure 15.17. Advantages of an *in situ* TEM reliability testing showing monitoring of real-time device failure.

During off-state biasing, failure through the punch-through mechanism can take place due to the formation of an extended depletion layer (figure 15.10) at relatively high drain voltage. A tunneling electron from the gate to buffer layer can be captured at the surface of the device, which further reduces surface states. The reduction of surface states acts as a virtual gate and an extended depletion layer is formed near the gate. This extended depletion layer reduces the 2DEG and discontinuation in 2DEG gas can be realized near the gate area, as shown in section 15.3.1 (figure 15.10). With the increment of drain bias, the electron gains sufficiently high kinetic energy and finds an alternative pathway through the buffer layer, as shown by the red arrow in figure 15.18(b) due to the discontinuation in 2DEG. Thus, at relatively high drain voltage, the punch-through mechanism may initiate failure between the drain and gate region of the buffer layer, as shown in figure 15.18.

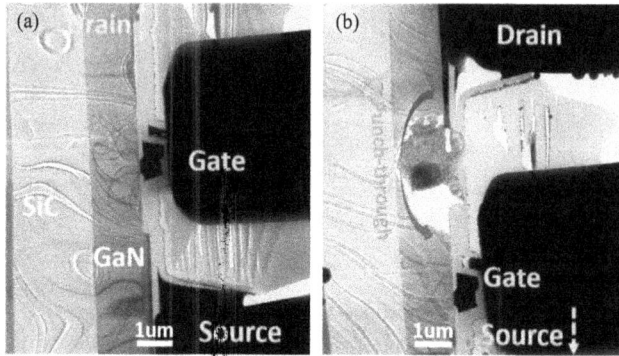

Figure 15.18. GaN HEMT reliability study: (a) before biasing and (b) after failure.

It is well known that both piezoelectric stress [28, 29] and hot electrons [20, 21, 26] play a dominant role in device degradation during off-state stressing. A typical off-state stressing and associated output characteristic curve is shown in figure 15.19. A significant level of degradation is observed at the gate electrode as shown in figure 15.19(b). The threshold voltage of this device is −3.1 V and during this study the gate voltage was kept constant at −5 V, which ensures off-state operating conditions. This device degrades at 23.5 V, as shown in figure 15.19(c). At this high drain voltage the piezoelectric stress is high enough to initiate physical degradation in the device layer. Degradation in the passivation layer both on the drain and source side is also noticeable in figure 15.19(b). At a high drain bias, lateral surface leakage may take place due to the surface hopping electron, as shown in figure 15.19(b). The effect of this surface hopping electron is reflected in the passivation layer degradation. Again, electrons from the source electrode may be injected at high drain bias, which can also degrade the source passivation layer accompanied by piezoelectric stress.

15.3.3.3 Failure mechanism of irradiated GaN HEMT in the off-state

As mentioned earlier, GaN HEMTs are an attractive choice for space applications. However, it also experiences a harsh environment at high altitudes due to energetic-particle irradiation. This energetic-particle irradiation can cause significant damage to the device layer. For example, vacancies, interstitials, and dislocations might appear due to the energetic-particle interactions with the device [97]. Thus, understanding of the irradiation damage and their effects on device performance is a vital issue. In this respect, *in situ* TEM experiments can provide useful information. Such a study is shown in figure 15.20(a)–(c). Irradiation can induce defects in the device layer, as shown in figure 15.20 by the yellow arrows and dotted box.

Figure 15.21(a) shows a TEM BF image of a HEMT after irradiation. The pink arrow shows a defective area in the device layer due to the irradiation damage before biasing. At 10 V drain bias, this area further degrades, as shown in figure 15.21(b). The change in contrast near the gate is also observed due to the high gate leakage, as marked by the yellow dashed rectangular region (figures 15.21(a) and (b)). At 10.2 V

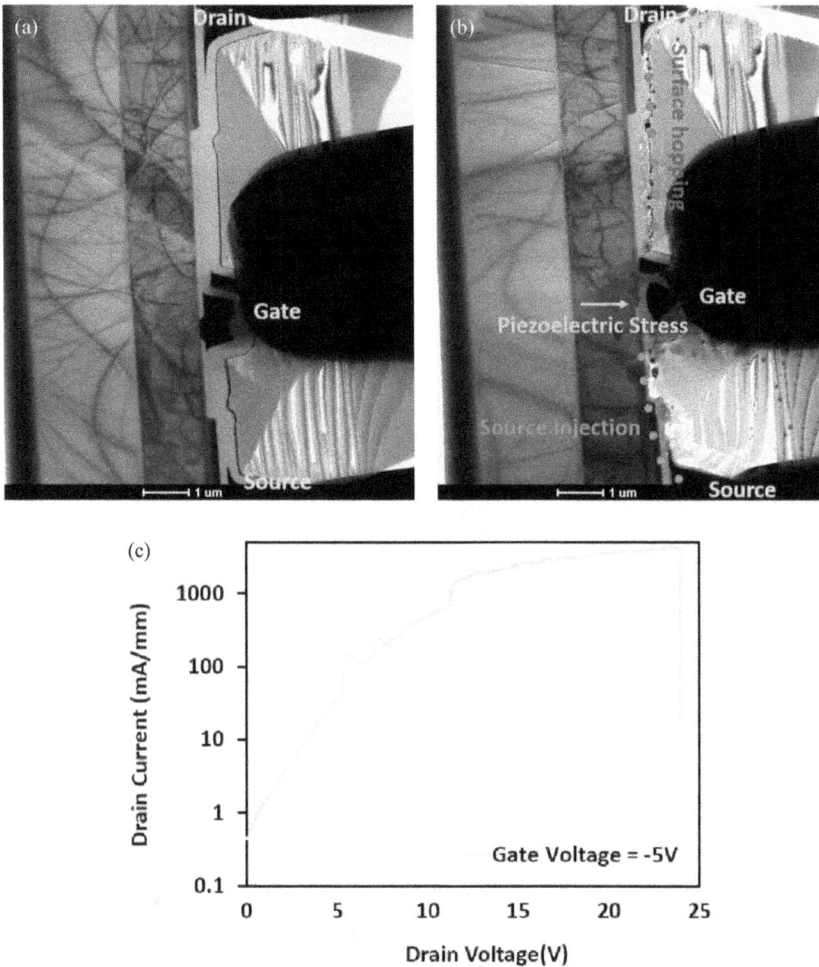

Figure 15.19. Direct evidence of surface hopping and source injection-induced failure in GaN HEMTs: (a) pristine specimen, (b) after failure, and (c) output response of a GaN HEMT in the off-state condition.

the device fails due to the high gate leakage current, which further creates permanent damage in the buffer layer near the gate area (figure 15.21(c)). Breaching between the passivation layer and the buffer layer is also visible (figure 15.21(c)). This could be attributed to the defect formation and surface roughness increment due to the ion irradiation at the interface [28, 29, 104, 105].

EDS analysis is a unique advantage of *in situ* TEM experiments which can aid researchers to understand chemical transformation during the operation of HEMT devices. Such an *in situ* study has been shown in figure 15.22. Figures 15.22(a)–(c) and 15.22(d)–(f) show EDS mapping before and after off-state stressing of the device, respectively. EDS mapping allows us to track chemical elements such as Ga and N atom (or any other elements) diffusion during the current/voltage stressing experiments. Due to the high leakage current N atoms can diffuse out from

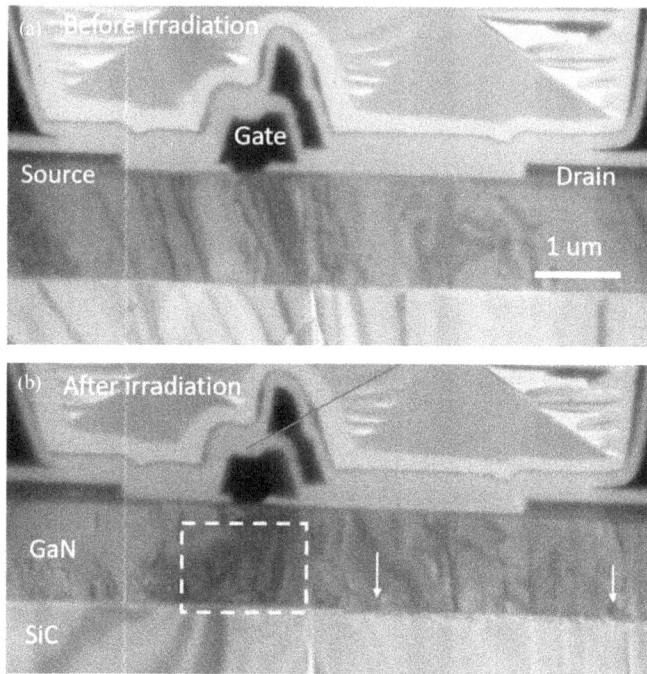

Figure 15.20. Irradiation-induced defects generation in the device layer. Reproduced with permission from [65]. Copyright 2019 Elsevier.

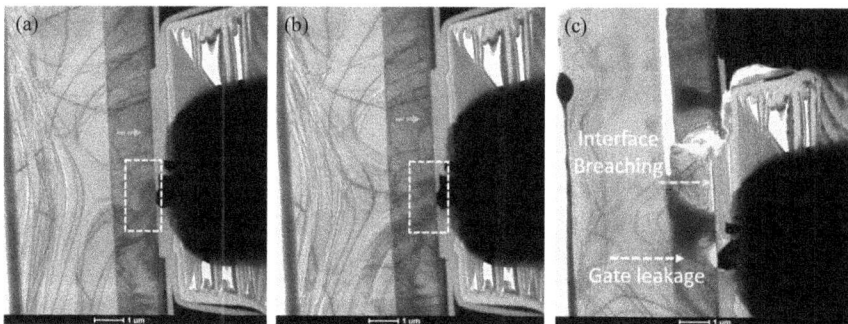

Figure 15.21. Off-state biasing of GaN HEMTs irradiated at 45 dpa: (a) before biasing, (b) biasing at a drain voltage of 10 V, and (c) failure at 10.2 V. Modified and reproduced with permission from [65]. Copyright 2019 Elsevier.

GaN and a Ga enriched metal pool can form in the device layer, as shown in figures 15.22(e) and 15.22(f). In addition to EDS mapping, high-resolution TEM imaging allows the identification of defect formation and measurement of the developed strains in the sample during operation. Such a study has been shown in figure 15.23(a), where the dislocation is shown by the yellow arrow in the GaN buffer layer. The geometric phase analysis (GPA) technique [106] also allows us to track the types and magnitude of strain in the sample. Local strain field mapping

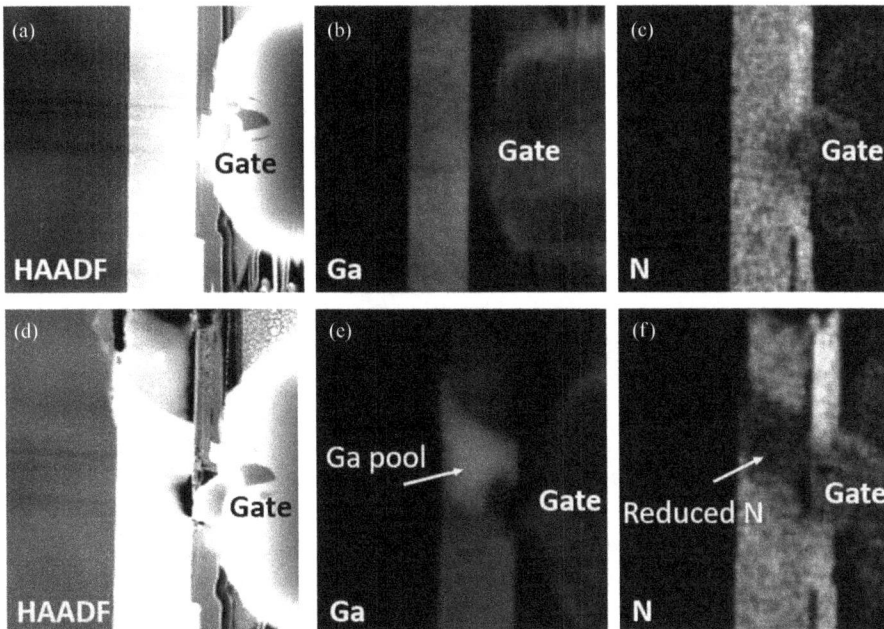

Figure 15.22. EDS analysis of an irradiated GaN HEMT sample: (a)–(c) before biasing and (d)–(f) after biasing.

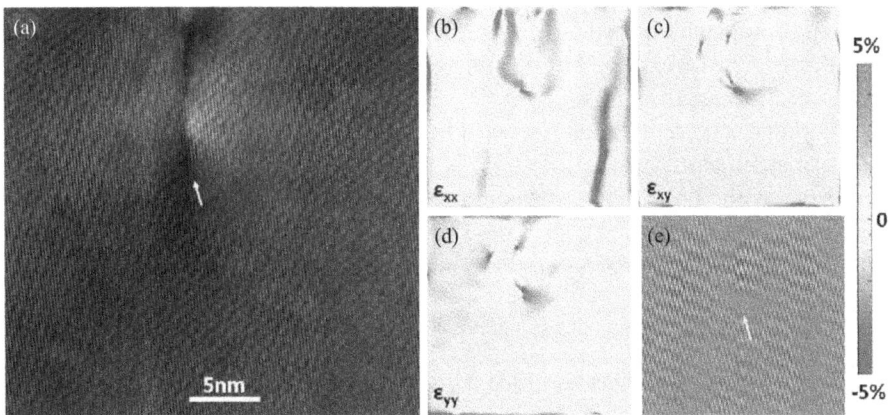

Figure 15.23. Dislocation analysis of a HRTEM image of a GaN HEMT: (a) strain analysis using GPA, (b) normal strain along the x-direction, (c) shear strain in the x–y plane, (d) normal strain in the y-direction, and (e) Bragg image simulation showing dislocation. The scale bar for all images is the same as that shown in figure 15.18(a).

allows us to quantify both normal and shear strain as shown in figure 15.23(b)–(d). Strain mapping indicates that dislocations are associated with both tensile and compressive strain and the highest magnitude of this strain is ±5%.

It can be concluded that an *in situ* TEM study could provide both qualitative and quantitative information during real-time operation of the device. Unlike post-failure analysis, it can also pinpoint the exact failure mechanism, point of failure initiation, and propagation directions in the device. This unique technique will be an advanced addition to the reliability community. However, more studies and correlations need to be developed since the geometry and boundary conditions of the electron-transparent samples are different to the bulk scale devices.

15.4 Future directions

Traditional reliability studies of materials and devices consider electron microscopy as an *ex situ* tool. Recent *in situ* TEM studies are shifts from this conventional trend and can provide quantitative as well as qualitative data at the same time. However, the preparation of an integrated electron-transparent GaN HEMT sample for *in situ* TEM study is challenging and requires special attention. Care should be taken to avoid any contamination and beam damage during sample preparation. Lower accelerating voltage and current during milling and cleaning may alleviate this problem. During a TEM experiment e-beam exposure from the TEM source could also affect the results. Thus, care should be taken to minimize the dose rate and exposure time. Another way to completely avoid e-beam exposure is to close the column valve/shutter during biasing. Strain engineering in electronic devices and semiconductor materials are always intriguing. Future work can be performed by applying strain in the electron-transparent thin films and tracking their subsequent response, as shown in figure 15.24.

As mentioned earlier, an electron-transparent thin film may induce residual stress in the thin sample which further introduces bend contours. Another viable option to perform *in situ* TEM experiments could be cutting a small window at the region of interest from a thicker sample, as shown in figure 15.25.

In summary, due to the geometry and dimensions of electron-transparent GaN HEMTs, a significant amount of study is required before making any quantitative comparison with bulk scale measurements. However, the failure modes and mechanisms observed during *in situ* TEM studies are in good agreement with *ex situ* experimental predictions.

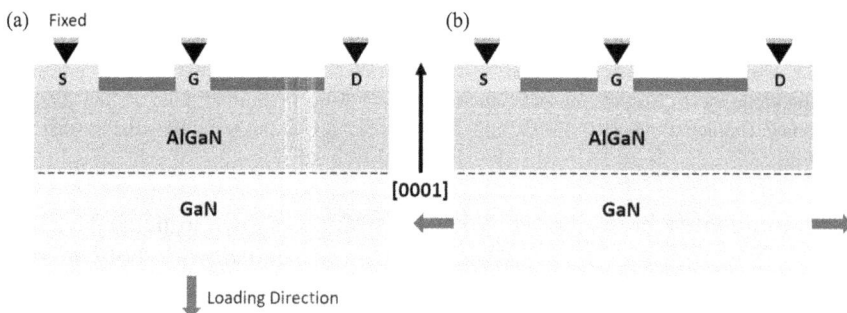

Figure 15.24. Schematic showing two different loading directions for future study.

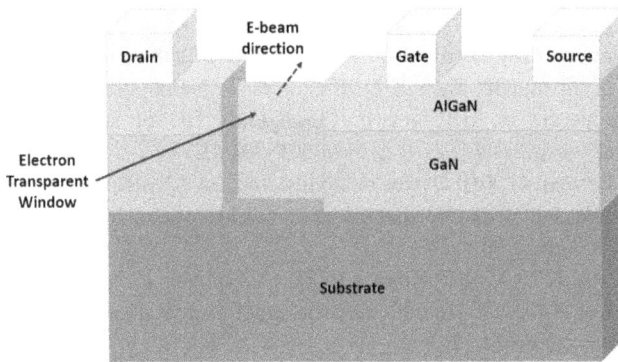

Figure 15.25. Schematic showing an electron-transparent window to avoid any residual stress effects in a future AlGaN/GaN HEMT reliability study.

Acknowledgement

We acknowledge support from the National Science Foundation through award #Civil, Mechanical and Manufacturing Innovation (CMMI) 1760931 and UES Inc.

References

[1] Mimura T 2002 The early history of the high electron mobility transistor (HEMT) *IEEE T. Microw. Theory* **50** 780–2

[2] Khan M A, Kuznia J N, Olson D T, Schaff W J, Burm J W and Shur M S 1994 Microwave performance of a 0.25 μM gate AlGaN/GaN heterostructure field-effect transistor *Appl. Phys. Lett.* **65** 1121–3

[3] Mishra U K, Parikh P and Wu Y F 2002 AlGaN/GaN HEMTs—an overview of device operation and applications *Proc. IEEE* **90** 1022–31

[4] Chow T P and Tyagi R 1994 Wide bandgap compound semiconductors for superior high-voltage unipolar power devices *IEEE Trans. Electron. Dev.* **41** 1481–3

[5] Wu Y, Saxler A, Moore M, Smith R P, Sheppard S, Chavarkar P M, Wisleder T, Mishra U K and Parikh P 2004 30-W/mm GaN HEMTs by field plate optimization *IEEE Electron Device Lett.* **25** 117–9

[6] Kanamura M, Kikkawa T and Joshin K 2004 A 100-W high-gain AlGaN/GaN HEMT power amplifier on a conductive n-SiC substrate for wireless base station applications *IEDM Technical Digest. IEEE Int. Electron Devices Meeting (13–15 December 2004)* pp 799–802

[7] Wu Y, Moore M, Wisleder T, Chavarkar P M, Mishra U K and Parikh P 2004 High-gain microwave GaN HEMTs with source-terminated field-plates *IEDM Technical Digest. IEEE Int. Electron Devices Meeting (13–15 December 2004)* pp 1078–9

[8] Joshin K, Kikkawa T, Hayashi H, Maniwa T, Yokokawa S, Yokoyama M, Adachi N and Takikawa M 2003 A 174 W high-efficiency GaN HEMT power amplifier for W-CDMA base station applications *IEEE Int. Electron Devices Meeting (8–10 December 2003)* pp 12.16.11–3

[9] Selvaraj S L, Watanabe A, Wakejima A and Egawa T 2012 1.4-kV breakdown voltage for AlGaN/GaN high-electron-mobility transistors on silicon substrate *IEEE Electron Device Lett.* **33** 1375–7

[10] Wang X *et al* 2007 AlGaN/AlN/GaN/SiC HEMT structure with high mobility GaN thin layer as channel grown by MOCVD *J. Cryst. Growth* **298** 835–9

[11] Nakamura N, Furuta K, Shen X Q, Kitamura T, Nakamura K and Okumura H 2007 Electrical properties of MBE-grown AlGaN/GaN HEMT structures by using 4H-SiC (0001) vicinal substrates *J. Cryst. Growth* **301–302** 452–6

[12] Rao H and Bosman G 2010 Device reliability study of AlGaN/GaN high electron mobility transistors under high gate and channel electric fields via low frequency noise spectroscopy *Microelectron. Reliab.* **50** 1528–31

[13] Chen W W, Ma X H, Hou B, Zhao S L, Zhu J J, Zhang J C and Hao Y 2014 Reliability investigation of AlGaN/GaN high electron mobility transistors under reverse-bias stress *Microelectron. Reliab.* **54** 1293–8

[14] Cheney D J *et al* 2013 Reliability studies of AlGaN/GaN high electron mobility transistors *Semicond. Sci. Tech.* **28** 074019

[15] Liu W, Kang J H, Sarkar D, Khatami Y, Jena D and Banerjee K 2013 Role of metal contacts in designing high-performance monolayer n-type WSe$_2$ field effect transistors *Nano Lett.* **13** 1983–90

[16] Killat N *et al* 2013 Reliability of AlGaN/GaN high electron mobility transistors on low dislocation density bulk GaN substrate: implications of surface step edges *Appl. Phys. Lett.* **103** 110321

[17] Wilson A F, Wakejima A and Egawa T 2013 Step-stress reliability studies on AlGaN/GaN high electron mobility transistors on silicon with buffer thickness dependence *Appl. Phys. Express* **6** 45

[18] Yang L, Hu G Z, Hao Y, Ma X H, Quan S, Yang L Y and Jiang S G 2010 Electric-stress reliability and current collapse of different thickness SiN$_x$ passivated AlGaN/GaN high electron mobility transistors *Chinese Phys.* B **19** 101

[19] Zeng C *et al* 2015 Investigation of abrupt degradation of drain current caused by under-gate crack in AlGaN/GaN high electron mobility transistors during high temperature operation stress *J. Appl. Phys.* **118** 123101

[20] Joh J, Gao F, Palacios T and del Alamo J A 2010 A model for the critical voltage for electrical degradation of GaN high electron mobility transistors *Microelectron. Reliab.* **50** 767–73

[21] Kim H, Tilak V, Green B M, Smart J A, Schaff W J, Shealy J R and Eastman L F 2001 Reliability evaluation of high power AlGaN/GaN HEMTs on SiC substrate *Phys. Status Solidi* A **188** 203–6

[22] Sozza A *et al* 2005 Evidence of traps creation in GaN/AlGaN/GaN HEMTs after a 3000 h on-state and off-state hot-electron stress *IEEE Int. Electron Devices Meeting, 2005. IEDM Technical Digest (5 December 2005)* pp 4–593

[23] Lee C *et al* 2005 Effects of AlGaN/GaN HEMT structure on RF reliability *Electron. Lett.* **41** 155–7

[24] Joh J and Alamo J A d 2006 Mechanisms for electrical degradation of GaN high-electron mobility transistors *2006 Int. Electron Devices Meeting (11–13 December 2006)* pp 1–4

[25] Joh J W and del Alamo J A 2008 Critical voltage for electrical degradation of GaN high-electron mobility transistors *IEEE Electron. Device Lett.* **29** 287–9

[26] Meneghesso G, Meneghini M, Stocco A, Bisi D, de Santi C, Rossetto I, Zanandrea A, Rampazzo F and Zanoni E 2013 Degradation of AlGaN/GaN HEMT devices: role of reverse-bias and hot electron stress *Microelectron. Eng.* **109** 257–61

[27] Kuball M, Ťapajna M, Simms R J T, Faqir M and Mishra U K 2011 AlGaN/GaN HEMT device reliability and degradation evolution: Importance of diffusion processes *Microelectron. Reliab.* **51** 195–200

[28] Cullen D A, Smith D J, Passaseo A, Tasco V, Stocco A, Meneghini M, Meneghesso G and Zanoni E 2013 Electroluminescence and transmission electron microscopy characterization of reverse-biased AlGaN/GaN devices *IEEE Trans. Device Mater. Res.* **13** 126–35

[29] Chang C *et al* 2011 Electric-field-driven degradation in off-state step-stressed AlGaN/GaN high-electron mobility transistors *IEEE Trans. Device Mater. Reliab.* **11** 187–93

[30] Choi S, Heller E, Dorsey D, Vetury R and Graham S 2013 The impact of mechanical stress on the degradation of AlGaN/GaN high electron mobility transistors *J. Appl. Phys.* **114** 124177

[31] Harima H 2002 Properties of GaN and related compounds studied by means of Raman scattering *J. Phys. Condens. Mat.* **14** R967–93

[32] Hellman E S 1998 The polarity of GaN: a critical review *MRS Internet J. Nitride Semicond. Res.* **3** e11

[33] Sumiya M and Fuke S 2004 Review of polarity determination and control of GaN *MRS Internet J. Nitride Semicond. Res.* **9** 56

[34] Bernardini F, Fiorentini V and Vanderbilt D 1997 Spontaneous polarization and piezo-electric constants of III–V nitrides *Phys. Rev. B* **56** 10024–7

[35] Leszczynski M, Suski T, Teisseyre H, Perlin P, Grzegory I, Jun J, Porowski S and Moustakas T D 1994 Thermal-expansion of gallium nitride *J. Appl. Phys.* **76** 4909–11

[36] Leszczynski M, Suski T, Perlin P, Teisseyre H, Grzegory I, Bockowski M, Jun J, Porowski S and Major J 1995 Lattice-constants, thermal-expansion and compressibility of gallium nitride *J. Phys. D: Appl. Phys.* **28** A149–53

[37] Leszczynski M *et al* 1996 Lattice parameters of gallium nitride *Appl. Phys. Lett.* **69** 73–5

[38] Levinshtein M E, Michael S L R and Shur S 2001 *Properties of Advanced Semiconductor Materials: GaN, AlN, InN, BN, SiC, SiGe* (New York: Wiley)

[39] Powell R C, Lee N E, Kim Y W and Greene J E 1993 Heteroepitaxial wurtzite and zinc-blende structure GaN grown by reactive-ion molecular-beam epitaxy: growth kinetics, microstructure, and properties *J. Appl. Phys.* **73** 189–204

[40] Wright A F and Nelson J S 1995 Consistent structural properties for Aln, Gan, and Inn *Phys. Rev. B* **51** 7866–9

[41] Yan W S, Zhang R, Xiu X Q, Xie Z L, Han P, Jiang R L, Gu S L, Shi Y and Zheng Y D 2007 Phenomenological model for the spontaneous polarization of GaN *Appl. Phys. Lett.* **90**

[42] Ibbetson J P, Fini P T, Ness K D, DenBaars S P, Speck J S and Mishra U K 2000 Polarization effects, surface states, and the source of electrons in AlGaN/GaN hetero-structure field effect transistors *Appl. Phys. Lett.* **77** 250–2

[43] Binari S C, Ikossi K, Roussos J A, Kruppa W, Park D, Dietrich H B, Koleske D D, Wickenden A E and Henry R L 2001 Trapping effects and microwave power performance in AlGaN/GaN HEMTs *IEEE Trans. Electron. Dev.* **48** 465–71

[44] Rivera C and Munoz E 2009 Erratum: 'The role of electric field-induced strain in the degradation mechanism of AlGaN/GaN high-electron-mobility transistors' (vol 94, 053501, 2009) *Appl. Phys. Lett.* **95** 145101

[45] Sarua A, Ji H, Kuball M, Uren M J, Martin T, Nash K J, Hilton K P and Balmer R S 2006 Piezoelectric strain in AlGaN/GaN heterostructure field-effect transistors under bias *Appl. Phys. Lett.* **88** 103502

[46] Jogai B 2003 Influence of surface states on the two-dimensional electron gas in AlGaN/GaN heterojunction field-effect transistors *J. Appl. Phys.* **93** 1631–5

[47] Gurnett K and Adams T 2006 Native substrates for GaN: the plot thickens *III-Vs Rev.* **19** 39–41

[48] Hoel V, Defrance N, Jaeger J C D, Gerard H, Gaquiere C, Lahreche H, Langer R, Wilk A, Lijadi M and Delage S 2008 First microwave power performance of AlGaN/GaN HEMTs on SopSiC composite substrate *Electron. Lett.* **44** 238–9

[49] Eastman L F *et al* 2001 Undoped AlGaN/GaN HEMTs for microwave power amplification *IEEE Trans. Electron. Dev.* **48** 479–85

[50] Chowdhury S and Biswas D 2015 Impact of varying buffer thickness generated strain and threading dislocations on the formation of plasma assisted MBE grown ultra-thin AlGaN/GaN heterostructure on silicon *AIP Adv.* **5** 057149

[51] Lo C-F *et al* 2012 Effect of buffer layer structure on electrical and structural properties of AlGaN/GaN high electron mobility transistors *J. Vac. Sci. Technol.* B **30** 011205

[52] Lee H P, Perozek J, Rosario L D and Bayram C 2016 Investigation of AlGaN/GaN high electron mobility transistor structures on 200-mm silicon (111) substrates employing different buffer layer configurations *Sci. Rep.* **6** 37588

[53] Cai Y, Zhu C, Jiu L, Gong Y, Yu X, Bai J, Esendag V and Wang T 2018 Strain analysis of GaN HEMTs on (111) silicon with two transitional $Al_xGa_{1-x}N$ layers *Materials* **11** 1968

[54] Rossetto I *et al* 2016 Study of the stability of e-mode GaN HEMTs with p-GaN gate based on combined DC and optical analysis *Microelectron. Reliab.* **64** 547–51

[55] Fontsere A, Perez-Tomas A, Placidi M, Baron N, Chenot S, Moreno J C and Cordier Y 2013 Bulk temperature impact on the AlGaN/GaN HEMT forward current on Si, sapphire and free-standing GaN *ECS Solid State Lett.* **2** P4–7

[56] Borysiuk J, Sobczak K, Wierzbicka A, Jezierska E, Klosek K, Sobanska M, Zytkiewicz Z R and Lucznik B 2014 Measurements of strain in AlGaN/GaN HEMT structures grown by plasma assisted molecular beam epitaxy *J. Cryst. Growth* **401** 355–8

[57] Gao F, Tan S C, del Alamo J A, Thompson C V and Palacios T 2014 Impact of water-assisted electrochemical reactions on the OFF-state degradation of AlGaN/GaN HEMTs *IEEE Trans. Electron. Dev.* **61** 437–44

[58] Whiting P G, Holzworth M R, Lind A G, Pearton S J, Jones K S, Liu L, Kang T S, Ren F and Xin Y 2017 Erosion defect formation in Ni-gate AlGaN/GaN high electron mobility transistors *Microelectron. Reliab.* **70** 32–40

[59] Sasaki H, Hisaka T, Kadoiwa K, Oku T, Onoda S, Ohshima T, Taguchi E and Yasuda H 2018 Ultra-high voltage electron microscopy investigation of irradiation induced displacement defects on AlGaN/GaN HEMTs *Microelectron. Reliab.* **81** 312–9

[60] Sasangka W A, Syaranamual G J, Gao Y, Made R I, Gan C L and Thompson C V 2017 Improved reliability of AlGaN/GaN-on-Si high electron mobility transistors (HEMTs) with high density silicon nitride passivation *Microelectron. Reliab.* **76–77** 287–91

[61] Gan C L, Sasangka W A and Thompson C V 2019 Electrochemical oxidation in AlGaN/GaN-on-Si high electron mobility transistors *2019 Electron Devices Technology and Manufacturing Conf., EDTM 2019* pp 71–3

[62] Sasangka W A, Gao Y, Gan C L and Thompson C V 2018 Impact of carbon impurities on the initial leakage current of AlGaN/GaN high electron mobility transistors *Microelectron. Reliab.* **88–90** 393–6

[63] Whiting P G, Rudawski N G, Holzworth M R, Pearton S J, Jones K S, Liu L, Kang T S and Ren F 2017 Nanocrack formation in AlGaN/GaN high electron mobility transistors utilizing Ti/Al/Ni/Au ohmic contacts *Microelectron. Reliab.* **70** 41–8

[64] Islam Z, Haque A and Glavin N 2018 Real-time visualization of GaN/AlGaN high electron mobility transistor failure at off-state *Appl. Phys. Lett.* **113** 183102

[65] Islam Z, Paoletta A L, Monterrosa A M, Schuler J D, Rupert T J, Hattar K, Glavin N and Haque A 2019 Heavy ion irradiation effects on GaN/AlGaN high electron mobility transistor failure at off-state *Microelectron. Reliab.* **102** 113493

[66] Lei B *et al* 2018 Direct observation of semiconductor–metal phase transition in bilayer tungsten diselenide induced by potassium surface functionalization *ACS Nano* **12** 2070

[67] Baoming W, Zahabul I, Aman H, Kelson C, Michael S, Eric H and Nicholas G 2018 *In situ* transmission electron microscopy of transistor operation and failure *Nanotechnology* **29** 31LT01

[68] Hummingbird Scientific 2004 Hummingbird Scientific (accessed 2 Feb 2020) http://hummingbirdscientific.com/

[69] Protochips 2002 Protochips (accessed 2 Feb 2020) https://protochips.com/products/poseidon-select/

[70] Islam Z, Haque A and Glavin N 2018 Real-time visualization of GaN/AlGaN high electron mobility transistor failure at off-state *Appl. Phys. Lett.* **113** 183102

[71] Wang B M, Islam Z, Haque A, Chabak K, Snure M, Heller E and Glavin N 2018 *In situ* transmission electron microscopy of transistor operation and failure *Nanotechnology* **29** 31LT01

[72] Haque M A and Saif M T A 2004 Deformation mechanisms in free-standing nanoscale thin films: a quantitative *in situ* transmission electron microscope study *Proc. Natl Acad. Sci. USA* **101** 6335–40

[73] Meneghesso G, Verzellesi G, Danesin F, Rampazzo F, Zanon F, Tazzoli A, Meneghini M and Zanoni E 2008 Reliability of GaN high-electron-mobility transistors: state of the art and perspectives *IEEE Trans. Device Mater. Reliab.* **8** 332–43

[74] Joh J, del Alamo J A, Langworthy K, Xie S J and Zheleva T 2011 Role of stress voltage on structural degradation of GaN high-electron-mobility transistors *Microelectron. Reliab.* **51** 201–6

[75] Nigam A, Bhat T N, Rajamani S, Dolmanan S B, Tripathy S and Kumar M 2017 Effect of self-heating on electrical characteristics of AlGaN/GaN HEMT on Si(111) substrate *AIP Adv.* **7** 085015

[76] Vitanov S, Palankovski V, Maroldt S and Quay R 2010 High-temperature modeling of AlGaN/GaN HEMTs *Solid-State Electronics* **54** 1105–12

[77] Chiu H C, Peng L Y, Wang H Y, Wang H C, Kao H L, Chien F T, Lin J C, Chang K J and Cheng Y C 2016 Temperature dependency and reliability of through substrate via InAlN/GaN high electron mobility transistors as determined using low frequency noise measurement *Jpn J. Appl. Phys.* **55** 143501

[78] Wilson A F, Wakejima A and Egawa T 2013 Origin and appearance of defective pits in the gate–drain region during reliability measurements of AlGaN/GaN high-electron-mobility transistors on Si *Appl. Phys. Express* **6** 122

[79] Singhal S, Roberts J C, Rajagopal P, Li T, Hanson A W, Therrien R, Johnson J W, Kizilyalli I C and Linthicum K J 2006 GaN–ON–Si failure mechanisms and reliability improvements *2006 IEEE Int. Reliability Physics Symp. Proc. (26–30 March 2006)* pp 95–8

[80] Cheney D J, Douglas E A, Liu L, Lo C-F, Gila B P, Ren F and Pearton S J 2012 Degradation mechanisms for GaN and GaAs high speed transistors *Materials* **5** 2498–520

[81] Chou Y C *et al* 2004 Degradation of AlGaN/GaN HEMTs under elevated temperature lifetesting *Microelectron. Reliab.* **44** 1033–8

[82] Meneghesso G, Rampazzo F, Kordos P, Verzellesi G and Zanoni E 2006 Current collapse and high-electric-field reliability of unpassivated GaN/AlGaN/GaN HEMTs *IEEE Trans. Electron. Dev.* **53** 2932–41

[83] Coffie R, Chen Y, Smorchkova I P, Heying B, Gambin V, Sutton W, Chou Y, Luo W, Wojtowicz M and Oki A 2007 Temperature and voltage dependent RF degradation study in AlGaN/GaN HEMTs *2007 IEEE Int. Reliability Physics Symp. Proc. 45th Annual (15–19 April 2007)* pp 568–9

[84] Min-Woo H, Seung-Chul L, Joong-Hyun P, Jin-Cherl H, Kwang-Seok S and Min-Koo H 2006 Silicon dioxide passivation of AlGaN/GaN HEMTs for high breakdown voltage *2006 IEEE Int. Symp. on Power Semiconductor Devices and IC's (4–8 June 2006)* pp 1–4

[85] Kumar S, Kumar V and Islam A 2016 Characterisation of field plated high electron mobility transistor *2016 Int. Conf. on Microelectronics, Computing and Communications (MicroCom) (23–25 January 2016)* pp 1–3

[86] Meneghini M, Stocco A, Ronchi N, Rossi F, Salviati G, Meneghesso G and Zanoni E 2010 Extensive analysis of the luminescence properties of AlGaN/GaN high electron mobility transistors *Appl. Phys. Lett.* **97** 154101

[87] Meneghini M, Hilt O, Wuerfl J and Meneghesso G 2017 Technology and reliability of normally-off GaN HEMTs with p-type gate *Energies* **10** 153

[88] Zhang Y, Feng S, Wang L, Ji Y, Han X, Shi L and Zhao Y 2015 Measuring temperature in GaN-based high electron mobility transistors by cathodoluminescence spectroscopy *Semicond. Sci. Tech.* **30** 055016

[89] Isobe Y *et al* 2017 Defect analysis in GaN films of HEMT structure by cross-sectional cathodoluminescence *J. Appl. Phys.* **121** 235703

[90] Cheney D, Deist R, Navales J, Gila B, Ren F and Pearton S 2012 Determination of the reliability of AlGaN/GaN HEMTs through trap detection using optical pumping *2012 IEEE Compound Semiconductor Integrated Circuit Symp. (CSICS) (La Jolla, CA)* pp 1–4

[91] Kim J, Freitas J A, Mittereder J, Fitch R, Kang B S, Pearton S J and Ren F 2006 Effective temperature measurements of AlGaN/GaN-based HEMT under various load lines using micro-Raman technique *Solid-State Electronics* **50** 408–11

[92] Tadjer M J, Raad P E, Komarov P L, Hobart K D, Feygelson T I, Koehler A D, Anderson T J, Nath A, Pate B and Kub F J 2018 Electrothermal evaluation of AlGaN/GaN membrane high electron mobility transistors by transient thermoreflectance *IEEE J. Electron. Dev.* **6** 922–30

[93] Baczkowski L, Carisetti D, Jacquet J, Kendig D, Vouzelaud F and Gaquiere C 2014 Thermal characterization of high power AlGaN/GaN HEMTs using infra red microscopy and thermoreflectance *20th Int. Workshop on Thermal Investigations of ICs and Systems (24–26 Sept. 2014)* pp 1–6

[94] Graff A, Simon-Najasek M, Poppitz D and Altmann F 2018 Physical failure analysis methods for wide band gap semiconductor devices *2018 IEEE Int. Reliability Physics Symp. (IRPS) (11–15 March 2018)* pp 3B.2-1–8

[95] Syaranamual G J, Sasangka W A, Made R I, Arulkumaran S, Ng G I, Foo S C, Gan C L and Thompson C V 2016 Role of two-dimensional electron gas (2DEG) in AlGaN/GaN

high electron mobility transistor (HEMT) ON-state degradation *Microelectron. Reliab.* **64** 589–93

[96] Makaram P, Joh J, del Alamo J A, Palacios T and Thompson C V 2010 Evolution of structural defects associated with electrical degradation in AlGaN/GaN high electron mobility transistors *Appl. Phys. Lett.* **96** 233509

[97] Galloway K F and Schrimpf R D 1990 MOS device degradation due to total dose ionizing radiation in the natural space environment: a review *Microelectron. J.* **21** 67–81

[98] Cheney D J *et al* 2013 Reliability studies of AlGaN/GaN high electron mobility transistors *Semicond. Sci. Tech.* **28** 074019

[99] Marcon D, Meneghesso G, Wu T, Stoffels S, Meneghini M, Zanoni E and Decoutere S 2013 Reliability analysis of permanent degradations on AlGaN/GaN HEMTs *IEEE Trans. Electron. Dev.* **60** 3132–41

[100] del Alamo J A and Joh J 2009 GaN HEMT reliability *Microelectron. Reliab.* **49** 1200–6

[101] Vassiliou B and Wilde F G 1957 A hexagonal form of silicon nitride *Nature* **179** 435–6

[102] Ganchenkova M G and Nieminen R M 2006 Nitrogen vacancies as major point defects in gallium nitride *Phys. Rev. Lett.* **96** 196402

[103] Usman M, Nazir A, Aggerstam T, Linnarsson M K and Hallén A 2009 Electrical and structural characterization of ion implanted GaN *Nucl. Instrum. Methods Phys. Res.* B **267** 1561–3

[104] Gu W, Xu X, Zhang L, Gao Z, Hu X and Zhang Z 2018 Study on neutron irradiation-induced structural defects of GaN-based heterostructures *Crystals* **8** 198

[105] Pearton S J, Ren F, Patrick E, Law M E and Polyakov A Y 2016 Review-ionizing radiation damage effects on GaN devices *ECS J. Solid State Sci.* **5** Q35–60

[106] Hÿtch M J, Snoeck E and Kilaas R 1998 Quantitative measurement of displacement and strain fields from HREM micrographs *Ultramicroscopy* **74** 131–46

IOP Publishing

Wide Bandgap Semiconductor-Based Electronics

Fan Ren and Stephen J Pearton

Chapter 16

Vertical GaN-on-GaN power devices

Houqiang Fu, Kai Fu and Yuji Zhao

This chapter reviews the recent progress in vertical GaN-on-GaN power devices including p–n diodes, Schottky barrier diodes (SBDs), advanced rectifiers, and normally-off transistors. For GaN power diodes, various edge termination techniques have been demonstrated to boost device breakdown voltages, such as ion implantation, field plates, bevel mesa, and plasma-based edge termination. Materials engineering in drift and buffer layers can also help improve device performance. In addition, it is found that screw-type dislocations play a vital role in the reverse leakage of GaN power diodes. Although GaN advanced power rectifiers such as junction barrier Schottky (JBS) diodes have been demonstrated, their performance is still inferior to conventional GaN power diodes partly due to the immature status of GaN technology in ion implantation, etching, and regrowth. Four types of normally-off vertical GaN transistors are discussed: current aperture vertical electron transistors (CAVETs), trench MOSFETs, junction FETs (JFETs), and FinFETs. Their advantages and disadvantages are examined. Finally, the recent development of selective area doping via regrowth for GaN power devices is presented.

16.1 Introduction

Wide bandgap semiconductor gallium nitride (GaN) has garnered considerable interest for efficient power conversion applications [1–8]. Due to its large bandgap (3.4 eV), high critical electric field (3.5 MV cm^{-1}), high saturation velocity (3×10^7 cm s^{-1}), and high Baliga's figure of merit (1000 times larger than Si), GaN power devices are promising to push the power electronics roadmap beyond the state-of-the-art Si technology. At the device level, GaN devices can achieve lower on-resistance (R_{on}) and a higher breakdown voltage (BV) than their Si counterparts, as indicated in figure 16.1(a). Therefore, GaN devices are expected to considerably cut down power conversion losses, in particular at high operating frequencies [2]. In the early development stage, most GaN power devices were heteroepitaxially grown on

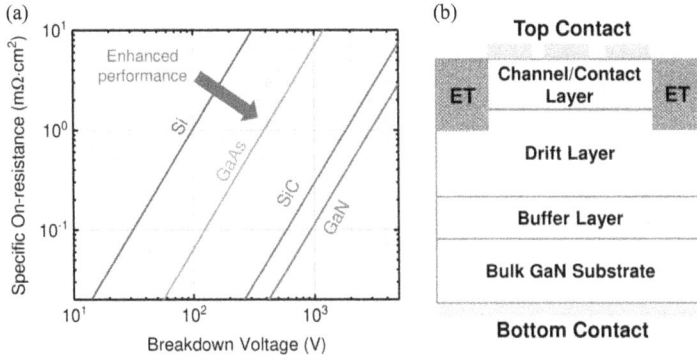

Figure 16.1. (a) Specific on-resistance versus breakdown voltage for different semiconductors. (b) Abstracted representation of vertical GaN-on-GaN power devices with different regions.

lattice-mismatched foreign substrates such as SiC and sapphire (later also on Si [5]). However, these devices suffered from high defect densities (e.g. 10^9 cm^{-2} on sapphire substrates) and relatively thin epilayer thickness, leading to limited device performance including large leakage currents and low BV.

With the availability of heavily doped bulk GaN substrates, vertical GaN-on-GaN power devices have shown significant performance enhancement [3, 9]. Figure 16.1(b) shows the schematics of vertical GaN-on-GaN power devices containing four regions: the channel/contact layer, edge termination, drift layer, and buffer layer. The benefits these vertical devices can offer are multifold. On the one hand, homoepitaxial growth can enable a drastic reduction in defect densities by three to five orders of magnitude (10^4–10^6 cm^{-2}) and the growth of very thick epilayers to achieve a high BV. On the other hand, compared to lateral devices, the vertical geometry can handle higher currents and voltages without enlarging the device area, and also offers other advantages including smaller chip size, immunity to surface-related issues, better heat dissipation, and better scalability [1, 10]. In this chapter, we will review the recent progress in vertical GaN-on-GaN power devices including p–n diodes, Schottky barrier diodes (SBDs), advanced rectifiers, and normally-off transistors. Selective area doping for GaN power devices will also be discussed. All the devices discussed below are grown using metal–organic chemical vapor deposition (MOCVD) on bulk GaN substrates unless specified otherwise.

16.2 Vertical GaN p–n diodes

Most research efforts in vertical GaN p–n diodes have been dedicated to enhancing the BV. From the growth perspective, the BV can always be enhanced by increasing the thickness and reducing the net doping concentration of device drift layers. Furthermore, edge termination is also indispensable for high voltage devices because it can prevent the premature breakdown at the device edge by alleviating the electric field crowding effects [11]. This section will focus on edge termination techniques for vertical GaN p–n diodes and explore possible reverse leakage mechanisms of the devices.

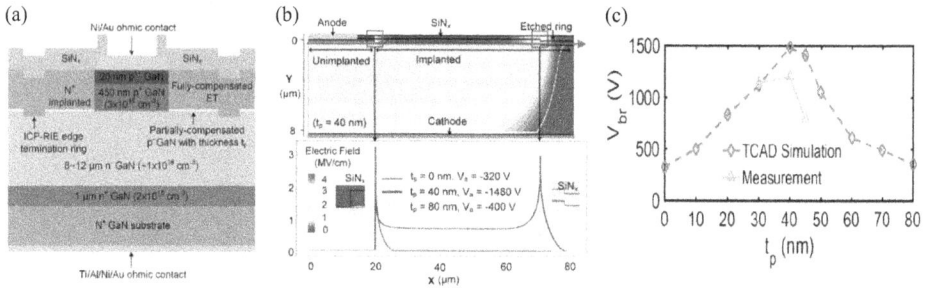

Figure 16.2. (a) Schematics of vertical GaN p–n diodes with ion-implantation based partially compensated edge termination. (b) Simulated electric field profile along the edge termination for different partial compensation layer thicknesses. (c) Simulated and measured breakdown voltages as a function of partial compensation layer thickness. Reproduced with permission from [21]. Copyright 2018 AIP Publishing.

16.2.1 Ion implantation

Ion implantation is one of the most widely used edge termination techniques for high voltage devices. It is a very mature technique in Si and SiC power devices [12], while it still lags behind in GaN technology. Nevertheless, ion implantation has been used either in the junction termination extension [13–17] or in the isolation of GaN devices by inducing material damage via ion bombardments to create mid-gap defects [18–21]. The latter can produce highly resistive layers at the edge to promote the lateral distribution of electric fields. Kizilyalli *et al* [17] used a two-step implantation process to fabricate the junction termination extension with a distance of 4.5 times of the drift layer thickness. This can result in smooth equipotential contours at the device edges and reduced the edge electric fields, as evidenced by the simulated electrostatic potential of the devices [17]. In addition, it was also revealed that the substrate orientation plays an important role in the device reverse leakage performance and reliability. This is because large hexagonal hillocks can easily form on on-axis (0001) plane growth by MOCVD. The optimal orientation is a nominal (0001) plane with a slight inclination toward the *m*-plane with several tenths of 1°. A large improvement in reverse leakage currents and BV was observed by utilizing the optimal substrate orientation. In addition, these devices also demonstrated avalanche breakdown capability. With a 40 μm drift layer and a net doping concentration of 2–5 × 10^{15} cm^{-3} in the drift layer, Kizilyalli *et al* [17] achieved a BV of 4 kV and an R_{on} of 2.8 mΩ cm^2.

Another version of this technology is the partially compensated ion-implantation edge termination. Figure 16.2(a) shows the schematics of vertical GaN p–n diodes with this edge termination [21]. The fabrication process used a triple-step nitrogen ion-implantation process to partially compensate p-GaN by creating damage. The implant depth can be tuned by adjusting the highest ion-implantation energy. The thickness of the remaining p-GaN, or the partial compensation layer, is a critical parameter for the BV. As shown in figure 16.2(b), when the p-GaN is fully compensated, the peak electrical field only occurred at the edge between the p–n junction and the implantation region. With increasing thickness of the partial compensation layer, two electric field peaks were observed at the edge of the p–n junction and the shallow etched mesa, respectively, with reduced electric fields. When this layer is too thick beyond the optimal thickness, the electric field peaked at the etched mesa because the partial compensation layer was not fully depleted. The width of the edge termination should exceed the drift layer thickness to avoid affecting the BV due to increased surface electric fields, similar to the reduced surface field (RESURF) method. Figure 16.2(c) presents the change of BV as the partial compensation layer thickness varies. Using this edge termination, Wang et al [21] demonstrated vertical GaN p–n diodes (a 12 μm drift layer) with a BV of 1.68 kV and an R_{on} of 0.15 mΩ cm^2.

16.2.2 Beveled field plate

Field plates are another very popular edge termination technique for GaN power p–n diodes [22–24]. When negatively biased (usually by connecting directly to the anodes), field plates can enlarge the extension of the depletion region along the surface, and thus reduce the edge electric field and increase the BV [11]. Nomoto et al [22] compared the performance of four GaN p–n diodes with different field plates. The four types of devices were: no field plate (NFP), no field plate with spin-on-glass (SOG) passivation (SOG-NFP), a long field plate with a steep mesa (LFP), and a long field plate with a beveled mesa (BLFP). The long field plates extended beyond the mesa bottom edge by 20 μm. With long field plates, the devices with a steep mesa showed a significant increase in BV from ~1–1.4 kV to ~3 kV. With both long field plates and beveled mesas, the BV was further increased to >3.2 kV with much improved BV uniformity with respect to the device size. It should also be noted that the length of the field plate is an important design parameter and needs to be optimized. In addition, all the devices showed similar forward I–V characteristics. This means the addition of field plates does not degrade the device forward performance, which is also widely observed in other edge termination techniques. Furthermore, avalanche breakdown was also observed in the devices. With a 32 μm drift layer and beveled field plates, a combination of 3.5 kV/1 mΩ cm^2 was achieved [22].

16.2.3 Mesa termination

Mesa termination is also a very common method to isolate GaN power devices. It is usually realized via dry etching by removing part of the materials at the device edge. Normally, a mesa termination with a 90° angle should be avoided due to strong electric field crowding effects, leading to reduced BV. However, Fukushima et al [25, 26] showed that using a deep steep mesa could actually relax the electric field at the edge. Figure 16.3(a) presents the simulated electric field distributions of GaN p–n diodes with different mesa depths [25]. It was clearly shown that with a mesa depth of 5–10 μm, the devices (a drift layer of 10 μm) achieved a uniform electric field distribution across the entire device. During the device fabrication, the dry etching was performed by inductively coupled plasma-reactive ion (ICP-RIE) etching with pure Cl$_2$ gas due to its good vertical directionality. To etch such a deep mesa, a hard mask consisting of Ni/Ti/Ni/Ti/Ni with a total thickness of

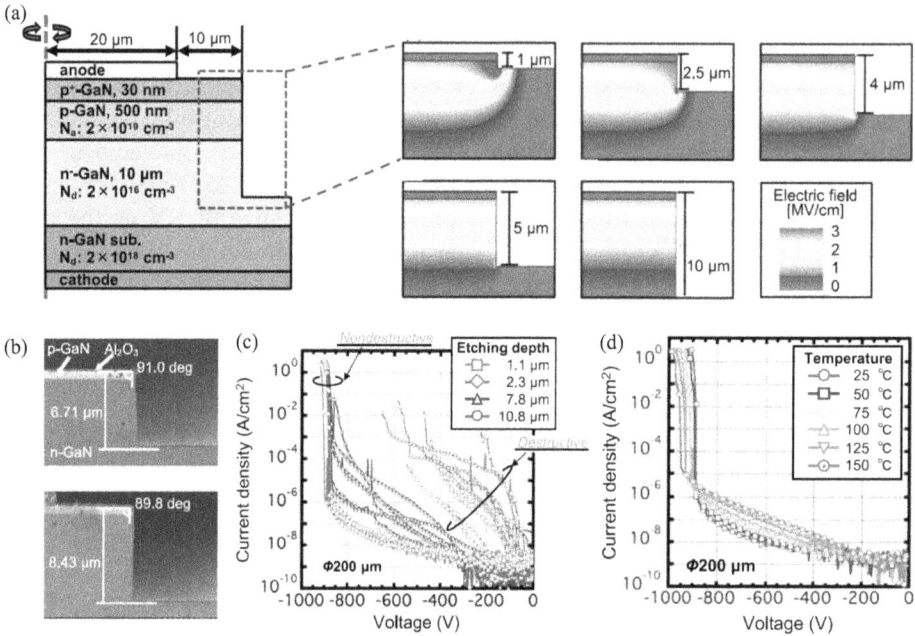

Figure 16.3. (a) Schematics of vertical GaN p–n diodes and simulated electric field distributions with different mesa depths (1, 2.5, 4, 5, and 10 μm) at −800 V. (b) Cross-sectional SEM images of GaN p–n diodes with mesa depths of 6.7 and 8.4 μm. (c) Reverse *I–V* characteristics of GaN p–n diodes with different mesa depths. (d) Temperature-dependent reverse *I–V* characteristics of GaN p–n diodes with a mesa depth of 10.8 μm. Reproduced with permission from [25]. Copyright 2019 The Japan Society of Applied Physics.

300 nm was used. The multilayer mask design was employed to prevent the cracking of thick Ni films. Figure 16.3(b) shows SEM images of etched vertical mesas with a depth of 6.7 μm and 8.4 μm, respectively. With shallow mesas (e.g. 1.1 and 2.3 μm), the reverse leakage was large, and the breakdown was destructive with a low BV of 450–600 V (figure 16.3(c)). The device performance also showed a large variation. With deep mesas (e.g. 7.8 and 10.8 μm), the reverse leakage was reduced by four orders of magnitude and the BV was enhanced to 880 V with very small variations. More interestingly, the breakdown was non-destructive. To further confirm whether it was an avalanche breakdown, temperature-dependent reverse *I–V* characteristics were measured, figure 16.3(d). A positive temperature-dependence of BV was observed, a signature of avalanche breakdown. In short, having a 13 μm drift layer and deep mesa termination, the GaN p–n diodes showed 880 V/0.8 mΩ cm^2 with avalanche capability [25].

In mesa termination for GaN devices, a more frequently used geometry is the beveled mesa, or negative beveled mesa, which is defined as one where more materials are removed from the heavily doped side than the lightly doped side [11]. In GaN p–n power diodes, this often means more materials are removed in the usually heavily doped p-GaN. The depletion region width along the negative beveled mesa sidewall is actually smaller than the depletion width in the bulk.

Therefore, it is counterintuitive to use negative beveled mesas for power devices. However, Maeda *et al* [27, 28] showed that it was still possible to reduce the surface electric field by using a sufficiently small negative bevel angle in combination with a relatively lightly doped p-GaN. With decreasing bevel angle θ and the acceptor concentration N_a to donor concentration N_d ratio (N_a/N_d), the breakdown electric field increases. When $\theta = 10°$ and N_a/N_d is less than 4, 95% of the critical electric field can be achieved. Using these conditions, simulations [28] revealed that the depletion region width along the surface was wider and the peak electric field was moved into the bulk, leading to alleviated electric field crowding at the device edge. Experimentally, the shallow bevel angle can be formed by using a thick photoresist mask with a post-exposure bake. Maeda *et al* [28] successfully realized a mesa angle of 11° using this method. The BV and breakdown electric fields of fabricated devices increased with decreasing Mg acceptor concentrations, which is consistent with the simulation results. The devices showed a BV of 400–500 V with 5 μm n-GaN and 2 μm p-GaN. In addition, avalanche breakdown was also observed in these devices. This technique is useful for relatively low BV (<1 kV) and some fundamental breakdown research of GaN, such as extracting impact ionization coefficients of GaN as in [29, 30].

16.2.4 Plasma-based edge termination

Hydrogen-plasma-based edge termination is a low-temperature, low-damage, low-cost, and easy-to-implement edge termination technology compared to ion-implantation edge termination [31–33]. Ion implantation usually requires a very high temperature (>1000 °C) annealing process to active implanted atoms and/or recover material damage. However, this high temperature process is highly undesirable since it can generate killer defects and degrade the device surface, which result in performance degradation and reliability issues [34–36]. In addition, ion implantation is also a complicated process with relatively high costs. In contrast, the plasma-based edge termination can be easily performed in the ICP tool (which is ubiquitous in nanofabrication centers) with an annealing temperature of only 400 °C, considerably reducing costs and simplifying processes [31–33]. The physical mechanism behind this method is that hydrogen can strongly bond with Mg acceptors in p-GaN to form thermally stable neutral Mg–H complexes that effectively passivate p-GaN into highly resistive GaN [37, 38] to serve as edge termination. This passivated p-GaN was also thermally stable [32]. The hydrogen plasma treatment has also been used to fabricate normally-off p-GaN/AlGaN/GaN high electron mobility transistors (HEMTs) [39–41].

Figure 16.4(a) shows the edge termination formation process where the anode works as a self-aligned mask. It requires both ICP and rapid thermal annealing (RTA) treatments. The cross-sectional SEM images of the devices before and after RTA treatment are shown in figures 16.4(b) and (c), respectively [32]. Without RTA treatment, the exposed p-GaN not covered by the anode showed a similar secondary electron (SE) contrast as the unexposed p-GaN layer under the anode. However, with RTA treatment, the sample had a very different cross-sectional image. The

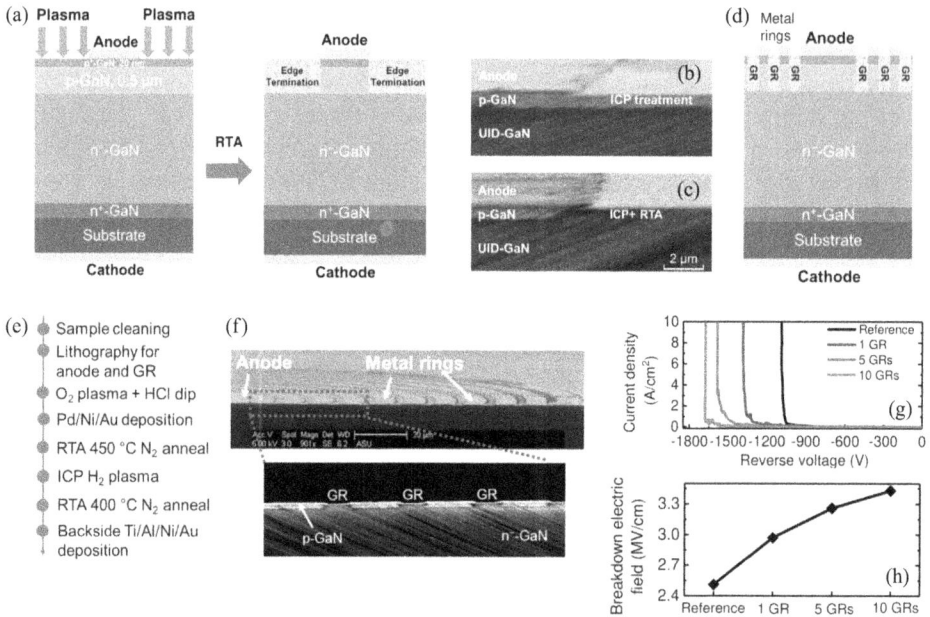

Figure 16.4. (a) Schematics of GaN p–n diodes with ICP plasma treatment and RTA treatment. Cross-sectional SEM images of GaN p–n diodes (b) with only ICP treatment and (c) with both ICP and RTA treatments. (d) Schematics of GaN p–n diodes with guard rings. (e) Fabrication process of the p–n diodes with guard rings. (f) SEM images of the p–n diodes with guard rings. (g) Reverse breakdown measurements and (h) breakdown electric fields of the p–n diodes with different numbers of guard rings. (a), (d)–(h) Reproduced with permission from [33]. Copyright 2019 IEEE. (b)-(c) Reproduced with permission from [32]. Copyright 2019 The Japan Society of Applied Physics.

exposed p-GaN outside the anode exhibited a much darker SE contrast, similar to the underlying unintentionally doped (UID) GaN layer, indicating the passivation of Mg acceptors and p-GaN. More details about the SE contrasts of p-GaN, UID-GaN, and n-GaN can be found in [42]. These results indicate that the ICP treatment alone can only deposit hydrogen atoms near the p-GaN surface, and a subsequent RTA treatment can thermally drive down these atoms into p-GaN to fully passivate it.

An advanced version of this technique is guard ring (GR) structures (figure 16.4(d)) where metal rings are deposited simultaneously with the anode to serve as masks [33]. The fabrication process is shown in figure 16.4(e). The cross-sectional SEM image of the fabricated devices is shown in figure 16.4(f), where the GR structure was clearly invisible. The BV and breakdown electric field increased with the increasing number of GRs due to more laterally spread electric fields. Without field plates or passivation, the GaN p–n diodes (9 μm drift layer) with plasma-based GRs showed BV/R_{on} of 1.70 kV/0.65 mΩ cm^2 [33], which are close to the GaN theoretical limit. These results indicate that plasma-based edge termination can serve as an effective edge termination technology for kV-class GaN p–n diodes. In addition, Ohta *et al* [43] demonstrated a mesa-based guard ring structure for 5 kV GaN p–n diodes. Figure 16.5 presents the benchmark plot for recently reported

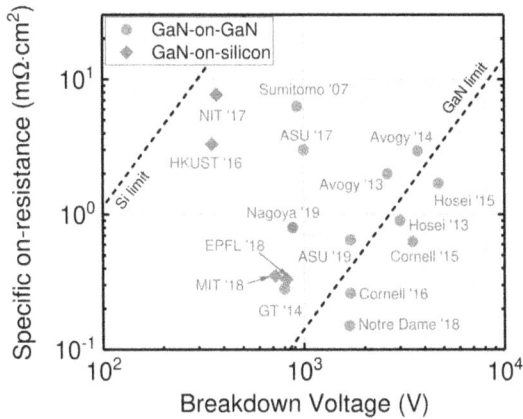

Figure 16.5. Benchmark plot of vertical GaN p–n diodes on bulk GaN and silicon substrates.

vertical GaN p–n diodes on bulk GaN and silicon substrates. The references for the devices can be found in [5, 21, 25, 31–33].

16.2.5 Leakage mechanism

Although the use of bulk GaN substrates has reduced the defect density of GaN devices to 10^4–10^6 cm^{-2}, these values are still much higher than those of Si and SiC devices. Therefore, it is very important to study the effect of defects on the device performance of GaN p–n diodes and identify the reverse leakage mechanism. Usami *et al* [44] used emission microscopy to observe the light emission from the leakage spots in GaN p–n diodes at a high reverse bias. As shown in figure 16.6(a), the emission microscopy image exhibited a dot-like emission pattern with a density of 4.8×10^5 cm^{-2}. Similar emission pattern was also observed in the forward bias. Each emission spot corresponded to a leakage spot. And devices with lower leakage currents showed fewer leakage spots. These results indicate that the high density of leakage spots is very likely to be the main cause of the large reverse currents and high ideality factors observed in vertical GaN-on-GaN p–n diodes [44].

Then cathodoluminescence (CL) was used to correlate these leakage spots with dislocations [44]. The dislocations appear as dark spots in CL due to their non-radiative nature. The density of dark spots of the drift layer in CL was ~ 1.3×10^7 cm^{-2}, which is close to the dislocation density of the bulk GaN substrates (1.0×10^7 cm^{-2}). This density was much higher than the leakage spot density. By comparing the emission microscopy and CL images, Usami *et al* [44] found that all the leakage spots corresponded to the CL dark spots and the rest of the dislocations did not contribute to leakage currents. Additionally, KOH etching revealed three types of pits (small, medium, and large) in figure 16.6(b), where only medium pits coincided with the observed leakage spots.

To identify the nature of the dislocations contributing to the leakage current, the dislocations under the medium etch pits were investigated by the large-angle

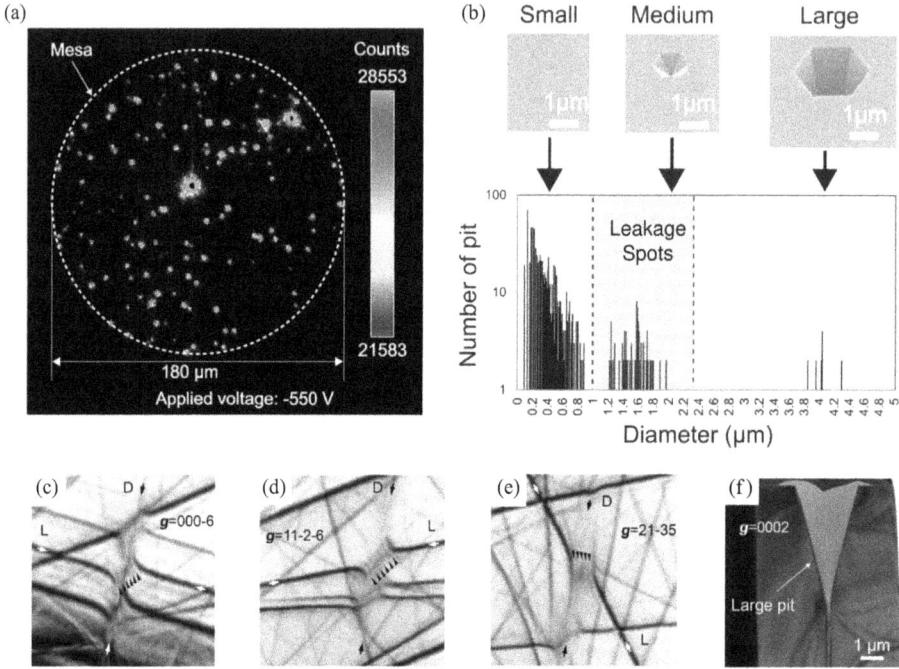

Figure 16.6. (a) Emission microscopy image of GaN p–n diodes at –550 V. Note that the large emission spot near the center of the mesa is caused by the contact of the probe. (b) Shapes and distributions of three different etch pits. Bright-field LACBED patterns obtained from the areas under the leakage spots (i.e. the medium etch pit) at different reciprocal lattice vectors of (c) [00–6], (d) [11–2–6], and (e) [21–35]. D indicates the dislocation and L indicates the Laue reflection lines. (f) A cross-sectional TEM image of a large pit. Reproduced with permission from [44]. Copyright 2018 AIP Publishing.

convergent-beam electron diffraction (LACBED) method in figures 16.6(c)–(e) [44]. Under three different reciprocal lattice vectors, the Burgers vector for the leakage-related dislocations was calculated to be [0001], which is close-core pure screw dislocations. In addition, the large etch pits in figure 16.6(f) were nanopipes or inversion domain with two parallel dislocation lines. The small etch pits should consist of mixed dislocations, and other dislocations not shown in the KOH etching may be edge dislocations. In short, the reverse leakage in vertical GaN-on-GaN p–n diodes is mainly related to $1c$ pure screw dislocations [44].

In addition, Rackauskas *et al* [45] observed abnormal features in the current transient of vertical GaN p–n diodes under reverse bias, and they attributed this phenomenon to impurity band conduction along dislocations. Tsou *et al* [46] proposed two different leakage mechanisms for vertical GaN p–n diodes at low and high temperatures: variable-range-hopping (VRH) conduction and multi-step electron transition. A theoretical study using density functional theory (DFT) confirmed that screw dislocations can form leakage paths in devices [47]. However, due to the relatively low density of screw dislocations in GaN-on-GaN devices, Usami *et al* [44] pointed out that VRH conduction was not likely to be the major leakage mechanism at room temperature. Therefore, the leakage mechanism of vertical GaN-on-GaN p–n diodes and its interplay with screw dislocations demand further investigation.

16.3 Vertical GaN Schottky barrier diodes

Vertical GaN SBDs have also been widely investigated due to some unique device advantages compared with GaN p–n diodes such as low turn-on voltage (V_{on}), fast switching due to the lack of minority carriers, and low-power conversion losses [48, 49]. However, the reported GaN SBDs still exhibit a lower BV and larger reverse leakage currents than GaN p–n diodes due to the Schottky contacts. In this section, we will discuss both materials engineering in drift and buffer layers and device engineering in edge termination to improve device performance, and possible leakage mechanisms for vertical GaN SBDs.

16.3.1 Carbon doping in the drift layer

Carbon doping is an important topic in GaN materials and devices [50–52]. Carbon can occupy either the Ga site to form shallow donor C_{Ga}, or the N site to form deep acceptor C_N. It has been shown that incorporating carbon into GaN buffer layers as deep acceptors can enhance the BV of GaN HEMTs [52]. Cao *et al* [48] studied the effect of carbon doping in the drift layer on the device performance of vertical GaN SBDs. They achieved a wide range of carbon concentrations in GaN as identified by secondary ion mass spectrometry (SIMS) from 3×10^{19} cm^{-3} (sample A), to 8×10^{17} cm^{-3} (sample B), to 8×10^{16} cm^{-3} (sample C), to $< 3 \times 10^{15}$ cm^{-3} (sample D), by increasing the growth pressure and V–III ratio. Atomic force microscopy (AFM) images (figure 16.7(a)–(d)) show that sample A had the roughest surface with large atomic steps and a root-mean-square (RMS) roughness of >1 nm. The rest of the epilayers with lower carbon concentrations showed very smooth surfaces with an RMS roughness of ~0.1 nm. According to x-ray rocking curve measurements, samples C and D had better material quality with lower defect densities.

For the forward *I–V* characteristics (figure 16.7(e)) only sample D showed high on-currents, indicating there are few free electrons when the carbon concentration is high. For the reverse *I–V* characteristics (figures 16.7(f)–(i)), sample D showed the lowest leakage currents and highest BV. These results suggest that carbon concentrations may affect the Fermi level in GaN and Schottky barrier heights. High carbon concentrations in samples A–C can result in a more insulating GaN film with a Fermi level closer to the mid-gap. Three leakage paths may exist: (i) thermionic emission over the Schottky barrier, (ii) direct tunneling through the Schottky barrier, and (iii) percolation leakage through electrons hopping assisted by the mid-gap states. The third leakage path was significantly suppressed in sample D due to a low carbon concentration, while the other two paths depend on the Schottky barrier height and width. In short, low carbon concentration in the drift layer can help improve the device performance of GaN SBDs. Using the growth conditions of sample D, Cao *et al* [48] demonstrated a GaN SBD (6 μm drift layer) with a BV of >800 V, an R_{on} of 5 mΩ cm^2, and a V_{on} of 0.77 V.

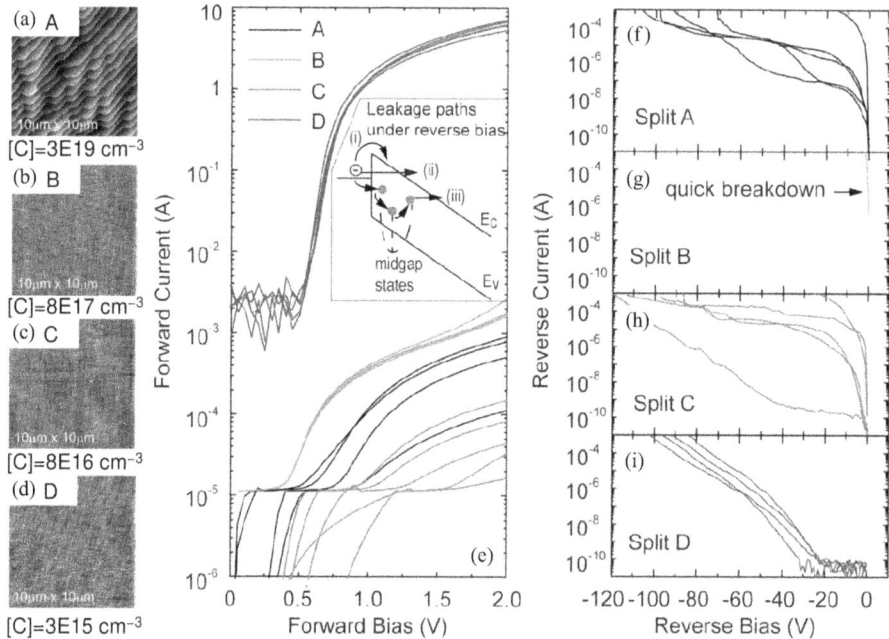

Figure 16.7. (a)-(d) AFM images of GaN epilayers with different carbon concentrations. The height scale is 12 nm for sample A and 1.2 nm for other samples. (b) Forward I–V characteristics of GaN SBDs fabricated on samples A, B, C, and D. The inset shows three possible leakage paths at reverse bias. (f)–(i) Reverse I–V characteristics of the four GaN SBDs. The GaN layer is 1 μm thick. For a quick turnaround, these devices were fabricated using a large 1.8 × 1.8 mm^2 Ni/Au pad as the Schottky contact and Ti/Au at the back of the substrates as the ohmic contact without isolation. Reproduced with permission from [48]. Copyright 2016 AIP Publishing.

16.3.2 Double drift layer

In the design of GaN SBDs, there is a trade-off between BV and R_{on}. While low R_{on} requires highly doped drift layers, high BV needs low doping in drift layers. To solve this problem, the double-drift-layer (DDL) design was introduced (figure 16.8(a)). With a well-controlled MOCVD growth process, the doping concentration of the top drift layer was reduced to suppress the peak electric field at the metal–GaN interface and increase the BV. The bottom drift layer was moderately doped to achieve low R_{on} to reduce power losses. Fu *et al* [49] compared the device performance of GaN SBDs with a single drift layer (SDL) and DDL, whose device structures are shown in figure 16.8(b). Detailed comparisons to GaN SBDs with a an SDL indicated that the DDL design did not degrade the device forward characteristics (figures 16.8(c) and (d)). The GaN SBDs with DDL exhibited a record low V_{on} of 0.59 V, an ultra-low R_{on} of 1.65 mΩ cm^2, a near unity ideality factor of 1.04, a high on–off ratio of $\sim 10^{10}$, and a high electron mobility of 1045.2 cm^2 V^{-1} s^{-1}. At reverse bias, the BV of GaN SBDs with DDL were also considerably enhanced compared to devices with as SDL (figure 16.8(e)). With proper design of thickness

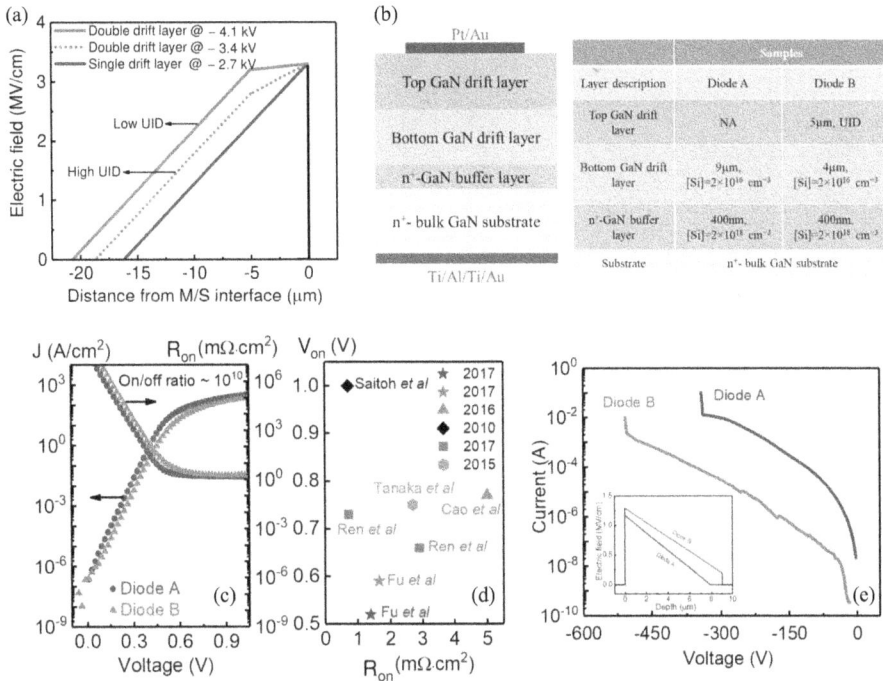

Figure 16.8. (a) Simulated electric field profiles using the one-dimensional Poisson's equation for GaN SBDs with a single drift layer and double drift layers where the top UID-GaN drift layers have low or high net doping concentrations. (b) Schematics of the GaN SBDs with double drift layers (left) and the details about each layer for the devices with single and double drift layers (right). (c) Forward current density and specific on-resistance as a function of voltage for the two SBDs in a semi-log scale. (d) Comparison of turn-on voltages and specific on-resistances of vertical GaN-on-GaN SBDs. (e) Reverse I–V characteristics of the two SBDs. The inset shows the electric field profiles of the two SBDs. Reproduced with permission from [49]. Copyright 2017 AIP Publishing.

and doping concentrations in each sub-drift-layer and controllable MOCVD growth, multi-drift-layer design [53] is expected to further improve the device performance. In short, the DDL design can balance the trade-off between BV and R_{on}, providing optimal overall device performance [49].

16.3.3 Effect of buffer layer thickness

Buffer layers also play a critical role in device performance. Fu *et al* [9] investigated the effect of buffer layer thickness on vertical GaN SBDs. The device structure and details of each layer are shown in figures 16.9(a) and (b), respectively. The devices consist of an n^+-GaN buffer layer ([Si] = 2×10^{18} cm^{-3}) with various thicknesses (20 nm, 100 nm, 400 nm, and 400 nm for SBD1, SBD2, SBD3, and SBD4, respectively), and a 9 μm thick UID or lightly doped ([Si] = 2×10^{16} cm^{-3} only for SBD4) drift layer. Very different forward I–V characteristics were observed on these devices in terms of on-currents and ideality factors. This indicates that buffer layer thickness can impact the electrical properties of drift layers. As shown in

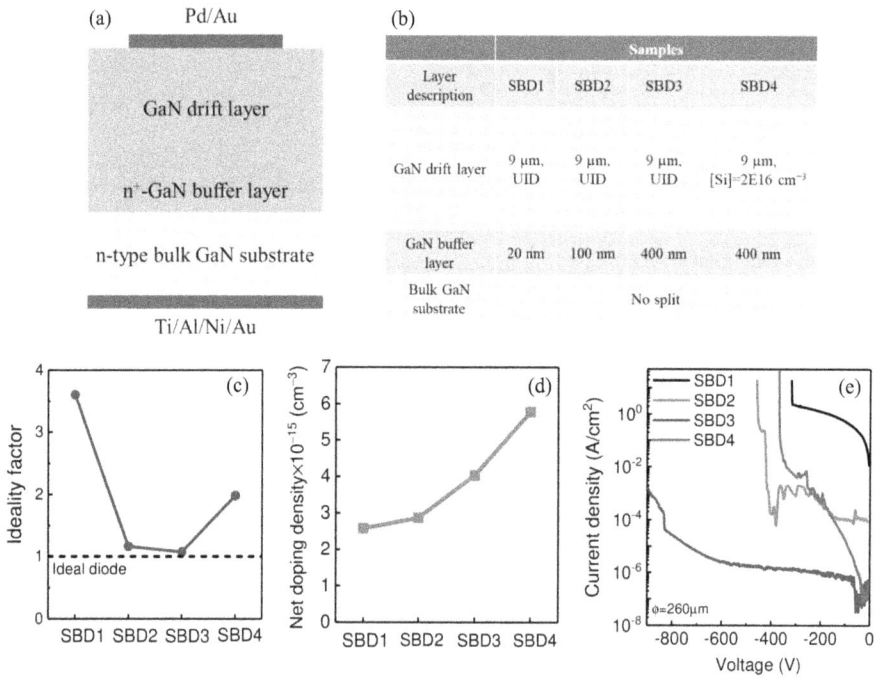

Figure 16.9. (a) Schematics of GaN SBDs. (b) Details of each layer for four samples with different buffer layer thicknesses. (c) Ideality factors of the four SBDs. (d) Net doping concentrations in the drift layers of the four SBDs extracted from C–V measurements. (e) Reverse I–V characteristics of the four SBDs. Reproduced with permission from [9]. Copyright 2017 IEEE.

figure 16.9(c), thicker UID drift layers resulted in smaller ideality factors, suggesting that thicker buffer layers may improve the material quality of the drift layer. In addition, higher drift layer doping concentration led to larger ideality factors at the same buffer layer thickness. The net doping concentration of the drift layer was extracted from C–V measurements according to methods in [54–57]. It was found the drift layer net doping concentration decreased with decreasing buffer layer thickness (figure 16.9(d)), which is possibly related to charged defects [58]. Figure 16.9(e) showed that the BV and breakdown electric field were also enhanced by thicker buffer layers, while the high doping concentration in SBD4 reduced the BV. A thicker buffer layer causes two competing effects: (i) it can improve the material quality that will enhance the BV and (ii) it can also result in a higher doping concentration in the drift layer that will reduce the BV. The fact that the thickest buffer layer still leads to the highest BV indicates that the material quality of a drift layer is more important in achieving a higher BV [9]. This should hold true for other high voltage GaN vertical devices.

16.3.4 Edge termination

The edge termination techniques discussed in the previous section can also be applied to GaN SBDs such as field plates, mesa, and the formation of a highly

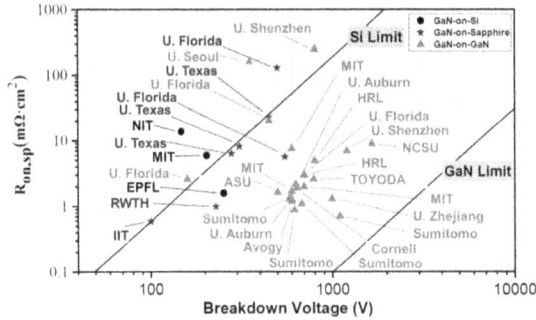

Figure 16.10. Benchmark plot of vertical GaN SBDs on bulk GaN, Si, and sapphire substrates. Reproduced from [62]. CC BY 4.0.

resistive region at the device edge. Recently, Han *et al* [59, 60] demonstrated two novel edge termination techniques for vertical GaN SBDs. The first one is incorporating fixed negative charges at the edge to achieve a smoother equipotential contours [59]. Due to their strong electronegativity, fluorine ions served as negative fixed charges in GaN. The process was realized by fluorine ion implantation, after which post-implantation annealing (PIA) was further applied to the devices. The fluorine-implanted termination with the PIA treatment increased the BV from ~260 V for unterminated devices to ~800 V. The PIA also reduced reverse leakage and improved the BV by recovering implantation damage. The other method is nitridation-based termination (NT) [60], which was realized by nitrogen plasma in plasma-enhanced chemical vapor deposition (PECVD). The GaN SBDs with NT showed a much lower leakage and higher BV of ~ 1 kV compared with the ~300 V of unterminated devices. Ultraviolet photoelectron spectroscopy (UPS) spectra revealed that nitridation shifted the Fermi level downward by 0.68 eV. As a result, the simulation showed the NT increased the barrier height/width at the junction edge that reduced the edge leakage [60]. In addition, Han *et al* [59] and Cao *et al* [61] showed that inserting a thin AlGaN tunneling layer reduced the reverse leakage current and the V_{on}. Figure 16.10 shows the benchmark plot of vertical GaN SBDs on bulk GaN, silicon, and sapphire substrates [62].

16.3.5 Leakage mechanism

Sang *et al* [63] identified the leakage current paths in vertical GaN SBDs using photon emission microscopy (PEM) with high spatial resolution and sensitivity to locate leakage positions. They used a thin and semi-transparent Schottky contact so that they can precisely pinpoint the leakage position from the front side. As shown in figures 16.11(a)–(d), two types of pits were observed: the polygonal and hexagonal pits, where only the former contributed to the leakage. The large reverse leakage in the reverse *I–V* characteristics of devices with polygonal pits also confirmed this (figure 16.11(e)). Wet KOH etching revealed the correlation between the pit shape and dislocations [64]. The non-leaky hexagonal pits were related to pure edge-type dislocations, and the leaky polygonal pits were due to the screw- and mixed-type

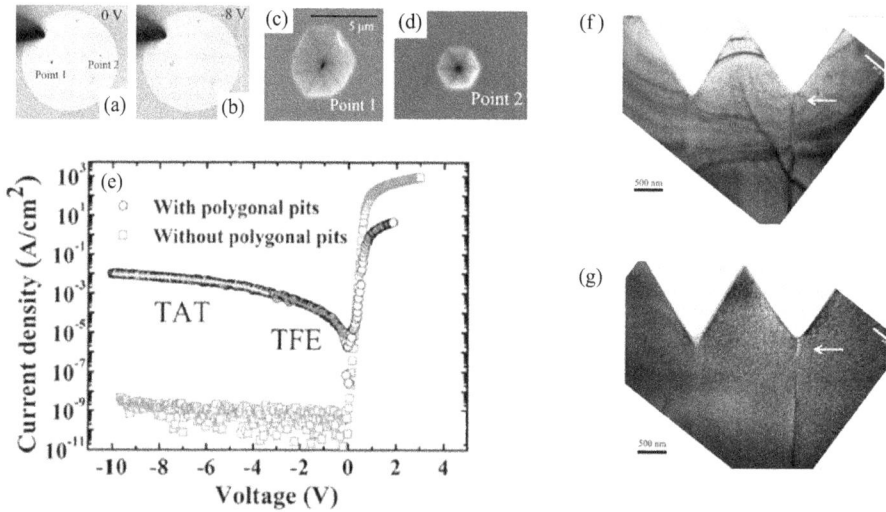

Figure 16.11. Photon emission microscopy mapping of vertical GaN SBDs at (a) zero bias and (b) −8 V from the front side of the devices with a semi-transparent Schottky contact. SEM images of (c) point 1 (polygonal pit) and (d) point 2 (hexagonal pit). (e) *I–V* characteristics of GaN SBDs with and without polygonal pits. Cross-sectional bright-field TEM images of the polygonal pits under reciprocal lattice vectors of (f) [0002] and (g) [11–20]. Reproduced with permission from [63]. Copyright 2017 AIP Publishing.

dislocations. Therefore, dislocations with a screw-type component are the major leakage pathways for vertical GaN SBDs [63].

The polygonal pits were further investigated by TEM under different reciprocal lattice vectors (figures 16.11(f) and (g)). The polygonal pits originated from the GaN epilayers during the MOCVD growth, and consisted of two V-shaped pits with screw- and mixed-type dislocations. From the energy dispersive x-ray spectroscopy (EDS) analysis, a large amount of carbon impurities accumulated to the dislocations. Optimizing MOCVD growth conditions to reduce carbon impurities may help reduce the density of polygonal pits. Finally, the reverse *I–V* characteristics of devices with polygonal pits in figure 16.11(e) can be fitted by the thermionic field emission (TFE) model at low reverse bias and trap assisted tunneling (TAT) model at high reverse bias. The traps inside the polygonal pits may play an important role in the tunneling process. In short, the initial large reverse leakage current of vertical GaN SBDs was caused by polygonal pits in the epilayers, which originated from the accumulation of carbon impurity to the dislocations with a screw-type component [63].

16.4 Vertical GaN advanced power rectifiers

Advanced power rectifiers can balance the trade-offs in conventional p–n diodes and SBDs, such as between on-state voltage drop, and reverse recovery currents and BV [65]. They usually involve more complicated fabrication processes, but can provide better performance. In this section, we will discuss recent demonstrations of vertical GaN merged p–n/Schottky (MPS) rectifiers [66, 67], junction barrier

Schottky (JBS) rectifiers [68], and trench metal–insulator–semiconductor (MIS) barrier Schottky (TMBS) rectifiers [69, 70].

16.4.1 Vertical GaN MPS rectifiers

In MPS rectifiers, the anode metal contacts form p–n junctions in some regions and Schottky junctions in the remaining regions. MPS rectifiers have a smaller on-state voltage drop and a smaller reverse recover current than p–n diodes [65], and a better surge current capability than JBS [66]. Figure 16.12(a) shows the schematics of a demonstrated GaN MPS rectifier with concentric patterns [66]. It was fabricated from a standard vertical GaN p–n diode structure. By dry etching, part of the p-GaN was removed to expose the underlying n-GaN region. Then the anode metals formed ohmic contacts with the p-GaN and Schottky contacts with the exposed n-GaN. Two turn-on processes were observed in the device: first the turn-on of the Schottky junction then the turn-on of the p–n junction (figure 16.12(b)). After the device was fully turned on, the on-currents increased with increasing p-GaN region width due to more minority carrier injection. In addition, the reverse leakage current decreased, and the BV increased with increasing p-GaN regions (figure 16.12(c)). Finally, Hayashida *et al* [66] demonstrated a vertical GaN MPS (18 μm drift layer) with a BV of 2 kV and an R_{on} of 1.7 mΩ cm^2. Moreover, Li *et al* [67] designed

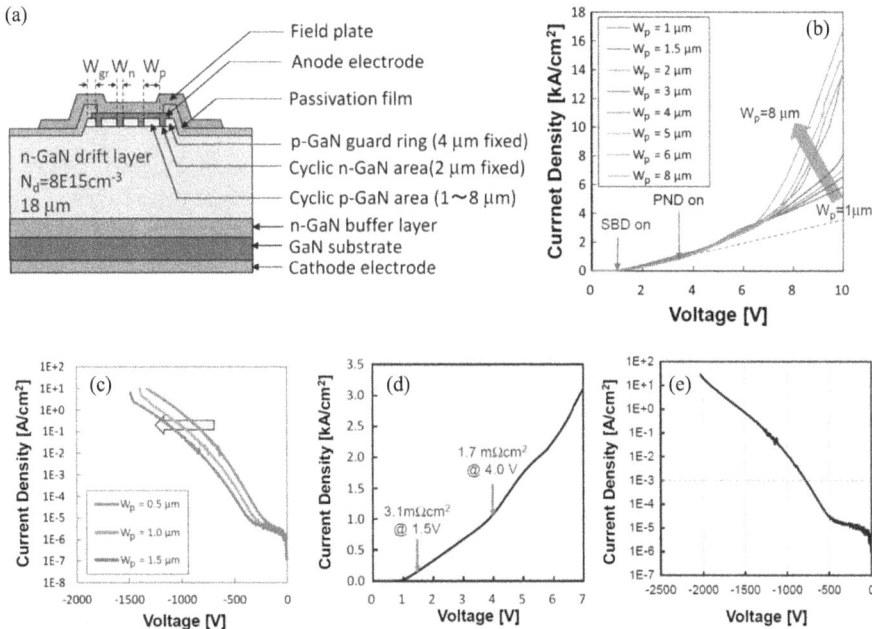

Figure 16.12. (a) Schematics of GaN MPS rectifiers with concentric circular patterns. (b) Forward *I–V* characteristics of concentric GaN MPS rectifiers. (c) Reverse *I–V* characteristics of GaN MPS rectifiers with different cyclic p-GaN widths. (d) Forward and (e) reverse *I–V* characteristics of GaN MPS rectifiers with a cyclic p-GaN width of 2 μm, a cyclic n-GaN width of 6 μm, and a guard ring width of 10 μm. Reproduced with permission from [66]. Copyright 2017 The Japan Society of Applied Physics.

another type of GaN MPS rectifier with circular trenches for Schottky contacts. Based on simulations, the surface electric field under the Schottky contact in MPS rectifiers was significantly reduced (i.e. the RESURF effect) compared to the conventional SBDs due to the barrier height difference between p-GaN and the Schottky metal. The GaN MPS rectifiers showed BVs comparable to GaN p–n diodes. The leakage current decreased with smaller trench sizes.

16.4.2 Vertical GaN JBS rectifiers

JBS rectifiers have a very similar structure to that of the MPS rectifier, except that the anode metal contacts do not form ohmic contacts to the p-GaN region. Therefore, there is no minority carrier injection in JBS rectifiers, making it faster than MPS rectifiers. Additionally, the p–n junction around the Schottky contact can create a potential barrier to shield the Schottky contact against high electric fields, reducing the reverse leakage current and increasing BV compared to SBDs [65]. Zhang et al [68] demonstrated planar vertical GaN JBS rectifiers using ion implantation. It can either be realized by Mg implantation into n-GaN to form p-wells, or by Si implantation into p-GaN to form n-wells. The two fabricated JBS rectifiers showed similar low V_{on} to SBDs, confirming the success of the ion-implantation method. Compared to conventional GaN SBDs, Mg-implanted JBS rectifiers showed larger R_{on} due to larger channel resistance and spread resistance. The former was caused by a smaller n-GaN area for the forward conduction and the latter by non-uniform current distributions. Compared with Mg-implanted devices, Si-implanted SBDs and JBS rectifiers showed even higher R_{on} due to low mobility in the implanted region. R_{on} and the forward voltage of GaN JBS rectifiers decreased with increasing n-well width and decreasing p-well width. As expected, the BV of GaN JBS rectifiers was between that of SBDs and p–n diodes. With increasing p-well width and decreasing n-well width, the BV increased due to the larger area of the p-GaN region.

16.4.3 Vertical GaN TMBS rectifiers

Similar to JBS rectifiers, the basic idea of TMBS rectifiers is also to reduce the electric field under the Schottky contact to suppress reverse leakage and increase the BV. It utilizes an MIS structure within trenches that are etched around the Schottky contact. Under reverse bias, the depletion regions between the trenches produce a potential barrier under the Schottky contact to shield it from high electric fields [65]. Zhang et al [69] fabricated vertical GaN TMBS rectifiers with implanted trench field rings (FR). Compared to conventional SBDs, TMBS showed lower reverse leakage current and higher BV. The additional of FRs further improved the performance. Furthermore, the geometry of trench structures also plays a critical role, such as the mesa width and the trench shape. A narrower mesa resulted in smaller reverse leakage currents and a higher BV due to stronger MIS depletion effects in the mesa as confirmed by simulations [69]. As shown in figure 16.13(a), different rounded trench shapes can be achieved by varying the ICP etching conditions combined with

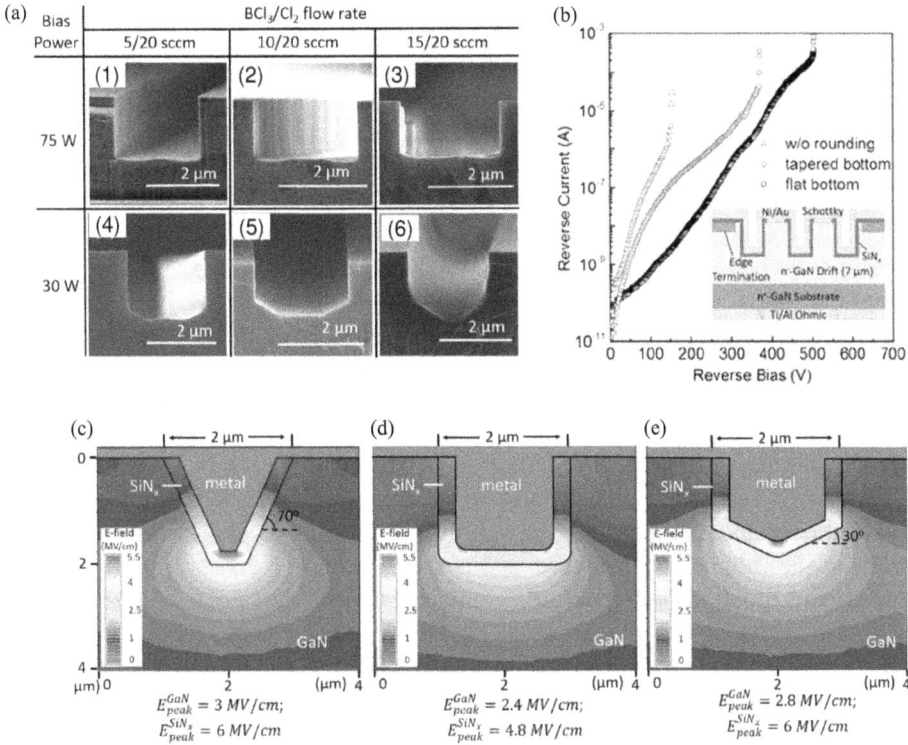

Figure 16.13. (a) Cross-sectional SEM images of trench structures in GaN with different ICP etching conditions. (b) Reverse I–V characteristics of TMBS rectifiers with non-rounded trenches, rounded flat-bottom trenches, and rounded tapered-bottom trenches in a semi-log scale. Simulated electric field distributions in the (c) non-rounded trench, (d) rounded flat-bottom trench, and (e) rounded tapered-bottom trench at −600 V. Reproduced with permission from [70]. Copyright 2017 AIP Publishing.

TMAH wet etching and piranha cleaning [70]. Figure 16.13(b) compares the reverse I–V characteristics of TMBS rectifiers with flat-bottom rounded trenches, tapered-bottom rounded trenches, and non-rounded trenches, where the first type of device had the best performance. The simulations in figures 16.13(c) and (e) reveal that rounded flat-bottom trenches showed a rather uniform electric field distribution with the lowest peak electric field among the three types of trenches. The benchmark plot for the demonstrated vertical GaN JBS, MPS, and TMBS rectifiers can be found in [69].

16.5 Normally-off vertical GaN power transistors

Normally-off operation is highly sought after in the development of vertical GaN power transistors due to its fail-safe feature and simple driver circuits. In this section, we will discuss different GaN transistor structures to achieve normally-off operation including current aperture vertical electron transistors (CAVETs) [71–74], trench MOSFETs [75–80], junction FETs (JFETs) [81–84], and FinFETs [85, 86].

16.5.1 Vertical GaN CAVETs

In GaN CAVETs, the AlGaN/GaN heterostructure produces a two-dimensional electron gas (2DEG) channel in the source access region. Electrons first flow laterally through the 2DEG channel and then vertically through the aperture formed by current blocking layers (CBLs) [1], [87]. Chowdhury *et al* [71] demonstrated the first normally-off GaN CAVETs on sapphire with a threshold voltage (V_{th}) of 0.6 V by exposing devices to CF_4 plasma. The CBLs used were Mg-implanted p-GaN. However, implanted CBLs suffered from Mg out diffusion during the subsequent high temperature channel regrowth. MOCVD grown p-GaN was also investigated as an option for CBLs. Ni *et al* [72] reported p-GaN gated GaN CAVETs with a V_{th} of 0.5 V, a BV of 1.5 kV, and an R_{on} of 2.2 mΩ cm^2. The CBLs were *in situ* grown p-GaN by MOCVD. It should also be noted the buried p-GaN CBL were shorted to the source contacts to avoid body effects.

To further increase the V_{th}, Shibata *et al* [73] fabricated p-GaN gated GaN trench CAVETs with a slanted channel. Band diagram analysis showed that the conduction band of the slanted channel shifted away from the Fermi level, indicating fewer free carriers in the channel. This is because the slanted channel on semipolar planes had reduced the polarization charges compared with the polar (0001) *c*-plane. Simulations revealed that with increasing tilted angle θ from the *c*-plane, the sheet carrier concentration in the channel decreased and the V_{th} increased. This trend is consistent with the fact that the polarization charges in GaN also decreased with increasing θ. As a result, the fabricated GaN CAVETs with slanted channels showed a V_{th} of 2.5 V compared to 1.0 V for those without slanted channels. Moreover, the carbon-doped GaN layer (HBL) can significantly suppress the punch-through current and increase the BV. With all these improvements, the GaN trench CAVETs showed a V_{th} of 2.5 V, a BV of 1.7 kV, and an R_{on} of 1.0 mΩ cm^2 [73]. In addition, Ji *et al* [74] also demonstrated a normally-off MIS gate GaN trench CAVETs.

16.5.2 Vertical GaN trench MOSFETs

Figure 16.14(a) shows the schematics of demonstrated vertical GaN trench MOSFETs. The structure is very similar to the well-studied Si trench MOSFETs. Without gate bias, there is no current conduction path due to the two back-to-back p–n junctions from the source to the drain, so they are normally-off devices. It should be noted that the source contact is also shorted to the p-GaN body. Oka *et al* [75] first demonstrated GaN trench MOSFETs with a BV of 1.6 kV, an R_{on} of 12.1 mΩ cm^2, and a V_{th} of 7 V. However, the R_{on} was too large. To solve this issue, they increased the doping concentration and reduced the thickness of the n$^-$-GaN drift layer [76]. In addition, they also employed a hexagonal trench gate layout (figure 16.14(b)) to increase the gate width per unit area to reduce R_{on}. As a result, a GaN trench MOSFET with a BV of 1.2 kV, an R_{on} of 1.8 mΩ cm^2, and a V_{th} of 3.5 V was successfully achieved [76]. The drain current was increased by 3.3 times compared to the first GaN trench MOSFETs. The reduction of V_{th} was also related to the increased drain current. One thing interesting to mention is that the obtained V_{th} was

(a)

(b)

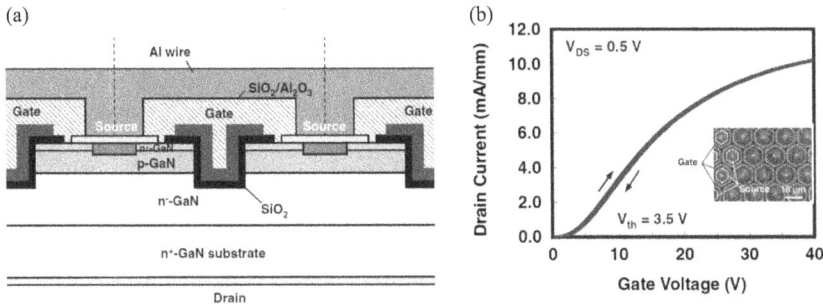

Figure 16.14. (a) Schematics of GaN trench MOSFETs. (b) Transfer characteristics of the trench MOSFETs. The inset shows the fabricated devices with a hexagonal layout. Reproduced with permission from [76]. Copyright 2015 The Japan Society of Applied Physics.

much smaller than the theoretical values (which are tens of volts). Oka *et al* [76] attributed this discrepancy to the fixed positive charges at the interface between the trench and the gate dielectric possibly due to residue impurities and/or nitrogen vacancies caused by ICP etching.

However, there are two challenges associated with the conventional GaN trench MOSFETs. The first one is the difficulty of forming good body contacts to the p-GaN channel layer after plasma etching. Without this contact, a high drain bias can lead to a non-zero positive potential in the p-GaN, thus reducing the V_{th} and increasing leakage currents [77]. Li *et al* [77] developed selective area regrowth of the n^+-GaN source layer to avoid the plasma etching of the p-GaN channel. They realized a very high V_{th} of 4.8 V. Another big challenge in GaN trench MOSFETs is the low channel electron mobility in the p-GaN inversion layer due to impurity and interface scattering. As a result, a very high gate bias (e.g. 40 V in [76]) needs to be applied to obtain a decent drain current, which is not practical for real life applications. To tackle this challenge, another type of GaN trench MOSFET was demonstrated, i.e. an *in situ* oxide, GaN interlayer based vertical trench MOSFET (OG-FET) [78–80]. A thin UID-GaN interlayer served as the channel with a much higher channel electron mobility due to lower doping concentration and weaker interface scattering with the *in situ* oxide. Ji *et al* [79] demonstrated a GaN OG-FET with a BV of 1.4 kV, an R_{on} of 2.2 mΩ cm^2, and a V_{th} of 4.7 V.

16.5.3 Vertical GaN JFETs

JFETs are another type of transistor that can also offer normally-off operations. Si and SiC JFETs have already been demonstrated, however, the development of GaN JFETs is still in its early stage. There are two types of GaN JFETs: lateral channel JFETs (figure 16.15(a)) and vertical channel JFETs (figure 16.15(b)). Ji *et al* [81] performed a comprehensive simulation study on GaN JFETs with a BV of 1.2 kV and low R_{on}. Kizilyalli *et al* [82] experimentally demonstrated a GaN lateral channel JFET. However, it was a normally-on device with a V_{th} lower than −20 V. Yang *et al* [83] showed GaN vertical channel JFETs by selective area regrowth of a p-GaN gate by MOCVD. The SEM image of the device cross section is shown in

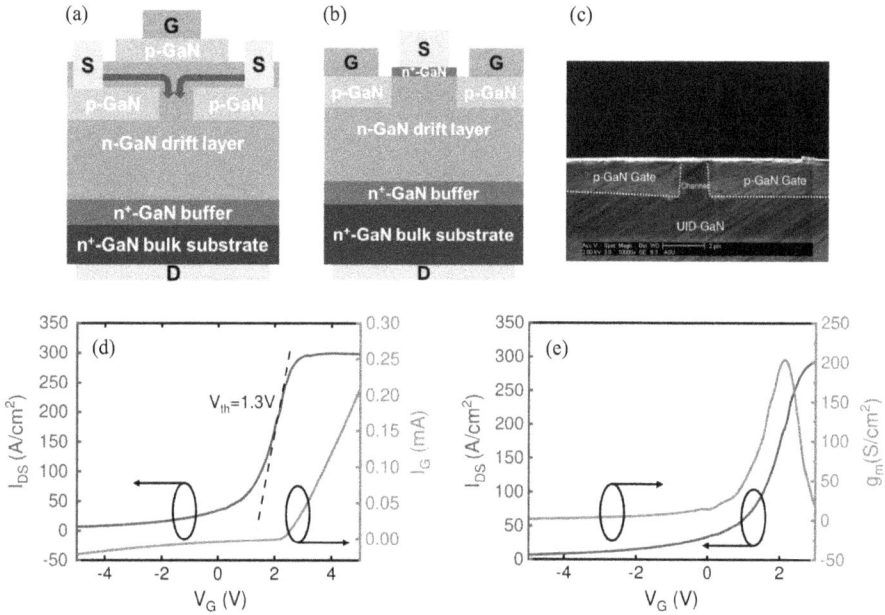

Figure 16.15. Schematics of (a) GaN lateral channel JFETs and (b) GaN vertical channel JFETs. (c) A cross-sectional SEM image of GaN vertical channel JFETs. (d) Transfer characteristics and gate leakage of fabricated GaN vertical channel JFETs at a drain bias of 10 V. (e) Transconductance as a function of gate voltage at a drain bias of 10 V.

figure 16.15(c). The devices showed a normally-off operation with a V_{th} of 1.3 V (figure 16.15(d)) and a decent transconductance (figure 16.15(e)). It should be noted that the crystal orientation of the channel sidewall plays a central role in the etching and regrowth processes. However, the BV of the devices was still very low. Kotzea *et al* [84] showed similar devices on sapphire substrates but with a normally-on operation. Future improvements are needed to develop kV-class normally-off vertical GaN JFETs.

16.5.4 Vertical GaN FinFETs

Sun *et al* [85] demonstrated vertical GaN FinFETs with only n-type GaN epilayers. When the fin width is small enough (usually sub-micron), the work function difference between the gate metal and GaN can deplete electrons in the channel from both sides of the fin, resulting a normally-on operation [85]. Due to the narrow width of the fin, electron beam lithography was used to define the fin region. During the etching of the fin, ICP etching usually results in a slanted sidewall, and a subsequent TMAH etching is critical to obtain vertical fins. Crystal orientation of the fin determines the sidewall roughness after etching. A positive V_{th} was obtained and it increased with decreasing fin width due to stronger depletion effects of the gate. Zhang *et al* [86] demonstrated a GaN FinFET with a BV of 1.2 kV, an R_{on} of 0.2 mΩ cm^2 and a V_{th} of 1 V. Large-area GaN FinFETs with over 600 fins showed a current of 10 A and a decent BV of 800 V.

The benchmark plot of vertical GaN-on-GaN power transistors can be found in [79, 80]. Finally, we compare the advantages and disadvantages of these normally-off vertical GaN transistors. GaN CAVETs have a very high channel mobility due to the 2DEG channel. However, CAVETs have complicated device structures and high requirements for growth. First, to achieve normally-off devices, special structures are needed for the depletion part of the channel such as the p-GaN gate and the slanted channel. Second, it is very difficult to realize good p-GaN CBL. Third, the regrowth of a high quality AlGaN/GaN channel is needed. Traditional GaN trench MOSFETs do not involve regrowth and are easier to grow, and have a very large V_{th} of >3 V. However, the channel mobility is very low and a very high gate bias is needed to achieve an acceptable drain current. In addition, forming good ohmic contacts to the buried p-GaN channel is also difficult. Although the proposed OG-FETs can improve the channel mobility, they require the regrowth of the UID-GaN channel. GaN JFETs can realize normally-off operation and do not need gate dielectrics. However, the gate bias cannot exceed ~3.5 V in order to prevent the turn-on of parasitic p–n junctions. In addition, the regrowth of a p-GaN gate is still challenging. GaN FinFETs do not need p-GaN and regrowth. However, the fabrication process is very complicated and requires electron beam lithography.

16.6 Selective area doping

Selective area doping is critical for realizing the full potential of GaN power electronics. The purpose of selective area doping is to form laterally patterned p–n junctions (figure 16.16(a)), which are the basis for a variety of advanced GaN power devices, such as JBS or merged MPS rectifiers, vertical JFETs, and superjunctions. In Si and SiC technology, selective area doping has already been realized by ion implantation. However, ion implantation is still very challenging for GaN, in particular for p-GaN mainly due to two reasons. First, ion implantation requires a subsequent thermal annealing process to activate implanted atoms and recover the crystal damage caused by the ion bombardments. The annealing temperatures usually need to go over 1200 °C. The problem is that GaN begins to decompose at ~900 °C. Second, although the researchers have tried to use capping layers such as AlN, multicycle rapid thermal annealing [88, 89], or ultra-high pressure [90] to alleviate GaN decomposition at high temperatures, the hole concentration or conductivity of the implanted p-GaN is still low and far from satisfactory from the perspective of power devices.

In addition to ion implantation, selective area growth or regrowth is regarded as one of the most promising alternative methods for selective area doping [91–94]. However, regrown p–n junctions usually suffer from large reverse leakage currents, which is likely due to surface contaminations at the regrowth interface such as impurities Si, O, and carbon, as revealed by SIMS [91], and ICP etching damage. Planar regrown p–n junctions were used as a test vehicle to obtain fundamental knowledge of the MOCVD regrowth. Figure 16.16(b) shows the etch-then-regrow process. Fu *et al* [92] evaluated the effects of ICP etching and surface treatments before regrowth on the performance of regrown GaN p–n junctions.

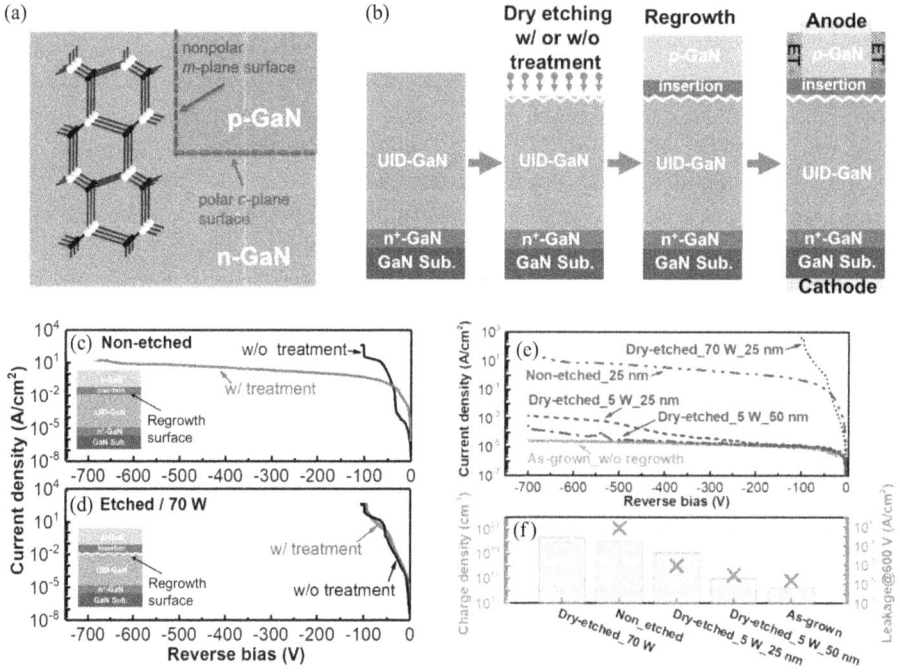

Figure 16.16. (a) Schematics of laterally patterned p–n junctions by selective area doping. (b) The growth and fabrication processes of GaN regrown p–n junctions. Reverse I–V characteristics of (c) a non-etched sample and (d) etched sample with or without surface treatments. (e) Reverse I–V characteristics of non-etched sample and etched samples with different ICP etching powers and insertion layer thicknesses. Charge density at the regrowth interface (histogram) and leakage current (line-shape) at -600 V for the five samples in (e). Reproduced with permission from [92]. Copyright 2019 IEEE.

Two samples were co-loaded into the MOCVD reactor for regrowth without any surface treatments: the non-etched sample and the etched sample with an ICP etching power of 70 W. Both samples showed large reverse leakage currents, indicating that the regrowth surface either with or without ICP etching was not ideal. Then, a combination of UV–ozone and acid surface treatments was applied to the two samples prior to regrowth. The non-etched sample exhibited a huge improvement in the reverse leakage current (figure 16.16(c)). However, these treatments had little impact on the etched sample (figure 16.16(d)). A hypothesis was proposed to explain this stark difference: for the etched sample, the etching damages are too severe to be recovered by the applied surface treatments. To confirm this hypothesis, ICP power was lowered to 5 W to reduce the etching damage. It is clear in figure 16.16(e) that the low-power etching dramatically reduced the reverse leakage currents. In addition, the thin insertion layer also helped further reduce the reverse leakage by moving the junction away from the regrowth interface. The etched sample with an ICP etching power of 5 W and a 50 nm insertion layer showed a reverse leakage current that was lower than the non-etched

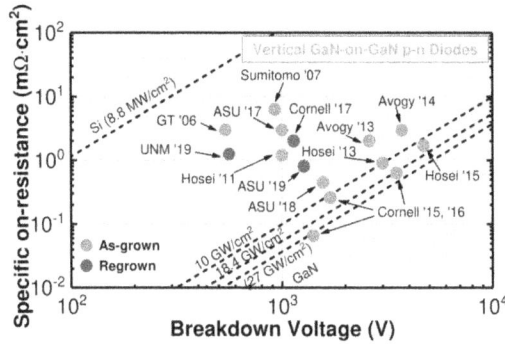

Figure 16.17. Benchmark plot for as-grown and regrown vertical GaN-on-GaN p–n diodes. Reproduced with permission from [92]. Copyright 2019 IEEE.

sample and similar to the as-grown sample. These results indicate that low-power slow etching combined with proper surface treatments is very effective for MOCVD regrowth.

Charge density at the regrowth interface was extracted by C–V measurements. Figure 16.16(f) compares the charge densities at the regrowth interface and reverse leakage currents of the five samples in figure 16.16(e). The two parameters shared the similar trend: the higher the interface charge density, the larger the leakage current. The as-growth sample had a low and constant charge distribution on the order of $\sim10^{16}$ cm^{-3}. The regrown sample showed a peak charge density at the regrowth interface on the order of 10^{17}–10^{20} cm^{-3}. These charges can create a large electric field and thin effective barrier at the regrowth interface for tunneling, leading to large leakage currents in regrown p–n junctions. Low-power etching can effectively reduce the interface charges. However, for some device structures with deep trenches and mesas, low-power etching is not always practical due to very low etching rate. To solve this issue, a multi-step etching was developed with decreasing ICP etching powers from 70 W to 35 W to 5 W to 2 W. This recipe provided not only reasonable etching rates, but also a high BV of over 1.2 kV. The sample with the multi-step etching also exhibited good forward rectifying behaviors, an on–off ratio of $\sim10^{10}$, and an R_{on} of 0.8 mΩ cm^2. Some other interesting studies in the MOCVD regrowth processes can be found in [95–98].

Figure 16.17 shows the benchmark plot for the as-grown and regrown vertical GaN-on-GaN p–n diodes [92]. Hu *et al* [99] demonstrated a regrown p–n junction using MBE with a BV of 1.1 kV and an R_{on} of 3.9 mΩ cm^2. Monavarian *et al* [100] showed a regrown nonpolar *m*-plane p–n junction with a BV of 540 V and an R_{on} of 1.7 mΩ cm^2. The Baliga's figure of merit of the regrown p–n diodes by Fu *et al* [92] was 2.0 GW cm^{-2}, which is even comparable to some of the reported as-grown p–n diodes. With future improvements in ICP etching, surface treatments, and device fabrication, the performance of GaN regrown p–n junctions is expected to further move towards the GaN limit.

16.7 Conclusion

The recent progress of vertical GaN-on-GaN power devices has been reviewed, including p–n diodes, SBDs, advanced rectifiers, and transistors. For vertical GaN p–n diodes, various edge termination techniques were demonstrated to reduce the electric field crowding effects and enhance the BV, such as ion implantation, field plates, bevel mesa, and plasma-based edge termination. For vertical GaN SBDs, researchers tried to improve the device performance from both materials and device engineering. Effects of carbon doping in drift layer and buffer layer thickness were studied. A double-drift-layer design proved to be effective in balancing the trade-off of R_{on} and BV. Some novel edge termination techniques were also demonstrated including nitridation and fluorine-implantation based edge termination. In addition, it was found that the reverse current leakage of both vertical GaN p–n diodes and SBDs was closely related to the screw-type dislocations in the drift layer. Further investigations are required to reveal the leakage mechanism and the role of screw-dislocations in the mechanism. The performance of reported vertical GaN advanced power rectifiers is still inferior to p–n diodes and SBDs. This is partly due to the immature status of GaN technology in some areas such as ion implantation, etching, and selective area doping. Four types of normally-off vertical GaN transistors were demonstrated: CAVETs, trench MOSFETs, JFETs, and FinFETs. Each type of device has its own advantages and disadvantages. Finally, selective area doping remains a huge hurdle for the development of GaN power devices. Selective area regrowth led to some decent regrown p–n junctions by optimizing etching and using proper surface treatments. Realizing high performance laterally patterned p–n junctions (either by ion implantation or regrowth) is still a critical on-going topic for GaN-on-GaN power devices.

References

[1] Chowdhury S and Mishra U K 2013 *IEEE Trans. Electron. Dev.* **60** 3060–6
[2] Chowdhury S, Swenson B L, Wong M H and Mishra U K 2013 *Semicond. Sci. Technol.* **28** 074014
[3] Kizilyalli I C, Edwards A P, Aktas O, Prunty T and Bour D 2015 *IEEE Trans. Electron. Dev.* **62** 414–22
[4] Amano H *et al* 2018 *J. Phys. D: Appl. Phys.* **51** 163001
[5] Zhang Y, Dadgar A and Palacios T 2018 *J. Phys. D: Appl. Phys.* **51** 273001
[6] Tsao J Y *et al* 2018 *Adv. Electron. Mater.* **4** 1600501
[7] Huang X *et al* 2019 *Nanotechnology* **30** 215201
[8] Montes J, Yang C, Fu H, Yang T H, Fu K, Chen H, Zhou J, Huang X and Zhao Y 2019 *Appl. Phys. Lett.* **114** 162103
[9] Fu H, Huang X, Chen H, Lu Z, Zhang X and Zhao Y 2017 *IEEE Electron Dev. Lett.* **38** 763–6
[10] Fu H *et al* 2018 *Appl. Phys. Express* **11** 111003
[11] Baliga B J 2008 *Fundamentals of Power Semiconductor Devices* (New York: Springer) ch 3, pp 91–155
[12] Sung W, Brunt E V, Baliga B J and Huang A Q 2011 *IEEE Electron. Dev. Lett.* **32** 880–2

[13] Laroche J, Ren F, Baik K, Pearton S, Shelton B and Peres B 2005 *J. Electrochem. Mater* **34** 370–4

[14] Kizilyalli I C, Edwards A P, Nie H, Disney D and Bour D 2013 *IEEE Trans. Electron. Dev.* **60** 3067–70

[15] Nie H, Edwards A P, Disney D R, Brown R J and Kizilyalli I C 2014 *US Patent* 8716716

[16] Kizilyalli I C, Edwards A P, Nie H, Bour D, Prunty T and Disney D 2014 *IEEE Electron Device Lett.* **35** 247–9

[17] Kizilyalli I C, Prunty T and Aktas O 2015 *IEEE Electron Device Lett.* **36** 1073–5

[18] Ozbek A M and Baliga B J 2011 *IEEE Electron Device Lett.* **32** 300–2

[19] Dickerson J R *et al* 2016 *IEEE Trans. Electron. Dev.* **63** 419–25

[20] Wang J, Cao L, Xie J, Beam E, McCarthy R, Youtsey C and Fay P 2017 *2017 IEEE International Electron Devices Meeting (IEDM) (San Francisco, CA)* pp 9.6.1–9.6.4

[21] Wang J, Cao L, Xie J, Beam E, McCarthy R, Youtsey C and Fay P 2018 *Appl. Phys. Lett.* **113** 023502

[22] Nomoto K *et al* 2015 *2015 IEEE International Electron Devices Meeting (IEDM) (Washington, DC)* pp 9.7.1–9.7.4

[23] Nomoto K, Song B, Hu Z, Zhu M, Qi M, Kaneda N, Mishima T, Nakamura T, Jena D and Xing H G 2016 *IEEE Electron. Dev. Lett.* **37** 161–4

[24] Hu Z, Nomoto K, Song B, Zhu M, Qi M, Pan M, Gao X, Protasenko V, Jena D and Xing H G 2015 *Appl. Phys. Lett.* **107** 243501

[25] Fukushima H, Usami S, Ogura M, Ando Y, Tanaka A, Deki M, Kushimoto M, Nitta S, Honda Y and Amano H 2019 *Appl. Phys. Express* **12** 026502

[26] Fukushima H, Usami S, Ogura M, Ando Y, Tanaka A, Deki M, Kushimoto M, Nitta S, Honda Y and Amano H 2019 *Jpn J. Appl. Phys.* **58** SCCD25

[27] Maeda T, Narita T, Ueda H, Kanechika M, Uesugi T, Kachi T, Kimoto T, Horita M and Suda J 2018 *2018 IEEE International Electron Devices Meeting (IEDM) (San Francisco, CA)* pp 30.1.1–30.1.4

[28] Maeda T, Narita T, Ueda H, Kanechika M, Uesugi T, Kachi T, Kimoto T, Horita M and Suda 2019 *IEEE Electron. Dev. Lett.* **40** 941–4

[29] Cao L, Wang J, Harden G, Ye H, Stillwell R, Hoffman A J and Fay P 2018 *Appl. Phys. Lett.* **112** 262103

[30] Ji D, Ercan B and Chowdhury S 2019 *Appl. Phys. Lett.* **115** 073503

[31] Fu H, Fu K, Huang X, Chen H, Baranowski I, Yang T H, Montes J and Zhao Y 2018 *IEEE Electron. Dev. Lett.* **39** 1018–21

[32] Fu H *et al* 2019 *Appl. Phys. Express* **12** 051015

[33] Fu H *et al* 2019 *IEEE Electron Device Lett.* **41** 127–30

[34] Ghandi R, Buono B, Domeij M, Malm G, Zetterling C M and Östling M 2009 *IEEE Electron. Dev. Lett.* **30** 1170–2

[35] Ghandi R, Buono B, Domeij M, Zetterling C M and Östling M 2011 *IEEE Trans. Electron. Dev.* **58** 2665–9

[36] Salemi A, Elahipanah H, Jacobs K, Zetterling C M and Östling M 2018 *IEEE Electron Device Lett.* **39** 63–6

[37] Nakamura S, Iwasa N, Senoh M and Mukai T 1992 *Jpn J. Appl. Phys.* **31** 1258–66

[38] Okamoto Y, Saito M and Oshiyma A 1996 *Jpn J. Appl. Phys.* **35** L807–9

[39] Hao R *et al* 2016 *Appl. Phys. Lett.* **109** 152106

[40] Hao R *et al* 2017 *IEEE Electron Device Lett.* **38** 1567–70

[41] Hao R *et al* 2018 *IEEE Trans. Electron. Dev.* **65** 1314–20

[42] Alugubelli S R, Fu H, Fu K, Liu H, Zhao Y and Ponce F A 2019 *J. Appl. Phys.* **126** 015704

[43] Ohta H, Hayashi K, Horikiri F, Yoshino M, Nakamura T and Mishima T 2018 *Jpn J. Appl. Phys.* **57** 04FG09

[44] Usami S *et al* 2018 *Appl. Phys. Lett.* **112** 182106

[45] Rackauskas B, Dalcanale S, Uren M J, Kachi T and Kuball M 2018 *Appl. Phys. Lett.* **112** 233501

[46] Tsou C W, Ji M H, Bakhtiary-Noodeh M, Detchprohm T, Dupuis R D and Shen S C 2019 *IEEE Trans. Electron. Dev.* **66** 4273–8

[47] Belabbas I, Chen J and Nouet G 2014 *Comput. Mater. Sci.* **90** 71–81

[48] Cao Y, Chu R, Li R, Chen M, Chang R and Hughes B 2016 *Appl. Phys. Lett.* **108** 062103

[49] Fu H, Huang X, Chen H, Lu Z, Baranowski I and Zhao Y 2017 *Appl. Phys. Lett.* **111** 152102

[50] Seager C H, Wright A F, Yu J and Gotz W 2002 *J. Appl. Phys.* **92** 6553

[51] Lyons J L, Janotti A and Van de Walle C G 2014 *Phys. Rev.* B **89** 035204

[52] Uren M J, Casar M, Gajda M A and Kuball M 2014 *Appl. Phys. Lett.* **104** 263505

[53] Ohta H, Kaneda N, Horikiri F, Narita Y, Yoshida T, Mishima T and Nakamura T 2015 *IEEE Electron. Dev. Lett.* **36** 1180–2

[54] Fu H, Baranowski I, Huang X, Chen H, Lu Z, Montes J, Zhang X and Zhao Y 2017 *IEEE Electron. Dev. Lett.* **38** 1286–9

[55] Fu H, Chen H, Huang X, Baranowski I, Montes J, Yang T H and Zhao Y 2018 *IEEE Trans. Electron. Dev.* **65** 3507–13

[56] Yang T, Fu H, Huang X, Montes J, Baranowski I, Fu K and Zhao Y 2019 *J. Semicond.* **40** 012801

[57] Montes J, Yang T H, Fu H, Chen H, Huang X, Fu K, Baranowski I and Zhao Y 2019 *IEEE Trans. Nucl. Sci.* **66** 91–6

[58] Cherns D and Jiao C G 2001 *Phys. Rev. Lett.* **87** 205504

[59] Han S, Yang S and Sheng K 2019 *IEEE Electron. Dev. Lett.* **40** 1040–3

[60] Han S, Yang S and Sheng K 2018 *IEEE Electron. Dev. Lett.* **39** 572–5

[61] Cao Y, Chu R, Li R, Chen M and Williams A J 2016 *Appl. Phys. Lett.* **108** 112101

[62] Sun Y, Kang X, Zheng Y, Lu J, Tian X, Wei K, Wu H, Wang W, Liu X and Zhang G 2019 *Electronics* **8** 575

[63] Sang L *et al* 2017 *Appl. Phys. Lett.* **111** 1221102

[64] Lu L *et al* 2008 *J. Appl. Phys.* **104** 123525

[65] Baliga B J 2009 *Advanced Power Rectifier Concepts* (Berlin: Springer)

[66] Hayashida T, Nanjo T, Furukawa A and Yamamuka M 2017 *Appl. Phys. Express* **10** 061003

[67] Li W, Nomoto K, Pilla M, Pan M, Gao X, Jena D and Xing H G 2017 *IEEE Trans. Electron. Dev.* **64** 1635–41

[68] Zhang Y *et al* 2017 *IEEE Electron Device Lett.* **38** 1097–100

[69] Zhang Y, Sun M, Liu Z, Piedra D, Pan M, Gao X, Lin Y, Zubair A, Yu L and Palacios T 2016 *2016 IEEE International Electron Devices Meeting (IEDM) (San Francisco, CA)* pp 10.2.1–10.2.4

[70] Zhang Y, Sun M, Liu Z, Piedra D, Hu J, Gao X and Palacios T 2017 *Appl. Phys. Lett.* **110** 193506

[71] Chowdhury S, Swenson B L and Mishra U K 2008 *IEEE Electron Device Lett.* **29** 543–5

[72] Nie H, Diduck Q, Alvarez B, Edwards A P, Kayes B M, Zhang M, Ye G, Prunty T, Bour D and Kizilyalli I C 2014 *IEEE Electron Device Lett.* **35** 939–41

[73] Shibata D, Kajitani R, Ogawa M, Tanaka K, Tamura S, Hatsuda T, Ishida M and Ueda T 2016 *2016 IEEE International Electron Devices Meeting (IEDM) (San Francisco, CA)* pp 10.1.1–10.1.4

[74] Ji D, Laurent M A, Agarwal A, Li W, Mandal S, Keller S and Chowdhury S 2017 *IEEE Trans. Electron. Dev.* **64** 805–8

[75] Oka T, Ueno Y, Ina T and Hasegawa K 2014 *Appl. Phys. Express* **7** 021002

[76] Oka T, Ina T, Ueno Y and Nishii J 2015 *Appl. Phys. Express* **8** 054101

[77] Li R, Cao Y, Chen M and Chu R 2016 *IEEE Electron. Dev. Lett.* **37** 1466–9

[78] Gupta C, Lund C, Chan S H, Agarwal A, Liu J, Enatsu Y, Keller S and Mishra U K 2017 *IEEE Electron. Dev. Lett.* **38** 353–5

[79] Ji D, Gupta C, Chan S H, Agarwal A, Li W, Keller S, Mishra U K and Chowdhury S 2018 *2018 IEEE International Electron Devices Meeting (IEDM) (San Francisco, CA)* pp 19.7.1–19.7.4

[80] Ji D, Gupta C, Agarwal A, Chan S H, Lund C, Li W, Keller S, Mishra U K and Chowdhury S 2018 *IEEE Electron. Dev. Lett.* **39** 711–4

[81] Ji D and Chowdhury S 2015 *IEEE Trans. Electron. Dev.* **62** 2571–8

[82] Kizilyalli I C and Aktas O 2015 *Semicond. Sci. Technol.* **30** 124001

[83] Yang C *et al* 2019 *IEEE Trans. Electron. Dev.* Unpublished

[84] Kotzea S, Debald A, Heuken M, Kalisch H and Vescan A 2018 *IEEE Trans. Electron. Dev.* **65** 5329–36

[85] Sun M, Zhang Y, Gao X and Palacios T 2017 *IEEE Electron Device Lett.* **38** 509–12

[86] Zhang Y, Sun M, Piedra D, Hu J, Liu Z, Lin Y, Gao X, Shepard K and Palacios T 2017 *2017 IEEE Compound Semiconductor Integrated Circuit Symposium (CSICS) (Miami, FL)* pp 1–3

[87] Ben-Yaacov I, Seck Y K, Mishra U K and DenBaars S P 2004 *J. Appl. Phys.* **95** 2073

[88] Feigelson B N, Anderson T J, Abraham M, Freitas J A, Hite J K, Eddy C R and Kub F J 2012 *J. Cryst. Growth* **350** 21–6

[89] Greenlee J D, Feigelson B N, Anderson T J, Hite J K, Hobart K D and Kub F J 2015 *ECS J. Solid State Sci. Technol.* **4** P382–6

[90] Sakurai H *et al* 2019 *Appl. Phys. Lett.* **115** 142104

[91] Fu K *et al* 2018 *Appl. Phys. Lett.* **113** 233502

[92] Fu K, Fu H, Huang X, Chen H, Yang T H, Montes J, Yang C, Zhou J and Zhao Y 2019 *IEEE Electron Device Lett.* **40** 1728–31

[93] Fu K *et al* 2019 Compound Semiconductor Week (CSW)

[94] Fu K *et al* 2020 *IEEE J. Electron Dev. Soc.* **8** 74–83

[95] Fu K *et al* 2019 *IEEE Electron. Dev. Lett.* **40** 375–8

[96] Fu K, Fu H, Huang X, Yang T H, Chen H, Montes J, Yang C, Zhou J and Zhao Y 2019 *Compound Semiconductor Week (CSW)*

[97] Liu H, Fu H, Fu K, Alugubelli S R, Su P Y, Zhao Y and Ponce F A 2019 *Appl. Phys. Lett.* **114** 082102

[98] Alugubelli S R, Fu H, Fu K, Liu H, Zhao Y, McCartney R and Ponce F A 2019 *Appl. Phys. Lett.* **115** 201602

[99] Hu Z, Nomoto K, Qi M, Li W, Zhu M, Gao X, Jena D and Xing H G 2017 *IEEE Electron. Dev. Lett.* **38** 1071–4

[100] Monavarian M *et al* 2019 *IEEE Electron. Dev. Lett.* **40** 387–90

Chapter 17

Electric-double-layer-modulated AlGaN/GaN high electron mobility transistors (HEMTs) for biomedical detection

Yu-Lin Wang and Chang-Run Wu

17.1 Introduction

In the age of the genome, biosensors have garnered research and market interest, paving the way for new biomedical technologies that focus on delivering high sensitivity detection, high throughput, and cost effectiveness. Field effect transistors (FETs) are potentially the most suitable candidates for biosensor implementation, as they are miniaturized, highly sensitive, and easy to batch produce, making them extremely cost-effective. There is a concerted effort in the field to reduce the dependency on laboratory based testing, as there is a growing demand for personal healthcare devices that can provide bed-side patient monitoring or home-care analysis. This can only be achieved if sensitive and reliable biosensors are designed and manufactured in keeping with the stringent demands of cost and complexity. FET biosensors can satisfy these deliverables but they have traditionally suffered from a major setback. In aqueous solutions, FET sensors suffer from charge screening effects which limit their use in high ionic strength solutions such as clinical samples [1–5]. This fundamental limitation has been tackled by researchers in various ways, such as performing test sample dilutions, filtering and washing, receptor bioengineering, and association with complex microfluidics. These are not desirable approaches for implementing a personal healthcare device which has to satisfy the cost, reliability, and accessibility constraints. There is a pressing demand for a biomarker sensing methodology that allows any user to perform laboratory-like testing in the convenience of their home.

In this chapter, we discuss the advent of electrical double layer (EDL) gated high electron mobility transistor (HEMT) based biosensors for protein detection in clinical samples. HEMT sensors have been previously used in wide ranging applications such as ion, chemical, gas, and biosensing [6–12]. Their chemical inertness, offering resistance to corrosive test solutions, mechanical and thermal capabilities, and highly sensitive electrical response make them particularly attractive for FET based biosensing applications. We have developed a new sensing methodology using EDL gating for HEMT sensors to directly detect proteins from complex media such as 1X PBS and human serum. The charge screening effect is overcome by fundamentally altering the sensor design to form a liquid-capacitor-like structure which is modulated by an external gate bias delivered as short duration DC voltage. The externally modulated electric field brings the advantage of high ionic strength to effectively modulate the HEMT channel current, which is highly sensitive to the concentration of the target protein. This sensing technique can be applied to all types of FET based sensors which employ affinity based reactions at the sensor surface to detect the concentration of target analyte in the test solution. We also developed a simple and robust sensor packaging technology to facilitate portable biosensing for potential applications in personal healthcare devices. Due to the standardized semiconductor manufacturing techniques, batch fabrication of EDL HEMT biosensors is quite easy and cost-effective. With the integration of cloud technology and data management, this diagnostic technique holds the potential to be an innovative Internet-of-things (IoT) based biosensor.

17.2 Fabrication of sensors

17.2.1 Fabrication of HEMT sensors

The AlGaN/GaN HEMT structure consists of a 3 μm thick undoped GaN buffer layer, a 150 Å thick undoped $Al_{0.25}Ga_{0.75}N$ layer, and a 10 Å thick undoped GaN cap layer. The AlGaN/GaN epi-wafer is grown by molecular beam epitaxy (MBE) on a silicon substrate, as shown in figure 17.1(a). The device active region is defined by inductively coupled plasma (ICP) etching with Cl_2/BCl_3 gases (35 sccm/35 sccm) under an ICP power of 300 W and an RF bias of 120 W at 2 MHz, as depicted in figure 17.1(b). Ohmic contacts are fabricated by deposition of a 200 Å thick Ti layer,

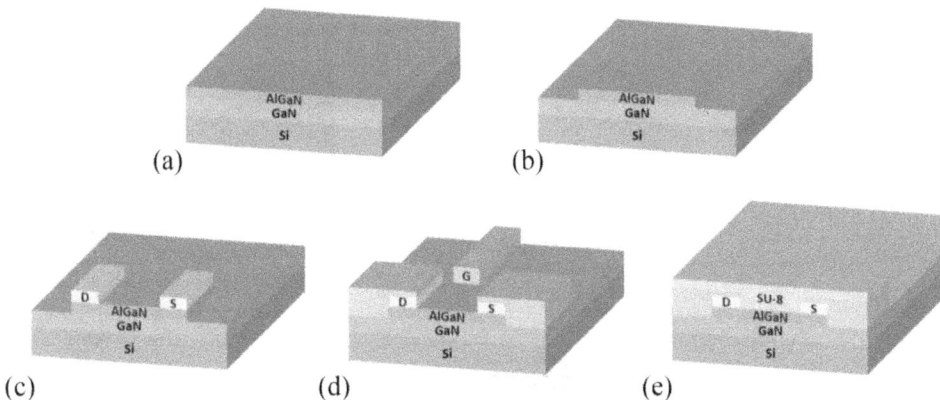

Figure 17.1. The schematic of the HEMT biosensor fabrication process: (a) structure of the epi-wafer, (b) defining the active area by ICP etching, (c) ohmic contact metal (Ti/Al/Ni/Au) deposited with E-beam and annealing, (d) final metal (Ti/Au) deposition with E-beam, and (e) passivation of the photoresist on the device.

a 400 Å thick Al layer, an 800 Å thick Ni layer, and a 1000 Å thick Au layer using an electron beam evaporator, followed by rapid thermal annealing at 850 °C for 45 s in N_2 environment. The ohmic contacts (60×60 μm^2) of the source and drain metals are separated by a 30 μm gap and the transistor's channel width is 50 μm, as illustrated in figure 17.1(c). Figure 17.1(d) shows a 1200 Å thick Au layer deposited as the final metal interconnects on the source and the drain, and as the gate electrode. A passivation layer was generated by spin coating a positive 2 μm thick photoresist (SU-8), followed by a typical photolithography process to create the openings on the gate electrode and on the transistor channel, as shown in figures 17.1(e) and 17.2(a). The gap between the openings on the gate electrode and on the transistor channel may include 65, 265 and 465 μm. The devices are then ready for surface modification and subsequent analyte detection.

17.2.2 Antibody and DNA aptamer immobilization

The gold gate electrode is functionalized with receptors (antibody or ssDNA aptamer) using the strong S–Au covalent binding. Disulfide bonds in the hinge region of the IgG molecule are cleaved using a moderate reducing agent such as 2-mercaptoethylamine (2-MEA) resulting in thiol terminated half IgG molecules binding on the gate electrode. Monoclonal antibodies (anti-CEA, anti-cTnI, and anti-NT-pro BNP) and the MEA mixture is incubated in room temperature on the device for 1.5 h followed by incubation at 4 °C for 12 h. Unbound half-IgG is washed away using PBS before detection of proteins. Thiol-modified DNA aptamers specific to HIV-1 RT and CRP are also used. The original aptamer solution is diluted in a TE buffer containing tris(2-carboxyethyl)phosphine (TCEP) in the molar ratio 1:1000 and heated up to 95 °C to separate the single stranded DNAs. After the mixture is cooled down, 5 μl is dropped on the device and incubated at room temperature for 24 h. The device is washed in TE buffer and PBS prior to measurements.

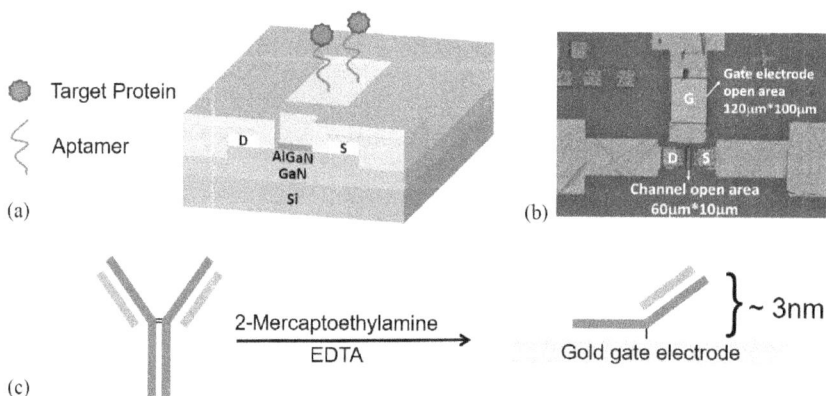

Figure 17.2. (a) A schematic diagram of a GaN HEMT biosensor. (b) Top view of the HEMT chip describing the sensing region. (c) IgG antibody immobilized with 2-MEA.

17.3 Principles and characteristics of EDL AlGaN/GaN HEMT sensors

EDL AlGaN/GaN HEMTs are designed as a separated gate electrode from the active channel of the HEMTs. The gate electrode and the active channel between the source and drain metals are all on the same plane. The AlGaN/GaN HEMTs are depletion mode n-channel field-effect transistors (FETs). When the gate electrode is applied with a positive bias, negative ions will accumulate on the surface of the gate electrode, in the opened area. At the same time, positive ions will also accumulate on the opening of the active channel, resulting in an increased two-dimensional electron concentration in the active channel of the HEMT, thus increasing its conductivity, as shown in figure 17.3(a). EDL forms on the gate metal and the surface of the channel as well. Thus, the solution on the top of the FET can be regarded as part of the gate dielectric of the FET. When the capacitance of the solution changes, the voltage drops in the solution and in the solid dielectric of the FET will both change, resulting in a drain current change, as depicted in figure 17.3(b).

The HEMT sensors shown in this chapter are operated under the linear region of conventional FETs. The drain current can be expressed as

$$I_d \cong \left(\frac{W\mu_n C}{L} \right)\left(V_g - V_t - \frac{1}{2}V_d \right)V_d, \tag{17.1}$$

where W and L represent the width and length of the gate region, C is the capacitance of the dielectric, μ_n is the electron mobility, and V_g, V_t, and V_d are the gate voltage, the threshold voltage, and the drain voltage, respectively. Because this device is a depletion mode FET, without applying any gate bias the device is always 'on' under a constant drain bias. The magnitude of drain current change depends on the transconductance of the FETs, shown as

$$g_m = \left(\frac{\partial I_d}{\partial V_g} \right)_{V_d}. \tag{17.2}$$

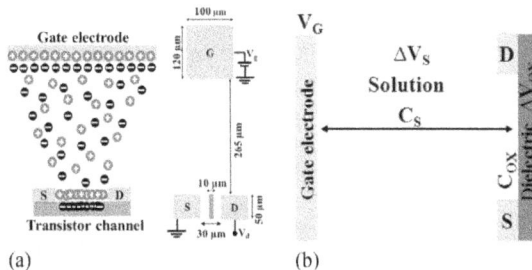

(a)   (b)

Figure 17.3. (a) Distribution of ions under a positive gate bias. Negative ions accumulate on the gate electrode surface and simultaneously positive ions accumulate on the active channel surface, increasing the output drain current. EDL formed on both the gate electrode surface and active channel constitute the gate dielectric of FET. Changes in sample solution capacitance thus directly modulate the drain current. (b) The applied gate voltage drops across the solution and the HEMT dielectric, based on the solution and dielectric capacitance, thus modulating the drain current.

As an FET has a larger transconductance, higher drain current can be induced at a certain ligand concentration. Therefore, an FET is expected to have a large g_m to create a large signal.

When V_g is applied to the gate electrode, the gate voltage drops through the solution (ΔV_s) and across the dielectric of the HEMT (ΔV_{ox}), as illustrated in figure 17.3(b). Because our HEMTs are depletion mode FETs, which have very high conductivity in the active channel, we can ignore the drop of V_g in the channel to simplify the model and assume that V_g only drops across the solution and the dielectric, and then it becomes grounded in the channel with the source terminal when V_d is applied to the drain terminal. By assuming that, the voltage drop of V_g can be divided as the voltage drop in solution, ΔV_s, and the voltage drop in the dielectric, ΔV_{ox}, as shown in equation (17.3). We assume that in the solution, when V_g is applied, the capacitance of the solution C_s is generated due to the formation of EDL, as shown in figure 17.3(a). By taking advantage of the linear relationship between the voltage drop and impedance, the voltage drop in the dielectric can be shown as in equation (17.4), which indicates a larger solution capacitance leading to a larger effective V_g applied on the dielectric of the FET, thus causing a larger drain current increase. For the application of biosensors, the detection of this FET-based sensor is based on the change of the solution capacitance caused by target protein binding with the antibody on the gate electrode opening. Obviously, the immobilization of the antibody or aptamer can also be monitored by the change of C_s, resulting in the change of the drain current gain. We can also conclude that in a higher ionic strength solution, the current gain is larger due to the larger solution capacitance, which is consistent with the explanation of ion density in EDL.

$$V_g = \Delta V_s + \Delta V_{ox} \tag{17.3}$$

$$\Delta V_{ox} = \frac{\frac{1}{j\omega C_{ox}}}{\frac{1}{j\omega C_{ox}} + \frac{1}{j\omega C_s}} \times V_g = \frac{C_s}{C_{ox} + C_s} \times V_g, \tag{17.4}$$

where j and ω are the current and angular frequency, respectively, ΔV_s and ΔV_{ox} are the gate voltage drops in the solution and in dielectric, respectively, and C_s and C_{ox} are the solution capacitance and the dielectric capacitance, respectively.

The EDL AlGaN/GaN HEMTs with a typical dimension that has a gap between the gate electrode opening and the active channel of 265 μm and a gate opening area of 100 μm $\times$ 120 μm, as shown in figure 17.3(a), were first measured in different ionic strength solutions, including de-ionized (DI) water, 0.01X PBS, 0.1X PBS, and 1X PBS in the time domain. The HEMT was biased at $V_{ds} = 2$ V and $V_{gs} = 0$ V in the beginning for 2 μs and then V_g was changed to 0.5 V and maintained for 50 μs at $V_{ds} = 2$ V, resulting in increased drain current. The increased drain current then quickly relaxes and becomes steady. The solution with the highest ionic strength causes the highest steady drain current, after V_g is applied. The DI water causes little current change when V_g is applied, as shown in figure 17.4(a). This phenomena can be easily explained as the high ionic strength solution causes larger EDL

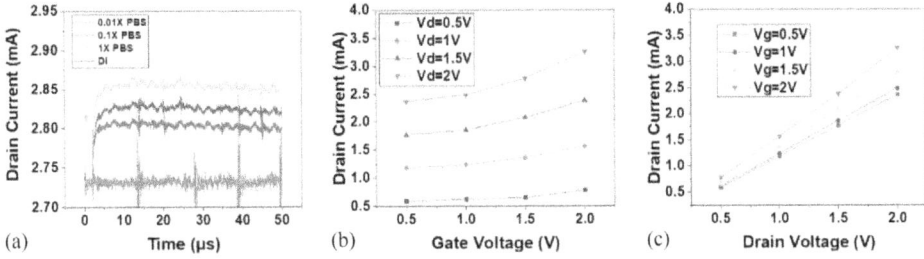

Figure 17.4. (a) Different sample solutions containing low to high ion concentrations are tested at $V_{ds} = 2$ V and $V_{gs} = 0.5$ V for 50 μs: de-ionized water, 0.01 X PBS, 0.1 X PBS, and 1 X PBS (physiological salt concentration); higher ionic strength leads to increased charge accumulation in the active channel leading to larger drain current. (b) The typical HEMT structure is biased at different drain voltages and (c) gate voltages. Figures (b) and (c) show the linear operation of the FET, in which the transconductance gain is proportional to the applied V_d.

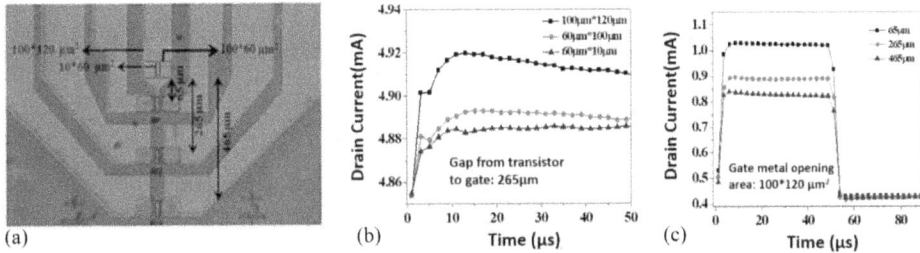

Figure 17.5. (a) Top view of an EDL HEMT depicting the different geometries designed. (b) A larger gate opening area leads to an increased drain current change. (c) Increased gap distances lead to decreased drain current change.

capacitance, resulting in larger conductivity of the solution, leading to less gate voltage drop in the liquid. Thus, the remaining gate voltage on the HEMT is higher and a larger drain current is measured, compared to that of the DI water. The current change can be regarded as the average 'gain', which is an important index (transconductance, $g_m = dI_d/dV_g$) indicating the amplification capability of FET. Here, we simply use 'current gain' to represent the current change after V_g is applied.

The EDL AlGaN/GaN HEMTs were measured in different bias conditions, including $V_d = 0.5$ V~2 V and $V_g = 0.5$ V~2 V, in 1X PBS, with a typical dimension of the HEMT that has a gap between the gate electrode opening and the active channel of 265 μm, and a gate opening area of 120 μm × 100 μm, as shown in figures 17.4(b) and (c). The results show that in such a bias condition, the FET is working in the linear region and the g_m is proportional to V_d, which satisfy the typical FET current–voltage model as shown as in equations (17.1) and (17.2).

Figure 17.5(a) shows three different gaps and opening areas on the gate metal of the EDL HEMTs. Three different gate electrode openings (length × width = $L \times W$), including $L \times W = 120$ μm × 100 μm, 60 μm × 100 μm, and 60 μm × 10 μm with the same gap, 265 μm, between the gate electrode opening and the active channel of the

AlGaN/GaN HEMT, are measured and compared at $V_d = 2$ V and $V_g = 0.5$ V, as shown in figure 17.5(b). The HEMT with a larger gate electrode opening has a higher current gain. Basically, this is attributed to the larger EDL capacitance for the larger opening area. However, the current gain is not always linearly proportional to the opening area. For example, with the same active channel, the device ($L \times W = 120$ μm $\times$ 100 μm) with twice the length of another device ($L \times W = 60$ μm $\times$ 100 μm), has nearly twice current gain. However, if a device ($L \times W = 60$ μm $\times$ 100 μm) has a width ten times larger than another one ($L \times W = 60$ μm $\times$ 10 μm), the current gain only increases by 20%, as depicted in figure 17.5(b). On the other hand, in another device, if the gate electrode opening is fixed at $L \times W = 120$ μm $\times$ 100 μm and the gaps between the gate electrode opening and the active channel are designed as 65, 265, and 465 μm, respectively, the current gains measured at $V_d = 2$ V and $V_g = 0.5$ V for these three gaps show a reversely proportional relationship, as illustrated in figure 17.5(c). This result can be explained as the gap decreases, the impedance also decreases, which causes less gate voltage drop in the solution, leading to larger drain current.

17.4 Beyond the Debye length for protein detection in physiological samples

17.4.1 Protein detection in 1X PBS and human serum

Post-receptor immobilization, the device is first tested for target proteins in a standard buffer solution with physiological salt concentration, which is 1X PBS, followed by testing in clinical human serum samples. To mimic the physiological conditions of human serum, 1% bovine serum albumin (BSA) is prepared in 1X PBS, which is used as the baseline sample (a zero concentration of target proteins) and varying concentrations of target proteins are prepared by diluting them in 1% BSA from the purified protein stock. Prior to electrical measurements, each of the test samples, including the baseline, is incubated on the device for 5 min, under ambient conditions. Post-sample incubation, a short duration (100 μs) gate pulse is applied to the HEMT, which is operated under linear region (V_d proportional to I_d). The response from 1% BSA with zero target protein concentration is recorded as the baseline, as shown in figure 17.6(a). The current gain of the sensor can be calculated as the difference in drain current before and after the application of the pulsed gate bias. This drain current response is a purely non-faradaic response which is marked by the absence of redox reactions at the sensor surface. This is verified by measuring the gate leakage current, which displays a pattern of quick relaxation of current to zero, indicating the absence of a concentration gradient induced electrical resistance from redox reactions. Since our EDL HEMT sensor displays a non-faradaic response, we can primarily exclude the resistive component and adopt a purely capacitive model to describe the sensor behavior. After establishing the baseline current gain response of the sensor, target proteins are introduced in varying concentrations and each of the samples is incubated on the device for 5 min prior to electrical testing. The subsequent current gain response of the sensor can be quantitatively correlated to the concentration of the target protein present in the test

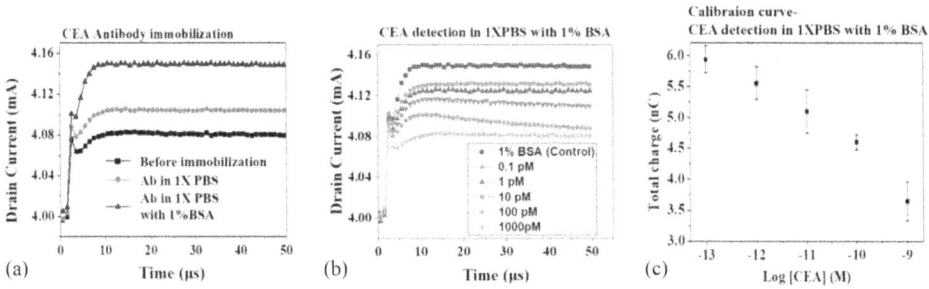

Figure 17.6. (a) Successful antibody immobilization (anti-CEA) is depicted by the time domain drain current gain before and after immobilization. The sensor was measured (in a bare gold electrode) before immobilization of the antibody, with the antibody in 1X PBS, and with the antibody in 1X PBS with 1%. (b) CEA concentration dependent drain current decrease. (c) CEA calibration curve. The total charge versus CEA concentration in 1X PBS with 1% BSA. The measured CEA concentrations are 100 fM, 1 pM, 10 pM, 100 pM, and 1 nM.

Table 17.1. The effective gate voltage changes show no correlation with the protein's net charge [13–15].

Proteins	CRP	CEA	NT-proBNP	HIV-1RT
Isoelectric point (pI)	7.4	4.7	8.5	8~9
Net charge of protein at $1 \times$ PBS	Neutral	Negative	Positive	Positive
Gate potential change caused by increasing protein concentration	+	—	—	—

solution, as shown in figure 17.6(b). The calibration curves for each of the target proteins can be set up by plotting total charge with respect to the protein concentration, as depicted in figure 17.6(c). The sensor displays a wide dynamic range and high sensitivity to the target protein, demonstrating the ability to detect biomarkers without the limitations of Debye length in high ionic strength solutions.

We observe an interesting phenomenon while surveying the calibration curves of various proteins. The current gain increases for increasing CRP protein concentration, while cTnI, NT-proBNP, and fibrinogen display an inverse trend, which is decreasing current gain for increasing protein concentration. This protein dependent behavior of the sensor response curves cannot be attributed to the net charge of each protein (table 17.1), as an inspection of their isoelectric points suggests that the net charge is pH = 7.4 test conditions [16, 17]. The charge-based FET biosensor mechanism of signal transduction has also been disputed in various other studies, even demonstrating the detection of uncharged species using FET devices [18–20]. We further examined this behavior by testing proteins using biphasic gate pulses as the input, where the drain current response is recorded for positive and negative cycles of gate bias. As seen from figure 17.7(a), the sensor response is not symmetric across positive and negative gate bias cycles, with the current gain increasing with increasing CRP concentration, in both the positive and negative directions. This response can experimentally rule out the charge based biosensing model previously

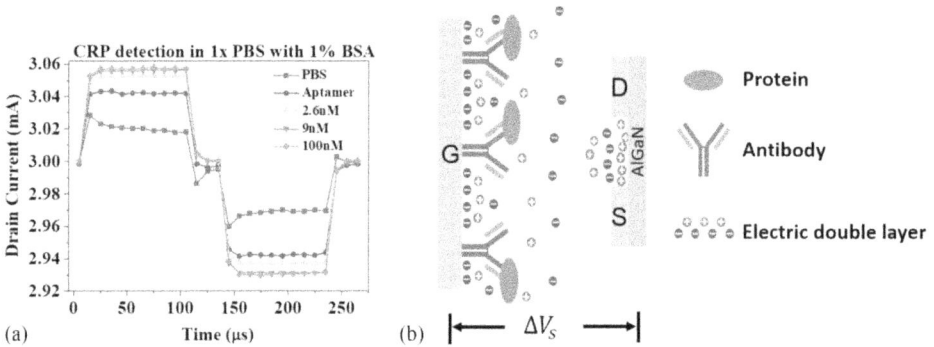

Figure 17.7. (a) Sensor response in dual gate bias (positive and negative 0.5 V V_{gs}). CRP detection in the time domain for 2.6 nM, 9 nM, and 100 nM. (b) Ion distribution in EDL in the solution.

described for FET sensors. The true physiochemical nature of FET sensors may be attributed to the change in capacitance brought about by the receptor–ligand interactions occurring at the sensor surface. The EDL undergoes charge redistribution or local charge perturbations as a result of the specific electrostatic interactions of the receptor and target proteins which changes the overall solution capacitance C_s, as described in equation (17.4) and shown in figure 17.7(b). The solution capacitance further modulates the transistor dielectric capacitance and hence the potential across it, thus gating the drain current of the HEMT accordingly. This change in capacitance brought about by the biochemical interactions cannot be linearly quantified from the sole consideration of the net electrical charge of the target protein. The receptor orientation on the sensor surface, the localized interaction of the myriad functional groups within the bound receptor–ligand moiety and the electrostatic attraction/repulsion of the dissociated salts present in the complex test media are all contributing factors in determining the overall capacitive behavior of the sensor.

The characteristic sensor response pertaining to each of the target proteins is also displayed in the tests conducted in human serum samples [21–23]. Figures 17.8(a)–(c) show the drain current response and the corresponding calibration curves for fibrinogen, cTnI, and NT-proBNP, respectively. The untreated human serum samples are obtained from clinical patients and the concentrations of target proteins are determined using laboratory standard instruments. The serum samples are also incubated on the sensor surface for 5 min prior to electrical measurements, similar to the assay protocols followed for buffer solution tests. Figures 17.8(d)–(f) shows the total charge versus protein concentration of fibrinogen, cTnI, and NT-proBNP, respectively. As seen from figure 17.8, the sensors were measured with Agilent B1530/B1500 and show very high sensitivity to the target proteins, even in the extremely complex and harsh medium of human serum. Apart from demonstrating robustness, these tests also show that EDL HEMT biosensors are also selective in nature. This key feature can be attributed to the sensing mechanism. During the short incubation period (5 min) the specific interactions between the receptor and the target proteins, which possess higher binding affinity and reaction rates, dominate

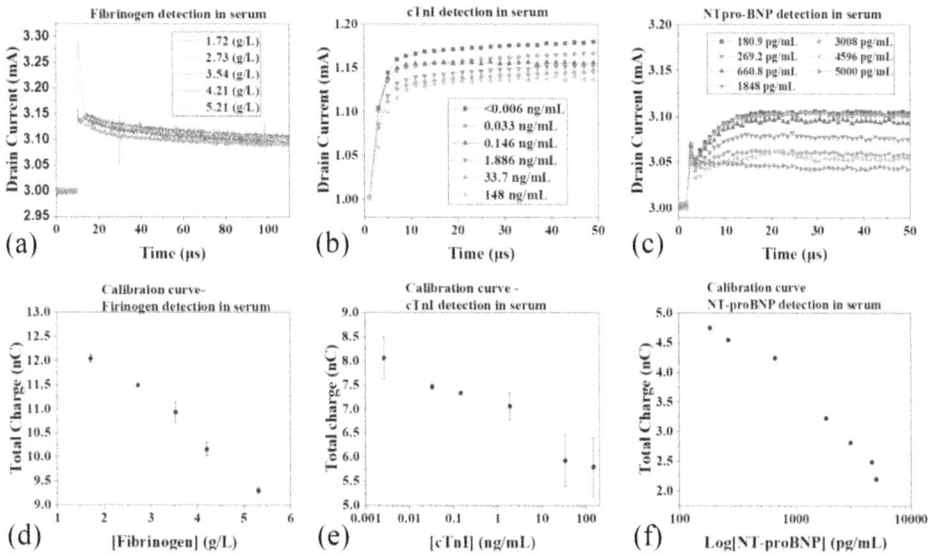

Figure 17.8. Drain current versus time graph for detection of (a) fibrinogen, (b) cTnI, and (c) NT-proBNP in clinical human serum samples. (d)–(f) sensor calibration curves formed by total charge versus protein concentration for the detection of fibrinogen, cTnI, and NT-proBNP, respectively, in clinical human serum samples.

over the non-specific interactions and thus few non-specific interactions contribute to the background signals. The incubation period is long enough to saturate the receptor–ligand binding but is not sufficient for non-specific interactions to contribute significantly to the overall change in solution capacitance. In this manner, our sensing technique provides sensitivity and selectivity to target species, without using complicated assay protocols involving extra reagents or wash/dilution steps. This type of assay is easier to translate to home-care and point-of-care diagnostics for biomarker panels and rapid screening tests.

17.4.2 Tunable and amplified sensitivity

The most important feature of our antibody/aptamer functionalized EDL HEMT sensor is the detection of proteins beyond the limits of Debye screening length. The sensor not only facilitates the detection of the proteins in extremely high ionic solutions, but does so with high sensitivity. This fundamental improvement in sensing warranted further studying of the sensing mechanism. The design of the sensor structure can essentially be described as a two plate capacitor with a liquid dielectric or in other words, a liquid capacitor. From the equation of capacitance, we know the linear inverse dependence of the distance/length of the charged plates/dielectric, respectively, on the overall capacitance. Although this linear distance dependence is true for solid capacitors and not for liquids, we explored the distance effect in our sensors by modulating the spacing between the functionalized gate electrode and the transistor active area. The fixed spacings used in the study were 65, 500, 1000, 3000, 5000, and 10 000 μm between the gate electrode and the transistor.

First the sensor readings are obtained by testing NT-proBNP in 4% BSA solution prepared in 1X PBS. The test solution containing the protein is dropped on the sensor surface and incubated for 5 min prior to electrical measurements. Since varying spacings are provided between the gate electrode and the transistor active area, the drain current response obtained from the EDL HEMT displays a strong correlation to the applied spacing. As seen from figure 17.9, the least spacing (65 μm) provides the largest sensor response, which is the change in current gain. Correspondingly, the largest spacing (10 000 μm) provides the least change in current gain to the applied protein concentration. This empirically establishes the relationship of sensitivity to the spacing between gate electrode and transistor active area, or in other words, the distance between the charged surfaces of the liquid capacitor.

Historically, the sensitivity of FET-based sensors was expressed in terms of change in the potential over concentration of the targeted species. To develop a quantitative comparison, we convert the current gain response to potential change

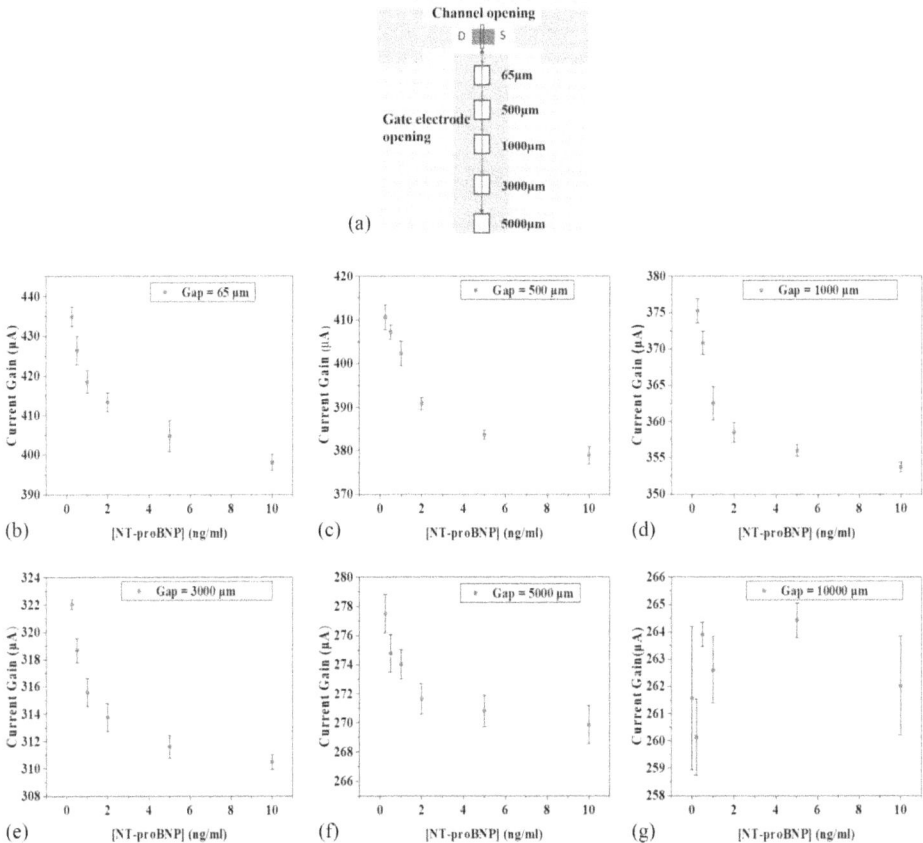

Figure 17.9. (a) Sensor with different spacings between the channel and gate electrode. (b)–(g) NT-proBNP detection in albumin containing buffer in different gaps of 65, 500, 1000, 3000, 5000, and 10 000 μm, respectively ($n = 3$).

by following the I–V response of the sensor described in section 17.4.3. The current gain of the sensor is recorded for varying applied gate bias voltages, which is then used to re-calibrate the gain versus protein concentration plots in terms of change in applied gate bias. The effective V_g for equivalent drain current response can be plotted with respect to the protein concentration, thus developing a calibration curve to deliver sensitivity in terms of V_g per/unit concentration. As seen from figure 17.10, the calibration curves depicting effective V_g over unit protein concentration demonstrates a wide dynamic range and can be expressed in the logarithmic scale to view the linear response. Figures 17.10(a)–(f) correspond to effective V_g versus log concentration curves for varying spacings of 65, 500, 1000, 3000, 5000, and 10 000 μm, respectively. By using linear regression on the calibration curves, we can obtain the sensitivities as 80.54, 75.54, 56.26, 33.53, and 25.38 mV/decade concentrations for spacings of 65, 500, 1000, 3000, and 5000 μm, respectively. At the extremely large spacing of 10 000 μm we observe no sensor response correlation to the protein

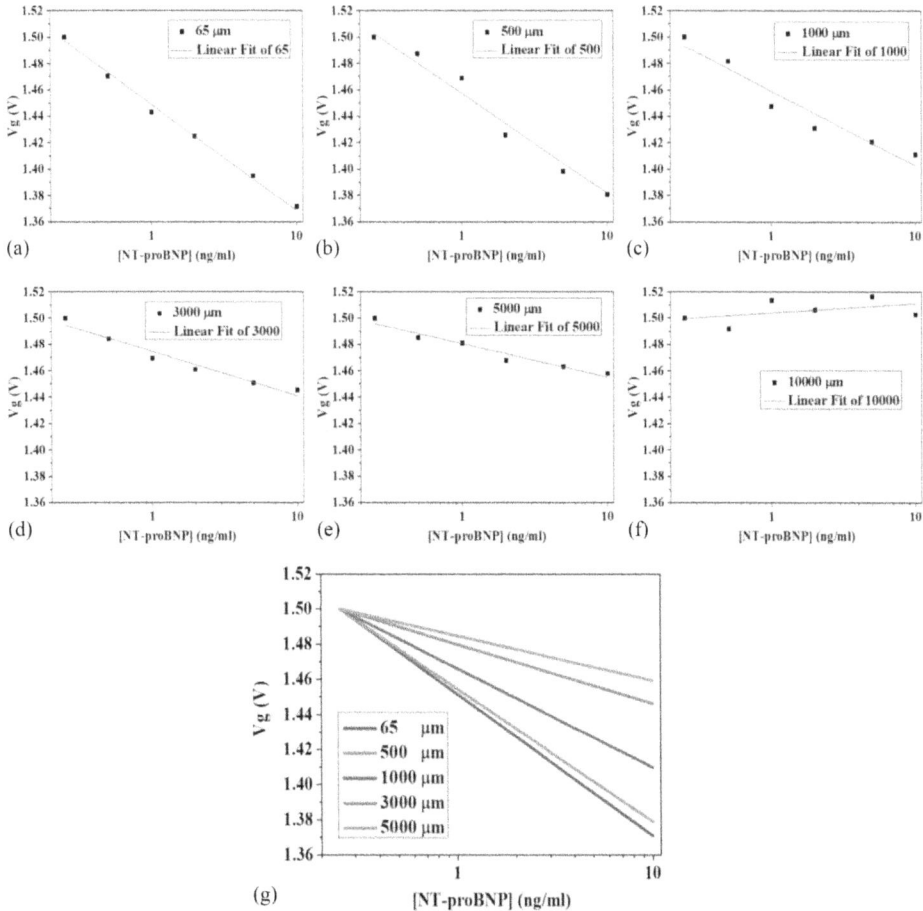

Figure 17.10. (a)–(f) V_g versus NT-proBNP concentration curves for different gaps of 65, 500, 1000, 3000, 5000, and 10 000 μm, respectively. (g) Consolidated V_g versus NT-proBNP concentration curves.

concentration, which marks the upper limit for detection. For traditional FET biosensors, the reference electrode (if there is one) is not applied with a biasing voltage or provided with a very low voltage. Additionally, the reference electrodes are also arbitrarily positioned and not placed at controlled distances away from the transistor active area, as is the case of our sensor [24]. This design, used by traditional FET sensors, can be comparable to our largest spacing design of 10 000 μm, which shows very minimal changes in current gain response, leading to insignificant sensitivity. This change in figure 17.10(f), is indistinguishable and the potential changes occurring at the binding sites are not sensed by the transistor. Similarly, in traditional FET sensors in the absence of external gating bias the strong charge screening in high ionic strength media screens off the potential changes resulting from ligand binding, rendering the FET sensor incapable of target sensing in physiological solutions. The results seen from table 17.2 demonstrate the sensitivity dependence on the applied spacing. The larger gate voltage develops over the smaller spacings and produces a larger change in drain current signal, leading to higher sensitivity. In other words, the EDL HEMT sensitivity is tunable in design and can be amplified by modulating the spacing between the gate electrode and transistor, and the applied gate bias voltage. The optimal combination of these factors can be explored to set the specification of each protein detection assay, as warranted by the purpose and application of the test.

At the advent of FET sensors, Bergveld proposed a quantitative model for an MOSFET-based pH sensor using Boltzmann distribution and EDL theory to estimate the ideal sensitivity, as 59.2 mV/decade [9]. This predicted ideal sensitivity for FET sensors is exactly the same as the ideal sensitivity described in the Nernst equation, used for potentiometric techniques [25–27]. This ideal sensitivity described for ionic selective FET sensors was studied in semi-quantitative models developed by our group for lead and mercury ion-selective FET sensors [27, 28]. It was shown that our sensors can obtain sensitivity higher than the ideal Nernst sensitivity, using a similar separated gate electrode sensing structure as adopted for our protein EDL HEMT sensors. From the results in table 17.2, we have seen that for protein sensors, the sensitivity can be as high as 80 mV/decade, which is much higher than the sensitivities reported in the literature pertaining to FET sensors, which are often in the range of 10–30 mV/decade [29–32]. Therefore, using increased gate bias and shortened gaps, for sensitivity tuning of EDL HEMTs we can amplify the sensor response without using extensive sample pre-treatments in clinical samples.

Table 17.2. Sensitivity values corresponding to different gaps between the channel and electrode.

Gap	65 μm	500 μm	1000 μm	3000 μm	5000 μm
Sensitivity	80.54 mV/decade	75.54 mV/decade	56.26 mV/decade	33.53 mV/decade	25.38 mV/decade
R-square	0.994	0.974	0.920	0.950	0.940

17.4.3 Portable devices for personal healthcare

The major advantages of our EDL HEMT technology include high sensitivity, low sample volume requirement, speedy detection, and cost effectiveness. The lack of sample pre-treatment techniques such as dilution, filtering, and washing renders the use of our sensor quite simple and easy to follow for anyone without prior training. To make this sensor technology easy to translate to clinical applications such as personal healthcare, we developed an integrated HEMT sensor, complete with a packaged microsensor and a portable sensor measurement unit [33, 34]. The sensor packaging technology was developed to use minimal interconnections and seamless integration to the portable measurement unit. The HEMT sensor with a die area of 1 mm^2 was transferred to epoxy resin using a PDMS master and cured using a two-step thermal process. Once the epoxy cures, it forms the substrate onto which the HEMT sensor is embedded. Metal interconnections are formed using photolithography and e-beam evaporation, which are in-plane and very easy to fabricate. This technique eliminates the need for complex thru-VIAs and wire-bondings. Additionally, the epoxy substrate can be patterned as a micro-SD card onto which the HEMT sensor is embedded, as shown in figure 17.11. In this manner, a standard

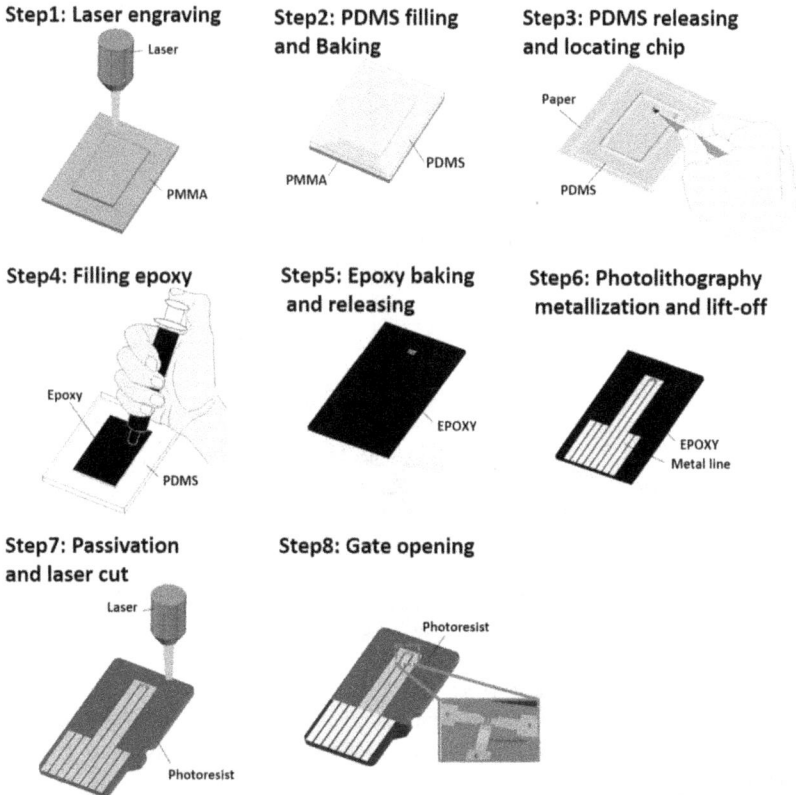

Figure 17.11. Schematic illustration of steps involved in packaging AlGaN/GaN HEMT devices in epoxy substrate.

micro-SD card reader and associated circuitry can be used to measure the sensor response. The packaged sensor is designed as a disposable sensor that can be measured using a portable measurement unit, shown in figure 17.12. This is similar to the current OTC available glucose meters and sensing chips which are the golden standard in personal healthcare. Another major challenge in developing a portable biosensor is sample delivery. Since biomarker detection has to be carried out using complex and harsh medium such as human serum, the sampling techniques often have to rely on complex microfluidics which add to the cost and bulk of the sensor measurement unit. We developed a simple capillary based microfluidic channel which transports extremely small sample volumes, roughly $\sim$2.5 μl from the inlet to the HEMT sensing region, similar to the wicking action of lateral flow assays. Figure 17.13 describes the micro-capillary fabrication and application. Together with the capillary, the packaged sensor can be used with the standalone measurement unit with the click of a button by anyone without the supervision of laboratory trained personnel. Such a sensor is highly attractive for home-care and non-clinical use where quantitative biomarker analyses are required but without the added complex instrumentation and cost. Figure 17.14 shows the detection of cTnI in human serum samples, carried out using the portable HEMT biosensor [22]. These results demonstrate the promising capability of our sensor methodology in developing rapid screening techniques for personal healthcare.

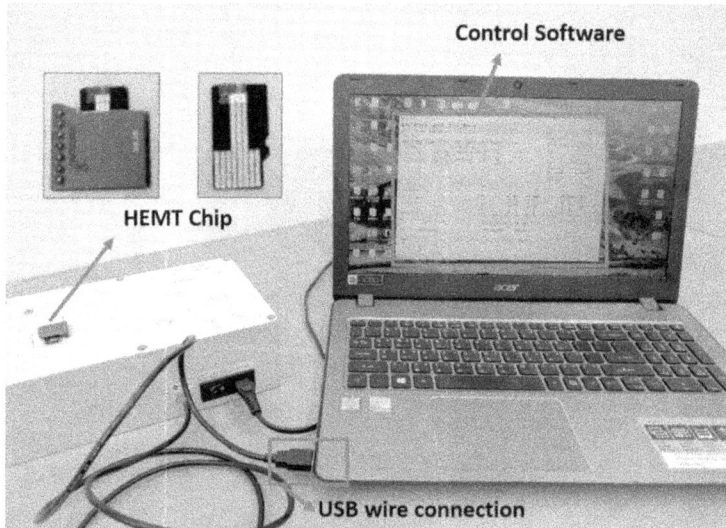

Figure 17.12. The portable measurement device with USB interface to connect to a laptop. The software installed in the laptop provides GUI to control the measurement parameters and display the test results. The inset shows the packaged AlGaN/GaN HEMT chip connected to the micro-SD card reader which is mounted on the portable device.

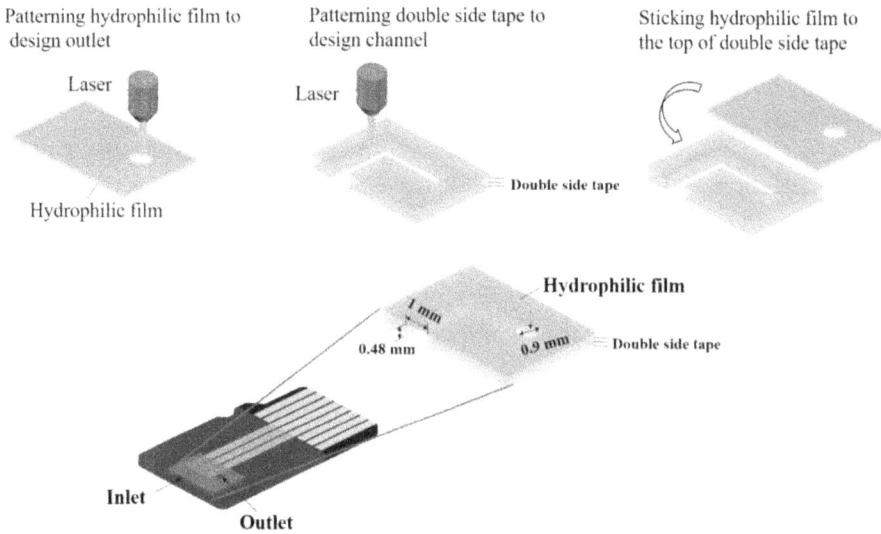

Figure 17.13. Schematic representation of design of the capillary microchannel. The hydrophilic film and stack of double-sided tapes are patterned using a laser drill and stuck together to form the capillary microchannel which is then placed on the packaged chip such that the AlGaN/GaN HEMT device is positioned inside the capillary channel.

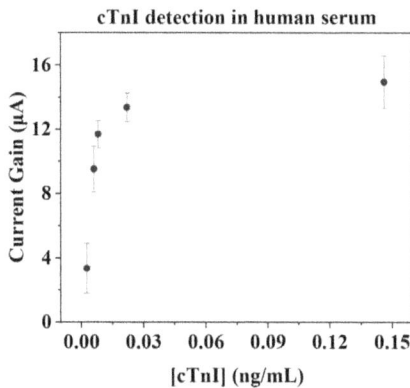

Figure 17.14. Current gain versus concentration graph obtained from a handheld device for the detection of troponin I in clinical human serum samples, including 0.0026, 0.006, 0.008, 0.022, and 0.146 ng ml^{-1}, respectively, obtained using the portable biosensor system.

17.5 Summary

In this chapter, we present the design, fabrication, and characterization of EDL HEMT biosensors and demonstrate their application in personal healthcare. The HEMT sensors are designed to overcome the fundamental limitations of traditional FET biosensors, which is Debye screening, and offer direct sensing of proteins in clinical samples. The fundamental differences of our sensor technology and traditional FET sensors are explored and experimental analyses are used to demonstrate why the present methodology can overcome the charge screening effect in high ionic strength solutions. The sensor packaging technology also integrates the features of ease and robustness to provide on-site marker detection without the hassle of complex instrumentation and excessive cost. This sensor technology can be applied to all types of FET-based sensors and demonstrates promising potential for commercialization and clinical applications.

References

[1] Chen K-I, Li B-R and Chen Y-T 2011 Silicon nanowire field-effect transistor-based biosensors for biomedical diagnosis and cellular recording investigation *Nano Today* **6** 131–54

[2] Lee C-S, Kim S K and Kim M 2009 Ion-sensitive field-effect transistor for biological sensing *Sensors* **9** 7111–31

[3] Lin S-P, Pan C-Y, Tseng K-C, Lin M-C, Chen C-D, Tsai C-C, Yu S-H, Sun Y-C, Lin T-W and Chen Y-T 2009 A reversible surface functionalized nanowire transistor to study protein–protein interactions *Nano Today* **4** 235–43

[4] Stern E, Vacic A, Rajan N K, Criscione J M, Park J, Ilic B R, Mooney D J, Reed M A and Fahmy T M 2010 Label-free biomarker detection from whole blood *Nat. Nanotechnol.* **5** 138

[5] Zheng G, Patolsky F, Cui Y, Wang W U and Lieber C M 2005 Multiplexed electrical detection of cancer markers with nanowire sensor arrays *Nat. Biotechnol.* **23** 1294

[6] Arya S K, Wong C C, Jeon Y J, Bansal T and Park M K 2015 Advances in complementary-metal–oxide–semiconductor-based integrated biosensor arrays *Chem. Rev.* **115** 5116–58

[7] Bausells J, Carrabina J, Errachid A and Merlos A 1999 Ion-sensitive field-effect transistors fabricated in a commercial CMOS technology *Sensors Actuators* B **57** 56–62

[8] Bergveld P 1970 Development of an ion-sensitive solid-state device for neurophysiological measurements *IEEE Trans. Biomed. Eng.* **17** 70–1

[9] Bergveld P 2003 Thirty years of ISFETOLOGY: what happened in the past 30 years and what may happen in the next 30 years *Sensors Actuators* B **88** 1–20

[10] Chu B H, Chang C, Kroll K, Denslow N, Wang Y-L, Pearton S, Dabiran A, Wowchak A, Cui B and Chow P 2010 Detection of an endocrine disrupter biomarker, vitellogenin, in largemouth bass serum using AlGaN/GaN high electron mobility transistors *Appl. Phys. Lett.* **96** 013701

[11] Cui Y and Lieber C M 2001 Functional nanoscale electronic devices assembled using silicon nanowire building blocks *Science* **291** 851–3

[12] Wang Y-L, Chu B, Chen K, Chang C, Lele T, Tseng Y, Pearton S, Ramage J, Hooten D and Dabiran A 2008 Botulinum toxin detection using AlGaN/GaN high electron mobility transistors *Appl. Phys. Lett.* **93** 262101

[13] Tsujimoto M, Inoue K and Nojima S 1983 Purification and characterization of human serum C-reactive protein *J. Biochem.* **94** 1367–73

[14] Casey B J and Kofinas P 2008 Selective binding of carcinoembryonic antigen using imprinted polymeric hydrogels *J. Biomed. Mater. Res.* A **87** 359–63

[15] Davis A J, Carr J M, Bagley C J, Powell J, Warrilow D, Harrich D, Burrell C J and Li P 2008 Human immunodeficiency virus type-1 reverse transcriptase exists as post-translation-ally modified forms in virions and cells *Retrovirology* **5** 115

[16] Burnette R R and Ongpipattanakul B 1987 Characterization of the permselective properties of excised human skin during iontophoresis *J. Pharm. Sci.* **76** 765–73

[17] Shah B, Surti N and Misra A 2011 Other routes of protein and peptide delivery: transdermal, topical, uterine, and rectal *Challenges in Delivery of Therapeutic Genomics and Proteomics* (Amsterdam: Elsevier) pp 623–71

[18] Chen R J, Bangsaruntip S, Drouvalakis K A, Kam N W S, Shim M, Li Y, Kim W, Utz P J and Dai H 2003 Noncovalent functionalization of carbon nanotubes for highly specific electronic biosensors *Proc. Natl Acad. Sci.* **100** 4984–9

[19] Chen R J, Choi H C, Bangsaruntip S, Yenilmez E, Tang X, Wang Q, Chang Y-L and Dai H 2004 An investigation of the mechanisms of electronic sensing of protein adsorption on carbon nanotube devices *JACS* **126** 1563–8

[20] Tsai C-C, Chiang P-L, Sun C-J, Lin T-W, Tsai M-H, Chang Y-C and Chen Y-T 2011 Surface potential variations on a silicon nanowire transistor in biomolecular modification and detection *Nanotechnology* **22** 135503

[21] Chu C-H, Sarangadharan I, Regmi A, Chen Y-W, Hsu C-P, Chang W-H, Lee G-Y, Chyi J-I, Chen C-C and Shiesh S-C 2017 Beyond the Debye length in high ionic strength solution: direct protein detection with field-effect transistors (FETs) in human serum *Sci. Rep.* **7** 5256

[22] Sarangadharan I, Regmi A, Chen Y-W, Hsu C-P, Chen P-c, Chang W-H, Lee G-Y, Chyi J-I, Shiesh S-C and Lee G-B 2018 High sensitivity cardiac troponin I detection in physiological environment using AlGaN/GaN high electron mobility transistor (HEMT) biosensors *Biosens. Bioelectron.* **100** 282–9

[23] Tai T-Y, Sinha A, Sarangadharan I, Pulikkathodi A K, Wang S-L, Shiesh S-C, Lee G-B and Wang Y-L 2018 Aptamer-functionalized AlGaN/GaN high-electron-mobility transistor for rapid diagnosis of fibrinogen in human plasma *Sensors Mater.* **30** 2321–31

[24] Tai T-Y, Sinha A, Sarangadharan I, Pulikkathodi A K, Wang S-L, Lee G-Y, Chyi J-I, Shiesh S-C, Lee G-B and Wang Y-L 2019 Design and demonstration of tunable amplified sensitivity of AlGaN/GaN high electron mobility transistor (HEMT)-based biosensors in human serum *Anal. Chem.* **91** 5953–60

[25] Guidelli E J, Guerra E M and Mulato M 2012 V_2O_5/WO_3 mixed oxide films as pH-EGFET sensor: sequential re-usage and fabrication volume analysis *ECS J. Solid State Sci. Technol.* **1** N39–44

[26] Kaisti M 2017 Detection principles of biological and chemical FET sensors *Biosens. Bioelectron.* **98** 437–48

[27] Chen Y-T, Sarangadharan I, Sukesan R, Hseih C-Y, Lee G-Y, Chyi J-I and Wang Y-L 2018 High-field modulated ion-selective field-effect-transistor (FET) sensors with sensitivity higher than the ideal Nernst sensitivity *Sci. Rep.* **8** 8300

[28] Chen Y-T, Hseih C-Y, Sarangadharan I, Sukesan R, Lee G-Y, Chyi J-I and Wang Y-L 2018 Beyond the limit of ideal Nernst sensitivity: ultra-high sensitivity of heavy metal ion detection with ion-selective high electron mobility transistors *ECS J. Solid State Sci. Technol.* **7** Q176–83

[29] Caras S and Janata J 1980 Field effect transistor sensitive to penicillin *Anal. Chem.* **52** 1935–7

[30] Miyahara Y, Moriizumi T and Ichimura K 1985 Integrated enzyme FETs for simultaneous detections of urea and glucose *Sensors Actuators* **7** 1–10

[31] Poghossian A 1997 Method of fabrication of ISFET-based biosensors on an $Si–SiO_2–Si$ structure *Sensors Actuators* B **44** 361–4

[32] Song K-S, Sakai T, Kanazawa H, Araki Y, Umezawa H, Tachiki M and Kawarada H 2003 Cl^- sensitive biosensor used electrolyte-solution-gate diamond FETs *Biosens. Bioelectron.* **19** 137–40

[33] Chen P-C, Chen Y-W, Sarangadharan I, Hsu C-P, Chen C-C, Shiesh S-C, Lee G-B and Wang Y-L 2017 Field-effect transistor-based biosensors and a portable device for personal healthcare *ECS J. Solid State Sci. Technol.* **6** Q71–6

[34] Hsu C-P, Chen P-C, Pulikkathodi A K, Hsiao Y-H, Chen C-C and Wang Y-L 2017 A package technology for miniaturized field-effect transistor-based biosensors and the sensor array *ECS J. Solid State Sci. Technol.* **6** Q63–7

IOP Publishing

Wide Bandgap Semiconductor-Based Electronics

Fan Ren and Stephen J Pearton

Chapter 18

Irradiation effects on high aluminum content AlGaN channel devices

Patrick Carey, Fan Ren, Jinho Bae, Jihyun Kim and Stephen Pearton

18.1 Introduction

Space bound devices are subject to significant environmental damage due to highly energetic charged particles. The dominant semiconductors of choice for extraterrestrial applications are SiC and GaN due to their intrinsic radiation hardness with mean displacement energies of 21 and 19.8 eV, respectively. One of the key engineering challenges in the design of a spacecraft is walking the fine line of sufficiently protecting the electronic devices and the added mass of the shielding. As more and more space bound devices are made of wide bandgap semiconductors rather than GaAs or Si, less wasted mass can be spent on shielding or the lifetime of the devices can be greatly extended by over 10× with the same shielding [1–11].

For communications system satellites within the low earth orbit the primary source of radiation is from the Van Allen radiation belts. The Van Allen belts originate from the strong magnetic field lines emanating from the Earth's core which create a toroidal circuit that traps low energy protons and electrons. The Van Allen belts were first discovered by accident as the Geiger counters on board spiked and reached their maximum reading [12]. The Van Allen belts are not uniform above the Earth and do have local hot spots. Galactic cosmic rays (GCRs) are another source of high energy particles with a composition of 87% hydrogen, 12% helium, and 1% heavy ions/neutrons/x-rays. GCRs originate from outside the protective heliosphere. The Voyager I and Voyager II probes have given the first glimpses of how the cold interstellar winds interact with the hot plasma of our solar wind and what lies beyond the influence of the Sun, and are able to measure GCRs free from our local influences. The third key source of radiation is solar particle events (SPEs). The sun regularly ejects energetic particles and these particles can reach the Earth within minutes. The collision of these particles with the dense atmosphere of Earth creates plasma and visual light that can be observed, the *aurora borealis* and *aurora australis*. The sun undergoes an 11 year cycle of four inactive years followed by seven active years with fluences above 5×10^7 particles/cm^2 at energies >10 MeV [13]. The last truly devastating ejection from the Sun occurred in 1859, the Carrington Event [14], when a coronal mass ejection induced the largest geomagnetic storm in recorded history, with miners in the Colorado Rockies waking in the middle of the night thinking it was daytime due to the intense brightness from the generated plasma.

To improve the longevity of extraterrestrial devices, a shift to ultra-wide bandgap materials may be useful in some applications. A shift to AlGaN devices drives up the bandgap and critical electric field of the material. For example, early results using a lateral metal–oxide–semiconductor field effect transistor (MOSFET) composed of $Al_{0.7}Ga_{0.3}N$ achieved an average critical field of 3.6 MV cm^{-1}, already surpassing the limitations of SiC and GaN [15]. For a $Al_{.085}Ga_{0.15}N/Al_{0.7}Ga_{0.3}N$ high electron mobility transistor (HEMT), high temperature operation up to 500°C in ambient atmosphere has been demonstrated with ideal gate-lag performance at 100 kHz [16]. This same HEMT was able to achieve a critical electric field of 2.6 MV cm^{-1} [17]. Early small signal results for this HEMT are promising with an f_t 28.4 GHz and f_{max} = 18.5 GHz and load-pull measurements at 3 GHz realizing an output power efficiency of 0.38 W mm^{-1} [18]. The third type of device that combines some of the benefits of both types of structures is a polarization doped field effect transistors (POLFET), where a continuous grading of the channel from low aluminum content to high aluminum content creates a strong polarization field and induces a near surface electron channel. This avoids the some of the difficulties associated with forming planar ohmic contacts to such highly insulative AlGaN layers [17, 19–24]. The other benefit of the POLFET structure is the removal of impurity scattering, potentially improving the high frequency performance of its HEMT counterpart.

In this chapter a discussion of the radiation response of high aluminum content AlGaN POLFET devices to proton and alpha irradiation is presented. Key computational modeling of irradiation effects will be used to describe the expected changes in device performance followed by the experimental results of such irradiation experiments.

18.2 SRIM modeling

There are two primary interactions that an incident particle can have with a given target: electronic and nuclear. A highly energetic charged particle will lose the majority of its energy through electronic (Coulombic) stopping, where energy is lost through inelastic energy transfer to electrons by ionizing and exciting the target atoms and the ion itself. The charged particle can be relatively assumed to traverse in a straight line as the likelihood of a collision with the target material nuclei is small. Only once sufficient Coulombic energy has been lost is the likelihood of the knock-on collision of the incident particle and the nuclei of a target atom reasonable to consider as straggle may begin to occur.

The primary workhorse for stopping range in motion (SRIM) modeling of incident particles and their Coulombic energy loss is via the Bethe equation [25],

$$S = \frac{4\pi e^4 Z_2}{m_e v^2} Z_1^2 \left[\ln\left(\frac{2mv^2}{\langle I \rangle}\right) - \ln(1 - \beta) - \beta^2 \right], \tag{18.1}$$

where S is the stopping cross section, Z_1 is the particle atomic number, Z_2 is the target atomic number, e is the charge of an electron, m_e is the mass of an electron, v is the particle velocity, $\langle I \rangle$ is the mean excitation energy per electron, and β is the relative particle velocity (v/c, where c is the speed of light). There have been many works since the original formulation which have gone on to offer corrections for a variety of factors that were not accounted for originally to bring the equation in closer alignment with experimental values; these have been summarized well in other works [26]. For our purposes of providing a brief insight into material comparisons rather than a direct relation to experimental data, these corrections are not necessary and would not offer more than a 10% correction at the energies used in the following experiments [27]. Elastic energy loss through collision of atomic nuclei is the primary generator of lattice displacements and amorphization through phonon energy dissipation. The nuclear energy loss can be determined if one knows the interatomic potential energy between the incident ion and the target. The Ziegler–Biersack–Littmark (ZBL) potential is a reasonable fit for the simple quantum calculation of interatomic potentials. The ZBL screening potential provided an analytical expression for the dimensionless screening potential ($\Phi(x)$) for arbitrary atom pairs,

$$\Phi(x) = 0.1818e^{-3.2x} + 0.5099e^{-0.9432x} + 0.2802e^{-0.4028x} + 0.02817e^{-0.2016x}, \tag{18.2}$$

where $\Phi(x)$ is a function of the reduced atomic radius (x) which is the atomic radius (r) normalized to the atomic scaling length (a_u), $x = \frac{r}{a_u}$. The atomic scaling length was described by ZBL as

$$a_u = \frac{0.8853 a_0}{Z_1^{0.23} + Z_2^{0.23}}, \tag{18.3}$$

where a_0 is the Bohr radius, and Z_1 and Z_2 are the atomic numbers of the ion and target atom, respectively.

The largest drawback with the use of SRIM is that all layers are treated as amorphous so the effects of channeling in crystalline materials cannot be accounted for. In particular with heterostructures, the damage cascades with ion implantation profiles may not be accurate. However, they can still provide important material insights into homogeneous layers.

For Coulombic energy loss, the effects are greater in high Z materials due to the higher electron density (figure 18.1).

The electron density for a material can be approximated by the expression

$$n = \frac{N_A Z \rho}{M}, \tag{18.4}$$

where N_A is Avogadro's number, ρ is the density of the material, and M is the molecular mass. For GaN and AlN we find approximate electron densities of

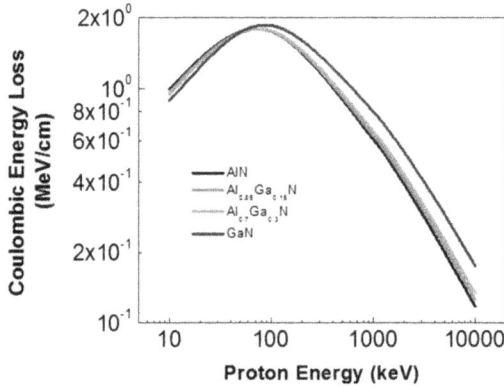

Figure 18.1. Coulombic energy loss as a function of proton energy in GaN to AlN. Copyright IOP Publishing. Reproduced with permission. All rights reserved.

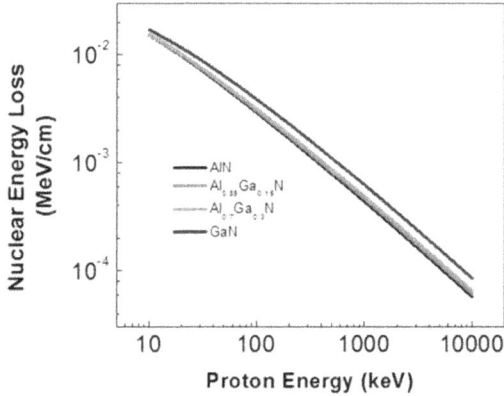

Figure 18.2. Nuclear energy loss as a function of proton energy in GaN to AlN. Copyright IOP Publishing. Reproduced with permission. All rights reserved.

1.7×10^{24} cm^{-3} and 9.6×10^{23} cm^{-3}, respectively. This implies a lower total ionization dose would be absorbed by materials closer to AlN than GaN. From SRIM, we find a 40% reduction in electronic stopping power at 10 MeV, which is near the average proton energy within the Van Allen radiation belts. The same trend is observed when considering nuclear energy loss (figure 18.2).

While the electron density does affect the shielding from nuclear interactions, knowledge of the bond strength also confirms this observation where the bond strength of Ga–N and Al–N are 8.92 eV/atom and 11.52 eV/atom, respectively. This further suggests a preference for highly alloyed AlGaN for radiative environments.

While protons do account for the majority of energetic particles in extraterrestrial applications, alpha particles (helium) make up the next highest portion. For a fixed energy we can determine a key rough proportionality from equation (18.1). For example, at a fixed energy of an ion a helium atom would have a 2× greater velocity than a hydrogen atom, $E = 1/2 \ mv^2$. This would lead to an approximately four-fold

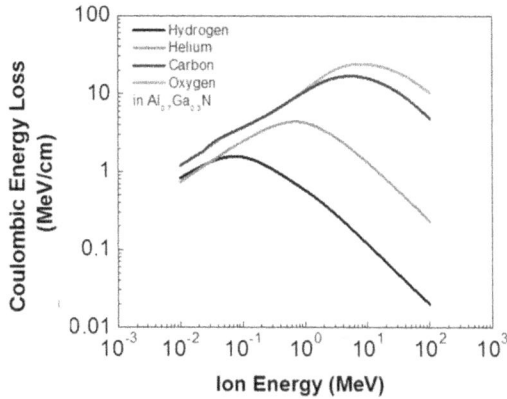

Figure 18.3. Coulombic energy loss as a function of ion energy in $Al_{0.7}Ga_{0.3}N$.

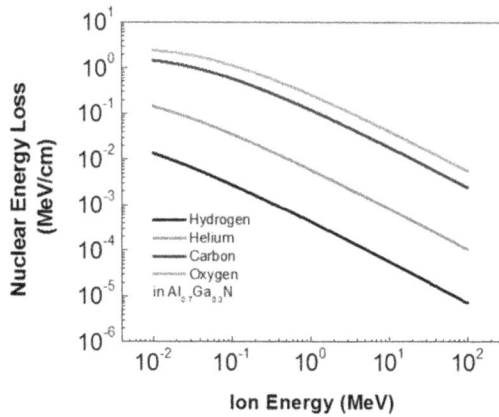

Figure 18.4. Nuclear energy loss as a function of ion energy in $Al_{0.7}Ga_{0.3}N$.

higher electronic stopping power for a material with incident helium versus hydrogen, figure 18.3.

A slightly larger than four-fold increase is observed within the SRIM calculation as the other terms do contribute. Two other elements are also provided for comparison: carbon and oxygen. As elements become heavier the maximum energy for coulombic energy loss becomes higher, due to the inverse square relation of the velocity to the coulombic energy loss. Below the maximum the energy loss begins to be dominated by nuclear stopping, which has a linear relation in a log–log scale, figure 18.4.

As the incident ions become more massive the scattering cross section becomes significantly larger and the likelihood of creating a defect is increased. While heavy ions can produce more damage, they are easier to shield a device from due to an increase in interactions with any material they pass through.

The next consideration is the types of defects which will be most prone to form. The key parameter in this case is the displacement energy, E_d, of the atoms in the crystal structure. Displacement energies in a crystal structure are directionally dependent and this can skew the type of defects formed.

In GaN, *ab initio* molecular dynamic (AIMD) simulations found the minimum E_d to be 39 eV and 17 eV for gallium and nitrogen atoms, respectively, along the $\langle\bar{1}010\rangle$ direction [28]. The gallium defect forms a gallium vacancy and gallium octahedral interstitial. The nitrogen defect forms a nitrogen vacancy and a tilted nitrogen–nitrogen dumbbell in the $\langle\bar{1}010\rangle$. If we then look along the [0001] direction a nitrogen vacancy requires 78 eV to form along with a nitrogen–nitrogen dumbbell.

In wurtzite AlN, AIMD found the minimum E_d to be 55 eV and 19 eV for aluminum and gallium, respectively, in the $[\bar{1}\bar{1}20]$ direction [29]. The aluminum defect forms an aluminum vacancy and aluminum octahedral interstitial. The nitrogen forms a tilted N dumbbell in the $[\bar{1}\bar{1}20]$ direction, but not a nitrogen vacancy. To form a nitrogen vacancy and dumbbell 44 eV is required.

In both binary cases, the nitrogen defect requires less energy to form and would be expected to be the dominant defect type under radiation. Most probably, with the formation of a majority p-type defect an induced loss of electrical carriers in n-channel devices would be expected.

18.3 Device fabrication overview

The POLFET structures were grown at low pressure (75 torr) using metal–organic vapor phase epitaxy (MOVPE) on sapphire substrates using trimethylgallium, trimethylammonia, and ammonia as precursors. First, a 2.3 μm thick AlN nucleation and buffer layer were grown on the (0001) *c*-plane sapphire misoriented by 0.2° toward the *m*-plane. Then a 0.25 μm unintentionally doped (UID) $Al_{0.7}Ga_{0.3}N$ was grown, followed by the linearly graded 110 nm $Al_xGa_{1-x}N$ from $x = 0.7$–0.85. The molar flux of the group III precursors were maintained at a constant value as the compositions was varied. Contactless measurement of the sheet resistance yielded 5500 Ω/□, which combined with CV measurement gives a carrier density, n_s, of 5.4×10^{12} cm^{-2}. Circular POLFETs were fabricated with a gate length of 2 μm, and symmetric source–drain to gate spacing of 3 μm. The gate circumference was 660 μm. Planar Zr/Al/Mo/Au ohmic contacts were deposited and subsequently annealed ($\rho_c = 1.1 \times 10^{-3}$ Ω cm^2). The gate was formed by deposition of Ni/Au into an opening on the 100 nm thick SiN dielectric.

18.4 Proton irradiation

Proton irradiation was performed using a cyclotron and two doses were tested at 10 MeV, 1×10^{14} cm^{-2}, and 3×10^{14} cm^{-2}. In figure 18.5 the drain current–voltage characteristics are shown for the POFLET devices pre- and post-irradiation. Drain saturation current was reduced by 25% and 45% at fluences of 1×10^{14} cm^{-2} and 3×10^{14} cm^{-2}, respectively. In the mobility dominated regime, below the knee voltage, the slope of *I–V* curves has been reduced, indicating a loss of electron mobility due to the removal of carriers.

I_d–V_g and g_m are shown in figure 18.6 with a reduction in transconductance of 19% and 47% at fluences of 1×10^{14} cm^{-2} and 3×10^{14} cm^{-2}, respectively. A positive

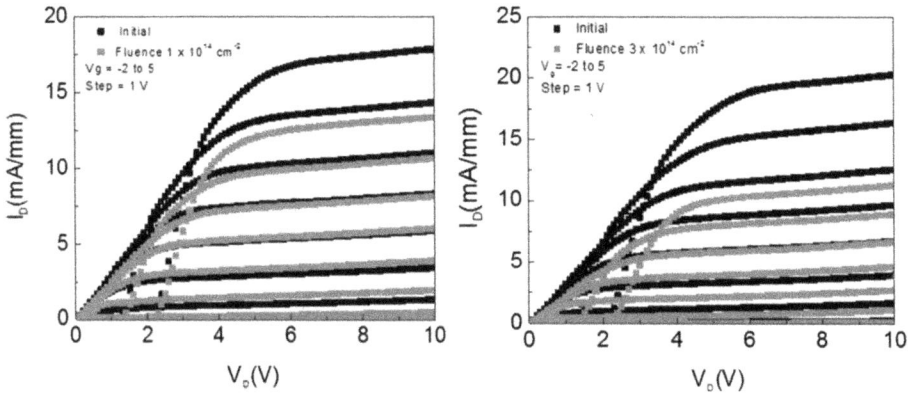

Figure 18.5. Current–voltage characteristics in the reference state and after proton irradiation. Copyright IOP Publishing. Reproduced with permission. All rights reserved.

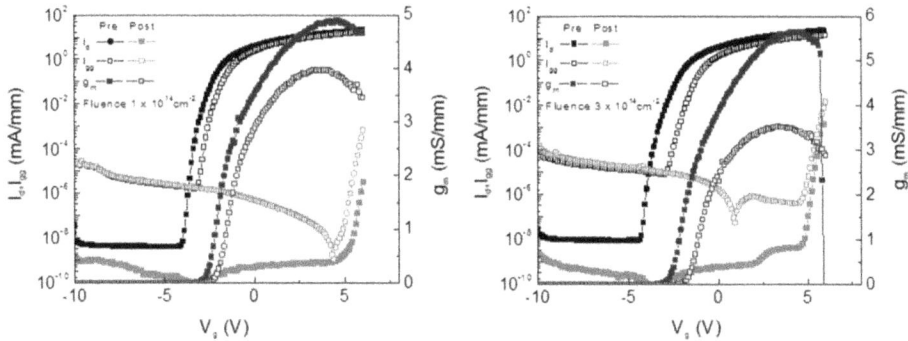

Figure 18.6. Transfer characteristics before and after proton irradiation. Copyright IOP Publishing. Reproduced with permission. All rights reserved.

shift was noted in the threshold voltage, further indicating a loss of carrier concentration. Using a relation that had previously been derived for GaN HEMT devices to describe the carrier density in the two-dimensional electron gas (2DEG) under strong inversion, we can approximate the carrier density [30]

$$n_s = \frac{2\varepsilon q D}{\varepsilon + 2q^2 D d}(V_{gs} - V_{th}),\qquad(18.5)$$

where ε is the dielectric constant of the barrier layer, V_{gs} is the applied gate to source voltage under strong inversion, V_{th} is the threshold voltage, and D is the conduction band density of states where $D = 4\pi m^*/h^2$, m^* is the effective electron mass and h is Planck's constant. The challenge with the application of this equation is that with a POLFET, a 3D electron slab (3DES) is formed rather than a 2DEG. The 3DES lies 55 nm beneath the surface of the device and encompasses the bottom 55 nm of the channel layer [21]. As the barrier layer is graded there is not a fixed value for the dielectric constant, so the dielectric constant was integrated across the barrier layer given known values for GaN and AlN. This gave in the reference state a carrier density of 5.97×10^{12} cm^{-2}, which compares very well to the contactless mercury probe CV measurement of 5.4×10^{12} cm^{-2}. For the two irradiated samples, the low and high dose gave carrier densities of 4.82×10^{12} cm^{-2} and 4.32×10^{12} cm^{-2}, respectively. The carrier removal rates were calculated using

$$R_c = \frac{n_{s0} - n_s}{\Phi},$$

(18.6)

where Φ is the proton fluence, n_{s0} is the initial carrier concentration, and n_s is the irradiated carrier concentration. The values were normalized to volume density to place them in units of cm^{-1} rather than being unitless for comparison to other irradiation studies. The removal rate as the energy is fixed is expected to be the same for both samples, as the results are normalized by the dose. However, carrier removal rates at the low and high dose were found to differ slightly, 1040 cm^{-1} and 515 cm^{-1}, respectively. This can be attributed to the effects of channeling of irradiation particles through the devices along with the small device-to-device variation.

Under irradiation the devices underwent significant degradation of the leakage current. This can be correlated to simultaneous changes in the gate contact performance. In figure 18.7 the gate diode current–voltage characteristics are presented.

Using the thermionic emission model, the Schottky barrier height (SBH) and ideality factor can be extracted:

$$I = I_0 \exp\left(\frac{qV}{nkT}\right)\left[1 - \exp\left(-\frac{qV}{kt}\right)\right]$$

(18.7)

$$I_0 = AA^{**}T^2 \exp\left(-\frac{q\phi_b}{kT}\right),$$

(18.8)

where I_0 is the reverse saturation current, V is the applied voltage, n is the ideality factor, A is effective diode area, A^{**} is the Richardson constant, k is the Boltzmann constant, and ϕ_b is the SBH. The SBH and ideality factor are presented in table 18.1.

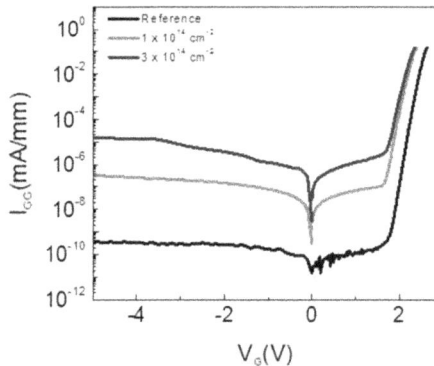

Figure 18.7. Gate I–V of the reference and irradiated POLFETs. Copyright IOP Publishing. Reproduced with permission. All rights reserved.

Table 18.1. Schottky barrier height and ideality factor extracted via the thermionic emission model.

Sample	n	SBH (eV)
Reference 1×10^{14} cm^{-2}	1.37	2.23
Irradiated 1×10^{14} cm^{-2}	1.58	1.84
Reference 3×10^{14} cm^{-2}	1.3	2.34
Irradiated 3×10^{14} cm^{-2}	1.88	1.61

Consistent with the drain and gate leakage current, the barrier height demonstrates a 0.39 eV and 0.73 eV reduction at the doses of 1×10^{14} cm^{-2} and 3×10^{14} cm^{-2}, respectively. With a reduction in barrier height, the expected shift in threshold voltage is to the negative, but that does not account for changes in charged traps within the graded layer which may shift the threshold to the positive.

Charged traps near the electron channel screen and decrease the carrier concentration. Additionally, the generation of new charged traps can significantly hinder the medium and high frequency operation by carrier scattering from radiation induced defects. To assess the changes under pulsed conditions, the effects of generation of traps and the formation of a virtual gate were examined using gate-lag measurements performed at 100 kHz and 10%–50% duty. The testing was performed on the POLFET devices alongside SiN$_x$ GaN MISHEMT devices for comparison, as shown in figure 18.8. The POLFET devices, outside of a small reduction in the overall current from a loss of carriers, demonstrate no change in pulsed characteristics, while the GaN HEMT devices' pulsed current is severely degraded. The carrier scattering effects will be much more pronounced within the 2DEG of the GaN HEMT, as the electrons may end up in the buffer or barrier layer and contribute to the formation of a virtual gate. However, in the POLFET device, as the 3DES has a thickness of approximately 55 nm, the scattering events are less likely for the electron to end outside the channel and not contribute to the forward current. The GaN-based devices exhibited the normal trend of smaller pulse width leading to lower current in switching performance, as there is less time for de-trapping of electrons to occur. The POLFET devices exhibited a slight but opposite trend of shorter pulse width giving a higher current. With the device being switched on for only 1–5 μs and remaining off for 9–5 μs, the device may not have enough time to reach steady state operating temperature. A longer pulse width leads to increased device heating and channel resistance. The increased time to steady state is due to the poor thermal conductivity of the ternary AlGaN graded channel layer, which experiences heightened effects of phonon–phonon and point-defect scattering over a GaN channel [31].

18.5 Alpha irradiation

Alpha irradiation was performed using a cyclotron and two doses were tested at 18 MeV: 1×10^{13} cm^{-2} and 3×10^{13} cm^{-2}. In figure 18.9 the drain current–voltage characteristics are shown for the POFLETs under the reference state and respective

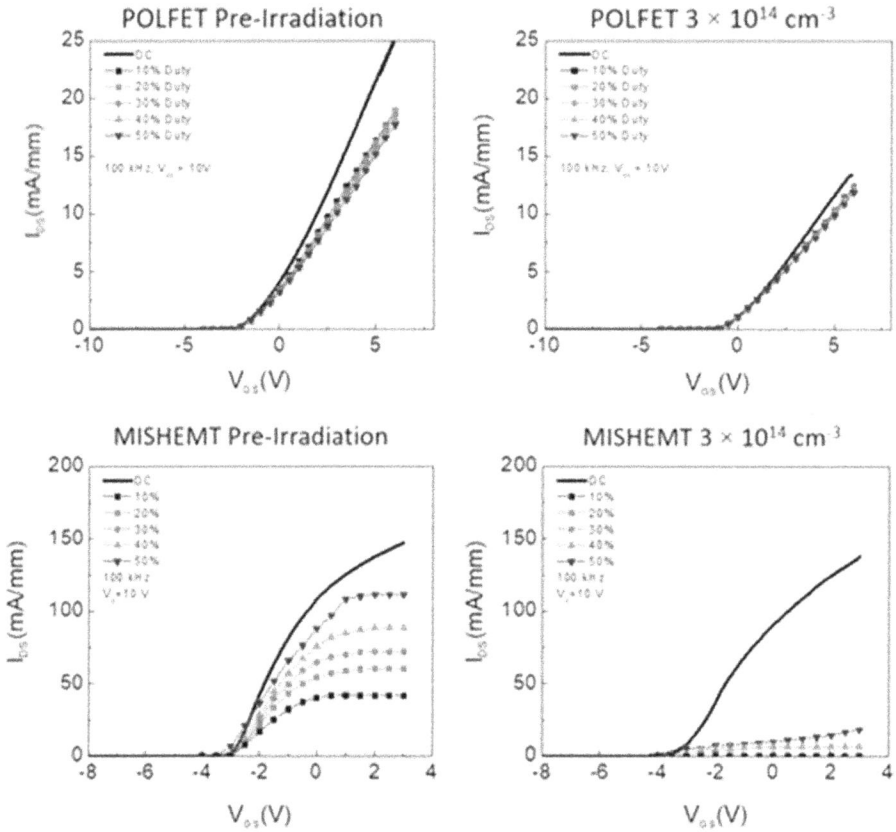

Figure 18.8. Gate-lag measurement at 100 kHz of an AlGaN POLFET and a GaN MISHEMT. Copyright IOP Publishing. Reproduced with permission. All rights reserved.

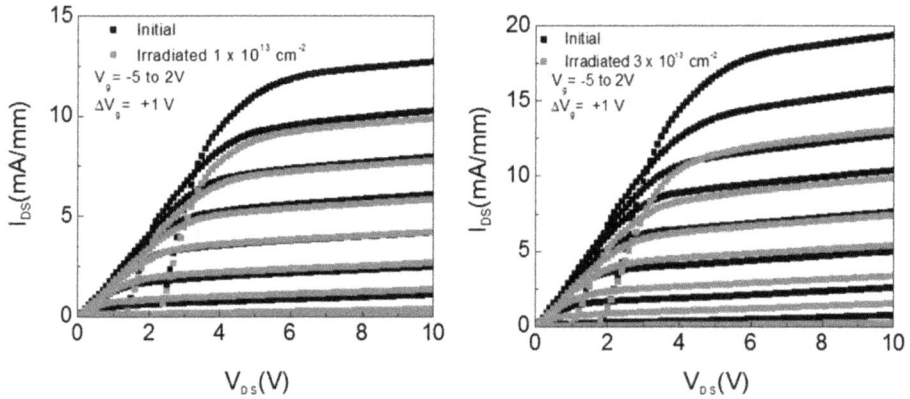

Figure 18.9. DC I–V curves for alpha irradiated POLFETs.

low and high dose. The forward current was reduced by approximately 23% and 33% for the low and high dose, respectively. The dosage for the alpha irradiation is one order of magnitude less than that of the proton irradiation, but produced similar reductions in forward current. This is due to the approximate one order of magnitude increase in both coulombic and nuclear stopping for helium ions as opposed to hydrogen, figures 18.3 and 18.4. The DC forward current degradation is again due to the removal of electron carriers and the generation of positively charged traps. The slope of the individual output curves prior to the knee voltage is reduced, indicating the loss of electron mobility and carrier concentration.

The transfer characteristics are presented in figure 18.10 and demonstrate a threshold voltage shift to the right, indicative of the channel opening slower due to the presence of positively charged defects, such as nitrogen vacancies. The subthreshold swing was not noted to change in any device before or after irradiation and stayed within the 80–90 mV/dec range. The transconductance was reduced by 19% and 18% at low and high doses, respectively. While this shift did not conform to the expectation that higher dosage leads to more damage, this may be due to channeling within the crystalline structure, as most of the damage will be deposited deep within the substrate as the particles slow down.

An interesting result was that the alpha irradiation did not induce a large degradation of the gate and drain leakage currents. Only at the highest fluence of 3×10^{13} cm^{-2} was any amount of change noted, indicating that the defects generated within the device are probably not conducting and are insulating only. The carrier concentration was extracted to further observe this change using equation (18.5). The reference state of the devices was found to have an average carrier concentration within the 3DES of 6.54×10^{12} cm^{-2}, which is slightly higher than the value extracted by the contactless mercury probe CV, 5.4×10^{12} cm^{-2}. This small difference may originate from an error in the assumption of the average electron mass within the 3DES and non-uniformity across the wafer. Table 18.2 presents the extracted carrier concentrations and removal rates for the fluences of 1×10^{13} cm^{-2} and 3×10^{13} cm^{-2}.

The removal rates of 7100 cm^{-1} and 2520 cm^{-1} for the low and high dose do differ again, due to effects of channeling and spatial clustering of defects.

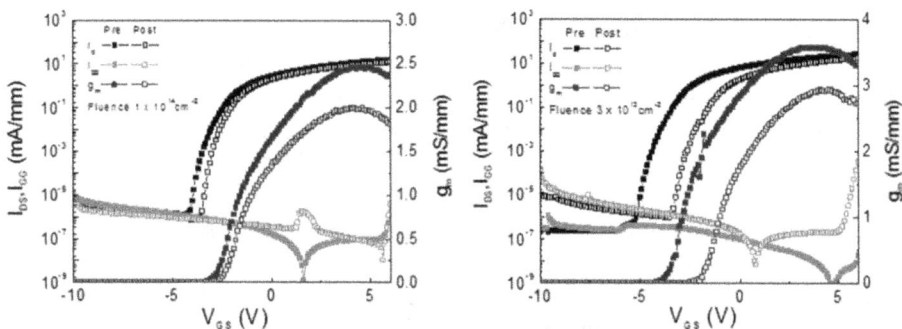

Figure 18.10. Transfer characteristics of alpha irradiated POLFETs.

Table 18.2. Extracted carrier concentrations and carrier removal rates for alpha irradiation of POLFETs.

Sample	n_s (cm^{-2})	% Carrier reduction	Carrier removal rate (cm^{-1})
Reference 1×10^{13} cm^{-2}	6.55×10^{12}	—	—
Irradiated 1×10^{13} cm^{-2}	5.77×10^{12}	11.9	7100
Reference 3×10^{13} cm^{-2}	6.51×10^{12}	—	—
Irradiated 3×10^{13} cm^{-2}	5.68×10^{12}	12.8	2520

Table 18.3. Schottky barrier height and ideality factor of alpha-irradiated POLFETs.

Sample	n	SBH (eV)
Reference 1×10^{13} cm^{-2}	1.69	1.87
Irradiated 1×10^{13} cm^{-2}	2.96	1.65
Reference 3×10^{13} cm^{-2}	1.58	1.89
Irradiated 3×10^{13} cm^{-2}	3.10	1.25

Gate characteristic *I–V* were used to extract information on barrier height lowering via the thermionic emission model, equations (18.7) and (18.8). The extracted Schottky barrier heights and ideality factors are presented in table 18.3. The barrier height lowering of 0.22 eV and 0.64 eV at fluences of 1×10^{13} cm^{-2} and 3×10^{13} cm^{-2}, respectively, is consistent with the observed degradation in drain leakage current presented in the transfer characteristics, shown in figure 18.10.

Gate-lag measurements were performed to evaluate the medium frequency power switching of the devices and observe any effects of the formation of a virtual gate. The devices were operated at 100 kHz and 10%–50% duty with a drain bias of +10 V. Figure 18.11 shows the gate lag for the POFLET and a reference SiN$_x$/ AlGaN/GaN MISHEMT that was irradiated simultaneously. The presence of a virtual gate, leading to a surface depletion region, caused a significant reduction in drain current in the reference MISHEMT sample. Scattering within the channel from irradiation induced defects and virtual gate formation are significant under switching operation. However, the POLFET sample shows essentially no change other than an overall reduction in current from the lower carrier concentration. This performance is unique—while scattering effects still occur, they are primarily confined within the 55 nm 3DES and the electrons do not necessarily leave the electron channel and become trapped to form a virtual gate.

When comparing the carrier removal rates for AlGaN POLFETs to their GaN counterparts, they do not appear to offer much improvement, see figure 18.12. The benefit comes in the performance leap for medium frequency switching applications, where GaN-based devices were entirely unusable, while the AlGaN POLFET had near ideal performance, see figures 18.8 and 18.11.

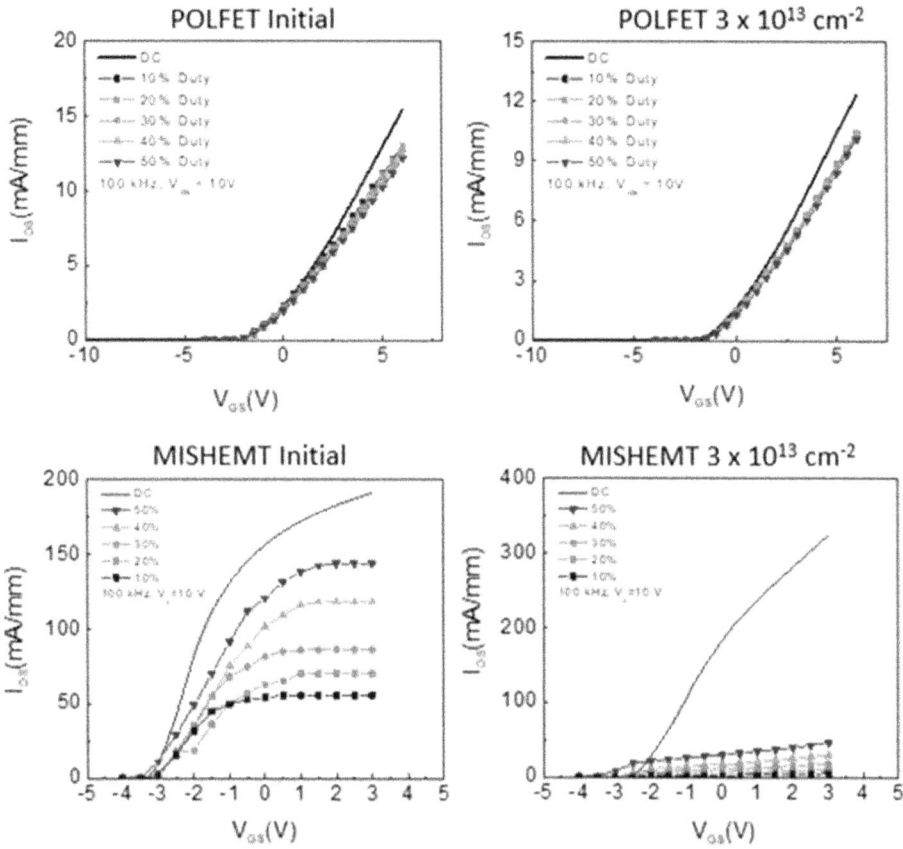

Figure 18.11. Gate lag of the POLFET and MISHEMT before and after alpha irradiation.

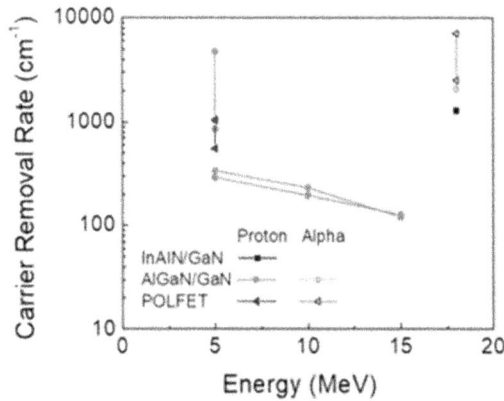

Figure 18.12. Comparison of GaN-based device carrier removal rates to AlGaN POLFETs [32–37].

18.6 Summary and conclusion

While AlGaN-based devices are still in their infancy with many of the simpler questions such as ohmic contact formation, passivation, and heat mitigation still being answered, their potential to grow and develop exponentially faster comes from the deep well of GaN knowledge. AlGaN will probably not surpass GaN for high frequency operation, but it may provide a better avenue for medium to low frequency power switching due to the ultra-wide bandgap and high critical break-down field.

References

[1] Anderson T J *et al* 2017 Impact of 2 MeV proton irradiation on the large-signal performance of Ka-band GaN HEMTs *ECS J. Solid State Sci. Technol.* **6** S3110–3

[2] Anderson T J *et al* 2016 Hyperspectral electroluminescence characterization of OFF-state device characteristics in proton irradiated high voltage AlGaN/GaN HEMTs *ECS J. Solid State Sci. Technol.* **5** Q289–93

[3] Gallagher J C *et al* 2017 Effect of surface passivation and substrate on proton irradiated AlGaN/GaN HEMT transport properties *ECS J. Solid State Sci. Technol.* **6** S3060–2

[4] Greenlee J D *et al* 2015 Degradation mechanisms of 2 MeV proton irradiated AlGaN/GaN HEMTs *Appl. Phys. Lett.* **107** 083504

[5] Pearton S J, Ren F, Patrick E, Law M E and Polyakov A Y 2015 Review—ionizing radiation damage effects on GaN devices *ECS J. Solid State Sci. Technol.* **5** Q35–60

[6] Weaver B D *et al* 2016 Editors' choice—on the radiation tolerance of AlGaN/GaN HEMTs *ECS J. Solid State Sci. Technol.* **5** Q208–12

[7] Koehler A D *et al* 2014 Proton radiation-induced void formation in Ni/Au-gated AlGaN/GaN HEMTs *IEEE Electron Device Lett.* **35** 1194–6

[8] Rostewitz M, Hirche K, Latti J and Jutzi E 2013 Single event effect analysis on DC and RF operated AlGaN/GaN HEMTs *IEEE Trans. Nucl. Sci.* **60** 2525–9

[9] Do T N T *et al* 2017 7–13 GHz MMIC GaN HEMT voltage-controlled-oscillators (VCOs) for satellite applications *12th European Microwave Integrated Circuits Conf. (EuMIC)*

[10] Yamasaki T *et al* 2010 A 68% efficiency, C-band 100W GaN power amplifier for space applications *2010 IEEE MTT-S Int. Microwave Symp.*

[11] Osawa K, Yoshikoshi H, Nitta A, Tanaka T, Mitani E and Satoh T 2016 Over 74% efficiency, L-band 200W GaN-HEMT for space applications *46th European Microwave Conf. (EuMC)*

[12] Van Allen J A, McIlwain C E and Ludwig G H 1959 Radiation observations with satellite 1958 ε *J. Geophys. Res.* **64** 271–86

[13] Feynman J, Spitale G, Wang J and Gabriel S 1993 Interplanetary proton fluence model: JPL 1991 *J. Geophys. Res.: Space Physics* **98** 13281–94

[14] Carrington R C 1859 Description of a singular appearance seen in the Sun on September 1, 1859 *Mon. Not. R. Astron. Soc.* **20** 13–5

[15] Bajaj S *et al* 2018 High Al-content AlGaN transistor with 0.5 A/mm current density and lateral breakdown field exceeding 3.6 MV cm^{-1} *IEEE Electron Device Lett.* **39** 256–9

[16] Carey P H C IV *et al* 2019 Operation up to 500°C of $Al_{0.85}Ga_{0.15}N/Al_{0.7}Ga_{0.3}N$ high electron mobility transistors *IEEE J. Electron Devices Soc.*

[17] Baca A G *et al* 2017 $Al_{0.85}Ga_{0.15}N/Al_{0.70}Ga_{0.30}N$ high electron mobility transistors with Schottky gates and large on/off current ratio over temperature *ECS J. Solid State Sci. Technol.* **6** Q161–5

[18] Baca A G *et al* 2018 RF performance of $Al_{0.85}Ga_{0.15}N/Al_{0.70}Ga_{0.30}N$ high electron mobility transistors with 80 nm gates *IEEE Electron Device Lett.* **40** 17–20

[19] Douglas E A *et al* 2017 Ohmic contacts to Al-rich AlGaN heterostructures *Phys. Status Solidi* A **214** 1600842

[20] Simon J, Protasenko V, Lian C, Xing H and Jena D 2010 Polarization-induced hole doping in wide-band-gap uniaxial semiconductor heterostructures *Science* **327** 60–4

[21] Armstrong A M *et al* 2018 Ultra-wide band gap AlGaN polarization-doped field effect transistor *Jpn J. Appl. Phys.* **57** 074103

[22] Armstrong A M *et al* 2019 AlGaN polarization-doped field effect transistor with compositionally graded channel from $Al_{0.6}Ga_{0.4}N$ to AlN *Appl. Phys. Lett.* **114** 052103

[23] Baca A G *et al* 2017 High temperature operation of $Al_{0.45}Ga_{0.55}N/Al_{0.30}Ga_{0.70}N$ high electron mobility transistors *ECS J. Solid State Sci. Technol.* **6** S3010–3

[24] Baca A G *et al* 2016 An AlN/Al0.85Ga0.15N high electron mobility transistor *Appl. Phys. Lett.* **109** 033509

[25] Bethe H 1932 Bremsformel für Elektronen relativistischer Geschwindigkeit *Z. Phys.* **76** 293–9

[26] Fano U 1963 Penetration of protons, alpha particles, and mesons *Ann. Rev. Nucl. Sci.* **13** 1–66

[27] Ziegler J F 1999 Stopping of energetic light ions in elemental matter *J. Appl. Phys.* **85** 1249–72

[28] Xiao H Y, Gao F, Zu X T and Weber W J 2009 Threshold displacement energy in GaN:Ab initiomolecular dynamics study *J. Appl. Phys.* **105** 123527

[29] Xi J, Liu B, Zhang Y and Weber W J 2018 *Ab initio* molecular dynamics simulations of AlN responding to low energy particle radiation *J. Appl. Phys.* **123** 054904

[30] Rashmi, Kranti A, Haldar S and Gupta R S 2002 An accurate charge control model for spontaneous and piezoelectric polarization dependent two-dimensional electron gas sheet charge density of lattice-mismatched AlGaN/GaN HEMTs *Solid-State Electronics* **46** 621–30

[31] Liu W and Balandin A A 2005 Thermal conduction in $Al_xGa_{1-x}N$ alloys and thin films *J. Appl. Phys.* **97** 073710

[32] Liu L *et al* 2013 Dependence on proton energy of degradation of AlGaN/GaN high electron mobility transistors *J. Vac. Sci. Technol.* B **31** 022201

[33] Liu L *et al* 2013 Impact of proton irradiation on DC performance of AlGaN/GaN high electron mobility transistors *J. Vac. Sci. Technol.* B **31** 042202

[34] Xi Y *et al* 2014 Effect of 5 MeV proton radiation on DC performance and reliability of circular-shaped AlGaN/GaN high electron mobility transistors *J. Vac. Sci. Technol.* B **32** 012201

[35] Ahn S *et al* 2015 Study on effect of proton irradiation energy in AlGaN/GaN metal–oxide semiconductor high electron mobility transistors *ECS Trans.* **69** 129–35

[36] Ahn S *et al* 2016 Effect of proton irradiation dose on InAlN/GaN metal–oxide semiconductor high electron mobility transistors with Al_2O_3 gate oxide *J. Vac. Sci. Technol.* B **34** 051202

[37] Fares C *et al* 2018 Effect of alpha-particle irradiation dose on SiN_x/AlGaN/GaN metal–insulator semiconductor high electron mobility transistors *J. Vac. Sci. Technol.* B **36** 041203

Part III

Zinc Oxide

IOP Publishing

Wide Bandgap Semiconductor-Based Electronics

Fan Ren and Stephen J Pearton

Chapter 19

BeMgZnO wide bandgap quaternary alloy semiconductor

Kai Ding, Vitaliy Avrutin, Natalia Izyumskaya, Ümit Özgür and Hadis Morkoç

The BeO–MgO–ZnO family is a wide bandgap oxide semiconductor material system offering a wide range of bandgap engineering and strain modulation for applications in ZnO-based electronic and optoelectronic devices. Compared to MgZnO and BeZnO ternaries, co-alloying of ZnO with BeO and MgO serves to enhance the incorporation rate of both Be and Mg while retaining the wurtzite phase and improving the stability, crystalline quality, and optical performance of the quaternary. This chapter provides the reader with a taste encompassing both theoretical and experimental studies of the structural, electrical, and optical properties, and also the application potential of the BeMgZnO quaternary system.

19.1 Introduction

Oxide semiconductor system based on zinc oxide (ZnO) has gained considerable attention as a platform for various electronic and optoelectronic devices such as field effect transistors (FETs), photodetectors, and solar cells, owing to its versatile properties including a tunable bandgap in a wide range (from the 3.37 eV of ZnO to the 10.6 eV of BeO) and a high electron saturation velocity (3.5×10^7 cm s^{-1} as predicted by theoretical simulations) [1–3]. A crucial factor in the realization of modern semiconductor devices is the creation of heterostructures, via energy band engineering, which are composed of stacks of multiple layers of different materials or the same material but with different compositions. The most relevant parameters that need to be taken into account in the design of a heterostructure are the valence- and conduction-band offsets between the individual layers and the lattice mismatch among them. Band offsets modulate carrier distributions and transport, while lattice mismatch alters atomic bonding, which can lead to piezoelectric strain, lattice imperfection, and wafer bending and even cracking. Figure 19.1 shows the bandgaps and in-plane lattice parameters for the ZnO related family of materials [1]. In

doi:10.1088/978-0-7503-2516-5ch19

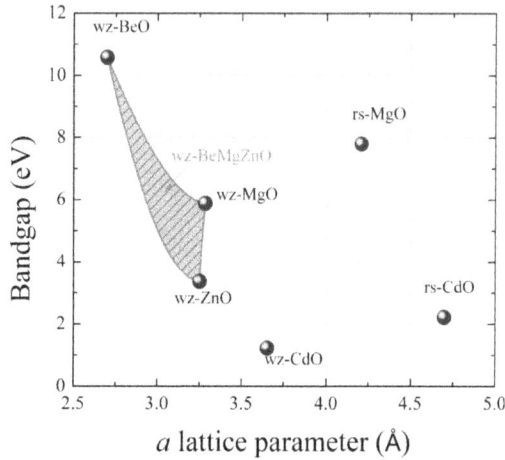

Figure 19.1. Bandgap versus in-plane lattice parameters for group-IIA oxides. The solid line and shaded area between binaries in the wurtzite phase correspond to wurtzite ternary and quaternary alloys, respectively.

analogy to the nitride semiconductor system, in which GaN can be alloyed with AlN and/or InN, energy band engineering in ZnO-based oxide semiconductor systems can be achieved by alloying ZnO with MgO, BeO, and/or CdO. However, unlike the nitride system, in which all binaries GaN, AlN, and InN possess the same wurtzite crystal structure, MgO and CdO are stable in the rocksalt structure, different to the wurtzite structure of ZnO [4, 5]. Due to the instability of MgO and CdO in the wurtzite structure, phase segregation becomes a major obstacle in MgZnO and CdZnO alloys with high Mg and Cd contents [6]. As an alternative to MgZnO, BeZnO was proposed for bandgap tuning as BeO is stable in the wurtzite structure, thus allowing modulation of the bandgap from 3.37 eV (ZnO) to 10.6 eV (BeO) [7]. Unfortunately, due to a large difference in covalent radii (1.22 Å for Zn, 0.96 Å for Be), BeZnO alloys with intermediate content of Be (from 35% to 70%) tend to segregate into high and low Be content phases. In order to overcome the aforementioned shortcomings of MgZnO and BeZnO ternaries, the BeMgZnO quaternary alloy has been considered, in which strain caused by the larger and smaller covalent radii of Mg (1.41 Å) and Be, respectively, can compensate for each other. This opens an avenue towards achieving wider bandgap materials with tunable strain sign and magnitude. Thus, these materials are highly desired for solar-blind UV detectors (cut-off wavelength <280 nm, i.e. E_g > 4.5 eV) [8], devices based on intersubband transitions (ISBTs) [9], and heterostructure field effect transistors (HFETs) with two-dimensional electron gas (2DEG) [10]. This chapter reviews both theoretical and experimental studies of the structural, electronic, and optical properties of the BeMgZnO quaternary system, and discusses its application potential with specific emphasis on HFETs. The organization of this chapter is as follows. In section 19.2, theoretical studies of the structural, chemical, and electrical properties of the BeMgZnO quaternary based on first-principles calculations are discussed. The growth methods employed to realize BeMgZnO quaternary materials, in particular using molecular beam epitaxy (MBE), are described in section 19.3.

This is followed by the characterization results of the BeMgZnO quaternary, with discussions of the determination of the quaternary composition in section 19.4. Finally, the application potential of this quaternary material system in the context of HFETs is discussed in section 19.5.

19.2 Theoretical studies

The stability of MgZnO, BeZnO, and BeMgZnO alloys based on calculations of their respective formation energies have been investigated by several groups [11, 12], making use of the first-principles density functional theory (DFT). Both the MgZnO and BeZnO ternary alloys were found to be unstable for high Mg and Be contents. However, the quaternary system BeMgZnO was predicted to be stable when the Be–Mg atom ratio is small. The formation energies of the wurtzite and rocksalt phases of the $Mg_xZn_{1-x}O$ alloys as a function of Mg content calculated by Su et al using the Vienna ab $initio$ simulation package (VASP) are shown in figure 19.2 (a) [11]. Clearly, the formation energy of the wurtzite $Mg_xZn_{1-x}O$ alloy increases linearly with Mg content x, whereas that of the rocksalt phase shows a decreasing linear relationship and a cross-over occurs at $x = 0.33$. Such tendencies indicate that low Mg content $Mg_xZn_{1-x}O$ alloys are more stable in the wurtzite phase while high

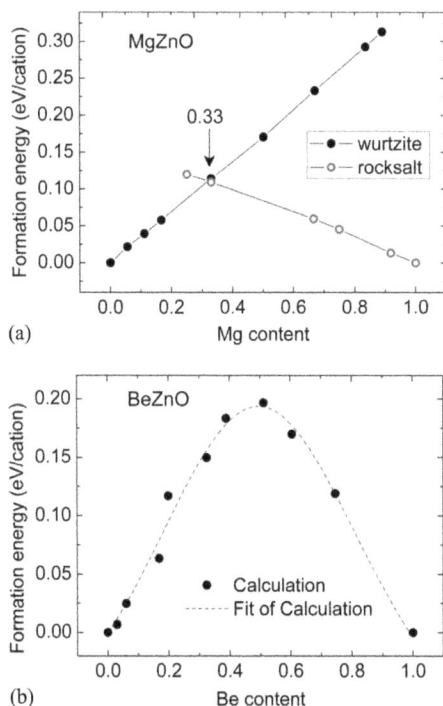

(a)

(b)

Figure 19.2. (a) Calculated formation energies of wurtzite and rocksalt $Mg_xZn_{1-x}O$ as a function of Mg content x. The phase transition point is found to be $x = 0.33$. (b) Calculated formation energies of $Be_xZn_{1-x}O$ as a function of Be content. $Be_xZn_{1-x}O$ with intermediate Be content has the highest formation energy. Reproduced with permission from [11]. Copyright 2014 American Chemical Society.

Mg content $Mg_xZn_{1-x}O$ alloys are more stable in the rocksalt phase, which is consistent with experimental findings. The formation energies of $Be_xZn_{1-x}O$ alloys as a function of Be content are shown in figure 19.2(b). $Be_xZn_{1-x}O$ alloys with Be content close to 0 or 1.0 are relatively stable, while $Be_xZn_{1-x}O$ alloys with intermediate Be content tend to segregate into low- and high-Be content phases due to their large formation energies, which can be ascribed to the large lattice mismatch between BeO and ZnO.

Compared with a relatively low Mg content of 0.33, where the wurtzite-to-rocksalt phase transition occurs in $Mg_xZn_{1-x}O$ ternary, the $Mg_xZn_{1-x}O$ ternary can readily maintain a wurtzite structure for Mg content as high as 0.83, as shown in figure 19.3(a). For the $Be_xMg_yZn_{1-x-y}O$ quaternary system, the wurtzite-to-rocksalt phase transition boundary for different Be and Mg contents is shown in figure 19.3(b). Above the boundary, denoted by the solid fitting line, the quaternary is stable in the rocksalt phase and below the boundary it is stable in the wurtzite phase. It is obvious that the incorporation of BeO into the quaternary can enhance the solubility of wurtzite MgO in the system when a small Be–Mg atomic ratio is used. The maximum Be–Mg ratio that can be used increases with increasing Be content. Figure 19.3(c) shows the formation energies of $Be_xMg_yZn_{1-x-y}O$ alloys with $y = 0$, $x + y = 0.33$, and $x + y = 0.50$. The energy saddle points at $x = 0.027$ for $x + y = 0.33$ and $x = 0.034$ for $x + y = 0.50$ suggest the optimum Be and Mg molar fractions for the most energetically favorable wurtzite phase $Be_xMg_yZn_{1-x-y}O$ alloys. The a-axis lattice compensation and four-fold coordination preference of the Be atom in $Be_xMg_yZn_{1-x-y}O$ were ascribed as the major origins for Be–Mg mutual stabilization in the hexagonal lattice [13]. The stabilization diagram of wurtzite or rocksalt phase $Be_xMg_yZn_{1-x-y}O$ alloys for different Be and Mg molar fractions is shown in figure 19.4, as calculated by Gorczyca et al [14].

Structural and electronic properties of the BeMgZnO quaternary system have been calculated for the entire compositional range by Toporkov et al [15]. The structural properties were calculated using Perdew–Burke–Ernzerhof (PBE) parameterization of the generalized gradient approximation (GGA) to the DFT, while the bandgaps were calculated using the exchange tuned Heyd–Scuseria–Ernzerhof (HSE06) hybrid functional with a fraction of exact exchange value of 0.375. Figures 19.5, 19.6, and 19.7 display the calculated a and c lattice parameters, and bandgaps of $Be_xMg_yZn_{1-x-y}O$ as a function of Be and Mg contents, respectively. Note that the in-plane lattice parameter of wurtzite MgO is very close to that of ZnO. On the other hand, the in-plane lattice parameter of BeO is substantially smaller than that of ZnO. Consequently, by choosing the proper Be and Mg contents, it is possible to achieve an in-plane lattice parameter larger or smaller than that of ZnO. The latter, for example, is very important in achieving tensile strain in the barrier layer of a Zn-polar BeMgZnO/ZnO heterostructure, which yields the proper sign of piezoelectric polarization and results in high 2DEG density at the interface. We will discuss the strain tuning in the BeMgZnO/ZnO interface for HFET application in section 19.4. As shown in figures 19.5 and 19.6, the bowing of the c lattice parameter is significantly larger than that of the a parameter due to the fact that the incorporation of Mg and Be has opposite effects on the in-plane lattice

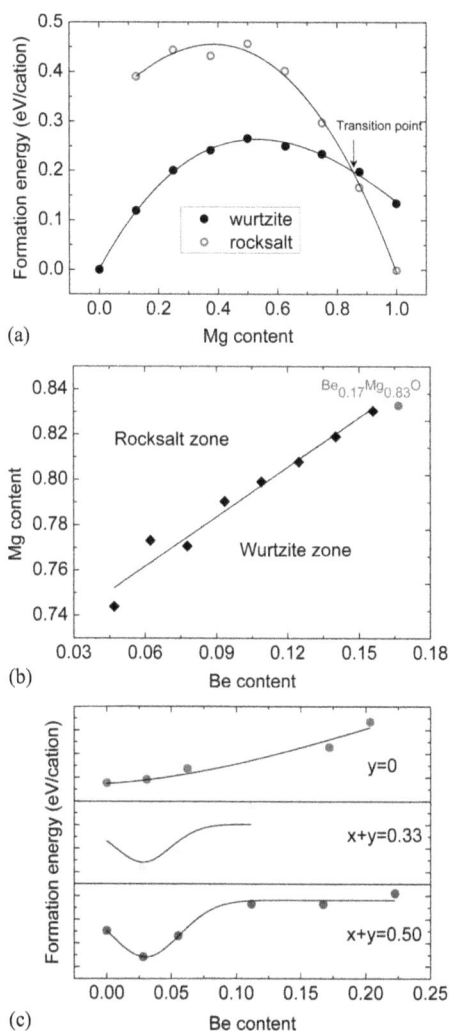

Figure 19.3. (a) Calculated formation energy of $Be_xMg_{1-x}O$ ternary alloy as a function of Mg content. (b) Calculated wurtzite-to-rocksalt phase transition points of $Be_xMg_yZn_{1-x-y}O$ quaternary as a function of Be content. (c) Calculated formation energy of wurtzite $Be_xMg_yZn_{1-x-y}O$ quaternary as a function of Be content. Reproduced with permission from [11]. Copyright 2014 American Chemical Society.

parameter (decreasing with Be, increasing with Mg), while the c parameter of BeMgZnO decreases with both increasing Be and Mg content. In addition to the shorter bond length between Be and O, stronger Be–O bonding is another reason for the decrease in the lattice constant when admixing Be to ZnO. The BeMgZnO alloy was found to exhibit stronger electron density localization than bulk ZnO. Particularly, the valence band maximum (VBM) orbitals localized on oxygen coordinated by Mg atoms tend to be more localized, compared to those coordinated by Zn atoms. The changes in the wavefunction with increasing Mg and Be contents are related to the bandgap bowing shown in figure 19.7. Bandgap broadening of the

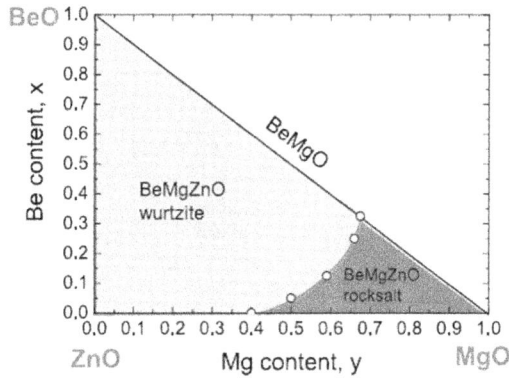

Figure 19.4. Stability diagram showing the composition (x, y) ranges in which the $Be_xMg_yZn_{1-x-y}O$ alloy is thermodynamically stable in the wurtzite or rocksalt phase. Reproduced with permission from [14]. Copyright 2016 AIP Publishing.

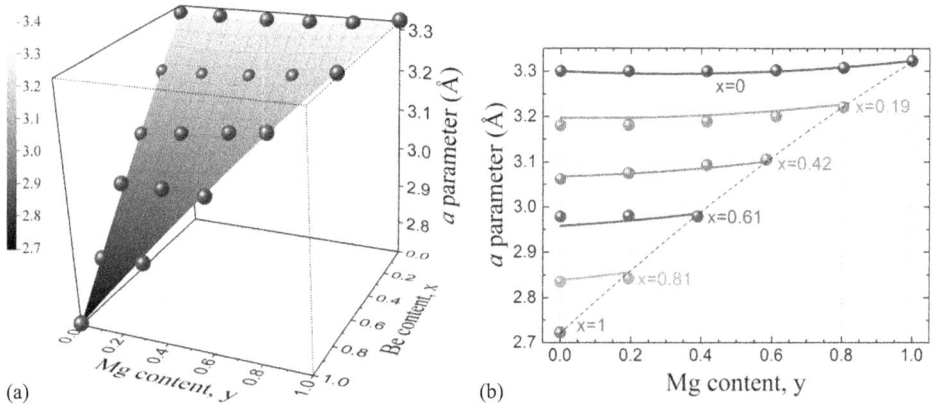

Figure 19.5. Calculated a lattice parameters of $Be_xMg_yZn_{1-x-y}O$ as a function of Be and Mg contents. Reproduced with permission from [15]. Copyright 2016 AIP Publishing.

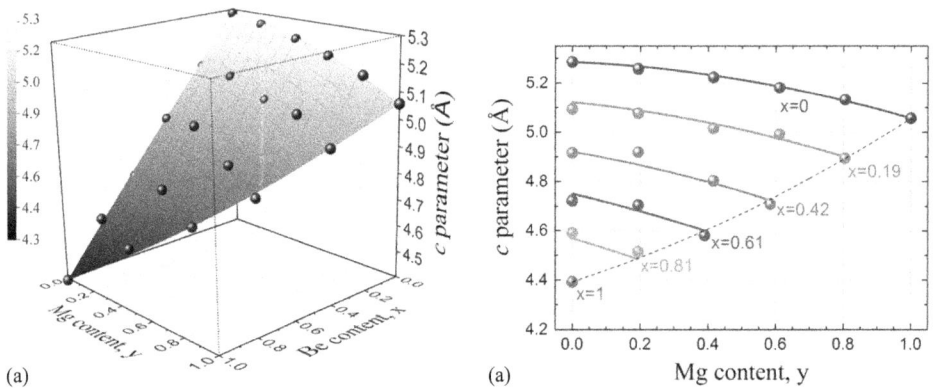

Figure 19.6. Calculated c lattice parameters of $Be_xMg_yZn_{1-x-y}O$ as a function of Be and Mg contents. Reproduced with permission from [15]. Copyright 2016 AIP Publishing.

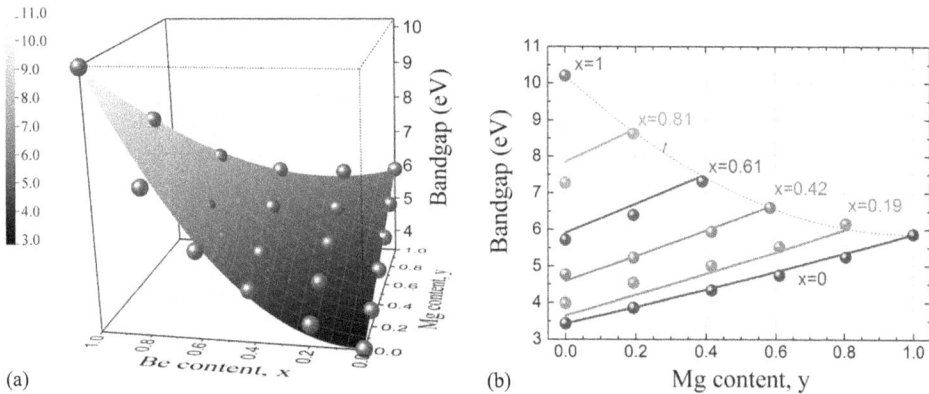

Figure 19.7. Calculated bandgaps of $Be_xMg_yZn_{1-x-y}O$ as a function of Be and Mg contents. Reproduced with permission from [15]. Copyright 2016 AIP Publishing.

BeMgZnO alloy was ascribed to the movement of the Zn4s states at conduction-band minimum (CBM) toward higher energies with the incorporation of Mg and Be [12].

19.3 Material growth

Only a few thin film growth approaches have been developed for BeMgZnO alloys, including MBE [15–18], pulsed laser deposition (PLD) [19–23], sputtering [24–26], and spin coating [27, 28], among which MBE is the most investigated method. Although direct growth of BeMgZnO on sapphire has been reported, for crystalline control most of the growth was carried out on ZnO templates with in the form of BeMgZnO/ZnO heterostructures. c-sapphire and GaN are the commonly used substrates due to their similar crystalline structures or close lattice parameters.

19.3.1 MBE of BeMgZnO

Typically for MBE of BeMgZnO, all the metal species Be, Mg, and Zn are evaporated from Knudsen cells, and radio frequency (RF)-activated radical oxygen gas is used as the O source. The growth progression is typically monitored *in situ* via reflection high-energy electron diffraction (RHEED). To avoid possible structural damage induced by low-energy ions generated by the oxygen plasma source [29], which is inherent to plasma-assisted MBE, an ion deflector with a negative bias is employed in some processes.

Ullah *et al* have investigated the crystalline structure and growth kinetics of O-polar BeMgZnO using MBE [30]. For growth on c-sapphire, O-polar ZnO templates with high temperature MgO and low temperature ZnO buffer layers and Ga-polar MOCVD grown GaN templates were used as substrates. These different substrates, using relatively well-developed buffering techniques, have been reported to provide high quality ZnO layers with smooth surfaces to meet the requirements for the growth of BeMgZnO quaternary based heterostructures. For growth on sapphire substrates, without the incorporation of Mg, $Be_{0.10}Zn_{0.90}O$ thin films exhibited a high density of stacking faults (SFs) and possible second-phase inclusions, as evidenced from dark field cross-sectional transmission electron microscopy (TEM). The dominant defects in O-polar $Be_{0.07}Mg_{0.30}Zn_{0.63}O$ films were threading dislocations (TDs) originating mostly from the ZnO/sapphire interface accompanied by compositional fluctuations, as shown in figure 19.8(a). The improved structural quality of the quaternary was confirmed using HRXRD, see figure 19.8(b), as evidenced by an order of magnitude higher intensity than that from the ternary, despite a large density of TDs.

Optimum Zn/(Be + Mg) ratios for obtaining single-crystal O-polar BeMgZnO were experimentally explored by growth on Ga-polar GaN/sapphire templates [30]. It was found that the suppression of second-phase formation at high substrate temperatures requires an accurate Zn/(Be + Mg) flux ratio. As shown in figure 19.9, a Zn/(Be + Mg) flux ratio of 3.9, which is suitable for growth at 450 °C, leads to the formation of the second phase while growing at 475 °C and 500 °C. Due to the reduction of spacing between RHEED reflection spots, the appearance of the second phase was attributed to Mg-rich MgZnO. To suppress its formation, higher Zn/(Be + Mg) flux ratios were needed along with increased substrate temperature.

Figure 19.10 shows a high angle annular dark field (HAADF) STEM image of the interface of an optimized O-polar $Be_{0.06}Mg_{0.33}ZnO/ZnO$ heterostructure. Under the observation conditions used, the contrast intensities of atomic columns are roughly proportional to $Z^{1.7}$, where Z is the atomic number of the elements. The BeMgZnO atomic columns are substantially darker than atomic columns in the ZnO layer. Because the lattice structures of BeMgZnO and ZnO are the same (wurtzite), the HAADF image in figure 19.10 represents the chemical contrast related to heavier (Zn in ZnO) and lighter elements (a mixture of Be, Mg, and Zn elements in the Zn sublattice of BeMgZnO). No misfit dislocations were found at the $Be_{0.06}Mg_{0.33}ZnO–ZnO$ interface for the investigated interface length of ~200 nm,

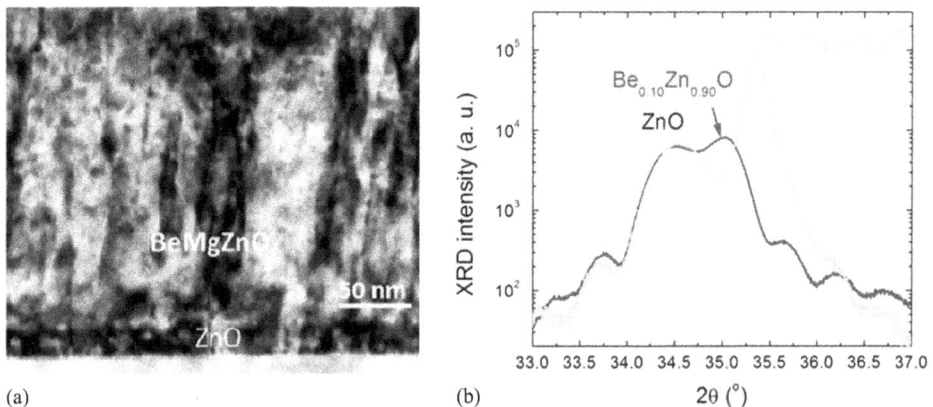

Figure 19.8. (a) Cross-sectional TEM image of O-polar $Be_{0.07}Mg_{0.30}Zn_{0.63}O/ZnO$ heterostructure grown on c-sapphire showing a large density of TDs (10^{10} cm^{-2}). (b) HRXRD of a 2θ–ω scan of (0002) reflection from $Be_{0.10}Zn_{0.90}O/ZnO$ and $Be_{0.07}Mg_{0.30}Zn_{0.63}O/ZnO$ heterostructures. Reproduced with permission from [30]. Copyright 2017 SPIE.

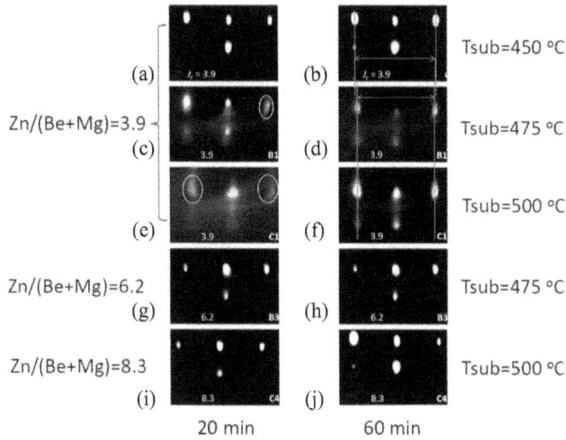

Figure 19.9. RHEED patterns taken along the [1$\bar{1}$00] azimuthal direction for Zn-polar BeMgZnO films with different Zn/(Be + Mg) ratios and substrate temperatures and growth times of 20 and 60 min. Reproduced with permission from [30]. Copyright 2017 SPIE.

Figure 19.10. HAADF STEM image of an O-polar Be$_{0.06}$Mg$_{0.33}$ZnO–ZnO interface taken along the [2$\bar{1}$10] zone axis. Courtesy of Dr Alexander V Kvit of the University of Wisconsin-Madison.

indicating that the density of misfit dislocations at this Be$_{0.06}$Mg$_{0.33}$ZnO–ZnO interface is substantially less than 5×10^4 cm^{-1} due to a small lattice distortion resulting from the appropriate Be–Mg content ratio.

The incorporation rate of Be and Mg atoms into the BeMgZnO alloy depends on the reactive oxygen to metal flux ratio due to growth kinetics [16]. When sufficiently high diffusion mobility of the metal adatoms is provided (by a sufficiently high substrate temperature), the respective formation enthalpies of the metal–oxygen binaries govern the formation of the metal–oxygen bonds. Compared with Be and Mg, Zn has a much higher equilibrium vapor pressure at typical growth temperatures and, therefore, the Zn metal adatoms are more likely to re-evaporate from the growth surface. For a given set of metal fluxes, Toporkov *et al* [16] have found that a reduction in the oxygen to metal flux ratio from 1.0 to 0.6 increased the bandgap of the BeMgZnO alloy from 4.0 eV to 4.5 eV and decreased the c lattice parameter from 5.08 Å to 5.02 Å, corresponding to the change of the compositions from $Be_{0.07}Mg_{0.21}Zn_{0.72}O$ to $Be_{0.10}Mg_{0.34}Zn_{0.56}O$. The oxygen deficiency on the growing surface under such a metal rich growth condition gave rise to a competition between the metal adatoms for the available reactive oxygen, resulting in a reduced incorporation rate of Zn on the lattice sites. Under certain MBE conditions, for BeZnO the incorporation of Be into the ZnO lattice is saturated at a Be content of 0.14 with a BeZnO bandgap of 3.57 eV [31]. Co-doping of Be and Mg has been shown to enhance the incorporation of both Be and Mg, leading to a much larger bandgap of 5.11 eV in $Be_{0.08}Mg_{0.52}Zn_{0.40}O$ with better structural quality compared to ternary BeZnO with similar Be content [31].

19.3.2 Other growth methods

Among other methods for BeMgZnO alloy growth are PLD, sputtering, and spin coating [19–28]. Growth of the BeMgZnO quaternary has been carried out using PLD with either a single BeMgZnO ceramic target or a combination of MgZnO and BeO targets [20–22]. The effect of oxygen partial pressure on the incorporation of Be and Mg was found to be similar to that for the growth using MBE in that both Be and Mg contents decreased with the increase of oxygen pressure in the chamber. Again, consistent with observations in MBE, a large increase in the solid solubility of Mg in the films via Be–Mg co-incorporation was found. Figure 19.11 shows the evolution of the *c*-axis lattice parameter and FWHM of XRD (0002) reflection in BeMgZnO films as a function of the number of the BeO ablation pulse, with a fixed pulse number of 25 on the $Zn_{0.8}Mg_{0.2}O$ target in each cycle [20]. The *c*-axis lattice parameter value reduces linearly with increasing BeO pulse number, i.e. Be content

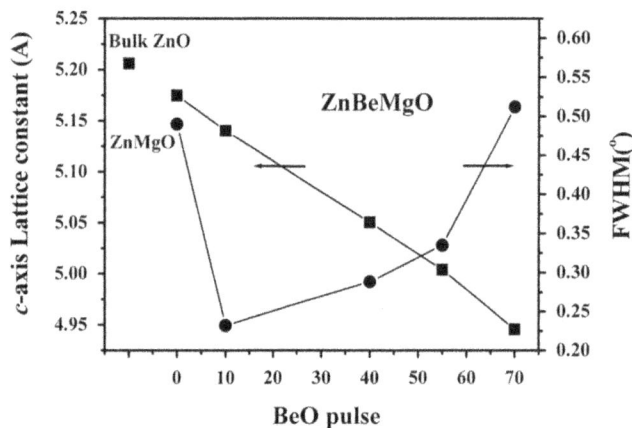

Figure 19.11. The *c*-axis lattice parameter and FWHM values of XRD (0002) reflection in ZnBeMgO films by pulsed laser deposition (PLD) as a function of the number of BeO pulses, with a fixed pulse number of 25 on the $Zn_{0.8}Mg_{0.2}O$ target in each cycle. Reproduced with permission from [20]. Copyright 2008 AIP Publishing.

in the film. The FWHM value of XRD (0002) reflection dropped significantly from 0.49° to 0.23° when Be was incorporated using ten BeO pulses. With further increase of the BeO pulse number, the FWHM value gradually increased due to the loss of Be–Mg mutual stabilization when excess Be is incorporated.

Continuing with deposition methods, radio frequency (RF) magnetron sputtering has also been employed for depositing BeMgZnO films with and without Ga doping [24–26]. A single target made from high purity powders of ZnO, MgO, and BeO was used as precursor for BeMgZnO deposition and Ga_2O_3 powder was mixed, as needed, into the target for Ga-doped BeMgZnO deposition. It was found that the electrical resistivity of the films substantially increased as the substrate temperature increased and as the oxygen gas was introduced to the Ar plasma. The optimized $Zn_{0.88}Mg_{0.05}Be_{0.03}Ga_{0.04}O$ film showed a room temperature resistivity of 1.6×10^{-3} Ω cm and an E_g of 3.75 eV, grown at room temperature.

Sol–gel spin-coating is another method that has been investigated for BeZnO and BeMgZnO thin film growth [28]. Zinc acetate dihydrate $(CH_3COO)_2Zn \cdot 2H_2O$, magnesium acetate dihydrate $(CH_3COCHC(O)CH_3)_2Mg \cdot 2H_2O$, and beryllium acetylacetonate $(CH_3COCH=C(O-)CH3)_2Be$ were used as the precursors of Zn, Mg, and Be, respectively. 2-methoxyethanol $(H_3OCH_2CH_2OH)$ and monoethanol-amine $NH_2CH_2CH_2OH$ were used as the solvent and stabilizer, respectively. The deposited films on sapphire substrates, however, consisted of irregularly shaped grains with a lateral size of 30–60 nm.

19.4 Compositional and optical characterizations of BeMgZnO

The accurate determination of Be and Mg contents in the BeMgZnO alloy is extremely challenging due to the lack of appropriate experimental tools for the detection of the light Be element with satisfactory accuracy. Element analytical techniques such as secondary ion mass spectroscopy (SIMS) for detecting Be requires a standard sample and suffers from artifacts caused during sample sputtering. X-ray photoelectron spectroscopy (XPS) also necessitates similar problems such as the need for surface sputtering and is limited by the uncertainty of the photoelectric cross-section and the inelastic mean free path (IMFP) of the photo-electrons from each element of the alloy.

Using a combination of nondestructive Rutherford backscattering spectrometry with a 1 MeV He^+ analyzing ion beam and non-Rutherford elastic backscattering experiments with 2.53 MeV energy protons, Zolnai *et al* have demonstrated that an atomic composition uncertainty of less than 1–2 atom % can be achieved for the BeZnO and BeMgZnO systems [32]. The main reason for accurate determination of Be with less uncertainty originates from the strong resonance with an enhancement factor of ~60 in the elastic backscattering cross-section of Be for proton energies around 2.5 MeV [33]. Using such ion beam analytical techniques, compositions, and layer thicknesses of wurtzite BeZnO/ZnO and BeMgZnO/ZnO heterostructures grown on c-sapphires with 7–19 atom % Be content were determined with less than 1–2 atom % uncertainty [32].

On the optical characterization side, potential fluctuations and exciton localization in the BeMgZnO alloy have been investigated by Toporkov *et al* in MBE grown oxygen-polar BeMgZnO thin films on GaN templates [34]. Exciton localization depth values determined from time resolved photoluminescence (TRPL) were found to increase with increasing Be–Mg ratio, from 98 meV for $Be_{0.04}Mg_{0.17}Zn_{0.79}O$, to 173 meV for $Be_{0.10}Mg_{0.25}Zn_{0.65}O$, and to 268 meV for $Be_{0.11}Mg_{0.15}Zn_{0.74}O$, as shown in figure 19.12. A similar correlation was observed with the temporal redshift of PL peak positions, which was ascribed to potential fluctuations and decrease of band filling effect in the localized states with time. As shown in figure 19.13, the redshift values at 15 K were 8 meV, 42 meV, and 55 meV for $Be_{0.04}Mg_{0.17}Zn_{0.79}O$, $Be_{0.10}Mg_{0.25}Zn_{0.65}O$, and $Be_{0.11}Mg_{0.15}Zn_{0.74}O$, respectively.

To confirm the presence of localized states, temperature dependent PL measurements have been performed for the $Be_{0.04}Mg_{0.17}Zn_{0.79}O$ and $Be_{0.11}Mg_{0.15}Zn_{0.74}O$ samples. As shown in figure 19.14, an S-shaped behavior of PL peak position as a function of temperature was observed for the $Be_{0.04}Mg_{0.17}Zn_{0.79}O$ sample, which evidences the presence of localized states. The extent of localization was determined to be 22 meV, about 1/5 of the localization depth value determined from TRPL (98 meV). For sample $Be_{0.11}Mg_{0.15}Zn_{0.74}O$, no S-shaped behavior was observed in the temperature range due to its much higher degree of localization (268 meV determined from TRPL) as temperatures higher than 300 K are needed to delocalize the carriers.

Improved PL performance with much higher band edge emission intensity to defect emission intensity ratio in the BeMgZnO alloy via Be incorporation has been reported by Su *et al* and Chen *et al* [17, 35]. As shown in figure 19.15, without Be incorporation the PL spectrum of MgZnO film consisted of near-band-edge emission at 360 nm and a wide defect-related visible emission band with comparable intensities. After alloying MgZnO with BeO, the defect-related emission was significantly suppressed. A theoretical calculation based on density functional theory revealed that the formation energy of oxygen vacancy V_O defects in BeMgZnO was increased compared to that in MgZnO, meaning that V_O defects have less possibility to form in BeMgZnO. Hence a reduced V_O density was explained as the reason for the defect-related emission in BeMgZnO [35]. In addition, the background electron concentration has also been reduced from 1.0×10^{17} cm^{-3} in MgZnO to 3.9×10^{16} cm^{-3} in BeMgZnO by Be incorporation in the alloy.

19.5 Applications of BeMgZnO

Applications of the BeMgZnO quaternary are not well developed partially due to the limitation imposed by the doping asymmetry problem in ZnO-based materials [36]. The unavailability of reliable p-type ZnO prevents applications of ZnO-based materials in light-emitting diodes and potential laser diodes, where BeMgZnO could otherwise find its position. However, researchers have attempted to employ BeMgZnO for devices in which p-type ZnO is not required, such as photodetectors and HFETs.

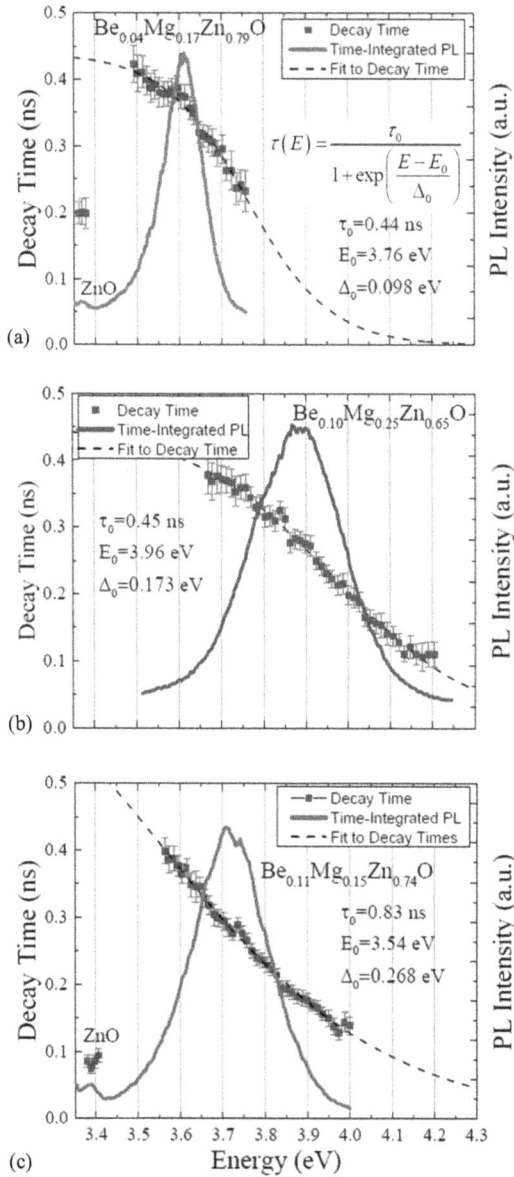

Figure 19.12. PL decay time dependence on the emission energy at 15 K and time integrated PL for samples (a) $Be_{0.04}Mg_{0.17}Zn_{0.79}O$, (b) $Be_{0.10}Mg_{0.25}Zn_{0.65}O$, and (c) $Be_{0.11}Mg_{0.15}Zn_{0.74}O$. Reproduced with permission from [34]. Copyright 2016 SPIE.

19.5.1 BeMgZnO/ZnO HFETs

One of the attractive applications of the BeMgZnO alloy is ZnO-based HFETs, owing to the capability of polarization engineering via Be and Mg content modulation in the barrier. In order to achieve competitive ZnO-based HFETs the

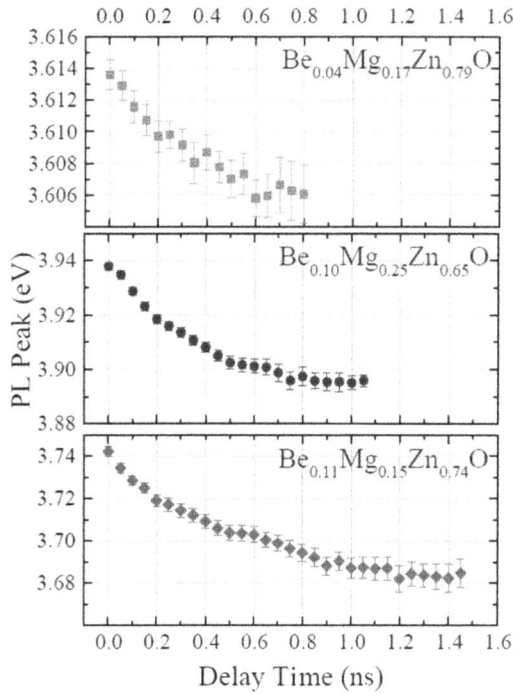

Figure 19.13. Temporal dependence of PL peak position (at 15 K) of O-polar BeMgZnO samples grown on GaN. Reproduced with permission from [34]. Copyright 2016 SPIE.

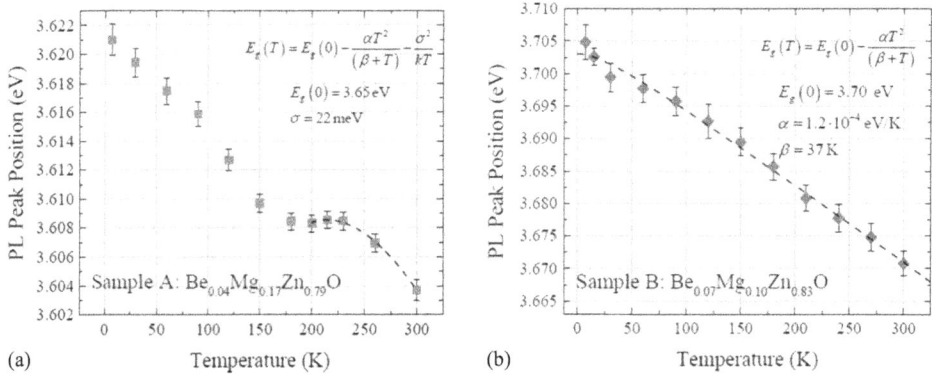

Figure 19.14. Temperature dependence of PL peak position for (a) $Be_{0.04}Mg_{0.17}Zn_{0.79}O$ and (b) $Be_{0.11}Mg_{0.15}Zn_{0.74}O$ grown on GaN. Reproduced with permission from [34]. Copyright 2016 SPIE.

following conditions must be satisfied: high quality of the ZnO layer that will serve as a medium for 2DEG layer and optimized growth conditions for the barrier layer that will provide proper spontaneous and piezoelectric polarization for the formation of 2DEG, low density of charge defects in the barrier, low specific contact resistivity ohmic contacts for the source and drain electrodes, and large barrier height and low leakage current for the gate electrode.

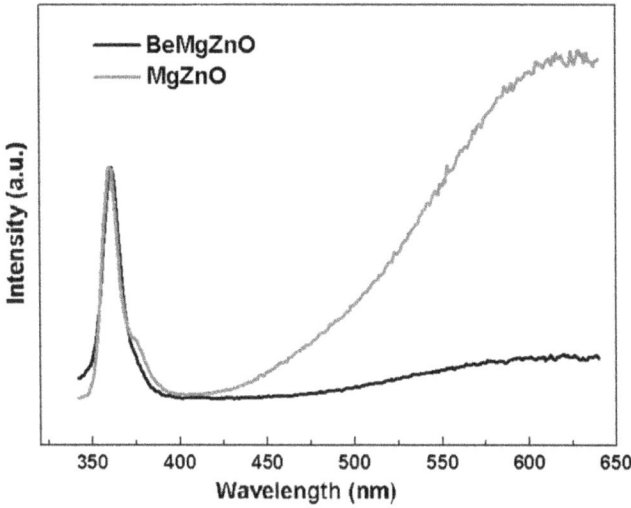

Figure 19.15. Room temperature PL spectra of MgZnO and BeMgZnO thin films grown by MBE on c-sapphire substrates. Reproduced with permission from [35]. Copyright 2015 Elsevier.

19.5.1.1 BeMgZnO versus MgZnO barrier

The conventionally used barrier material, MgZnO ternary, necessitates a very high Mg content (>40%) and growth at low substrate temperatures (300 °C or lower) [37, 38], and as such these structures are apt to degrade under high power operation and during thermal treatment, even if the unwanted charge density in the barrier layer is sufficiently low for efficient gate modulation. To circumvent this obstacle, BeMgZnO was proposed as the barrier, in which the strain sign in the barrier can be switched from compressive to tensile through the incorporation of Be, making the spontaneous and piezoelectric polarizations additive [18].

Figures 19.16(a) and (b) show schematics depicting the formation of 2DEG and the directions of the piezoelectric and spontaneous polarization vectors for Zn-polar MgZnO/ZnO and BeMgZnO/ZnO heterostructures, respectively. The 2DEG layer represented by a dashed line forms inside the ZnO near the barrier–ZnO interface. The spontaneous polarization (P_{SP}) vector associated with the partial ionicity of chemical bonds in ZnO-based materials is always orientated in the [000$\bar{1}$] direction. The direction of the strain-induced piezoelectric polarization (P_{pe}) depends on the sign of strain. Figure 19.16(c) shows a contour plot illustrating that a few percent of Be is sufficient to change the strain sign in the BeMgZnO barrier. As a result, BeMgZnO and MgZnO barriers in the Zn-polar heterostructures are under tensile and compressive biaxial strain, respectively. This means that the incorporation of Be into wurtzite lattice changes the direction of piezoelectric polarization through switching the strain sign from compressive to tensile, making the piezoelectric and spontaneous polarizations to be additive in BeMgZnO/ZnO HFETs. In this case, the polarization-induced sheet charge density σ at the BeMgZnO–ZnO interface is described by $|\sigma| = |P_{sp}(Be_xMg_yZn_{1-x-y}O) + P_{pe}(Be_xMg_yZn_{1-x-y}O) - P_{sp}(ZnO)|$,

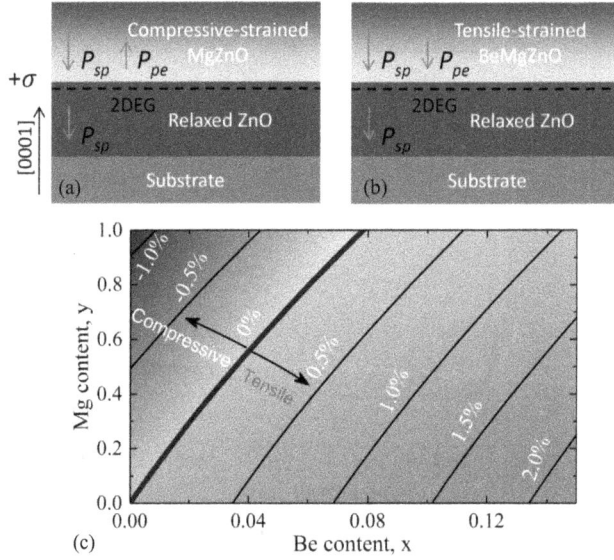

Figure 19.16. Schematics of the polarization-induced 2DEG channels and the directions of the spontaneous and piezoelectric polarization vectors for Zn-polar (a) MgZnO/ZnO and (b) BeMgZnO/ZnO heterostructures. (c) Contour plot of strain in the $Be_xMg_yZn_{1-x-y}O$ layer on ZnO as a function of Be and Mg contents. The negative and the positive signs represent the compressive and tensile strain, respectively. Reproduced with permission from [18]. Copyright 2017 AIP Publishing.

with a negative $P_{pe}(Be_xMg_yZn_{1-x-y}O)$ value instead of a positive value in the case of MgZnO/ZnO heterostructures.

Simple calculations, using the Vegard's law for the lattice parameters coupled with linear interpolation for the estimation of piezoelectric and elastic constants and spontaneous polarization values [39], show that using BeMgZnO as the barrier with a small amount of Be (<5%) and relatively low Mg content (~30%), one can achieve a polarization sheet charge density of $\sim10^{13}$ cm^{-2}, which otherwise would require very high Mg content (~60%) in the MgZnO/ZnO HFET analogue, as denoted by the dashed lines in figure 19.17. Utilizing this approach, high 2DEG densities near and above the plasmon-LO phonon resonance ($\sim7 \times 10^{12}$ cm^{-2}) in BeMgZnO/ZnO heterostructures with a Mg content below 30% and a Be content of only 2%~3% was experimentally realized [18]. The highest 2DEG concentration of 1.2×10^{13} cm^{-2} was achieved in the BeMgZnO/ZnO heterostructure with 3% Be and 41% Mg.

19.5.1.2 Zn-polar versus O-polar
Basal plane sapphire, c-sapphire, is the most commonly used substrate for ZnO-based materials. Due to the large in-plane lattice mismatch between c-sapphire and ZnO (18%), MgO buffer techniques have been developed for the growth of good quality ZnO on sapphire substrates [40]. Generally Zn-polar growth occurs when the thickness of the MgO buffer is more than 3 nm and O-polar growth occurs when the MgO layer is less than 2 nm thick. Although both Zn- and O-polar ZnO can be obtained via control of the MgO buffer thickness, Zn-polar ZnO usually exhibits

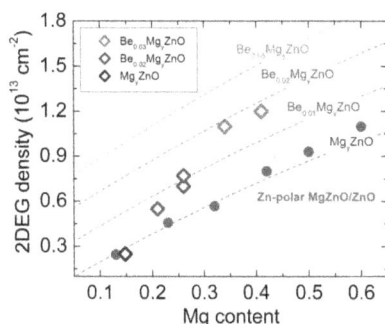

Figure 19.17. 2DEG densities in Zn-polar BeMgZnO/ZnO heterostructures compared with those in Zn-polar MgZnO/ZnO heterostructures. The dashed lines are calculated polarization sheet charge densities for BeMgZnO/ZnO heterostructures with 0%, 1%, 2%, and 3% Be content. Reproduced with permission from [18]. Copyright 2017 AIP publishing.

worse structural quality. This is because when the MgO buffer is thicker than 1 nm, the growth mode of MgO changes from two-dimensional to three-dimensional. Therefore, it is difficult to grow Zn-polar ZnO with both low background electron concentration and a smooth surface on the c-sapphire, both of which are critical to the performance of BeMgZnO/ZnO HFETs. If an O–ZnO growth orientation is adopted, the growth sequence of the ZnO channel and BeMgZnO barrier needs to be reversed, i.e. ZnO must be grown on BeMgZnO [30]. The issue with such a growth sequence is that the BeMgZnO layer tends to degrade during the high temperature ZnO growth step.

To avoid low quality Zn-polar ZnO and barrier degradation with O-polar designs, Zn-polar growth on a Ga-polar GaN template has been proposed as an alternative avenue for achieving BeMgZnO/ZnO HFETs [41]. The polarity of MBE grown ZnO on GaN was controlled by tuning the VI–II ratio during ZnO nucleation. It was found that following the Zn pre-exposure treatment, ZnO layers grown on GaN exhibit Zn-polarity when nucleated with low VI–II ratios (<1.5), while those nucleated with VI–II ratios above 1.5 exhibit O-polarity. Figure 19.18 shows the XRD triple-axis $2\theta-\omega$ scan and atomic force microscopy (AFM) image of a typical Zn-polar $Be_{0.02}Mg_{0.21}ZnO/ZnO$ heterostructure with a $Be_{0.02}Mg_{0.21}ZnO$ thickness of 50 nm grown on metal–organic chemical vapor deposition (MOCVD)-grown GaN templates using MBE [18]. The three peaks at 34.46°, 34.54°, and 34.71° correspond to (0002) reflections of ZnO, GaN, and $Be_{0.02}Mg_{0.21}ZnO$, respectively. The tensile biaxial strain in the ZnO layer is an indication of Zn-polarity of the heterostructure. Starting with a GaN template that had a root mean square (RMS) surface roughness of 0.28 nm for a 5 μm × 5 μm area scan, the roughness increased to 0.35 nm after the growth of a 300 nm thick high temperature ZnO layer. It further increased to 0.65 nm after the growth of the BeMgZnO barrier layer, featuring hexagonal pits on the surface.

High 2DEG density ranging from mid-10^{12} cm^{-2} to above 1×10^{13} cm^{-2} with a high electron mobility of above 200 cm^2 V^{-1} s^{-1} has been reported in Zn-polar BeMgZnO/ZnO heterostructures on GaN templates [18]. However, it was noticed

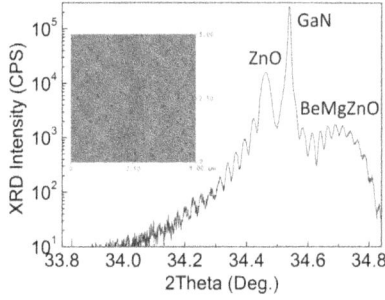

Figure 19.18. An XRD triple-axis 2θ–ω scan of a typical Zn-polar $Be_{0.02}Mg_{0.21}ZnO/ZnO$ heterostructure with a $Be_{0.02}Mg_{0.21}ZnO$ thickness of 50 nm. Inset: 5 μm × 5 μm atomic force microscopy (AFM) image of the heterostructure with RMS roughness of 0.65 nm. Reproduced with permission from [18]. Copyright 2017 AIP publishing.

that there exists a non-negligible parallel conduction contribution with a sheet electron density of ~2.4×10^{12} cm^{-2} in addition to the 2DEG channel, which is deemed detrimental to the device performance. One possible origin may be the spontaneous polarization difference between GaN (−0.029 C m^{-2}) and ZnO (−0.053 C m^{-2}). This parallel conduction caused reconsideration of direct growth of Zn-polar heterostructures on c-sapphire with improved quality. By tuning the growth temperature and thickness of the MgO buffer and the subsequent LT-ZnO buffer layer, Ding *et al* have found that it is possible to achieve highly resistive Zn-polar ZnO with smooth morphology directly on sapphire [42]. As shown in figure 19.19, the Zn-polar ZnO on c-sapphire exhibits similar features such as that on the GaN template, with a slightly higher RMS roughness of 1.2 nm. Such ZnO templates should be suitable for HFETs devices.

19.5.1.3 Effects of growth parameters

The electrical properties of BeMgZnO/ZnO heterostructures depend on the BeMgZnO barrier growth parameters including substrate temperature, O_2 flow rate, and barrier thickness [43]. As mentioned in section 19.3, a high substrate temperature and a low VI–II ratio are favorable for Mg and Be incorporation, resulting in a relatively high electron density but a low electron mobility due to enhanced interface/alloy scattering. As shown in figure 19.20, capacitance–voltage (C–V) depth profiling for $Be_{0.02}Mg_{0.26}ZnO/ZnO$ heterostructures with different barrier thicknesses indicates that a $Be_{0.02}Mg_{0.26}ZnO$ barrier with thickness slightly below 20 nm is sufficiently thick for generating 2DEG but with poor electron confinement. Good confinement can be obtained with barriers thicker than 20 nm.

The barrier thickness dependence of the electronic properties of the $Be_{0.02}Mg_{0.26}ZnO/ZnO$ heterostructures has been investigated by Hall-effect measurements at liquid nitrogen temperature (77 K). As can be seen in figure 19.21, the sheet electron density decreases as the BeMgZnO thickness is increased from 20 to 70 nm, and electron mobility reaches its maximum for a barrier thickness of 30 nm. The observed behavior is very similar to that of the 2DEG in AlGaN/GaN heterostructures [44, 45], such observed characteristics of which have been attributed

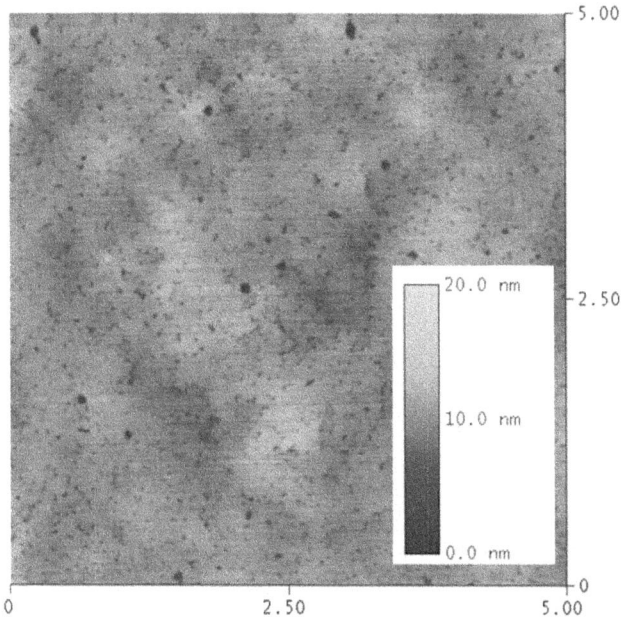

Figure 19.19. A 5 μm × 5 μm AFM image of Zn-polar ZnO directly grown on a c-sapphire substrate by MBE with an RMS roughness of 1.2 nm.

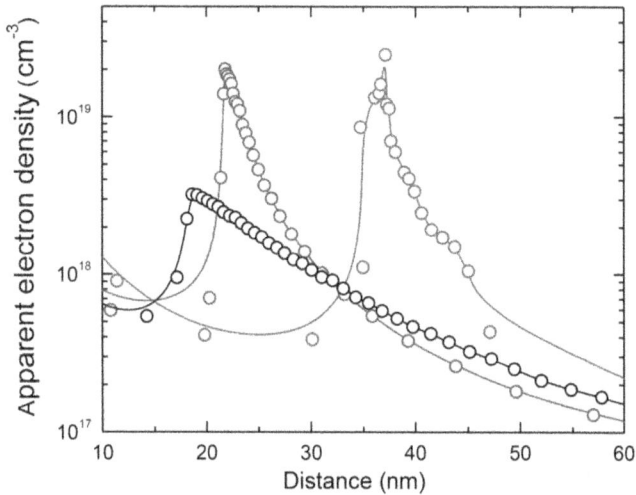

Figure 19.20. Capacitance–voltage (C–V) depth profiling for $Be_{0.02}Mg_{0.26}ZnO/ZnO$ heterostructures with different barrier thicknesses. Reproduced with permission from [43]. Copyright 2019 SPIE.

to the distribution of 2DEG in the channel and relaxation of the tensile strain in the barrier.

Figure 9.22 shows the room temperature electron sheet density and electron mobility of BeMgZnO/ZnO heterostructures on GaN templates with a BeMgZnO

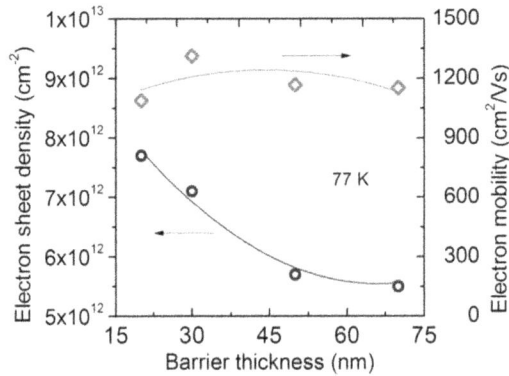

Figure 19.21. 77 K electron sheet density and electron mobility of $Be_{0.02}Mg_{0.26}ZnO/ZnO$ heterostructures with different barrier thicknesses. The solid lines are guides for the eye. Reproduced with permission from [43]. Copyright 2019 SPIE.

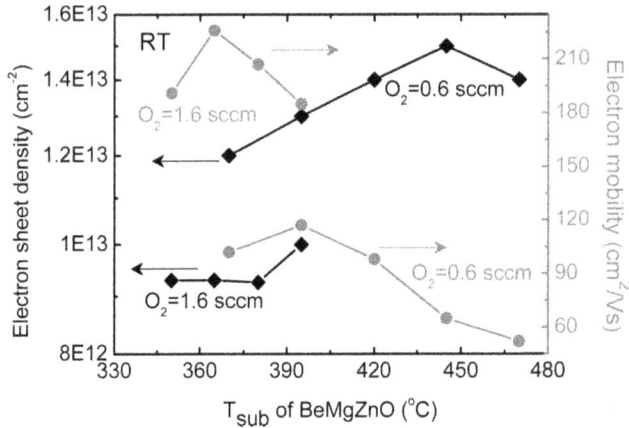

Figure 19.22. Room temperature electron sheet density and electron mobility of BeMgZnO/ZnO heterostructures grown with different substrate temperatures and O_2 flow rates for the BeMgZnO barrier layer.

barrier layer grown at substrate temperatures and oxygen flow rates varying from 350 °C to 470 °C, and 0.6 to 1.6 sccm, respectively. The BeMgZnO barrier thickness for all the samples is about 30 nm. For an O_2 flow rate of 0.6 sccm, the electron density increases linearly from 1.2×10^{13} cm^{-2} to 1.5×10^{13} cm^{-2} with increasing substrate temperature in the range from 370 °C to 450 °C, but reduces with further temperature increase. When the O_2 flow rate is increased to 1.6 sccm, the electron density decreases to less than 1.0×10^{13} cm^{-2} in the substrate temperature range of 350 °C to 395 °C. This effect is attributed to relatively lower Mg and Be content under more O_2-rich conditions caused by variation in Zn incorporation [16] and to some extent to smaller bulk contribution from the barrier layer with an optimal VI–II ratio (~5.3). The evolution of the electron mobility illustrated in figure 19.22 shows that lower substrate temperatures (350 °C–400 °C) and a high O_2 flow rate of 1.6 sccm are beneficial for obtaining high electron mobility in the heterostructures.

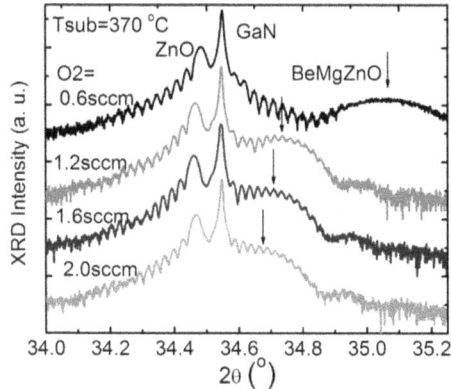

Figure 19.23. HRXRD triple-axis 2θ–ω scans of BeMgZnO/ZnO heterostructures grown with oxygen flow rates varied from 0.6 to 2.0 sccm. Reproduced with permission from [43]. Copyright 2019 SPIE.

This improvement may be a result of reduced interface/alloy scattering from reduced lattice disorder induced by lower Mg and Be content. Figure 19.23 shows the HRXRD triple-axis 2θ–ω scans of BeMgZnO/ZnO heterostructures grown with oxygen flow rates from 0.6 to 2.0 sccm at a substrate temperature of 370 °C for the BeMgZnO barrier layer. It can be noted that as the O_2 flow rate increases, the BeMgZnO reflections shift to lower angles, indicating decrease of the Mg and Be content in the quaternary alloys, while the full width at half maximum (FWHM) decreases due to improvement in crystalline quality and reduced lattice disorder.

19.5.1.4 Schottky barrier fabrication

Fabrication of Schottky barriers on BeMgZnO alloy has been reported by Ullah *et al* [46], and the detailed fabrication process has been visually demonstrated by Ding *et al* [47]. The ohmic contact was formed by electron beam evaporation of Ti/Au (30 nm/50 nm) followed by rapid thermal annealing (RTA) at 550 °C for 10 s in a nitrogen ambient. The specific ohmic contact resistivity was measured to be ~2.25×10^{-5} Ω cm^2. A Ag metal layer of 50 nm thickness was then deposited as the Schottky metal preceded with remote oxygen plasma of the BeMgZnO surface for 5 min with an RF power of 50 W and an oxygen flow of 35 sccm. Figure 19.24 shows the current–voltage (I–V) characteristics measured at room temperature for four representative Ag/Be$_{0.02}$Mg$_{0.26}$ZnO/ZnO Schottky diodes with a Schottky area of 1.1×10^{-4} cm^2 on the same wafer. The forward current increases exponentially with applied voltage up to 0.75 V, beyond which the voltage drop across the series resistance becomes apparent. The highest Schottky barrier height of $\Phi_{ap} = 1.07$ eV was attained with an ideality factor of $n = 1.22$. Rectification ratios of about 1×10^8 were achieved using the current values measured at $V = \pm 2$ V.

19.5.2 Other applications

Metal–semiconductor–metal (MSM) structured solar-blind UV detectors based on Be$_{0.17}$Mg$_{0.54}$Zn$_{0.29}$O have been demonstrated [11]. A detector showed a dark current

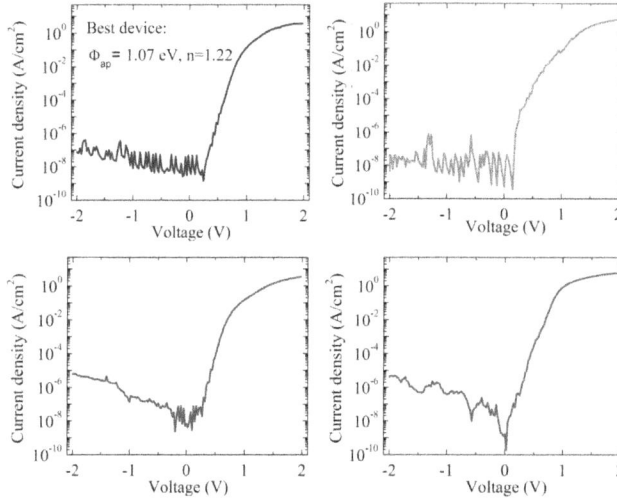

Figure 19.24. Typical I–V characteristics of four representative Ag/Be$_{0.02}$Mg$_{0.26}$ZnO/ZnO Schottky diodes measured at room temperature. The similarity of the four I–V curves indicates the high in-wafer uniformity of the sample. Reproduced with permission from [47]. Copyright 2018 JoVE.

of 9.3 pA under 5 V bias, a peak responsivity of 18 mA W^{-1} at 241 nm, and a cut-off wavelength of 267 nm. The high rejection ratio value of 21 (response at peak/ response at 280 nm) was much higher than that in reported MgZnO-based detectors. Heterojunction solar-blind UV detectors based on n-type BeMgZnO grown on p-type Si have also been reported [48]. For this work, a ceramic Zn$_{0.7}$Be$_{0.1}$Mg$_{0.2}$O target was used as the precursor for PLD growth of the n-type BeMgZnO. Under zero bias, the detector exhibited a cut-off wavelength of 280 nm, a peak responsivity of 110 mA W^{-1} at 270 nm, a rise time of 20 ns, and a fall time of 250 μs.

Be$_{0.083}$Mg$_{0.047}$Zn$_{0.870}$O film deposited by RF magnetron sputtering using three targets of MgO, Be, and Zn was employed as the active layer of ultraviolet (UV) light-emitting diodes (LEDs) with heterostructures such as p-GaN/i-MgBeZnO/n-MgZnO:Al and p-ZnO/(i-MgZnO/i-MgBeZnO, three pairs)/n-MgZnO:Al [49, 50]. The peak emission wavelength of the UV LEDs ranged from 352.8 nm to 362.6 nm as the injection current increased from 5 mA to 80 mA.

19.6 Summary

In conclusion, we have reviewed recent investigations of the BeMgZnO quaternary wide bandgap material system, which provides better bandgap engineering for ZnO-based structures, compared to MgZnO and BeZnO ternaries. The a-axis lattice compensation between BeO and MgO, and four-fold coordination preference of the Be atom in Be$_x$Mg$_y$Zn$_{1-x-y}$O lead to Be–Mg mutual stabilization in the hexagonal lattice, resulting in an extended wurtzite-to-rocksalt phase transition boundary, and improved stability, crystalline quality, and optical properties. The capability of strain modulation with a small amount of Be makes BeMgZnO an attractive barrier material for HFETs with high 2DEG density and electron mobility. The electrical properties of MBE grown BeMgZnO/ZnO heterostructures have shown some prospects in applications of HFETs, while further explorations are needed to demonstrate the merit of high electron saturation velocity in ZnO.

References

[1] Özgür Ü, Alivov Y I, Liu C, Teke A, Reshchikov M, Doğan S, Avrutin V, Cho S-J and Morkoç H 2005 *J. Appl. Phys.* **98** 11

[2] Son D-Y, Im J-H, Kim H-S and Park N-G 2014 *J. Phys. Chem.* C **118** 16567

[3] Liu K, Sakurai M and Aono M 2010 *Sensors* **10** 8604

[4] Morkoç H 2013 *Nitride Semiconductors and Devices* (Berlin: Springer)

[5] Zhu Y, Chen G, Ye H, Walsh A, Moon C and Wei S-H 2008 *Phys. Rev.* B **77** 245209

[6] Singh A, Vij A, Kumar D, Khanna P, Kumar M, Gautam S and Chae K 2013 *Semicon. Sci. Tech* **28** 025004

[7] Ryu Y, Lee T, Lubguban J, Corman A, White H, Leem J, Han M, Park Y, Youn C and Kim W 2006 *Appl. Phys. Lett.* **88** 052103

[8] Zheng Q, Huang F, Ding K, Huang J, Chen D, Zhan Z and Lin Z 2011 *Appl. Phys. Lett.* **98** 221112

[9] Le Biavan N, Hugues M, Montes Bajo M, Tamayo-Arriola J, Jollivet A, Lefebvre D, Cordier Y, Vinter B, Julien F-H and Hierro A 2017 *Appl. Phys. Lett.* **111** 231903

[10] Sasa S, Maitani T, Furuya Y, Amano T, Koike K, Yano M and Inoue M 2011 *Phys. Stat. Sol.* A **208** 449

[11] Su L, Zhu Y, Yong D, Chen M, Ji X, Su Y, Gui X, Pan B, Xiang R and Tang Z 2014 *ACS Appl. Mater. Interfaces* **6** 14152

[12] Su X, Si P, Hou Q, Kong X, Cheng W and Tang Z 2009 *Physica* B **404** 1794

[13] Zhu Y, Su L, Chen M, Su Y, Ji X, Gui X and Tang Z 2015 *J. Alloys Compd* **631** 355

[14] Gorczyca I, Teisseyre H, Suski T, Christensen N and Svane A 2016 *J. Appl. Phys.* **120** 215704

[15] Toporkov M, Demchenko D, Zolnai Z, Volk J, Avrutin V, Morkoç H and Özgür Ü 2016 *J. Appl. Phys.* **119** 095311

[16] Toporkov M, Ullah M, Demchenko D, Avrutin V, Morkoç H and Özgür Ü 2017 *J. Cryst. Growth* **467** 145

[17] Su L, Zhu Y, Zhang Q, Chen M, Wu T, Gui X, Pan B, Xiang R and Tang Z 2013 *Appl. Surf. Sci.* **274** 341

[18] Ding K, Ullah M, Avrutin V, Özgür Ü and Morkoç H 2017 *Appl. Phys. Lett.* **111** 182101

[19] Yang R, Li M, Chang G, Lu Y and He Y 2018 *Mater. Sciences* **8** 559

[20] Yang C, Li X, Gu Y, Yu W, Gao X and Zhang Y 2008 *Appl. Phys. Lett.* **93** 112114

[21] Yang M L R, Chang G, Lu Y and He Y 2018 *Mater. Sciences* **8** 571–5

[22] Yang C, Li X, Gao X, Cao X, Yang R and Li Y 2010 *J. Cryst. Growth* **312** 978

[23] Panwar N, Liriano J, Puli V S and Katiyar R S 2010 *American Physical Society, APS March Meeting 2010, March 15-19* abstract id. D32.003

[24] Park J-H, Cuong H B, Jeong S-H and Lee B-T 2014 *J. Alloys Compd* **615** 126

[25] Cuong H B and Lee B-T 2015 *Appl. Surf. Sci.* **355** 582

[26] Cuong H B, Lee C-S and Lee B-T 2014 *Thin Solid Films* **573** 95

[27] Panwar N, Liriano J and Katiyar R S 2011 *(Symposium MM - Transparent Conducting Oxides and Applications)* **1315** mrsf10-1315-mm05-03-f05-03

[28] Vettumperumal R, Kalyanaraman S and Thangavel R 2013 *J. Sol-Gel Sci. Technol.* **68** 334

[29] Avrutin V, Reshchikov M, Nie J, Izyumskaya N, Shimada R, Özgür Ü, Foreman J, Everitt H, Litton C and Morkoç H 2008 *Proc. SPIE* **6895** 68950Y

[30] Ullah M, Toporkov M, Avrutin V, Özgür Ü, Smith D and Morkoç H 2017 *Proc. SPIE* **10105** 101050

[31] Toporkov M, Avrutin V, Okur S, Izyumskaya N, Demchenko D, Volk J, Smith D, Morkoç H and Özgür Ü 2014 *J. Cryst. Growth* **402** 60

[32] Zolnai Z, Toporkov M, Volk J, Demchenko D O, Okur S, Szabó Z, Özgür Ü, Morkoç H, Avrutin V and Kótai E 2015 *Appl. Surf. Sci.* **327** 43

[33] Khoshman J, Ingram D and Kordesch M 2008 *J. Non-Cryst. Solids* **354** 2783

[34] Toporkov M, Ullah M B, Hafiz S, Nakagawara T, Avrutin V, Morkoç H and Özgür Ü 2016 *Proc. SPIE* 9749 974910

[35] Chen S, Pan X, Li Y, Chen W, Zhang H, Dai W, Ding P, Huang J, Lu B and Ye Z 2015 *Phys. Lett.* A **379** 912

[36] Avrutin V, Silversmith D J and Morkoç H 2010 *Proc. IEEE* **98** 1269

[37] Koike K, Hama K, Nakashima I, Takada G-y, Ogata K-i, Sasa S, Inoue M and Yano M 2005 *J. Cryst. Growth* **278** 288

[38] Du X, Mei Z, Liu Z, Guo Y, Zhang T, Hou Y, Zhang Z, Xue Q and Kuznetsov A Y 2009 *Adv. Mater.* **21** 4625

[39] Ullah M B, Avrutin V, Nakagawara T, Hafiz S, Altuntaş I, Özgür Ü and Morkoç H 2017 *J. Appl. Phys.* **121** 185704

[40] Kato H, Miyamoto K, Sano M and Yao T 2004 *Appl. Phys. Lett.* **84** 4562

[41] Ullah M B, Avrutin V, Li S Q, Das S, Monavarian M, Toporkov M, Özgür Ü, Ruterana P and Morkoç H 2016 *Phys. Stat. Sol. (RRL)* **10** 682

[42] Ding V A K, Izyumskaya N, Özgür Ü and Morkoç H 2020 *J. Vacuum Sci. Technol.* A **38** 023408

[43] Ding K, Avrutin V, Izyumskaya N, Özgür Ü, Morkoç H, Šermukšnis E and Matulionis A 2019 *Proc. SPIE* 10919 1091917

[44] Smorchkova I, Elsass C, Ibbetson J, Vetury R, Heying B, Fini P, Haus E, DenBaars S, Speck J and Mishra U 1999 *J. Appl. Phys.* **86** 4520

[45] Ibbetson J P, Fini P, Ness K, DenBaars S, Speck J and Mishra U 2000 *Appl. Phys. Lett.* **77** 250

[46] Ullah M B, Ding K, Nakagawara T, Avrutin V, Özgür Ü and Morkoç H 2018 *Phys. Status Solidi Lett.* **12** 1700366

[47] Ding K, Avrutin V, Izioumskaia N, Ullah M B, Özgür Ü and Morkoç H 2018 *J. Vis. Exp.* **140** e58113

[48] Yang C, Li X, Yu W, Gao X, Cao X and Li Y 2009 *J. Phys. D: Appl. Phys.* **42** 152002

[49] Lee H-Y, Chang H-Y, Lou L-R and Lee C-T 2013 *IEEE Photonics Technol. Lett.* **25** 1770

[50] Lee C-T and Chang H-Y 2014 *Proc. SPIE* 8987 89871

Part IV

Boron Nitride

IOP Publishing

Wide Bandgap Semiconductor-Based Electronics

Fan Ren and Stephen J Pearton

Chapter 20

Growth and properties of hexagonal boron nitride (*h*-BN) based alloys and quantum wells

Q W Wang, J Li, J Y Lin and H X Jiang

20.1 Introduction and unique properties of *h*-BN

Due to its extraordinary physical properties, layered structure, quasi-2D nature, and bandgap of about 6.5 eV, hexagonal boron nitride (*h*-BN) possesses promising rich novel properties. Currently, AlN is the best known deep UV semiconductor material with a comparable bandgap (~6.1 eV) to *h*-BN. Several unique features of *h*-BN in comparison to AlN are of immense interest for novel device exploration and can be summarized as follows.

Ability to enable novel flexible devices with versatile form factors. Significant progress in epitaxial growth using metal–organic chemical vapor deposition (MOCVD) for producing thick *h*-BN epilayers has been made in recent years [1–7]. When the thickness exceeds ~30 μm, after growth and during cooling down an *h*-BN epilayer tends to automatically separate from the sapphire substrate due to its unique hexagonal (layered) structure and the difference in thermal expansion coefficients between the *h*-BN and sapphire substrate [1–7]. As shown in figure 20.1, free-standing *h*-BN epilayers can be sliced into varying shapes and tiled together to form arrays, and are flexible with good conformability and transferability. Devices based on free-standing *h*-BN epilayers can be attached to rigid, flat, or curved surfaces. As such, these materials potentially offer a disruptive platform to design a wide range of photonic and electronic devices in flexible form factors.

High optical emission efficiency. As shown in figure 20.2(a), in comparison with AlN, the full width at half maximum (FWHM) of the x-ray diffraction (XRD) rocking curve of the *h*-BN (0002) diffraction peak is ~380 arcsec [8], which is 5–8 times broader than that of high quality AlN epilayers deposited on *c*-plane sapphire (around 60 arcsec) [9]. The FWHM of the XRD rocking curve of *h*-BN is comparable to those of typical GaN epilayers deposited on *c*-plane sapphire (around 290 arcsec) [10]. In III-nitride materials, it is well known that the FWHM of XRD

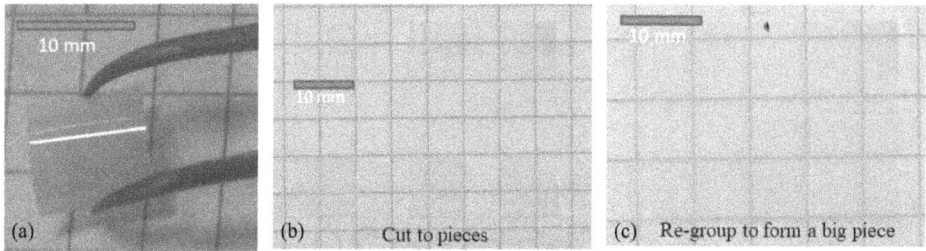

Figure 20.1. Optical images of free-standing (FS) *h*-BN epilayers of 50 μm in thickness: (a) an FS *h*-BN epilayer displays flexibility, (b) an FS *h*-BN epilayer cut into small pieces, and (c) small pieces of FS *h*-BN epilayers grouped to form a large array.

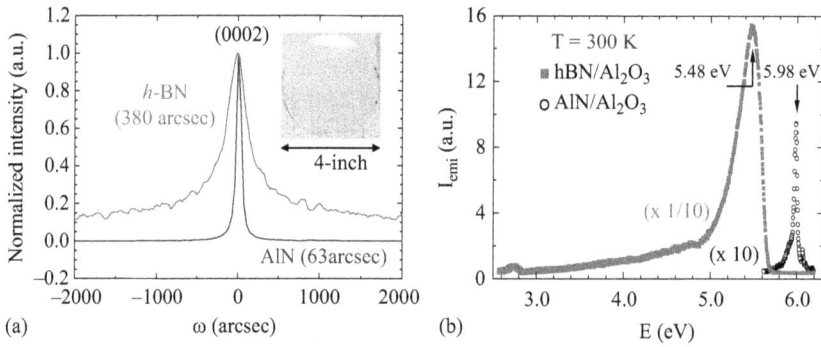

Figure 20.2. (a) Comparison of XRD rocking curves of the (0002) reflection peaks between *h*-BN and AlN epilayers (after [8]). (b) Comparison of DUV PL spectra between *h*-BN and AlN epilayers measured at 300 K (after [13] and [14]).

rocking curves is correlated with the density of dislocations [9–12]. In the case of *h*-BN with a layered structure, the FWHM could also reflect the presence of misalignment between layers, turbostratic (t-)phase layers, or variations in the c-plane orientation of individual layers within *h*-BN, stacking faults, and native and point defects. However, despite the fact that the crystalline quality of *h*-BN is poorer than that of AlN, the photoluminescence (PL) emission intensity of the band-edge transition in *h*-BN epilayers is generally more than two orders of magnitude higher than that of AlN epilayers with a comparable thickness [13, 14]. Note that the PL spectra shown in figure 20.2(b) are multiplied by a factor of 1/10 for *h*-BN and by 10 for AlN, respectively. These results imply that *h*-BN deep UV devices could potentially be even more efficient than AlN. This high emission efficiency is in part due to the quasi-2D nature of *h*-BN [13].

TE versus TM mode. It is well known that the band-edge emission in AlN and high Al-content AlGaN is dominated by the transverse-magnetic (TM) polarization mode due to the unique band structure of AlN [15, 16]. However, the TE mode is generally preferred for optoelectronic device applications because this mode allows for surface emitting light emitting diodes (LEDs) and is associated with a lower threshold and higher optical gain and lasing beam quality for laser diodes (LDs).

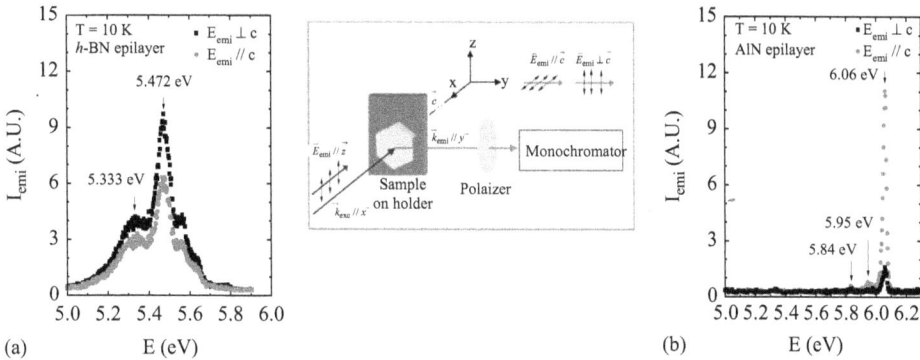

Figure 20.3. Comparison of polarization-resolved low temperature (10 K) band-edge PL spectra between (a) the h-BN epilayer and (b) AlN epilayer, with the emission polarization parallel ($E_{emi}//c$) and perpendicular ($E_{emi}\perp c$) to the c-axis (after [13] and [14]). The PL measurement configuration is depicted in the inset and the excitation laser line is polarized in the direction perpendicular to the c-axis ($E_{exc}\perp c$).

Figure 20.3 compares the polarization-resolved band-edge PL emission spectra of h-BN and AlN epilayers measured at 10 K [13, 14]. The PL emission spectral lineshape for h-BN shown in figure 20.3(a) for the configuration with emission polarization along the crystal c-axis ($E_{emi}//c$) is observed to be very similar to that in the ($E_{emi}\perp c$) configuration. However, the emission intensity is about 1.7 times stronger in the ($E_{emi}\perp c$) configuration, which is in sharp contrast to the polarization-resolved PL spectra of AlN shown in figure 20.3(b). However, the emission polarization of h-BN is the same as that of GaN [16–18]. The results clearly reveal that the band-edge emission in h-BN is predominantly transverse-electric (TE) polarized and agree with theoretical calculation results [13]. Due to the combination of the TE mode, large exciton binding energy, and 2D nature, h-BN light emitting devices are expected to be very efficient and h-BN based deep UV lasers should perform better than AlN based deep UV lasers.

2D versus 3D. Having a layered structure, h-BN is a quasi-2D material, whereas AlN is a 3D material. It was demonstrated that the strong optical transitions in h-BN partly originates from the unusually strong $p \rightarrow p$-like transitions due to its quasi-2D nature, giving rise to a very high density of states near the band edge [13]. Furthermore, the above band-edge optical absorption coefficient is around 7×10^5 cm^{-1} for h-BN [19, 20] in comparison to 2×10^5 cm^{-1} for AlN [21, 22]. This is also due to the quasi-2D nature of h-BN in which each layer absorbs 2.3% for the above bandgap photons [19].

Improved p-type conductivity control over AlN. Although Al-rich AlGaN ternary alloys and AlN have been the default choice for the development of LEDs and LDs operating at wavelengths below 300 nm, the highly resistive nature of p-layers leads to very low hole injection efficiency and hence low quantum efficiency (QE). The poor p-type conductivity of Al-rich $Al_xGa_{1-x}N$ alloys is the major obstacle that limits the QE of deep ultraviolet (DUV) light emitting devices. This problem is due to the deepening of the Mg acceptor level in $Al_xGa_{1-x}N$ with increasing x, from about 170 meV ($x = 0$) to 530 meV ($x = 1$) [23–26]. Since the free hole concentration

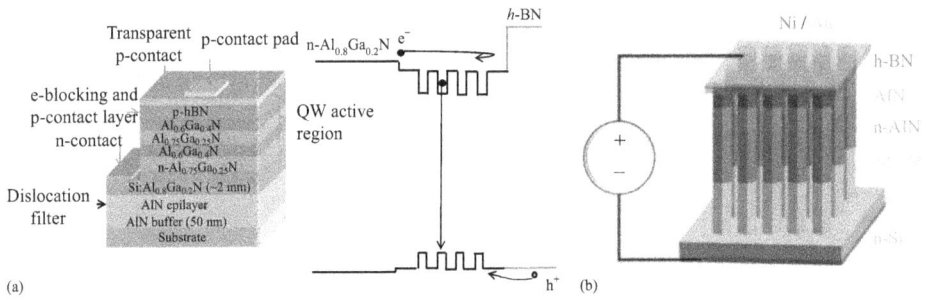

Figure 20.4. (a) Schematic of a DUV LED layer structure incorporating p-type h-BN. The wide bandgap p-type *h*-BN serves as a natural electron-blocking (e-blocking), hole injection and p-type contact layer. The approach is based on *h*-BN's unique band-edge alignment with AlGaN, improved p-type conductivity over AlN and transparency to DUV photons (after [23] and [29]). (b) A 210 nm deep UV LED fabricated from p-type *h*-BN/n-AlN nanowire heterostructure. Reproduced with permission from [30]. Copyright 2017 American Chemical Society.

depends exponentially on the acceptor energy level (E_a), an E_a value around 500 meV translates to only 1 free hole for roughly every 2 billion (2×10^9) incorporated Mg impurities (at 300 K). This causes an extremely low free hole injection efficiency into the quantum well (QW) active region and is a major obstacle for the realization of high-performance AlGaN-based DUV emitters. Significant advances in the QE of DUV emitters will require the exploitation of disruptive device concepts. It was shown that a much lower p-type resistivity can be achieved in *h*-BN than in AlN. A p-type resistivity of ~2 Ω cm at 300 K by Mg doping has been demonstrated in *h*-BN [23, 27, 28], which is about five to six orders of magnitude lower than what has been possible for Mg doped AlN (>10^5 Ω cm) [24]. A p-BN/n-AlGaN heterostructure, as schematically shown in figure 20.4(a) [23, 29], has been explored for deep UV device applications and can be further improved to tackle the *p*-type conductivity issue in DUV emitters based on high Al-content AlGaN. More recently, very promising results have been demonstrated for AlGaN nanowire based DUV LED structures, as illustrated in figure 20.4(b), by utilizing *h*-BN as a p-type current injection layer [30].

Huge exciton binding energy. The calculated exciton binding energy in 2D *h*-BN is huge and ranges from 0.71 to 0.76 eV [13, 31–33]. Experimentally, from the temperature dependence of the exciton decay lifetime, an exciton binding energy of ~740 meV and a small exciton Bohr radius of ~8 Å have been indirectly deduced [34]. Photocurrent excitation spectroscopy has also been utilized to directly probe the fundamental band parameters of *h*-BN. As shown in figure 20.5(a), transitions in a photoexcitation spectrum obtained at a fixed bias voltage corresponding to the direct band-to-band, free excitons, and impurity bound excitons have been directly observed. From the observed transition peak positons, the room temperature bandgap ($E_g \sim 6.42$ eV) and binding energy of excitons ($E_x \sim 0.73$ eV) were directly measured and a band diagram has been constructed as shown in figure 20.5(b) [35]. This exciton binding energy is about one order of magnitude larger than the well-known very large exciton binding energy in AlN [36–39] and is likely the largest

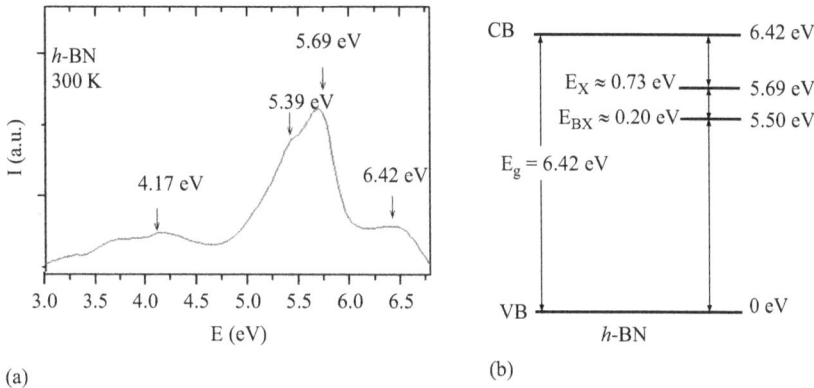

Figure 20.5. (a) Room temperature photocurrent excitation spectrum of a metal–semiconductor–metal detector consisting of micro-strip interdigital fingers (6 μm/9 μm of width/spacing) fabricated from an *h*-BN epilayer measured at a bias voltage = 100 V applied between detector fingers. (b) Energy band diagram including the room temperature energy bandgap (E_g), the binding energy of a free exciton (E_x), and the binding energy of an impurity bound exciton (E_{bx}) in *h*-BN epilayers constructed from the photocurrent excitation spectrum shown in (a), neglecting possible effects due to strain (after [35]).

in inorganic 2D semiconductors. This huge binding energy is expected to have large consequences on its applications as light emitting devices.

Low electron affinity. *h*-BN has a very low or even negative electron affinity, similar to diamond and appears to be promising as a high-performance electron emitter material with potential advantages of high power/power density and low transverse emissivity [20].

Ideal material for h-BN/graphene heterostructure applications. Due to its similar in-plane lattice constant to graphene and chemical inertness and resistance to oxidation, *h*-BN has been established as an ideal material for the exploration of van der Waals heterostructures and devices with new physics and applications [40–43].

Possible host for single photon emitters. More recently, *h*-BN has emerged as a promising material to host single photon emitters. More specifically, room temperature stable single photon emissions in a broad spectral range from about 1.65 to 2.2 eV have been observed in tape exfoliated *h*-BN multilayers as well as in small *h*-BN powder and bulk crystals [44–50]. In contrast to the well-studied diamond material, in which it is generally accepted that NV centers are the origin of single photon emitters, the physical origin of single photon emitters in *h*-BN is still a topic of experimental and theoretical debate [44–51].

Material of choice for solid-state neutron detectors. As shown in figure 20.6(a), the isotope B-10 is among a few isotopes that can interact with neutrons with a large capture cross-section (only second to ^{3}He) and has a capture cross-section of about 3840 barns for thermal neutrons (neutrons with an average energy = 25 meV) [52]. As a semiconductor, the density of B atoms in 100% ^{10}B-enriched BN which can interact with thermal neutrons is 5.5×10^{22}/cm^3, which is about 550 times higher than that in He-3 gas pressurized at 4 atm [1]. This unique property makes *h*-BN a

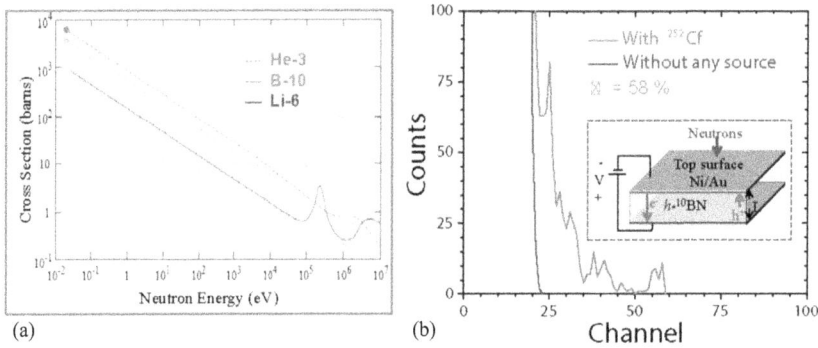

Figure 20.6. (a) Neutron capture cross sections as functions of the kinetic energy of neutrons for He-3, B-10, and Li-6. The green and orange dots indicate, respectively, the cross sections in He-3 and B-10 for thermal neutrons (neutrons with an average energy of 0.025 eV). (b) Nuclear reaction pulse height spectrum of a B-10 enriched *h*-BN detector with the device configuration shown in the inset under thermal neutron radiation. The neutron response was measured by placing the detector at 30 cm away from the ^{252}Cf source moderated by a one-inch thick high-density polyethylene (HDPE) moderator at a bias voltage of 200V for 15 min. Reproduced with permission from [3]. Copyright 2018 AIP Publishing.

highly promising material for the fabrication of solid-state neutron detectors [1–7]. As illustrated in figure 20.6(b), the recent achievement of free-standing B-10 enriched *h*-BN epilayers of large thicknesses (> 50 μm) has enabled the realization of thermal neutron detectors with an unprecedented record high detection efficiency among all solid-state detectors at 58% [3, 6]. In comparison to He-3 gas detectors, BN neutron detectors possess all the intrinsic advantages of semiconductor devices: light weight, compact size, fast response speed, ease of mass production via utilization of existing semiconductor manufacturing infrastructures, low cost for operation and maintenance, flexible form factors, and durability. The price of He-3 gas is also very high because He-3 gas is extremely rare on Earth. Neutron detectors have applications in many industries including special nuclear material tracking, nuclear hazard detection, life (water) searching in space, and geothermal and well logging. Other potential applications of neutron detectors include neutron scattering experiments for materials research, neutron cameras for neutron radiography for monitoring metal structures and parts, and for medical applications such as neutron imaging and boron neutron capture therapy for cancer treatment.

Despite its unique combination of superb properties, as with all other compound semiconductors, achieving the ability to tune the optoelectronic properties through alloying and heterojunction formation is highly desirable and will further expand the usefulness of *h*-BN. One way to achieve bandgap engineering in *h*-BN is to form alloys with GaN, namely $Ga_xB_{1-x}N$ alloys [53, 54]. Despite the fact that the equilibrium phases of GaN and BN are wurtzite and hexagonal, respectively, recent theoretical studies indicated that a 2D hexagonal layered structure is thermodynamically the most stable structure for few-layer nitride semiconductors (GaN, AlN, InN) [55–57] and the transition from layered structure to wurtzite structure occurs at eight layers for InN, ten for GaN, and 22 for AlN in the absence of strain [56]. An alternative approach to realizing *h*-BN based alloys is to consider the layer-structured hexagonal boron nitride carbon semiconductor alloys, h-(BN)C [58–61]. Both h-BN and graphite (C) are layer-structured materials with similar lattice parameters and crystalline structures. The in-plane a-lattice constant difference is only about 1.5% between graphite and h-BN. The h-(BN)$_{1-x}$(C$_2$)$_x$ alloy system thus holds the unique advantages of identical crystalline structure (hexagonal) and excellent matches in lattice constants, thermal expansion coefficients, and melting points throughout the entire alloy range. In this chapter, we summarize recent progress made in terms of MOCVD growth and the understanding of the basic properties of hexagonal (BN)C and BGaN alloys and heterostructures.

20.2 Prospects of h-BN-based alloys and heterostructures

With the recent progress in achieving h-BN at the wafer-scale using MOCVD growth, we are now provided with an ideal template to expand the epitaxial growth technique to produce novel 2D layer-structured h-BCN and h-BGaN alloys, heterostructures, and QWs. These new material systems are expected to possess different properties from bulk or epilayer structures with respect to the energy bandgaps, 2D transport and optical properties. Some of the basic properties that can be envisioned from h-BGaN based alloys include the following.

Bandgap variation in 2D semiconductors. The energy bandgap variations of these alloy systems are expected to be very large. Both hexagonal boron nitride (h-BN) and graphite (C) are layered materials with similar lattice parameters and crystalline structures. The in-plane a-lattice constant difference is only about 1.5% between graphite and h-BN, which provides the potential to synthesize layer-structured h-(BN)$_{1-x}$(C$_2$)$_x$ alloys. The h-(BN)$_{1-x}$(C$_2$)$_x$ alloy system thus holds the unique advantages of identical crystalline structure (hexagonal) and excellent matches in lattice constants, thermal expansion coefficients, and melting points throughout the entire alloy range. Moreover, as schematically shown in figure 20.7(a), this layer-structured alloy system potentially possesses an extremely large energy gap (E_g) variation from around 6.5 eV for h-BN to 0 for graphite. From an electrical properties perspective, this alloy system provides a large range of conductivity control from highly resistive semiconductors (undoped h-BN) to semi-metal (graphite). The h-(BN)$_{1-x}$(C$_2$)$_x$ alloys in the BN-rich side would complement III-nitride wide bandgap semiconductors as a layer-structured material system functioning in the deep UV spectral region. The h-(BN)$_{1-x}$(C$_2$)$_x$ alloys in the C-rich side have the potential to provide an unprecedented degree of freedom in the design of infrared (IR) detector and electronic devices. More specifically, the bandgap energy tunability in the C-rich side could offer IR detectors with cut-off wavelengths from the short wavelength IR (SWIR: 1–3 μm) to very long wavelength (VLWIR: 14–30 μm) range. Furthermore, C-rich h-(BN)$_{1-x}$(C$_2$)$_x$ alloys would address the major challenges facing the emerging 2D materials and open up new realms for novel physical properties and device exploration. On the other hand, with the energy gaps of h-BN and h-GaN of around 6.5 eV [34, 35] and 4.5 eV [62], respectively, this layer-structured h-BGaN alloy system appears to offer a large bandgap variation.

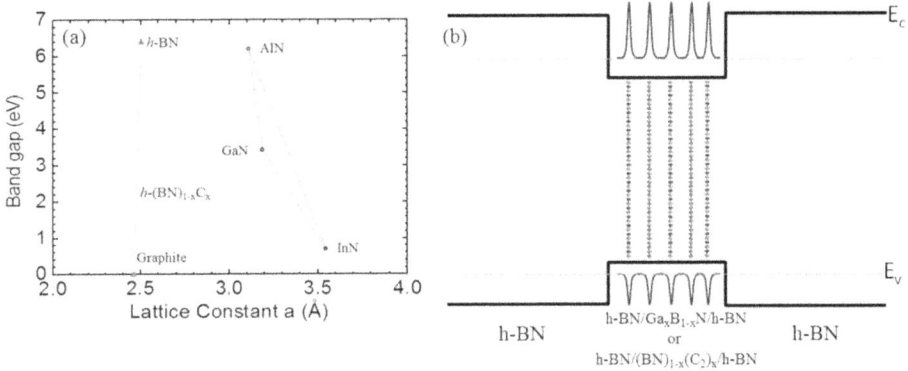

Figure 20.7. (a) Comparison of the energy bandgap variation of the a-lattice constant of h-(BN)C alloys with InGaAlN alloys. (b) Schematic of the band structure of a h-BN/h-Ga$_x$B$_{1-x}$N/h-BN or h-BN/h-(BN)$_{1-x}$(C$_2$)$_x$/h-BN QW with five well layers of h-Ga$_x$B$_{1-x}$N or h-(BN)$_{1-x}$(C$_2$)$_x$, in which carriers are confined in the h-Ga$_x$B$_{1-x}$N or h-(BN)$_{1-x}$(C$_2$)$_x$ layers in addition to the confinement provided by h-BN barriers.

The system also offers the possibility for conductivity control from the highly insulating semiconductor (undoped h-BN) to conductive GaN and therefore potentially enables the possibility of monolithic integration of the active layer, carrier injection layers, dielectric layer, and passivation layers via a single growth process.

Unique heterostructures and QWs. In h-BN/h-Ga$_x$B$_{1-x}$N/h-BN and h-BN/h-(BN)$_{1-x}$(C$_2$)$_x$/h-BN QWs, excitons are confined in the layers in addition to the confinement provided by h-BN barriers, as schematically illustrated in figure 20.7(b). This unique property stems from the fact that the wavefunctions of electrons and holes in these QWs do not extend through the whole well region as in those conventional QWs, since carriers and excitons are confined in the layers and there is a large potential barrier between layers. This naturally occurring large quantum confinement may significantly enhance the collective dipole interaction between the excitons and photon fields in micro-size cavities.

Possibility for novel detector structures. For 2D layered structures, a single layer or a few layers are adequate for many applications. However, for certain applications such as photodetectors, tens of layers may be needed. Other than the quantum confinement effects resulting from these layered QWs, h-BN can also be used as separation layers for a single layer or a few layers of h-(BN)$_{1-x}$(C$_2$)$_x$ and h-BGaN by recognizing the fact that each layer only absorbs 2.3% of incoming light with energies above its bandgap. Due to the capability for bandgap engineering via alloying, if successfully realized, these structures have the potential to serve as a basic building blocks for the construction of emitters and detectors operating from the DUV to FIR and THz, full spectrum solar cells, and multi-spectral detectors.

Possibility for novel laser structures. Due to the photon–exciton interaction in semiconductors, there arise new elementary excitations, which are essentially mixed states of the photon and exciton: excitonic-polaritons. Polaritons can be utilized to design polariton lasers. In contrast to conventional semiconductor lasers, polariton lasers have very low threshold current because polariton lasing involves only the spontaneous emission of coherent light of an exciton–polariton condensate and does not rely on population inversion. In recent years, there has been intense interest in understanding polaritons and developing room temperature polariton lasers using wide bandgap semiconductors [63, 64]. In wide bandgap semiconductors, the exciton binding energy and oscillator strength are large enough such that polariton condensation and lasing at visible to ultraviolet wavelengths can survive at 300 K. Due to a very strong collective dipole interaction between the QW excitons and microcavity photon fields, even with a relatively low-Q cavity, the planar microcavity system is expected to feature a reversible spontaneous emission and thus normal-mode splitting into upper and lower branches of polaritons, spectrally separated by the Rabi splitting energy $\hbar\Omega$. The exciton oscillator strength f is written as

$$f = \frac{2m^*\omega}{\hbar}\,|\langle u_v\,|\,r\cdot e\,|\,u_c\rangle|^2\,\frac{v}{\pi a_B^3}. \tag{20.1}$$

Here $m^* = (m_e^{-1} + m_h^{-1})^{-1}$ is the effective mass of the exciton, m_e (m_h) is the effective mass of the electron (hole), ω is the frequency of photons, $|u_c\rangle$ and $|u_v\rangle$ are the electron and hole Bloch wavefunctions, V is the quantization volume, and a_B is the Bohr radius of the exciton. The binding energy (E_B) of the ground exciton state is

$$E_B = \frac{e^2}{2\varepsilon a_B}\ \text{or}\ E_B = \frac{h^2}{\pi^2\mu a_B^2}, \tag{20.2}$$

where ε is the crystal's dielectric constant, e is the electron electric charge, h is Planck's constant, μ is the exciton reduced mass $\mu = \varepsilon m_o a_o/a_B$, $a_o = 0.53$ Å is the Bohr radius of the hydrogen atom, and m_o is the mass of free electron. The ratio $V/\pi a_B^3$ reflects the oscillator strength enhancement due to enhanced electron–hole overlap in an exciton compared to an unbound electron and hole pair. This ratio is very large, i.e. the exciton oscillator strength is much larger in wide bandgap semiconductors which have larger exciton binding energies or smaller exciton Bohr radii than those in small bandgap semiconductors. The exciton binding energy of about 740 meV in h-BN is at least one order of magnitude larger whereas the exciton Bohr radius of ~8 Å in h-BN is much smaller than those in the other well-known wide bandgap semiconductors, GaN, ZnO, and AlN. It is further noted that the predicted exciton binding energy in a single h-BN layer (or h-BN single sheet) is as large as 2.1 eV [32]. Moreover, h-BN is also a natural hyperbolic material with a strong uniaxial anisotropy, in which the transverse and longitudinal permittivities (the real parts) have opposite signs [65] and hence it is likely to enable many novel applications. However, developing h-BN based alloys and QWs structures such as those shown in figure 20.7(b) is necessary to enable carrier injection from the barrier into the well regions and practical current injection devices.

20.3 Epitaxy growth and properties of *h*-BGaN alloys and QWs

BGaN and BAlN in the wurtzite (WZ) phase have attracted considerable interest for UV material and device applications, as these alloys potentially offer the ability of tuning the bandgap energy from the blue to deep UV spectral region as well as providing an improved lattice match with SiC substrates [66–69]. Significant progress has been made recently in realizing $B_xAl_{1-x}N$ epitaxial layers in single WZ phase by MOCVD [70, 71]. Similarly, a majority of theoretical studies have also focused on the WZ phase BGaN and BAlN alloys [72–74].

Theoretical and experimental studies on the synthesis and structural stability of BGaN alloys in the hexagonal phase are scarce. Most recently, an empirical bond-order potential (BOP) with the aid of *ab initio* calculations has been applied to investigate the structures and miscibility of $B_xAl_{1-x}N$ and $B_xGa_{1-x}N$ alloys [75, 76]. Table 20.1 shows the calculated equilibrium bond length r_e and the cohesive energy E_{coh} of BN and GaN for free-standing hexagonal (Hex), wurtzite (WZ), and zinc blende (ZB) structures. Figure 20.8(a) shows the cohesive energy differences between

Table 20.1. Calculated cohesive energy E_{coh} and equilibrium bond length r_e for Hex, WZ, and ZB structures in BN and GaN. Adapted from [76]. Copyright 2019 The Japan Society of Applied Physics.

Material	Structure	E_{coh} (eV atom^1)	r_e (Å)
BN	Hex	−6.944	1.48
	WZ	−6.796	1.57
	ZB	−6.813	1.57
GaN	Hex	−4.261	1.90
	WZ	−4.677	1.96
	ZB	−4.673	1.96

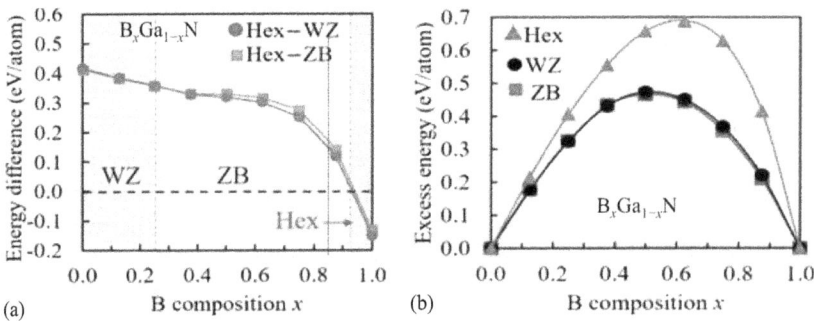

Figure 20.8. (a) Cohesive energy difference with respect to the cohesive energy of the hexagonal (Hex) structure for $B_xGa_{1-x}N$ alloys as a function of boron composition, *x*. Positive values indicate that the wurtzite (WZ) and zinc blende (ZB) structures are more stable than the Hex structure. (b) Excess energy of $B_xGa_{1-x}N$ alloys as a function of boron composition, *x*. The triangle, circle, and squares represent the excess energies of the Hex, WZ, and ZB structures. Reproduced with permission from [76]. Copyright 2019 The Japan Society of Applied Physics.

Hex and WZ structures as well as between Hex and ZB structures for $B_xGa_{1-x}N$ alloys as functions of boron composition x. The energy differences were calculated using $\Delta E(\text{Hex-WZ}) = E(\text{Hex}) - E(\text{WZ})$ and $\Delta E(\text{Hex-ZB}) = E(\text{Hex}) - E(\text{ZB})$, where $E(\text{Hex})$, $E(\text{WZ})$, and $E(\text{ZB})$ denote the cohesive energies of alloys with Hex, WZ, and ZB structures, respectively. As shown in figure 20.8(a), the calculated energy differences are positive for boron compositions <0.93, meaning that the formation of $B_xGa_{1-x}N$ alloys in the hexagonal phase in the range of boron compositions less than 0.93 (or Ga compositions greater than 7%) is energetically unfavorable. However, figure 20.8(a) shows that the calculated energy differences sharply become negative values and clearly revealed that for the boron composition $\geqslant 0.93$ (or for Ga compositions $\leqslant 7\%$) the hexagonal structure can be stabilized in free-standing (BN)-rich BGaN alloys [75, 76]. The authors attributed this effect to the small energy difference among Hex, WZ, and ZB structures in BN of ~0.13 eV atom^{-1}. Figure 20.8(b) shows the calculated excess energy ΔE_{ex} of the $B_xGa_{1-x}N$ alloy as a function of boron composition x, where the excess energy of the $B_xGa_{1-x}N$ alloy is given by $\Delta E_{ex} = E(B_xGa_{1-x}N) - \{xE(\text{BN}) + (1-x)E(\text{GaN})\}$. The calculated excess energies of $B_xGa_{1-x}N$ alloys in WZ and ZB structures, shown in figure 20.8(b), are lower than those in Hex structure. The calculation results shown in figure 20.8(b) therefore indicate that the miscibility of $B_xGa_{1-x}N$ alloys with hexagonal structure is lower than that with WZ and ZB structures due to the relaxation of atoms within in-plane directions in the hexagonal structure. In particular, the calculation results revealed that the ZB structure tends to stabilize over a wide boron composition range for $B_xGa_{1-x}N$ alloys [75, 76].

Experimentally, it has been observed that a few layers of ZnO grown on Ag(111) substrates indeed crystallize into a graphitic or hexagonal BN (h-BN) structure instead of the equilibrium wurtzite structure [77]. The synthesis of 2D GaN has been recently demonstrated via a migration-enhanced encapsulated growth technique utilizing epitaxial graphene [62]. However, the hexagonal phase has not been experimentally observed in free-standing III-nitride semiconductors (AlN, GaN, InN, and their alloys) or deposited on a substrate prior to recent works [53, 54]. Epitaxial growth of hexagonal phase (BN)-rich $Ga_xB_{1-x}N$ alloys ($h\text{-}Ga_xB_{1-x}N$) has been attempted only recently via an MOCVD technique by utilizing the h-BN epilayer as a template to promote the crystallization of BGaN alloys into the hexagonal phase [53, 54]. The results presented in the sub-sections below are based on these recent published results.

20.3.1 Epitaxial growth of $h\text{-}Ga_xB_{1-x}N$ alloys

For the epitaxial growth of hexagonal phase (BN)-rich $Ga_xB_{1-x}N$ alloys by MOCVD, triethylboron (TEB), ammonia (NH$_3$), and trimethylgallium (TMGa) were used as precursors for B, N, and Ga, respectively, and hydrogen as a carrier gas [53]. Prior to the deposition of h-BGaN, a 10 nm thick h-BN epilayer was first deposited on c-plane sapphire substrate at 1300 °C to serve as a template. Because GaN tends to decompose above 1200 °C, a growth temperature of 1225 °C was chosen for h-BGaN alloys in order to accommodate the incorporation of Ga while retaining a reasonable crystalline quality. The alloy composition was controlled via the variation of the TMGa flow rate (R_{TMG}) while keeping both the TEB and NH$_3$ flow rates constant.

The XRD θ–2θ scans for h-BGaN alloys grown under different TMGa flow rates (R_{TMG}) shown in figure 20.9(a) all exhibit diffraction peaks near 26°, which are close to that of the h-BN (0002) diffraction peak. The results thus confirmed that these alloys have been crystallized into the hexagonal phase. Consequently, the corresponding c-lattice constants of h-BGaN alloys can be calculated based on the measured θ–2θ peak positions as summarized in figure 20.9(b), revealing an onset point occurring at $R_{TMG} = 4$ sccm. Figure 20.9(b) indicate that the c-lattice constant of h-BGaN increases from 6.85 Å to 6.94 Å as R_{TMG} was increased from 0 to 4 sccm. No further increase in the c-lattice constant was noticeable when R_{TMG} was increased beyond 4 sccm. As the theoretically calculated a-lattice constant of single layer h-GaN is $a = 3.21$ Å [2], h-BGaN epilayers grown on h-BN templates are expected to undergo a compressive strain. Therefore, an increase in the inter-plane distance ($c/2$) is expected with the incorporation of Ga. Moreover, it interesting to note in figure 20.9(b) that the XRD θ–2θ peak intensity as a function of TMGa flow rate (R_{TMG}) exhibits an initial rapid decrease followed by almost no change, with the onset point also occurring at $R_{TMG} = 4$ sccm, corroborating the dependence of the c-lattice constant on R_{TMG}.

Since XRD θ–2θ scans provide information concerning the c-lattice constants and there are no available data for the c-lattice constant of h-GaN, the well-established Vegard's law cannot be applied to determine the composition of Ga in h-BGaN alloys. Instead, the Ga composition in h-Ga$_x$B$_{1-x}$N alloys were determined by x-ray photoelectron spectroscopy (XPS) measurements. Figure 20.9(c) plots the dependence of x in h-Ga$_x$B$_{1-x}$N alloys on R_{TMG} utilizing the Ga compositions obtained

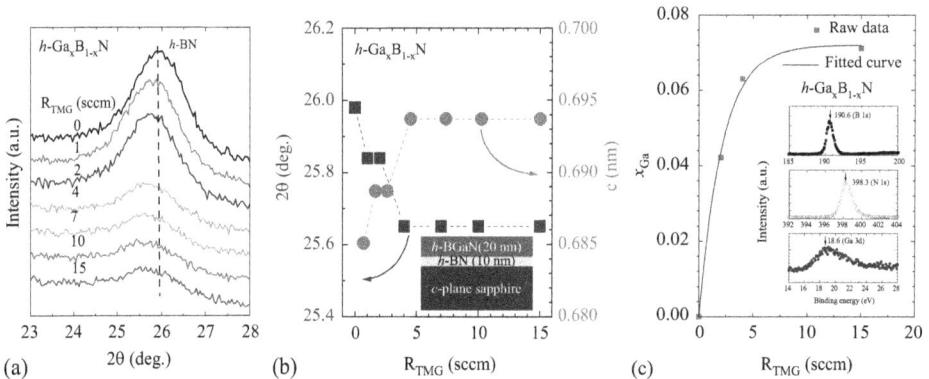

Figure 20.9. (a) The XRD (0002) diffraction peak intensity of h-BGaN alloys as a function of the R_{TMG}. (b) The XRD (0002) diffraction peak position (left axis) and the corresponding c-lattice constant (right axis) versus the TMGa flow rate, R_{TMG}. (c) The Ga composition (x) in h-Ga$_x$B$_{1-x}$N alloys obtained via XPS measurements versus R_{TMG}. The solid curve is the least squares fit of data with equation (20.3). Inset: XPS spectra of an h-BGaN sample grown at $R_{TMG} = 15$ sccm for B 1s, N 1s, and Ga 3d peaks, from which the alloy composition was determined (after [53]).

from XPS measurement data, which revealed that x in h-Ga$_x$B$_{1-x}$N sharply increases with R_{TMG} initially followed by a saturation with an onset occurring again at $R_{TMG} \sim 4$ sccm; whereas x versus R_{TMG} can be described by the following equation [53]:

$$x = x_0[1 - \exp(-R_{TMG}/A)], \qquad (20.3)$$

where x_0 denotes the saturation value of Ga composition and A describes an exponential dependence of x in h-Ga$_x$B$_{1-x}$N with R_{TMG}. The least squares fit between the experimentally measured Ga compositions with equation (20.3) yielded a saturation value for the Ga composition of $x_0 = (7.2 \pm 0.2)$ % and $A = 2.14 \pm 0.17$ sccm. The Ga composition saturation behavior shown in figure 20.9(c) is consistent with the observed onset behavior exhibited in the dependence of the XRD intensity and c-lattice constant on the TMGa flow rate shown in figure 20.9(b).

The observation of the onset point occurring at $R_{TMG} = 4$ sccm suggests that only a limited fraction of GaN, up to ~7.2%, can be incorporated into h-BN [53]. Since the temperature employed for the growth of h-BGaN alloys of 1225 °C is higher than the decomposition temperature of GaN, it is plausible to attribute the Ga composition saturation to GaN decomposition at high growth temperature [53, 54]. However, the calculation results shown in figure 20.8(a) clearly revealed that the hexagonal structure can be stabilized in free-standing (BN)-rich BGaN alloys only for a B composition greater than 93% or equivalently for a Ga composition $x \leqslant 7\%$ [75, 76]. In fact, the experimentally observed Ga saturation composition in h-B$_{1-x}$Ga$_x$N of $x_0 \sim 7.2\%$ [53] is in perfect agreement with the theoretical prediction of the critical Ga composition of ~7%, as illustrated in figure 20.8(a) [75, 76]. It is interesting to note that the theoretical prediction is made for free-standing h-B$_{1-x}$Ga$_x$N [75, 76], whereas the MOCVD grown layers were deposited on h-BN epilayer templates [53, 54]. Another interesting observation is that the theoretical calculation suggested that the use of AlN or GaN substrate promotes B$_{1-x}$Ga$_x$N alloys to crystalline into the ZB and WZ phase owing to the large cohesive energy of BN with the lattice constraint of the WZ-AlN or WZ-GaN substrate [75, 76]. This also explains the fact that experimentally the use of h-BN templates is necessary to promote the crystallization of h-B$_{1-x}$Ga$_x$N alloys into the hexagonal lattice [53, 54].

Since undoped h-BN is highly insulating and unintentionally doped GaN is usually conductive, one expects that the conductivity of the h-B$_{1-x}$Ga$_x$N alloy system increases with an increase in x. Indeed, the measured electrical resistivity values for h-BGaN were $\rho \approx 5 \times 10^{10}$, 5×10^6, and 2×10^6 $\Omega \cdot$ cm for $x = 0$, 0.045, and 0.07, respectively, corresponding to an enhancement in the electrical conductivity of four orders of magnitude as the Ga composition in h-B$_{1-x}$Ga$_x$N was increased from 0 to 0.07. This ability to control conductivity is expected to be very useful for certain device applications [53].

20.3.2 Growth of h-BGaN QWs and photoluminescence emission properties

With the baseline for the growth of h-BGaN epilayers established, the growth of an h-BN/Ga$_x$B$_{1-x}$N/BN QW was attempted [53]. As for the growth of h-BGaN epilayers, prior to the deposition of a 2 nm thick *h*-BGaN QW, a 10 nm thick *h*-BN epilayer was first deposited on a *c*-plane sapphire substrate at 1300 °C to serve as a template as well as the bottom barrier layer. To avoid GaN decomposition in the well region, the growth temperature of the well and the top *h*-BN barrier was set to be the same as that of the *h*-BGaN epilayers at 1225 °C. A 2 nm thick *h*-BGaN epilayer without the top *h*-BN barrier was also grown as a reference for direct comparison. The targeted GaN fraction in the well region and in the reference *h*-BGaN epilayer was ~4%. The layer structures of the *h*-BN/Ga$_x$B$_{1-x}$N/BN QW and the reference *h*-Ga$_x$B$_{1-x}$N epilayer are schematically shown in figure 20.10.

Figure 20.10 presents the low temperature photoluminescence spectra for the (a) *h*-BN/BGaN/BN QW and (b) *h*-BGaN reference epilayer. The dominant peaks around 4 eV are due to the recombination of a donor–acceptor pair and its phonon replicas involving the 200 meV longitudinal optical phonon mode [78–80], whereas the band-edge emission line near 5.55 eV in the *h*-BGaN alloy can be attributed to the recombination of bound excitons (or self-trapped excitons) [14, 78–81]. One of the distinctive features exhibited by the PL spectra of both the *h*-BN/BGaN/BN QW and *h*-BGaN reference epilayer is a pair of very sharp band-edge transition lines with an energy separation of exactly 170 meV (at 5.57 and 5.4 eV for *h*-BN/BGaN/BN QW and at 5.55 and 5.38 eV for the *h*-BGaN reference epilayer). This observed energy separation matches perfectly with the phonon energy of the E_{2g} symmetry vibration mode in *h*-BN ($\Delta\sigma - 1371$ cm^{-1}) [82]. The PL results thus suggest that the 5.38 eV (5.40 eV) emission line is the one-phonon replica of the 5.55 eV (5.57 eV) zero-phonon line. Since the E_{2g} symmetry vibration mode corresponds to the in-plane stretch of B and N atoms, the observation of a pronounced phonon replica line involving E_{2g} symmetry vibration mode corroborates the XRD results shown in

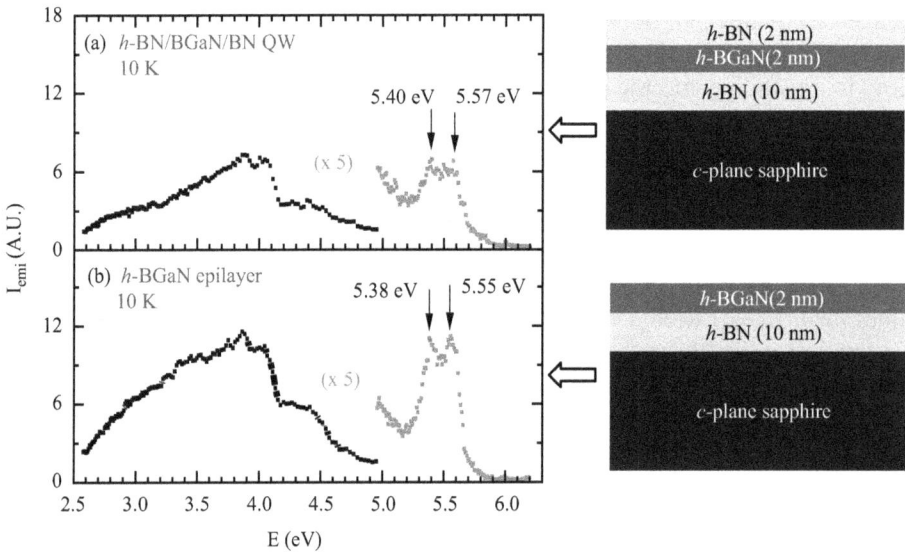

Figure 20.10. Low temperature (10 K) photoluminescence spectra for (a) *h*-BN/B$_{1-x}$Ga$_x$N/BN ($x \sim 0.04$) QW and (b) *h*- B$_{1-x}$Ga$_x$N ($x \sim 0.04$) epilayer. The layer structures are shown to the right. (After [53]).

figure 20.9(a) and further confirms that BN/BGaN/BN QW and BGaN epilayers have been crystallized into the hexagonal phase with a reasonable layered structure.

In comparison, the PL spectral line shapes of the 2 nm thick h-BGaN epilayer and 2 nm h-BGaN QW are quite similar. However, a careful inspection reveals that both the band-edge emission and its phonon replica lines in h-BGaN QW were slightly blue shifted by ~20 meV with respect to those in the h-BGaN epilayer. This seems to suggest that a layer-structured QW has been successfully synthesized. Moreover, the results revealed that the formation of h-BN/BGaN/BN QW induces a quantum confinement effect, as in the conventional QWs. However, due to the larger exciton binding energy (~0.7 eV) [31–35] and smaller exciton Bohr radius (~ 0.1 nm) [34] in h-BN (and hence in B-rich h-BGaN), the confinement effect in h-BGaN QW is expected to be less visible in PL emission spectra than in more conventional QWs (e.g. GaN/AlN) with the same well width. However, h-BN barrier layers are necessary to facilitate carrier injection. These layer structures and the results shown in figures 20.9 and 20.10 provide a foundation for the further development of h-BGaN QW-based optoelectronic devices.

20.3.3 Probing the critical thickness and phase separation effects in h-GaBN/BN heterostructures

Despite the fact that the equilibrium phases of GaN and BN are wurtzite and hexagonal, respectively, the results described in the previous sub-sections demonstrated the successful synthesis of $Ga_xB_{1-x}N$ alloys as well as h-$Ga_xB_{1-x}N$/BN QWs in the hexagonal phase using MOCVD, with a maximum incorporation of Ga composition up to ~7.2% at a growth temperature of 1225 °C [53]. This experimentally observed maximum Ga composition in h-$Ga_xB_{1-x}N$ agrees perfectly with the theoretical calculation results of ~7% [75, 76]. The agreement between experimental and calculation results seems to suggest that GaN decomposition at relatively high growth temperatures is not a factor which limits the maximum Ga composition in h-$Ga_xB_{1-x}N$, but rather it is the different equilibrium crystalline phases between BN (hexagonal) and GaN (wurtzite) which limits the incorporation of Ga in in h-$Ga_xB_{1-x}N$. It is desirable to find ways to produce h-GaBN alloys offering an extended range of variation in the bandgap as well as in the optical and electrical properties. However, this would require a better understanding of the basic properties of h-$Ga_xB_{1-x}N$/h-BN heterostructures. One of the fundamental issues of heterostructure epitaxy is the critical thickness. Another correlated and critical issue is the evolution of the phase separation with the layer thickness.

Two sets of h-GaBN samples with varying h-GaBN thicknesses deposited on h-BN epi-templates grown under the TMG flow rates of $R_{TMG} = 5$ and 20 sccm were synthesized for the investigation of the critical thickness (L_C) phenomenon in the h-GaBN/BN heterostructure system [54]. The layer structure is identical to that shown in the inset of figure 20.9(b). Based on the results shown in figure 20.9(c), the TMG flow rates employed were beyond the onset point of $R_{TMG} = 4$ sccm and hence the expected Ga contents in h-GaBN layers with varying thicknesses (d) should be greater than x_0 (≈ 0.072). The microscopy images of a set of samples grown under

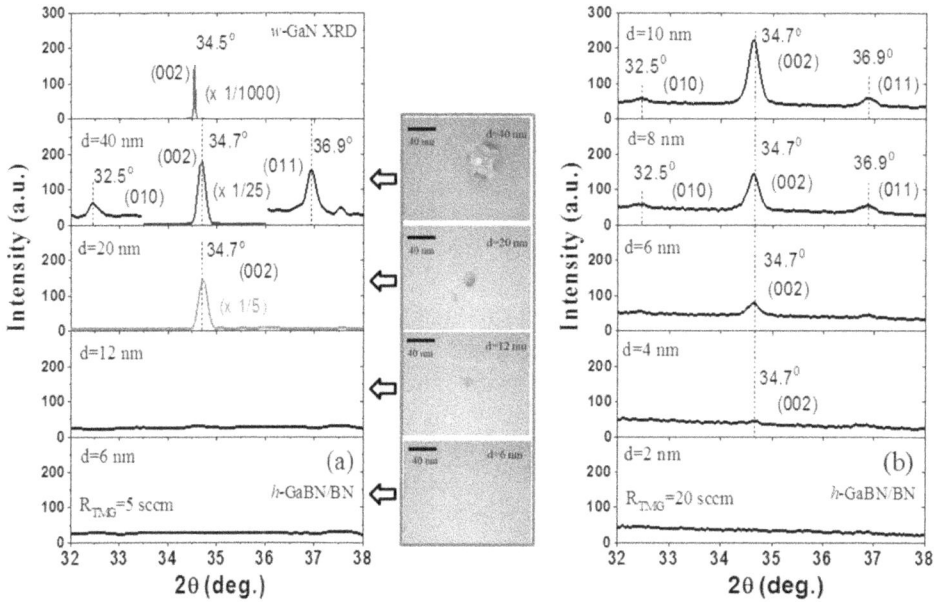

Figure 20.11. XRD θ–2θ scans of h-GaBN/BN heterostructure samples of different layer thickness grown under (a) $R_{TMG} = 5$ sccm and (b) $R_{TMG} = 20$ sccm. For reference, an XRD θ–2θ scan of the (002) peak for a wurtzite-GaN (w-GaN) epilayer is shown in the top panel of in (a). The inset contains microscopy images of h-GaBN alloy samples grown under $R_{TMG} = 5$ sccm with different thicknesses (after [54]).

$R_{TMG} = 5$ sccm are shown in the inset of figure 20.11(a) and the results indicate that the 6 nm thick h-GaBN layer sample exhibits a very smooth surface. However, the formation of extra complex features with sizes that increase with the layer thickness (d) is clearly observed in films with thicknesses >12 nm. For the film with $d = 40$ nm, the size of the extra features becomes quite large with a diameter as large as 50 μm. XRD θ–2θ scans were employed to assess the crystalline structures of both sets of samples grown under (a) $R_{TMG} = 5$ sccm and (b) 20 sccm with different thicknesses and the results are shown in figure 20.11, covering the spectral range of the wurtzite (w)-GaN (002) peak between 32° and 38°. An XRD θ–2θ scan of an MOCVD grown wurtzite (w)-GaN epilayer is included for reference in the top panel of figure 20.11(a), which exhibits the expected w-GaN (002) peak at 34.5°. As shown in figure 20.11(a), for the set of samples grown with $R_{TMG} = 5$ sccm, no w-GaN peak was observed for $d = 6$ and 12 nm. However, the w-GaN (002) peak was clearly resolved at about at 34.7° in thick layers with $d = 20$ nm and 40 nm. The shift in the (002) peak position from 34.5° for the w-GaN epilayer to 34.7° for the w-GaN domains inside h-GaBN alloys is most likely due to the presence of strain in w-GaN domains. As the thickness (d) further increases to 40 nm, in addition to the GaN (002) peak, other diffraction peaks corresponding to w-GaN (010) at 32.5° and GaN (011) at 36.9° were also observable. For the set of samples grown under $R_{TMG} = 20$ sccm, the w-GaN (002) peak was absent only in the sample with $d = 2$ nm and it starts to appear at $d = 4$ nm, as shown in figure 20.11(b). The (010) peak and (011) peak of the w-GaN phase were also observed as d further increases to above 6 nm. Compared with the XRD results of the reference w-GaN epilayer sample, the XRD results of h-GaBN alloys suggest that the extra features seen in figure 20.11 are related to the formation of the w-GaN phase inside the h-GaBN alloy matrix and they are only observable when the h-GaBN layer thickness is beyond 12 nm in samples grown under R_{TMG} = 5 sccm (figure 20.11(a)) and 2 nm in samples grown under R_{TMG} = 20 sccm (figure 20.11(b)).

Room temperature PL spectra for these two sets of h-GaBN/BN heterostructure samples have been measured and are plotted in figure 20.12 for samples grown under (a) R_{TMG} = 5 sccm and (b) R_{TMG} = 20 sccm. The PL spectrum of a standard w-GaN epilayer is also included in the top panel of figure 20.12(a) for reference, which exhibits a strong band-edge emission line in w-GaN at 3.39 eV. For films grown under R_{TMG} = 5 sccm, the band-edge emission line from the wurtzite phase GaN is absent for the sample with a thickness of d = 6 nm and a small peak near 3.34 eV starts to emerge for samples with d > 12 nm. In fact, this emission line at 3.34 eV becomes rather prominent in the sample with d = 40 nm. We believe that the new 3.34 eV line is the band-edge emission of w-GaN domains inside h-GaBN, which is redshifted with respect to 3.39 eV observed in w-GaN epilayers. It is reasonable to attribute the cause of this redshift in the PL emission peak in w-GaN domains with respect to that of the w-GaN epilayer to the same strain which produced a small shift

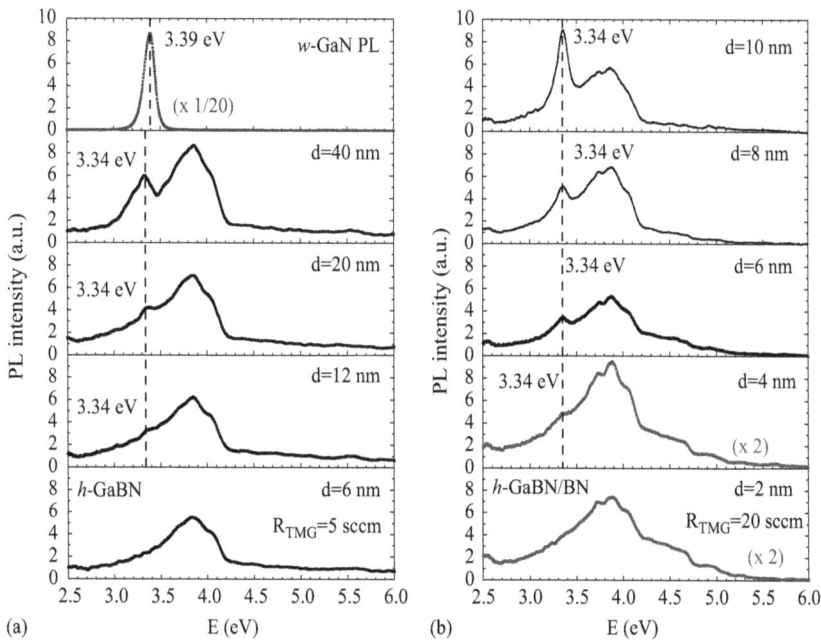

Figure 20.12. Room temperature photoluminescence spectra of h-GaBN/BN heterostructures of different layer thickness, d, for samples grown under (a) R_{TMG} = 5 sccm and (b) R_{TMG} = 20 sccm. For reference, the PL spectra of a wurtzite-GaN (w-GaN) epilayer is presented in the top panel of (a) (after [54]).

in the XRD (002) peak position of w-GaN domains in h-GaBN alloys with respect to that of w-GaN epilayer shown in figure 20.11.

For samples grown under $R_{TMG} = 20$ sccm, the emission line near 3.34 eV related to w-GaN domains was absent only in the sample with $d = 2$ nm and becomes the dominant emission line in the sample with $d = 10$ nm. PL results for both sets of samples show that the emission line near 3.34 eV related to w-GaN domains increases with the h-GaBN thickness (d) when d is beyond a critical thickness. Thus, the PL results collaborate the XRD results well. In addition to the band-edge emission line of w-GaN domains in h-GaBN observed at 3.34 eV, a broad peak near 3.9 eV is believed to be associated with a shallow donor to deep acceptor (DAP) transition in h-GaBN alloys, as its spectral features and peak position are very close to a DAP transition in the h-BN epilayers [80].

The XRD and PL results conclusively show that there exists a critical thickness below which $B_xGa_{1-x}N$ alloys of single hexagonal phase can be synthesized. One of the very surprising observations is the absence of the zinc blende (ZB) GaN domains inside the h-GaBN alloy matrix, despite the fact that the calculation results revealed that the ZB structure tends to be stabilized over a wide boron composition range for $B_xGa_{1-x}N$ alloys [75, 76]. Based on the calculation results, one expects a phase transition between hexagonal and zinc blende to occur when the layer thickness exceeds the critical thickness. However, the XRD and PL results shown in figures 20.11 and 20.12 clearly revealed that $B_xGa_{1-x}N$ alloy samples contain only W-GaN domains when their layer thicknesses exceed a critical thickness (L_C).

To determine the critical thickness (L_C) quantitatively, figure 20.13(a) plots the w-GaN (002) XRD peak intensity versus the layer thickness of h-GaBN alloys (d) for (a) $R_{TMG} = 5$ sccm and (b) $R_{TMG} = 20$ sccm. The intensity of the w-GaN (002) peak at 34.7° displays a power law dependence on d when d is beyond ~6 nm for $R_{TMG} = 5$ sccm and ~2 nm for $R_{TMG} = 20$ sccm. We believe that the results are indicative of a phase separation, meaning that w-GaN domains start to form inside layer-structured h-GaBN alloys when d is beyond a critical layer thickness (L_C). When the h-GaBN film thickness is less than L_C, no phase separation occurs. The relationship between the XRD w-GaN (002) peak intensity (I_{XRD}) and h-GaBN layer thickness d can be described by

$$I_{XRD}(d) = I_0(d - L_C)^n. \tag{20.4}$$

The least squares fit of data with equation (20.4) yielded values of $L_C = 6.0$ nm and 2.3 nm for the two sets of samples grown with $R_{TMG} = 5$ sccm and 20 sccm, respectively, with the other fitting parameters being $I_0 = 3.0$ (1.5), $L_C = 6.0$ nm (2.3 nm), and $n = 2.1$ (2.4) for $R_{TMG} = 5$ sccm (20 sccm). The results shown in figure 20.13 indicate that the critical thickness, L_C, depends on the total Ga composition in h-GaBN alloys, as expected. The w-GaN (002) peak intensity increases with L_C following a power law dependence with an exponent n being very close to 2, which is probably related to the fact that h-GaBN films are 2D in nature. The results suggest that the formation of w-GaN polycrystalline structures inside h-GaBN occurs when d is larger than L_C.

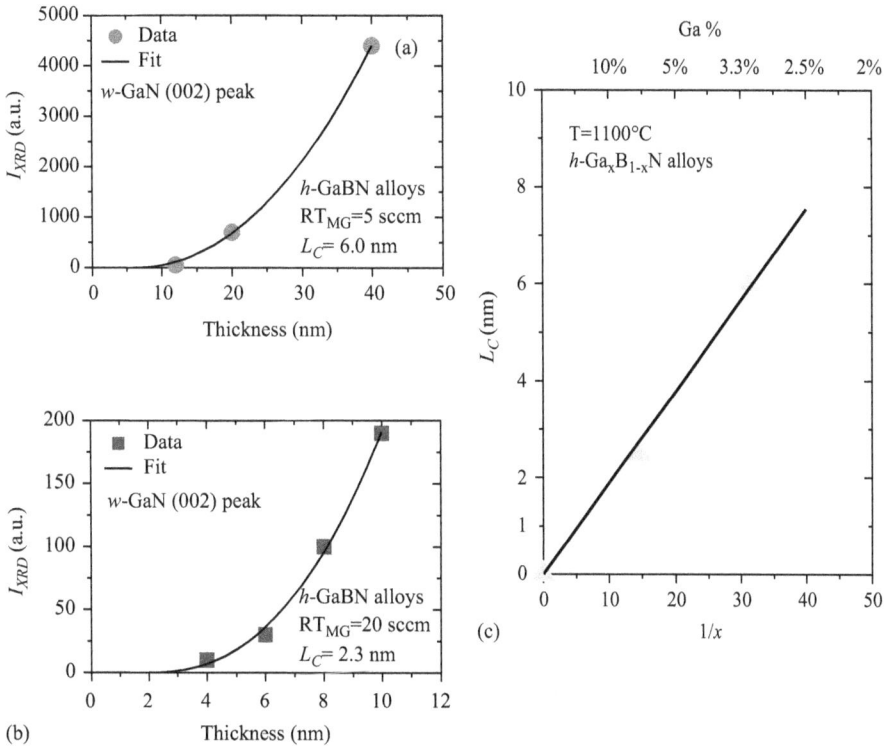

Figure 20.13. The wurtzite-GaN (w-GaN) (002) peak intensity versus h-GaBN alloy layer thickness for samples grown under (a) $R_{TMG} = 5$ sccm and (b) $R_{TMG} = 20$ sccm. The solid curves are the least squares fit of data with equation (20.4). (c) The Ga composition (x) dependence of the critical thickness (L_C) of h-$Ga_xB_{1-x}N$/BN heterostructures. The solid triangles show experimental data and the solid line is the least squares fit of data with equation (20.7) (after [54]).

Phase separation in ternary alloys $A_xB_{1-x}C$ is generally due to the internal elastic strain, caused by the lattice constant mismatch between two binary constituents AC and BC and the underlying epilayer template [83, 84]. The origin of phase separation observed in h-GaBN alloys differs from a ternary alloy of the same crystalline structure such as InGaN alloys. The calculation results shown in figure 20.8 predicted a phase transition from hexagonal to zinc blende structure in GaBN alloys when the Ga composition is greater than 7%. However, the experimental results shown in figures 20.11 and 20.12 indicate that a phase separation between hexagonal and wurtzite structures occurs at a critical layer thickness (L_C). Initially, the stable phase of BN in our growth condition is hexagonal, while that of GaN is wurtzite. This implies that a phase separation will occur when the Ga composition is above a critical value. Therefore, it is not a surprise to observe a phase separation in h-GaBN alloys. What is interesting, however, is the observation of the critical thickness in this layer-structured alloy system. It is well known that the interlayer interaction is very weak in 2D materials. However, the results shown in figures 20.11 and 20.12 indicate that interlayer interaction is strong enough to give rise to a critical thickness in h-GaBN alloys, beyond which phase separation is to set in. The critical thickness (L_C) of a lattice mismatched heterostructure can be calculated based on the strain energy [84, 85] and can be further simplified to the following relation [86]:

$$L_C(\text{nm}) \cong 0.1\frac{a_s}{f},\qquad(20.5)$$

where a_s is the lattice constant of the underlying template (h-BN) and f is the misfit between the subsequent epilayer (h-Ga$_x$B$_{1-x}$N layer) and the underlying template layer, which is defined by

$$f = \frac{a_e - a_s}{a_s} = \frac{b}{a_s}x.\qquad(20.6)$$

Here, a_e denotes the lattice constant of the h-Ga$_x$B$_{1-x}$N layer, which is proportional to the Ga composition, x. From equations (20.5) and (20.6) a simple linear relationship between the critical thickness L_C and Ga composition x in h-Ga$_x$B$_{1-x}$N alloys can be obtained as follows:

$$L_C\ (\text{nm}) = A\frac{1}{x},\qquad(20.7)$$

where A is a proportionality constant. The average Ga compositions in Ga$_x$B$_{1-x}$N alloys were determined by XPS measurements [54]. For instance, the measured average B, N, and Ga compositions in the sample with $d = 10$ nm grown under $R_{\text{TMG}} = 20$ sccm were 46.5%, 50.1%, and 3.4%, respectively, corresponding to a h-Ga$_x$B$_{1-x}$N alloy with $x = 0.068$. For the sample with $d = 6.0$ nm grown under $R_{\text{TMG}} = 5$ sccm, the average Ga composition in the h-Ga$_x$B$_{1-x}$N alloys deduced from the XPS measurements was $x = 0.034$. The experimentally measured L_C for these two samples with different Ga composition x are plotted in figure 20.13(c). Fitting of L_C versus $1/x$ with equation (20.7) by including the point at the origin provides a fitting parameter $A = 0.19$. It is interesting to note that the simplified relationship of equation (20.7), which was developed for 3D materials, appears to adequately describe the critical thickness phenomenon in layered heterostructures.

Based on the theoretical calculation results shown in figure 20.8 [75, 76] and experimental results for thick (20 nm) h-Ga$_x$B$_{1-x}$N alloys shown in figure 20.9 [53], the expected maximum Ga composition in single phase hexagonal h-Ga$_x$B$_{1-x}$N alloys is about 7.2%. However, the results shown in figure 20.13(c) means that it is possible to increase the Ga composition in h-GaBN alloys without phase separation by further reducing the h-GaBN layer thickness. For instance, if we consider the extreme case of $L_C = 1$ monolayer $= 0.33$ nm, the results shown in figure 20.13(c) imply that it is possible to synthesize h-Ga$_x$B$_{1-x}$N films with a Ga composition as high as $x = 0.55$ without phase separation. These results therefore provide the first order baseline to guide the further development of layered h-GaBN/BN heterostructure and QWs.

20.4 Epitaxy growth and properties of *h*-(BN)C semiconductor alloys

The *h*-BNC alloy system appears to possess the advantages of identical crystalline structure (hexagonal) and excellent matches in lattice constants (with an in-plane *a*-lattice constant mismatch of only about 1.5% between graphite and *h*-BN), thermal expansion coefficients, and melting points throughout the entire alloy range and potentially offers the possibility of synthesizing homogeneous *h*-BNC alloys in a wide range of compositions. However, it has been proven theoretically [87, 88] and experimentally [89–91] that synthesizing homogeneous *h*-BNC alloys is very challenging due to the strong inter-atomic bonds between B–N and C–C and the large difference in the bond energies, with respective bond energies of 4.0 eV (B–N) and 3.71 eV (C–C) compared with values of 2.83 eV for the C–N bond and 2.59 eV for the C–B bond [92, 93]. Consequently, atomic arrangements with C–C and B–N bonds are energetically favored over the ones with B–B and N–N bonds, leading to phase separation in *h*-BNC alloys [87, 88]. In general, the expression of $(BN)_{1-x}(C_2)_x$ for these alloys is used because C atoms tend to incorporate as C–C (C_2) pairs [92]. Atomic layers consisting of hybridized, randomly distributed domains of *h*-BN and C phases with compositions ranging from pure BN to pure graphene have been synthesized by various growth techniques [89–91], making the distributions of atoms in these alloy materials far from perfectly random.

Phase stability for a monolayer *h*-BNC alloy system was examined using Monte Carlo simulations and the cluster expansion technique based on first-principles calculations to serve as a foundation for understanding the phase stability of h-$(BN)_x(C_2)_{1-x}$ alloys [87]. The simulation results revealed several key finds [87]: (i) no stable intermediate phase exists between monolayer BN and graphene through construction of a ground-state diagram—the formation energies for all the possible atomic arrangements are positive; (ii) the atomic arrangements strongly favor neighboring B–N and C–C bonds and disfavor B–C and C–N atoms; (iii) no B–B and N–N atoms exist for these atomic arrangements; (iv) complete miscibility requires a very high growth temperature and the critical temperature for complete miscibility is around $T_c = 4500$ K; (v) random alloys can only exist for extremely small carbon or BN compositions; and (vi) lattice vibration enhances the solubility limits in (BN)-rich and C-rich h-$(BN)_x(C_2)_{1-x}$ alloys and also reduces the critical temperature of complete miscibility from $T_c = 4500$ K to $T_c = 3500$ K.

As with any new semiconductor material in the development stage, the ability to synthesize large wafers of homogenous alloys (instead of domains) is essential for the realization of technologically significant device applications. There was, however, experimental evidence for obtaining homogeneous h-$(BN)_x(C_2)_{1-x}$ alloys with $x = 0.5$ (BNC_2) [93, 94]. These BC_2N thin films were prepared using CVD with acetonitrile and boron trichloride and helium gas as a carrier gas to ensure forward flow of the reactants and deposited thin films on the cleaved surfaces of *h*-BN substrates at a substrate temperature of about ~850 °C [94]. The bandgap was examined using scanning tunneling microscopy (STM) and PL emission spectroscopy and the results are shown in figure 20.14. The results of the current–voltage (I–V) using STM measurements are shown in figure 20.14(a). The differential

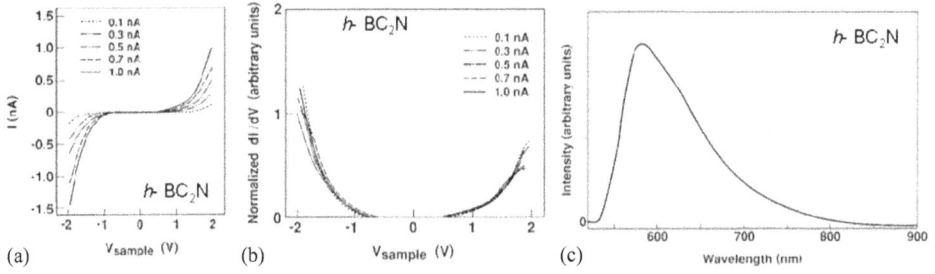

Figure 20.14. (a) Spectroscopic data obtained using STM for the cleaved surface of an h-$(BN)_{1-x}(C_2)_x$ alloy sample with $x = 0.5$ (BC_2N) and an Re tip. Each curve is acquired at constant tip–sample separation, which was controlled by tunneling currents of 0.1, 0.3, 0.5, 0.7, and 1.0 nA with a sample voltage of 2 V. (b) Normalized differential conductivity against sample bias voltage. (c) Photoluminescence spectrum taken from BC_2N thin film at a low temperature (4.2 K). Reproduced with permission from [94]. Copyright 1996 the American Physical Society.

conductivity, dI/dV shown in figure 20.14(b), deduced from the I–V characteristics provides a measurement of the surface density of states. A dI/dV above 0 V corresponds to the conduction band and that below 0 V corresponds to the valence band. From figure 20.14(b), the bandgap of h-BC_2N thin films was estimated to be about 2 eV [94]. A typical low temperature (4.2 K) PL spectrum of these h-BC_2N thin films is shown in figure 20.14(c). The PL emission spectra exhibit peak energies at 600 nm at room temperature and 580 nm at 4.2 K, corresponding to 2.07 and 2.14 eV, respectively. These values agree well with the bandgap energy of 2 eV estimated using STM [94]. The observation of the PL emission peak corresponding well to the measured bandgap implies that that BNC_2 is a direct bandgap semiconductor with an estimated bandgap energy $E_g = 2.0$ eV [94].

More recently, thin films of homogeneous h-$(BN)_{1-x}(C_2)_x$ alloys in both the BN-rich and C-rich sides have been synthesized using MOCVD by utilizing h-BN as a template [58–61]. Experimental results suggest that the critical carbon concentration (x_c) to form the homogeneous h-$(BN)_{1-x}(C_2)_x$ alloys in the BN-rich is about 3.2% and on the C-rich side is ~95% at a growth temperature of 1300 °C [58–60]. Furthermore, an enhancement of approximately 15 orders of magnitude in the electrical conductivity has been attained by increasing the carbon concentration (x) from 0 (h-BN) to 1 (graphite) [58–61]. These recent results are summarized in the sub-sections below.

20.4.1 BN-rich h-$(BN)_{1-x}(C_2)_x$ alloys

Epitaxial layers of h-$(BN)_{1-x}(C_2)_x$ alloys of about 60 nm in thickness were synthesized on sapphire (0001) substrates at 1300 °C using MOCVD. Prior to the growth of h-$(BN)_{1-x}(C_2)_x$ alloy, a 5 nm thick h-BN epilayer was first deposited on sapphire to serve as a template for the subsequent growth of the h-$(BN)_{1-x}(C_2)_x$ alloy. Triethylboron (TEB), ammonia (NH_3), and propane (C_3H_8) were used as the B, N, and C precursors, respectively, and hydrogen was used as a carrier gas. The carbon concentration was controlled by the C_3H_8 flow rates [58]. A pulsed MOCVD

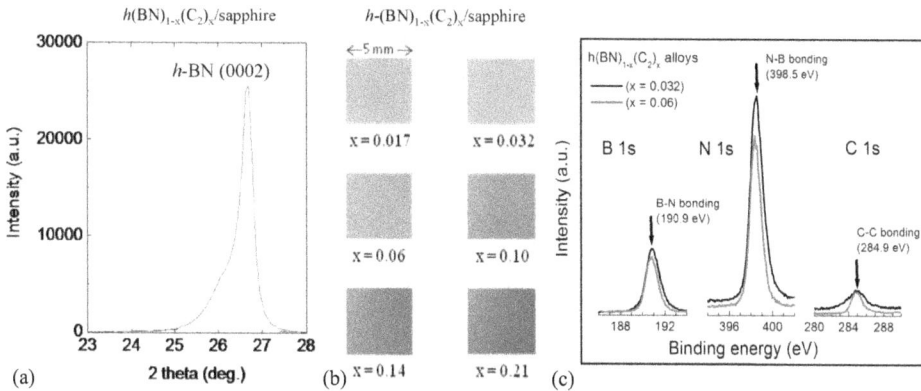

Figure 20.15. (a) An XRD θ–2θ scan around the h-BN (0002) diffraction peak of an h-BN epilayer deposited on sapphire substrate. (b) Optical micrographs (5 mm × 5 mm) of a set of h-BN-rich h-(BN)$_{1-x}$(C$_2$)$_x$ epilayers with different C concentrations deposited on sapphire substrate at the same temperature as the h-BN epilayer in (a). (c) XPS spectra of the B 1s, N 1s, and C 1s core levels for the h-(BN)$_{1-x}$(C$_2$)$_x$ epilayers with x = 0.032 and 0.06. Reproduced with permission from [58]. Copyright 2014 AIP Publishing.

epitaxial growth process was employed to grow the h-(BN)$_{1-x}$(C$_2$)$_x$ alloys [61]. Both TEB and C$_3$H$_8$ were transported together into the reactor and NH$_3$ was supplied separately with respective flow rates of 0.06 sccm and 3.0 standard liters per minute (SLM). Samples were grown under N-rich conditions, ensuring a very high V–III ratio. C$_3$H$_8$ flow rates were increased in small steps to obtain h-(BN)$_{1-x}$(C$_2$)$_x$ alloys with different carbon concentrations (x). The XRD θ–2θ scan patterns of h-(BN)$_{1-x}$(C$_2$)$_x$ samples have a similar spectral shape as those of the h-BN templates shown in figure 20.15(a). Figure 20.15(b) shows optical micrographs of a 5 mm × 5 mm area of h-(BN)$_{1-x}$(C$_2$)$_x$ epilayers with different carbon concentrations. XPS measurements were employed to determine the carbon concentrations. Figure 20.15(c) shows the XPS spectra of B 1s, N 1s, and C 1s core levels for two representative h-(BN)$_{1-x}$(C$_2$)$_x$ alloy samples with x = 0.032 and x = 0.06 [58]. Based on the bandgap value of h-BN ($\sim$6 eV) and the small values of x (x = 0 to 0.21), these h-(BN)$_{1-x}$(C$_2$)$_x$ epilayers should be transparent in the visible spectral range. However, only samples with x = 0, 0.017, and 0.032 are transparent. Samples with x = 0.06, 0.10, 0.14, and 0.21 appear dark and the darkness increases with an increase of x. We believe that this is due to the formation of C clusters in these alloys. The results shown in figure 20.15(b) seem to suggest that the critical C concentration, x_c, for homogeneous alloy formation or the solid solubility limit of C in h-BN is around 0.032 at a growth temperature of 1300 °C.

UV–visible optical absorption spectroscopy was employed to measure the bandgap variation with C incorporation. We have observed that the optical absorption edge of h-(BN)$_{1-x}$(C$_2$)$_x$ epilayers decreases with an increase of the C concentration, as expected. The absorption coefficients (α) were obtained from the absorption spectrum. The energy bandgap E_g values were estimated from the Tauc plot of the absorption coefficients [95]. Figure 20.16(a) plots α^2 as a function of the excitation photon energy for h-(BN)$_{1-x}$(C$_2$)$_x$ epilayers with x = 0.0, 0.017, 0.032, 0.06, 0.10, 0.14, and 0.21. E_g values were obtained from the intersections between

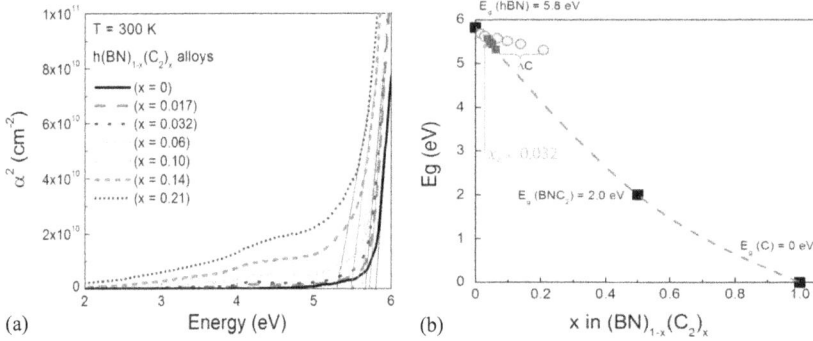

Figure 20.16. (a) Tauc plots of absorption coefficients of h-BN-rich h-(BN)$_{1-x}$(C$_2$)$_x$ epilayers with different C compositions. (b) Energy bandgap E_g versus C composition in h-(BN)$_{1-x}$(C$_2$)$_x$ alloys. Green circles are the measured data of E_g obtained from the optical absorption spectra shown in (a) and the total C concentrations from XPS. The dashed curve is the plot of equation (20.8), $E_g[h$-(BN)$_{1-x}$(C$_2$)$_x]$ = $(1 - x)E_g(h$-BN$) + xE_g(C) - b(1 - x)x$, where E_g (graphite) = 0 eV and E_g (BNC$_2$) = 2.0 eV [94] were used, providing a fitted value of b = 3.6 eV. Blue squares are the data points extrapolated from equation (20.8) representing the C concentrations in the homogeneous h-(BN)$_{1-x}$(C$_2$)$_x$ alloys (y), which is lower than the total C concentration in the samples (x). ΔC (= $x - y$) denotes the amount of excess C concentration in h-(BN)$_{1-x}$(C$_2$)$_x$ alloys with $x > x_c$ ($\approx$ 0.032) which ends up in the separated C phase. Reproduced with permission from [58]. Copyright 2014 AIP Publishing.

the straight lines and the horizontal axis to be 5.80, 5.70, 5.65, 5.60, 5.55, 5.47, and 5.30 eV for samples with x = 0.0, 0.017, 0.032, 0.06, 0.10, 0.14, and 0.21, respectively. The optical absorption edge appears to be strongly affected by the excitonic effects because the actual bandgap of h-BN is near 6.4 eV. Figure 20.16(b) plots the measured excitonic bandgap E_g versus x for h-(BN)$_{1-x}$(C$_2$)$_x$ epilayers that include the bandgap energies of h-BN (x = 0), graphite, and h(BNC$_2$) from figure 20.14 [94]. The dashed curve represents a fitting using the general equation for describing the bandgaps of semiconductor ternary alloys,

$$E_g[h\text{-}(BN)_{1-x}(C_2)_x] = (1 - x)E_g(h\text{-}BN) + xE_g(C) - b(1 - x)x, \qquad (20.8)$$

where x is the C composition in the h-(BN)$_{1-x}$(C$_2$)$_x$ alloys and b is the bowing parameter. E_g(C) = 0 is the energy bandgap of graphite. The fitted bowing parameter is b = 3.6 eV. The green filled circles are the measured C concentrations obtained from XPS and the corresponding E_g values obtained from the optical absorption spectra. The results show that the data for samples with x = 0.017 and 0.032 fit well with equation (20.8), suggesting the formation of homogeneous alloys.

Samples with x = 0.06, 0.10, 0.14, and 0.21 deviate significantly from equation (20.8). This is a result phase separation occurring in samples with $x > x_c$ ($\approx$0.032). In the phase separated materials, the XPS measures the total carbon concentrations, while the carbon concentration deduced from the bandgap variation of equation (20.8) represents the carbon concentration in the homogeneous h-(BN)$_{1-x}$(C$_2$)$_x$ alloy phase. Therefore, the deviation from equation (20.8) is a measure of the excess C concentration in the phase separated C clusters (graphite phase). In other words,

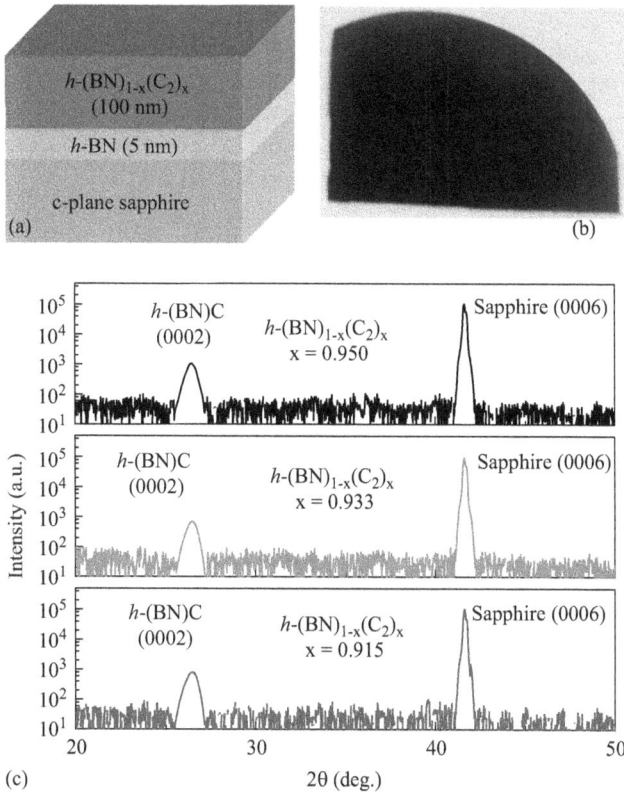

Figure 20.17. (a) Schematic of c-rich h-$(BN)_{1-x}(C_2)_x$ epilayers. (b) Optical micrograph of a h-$(BN)_{1-x}(C_2)_x$ alloy sample with $x = 0.95$. (c) XRD θ–2θ scans of C-rich h-$(BN)_{1-x}(C_2)_x$ epilayers grown on c-plane sapphire substrates under 1, 2, and 3 sccm NH_3 flow rates (after [59]).

h-$(BN)_{1-x}(C_2)_x$ alloys with $x > x_c$ (≈ 0.032) contain separated carbon domain. The actual carbon concentrations (y) in the homogeneous alloys is less than the total carbon concentration determined using XPS (x). The carbon (C) concentrations in the homogeneous alloy phase (y) are plotted as blue filled squares in figure 20.16(b), which are deduced from E_g measured by the optical absorption spectroscopy coupled with the use of equation (20.8). The C concentration difference between x and y is denoted as $\Delta C = x - y$ (indicated in figure 20.16 (b)), which measures the excess carbon concentrations in the separated carbon phases (or graphite phase) in samples with $x = 0.06$, 0.10, 0.14, and 0.21. However, at $x > x_c$ the phases of h-BN, graphite, and h-$(BN)_{1-y}(C_2)_y$ alloys with $y < x$ co-exist.

20.4.2 C-rich h-$(BN)_{1-x}(C_2)_x$ alloys

Epitaxial layers of h-$(BN)_{1-x}(C_2)_x$ alloys ($x \sim 0.92$ to 0.95) of about 100 nm in thickness, shown schematically in figure 20.17(a), were also synthesized on h-BN/c-plane sapphire templates [59, 61]. Triethylboron (TEB), ammonia (NH_3), and propane (C_3H_8) were used as the B, N, and C precursors, respectively. Samples were grown using nitrogen as a carrier gas at 1300 °C. The carbon compositions were controlled during MOCVD by varying the NH_3 flow rate and verified by XPS measurements, which revealed that x (the C mole fraction) decreases almost linearly with an increase in the NH_3 flow rate (sccm). This is due to the increase in BN fraction in the h-$(BN)_{1-x}(C_2)_x$ epilayers with an increase in N (NH_3 flow rate). Figure 20.17(b) shows the optical micrograph of an h-$(BN)_{1-x}(C_2)_x$ alloy sample with $x = 0.95$. Based on the bandgap value of graphite (zero gap), h-$(BN)_{1-x}(C_2)_x$ epilayers with large values of x should appear black under the visible light. Figure 20.17(c) shows XRD θ–2θ scans of h-$(BN)_{1-x}(C_2)_x$ samples with $x \sim$ 95.0%, 93.3%, and 91.5%, revealing a lattice constant of $c = 6.73$ Å for all samples, which closely matches a value of $c = 6.70$ Å for graphite [96]. No other diffraction peaks were observed, which confirms the hexagonal crystalline structure of the $(BN)_{1-x}(C_2)_x$ epilayers grown using MOCVD. The slight increase in the c lattice constant over graphite is due to the fact that these films were grown on a h-BN template and experience a 'compressive'-like strain in the c-plane.

Raman spectroscopy measurements were employed to probe the phase separation effect in C-rich h-$(BN)_{1-x}(C_2)_x$ alloys. Raman spectra of pure h-BN, graphite, and selected h-$(BN)C$ samples are shown in figure 20.18(a). The graphite spectrum exhibits the typical characteristic graphitic $E_{2g}(G)$ vibration peak at 1588 cm^{-1} and the defect induced D peak at 1345 cm^{-1} [97]. The spectra of the h-$(BN)_{1-x}(C_2)_x$ epilayers with $x < 0.95$ show that the G peak and the D peak are at the same positions as those in graphite. This implies that the compositions in these h-$(BN)C$ epilayers are phase separated with separate C–C domains. Moreover, the intensity and peak line widths of the E_{2g} modes of these samples are also like those of graphite. In contrast, the Raman spectrum of the h-$(BN)_{1-x}(C_2)_x$ alloy with $x = 0.95$ (black solid line) shows that the G peak shifted to 1576 cm^{-1} towards the

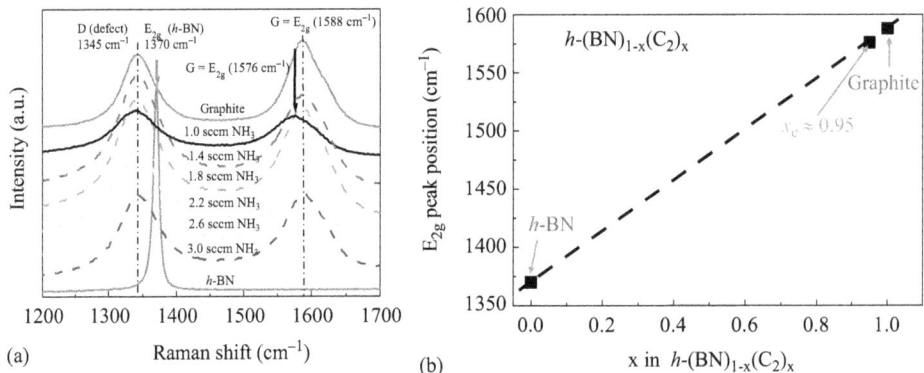

Figure 20.18. (a) Raman spectra of h-BN ($x = 0$), graphite ($x = 1$), and selected C-rich h-$(BN)_{1-x}(C_2)_x$ epilayers grown at different NH_3 flow rates. The G peak at 1576 cm^{-1} of the h-$(BN)_{1-x}(C_2)_x$ alloy with $x = 0.95$ (1 sccm NH_3 flow rate) shifts towards the sp^2 bonded h-BN vibrational peak (1370 cm^{-1}), which is a signature of the formation of homogeneous h-$(BN)C$ alloy. (b) E_{2g} vibration peak position versus x in C-rich h-$(BN)_{1-x}(C_2)_x$ showing peak positions for h-BN, graphite, and h-$(BN)_{1-x}(C_2)_x$ alloy with $x = 0.95$. The dashed line is a guide for the eyes (after [59]).

characteristic of a pure h-BN peak (1370 cm^{-1}). Figure 20.18(b) plots the measured E_{2g} peak position versus x in h-$(BN)_{1-x}(C_2)_x$ alloys using the measured values of h-BN, graphite, and h-$(BN)_{1-x}(C_2)_x$ (x = 0.95). Typically, the formation of homogeneous alloys leads to a shift in the Raman spectral peak that varies linearly according to the peak positions of the constituent atoms or binary compounds [98]. For homogeneous C-rich h-$(BN)_{1-x}(C_2)_x$ alloys, the Raman peak position is expected to shift towards lower frequency with a decrease of x. The observed Raman peak shift for h-$(BN)_{1-x}(C_2)_x$ alloy with $x = 0.95$ suggests that homogeneous C-rich h-$(BN)_{1-x}(C_2)_x$ alloys with $x \geqslant 0.95$ have been successfully synthesized.

It is interesting to note that a slight deviation from the ideal (1:1) stoichiometry ratio between B and N could have a strong influence on the conductivity type of C-rich h-$(BN)_{1-x}(C_2)_x$ [59]. This is because h-$(BN)_{1-x}(C_2)_x$ alloys are formed by incorporating group-III (B) and group-V (N) atoms into the group-IV (C) lattice, in which B and N may also serve as dopants. N could replace C to give rise to n-type conductivity and B could replace C to give rise to p-type conductivity. Figure 20.19(a) shows the room temperature Hall-mobility and carrier concentration of h-(BN)C epilayers as functions of the NH$_3$ flow rate employed during the MOCVD growth, which clearly revealed that the carrier type is p-type for samples synthesized under NH$_3$ flow rates below 2.1 sccm and is n-type for the samples synthesized under NH$_3$ flow rates above 2.1 sccm. The results thus indicate that when h-(BN)C alloys are synthesized under NH$_3$ flow rates below 2.1 sccm, there are fewer N atoms than B atoms in the materials, leading to a p-type conductivity. Supplying more N atoms to the reaction zone by increasing the NH$_3$ flow rate to above 2.1 sccm produces n-type materials because there are more N atoms than B atoms. This speculation is further verified by the XPS measurement results which revealed that samples synthesized under 1 and 2 sccm NH$_3$ flow rates have a lower N concentration than B, while the sample synthesized under a 3 sccm NH$_3$ flow rate has a higher N concentration than B [59, 61]. At NH$_3$ = 2.1 sccm, C-rich h-$(BN)_{1-x}(C_2)_x$ alloys have a 1:1 stoichiometry ratio between B atoms and N atoms, [B] = [N], a transition from p- to n-type conductivity occurs [59, 61].

The measured background carrier concentrations at room temperature for both p- and n-type C-rich h-$(BN)_{1-x}(C_2)_x$ epilayers are relatively high ($\sim 1.5 \times 10^{20}$ cm^{-3}) [59]. This is mostly due to the fact that the energy bandgaps of C-rich h-(BN)C alloys are small and the energy level of the N donors (B acceptors) in C-rich h-(BN)C alloys is expected to be very shallow or possibly even lie within the conduction (valence band). The measured electron and hole mobilities (μ_e or μ_h) and concentrations (n or p) are quite comparable and their dependence on the NH$_3$ flow rate are almost symmetric around NH$_3$ = 2.1 sccm. The results suggest that the effective masses of electrons and holes in C-rich h-$(BN)_{1-x}(C_2)_x$ alloys must be comparable, similar to the case of single sheet h-BN [34, 99]. In these alloys, the carrier mobility depends only on the carrier concentration near room temperature, whereas the room temperature mobility for both p- and n-type h-$(BN)_{1-x}(C_2)_x$ epilayers is ~ 15 cm^2 V^{-1} s^{-1}.

With the expectation of small bandgaps for C-rich h-$(BN)_{1-x}(C_2)_x$ alloys, variable temperature van der Pauw Hall-effect measurements were attempted to extract the

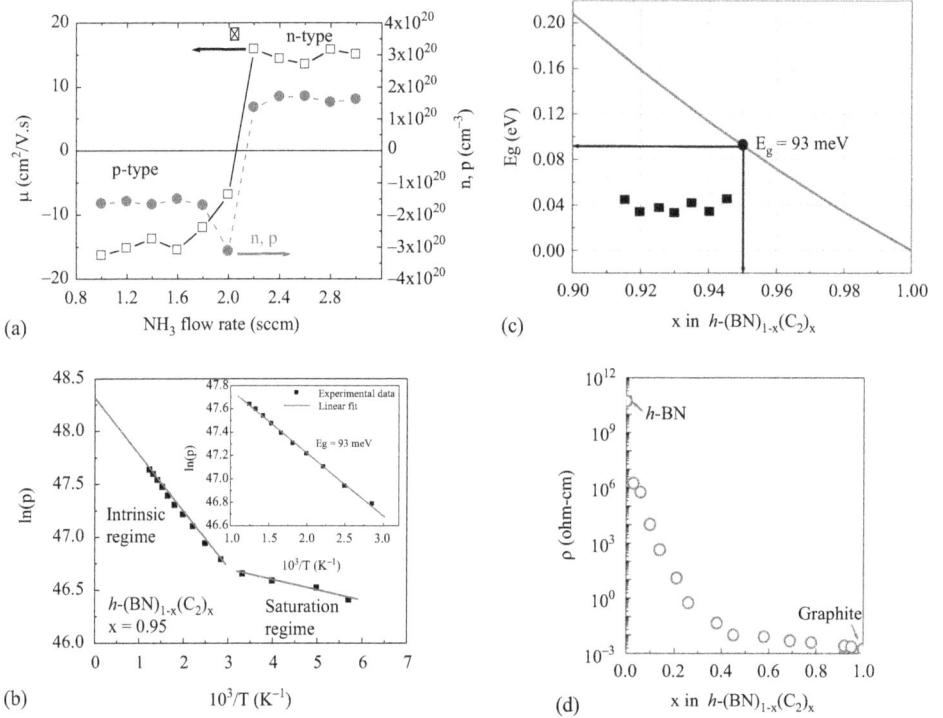

Figure 20.19. (a) Mobility and carrier concentration of C-rich h-$(BN)_{1-x}(C_2)_x$ epilayers grown under different NH_3 flow rates. The carrier type in the alloys is p-type for NH_3 flow rates below 2.1 sccm and changes to n-type for NH_3 flow rates above 2.1 sccm. (b) Temperature dependent hole concentration plotted in the scale of $\ln(p)$ versus $1/T$ for an h-$(BN)_{1-x}(C_2)_x$ sample with $x = 0.95$. The temperature dependence of the free carrier concentration shows a typical behavior of narrow gap semiconductors consisting of both saturation and intrinsic carrier conduction regime. The inset shows $\ln(p)$ versus $1/T$ plot for the intrinsic region, from which a bandgap $E_g \sim 93$ meV is obtained. (c) Plot of the bandgap energy of h-$(BN)_{1-x}(C_2)_x$ alloys versus carbon concentration (x) in the C-rich side obtained from equation (20.8) with E_g (h-BN) = 6.4 eV, E_g(C) = 0 eV, and the bowing parameter of $b = 4.8$ eV. The activation energy (or the bandgap) value of h-$(BN)_{1-x}(C_2)_x$ with $x = 0.95$ is marked as a solid circle on the plot of the bandgap variation of h-$(BN)_{1-x}(C_2)_x$ with x. Activation energies of h-$(BN)_{1-x}(C_2)_x$ epilayers with $x < 0.95$ are also shown as solid squares, which deviate from the plot of the bandgap variation of h-$(BN)_{1-x}(C_2)_x$ with x. (d) Room temperature resistivity (conductivity) of the h-$(BN)_{1-x}(C_2)_x$ epilayers with different C concentrations, including h-BN and graphite (after [58] and [59]).

bandgap from the intrinsic carrier conduction regime [59]. Figure 20.19(b) shows the temperature dependence of the free hole concentration (p) for the h-$(BN)_{1-x}(C_2)_x$ sample with $x = 0.95$ in the temperature range of 175–800 K. The dependence of the carrier concentration on temperature follows that of a very typical narrow bandgap semiconductor with two distinct regimes [100]. In the medium or low temperature region, nearly all acceptors are ionized, and the carrier concentration is nearly saturated and is independent of the temperature. In the high temperature region, the carrier concentration is predominantly intrinsic. The crossover from the saturation to the intrinsic conduction regime occurs around 350 K. The inset of figure 20.19(b) shows the Arrhenius plot of the carrier concentration in the intrinsic conduction regime, in which the hole concentration (p) and energy gap (E_g) can be expressed in terms of temperature as

$$p \propto \exp\left(-\frac{E_g}{2k_bT}\right), \tag{20.9}$$

where k_b is the Boltzmann constant. The fitted value of E_g obtained for the h-(BN)$_{1-x}$(C$_2$)$_x$ alloy ($x = 0.95$) is ~93 meV.

The bandgap value of ~93 meV obtained from the temperature dependent carrier concentration in the intrinsic conduction region shown in figure 20.19(b) can be further verified by comparing it with the expected value. The expected bandgap variation of homogeneous h-(BN)$_{1-x}$(C$_2$)$_x$ alloys with x follows the relationship described by equation (20.7) and is plotted in figure 20.19(c) for the C-rich side, where values of $E_g(h\text{-BN}) = 6.4$ eV, $E_g(\text{C}) = 0$ eV, and the bowing parameter $b = 4.8$ eV were used. Based on equation (20.7) and figure 20.19(c), a bandgap value of $E_g = 91$ meV is deduced for the h-(BN)$_{1-x}$(C$_2$)$_x$ alloy ($x = 0.95$), which agrees almost perfectly with the value of ~93 meV obtained from the temperature dependent carrier concentration presented in the inset of figure 20.19(b) [59]. This excellent agreement between the measured and expected bandgap values provides strong evidence that homogeneous h-(BN)$_{1-x}$(C$_2$)$_x$ alloys with $x \geqslant 0.95$ have been successfully synthesized using MOCVD at a growth temperature of 1300 °C, corroborating well with the conclusion drawn from Raman spectroscopy data shown in figure 20.18.

The temperature dependent free carrier concentration in h-(BN)$_{1-x}$(C$_2$)$_x$ epilayers with $x < 0.95$ has also been measured. The measured thermal activation energies for h-(BN)$_{1-x}$(C$_2$)$_x$ epilayers with $x < 0.95$ are plotted in the inset of figure 20.19(c) and vary from ~32 meV to ~45 meV. These values significantly deviate from the E_g versus x plot shown in figure 20.19(c), which implies that the carrier conduction is no longer intrinsic and that phase separation and formation of separate C–C and B–N domains occurred in the h-(BN)$_{1-x}$(C$_2$)$_x$ epilayers with $x < 0.95$.

Another unique features of the h-(BN)$_{1-x}$(C$_2$)$_x$ alloy system is that it offers the conductivity variation from highly insulating semiconductor (undoped h-BN) to semi-metal (graphite). Figure 20.19(d) shows the measured 300 K electrical resistivity for samples with different carbon concentrations. Both van der Pauw and I–V characteristic measurements were performed to compare the relative electrical resistivities of h-(BN)$_{1-x}$(C$_2$)$_x$ epilayers with varying C concentrations. The results shown in figure 20.19(b) has demonstrated that the electrical resistivity decreases by approximately 15 orders of magnitude when x increases from 0 to 1 [61]. Since graphite is a semi-metal and undoped h-BN is highly insulating, it is expected that when these two systems combine to form h-(BN)$_{1-x}$(C$_2$)$_x$ alloys, a large range of electrical conductivity control can be achieved and the electrical conductivity increases with an increase in the C composition. In the phase separated materials, the formation of C clusters further increases the electrical conductivity.

A more recent theoretical study using first-principles ab $initio$ calculations combined with a rigorous statistical approach based on cluster expansion was performed to study the effects of disorder, phase segregation, and composition fluctuations on $(BN)_{1-x}(C_2)_x$ monolayer alloy properties [88]. The calculations combined with the generalized quasichemical approximation to account for disorder effects have further confirmed that atomic arrangements with C–C and B–N bonds are energetically favored over the ones with B–B and N–N bonds [88], which further confirmed the simulation results by Yuge [87]. It was found that the most energetically favorable configurations are the atomic arrangements BNBNBNBN, CCBNBNBNBN, CCCCBNBN, CCCCCCBN, and CCCCCCCC. The calculated T–x phase diagram shown in figure 20.20(a) revealed a huge critical temperature $T_c = 5200$ K, which is comparable to the result of $T_c = 4500$ K calculated using Monte Carlo simulations, neglecting the lattice vibrations [87]. The calculation results indicated that significant contributions from clusters with B–B and N–N bonds may also occur only at growth temperatures close to the critical temperature. The phase diagram in figure 20.20(a) shows that for normal growth temperatures employed in CVD or MOCVD ($\sim$1000 °C) random alloys can only exist for extremely small C composition in the h-BN-rich side or small h-BN composition in the C-rich side and that carbon-rich alloys are more thermodynamically favored than h-BN-rich ones. More specifically, for the growth temperature employed in obtaining h-$(BN)_{1-x}(C_2)_x$ alloys shown in figures 20.15–20.19 of 1300 °C (or $T = 1573$ K), the calculation results predicted a very small carbon solubility in h-BN with a critical carbon composition at $x_c = 0.028$ and a h-BN solubility in graphene with a critical carbon composition at $x_c = 0.958$. These predication results

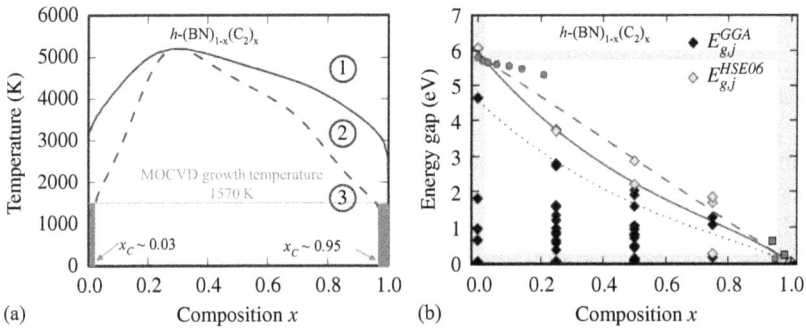

Figure 20.20. (a) The T–x phase diagram of h-$(BN)_{1-x}(C_2)_x$. The bimodal (spinodal) curve is represented by the full (dashed) line. The stable, metastable, and unstable regions in the T–x diagram are labeled by 1, 2, and 3, respectively. The horizontal line indicates the typical growth temperature employed in MOCVD or CVD growth and the vertical green shaded areas indicate the composition regions where the synthesis of homogeneous h-$(BN)_{1-x}(C_2)_x$ alloys has been realized at 1300 °C [58–60]. (b) The energy bandgap as a function of the composition for h-$(BN)_{1-x}(C_2)_x$ alloys obtained with the GGA functional (dotted black curve) and within the HSE06 approach with and without phase decomposition effects (dashed and solid blue lines, respectively). The vertical gray shades indicate the x compositions where the alloy is stable and the horizontal ones correspond to the energy gap tune range at the stable composition conditions. The black and light blue diamonds represent the energy bandgap obtained by GGA and HSE06 calculations for the nine most statistically relevant configurations with the lowest excess energies. The experimental data of [58] and [59] (figure 20.16 (b) and 20.19(c)) are represented by red circles and those of [90] are represented by red squares [90]. Reproduced with permission from [88]. Copyright 2017 the American Physical Society.

are in excellent agreement with the experimentally observed values of $x_c = 0.032$ on the h-BN-rich side shown in figures 20.15 and 20.16 and $x_c = 0.95$ observed on the C-rich side shown in figures 20.18 and 20.19 [58–61].

The same authors have also estimated the structural, electronic, and optical properties within the generalized quasichemical approximation (GQCA) formalism considering a typical growth temperature of $T = 1600$ K [88]. The simulation results verified that the lattice parameters of the h-(BN)$_{1-x}$(C$_2$)$_x$ alloys follow Vegard's law, $a(x) = a_{BN}(1 - x) + a_C x$, $a_C = 2.47$ Å, and $a_{BN} = 2.51$ Å and that there is no appreciable layer buckling [88].

Figure 20.20(b) shows the simulated energy bandgap as a function of the carbon composition (x) for h-(BN)$_{1-x}$(C$_2$)$_x$ alloys obtained using the generalized gradient approximation (GGA) functional approach as well as the Heyd, Scuseria, and Ernzerhof (HSE06) hybrid functional approach [101] with and without phase decomposition effects, by restricting the electronic structure calculations using the hybrid functional HSE06 to nine most statistically relevant configurations with the lowest excess energies [88]. By assuming that the growth conditions are controlled so that domain sizes are comparable with the considered clusters, including only the composition fluctuation effect, the simulated energy bandgap using the HSE06 approach was shown to vary between $E_g(h\text{-BN}) = 6.06$ eV and $E_g(C) = 0$ eV, following equation (20.7). However, the authors introduced a carbon composition dependent bowing parameter, $b(x)$. The calculated bowing parameter is $b(x) = (5.6 - 4.9x)$ eV [88], which agrees reasonably well with experimental measured values of 3.6 eV in the h-BN-rich side shown in figure 20.16(b) [58] and 4.8 eV in the C-rich side shown in figure 20.19(c) [59].

20.5 Concluding remarks

Phase separation appears to be a critical issue towards the development of h-BN based alloys. Phase separation is a long outstanding issue for InGaN alloys and remains to be solved. Although not much information is available on the phase separation issue in h-Ga$_x$B$_{1-x}$N and h-(BN)$_{1-x}$(C$_2$)$_x$ alloys, the differences between InGaN, h-Ga$_x$B$_{1-x}$N, and h-(BN)$_{1-x}$(C$_2$)$_x$ alloys in terms of phase separation can be briefly summarized as below. Phase separation in InGaN alloys is mainly caused by the difference in the in-plane a-lattice constants between InN ($a = 0.3544$ nm) and GaN ($a = 0.3189$ nm), whereas the difference in bond energies between B–N (4.0 eV), C–C (3.71 eV), C–N (2.83 eV), and C–B (2.59 eV) is the main cause of the phase separation in h-(BN)$_{1-x}$(C$_2$)$_x$ alloys. On the other hand, phase separation occurs in h-GaBN because the stable crystal phases of BN and GaN are hexagonal and wurtzite, respectively.

In addition to the lattice constant mismatch between InN and GaN, the growth temperature mismatch between InN and GaN makes the phase separation issue in InGaN even more difficult to overcome [103]. Theoretically, increasing the growth temperature will reduce the miscibility gap [83, 84, 87, 88], but InN decomposes at high temperatures ($T > 700$ °C). While the crystalline quality of InGaN alloys can be improved by employing a low growth rate in the low In-content regime [102], growth conditions farther away from the thermodynamic equilibrium (such as high growth rates) are helpful to promote the formation of single phase InGaN alloys in the theoretically predicted miscibility gap region (the middle range of the alloy composition) [103].

Phase separation is also a significant challenge to the synthesis of homogeneous h-BN based alloys. However, by combining the results of (BN)-rich h-$(BN)_{1-x}(C_2)_x$ and C-rich h-$(BN)_{1-x}(C_2)_x$ alloys [58–61], we can conclude that h-$(BN)_{1-x}(C_2)_x$ epilayers with $0.032 < x < 0.95$ are most likely phase separated at a growth temperature of 1300 °C. Recent theoretical calculations performed on monolayer h-$(BN)_{1-x}(C_2)_x$ suggests a low solubility of C in BN and BN in graphene at a growth temperature of 1300 °C [87, 88] and shows that the miscibility gap decreases with an increase in the growth temperature [87, 88]. The same trend is expected for h-GaBN and $h(BN)_{1-x}(C_2)_x$ alloys. The phase diagram for h-$(BN)_{1-x}(C_2)_x$ alloys shown in figure 20.20(a) reconstructed from combining the experimental [58, 59] and calculation [88] results can serve as a guideline for the further development of homogeneous h-$(BN)_{1-x}(C_2)_x$ alloys. Due to the excellent matches in lattice constants, thermal expansion coefficients, and melting temperatures throughout the entire alloys, it is expected that the alloy miscibility gap in $h(BN)_{1-x}(C_2)_x$ alloys can be reduced or completely removed by increasing the growth temperature. This could be a huge advantage over InGaN in which InN decomposes at high temperatures and a high growth temperature cannot be utilized to close the miscibility gap. It is also expected that the crystalline quality of h-BN based alloys will be improved at higher growth temperatures. However, theoretical insights concerning possible approaches to overcome the phase separation in h-$Ga_xB_{1-x}N$ and h-$(BN)_{1-x}(C_2)_x$ alloys and the composition dependent electronic properties of epitaxial films are still needed to guide material synthesis.

Acknowledgement

The research at Texas Tech University (TTU) on h-$Ga_xB_{1-x}N$ is supported by the US Army Research Office (ARO) (No. W911NF-16–1–0268). The work on h-$(BN)_{1-x}(C_2)_x$ was supported by the National Science Foundation (NSF-DMR-1206652). The development of h-BN material and device technologies at TTU has been supported by DOE ARPA-E (DE-AR0000964), NSSA SSAA program (DE-NA0002927), and (DE-FG02–09ER46552); DHS ARI Program (No. 2011-DN077-ARI048); and NSF (DMR-1206652) and (ECCS-1402886). H X Jiang and J Y Lin are grateful to the AT&T Foundation for the support of Ed Whitacre and Linda Whitacre endowed chairs.

References

[1] Maity A, Doan T C, Li J, Lin J Y and Jiang H X 2016 Realization of highly efficient hexagonal boron nitride neutron detectors *Appl. Phys. Lett.* **109** 072101

[2] Maity A, Grenadier S J, Li J, Lin J Y and Jiang H X 2017 Toward achieving flexible and high sensitivity hexagonal boron nitride neutron detectors *Appl. Phys. Lett.* **111** 033507

[3] Maity A, Grenadier S J, Li J, Lin J Y and Jiang H X 2018 Hexagonal boron nitride neutron detectors with high detection efficiencies *J. Appl. Phys.* **123** 044501

[4] Grenadier S J, Maity A, Li J, Lin J Y and Jiang H X 2018 Origin and roles of oxygen impurities in hexagonal boron nitride epilayers *Appl. Phys. Lett.* **112** 162103

[5] Maity A, Grenadier S J, Li J, Lin J Y and Jiang H X 2019 Effects of surface recombination on the charge collection in *h*-BN neutron detectors *J. Appl. Phys.* **125** 104501

[6] Maity A, Grenadier S J, Li J, Lin J Y and Jiang H X 2019 High sensitivity hexagonal boron nitride lateral neutron detectors *Appl. Phys. Lett.* **114** 222102

[7] Grenadier S J, Maity A, Li J, Lin J Y and Jiang H X 2019 Lateral charge carrier transport properties of B-10 enriched hexagonal BN thick epilayers *Appl. Phys. Lett.* **115** 072108

[8] Jiang H X and Lin J Y 2017 Review—hexagonal boron nitride epilayers: growth, optical properties and device applications *ECS J. Solid State Sci. Technol.* **6** Q3012

[9] Pantha B N, Dahal R, Nakarmi M L, Nepal N, Li J, Lin J Y, Jiang H X, Paduano Q S and Weyburne D 2007 Correlation between optoelectronic and structural properties and epilayer thickness of AlN *Appl. Phys. Lett.* **90** 241101

[10] Nakamura S, Fasol G and Pearton S J 2000 *The Blue Laser Diode: The Complete Story* (New York: Springer)

[11] Kaganer V M, Brandt O, Trampert A and Ploog K H 2005 X-ray diffraction peak profiles from threading dislocations in GaN epitaxial films *Phys. Rev.* B **72** 045423

[12] Lee S R, West A M, Allerman A A, Waldrip K E, Follstaedt D M, Provencio P P, Koleske D D and Abernathy C R 2005 Effect of threading dislocation on the Bragg peak linewidths of GaN, AlGaN and AlN heterolayers *Appl. Phys. Lett.* **86** 241904

[13] Huang B, Cao X K, Jiang H X, Lin J Y and Wei S H 2012 Origin of the significantly enhanced optical transitions in layered boron nitride *Phys. Rev.* B **86** 155202

[14] Majety S, Cao X K, Li J, Dahal R, Lin J Y and Jiang H X 2012 Band-edge transitions in hexagonal boron nitride epilayers *Appl. Phys. Lett.* **101** 051110

[15] Li J, Nam K B, Nakarmi M L, Lin J Y, Jiang H X, Carrier P and Wei S-H 2003 Band structure and fundamental optical transitions in wurtzite AlN *Appl. Phys. Lett.* **83** 5163

[16] Nam K B, Li J, Nakarmi M L, Lin J Y and Jiang H X 2004 Unique optical properties of AlGaN alloys and related ultraviolet emitters *Appl. Phys. Lett.* **84** 5264

[17] Chen G D, Smith M, Lin J Y, Jiang H X, Wei S-H, Asif Khan M and Sun C J 1996 Fundamental optical transitions in GaN *Appl. Phys. Lett.* **68** 2784

[18] Li X H *et al* 2015 Demonstration of transverse-magnetic deep-ultraviolet stimulated emission from AlGaN multiple-quantum-well lasers grown on a sapphire substrate *Appl. Phys. Lett.* **106** 041115

[19] Li J, Majety S, Dahal R, Zhao W P, Lin J Y and Jiang H X 2012 Dielectric strength, optical absorption, and deep ultraviolet detectors of hexagonal boron nitride epilayers *Appl. Phys. Lett.* **101** 171112

[20] Sugino T, Tanioka K, Kawasaki S and Shirafuji J 1997 Characterization and field emission of sulfur-doped boron nitride synthesized by plasma-assisted chemical vapor deposition *Jpn J. Appl. Phys.* **36** L463

[21] Ioffe Institute, Russia (accessed 17 Jan 2020) http://ioffe.ru/SVA/NSM/Semicond/

[22] Rumyantsev S L, Levinshtein M E, Jackson A D, Mohammmad S N, Harris G L, Spencer M G and Shur M S 2001 *Properties of Advanced Semiconductor Materials GaN, AlN, InN, BN, SiC, SiGe* ed M E Levinshtein, S L Rumyantsev and M S Shur (New York: Wiley) pp 67–92

[23] Jiang H X and Lin J Y 2014 Hexagonal boron nitride for deep ultraviolet photonic devices *Semicond. Sci. Technol.* **29** 084003

[24] Nakarmi M L, Kim K H, Khizar M, Fan Z Y, Lin J Y and Jiang H X 2005 Electrical and optical properties of Mg-doped $Al_{0.7}Ga_{0.3}N$ alloys *Appl. Phys. Lett.* **86** 092108

[25] Taniyasu Y, Kasu M and Makimoto T 2006 An aluminium nitride light-emitting diode with a wavelength of 210 nanometres *Nature* **441** 325

[26] Nakarmi M L, Nepal N, Lin J Y and Jiang H X 2009 Photoluminescence studies of impurity transitions in Mg-doped AlGaN alloys *Appl. Phys. Lett.* **94** 091903

[27] Dahal R, Li J, Majety S, Pantha B N, Cao X K, Lin J Y and Jiang H X 2011 Epitaxially grown semiconducting hexagonal boron nitride as a deep ultraviolet photonic material *Appl. Phys. Lett.* **98** 211110

[28] Majety S, Li J, Cao X K, Dahal R, Pantha B N, Lin J Y and Jiang H X 2012 Epitaxial growth and demonstration of hexagonal BN/AlGaN p–n junctions for deep ultraviolet photonics *Appl. Phys. Lett.* **100** 061121

[29] Jiang H X *et al* 2015 Structures and devices based on boron nitride and boron nitride-III-nitride heterostructures *US Patent* #9093581

[30] Laleyan D A, Zhao S, Woo S Y, Tran H N, Le H B, Szkopek T, Guo H, Botton G A and Mi Z 2017 AlN/*h*-BN heterostructures for Mg dopant-free deep ultraviolet photonics *Nano Lett.* **17** 3738

[31] Arnaud B, Lebe'gue S, Rabiller P and Alouani M 2006 Huge excitonic effects in layered hexagonal boron nitride *Phys. Rev. Lett.* **96** 026402
Arnaud B, Lebe'gue S, Rabiller P and Alouani M 2008 *Phys. Rev. Lett.* **100** 189702

[32] Wirtz L, Marini A and Rubio A 2006 Excitons in boron nitride nanotubes: dimensionality effects *Phys. Rev. Lett.* **96** 126104

[33] Wirtz L, Marini A, Gruning M, Attaccalite C, Kresse G and Rubio A 2008 Comment on huge excitonic effects in layered hexagonal boron nitride *Phys. Rev. Lett.* **100** 189701

[34] Cao X K, Clubine B, Edgar J H, Lin J Y and Jiang H X 2013 Two-dimensional excitons in three-dimensional hexagonal boron nitride *Appl. Phys. Lett.* **103** 191106

[35] Doan T C, Li J, Lin J Y and Jiang H X 2016 Bandgap and exciton binding energies of hexagonal boron nitride probed by photocurrent excitation spectroscopy *Appl. Phys. Lett.* **109** 122101

[36] Nam K B, Li J, Nakarmi M L, Lin J Y and Jiang H X 2003 Deep ultraviolet picosecond time-resolved photoluminescence studies of AlN epilayers *Appl. Phys. Lett.* **82** 1694

[37] Chen L *et al* 2004 Band-edge exciton states in AlN single crystals and epitaxial layers *Appl. Phys. Lett.* **85** 4334

[38] Silveira E, Freitas J A Jr, Kneissl M, Treat D W, Johnson N M, Slack G A and Schowalter L J 2004 Near-bandedge cathodoluminescence of an AlN homoepitaxial *Appl. Phys. Lett.* **84** 3501

[39] Dvorak M, Wei S-H and Wu Z 2013 Origin of the variation of exciton binding energy in semiconductors *Phys. Rev. Lett.* **110** 016402

[40] Dean C R *et al* 2010 Boron nitride substrates for high-quality graphene electronics *Nat. Nanotechnol.* **5** 722

[41] Britnell L *et al* 2012 Field-effect tunneling transistor based on vertical graphene heterostructures *Science* **335** 947

[42] Dean C, Young A F, Wang L, Meric I, Lee G-H, Watanabe K, Taniguchi T, Shepard K, Kim P and Hone J 2012 Graphene based heterostructures *Solid State Commun.* **152** 1275

[43] Geim A K and Grigorieva I V 2013 Van der Waals heterostructures *Nature* **499** 419

[44] Tran T T, Elbadawi C, Totonjian D, Lobo C J, Grosso G, moon H, Englund D R, Ford M J, Aharonovich I and Toth M 2016 Robust multicolor single photon emission from point defects in hexagonal boron nitride *ACS Nano.* **10** 7331

[45] Tran T T, Bray K, Ford M J, Toth M and Aharonovich I 2016 Quantum emission from hexagonal boron nitride monolayers *Nat. Nanotechnol.* **11** 37

[46] Martínez L J, Pelini T, Waselowski V, Maze J R, Gil B, Cassabois G and Jacques V 2016 Efficient single photon emission from a high-purity hexagonal boron nitride crystal *Phys. Rev.* B **94** 121405(R)

[47] Jungwirth N R, Calderon B, Ji X, Spencer M G, Flatté M E and Fuchs G D 2016 Temperature dependence of wavelength selectable zero-phonon emission from single defects in hexagonal boron nitride *Nano Lett.* **16** 6052

[48] Chejanovsky N *et al* 2016 Structural attributes and photodynamics of visible spectrum quantum emitters in hexagonal boron nitride *Nano Lett.* **16** 7037

[49] Exarhos A L, Hopper D, Grote R R, Alkauskas A and Bassett L C 2017 Optical signatures of quantum emitters in suspended hexagonal boron nitride *ACS Nano.* **11** 3328

[50] Koperski M, Nogajewski K and Potemski M 2018 Single photon emitters in boron nitride: more than a supplementary material *Opt. Commun.* **411** 158

[51] Weston L, Wickramaratne D, Mackoit M, Alkauskas A and Van de Walle C G 2018 *Phys. Rev.* B **97** 214104

[52] Knoll G F 2010 *Radiation Detection and Measurement* 4th edn (New York: Wiley)

[53] Wang Q W, Uddin M R, Du X Z, Li J, Lin J Y and Jiang H X 2019 Synthesis and photoluminescence properties of hexagonal BGaN alloys and quantum wells *Appl. Phys. Expr.* **12** 011002

[54] Wang Q W, Li J, Lin J Y and Jiang H X 2019 Critical thickness of hexagonal GaBN/BN heterostructures *J. Appl. Phys.* **125** 205703

[55] Freeman C L, Claeyssens F, Allan N L and Harding J H 2006 Graphitic nanofilms as precursors to wurtzite films: theory *Phys. Rev. Lett.* **96** 066102

[56] Wu D X, Lagally M G and Liu F 2011 Stabilizing graphitic thin films of wurtzite materials by epitaxial strain *Phys. Rev. Lett.* **107** 236101

[57] Kecik D, Onen A, Konuk M, Gürbüz E, Ersan F, Cahangirov S, Aktürk E, Durgun E and Ciraci S 2018 Fundamentals, progress and future directions of nitride-based semiconductors and their composites in two-dimensional limit: a first-principles perspective to recent synthesis *Appl. Phys. Rev.* **5** 011105

[58] Uddin M R, Majety S, Li J, Lin J Y and Jiang H X 2014 Layer-structured hexagonal (BN) C semiconductor alloys with tunable optical and electrical properties *J. Appl. Phys.* **115** 093509

[59] Uddin M R, Li J, Lin J Y and Jiang H X 2015 Carbon-rich hexagonal (BN)C alloys *J. Appl. Phys.* **117** 215703

[60] Uddin M R, Doan T C, Li J, Ziemer K S, Lin J Y and Jiang H X 2014 Electrical transport properties of (BN)-rich hexagonal (BN)C semiconductor alloys *AIP Adv.* **4** 087141

[61] Uddin M R 2016 Epitaxial growth and characterization of hexagonal boron nitride carbon semiconductor alloys *PhD Dissertation* Texas Tech University, Lubbock, TX

[62] Al Balushi Z Y *et al* 2016 Two-dimensional gallium nitride realized via graphene encapsulation *Nat. Mater.* **15** 1166

[63] Christopoulos S, Hogersthal G B, Grundy A J D, Lagoudakis P G, Kavokin A V and Baumberg J J 2007 Room-temperature polariton lasing in semiconductor microcavities *Phys. Rev. Lett.* **98** 126405

[64] Bhattacharya P, Frost T, Deshpande S, Baten M Z, Hazari A and Das A 2014 Room temperature electrically injected polariton laser *Phys. Rev. Lett.* **112** 236802

[65] Caldwell J D *et al* 2014 Sub-diffractional volume-confined polaritons in the natural hyperbolic material hexagonal boron nitride *Nat. Commun.* **5** 5221

[66] Honda T, Kurimoto M, Shibata M and Kawanishi H 2000 Excitonic emission of BGaN grown on (0001) 6H–SiC by metal–organic vapor-phase epitaxy *J. Lumin.* **87** 1274

[67] Wei C H, Xie Z Y, Edgar J H, Zeng H C, Lin J Y, Jiang H X, Chaudhuri J, Ignatiev C and Braski D N 2000 MOCVD growth of GaBN on 6H–SiC (0001) substrates *J. Electron. Mater.* **29** 452

[68] Polyakov A Y, Shin M, Qian W and Skowronski M 1997 Growth of AlBN solid solutions by organometallic vapor-phase epitaxy *J. Appl. Phys.* **81** 1715

[69] Akasakaa T and Makimoto T 2006 Flow-rate modulation epitaxy of wurtzite AlBN *Appl. Phys. Lett.* **88** 041902

[70] Li X *et al* 2015 MOVPE grown periodic AlN/BAlN heterostructure with high boron content *J. Cryst. Growth* **414** 119

[71] Li X, Wang S, Liu H, Ponce F A, Detchprohm T and Dupuis R D 2017 100-nm thick single-phase wurtzite BAlN films with boron contents over 10% *Phys. Status Solidi* B **254** 1600699

[72] Zhang M and Li X 2017 Structural and electronic properties of wurtzite $B_xAl_{1-x}N$ from first-principles calculations *Phys. Status Solidi* B **254** 1600749

[73] Shen J X, Wickramaratne D and Van de Walle C G 2017 Band bowing and the direct-to-indirect crossover in random BAlN alloys *Phys. Rev. Mater.* **1** 065001

[74] Turiansky M E, Shen J X, Wickramaratne D and Van de Walle C G 2019 First-principles study of bandgap bowing in BGaN alloys *J. Appl. Phys.* **126** 095706

[75] Hasegawa Y, Akiyama T, Pradipto A M, Nakamura K and Ito T 2018 Empirical interatomic potential approach to the stability of graphitic structure in BAlN and BGaN alloys *J. Cryst. Growth* **504** 13

[76] Hasegawa Y, Akiyama T, Pradipto A M, Nakamura K and Ito T 2019 Theoretical investigations on the structural stability and miscibility in BAlN and BGaN alloys: bond-order interatomic potential calculations *Jpn J. Appl. Phys.* **58** SCCB21

[77] Tusche C, Meyerheim H L and Kirschner J 2007 Observation of depolarized ZnO (0001) monolayers: formation of unreconstructed planar sheets *Phys. Rev. Lett.* **99** 026102

[78] Museur L and Kanaev A 2008 Near band-gap photoluminescence properties of hexagonal boron nitride *J. Appl. Phys.* **103** 103520

[79] Silly M G, Jaffrennou P, Barjon J, Lauret J-S, Ducastelle F, Loiseau A, Obraztsova E, Attal-Tretout B and Rosencher E 2007 Luminescence properties of hexagonal boron nitride: cathodoluminescence and photoluminescence spectroscopy measurements *Phys. Rev.* B **75** 085205

[80] Du X Z, Li J, Lin J Y and Jiang H X 2015 The origin of deep-level impurity transitions in hexagonal boron nitride *Appl. Phys. Lett.* **106** 021110

[81] Watanabe K, Taniguchi T and Kanda H 2004 Direct-bandgap properties and evidence for ultraviolet lasing of hexagonal boron nitride single crystal *Nature Mater.* **3** 404

[82] Nemanich R J, Solin S A and Martin R M 1981 Light scattering study of boron nitride microcrystals *Phys. Rev.* B **23** 6348

[83] Ho I H and Stringfellow G B 1996 Solid phase immiscibility in GaInN *Appl. Phys. Lett.* **69** 2701

[84] Karpov S Y 1998 Suppression of phase separation in InGaN due to elastic strain MRS *Internet J. Nitride Semicond. Res.* **3** 6

[85] Van der Merwe J H 1962 Crystal interface. Part II. finite overgrowths *J. Appl. Phys.* **34** 123

[86] People R and Bean J C 1985 Calculation of critical layer thickness versus lattice mismatch for Ge_xSi_{1-x}/Si strained-layer heterostructures *Appl. Phys. Lett.* **47** 322

[87] Yuge K 2009 Phase stability of boron carbon nitride in a heterographene structure: a first-principles study *Phys. Rev.* B **79** 144109

[88] Guilhon I, Marques M and Teles L K 2017 Optical absorbance and band-gap engineering of $(BN)_{1-x}(C_2)_x$ two-dimensional alloys: phase separation and composition fluctuation effects *Phys. Rev.* B **95** 035407

[89] Ci L *et al* 2010 Atomic layers of hybridized boron nitride and graphene domains *Nature Mater* **9** 430

[90] Chang C K *et al* 2013 Band gap engineering of chemical vapor deposition graphene by *in situ* BN doping *ACS Nano* **7** 1333

[91] Gong Y *et al* 2014 Direct chemical conversion of graphene to boron- and nitrogen- and carbon-containing atomic layers *Nat. Commun.* **5** 3193

[92] Bahandary S and Sanyal B 2012 *Graphene-Boron Nitride Composite: A Material with Advanced Functionalities* (Rijeka: InTech)

[93] Nozaki H and Itoh S 1996 Structural stability of BC_2N *J. Phys. Chem. Solids* **57** 41

[94] Watanabe M O, Itoh S, Sasaki T and Mizushima K 1996 Visible-light-emitting layered BC_2N semiconductor *Phys. Rev. Lett.* **77** 187

[95] Hoffman D M, Doll G L and Eklund P C 1984 Optical properties of pyrolytic boron nitride in the energy range 0.05–10 eV *Phys. Rev.* B **30** 6051

[96] Dresselhaus M S and Dresselhaus G 2002 Intercalation compounds of graphite *Adv. Phys.* **51** 1

[97] Reich S and Thomsen C 2004 Raman spectroscopy of graphite *Phil. Trans. R. Soc. Lond.* A **362** 2271

[98] Meher S R, Biju K P and Jain M K 2011 Raman spectroscopic investigation of phase separation and compositional fluctuations in nanocrystalline $In_xGa_{1-x}N$ thin films prepared by modified activated reactive evaporation *Phys. Status Solidi* A **208** 2655

[99] Semenoff G W 1984 Condensed-matter simulation of a three-dimensional anomaly *Phys. Rev. Lett.* **53** 2449

[100] Razeghi M 2009 *Fundamentals of Solid State Engineering* (Berlin: Springer) ch 7

[101] Heyd J, Scuseria G E and Ernzerhof M 2003 Hybrid functionals based on a screened Coulomb potential *J. Chem. Phys.* **118** 8207
Heyd J, Scuseria G E and Ernzerhof M 2006 *J. Chem. Phys.* **124** 219906

[102] Nakamura S and Fasol G 1997 *The Blue Laser Diode* (Berlin: Springer) pp 201–60

[103] Pantha B N, Li J, Lin J Y and Jiang H X 2010 Evolution of phase separation in In-rich InGaN alloys *Appl. Phys. Lett.* **96** 232105

Part V

Diamond

IOP Publishing

Wide Bandgap Semiconductor-Based Electronics

Fan Ren and Stephen J Pearton

Chapter 21

Recent advances in SiC/diamond composite devices

Debarati Mukherjee, Miguel Neto, Filipe J Oliveira, Rui F Silva, Luiz Pereira, Shlomo Rotter and Joana C Mendes

SiC and diamond are widely considered suitable materials for providing optimal solutions for the high power requirements of modern electronics. Being wide bandgap (WBG) materials, SiC and diamond are perfect candidates for high power/high temperature operation applications and find potential uses in many of today's technological needs, from consumer electronics, avionics, and vehicular transport to renewable energy technologies. In the last decades impressive progress has been recorded in the fabrication of SiC substrates. Six inch SiC wafers can be currently purchased from a few manufacturers, however at a price considerably higher than for silicon (Si) wafers. Nevertheless, SiC is regarded as the semiconductor of choice in applications where Si electronics has reached its performance limits, such as the electric vehicle industry. In addition, diamond devices with excellent performance have already been reported. However, the lack of large-area single crystalline substrates has hampered the effective implementation of diamond based electronics. This chapter presents a review of the field of SiC and CVD diamond electronic devices and discusses the current technological issues in both fields. In addition, it describes possible approaches to combine these two premier materials and presents some preliminary results related to the fabrication of diamond–SiC p–n heterojunctions, fabricated using the very versatile and economical method of hot filament (HF) CVD. Devices such as these are expected to perform optimally under high temperature/high power requirements.

21.1 Introduction

Wide bandgap (WBG) semiconductors have been a research topic for many decades because of their unique electronic and optical properties. With the upsurge of experimental and theoretical activities in recent years, the promise of their

doi:10.1088/978-0-7503-2516-5ch21

21-1

technological application has come to be realized [1]. As a consequence of their WBG, these materials have low intrinsic carrier concentrations and high breakdown voltages, which allows for high temperature and high power applications. Combined with the high thermal conductivity that some WBG materials show, these properties permit higher voltage and power handling capability, higher switching frequency, lower conduction voltage drop, higher junction temperature, and better radiation hardness—properties that promise a renaissance in power electronics, in particular for higher power applications, as well as for smart grid, renewable energy systems, data centres, motor drives, rail traction, and hybrid electric vehicles [2, 3].

Silicon carbide (SiC) and diamond are two such WBG materials with extreme properties. SiC electronics is already a reality. As an example, 24 SiC MOSFET modules are used in each of the inverters of the Tesla Model 3 electric car [4]. Diamond devices with excellent performance have also been reported. However, commercially available diamond based electronics are not yet a reality, due mostly to the lack of large-area electronic-grade diamond wafers. This chapter describes the current status of devices fabricated with these WBG materials and discusses the technological challenges each material is currently facing. The integration of diamond and SiC for the development of hybrid devices with improved character- istics is also discussed. In particular, the nucleation procedure proposed by Rotter [5] is shown to have a positive impact in the electrical characteristics of p-type diamond films deposited on SiC substrates, paving a new way for the fabrication of a new generation of diamond/SiC heterojunctions with improved characteristics.

21.2 Silicon carbide

Silicon carbide (SiC) and gallium nitride (GaN) are almost unanimously recognized as the best materials for the next generation of power electronic devices since they can exceed the performance of silicon (Si) devices in terms of power handling, maximum operating temperature, and conversion efficiency in power modules. The higher thermal conductivity and lower on-resistance of SiC, in particular, reflect the potential for high-density integration of SiC devices, making this material particularly attractive for high temperature and high power industrial and research applications. Recent technological developments and the scale-up of the seeded-sublimation growth technique as well as the epitaxial methods that allow the commercial production of single crystal SiC wafers have contributed largely to the generalized interest in this material. Six inch n-type, p-type, and semi-insulating SiC wafers can be commercially obtained from a few manufacturers. However, they are still more expensive and show higher densities of crystalline defects than the commonplace Si and GaAs wafer standards. Further improvements in deposition and growth techniques are expected to improve the wafer surface quality and lower the wafer cost.

Among the several possible SiC polytypes, three are suitable for device fabrica- tion: 3C, 4H, and 6H. 4H-SiC shows a higher mobility [6] and may be preferable for electronic applications. The unique properties of the 4H-SiC include a high break- down field, large bandgap, high electron saturation velocity, high thermal con- ductivity, and reasonably high electron mobility (table 21.1). SiC is an indirect

Table 21.1. Selected electronic properties of Si, major SiC polytypes, GaN [8], and natural and CVD diamond [9].

Property	Si	3C-SiC	4H-SiC	6H-SiC	GaN	Natural diamond	CVD diamond
Bandgap (eV)	1.1	2.3	3.2	3	3.4	5.47	5.47
Breakdown field ($\times 10^6$ V cm^{-1})	0.6	1.8	//c-axis: 3.0 ⊥c-axis: 2.5	//c-axis: 3.2 ⊥c-axis: >1	2–3	10	10
Electron mobility (cm^2 V^{-1} s^{-1})	1200	750	//c-axis: 800 ⊥c-axis: 800	//c-axis: 60 ⊥c-axis: 400	900	200–2800	4500
Hole mobility (cm^2 V^{-1} s^{-1})	420	40	115	90	200	1800–2100	3800
Electron saturation velocity ($\times 10^7$ cm s^{-1})	1	2.5	2	2	2.5	2	2
Hole saturation velocity ($\times 10^7$ cm s^{-1})	n/a	n/a	n/a	n/a	n/a	0.8	0.8
Thermal conductivity (W cm^{-1} K^{-1})	1.5	3–5	3–5	3–5	1.3	22	22

bandgap material, which is a further advantage for operation under high temperature and high power density conditions. Thanks to all these properties, SiC can broadly exceed the ultimate performance reached by Si devices in terms of power handling and maximum operating temperature. In addition, the high temperature operation of SiC devices means a simplification of the bulky cooling units, which are often required in Si power applications. The potential of SiC electronics for demanding high power and high temperature applications is unquestionable. According to Yole specialists, the total SiC-based power device market is expected to grow steadily, reaching more than $1.5B in 2023 [7].

21.2.1 SiC power devices

SiC power devices have already asserted themselves as devices with unmatchable performance in a few areas. The on-resistance of SiC is considerably lower than that of Si, which directly improves the efficiency of SiC devices [10]. This translates into SiC power modules with improved power handling, maximum operating temperature, and conversion efficiency capabilities in comparison to the Si counterparts [11]. The replacement of Si with SiC technology can reduce the overall power system losses by 50% [12]. SiC-based power converters can operate efficiently in high temperature environments [13] and at high switching frequencies [14]. The applications of SiC power modules have been widely researched and published. Potential application areas range from smart grid systems [15] to optically activated SiC power transistors for pulsed-power application [16], modules for hybrid electric vehicles (HEVs), and plug-in HEVs (PHEVs) [17–19] in addition to various power electronics devices applying novel SiC power semiconductor modules [20]. Testing of a lightweight SiC power module for avionic applications with 540 V DC link voltage and a 6 kW output power with 100 kHz switching frequency has shown that the module can be successfully operated with high efficiency at high switching frequencies [21]. The testing results of high temperature SiC power modules with integrated SiC gate drivers for future high-density power electronics applications operating with a switching frequency of 200 kHz and peak output power of approximately 5 kW have already been presented [22]. The efficiency of the converter was evaluated experimentally and optimized by increasing the overdrive voltage on the SiC gate driver ICs and an overall peak efficiency of 97.7% was measured at 3.0 kW output and 95% at 500 kHz, with an output power of 2.1 kW, with no further optimization. Thanks to the high bandgap, SiC unipolar devices are also expected to replace Si bipolar devices in the blocking-voltage range ≈ 300–4500 V; for voltages above 4500 V, SiC bipolar devices may be considered [23].

SiC power devices exhibit fast switching with minimum reverse recovery (little current overshoot), when compared to Si devices [24]. SiC bipolar junction transistors (BJTs) operating at considerable power levels (300 V, >7 A) were able to switch at a speeds (20–80 ns) comparable to or faster than similar commercial Si unipolar devices. Furthermore, the overall switching speed remains almost unchanged at temperatures as high as 275 °C [25].

Different types of electronic devices have already been fabricated with SiC, such as Schottky barrier diodes with a maximum reverse voltage of 1400 V and forward current densities higher than 700 A cm^{-2} [26], junction-gate field-effect transistors (JFETs) [27, 28], metal–oxide–semiconductor field-effect transistors (MOSFETs) [29, 30] and ultra-violet (UV) photo detectors [31]. 1200 V SiC u-shaped MOSFETs (1200 V) are projected to have 15 times the current density of Si insulated-gate bipolar transistors (IGBTs) [26]. 4H-SiC Schottky diodes have already definitively entered the market as a result of which great effort is being devoted to fundamental studies on SiO$_2$–SiC interfaces for the optimization of vertical MOSFETs [32]. These 4H-SiC MOSFETs are being considered as excellent alternatives to the currently used Si IGBTs. These devices have been shown to enable lower power consumption at higher switching frequencies and at junction temperatures above 200 °C [33–35].

The radiation resistance of SiC devices has already been demonstrated, showing the potential of using SiC devices in applications which combine high temperature and radiation environments and where the use of Si and GaAs technologies is limited [36]. At the present moment, bipolar integrated circuits operating at 500 °C with improved radiation hardness have already been reported [37].

Overall, there has been a ten-fold improvement in SiC MOSFETs and power JFET voltage ratings from sub-kV breakdown to over 10 kV breakdown between 1997 and 2014 [38–43]. A similar trend with respect to breakdown voltage rating, at the same time optimizing specific on-resistance and current density capabilities has also been observed in the past few years [44–46].

More recently, Monolith Semiconductor partnered with XFab Inc., an automotive qualified Si manufacturing foundry, to develop and implement an advanced SiC power MOSFET process on their 150 mm diameter Si production line. The AMPERES project aimed to develop devices at a cost lower than $0.1/A [47]. As of June 2019, the actual cost is between US$0.40 and US$0.80, still 4–10 times higher than for Si based technology [48]. Utilizing existing Si foundries, often referred to as the 'foundry model' [49], has been seen to be the most effective cost-reduction path and may significantly reduce fabrication costs in the future.

21.2.2 Technological challenges

The increasing interest in SiC for power applications has played the role of a trigger for innovations in bulk and epitaxial growth technologies [23]. Since SiC does not melt at atmospheric pressure, SiC ingots are grown by a sublimation method at extremely high temperatures, above 2200 °C. Six-inch single crystalline SiC wafers are commercially available, however, the cost is prohibitively high compared to their Si counterparts and the dislocation density is still in the range of 10^3–10^4 cm^{-2}.

The impact of dislocations on the performance and reliability of SiC devices is an area of extensive research and development. Due to the limitations in fabrication technology with regards to SiC bulk material, SiC devices run the chance of suffering from a low inversion channel mobility (much lower that the theoretical bulk mobility value) [50] brought about by the presence of electrically active defects in the metal–oxide–SiC interface. These defects lead to a high density of interface traps near the conduction band edge [32, 51]. Electrically active defects of this nature, for example, the clustering of carbon (C) and/or intrinsic defects near the interface region, contribute to the non-ideality of the metal–oxide–semiconductor system [51]. Such interface states along with fixed charges in the oxide severely impact the properties of the MOSFET conduction channel. In order to improve the channel mobility, different processes for passivating the SiO_2–SiC interface states have been explored in the last decade, mainly consisting in thermal annealing in nitrogen-rich atmospheres (NO or N_2O). The typical annealing temperatures for these passivation processes are in the range of 1100 °C–1400 °C. The nitridation of the gate oxide produces a significant improvement in the channel mobility ($\approx$ 20–50 cm^2 V^{-1} s^{-1}), when compared to non-annealed gate oxides formed by dry oxidation processes (<5 cm^2 V^{-1} s^{-1}). The passivation mechanism behind this improvement of channel mobility consists of the saturation of dangling bonds or in the elimination of an excess of C at the interface itself [52]. Atomically resolved scanning-tunnelling microscopy (STM) images and low-energy electron diffraction (LEED) studies of the effect of annealing at high temperatures (>1600 °C) have shown the formation of step-bunched surfaces with a high RMS surface roughness [53]. This surface roughness was brought below 0.1 nm using a combination of a protective cap layer and subsequent mechanical polishing. The thickness of the cap layer was found to be proportional to the surface roughness. The reduction in net roughness has been proposed to be due to the inhibited mobility of surface atoms in the presence of the cap layer [53].

It is known that as the operating temperature increases, the changes in semiconductor characteristics such as intrinsic carrier concentration, p–n junction leakage current, thermionic leakage, and carrier mobility influence and degrade the devices' performance [54]. The performance degradation of 4H-SiC RF power metal–semiconductor FET (MESFETs) caused by self-heating and high operating temperature has been studied using two-dimensional electro-thermal simulations and DC and S-parameters measurements. The effects of different device layouts/designs and packaging materials/solutions on the junction temperature of SiC devices has been studied by Liu and co-workers [55]. The temperature was shown to profoundly influence device performance in terms of breakdown voltage, saturation current, power density, high frequency current gain and power gain, cut-off frequency, and maximum frequency of oscillation. As the operating temperature increased from 25 °C to 250 °C, a 38% reduction in drain saturation current and 29% and 41% drops in cut-off frequency and maximum frequency of oscillation, respectively, were observed.

The knowledge of the failure limits of a device with respect to defects and other factors is important in order to know the safety margin that must be kept in mind for the device application. From the viewpoint of system optimization, this directly impacts the robustness limits. Lutz *et al* [56] studied the impact of different parameters (such as bandgap, critical field strength, electron mobility, hole mobility, and thermal conductivity) on SiC power device failure mechanisms using the failure physics known from Si experimental results. They concluded that with well-designed SiC devices, some of the challenging and, thereby, limiting effects in Si, such as dynamic avalanche and second breakdown, do not occur with SiC. They also concluded that SiC devices can be designed to be very rugged in a static avalanche and that they withstand extreme high temperatures. A separate study by Fayyaz et al [57] presented a detailed functional and structural characterization of a commercial 1200 V SiC power MOSFET under single pulse and repetitive UIS test conditions to demonstrate robustness and avalanche ruggedness. The single pulse robustness test showed that the robustness level of SiC MOSFETs is hindered only by the structure and packaging related issues. In fact, optimization of wafer and packaging technologies is indispensable and its unavailability will hinder the development and utilization of devices in the 300 °C–600 °C temperature range.

The prime considerations for packaging technologies are the chemical, physical, and electrical stabilities of both the packaging and semiconductor materials, as well as the interfaces between the device and the packaging material itself. The matching of the thermal coefficients between the materials is also an important consideration to be kept in mind. Conventional metals used for integrated circuit (IC) packaging, such as Cu, Al, and Au/Ni-coated Kovar, undesirably oxidize at temperatures approaching 500 °C when atmospheric oxygen is present [54]. Hence high temperature operation may only be considered in a perfectly hermetically inert or vacuum environment. In case more than one metal or metallic material is used in the fabrication of electrical connections, one must also consider the undesirable metallic phases that may occur at higher temperatures for metals such as Al and Cu. These phases are deemed undesirable as they often result in the reduction of overall mechanical strength. During standard cycles of usage, devices are expected to encounter large temperature ranges (55 °C–600 °C). This pattern of operation may lead to an increase in the mismatch between the thermal expansion of the materials, placing in stress the WBG die itself. Hence, for proper device design and minimization of stress, the thermal characteristics of the different materials must ideally be matched. The bonding material must also be chemically and electrically compatible with the metallization on the wafer backside as well. However, obtaining matched materials that do not diffuse or undergo similar destructive processes at high temperatures in order to maintain the hermetically sealed environment necessary for operation is a difficult task. The need for innovative packaging materials and packaging design concepts naturally follows [54]. Aluminium nitride (AlN), with a thermal expansion close to that of SiC, was first proposed as an insulating non-crystalline ceramic substrate material for high temperature device packaging in the late 1990s [58]. Around the same time, thick-film materials were also proposed to be promising for use as substrate metallization in realizing 500 °C hybrid packaging [59]. More recently, a high temperature, wire-bondless power electronics module with double-sided cooling capability was successfully fabricated [60]. The module used a low-temperature co-fired ceramic (LTCC) substrate as the dielectric as well as a chip carrier. On this LTCC carrier, conducting vias were created to realize the interconnection, and the absence of a base plate altogether reduced the overall thermal resistance and improved the fatigue life as well. Nano silver paste was used to attach power devices to the substrate as well as to pattern the gate connection. The electrical measurements of a SiC MOSFET and SiC diode switching position demonstrated the functionality of the module [61].

21.3 Diamond

Like SiC, diamond is another WBG material with extreme properties (table 21.1). The values of its breakdown field, bandgap, electron saturation velocity, thermal conductivity, and electron mobility are larger than those in SiC, however, the chemical inertness of SiC is superior in comparison, since SiC remains stable in a high temperature (>700 °C) oxidizing ambient, whereas diamond graphitizes [62].

Diamond can be synthesized in the form of bulk crystals by high pressure high temperature (HPHT) methods. These crystals retain the properties of their naturally existing counterparts and are electrically insulating. Diamond films can also be deposited in the form of thin films by chemical vapour deposition (CVD), a technique that has already been thoroughly investigated in the decades since the first successful experiments by Eversole (1962) [70] and Deryagin *et al* [63]. Unlike the HPHT growth process, the CVD of diamond takes place at low pressures, where diamond is the metastable form of C [64] and involves the production of diamond from a gaseous precursor, typically methane, diluted in hydrogen (H). As methane molecules are heated by a high energy source, they form radicals whose C atoms bond with atoms at the substrate. During this process non-diamond bonds, such as graphite, are also created, however, they are etched away by the activated H atoms, leaving behind the sp^3 C bonds. The gas phase activation can be done using different approaches, hot filament (HF) and microwave plasma (MPCVD) being two of the most commonly used methods. Figure 21.1 represents a schematic view of a generic HFCVD system.

Diamond films can be deposited by CVD on HPHT diamond crystals (so-called homoepitaxial diamond films) as well as on non-diamond substrates (heteroepitaxial diamond films), such as Si, SiC, or silicon nitride (Si_3N_4), to name a few. However,

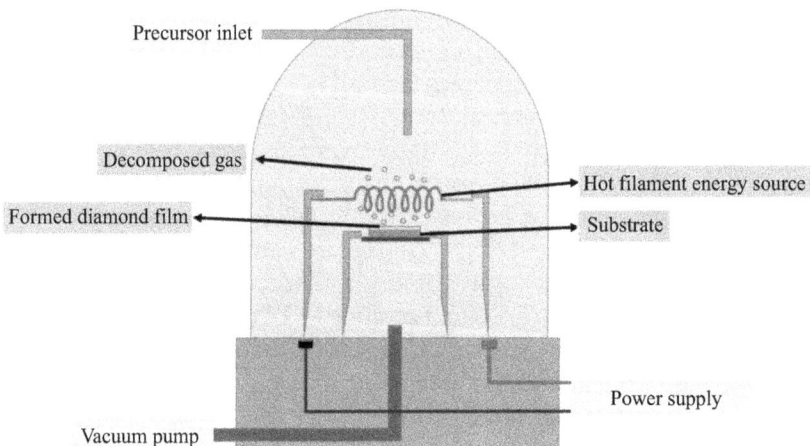

Precursor inlet

Decomposed gas

Formed diamond film

Hot filament energy source

Substrate

Power supply

Vacuum pump

Figure 21.1. Schematic representation of an HFCVD system.

for diamond to grow on a non-diamond substrate, the surface of the latter needs to be enriched with diamond nanoparticles before diamond growth takes place, otherwise only 10^2–10^3 cm^{-2} isolated diamond crystals are deposited [65].

Different methods can be used to this purpose, such as scratching the substrate with diamond or an abrasive grit [66] or ultrasonically (US) seeding the substrate in a suspension containing diamond particles [65]. As an example, seeding with a colloidal solution of detonation nanodiamond (DND particles) enables nucleation densities in excess of 10^{12} cm^2 [67–71]. The ζ potential of DND particles can be further changed by functionalizing the particles' surface [72] and suspensions with positive ζ potential have been used to grow diamond films on gallium nitride (GaN) substrates [73] and Au (gold) films [74]. Bias-enhanced nucleation (BEN) [75, 76] is another commonly used method that employs *in situ* surface bombardment in the gas phase by C species through the application of negative bias on the substrate, resulting in a C-rich surface that enhances diamond nucleation [77]. BEN has been used for growing highly oriented diamond films on multilayered iridium (Ir) substrates such as Ir/yttria-stabilized zirconia/Si [78] or Ir/SrTiO$_3$/Si [79] and for fabricating devices such as Schottky diodes [80], field-effect transistors [81], and mechanical microactuators [82]. Other seeding methods include dipping the substrate in a suspension containing diamond nanopowders [83, 84], ink jet printing [85–87], electrostatic self-assembly [88, 89], or exposing the substrate surface to diamond growth conditions for a short period before following with the seeding step [5]. The later method, in particular, has been used to coat mesa and trench structures, EEPROM integrated circuits and GaAs HEMT devices [90, 91], to grow diamond films on epitaxial SiC substrates [92], and to coat platinum electrodes [93].

Following the seeding step, the substrate is introduced in the deposition chamber. Upon exposure to growth conditions, small diamond seeds on the surface of the substrate grow three-dimensionally and eventually coalesce, creating a closed diamond film. As the thickness of the diamond film increases, crystals with the fastest growth rate progressively dominate (the van der Drift process [94]) and columnar growth appears [95]. Depending on the growth conditions (deposition time, temperature, composition of the gas phase) the diamond crystals can have different sizes, ranging from a few micrometres (poly or microcrystalline diamond (PCD/MCD)) to several nanometres (nanocrystaline diamond (NCD)). When the films are grown in hydrogen-deficient chemistries, the size of the grains can be as small as 3–4 nm (ultra-nanocrystalline diamond (UNCD)).

21.3.1 Doped diamond

Diamond is an electrically insulating WBG semiconductor, yet like other semiconductor materials it can be doped. An efficient doping is achieved when the impurity atoms that substitute C atoms in the lattice are electrically active, that is, contribute with electrons to the conduction band (n-type) or holes to the valence band (p-type). Impurities that do not occupy a substitutional site in the lattice are unwanted, since they may generate additional defects which can yield levels deep in the bandgap susceptible to compensate or trap the original useful carriers. The distribution of the impurities must also be uniform through the volume so there is no accumulation or segregation in certain sites.

N-type diamond films can be obtained by adding nitrogen (N) or phosphorous (P) impurities. As a consequence of the high activation energy of N dopants (1.7 eV), activation of carriers only happens for temperatures higher than 800 °C [96] and at room temperature N-doped diamond films behave like a lossy dielectric [97]. Despite being electrically inactive, the N level acts as a recombination centre in radiative transitions [98]. On the other hand, P atoms introduce a deep donor level 0.6 eV below the conduction band. Due to the large covalent radius of P atoms, as well as to the positive formation energy of the impurity, the efficiency of incorporation of P in the diamond lattice is very low [99]. The introduction of the P atoms in the diamond lattice is usually performed by CVD and its efficiency depends on the orientation of the crystal surface. In the (111) orientation it ranges between 5% [100] and 18% [101], decreasing to 0.026% in the (111) orientation [102]. The maximum carrier density also depends on the type of film. In the case of single crystal films, values as high as 10^{20} and 10^{18} cm^{-3} were obtained in the (111) [103] and (100) [102, 104] orientations, respectively. In the case of micro- and nanocrystalline diamond films, the maximum reported carrier concentrations are in the order of 10^{18} cm^{-3} [105, 106]. Due to the high activation energy, P-doped diamond films also have relatively high resistivity [107].

p-type characteristics can be obtained with H or boron (B) doping. The growth surface of diamond films deposited by CVD in H-rich chemistries is invariably terminated with H atoms. This induces the appearance of a p-type highly conductive layer at room temperature [108] on homoepitaxial as well as on heteroepitaxial CVD diamond films.

B is the most extensively studied dopant in CVD diamond and was identified as being an effective acceptor level in 1971 [109]. The doping of synthetic HPHT diamond crystals with B can be done during HPHT synthesis [110], via the gas phase [111], and through ion implantation [112]. p-type HPHT diamond substrates are commercially available, however, their size is limited to a few square millimetres, which limits their practical applications [113, 114]. On the other hand, ion implantation is a mature technique for Si, but because diamond is metastable, there might be conversion to graphite during the process or during post-process annealing. Nevertheless, it remains a widely investigated and used technique to obtain p-type diamond [112]. The doping of PCD films by B ion implantation [115, 116] has yielded results comparable to those obtained for single crystal type IIa diamonds (although with poorer electrical properties), indicating that the technology developed for implantation doping of single crystalline material is also applicable to the implantation doping of CVD diamond films.

The incorporation of B atoms from the gas phase during the CVD of diamond has been widely studied [95]. The efficiency of the process depends on the precursors used, on the deposition technique, and on the nature of the diamond film (PCD or homoepitaxial diamond). Different precursors have been introduced in the gas phase, such as a thin B rod [117], diborane, boron oxide dissolved in acetone, ethanol or methanol, trimethylborate, and trimethylboron, diborane and trimethylboron being the most usual [95]. Doped films can be grown using both the MPCVD and HFCVD techniques, however, the intrinsic nature of the CVD process influences the quality and growth rate of the deposited films. The lifetime τ of B-related molecules, calculated based on a step-flow motion model, was found to be ~13 times higher for HFCVD than for MPCVD. This attests to the longer migration length of B-related molecules in HFCVD than in MPCVD, which limits the incorporation of B in the latter. In fact, the higher flux of atomic H, the very factor that leads to higher growth rates of undoped diamond films in MPCVD, has been seen to hinder the growth rate of doped diamond films by causing surface etching [118]. As a consequence, as the concentration of B in the gas phase increases, the diamond deposition rate decreases with MPCVD. In contrast, it increases when HFCVD is used. MPCVD has also been known to lead to soot formation when preparing p-type layers with a high doping level [114]. This limits growth itself, and possibly becomes an origin of non-epitaxial components. In HFCVD, this problem is avoided even at high doping levels. As an example, films with a doping level of 10^{21} cm^{-3} and a resistivity of 1 m cm have been obtained with this technique [119]. For low doping levels and independently of the deposition technique, the incorporation of B atoms reduces the concentration of the planar and of some point defects, thus improving the quality of the polycrystalline films (with the exception of using trimethylborate as a precursor) [95]. However, B incorporation increases the concentration of unepitaxial and pyramidal crystallite in homoepitaxial films [120].

B atoms create acceptor states 0.37 eV above the valence band [109]. This value is much larger than the thermal voltage k_BT (where k_B is the Boltzmann constant and T is the temperature) at 300 K, which means that in films with low doping the levels are not activated at room temperature. In this case, the number of carriers (and hence the electrical conductivity) increases with increasing temperature. However, as the B concentration increases, the activation energy decreases and becomes null for concentrations around 10^{20} cm^{-3} [121]. Thanks to this metallic conductivity, the devices fabricated on highly doped films are fully activated at room temperature.

21.3.2 Diamond based devices

Different types of diamond devices have been fabricated with B-doped diamond films. Schottky devices with current densities of 100 A cm^{-2} [122] or operating at temperatures as high as 1000 °C [123] and p–n junctions operating up to 400 °C [124] have been reported. Diamond Schottky diodes have the potential of being an alternative to Si p–i–n and SiC Schottky diodes in power electronic circuits [125]. Lateral as well as vertical diamond emitter diodes with excellent field emission characteristics have been demonstrated. The vertical configuration, with an indented anode, exhibited high emission currents (>0.1 A). The lateral configuration with anode–cathode spacing <2 μA exhibited emission currents in the μA range [126]. Other devices, such as FETs, capacitors, and switches have worked up to 450 °C [84].

Since the p-type conductive characteristics of H-terminated diamond surfaces (CVD diamond films) were first reported in 1989 [127] and the application to an FET device was demonstrated in 1994 [128], different FET devices have been fabricated on high quality CVD diamond films. The radio frequency characterization of these devices shows values of maximum transition frequency (f_T) and maximum frequency of oscillation (f_{max}) higher than the highest values obtained for single crystalline diamond FETs [129]. Another study reported 0.2 μm gate diamond metal–insulator–semiconductor field-effect transistors (MISFETs) with 23 and 25 GHz cut-off and maximum frequency of oscillation, respectively [130]. Great improvements have also been reported in the output power density of diamond FETs whereby a maximum power gain of 10.9 dB and power added efficiency of 31.8% were obtained [131]. Geometry optimization can further enhance the performance of diamond MESFETs, for example by reducing the total length of the channel (the source–drain distance) at a constant gate–source distance the knee voltage could be decreased [132].

Another obvious application of this material is the use of diamond plates as carriers or heat spreaders for high power devices. In fact, thermal grades of diamond wafers are commercially available from a few vendors. However, they remain too expensive for standard utilization. With regards to power electronics, silver/diamond composites have been investigated as an effective packaging solution for GaN devices [133]. The viability of CVD diamond for electronic packaging applications has also been demonstrated for plastic-packaged GaAs devices [134] and more recently AlGaN/GaN HEMTs with over 7 W mm^{-1} output power density at 10 GHz [135].

CVD diamond films have been studied as a promising packaging material for micro-electromechanical systems (MEMS) [136–138]. In particular, for a given geometry, poly-carbon (poly-C) thin film packages withstand more differential pressure than any other material because of the material's high Young's modulus [137].

21.3.3 Technical challenges

The aforementioned properties raised a lot of expectations regarding the development of an all-diamond based electronics for extreme applications. However, and despite the generalized initial enthusiasm, the effective commercial development of diamond devices has been hampered by some technological constraints.

The first limiting issue is the typically low area of high quality/low defect density HPTH substrates, which lies in the range of a few square millimetres. More recently, large-area single crystal substrates, fabricated using the 'mosaic technique', became available on the market [139]. This growth technique involves the CVD of high quality single crystal diamond on a suitable substrate after which a planar crystal is fabricated using the lift-off process based on ion implantation. The final step is the lateral jointing of two or more single crystals, thus leading to the nomenclature 'mosaic'. Since this fabrication process requires the availability of a single large (inch sized) seed wafer, recently research on bulk diamond crystal growth techniques received a new impetus [140]. However, and although the technique can be used to fabricate large-area substrates, defect densities still need to be improved in order to turn these substrates into electronic-grade wafers [141].

When diamond films are deposited by CVD on HPHT crystals, the initial orientation of the substrate influences the synthesis of thick high quality intrinsic and doped diamond layers [140]. Due to the availability and relative ease of growing low defect films, the [1 0 0] orientation is still the most widely used. However, doping with phosphorous to obtain active n-type films is not straightforward and the efficiency of doping with B to obtain p-type films is limited. Growing the films on [1 1 0] orientation improves the efficiency of B doping. However the usable area in this case is reduced, requiring the improvement of the growth conditions. The [1 1 1] orientation is probably the most attractive for doping since it allows the growth of active n-type films with phosphorous and the achievement of the highest B concentrations. However, this orientation is prone to the formation of twins and macroscopic defects that restrain the maximum reachable thickness while the crystalline quality remains average. The growth of diamond films on alternative orientations remains an area of active research interest and the development of specific growth strategies on alternative orientations and substrates plays an important role in exploiting the properties of semiconducting diamond [140]. The optimization of the growth process is particularly important because the carrier mobilities, for example, are process dependent and increase with the diamond quality [142]. To achieve high blocking voltages and correspondingly high break-down electric fields, highly pure homoepitaxial films with very low defect concentration are required. So far only ultrahigh homoepitaxial films with a dopant concentration lower than $\sim 10^{13}$ cm^{-3} and around 10^{15} cm^{-3} for nitrogen and B doping, respectively, have been demonstrated [143].

Heteroepitaxial diamond films deposited on iridium and SiC substrates [140] do not face the size limitations that homoepitaxial diamond does, but two main challenges still need to be addressed. First, the dislocation density needs to be reduced. The current state-of-the-art of heteroepitaxial diamond growth on iridium substrates using bias-enhanced nucleation (BEN) starts with a dislocation density higher than 10^{10} cm^{-2}, a value that needs to be brought down to 10^6 cm^{-2} [144]. Second, it has to be demonstrated that diamond device elements really can compete with existing technologies based on other WBG materials (e.g. SiC and GaN). This may be done by fabricating devices on the best single crystal material available with the lowest dislocation density (close to zero) and then proceed with the fabrication of devices with identical performance on state-of-the-art heteroepitaxial diamond. In case it turns out that device fabrication requires the availability of crystals containing no dislocations or a maximum of several 100–1000 cm^{-2}, diamond heteroepitaxy on iridium substrates is probably outside the race for diamond electronics [144]. The heteroepitaxy of diamond on SiC remains promising, however, mainly because of the lower lattice mismatch between the lattice parameters of diamond and SiC than that between diamond and Si, the most commonly used substrate for the heteroepitaxial growth of diamond [145].

It can thus be said that before the large scale application of diamond in electronics takes place, diamond synthesis must overcome the limitations related to deficiencies in doping concentrations and a high density of impurities and defects—and, most important of all, at affordable prices. On the other hand, the exploitation of these defects and impurities for their use as quantum devices is an active area of research. The identification of more shallow dopants [113] and new quantum defects and effects [146, 147] are still open challenges. The deep impurity levels in diamond result in low free carrier densities and correspondingly high resistivity. Although many attempts to identify shallower dopant impurities have been made, these are still to be found. Hence, the ultimate in high voltage tolerant devices can be obtained, due to the high breakdown field of diamond, however, the low carrier density results in very large on-resistances and long drift layers. As a consequence, diamond devices have not demonstrated the theoretically predicted FOMs when compared to other power WBG semiconductors. In order to reduce the on-resistance, researchers have been designing diamond power devices for high temperature operation or utilizing the highly conductive hopping transport to achieve high doping levels [142].

In addition to the obvious challenges in the production steps, the charge transport characteristics and carrier dynamics also remain an under-explored issue. The electrical interfaces obtained with CVD diamond metals or other semiconductors have only been studied with respect to specific devices and an understanding of the underlying physics is still missing. With respect to power devices, predicting the electrical and thermal performance is a key step for their next-generation development. For the development of diamond power devices in particular, the implementation of the physical models of CVD diamond into a finite element simulator is mandatory.

21.4 Diamond/SiC composite devices

As was mentioned above, SiC is a material that already excels itself in high power/high temperature applications. In the case of hybrid and electric vehicles, for example, SiC components are often the only realistic technical solution. So the question that naturally arises is what added benefits can diamond bring to SiC electronics.

21.4.1 Thermal management

One of the obvious answers is improved thermal management. In fact, and even though diamond loses the race with SiC with respect to the availability of electronic-grade wafers at realistic prices, the optimization of the CVD deposition process has resulted in the production of thermal grade diamond plates (with a thermal conductivity in excess of 1800 W m^{-1} K^{-1}). These plates can be purchased from several vendors around the world and are available with different sizes, thickness and metallization schemes (if required). They are electrically insulating and this provides an extra advantage in high-density electronic packaging. As an example, SiC devices can be mounted directly on a diamond heat spreader, which causes a decrease in the junction temperature and allows for higher power operation and improved reliability for a given power [148].

21.4.2 Device passivation

Device passivation is another use of diamond. One of the important advantages of SiC for power device applications is its high electric breakdown field. However, in order to obtain real benefit from this property, the passivation materials should provide enough robustness to withstand the high electric fields in the areas of the device where electric field concentration occurs. Failure to do so results in the decrease of the breakdown voltage, as well as in the occurrence of irreversible and destructive electric breakdown. Different materials have been proposed to be used as passivation layers of SiC devices, such as aluminium nitride [149], porous SiC [150], polyimide [151], and aluminium oxide [152]. The electric breakdown field of diamond exceeds that of each of these materials and, as such, the passivation of the SiC surface with CVD diamond not only increases the breakdown voltage of a particular device but spreads any heat generated at local hotspots over the whole area of the film. However, in order to provide the effective passivation of the SiC electric defects, the interface between both these materials is required to be as intimate as possible. It is known that the nucleation surface of CVD diamond films typically has a lower quality than the growth surface. This fact is related to the higher concentration of grain boundaries in the former. These grain boundaries have a negative impact when device passivation or heat extraction are desired. By being saturated with non-sp^3 diamond bonds, they may provide channels through which the charge carriers can flow [153] and this increases the leakage currents and decreases the electric breakdown field. In addition, they induce phonon scattering and also compromise the thermal conductivity of the films [154, 155].

Mukherjee *et al* [92] studied the impact of the diamond deposition parameters in the quality of the diamond deposits at the SiC–diamond interface. They showed that the type of diamond particles in the seeding process has an impact on the film quality, and that lower deposition temperatures in conjunction with the novel nucleation procedure (NNP) favour the lateral growth of the diamond grains during the early stages of diamond growth, thus decreasing the grain boundary–grain ratio.

21.4.3 Diamond/SiC heterojunctions

Heterojunctions exhibit useful electronic properties because of the discontinuities in the local band structure at the interface. In order to have a useful heterojunction, the lattice constants of the two materials must be well matched, as in the case of 4H-SiC [156] and diamond [157], since lattice mismatch can introduce dislocations and result in interface states.

The first studies of the diamond–SiC interface were performed as early as the 1990s. They included theoretical and experimental studies of the interface formed as a result of heteroepitaxial MPCVD growth of diamond on β-SiC (cubic phase of SiC) [158–161] and on α-SiC (the 6H-SiC polytype in this case) [162], as well as of the growth mechanisms involved in the heteroepitaxy [163, 164]. It should be noted that these studies have not included any investigation into the electrical properties of these heterostructures.

The earliest reported SiC–diamond heterojunctions involved the deposition of diamond films using CVD on Si wafers and post-growth high temperature annealing to obtain a SiC layer at the interface [165] or Si wafers nucleated by a two-step process that involves the conversion of the Si surface to an epitaxial SiC layer, followed by bias-enhanced nucleation of diamond [166]. Buried SiC regions in diamond have been achieved by implanting diamond substrates at elevated temperatures with high fluences of Si, resulting in a buried layer with crystalline 3C-SiC domains in perfect epitaxial relation to the diamond substrate. Amorphization and graphitization were completely prevented by the elevated temperature during implantation [167, 168]. p–n heterojunctions have been fabricated by depositing n-[169] and p-type [170–172] NCD diamond films as an electrical contact on the SiC substrate. Rectification ratios of 10^7 were observed at room temperature and 10^4 at 570 K [169] for the n-type NCD/p-type SiC heterojunction. Heavily doped metallic p ++ NCD films have also been evaluated for application as high temperature surface passivating contacts on n-type SiC Schottky diodes [170]. The relative stability of the current–voltage characteristics with temperature and the negligible reverse leakage current levels (10^{-8} and up to 3×10^{-7} A cm^{-2} at −20 V) demonstrate that highly doped NCD is a versatile high temperature contact to 4H-SiC. The characterization of the lightly doped p-type NCD/n-type SiC heterojunction with temperature indicated near-ideal thermionic rectification. Transport was found to occur both by thermal excitation over the barrier and tunnelling through it at low forward bias and low SiC doping levels. These heterojunctions have demonstrated a curvature coefficient of 10^5 V^{-1}, the highest ever reported for a diode [172].

21.5 PCD/SiC heterojunctions

Although the NCD-SiC heterojunctions reported so far show a promising performance, no studies concerning the PCD/SiC heterojunction can be found in the literature. NCD is appropriate to be used as an optical coating [84, 173] and for MEMs and NEMs applications [174–176], and B-doped NCD has raised considerable interest as an electrochemical electrode [177–181]. However, NCD is not the best candidate for active electronic device applications [182]. Due to the small grain size, these films have a large concentration of grain boundaries that have a detrimental effect in the performance of electronic devices. On one hand, if the density of grain boundaries is high enough, the latter may provide channels through which the charge carriers can flow [153], increasing leakage currents and decreasing the electric breakdown field. On the other hand, the grain boundaries induce phonon scattering and this effect compromises the thermal conductivity of the diamond films [154, 155]. Finally, the grain boundaries reduce the carrier mobility of doped films considerably, which remains around 1 cm^2 V^{-1} s^{-1} for both NCD and UNCD films regardless of the doping concentration [182–184]. On the other hand, single crystal diamond and large grain size PCD show a pronounced increase in hole mobility as the B doping level decreases [182].

The promising electrical performance of PCD/SiC heterojunctions justifies a thorough understanding of all the parameters that influence their electrical properties. Technological challenges, such as the difficulty in relating the conduction mechanism in these heterojunctions to a pre-determined band structure, or the causal variety of defect levels present in CVD-grown diamond, justify fundamental experimental study for characterizing the device behaviour as well as the parameters that affect it. The electrical properties of the diamond layer (and hence of the heterojunction) depend on the crystal quality, on the diamond surface termination, on the metal–diamond junction characteristics, and on geometrical factors [132]. Another critical parameter is the diamond–SiC interface itself. The deposition of diamond films on foreign substrates requires a seeding procedure that enriches the substrate surface with diamond nanoparticles that grow and coalesce during the CVD cycle [185]. The density of these seeds, as well as the speed at which they grow and coalesce, are critical factors that determine the sp^2–sp^3 ratio at the interface. This, in turn, deeply influences the rectifying ratio, as well as the leakage currents at the interface [153]. In addition, the dependence of the resistivity of B-doped PCD films with B incorporation is very complex and the parameters such as concentration, location, and structure of B defects are known to affect the concentration of free holes in PCD films [186]. These characteristics in turn depend on the CVD deposition parameters, such as gas composition and pressure, substrate temperature, and seeding method.

21.5.1 Experimental details

A series of experiments was performed in order to understand how the seeding method and the HFVCD parameters impact the morphology and sheet resistance of the diamond films. To that end, a 4 inch SiC epitaxial wafer with a carrier concentration of 1×10^{16} cm^{-3} and a resistivity $\leqslant 0.025$ Ω cm purchased from Cree Inc. was diced into 5×5 mm^2 pieces. P-type B-doped PCD films were deposited on the top side of the samples using a custom made HFCVD system [187]; the temperature was measured by a thermocouple touching the back of a sacrificial SiC substrate.

The SiC samples were initially cleaned in an ultrasonic (US) bath with acetone (CH$_3$COCH$_3$) followed by ethanol (C$_2$H$_5$OH), 5 min each. Following this initial solvent cleanse, they were treated with ammonium hydroxide (NH$_4$OH), hydrogen peroxide (H$_2$O$_2$), and hydrofluoric (HF) and hydrochloric (HCl) acids. The details can be seen in table 21.2.

Half of the samples were exposed to diamond growth conditions prior to any seeding procedure. This step will be referred to as pre-treatment (PT) hereafter. The pre-treated samples and the remaining cleaned SiC substrates were then seeded using three different procedures: (i) 6 nm ND particles in a water based suspension (Diamecânica®), (ii) 0.2 g of 6–12 μm diamond grit (Saint Gobain ®) in 20 ml C$_2$H$_5$OH, and (iii) manual scratching method (by pressing the sample against a tissue with diamond particles) [187]. The B source was B$_2$O$_3$ diluted in C$_2$H$_5$OH, evaporated at a rate of 0.25 μl min^{-1}, and was dragged by argon (Ar) at different

Table 21.2. Cleaning procedure for SiC samples.

Cleaning step	Liquid	Duration (s)	Temperature (K)	Additional notes
SC1	i. 5 parts DI* H_2O ii. 1 part NH_4OH (28%) iii. 1 part H_2O_2 (30%)	300	353	• Sample placed in beaker, on a magnetic hot plate with a stirrer • Rinsed with DI* H_2O post treatment
Oxide clean	i. 1 part HF (49%) ii. 30 parts DI H_2O	15	Ambient	• Sample placed in Teflon beaker and stirred • Rinsed with DI* H_2O post treatment
SC2	i. 6 parts DI H_2O ii. 1 part HCl (37%) iii. 1 part H_2O_2 (30%)	600	353	• Sample placed in beaker, on a magnetic hot plate with a stirrer • Rinsed with DI* H_2O post treatment

*DI stands for deionized water

flow rates, CH_4/H_2 gas ratios, and system pressures. The substrate temperature was obtained by adjusting the filament temperature (within the range of 2173 to 2473 K) and the filament–substrate distance was maintained at 5 mm. Tungsten (W) filaments were used, as W is a carbide forming metal, a carburization step is required before stable PT or GC conditions can be obtained. To this end the HFCVD system was run with 5% CH_4/H_2 and filament temperature was set to 2673 K for 30 min. During this step, performed prior to PT (where applicable) and growth cycle (GC), the samples were kept at a much larger distance from the filament (>5 cm). Depositions were performed at 4.5 and 9 kPa. The deposition conditions are listed in table 21.3. The films were characterized with a Hitachi SU70 SEM.

In order to optimize the fabrication of the ohmic contacts, 100 nm thick gold (Au) and gold/palladium (Au/Pd) 2 mm diameter circular contacts 3 mm apart were deposited on the surface of film A3 using an SEM sputtering system (E5000) and a shadow metal mask. These metals were chosen because Au contacts deposited on H-terminated diamond surfaces show ohmic behaviour after annealing [108]. On the other hand, Pd is known to improve the adhesion between the contact and the diamond surface [188]. Following metallization, the sample underwent annealing in pure N_2 atmosphere at atmospheric pressure and at a temperature of 973 K for 60 min, after which the temperature was increased to 1173 K for further 30 min. The sample was cooled down to room temperature without any forced cooling in the

Table 21.3. Diamond films deposition conditions; samples not given PT are marked with ′.

Sample ID	Pre-treatment					Growth			
	CH_4/H_2 (%)	T^* (K)	Duration (min)	P^* (kPa)	Seeding	$CH_4/Ar/H_2$ (%)	T^* (K)	Duration (min)	P^* (kPa)
A1/A1′	1					1/8/100			
A2/A2′	2				ND	2/8/100			
A3/A3′	3	993		4.5		3/8/100	993		4.5
A4/A4′	3				scratch	3/8/100			
A5/A5′	2		90		grit	2/8/100		240	
A6/A6′	1					1/10/100			
B1/B1′	2	1023		9.0	ND	2/8/100	1023		9.0
B2/B2′	3					3/8/100			

$^*T$ stands for temperature and P for pressure.

same N_2 atmosphere. Current–voltage measurements were made using a Keithley 2400.

A second set of samples was prepared in order to further evaluate the impact of the PT and CH_4 ratio on the resistivity of the films. The cleaning procedure was the same as in the previous set and again half of the samples were given the PT prior to the seeding step. PT and no-PT samples were US seeded in a suspension with ND particles and the films were deposited with a higher Ar flow rate ($1.67 \times 10^{-7}\,m^3\,s^{-1}$) and varying the CH_4 ratio in the plasma according to the conditions listed in table 21.4.

$0.5 \times 5\,mm^2$ equally spaced AuPd contacts were sputtered on the surface of the second set of films using shadow masks and were further annealed according to the conditions described previously. The electrode layout can be seen in figure 21.2(a). Resistivity measurements were made using the four-point probe technique represented in figure 21.2(b) and a Keithley 2400.

21.5.2 Morphological characterization of the BDD films

The main parameters that were varied in this study were: (i) PT (carried out or not), (ii) type of seeding, (iii) Ar (consequently, rate of flow of dopant) and CH_4/H_2 ratio in plasma, and (iv) chamber pressure during deposition. Each of these factors profoundly influenced the characteristics of the respective films.

Figures 21.3(a) and (b) show the films deposited on samples A1′ and A5′ (seeded with ND particles and diamond grit, respectively) after 240 min of deposition at 4.5 kPa and without PT. The PCD films are closed and highly adherent and have an average crystal size of 600 nm. The films deposited on PT samples, as well as films from series A3, reveal similar trends and are not shown. On the other hand, the films deposited on samples seeded with the scratch method (A4′) without PT showed areas where the film was not yet fully closed (figure 21.3(d)). In contrast, the samples seeded with the scratch method and given PT show closed films (figure 21.3(c)). These results suggest that the US seeding with ND particles or grit is more efficient that the scratch method and that the PT has a beneficial effect when the substrates are seeded with the scratch method.

The SEM images of samples A2 and A5 (figure 21.4), both given PT and deposited in the same run, show the impact of seeding with different diamond suspensions. The diamond grains on sample A5 (seeded with diamond grit) are 0.5–1.5 μm in size (figure 21.4(a)), whereas sample A2 (seeded with ND) shows an NCD film (grain size < 200 nm) with some larger (1.5 μm) crystals (figure 21.4(b)). These larger crystals were not observed in the samples that were not given PT, so they can be related with spontaneous diamond nucleation during the PT step.

The impact of the deposition pressure on the quality of the diamond deposits can be assessed by comparing samples A3 and B2 (figure 21.5). The seeding was performed with ND and the deposition parameters were the same with exception of the deposition pressure which was 4.5 kPa for A3 and 9 kPa for B2. Again, SEM images of the two reveal stark differences. Growth at a lower pressure leads to significantly smaller crystal size, as can be seen by comparing the images at higher

Table 21.4. Deposition conditions of second set of BDD films; samples not given PT are marked with ′.

Sample ID	Pre-treatment				Seeding	Growth			
	CH_4/H_2 (%)	T^* (K)	Duration (min)	P^* (kPa)		$CH_4/Ar/H_2$ (%)	T^* (K)	Duration (min)	P^* (kPa)
BDD_1/BDD_1′						1/10/100			
BDD_2/BDD_2′						2/10/100			
BDD_3/BDD_3′	3	1023	90	9.0	ND	3/10/100	1023	240	9.0
BDD_4/BDD_4′						4/10/100			
BDD_5/BDD_5′						5/10/100			

* T stands for temperature and P for pressure.

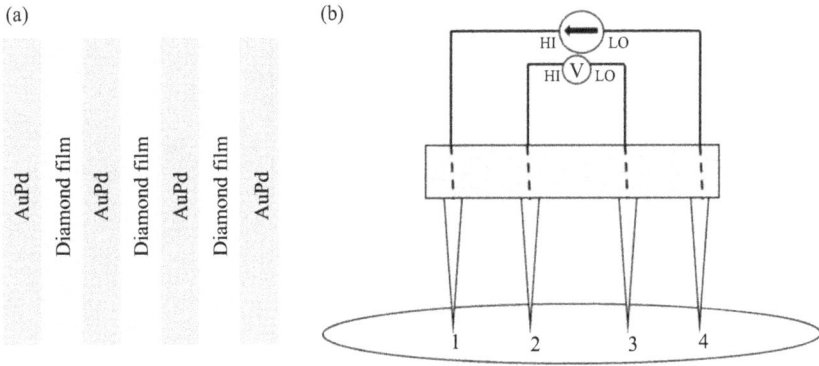

Figure 21.2. (a) Contacts used for four-point probe measurements. (b) Measurement set-up for the four-point probe technique.

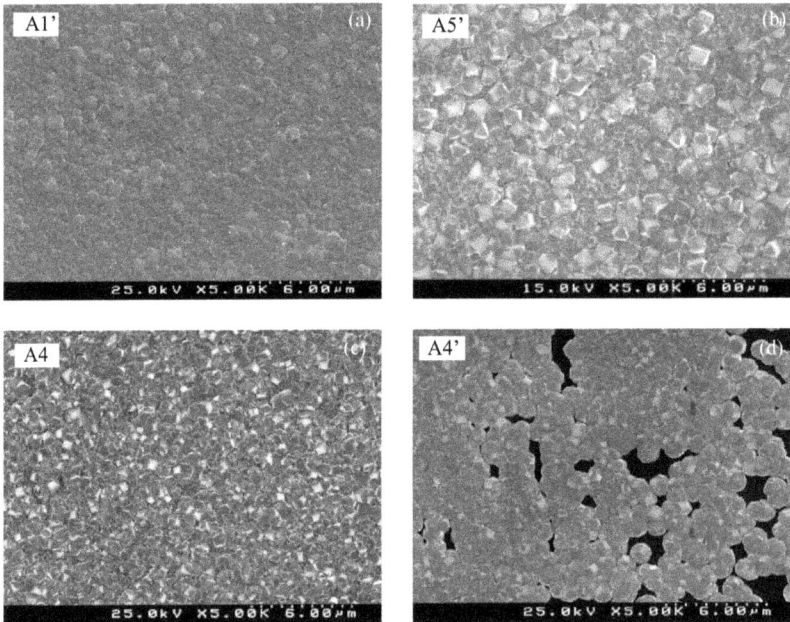

Figure 21.3. SEM images of samples (a) A1', (b) A5', (c) A4, and (d) A4'.

magnification in figures 21.5(a) and (b). As deposition pressure increases, so does the grain size, from a few nanometres at 4.5 kPa to a few microns at 9 kPa. The edges and corners of the diamond grains become more trenched, suggesting that growth happens in a larger number of steps, forming more step edges [189]. At high deposition pressures, the effect of the PT on the characteristics of the diamond deposits is also evident. Comparing samples B2 and B2' (PT and no-PT, respectively), it can be seen that the trenches observed in B2 are not present in B2' (figures 21.5(b) and (c)).

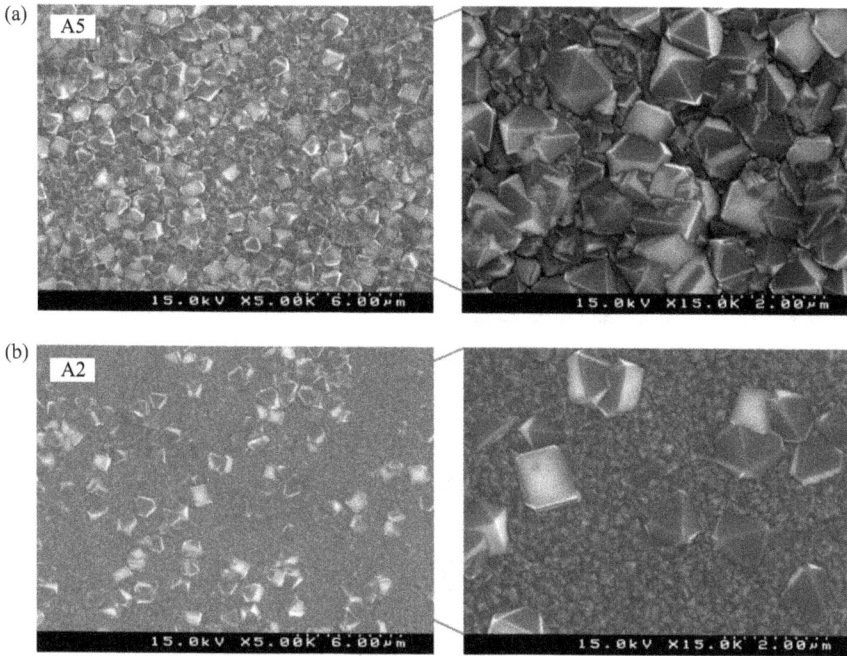

Figure 21.4. SEM images of samples (a) A5 and (b) A2.

The SEM images of samples B1 and B2 (figure 21.6) (grown with the same flow rate of Ar) show the impact of the CH_4 ratio on the film characteristics—a higher amount of CH_4 in the plasma leads to significantly smaller crystal size in sample B2, i.e. the grain size in sample B1 is $\approx$ 2–3 μm whereas in B2 it is smaller than 300 nm. This effect of decreasing crystal size with increasing CH_4 is enhanced in the presence of a chemically inert gas such as Ar. It has been suggested that the Ar in the gas mixture increases the degree of dissociation of CH_4, as a result of which the conditions in the HFCVD chamber are more likely to induce NCD growth [190, 191].

SEM images of samples A1 and A6 reveal the effect of the Ar ratio (and hence the flow rate of the dopant during the deposition process) on the film characteristics (figure 21.7). A higher Ar flow seems to promote the uniformity of the diamond deposits (figure 21.7(a)). Again the large isolated crystals that can be seen in the image are attributed to the spontaneous nucleation of diamond during the PT step.

The results above suggest that seeding with ND and depositing the diamond films with higher Ar flows improve the uniformness of the diamond deposits. In addition, the morphological impact of pre-treating the samples was most clear at 9 kPa. As a consequence, the seeding of the second set of samples was performed with ND and the films were deposited with 10% Ar and at a pressure of 9 kPa (varying only the CH_4/H_2 ratio).

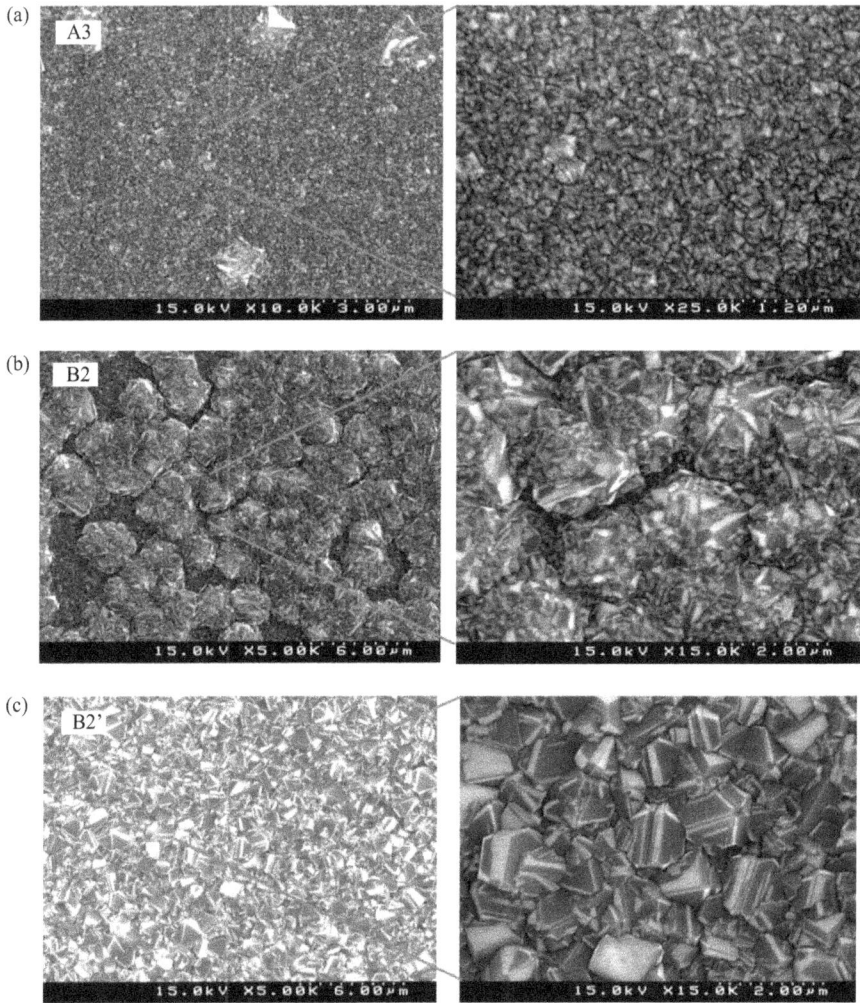

Figure 21.5. SEM images of samples (a) A3, (b) B2, and (c) B2′.

Figure 21.6. SEM images of samples (a) B1 and (b) B2.

Figure 21.7. SEM images of samples (a) A6 and (b) A1.

21.5.3 Electrical characteristics of the BDD films

The I–V characteristics of the as-sputtered and annealed Au and Au/Pd deposited on sample A3 are shown in figure 21.8. The resistance of both types of contacts decreases upon annealing, which translates into an increase of the current level. In addition, the current measured across the Au/Pd contacts is one order of magnitude higher than the current that flows across the Au contacts. The low current values (below mA level) are a consequence of the large electrode separation (3 mm).

The same type of contacts were deposited on the surface of the second set of films using the contact geometry represented in figure 21.2(a). By using four probes for thin film measurements, measurement errors due to probe resistance, spreading resistance under each probe, and contact resistance were eliminated. The sheet resistance of each film was then calculated from the I–V measurements and plotted in figure 21.9.

As the CH_4–H_2 ratio increases, the sheet resistance of the samples decreases (with the exception of the PT sample deposited with 3% CH_4/H_2). This result is related to the increase of the sp^2–sp^3 ratio in the films as a consequence of the CH_4 increase [153], since the non-sp^3 bonds provide an additional path through which current can flow. When comparing the PT and no-PT samples, it can be seen that the PT of the SiC substrate results in a lower sheet resistance, suggesting that the PT favours the incorporation of B atoms in the diamond film. It has been shown that exposing the substrate to diamond growth conditions prior to the seeding step creates a carbon film on the surface of the sample that enhances the lateral growth of the diamond grains in the early stages of deposition [5, 92]. This lateral growth decreases

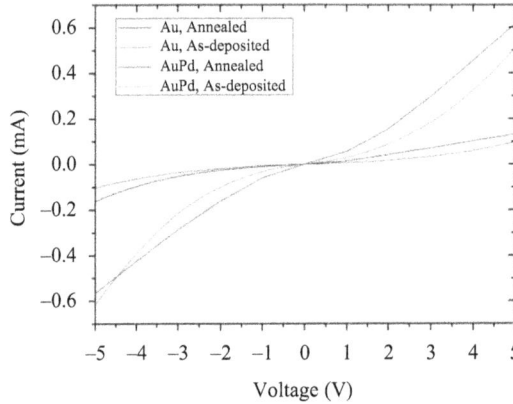

Figure 21.8. *I–V* curves of sample A3 with Au and Au/Pd ohmic contacts before and after annealing.

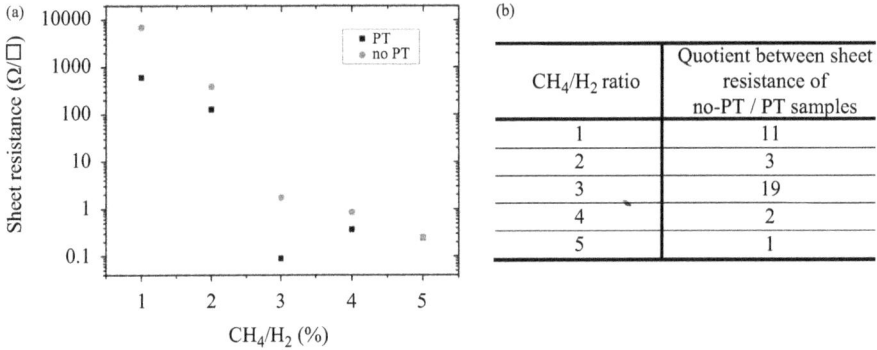

CH$_4$/H$_2$ ratio	Quotient between sheet resistance of no-PT / PT samples
1	11
2	3
3	19
4	2
5	1

Figure 21.9. (a) Sheet resistance of the PT and no-PT samples and (b) the ratio between the sheet resistance of PT and no-PT samples as a function of the CH$_4$–H$_2$ ratio.

the amount of grain boundaries in the film and increases the size of the crystals, in comparison to the sizes of the crystals grown without the PT step. The higher current levels obtained with the PT samples thus suggest that these larger diamond grains also favour the incorporation of B atoms in the PCD film.

The quotient between the sheet resistance of PT and no-PT samples (figure 21.9(b)) varies with the CH$_4$/H$_2$ content, however, this variation is not monotonic. For 1% CH$_4$/H$_2$, the sheet resistance of the no-PT sample is 11 times larger than the sheet resistance of the PT sample. This quotient decreases to 3, 2, and 1 for 2%, 4%, and 5% CH$_4$/H$_2$, respectively, suggesting that the beneficial effect of PT is reduced as the C supply from the gas phase increases. It has been observed that the lateral growth of the grains is more evident when the diamond films are deposited with low CH$_4$–H$_2$ ratios [92] so it is also expected that, for higher amounts of C from the gas phase, the enhancement of the lateral growth of the crystals is not so noticeable. As a consequence, the impact of pre-treating the sample decreases as the CH$_4$/H$_2$ content increases.

Surprisingly, the quotient between the sheet resistance of the no-PT and PT samples increases to 19 for 3% CH_4/H_2, meaning that the impact of pre-treating the sample is maximum under these conditions. The sheet resistance of the PT sample also reaches its minimum value, increasing to higher values for 4% and 5% CH_4/H_2. This apparently unexpected result is not attributed to experimental artefacts. Instead, it may reflect the complexity of the chemistry during the deposition. Even though the Ar flow (and consequently the rate at which the dopant atoms are supplied to the gas phase) was kept constant, it is known that the concentration of CH_4 in the chamber impacts not only the amount of non-sp^3 bonds but also the incorporation of B atoms in the diamond film [95] and hence the conductivity of the films. In addition, there is an extra supply of carbon atoms during the initial stages of diamond deposition from the PT film—which favours the lateral growth of the diamond grains and also impacts the incorporation of the B atoms. In the case of undoped PCD films, the lateral growth of the diamond grains is more evident when the supply of C from the gas phase is lower, so it would be expected that the impact of the PT was higher for lower CH_4/H_2. However, in the current case other chemical reactions occur in the deposition chamber due to the presence of the Ar and $B_2O_3^-$ and the enhancement of the lateral growth of the diamond grains may show a more complex dependence of the CH_4/H_2 content. In order to fully understand the dependence of the sheet resistance with the deposition conditions, future work will include the characterization of the doped films with Hall effect measurements (to determine the carrier concentration), as well as x-ray diffraction and transmission electron microscopy imaging at the SiC–diamond interface. Finally, the I–V response of the p-type diamond films/n-type SiC heterojunctions will also be evaluated.

21.6 Conclusions

SiC and diamond are two WBG materials that have received a lot of attention from the scientific community. SiC devices have already asserted themselves as a solid alternative for high power/high temperature electronic applications, such as the electric vehicle industry. Diamond, even though it has the ultimate properties such as an extremely high bandgap and thermal conductivity, has failed to conquer the market due mostly to the unavailability of large-area electronic-grade wafers.

Diamond and SiC can also be combined in order to fabricate hybrid devices. For example, the deposition of diamond films directly on the SiC surface may provide an alternative way of passivating SiC devices. SiC/diamond heterojunctions may be fabricated by growing p-type diamond films on n-type SiC substrates. NCD-SiC heterojunctions with 10^{-8} A cm^{-2} leakage current and a curvature coefficient of 10^5 V^{-1} have already been fabricated.

This chapter also presents results concerning the deposition of p-type PCD films on n-type SiC substrates and discusses the impact of the deposition parameters on the morphology and conductivity of the diamond films. In particular, the exposure of the SiC samples to deposition conditions prior to the seeding step is shown to have a positive impact on the conductivity of the diamond films.

References

[1] Armstrong K O, Das S and Cresko J 2017 Wide bandgap semiconductor opportunities in power electronics Oak Ridge National Laboratory ORNL/TM-2017/702 https://info.ornl.gov/sites/publications/Files/Pub104869.pdf

[2] Armstrong K O, Das S and Cresko J 2016 Wide bandgap semiconductor opportunities in power electronics *4th IEEE Work Wide Bandgap Power Devices Appl.* pp 259–64

[3] Rashid M H 2018 *Power Electronics Handbook* 4th edn (Amsterdam: Elsevier)

[4] Power P 2018 About the SiC MOSFETs modules in TESLA MODEL 3 (accessed 12 Dec 2019) https://www.pntpower.com/tesla-model-3-powered-by-st-microelectronics-sic-mosfets/

[5] Rotter S Z and Madaleno J C 2009 Diamond CVD by a combined plasma pretreatment and seeding procedure *Chem. Vap. Depos.* **15** 209–16

[6] Roschke M, Schwierz F, Paasch G and Schipanski D 2009 Evaluating the three common SiC polytypes for MESFET applications *Mater. Sci. Forum* **264–268** 965–8

[7] Rosina M 2018 GaN and SiC power device: market overview, remaining challenges *Semicond. Eur.* (Yole Développement) (accessed 17 Nov 2019) http://www1.semi.org/eu/sites/semi.org/files/events/presentations/02_MilanRosina_Yole.pdf

[8] Neudeck P G 2006 Silicon carbide technology *VLSI Handbook* 2nd edn (Boca Raton, FL: CRC Press) pp 5-1–34

[9] Wort C J H and Balmer R S 2008 Diamond as an electronic material *Mater. Today* **11** 22–8

[10] Okumura K, Hase N, Ino K, Nakamura T and Tanimura M 2012 Ultra low on-resistance SiC trench devices *Power Semicond.* **4** 22–5

[11] Shur M, Rumyantsev S L and Levinshtein M E 2006 *SiC Materials and Devices* (Singapore: World Scientific)

[12] Research A M 2019 *World Silicon Carbide (SIC) Market—Opportunities and Forecasts* pp 2019–26 (Yole Développement) (accessed 17 Nov 2019) https://alliedmarketresearch.com/silicon-carbide-market

[13] Funaki T, Balda J C, Junghans J, Kashyap A S, Mantooth H A, Barlow F, Kimoto T and Hikihara T 2007 Power conversion with SiC devices at extremely high ambient temperatures *IEEE Trans. Power Electron.* **22** 1321–9

[14] Abou-Alfotouh A M, Radun A V, Chang H R and Winterhalter C 2006 A 1-MHz hard-switched silicon carbide dc–dc converter *IEEE Trans. Power Electron.* **21** 880–9

[15] Colak I, Kabalci E, Fulli G and Lazarou S 2015 A survey on the contributions of power electronics to smart grid systems *Renew. Sustain. Energy Rev.* **47** 562–79

[16] Zhao F and Islam M M 2010 Optically activated SiC power transistors for pulsed-power application *IEEE Electron Device Lett.* **31** 1146–8

[17] Chinthavali M, Tolbert L M, Zhang H, Han J H, Barlow F and Ozpineci B 2010 High power SiC modules for HEVs and PHEVs *2010 Int. Power Electron. Conf.—ECCE Asia* **vol 6** pp 1842–8

[18] Zhang H, Tolbert L M and Ozpineci B 2011 Impact of SiC devices on hybrid electric and plug-in hybrid electric vehicles *IEEE Trans. Ind. Appl.* **47** 912–21

[19] Hamada K 2012 SiC device and power module technologies for environmentally friendly vehicles *7th Int. Conf. Integr. Power Electron. Syst.* pp 1–6

[20] Mino K, Yamada R, Kimura H and Matsumoto Y 2014 Power electronics equipments applying novel SiC power semiconductor modules *2014 Int. Power Electron. Conf. IPEC, Hiroshima* pp 1920–4

[21] Mills L, Castellazzi A, Lopez-Arevalo S, De D, Gurpinar E and Li J 2014 Testing of a lightweight SiC power module for avionic applications *7th IET International Conference on Power Electronics, Machines and Drives* 1.1.03

[22] Whitaker B *et al* 2014 High-temperature SiC power module with integrated SiC gate drivers for future high-density power electronics applications *2nd IEEE Work Wide Bandgap Power Devices Appl.* pp 36–40

[23] Kimoto T 2010 SiC technologies for future energy electronics *Dig. Tech. Pap.—Symp. VLSI Technol.* pp 9–14

[24] Kimoto T, Kawahara K, Niwa H, Okuda T and Suda J 2013 Junction technology in SiC for high-voltage power devices *Ext. Abstr. 13th Int. Work. Junction Technol. 2013, IWJT 2013* pp 54–7

[25] Sheng K, Yu L C, Zhang J and Zhao J H 2006 High temperature characterization of SiC BJTs for power switching applications *Solid. State. Electron.* **50** 1073–9

[26] Weitzel C E, Palmour J W, Carter C H, Moore K, Nordquist K J, Alien S, Thero C and Bhatnagar M 1996 Silicon carbide high-power devices *IEEE Trans. Electron. Dev.* **43** 1732–41

[27] Pan S, Li L and Chen Z 2011 Research of solar inverter based on silicon carbide JFET power device *Energy Procedia* **16** 1986–93

[28] Zhong X, Zhang L, Xie G, Guo Q, Wang T and Sheng K 2013 High temperature physical modeling and verification of a novel 4H-SiC lateral JFET structure *Microelectron. Reliab.* **53** 1848–56

[29] Othman D, Berkani M, Lefebvre S, Ibrahim A, Khatir Z and Bouzourene A 2012 Comparison study on performances and robustness between SiC MOSFET & JFET devices—abilities for aeronautics application *Microelectron. Reliab.* **52** 1859–64

[30] Othman D, Lefebvre S, Berkani M, Khatir Z, Ibrahim A and Bouzourene A 2013 Robustness of 1.2 kV SiC MOSFET devices *Microelectron. Reliab.* **53** 1735–8

[31] Zhu H, Chen X, Cai J and Wu Z 2009 4H-SiC ultraviolet avalanche photodetectors with low breakdown voltage and high gain *Solid. State. Electron.* **53** 7–10

[32] Roccaforte F, Giannazzo F and Raineri V 2010 Nanoscale transport properties at silicon carbide interfaces *J. Phys. D: Appl. Phys.* **43** 223001 1–19

[33] Millán J 2007 Wide band-gap power semiconductor devices *IET Circuits Devices Syst.* **1** 372–9

[34] Matocha K 2008 Challenges in SiC power MOSFET design *Solid. State. Electron.* **52** 1631–5

[35] Ryu S H *et al* 2011 3.7 mΩ-cm^2, 1500 V 4H-SiC DMOSFETs for advanced high power, high frequency applications *Proc. Int. Symp. Power Semicond. Devices ICs* pp 227–30

[36] McGarrity J M, McLean F B, DeLancey W M, Palmour J, Carter C, Edmond J and Oakley R E 1992 Silicon carbide JFET radiation response *IEEE Trans. Nucl. Sci.* **39** 1974–81

[37] Zetterling C M *et al* 2017 Bipolar integrated circuits in SiC for extreme environment operation *Semicond. Sci. Technol.* **32** 034002

[38] Shenoy J N, Cooper J A and Melloch M R 1997 High-voltage double-implanted power MOSFETs in 6H-SiC *IEEE Trans. Electron. Dev.* **18** 93–5

[39] Sugawara Y and Asano K 1998 1.4 kV 4H-SiC UMOSFET with low specific on-resistance *Proc. 10th Int. Symp. Power Semicond. Devices ICs ISPSD* pp 119–22

[40] Saha A and Cooper J A 2007 A 1-kV 4H-SiC power DMOSFET optimized for low ON-resistance *IEEE Trans. Electron. Dev.* **54** 2786–91

[41] Alok D, Arnold E, Egloff R, Barone J, Murphy J, Conrad R and Burke J 2001 4H-SiC RF power MOSFETs *IEEE Electron. Dev. Lett.* **22** 577–8

[42] Bellone S and Benedetto L D 2014 Design and performances of 4H-SiC bipolar mode field effect transistor (BMFETs) *IEEE Trans. Power Electron.* **29** 2174–9

[43] Anthony P, McNeill N and Holliday D 2014 High-speed resonant gate driver with controlled peak gate voltage for silicon carbide MOSFETs *IEEE Trans. Ind. Appl.* **50** 573–83

[44] Friedrichs P, Mitlehner H, Schörner R, Dohnke K O and Stephani D 2009 High-voltage modular switch based on SiC VJFETs—first results for a fast 4.5 kV/1.2 Ω configuration *Mater. Sci. Forum* **433–436** 793–6

[45] Zhang H-R C Q 2005 10 kV trench gate IGBTs on 4H-SiC *Proc. ISPSD '05–17th IEEE Int. Symp. Power Semicond. Devices ICs* pp 303–6

[46] Agarwal A K, Casady J B, Rowland L B, Seshadri S, Siergiej R R, Valek W F and Brandt C D 1997 700-V asymmetrical 4H-SiC gate turn-off thyristors (GTO's) *IEEE Electron Device Lett.* **18** 518–20

[47] Monolith SemiconductorTexas X-F, Institute R P and Center U T R 2014 A new model for wide bandgap semiconductor manufacturing in the US (accessed 17 Nov 2019) https://arpa-e.energy.gov/?q=impact-sheet/monolith-semiconductor-switches

[48] Adan A O, Tanaka D, Burgyan L L and Kakizaki Y 2019 The current status and trends of 1,200-V commercial silicon-carbide MOSFETs: deep physical analysis of power transistors from a designer's perspective *IEEE Power Electron. Mag.* **6** 36–47

[49] Horowitz K, Remo T and Reese S 2017 A manufacturing cost and supply chain analysis of SiC power electronics applicable to medium-voltage motor drives *Natl Renew. Energy Lab Technical Report NREL/TP-6A20-67694* (accessed 6 Dec 2019)

[50] Roccaforte F, Fiorenza P, Greco G, Vivona M, Lo Nigro R, Giannazzo F, Patti A and Saggio M 2014 Recent advances on dielectrics technology for SiC and GaN power devices *Appl. Surf. Sci.* **301** 9–18

[51] Afanas'ev V V, Ciobanu F, Dimitrijev S, Pensl G and Stesmans A 2004 Band alignment and defect states at SiC/oxide interfaces *J. Phys. Condens. Matter* **16** S1839–56

[52] Jamet P, Dimitrijev S and Tanner P 2001 Effects of nitridation in gate oxides grown on 4H-SiC *J. Appl. Phys.* **90** 5058–63

[53] Guy O J *et al* 2008 Investigation of the 4H-SiC surface *Appl. Surf. Sci.* **254** 8098–105

[54] Neudeck P G, Okojie R S and Chen L Y 2002 High-temperature electronics—a role for wide bandgap semiconductors? *Proc. IEEE* **90** 1065–76

[55] Liu W, Zetterling C and Ostling M 2004 Thermal-issues for design of high power SiC MESFETs *HDP '04. Proc. Sixth IEEE CPMT Conf. High Density Microsyst. Des. Packag. Compon. Fail. Anal.* pp 31–5

[56] Lutz J and Baburske R 2014 Some aspects on ruggedness of SiC power devices *Microelectron. Reliab.* **54** 49–56

[57] Fayyaz A, Yang L, Riccio M, Castellazzi A and Irace A 2014 Single pulse avalanche robustness and repetitive stress ageing of SiC power MOSFETs *Microelectron. Reliab.* **54** 2185–90

[58] Bratcher M and Whitworth R J Y B 1996 Aluminum nitride package for high-temperature applications *Proc. 3rd Int. High-Temperature Electron. Conf. (Albuquerque, NM)* pp 21–6

[59] Salmon J S and Palmer R W J M 1998 Thick film hybrid packaging techniques for 500 °C operation *Proc. 4th Int. High Temperature Electron. Conf. (Albuquerque, NM)* pp 103–8

[60] Liang Z, Wang F and Tolbert L 2014 Development of packaging technologies for advanced SiC power modules *2nd IEEE Work Wide Bandgap Power Devices Appl. WiPDA 2014* pp 42–7

[61] Zhang H, Ang S S, Mantooth H A and Krishnamurthy S 2013 A high temperature, double-sided cooling SiC power electronics module *2013 IEEE Energy Convers. Congr. Expo. ECCE 2013* pp 2877–83

[62] Senesky D G, Cheng K B, Pisano A P and Jamshidi B 2009 Harsh environment silicon carbide sensors for health and performance monitoring of aerospace systems: a review *IEEE Sens. J.* **9** 1472–8

[63] Derjaguin B V, Fedoseev D V, Lukyanovich V M, Spitzin B V, Ryabov V A and Lavrentyev A V 1968 Filamentary diamond crystals *J. Cryst. Growth* **2** 380–4

[64] Angus J C and Hayman C C 1988 Low-pressure, metastable growth of diamond and 'diamondlike' phases *Science* **241** 913–21

[65] Iijima S, Aikawa Y and Baba K 1990 Early formation of chemical vapor deposition diamond films *Appl. Phys. Lett.* **57** 2646–8

[66] Mitsuda Y, Kojima Y, Yoshida T and Akashi K 1987 The growth of diamond in microwave plasma under low pressure *J. Mater. Sci.* **22** 1557–62

[67] Williams O A, Douhéret O, Daenen M, Haenen K, Osawa E and Takahashi M 2007 Enhanced diamond nucleation on monodispersed nanocrystalline diamond *Chem. Phys. Lett.* **445** 255–8

[68] Kromka A, Potocký Š, Čermák J, Rezek B, Potměšil J, Zemek J and Vaněček M 2008 Early stage of diamond growth at low temperature *Diam. Relat. Mater.* **17** 1252–5

[69] Shenderova O, Hens S and McGuire G 2010 Seeding slurries based on detonation nanodiamond in DMSO *Diam. Relat. Mater.* **19** 260–7

[70] Butler J E and Sumant A V 2008 The CVD of nanodiamond materials *Chem. Vap. Depos* **14** 145–60

[71] Mallik A K, Mendes J C, Rotter S Z and Bysakh S 2014 Detonation nanodiamond seeding technique for nucleation enhancement of CVD diamond—some experimental insights *Adv. Ceram. Sci. Eng* **3** 36–45

[72] Krueger A and Lang D 2012 Functionality is key: recent progress in the surface modification of nanodiamond *Adv. Funct. Mater.* **22** 890–906

[73] Mandal S *et al* 2017 Surface zeta potential and diamond seeding on gallium nitride films *ACS Omega* **2** 7275–80

[74] Nicley S S, Drijkoningen S, Pobedinskas P, Raymakers J, Maes W and Haenen K 2019 Growth of boron-doped diamond films on gold-coated substrates with and without gold nanoparticle formation *Cryst. Growth Des.* **19** 3567–75

[75] Yugo S, Kanai T, Kimura T, Muto T, Yugo S, Kanai T, Kimura T and Muto T 2001 Generation of diamond nuclei by electric field in plasma chemical vapor deposition *Appl. Phys. Lett.* **58** 1036–8

[76] Stoner B R, Ma G H M, Wolter S D and Glass J T 1992 Characterization of bias-enhanced nucleation of diamond on silicon by invacuo surface analysis and transmission electron microscopy *Phys. Rev.* B **45** 11067–84

[77] Pecoraro S, Arnault J C and Werckmann J 2005 BEN-HFCVD diamond nucleation on Si (111) investigated by HRTEM and nanodiffraction *Diam. Relat. Mater.* **14** 137–43

[78] Mayr M, Stehl C, Fischer M, Gsell S and Schreck M 2014 Correlation between surface morphology and defect structure of heteroepitaxial diamond grown on off-axis substrates *Phys. Status Solidi Appl. Mater. Sci.* **211** 2257–63

[79] Lee K H *et al* 2016 Epitaxy of iridium on SrTiO₃/Si (001): a promising scalable substrate for diamond heteroepitaxy *Diam. Relat. Mater.* **66** 67–76

[80] Kawashima H, Noguchi H, Matsumoto T, Kato H, Ogura M, Makino T, Shirai S, Takeuchi D and Yamasaki S 2015 Electronic properties of diamond Schottky barrier diodes fabricated on silicon-based heteroepitaxially grown diamond substrates *Appl. Phys. Express* **8** 104103

[81] Kubovic M, Aleksov A, Schreck M, Bauer T, Stritzker B and Kohn E 2003 Field effect transistor fabricated on hydrogen-terminated diamond grown on SrTiO₃ substrate and iridium buffer layer *Diam. Relat. Mater.* **12** 403–7

[82] Kusterer J, Schmid P and Kohn E 2006 Mechanical microactuators based on nanocrystalline diamond films *New Diam. Front. Carbon Technol.* **16** 295–321

[83] Singh M K, Titus E, Madaleno J C, Cabral G and Gracio J 2008 Novel two-step method for synthesis of high-density nanocrystalline diamond fibers *Chem. Mater.* **20** 1725–32

[84] Bogdanowicz R, Sobaszek M, Ryl J, Gnyba M, Ficek M, Gołuński Ł, Bock W J, mietana M and Darowicki K 2015 Improved surface coverage of an optical fibre with nanocrystalline diamond by the application of dip-coating seeding *Diam. Relat. Mater.* **55** 52–63

[85] Fox N A, Youh M J, Steeds J W and Wang W N 2000 Patterned diamond particle films *J. Appl. Phys.* **87** 8187–91

[86] Chen Y C, Tzeng Y, Cheng A J, Dean R, Park M and Wilamowski B M 2009 Inkjet printing of nanodiamond suspensions in ethylene glycol for CVD growth of patterned diamond structures and practical applications *Diam. Relat. Mater.* **18** 146–50

[87] Sartori A F, Belardinelli P, Dolleman R J, Steeneken P G, Ghatkesar M K and Buijnsters J G 2019 Inkjet-printed high-Q nanocrystalline diamond resonators *Small* **15** 1803774

[88] Hees J, Kriele A and Williams O A 2011 Electrostatic self-assembly of diamond nano-particles *Chem. Phys. Lett.* **509** 12–5

[89] Yoshikawa T, Gao F, Zuerbig V, Giese C, Nebel C E, Ambacher O and Lebedev V 2016 Pinhole-free ultra-thin nanocrystalline diamond film growth via electrostatic self-assembly seeding with increased salt concentration of nanodiamond colloids *Diam. Relat. Mater.* **63** 103–7

[90] Rotter S 1998 Applications of conformal CVD diamond films *Isr. J. Chem.* **38** 135–40

[91] Kyatam S, Mukherjee D, Silva A, Alves L, Rotter S, Neto M, Oliveira F, Silva R, Neto H and Mendes J C 2019 CVD diamond films for thermal management applications *IEEE Conf. Microwaves, Commun. Antennas Electron. Syst. (Tel Aviv, Israel)*

[92] Mukherjee D, Oliveira F, Trippe S C, Rotter S, Neto M, Silva R, Mallik A K, Haenen K, Zetterling C-M and Mendes J C 2020 Deposition of diamond films on single crystalline silicon carbide substrates *Diam. Relat. Mater.* **101** 107625

[93] Fhaner M, Zhao H, Bian X, Galligan J J and Swain G M 2011 Improvements in the formation of boron-doped diamond coatings on platinum wires using the novel nucleation process (NNP) *Diam. Relat. Mater.* **20** 75–83

[94] Van der Drift A 1967 Evolutionary selection, a principle governing growth orientation in vapour-deposited layers *Philips Res. Rep.* **22** 267

[95] Deneuville A 2003 Boron doping of diamond films from the gas phase *Semiconductor Semimetals* ed C E Nebel and J Ristein (Amsterdam: Elsevier) pp 183–238

[96] Collins A T 1989 Diamond electronic devices—a critical appraisal *Semicond. Sci. Technol.* **4** 605–11

[97] Borst T H, Strobel S and Weis O 1995 High-temperature diamond p–n junction: B-doped homoepitaxial layer on N-doped substrate *Appl. Phys. Lett.* **67** 2651–3

[98] Vavilov V S, Gippius A A and Konorova E A 1985 *Electron and Optical Processes in Diamond* (Moscow: Moscow Izdatel Nauka)

[99] Miyazaki T, Kato H, Okushi H and Yamasaki S 2006 *Ab initio* energetics of phosphorus impurity in subsurface regions of hydrogenated diamond surfaces *Surf. Sci. Nanotechnol* **4** 124–8

[100] Yamamoto T, Janssens S D, Ohtani R, Takeuchi D and Koizumi S 2016 Toward highly conductive n-type diamond: incremental phosphorus-donor concentrations assisted by surface migration of admolecules *Appl. Phys. Lett.* **109** 1821021–5

[101] Kociniewski T *et al* 2006 N-type CVD diamond doped with phosphorus using the MOCVD technology for dopant incorporation *Phys. Status Solidi Appl. Mater. Sci.* **203** 3136–41

[102] Kato H, Makino T, Yamasaki S and Okushi H 2007 N-type diamond growth by phosphorus doping on (001)-oriented surface *J. Phys. D: Appl. Phys.* **40** 6189–200

[103] Temahuki N, Gillet R, Sallet V, Jomard F, Chikoidze E, Dumont Y, Pinault-Thaury M A and Barjon J 2017 New process for electrical contacts on (100) N-type diamond *Phys. Status Solidi Appl. Mater. Sci.* **214** 1700466

[104] Maida O, Tada S, Nishio H and Ito T 2015 Substrate temperature optimization for heavily-phosphorus-doped diamond films grown on vicinal (001) surfaces using high-power-density microwave-plasma chemical-vapor-deposition *J. Cryst. Growth* **424** 33–7

[105] Haenen K, Lazea A, Barjon J, D'Haen J, Habka N, Teraji T, Koizumi S and Mortet V 2009 P-doped diamond grown on (110)-textured microcrystalline diamond: growth, characterization and devices *J. Phys. Condens. Matter* **21** 364204

[106] Vlckova Zivcova Z, Frank O, Drijkoningen S, Haenen K, Mortet V and Kavan L 2016 N-type phosphorus-doped nanocrystalline diamond: electrochemical and *in situ* Raman spectroelectrochemical study *RSC Adv.* **6** 51387–93

[107] Gildenblat G S, Grot S A and Badzian A 1991 The electrical properties and device applications of homoepitaxial and polycrystalline diamond films *Proc. IEEE* **79** 647–68

[108] Kawarada H 1996 Hydrogen-terminated diamond surfaces and interfaces *Surf. Sci. Rep.* **26** 205–6

[109] Collins A T and Williams A W S 1971 The nature of the acceptor centre in semiconducting diamond *J. Phys. C: Solid State Phys.* **4** 1789–800

[110] Ekimov E A, Sidorov V A, Rakhmanina A V, Mel N N, Sadykov R A and Thompson J D 2006 High-pressure synthesis and characterization of superconducting boron-doped diamond *Sci. Technol. Adv. Mater.* **7** S2–6

[111] Kudryavtsev O S, Khomich A A, Sedov V S, Ekimov E A and Vlasov I I 2018 Fluorescence and Raman spectroscopy of doped nanodiamonds *J. Appl. Spectrosc.* **85** 295–9

[112] Prins J F 2003 Ion implantation of diamond for electronic applications *Semicond. Sci. Technol.* **18** S27–33

[113] Ohmagari S 2018 Doping and semiconductor characterizations *Power Electronics Device Applications of Diamond Semiconductors* ed S Koizumi, H U Pernot and M Suzuki (Cambridge: Woodhead Publishing) pp 99–189

[114] Yamada H 2019 Diamond *Single Crystals of Electronic Materials* ed R Fornari (Amsterdam: Elsevier) pp 331–50

[115] Fontaine F, Uzan-Saguy C, Philosoph B and Kalish R 1996 Boron implantation/*in situ* annealing procedure for optimal p-type properties of diamond *Appl. Phys. Lett.* **68** 2264–6

[116] Kalish R, Uzan-Saguy C, Samoiloff A, Locher R and Koidl P 1994 Doping of polycrystalline diamond by boron ion implantation *Appl. Phys. Lett.* **64** 2532–4

[117] Aleksov A, Vescan A, Kunze M, Gluche P, Ebert W, Kohn E, Bergmeier A and Dollinger G 2002 Diamond junction FETs based on δ-doped channels *Diam. Relat. Mater.* **8** 941–5

[118] Ohmagari S, Ogura M, Umezawa H and Mokuno Y 2017 Lifetime and migration length of B-related admolecules on diamond {100}-surface: comparative study of hot-filament and microwave plasma-enhanced chemical vapor deposition *J. Cryst. Growth* **479** 52–8

[119] Ohmagari S, Srimongkon K, Yamada H, Umezawa H, Tsubouchi N, Chayahara A, Shikata S and Mokuno Y 2015 Low resistivity p+ diamond (100) films fabricated by hot-filament chemical vapor deposition *Diam. Relat. Mater.* **58** 110–4

[120] Okushi H 2001 High quality homoepitaxial CVD diamond for electronic devices *Diam. Relat. Mater.* **10** 281–8

[121] Borst T H and Weis O 1995 Electrical characterization of homoepitaxial diamond films doped with B, P, Li and Na during crystal growth *Diam. Relat. Mater.* **4** 948–53

[122] Gurbuz Y, Kang W P, Davidson J L, Member S, Kerns D V and Zhou Q 2005 PECVD diamond-based high performance power diodes power electron *IEEE Trans.* **20** 1–10

[123] Vescan A, Daumiller I, Gluche P, Ebert W and Kohn E 1998 High temperature, high voltage operation of diamond Schottky diode *Diam. Relat. Mater.* **7** 581–4

[124] Aleksov A, Denisenko A and Kohn E 2000 Prospects of bipolar diamond devices *Solid. State. Electron.* **44** 369–75

[125] Brezeanu M *et al* 2007 Single crystal diamond M–i–P diodes for power electronics *IET Circuits, Devices Syst.* **1** 380–6

[126] Kang W P, Davidson J L, Wisitsora-At A, Wong Y M, Takalkar R, Holmes K and Kerns D V 2004 Diamond vacuum field emission devices *Diam. Relat. Mater.* **13** 1944–8

[127] Landstrass M I and Ravi K V 1989 Resistivity of chemical vapor deposited diamond films *Appl. Phys. Lett.* **55** 975–7

[128] Kawarada H, Aoki M and Ito M 1994 Enhancement mode metal–semiconductor field effect transistors using homoepitaxial diamonds *Appl. Phys. Lett.* **65** 1563–5

[129] Ueda K, Kasu M, Yamauchi Y, Makimoto T, Schwitters M, Twitchen D J, Scarsbrook G A and Coe S E 2006 Diamond FET using high-quality polycrystalline diamond with f_T of 45 GHz and f_{max} of 120 GHz *IEEE Electron Device Lett.* **27** 570–2

[130] Matsudaira H, Miyamoto S, Ishizaka H, Umezawa H and Kawarada H 2004 Over 20-GHz cutoff frequency submicrometer-gate diamond MISFETs *IEEE Electron Device Lett.* **25** 480–2

[131] Kasu M, Ueda K, Ye H, Yamauchi Y, Sasaki S and Makimoto T 2005 2 W/mm output power density at 1 GHz for diamond FETs *Electron. Lett.* **41** 1249

[132] Verona C, Ciccognani W, Colangeli S, Pietrantonio F D, Giovine E, Limiti E, Marinelli M and Verona-Rinati G 2015 Gate–source distance scaling effects in H-terminated diamond MESFETs *IEEE Trans. Electron. Dev.* **62** 1150–6

[133] Faqir M, Batten T, Mrotzek T, Knippscheer S, Massiot M, Buchta M, Blanck H, Rochette S, Vendier O and Kuball M 2012 Improved thermal management for GaN power electronics: silver diamond composite packages *Microelectron. Reliab.* **52** 3022–5

[134] Fabis P M 2002 The processing technology and electronic packaging of CVD diamond: a case study for GaAs/CVD diamond plastic packages *Microelectron. Reliab.* **42** 233–52

[135] Dumka D C, Chou T M, Faili F, Francis D and Ejeckam F 2013 AlGaN/GaN HEMTs on diamond substrate with over 7 W/mm output power density at 10 GHz *Electron. Lett.* **49** 1298–9

[136] Zhu X and Aslam D M 2006 CVD diamond thin film technology for MEMS packaging *Diam. Relat. Mater.* **15** 254–8

[137] Zhu X, Aslam D M and Sullivan J P 2006 The application of polycrystalline diamond in a thin film packaging process for MEMS resonators *Diam. Relat. Mater.* **15** 2068–72

[138] Su Q, Liu J, Wang L, Shi W and Xia Y 2006 Efficient CVD diamond film/alumina composite substrate for high density electronic packaging application *Diam. Relat. Mater.* **15** 1550–4

[139] Excellent Diamond Products (n.d.). http://d-edp.jp/en/technology-2.html (accessed 6 February 2020)

[140] Koizumim S, Umezawa H, Pernot J and Suzuki M (ed) 2018 Diamond wafer technologies for semiconductor device applications *Power Electronics Device Applications of Diamond Semiconductors* (Cambridge: Woodhead Publishing) pp 1–97

[141] Shikata S 2016 Single crystal diamond wafers for high power electronics *Diam. Relat. Mater.* **65** 168–75

[142] Yamasaki S and Nemanich R J 2018 Strategies for diamond power device applications *Power Electronics Device Applications of Diamond Semiconductors* ed S Koizumim, H Umezawa, J Pernot and M Suzuki (Cambridge: Woodhead Publishing) pp 200–18

[143] Teraji T 2018 Homoepitaxial growth of ultrahigh purity diamond films *Power Electronics Device Applications of Diamond Semiconductors* ed S Koizumim, H Umezawa, J Pernot and M Suzuki (Cambridge: Woodhead Publishing) pp 27–40

[144] Schreck M and Arnault J-C 2018 Heteroepitaxy of diamond on Ir/metal-oxide/Si substrates *Power Electronics Device Applications of Diamond Semiconductors* ed S Koizumim, H Umezawa, J Pernot and M Suzuki (Cambridge: Woodhead Publishing) pp 58–80

[145] Yaita J, Iwasaki T and Hatano M 2018 Heteroepitaxy of diamond on SiC *Power Electronics Device Applications of Diamond Semiconductors* ed S Koizumim, H Umezawa, J Pernot and M Suzuki (Cambridge: Woodhead Publishing) pp 81–98

[146] Magyar A, Hu W, Shanley T, Flatté M E, Hu E and Aharonovich I 2014 Synthesis of luminescent europium defects in diamond *Nat. Commun.* **5** 3523

[147] Isberg J, Gabrysch M, Hammersberg J, Majdi S, Kovi K K and Twitchen D J 2013 Generation, transport and detection of valley-polarized electrons in diamond *Nat. Mater.* **12** 760

[148] Saxler A W 2006 Silicon carbide on diamond substrates and related devices and methors *US Patent* 7033912 B2

[149] Harris C, Janzen E and Konstantinov A 1997 Semiconductor device having a passivation layer *US Patent* 5650638

[150] Harris C I, Konstantinov A O, Hallin C and Janzén E 2002 SiC power device passivation using porous SiC *Appl. Phys. Lett.* **66** 1501–2

[151] Zelmat S, Locatelli M L, Lebey T and Diaham S 2006 Investigations on high temperature polyimide potentialities for silicon carbide power device passivation *Microelectron. Eng.* **83** 51–4

[152] Hallén A, Usman M, Suvanam S, Henkel C, Martin D and Linnarsson M K 2014 Passivation of SiC device surfaces by aluminum oxide *IOP Conf. Ser.: Mater. Sci. Eng.* **56** 012007

[153] Mendes J C, Gomes H L, Trippe S C, Mukherjee D and Pereira L 2019 Small signal analysis of MPCVD diamond Schottky diodes *Diam. Relat. Mater.* **93** 131–8

[154] Liu W L, Shamsa M, Calizo I, Balandin A A, Ralchenko V, Popovich A and Saveliev A 2006 Thermal conduction in nanocrystalline diamond films: effects of the grain boundary scattering and nitrogen doping *Appl. Phys. Lett.* **89** 89–91

[155] Anaya J *et al* 2017 Simultaneous determination of the lattice thermal conductivity and grain/grain thermal resistance in polycrystalline diamond *Acta Mater.* **139** 215–25

[156] Persson C and Lindefelt U 1997 Relativistic band structure calculation of cubic and hexagonal SiC polytypes *J. Appl. Phys.* **82** 5496–508

[157] Paszkowicz W, Piszora P, ŁAsocha W, Margiolaki I, Brunelli M and Fitch A 2010 Lattice parameter of polycrystalline diamond in the low-temperature range *Acta Phys. Pol.* A **117** 323–7

[158] Zhu W, Wang X H, Stoner B R, Ma G H M, Kong H S, Braun M W H and Glass J T 1993 Diamond and β-SiC heteroepitaxial interfaces: a theoretical and experimental study *Phys. Rev.* B **47** 6529–42

[159] Kawarada H, Suesada T and Nagasawa H 1995 Heteroepitaxial growth of smooth and continuous diamond thin films on silicon substrates via high quality silicon carbide buffer layers *Appl. Phys. Lett.* **583** 583

[160] Suesada T, Nakamura N, Nagasawa H and Kawarada H 1995 Initial growth of heteroepitaxial diamond on Si(001) substrates via β-SiC buffer layer *Jpn J. Appl. Phys.* **34** 4898–904

[161] Kawarada H, Wild C, Herres N, Locher R, Koidl P and Nagasawa H 1997 Heteroepitaxial growth of highly oriented diamond on cubic silicon carbide *J. Appl. Phys.* **81** 3490–3

[162] Chang L, Yan J E, Chen F R and Kai J J 2000 Deposition of heteroepitaxial diamond on 6H-SiC single crystal by bias-enhanced microwave plasma chemical vapor deposition *Diam. Relat. Mater.* **9** 283–9

[163] Jiang X, Schiffmann K, Klages C P, Wittorf D, Jia C L, Urban K and Jäger W 1998 Coalescence and overgrowth of diamond grains for improved heteroepitaxy on silicon (001) *J. Appl. Phys.* **83** 2511–8

[164] Srikanth V V S S, Staedler T and Jiang X 2008 Structural and compositional analyses of nanocrystalline diamond/β-SiC composite films *Appl. Phys.* A **91** 149–55

[165] Humphreys T P, Hunn J D, Patnaik B K, Parikh N R, Malta D M and Das K 1993 Silicon carbind/diamond heterostructure rectifying contacts *Electronics Lett.* **29** 1332–4

[166] Stoner B R, Sahaida S R, Bade J P, Southworth P and Ellis P J 1993 Highly oriented, textured diamond films on silicon via bias-enhanced nucleation and textured growth *J. Mater. Res.* **8** 1334–40

[167] Weishart H, Heera V and Skorupa W 2005 N-type conductivity in high-fluence Si-implanted diamond *J. Appl. Phys.* **97** 103514

[168] Heera V, Fontaine F, Skorupa W, Pécz B and Barna Á 2000 Ion-beam synthesis of epitaxial silicon carbide in nitrogen-implanted diamond *Appl. Phys. Lett.* **77** 226–8

[169] Goto M, Amano R, Shimoda N, Kato Y and Teii K 2014 Rectification properties of n-type nanocrystalline diamond heterojunctions to p-type silicon carbide at high temperatures *Appl. Phys. Lett.* **104** 153113

[170] Tadjer M J *et al* 2007 Nanocrystalline diamond films as UV-semitransparent Schottky contacts to 4H-SiC *Appl. Phys. Lett.* **91** 2005–8

[171] Tadjer M J *et al* 2014 Thermionic-field emission barrier between nanocrystalline diamond and epitaxial 4H-SiC *IEEE Electron. Device Lett.* **35** 1173–5

[172] Tadjer M J *et al* 2010 On the high curvature coefficient rectifying behavior of nanocrystalline diamond heterojunctions to 4H-SiC *Appl. Phys. Lett.* **97** 5–8

[173] Yang W B, Lü F X and Cao Z X 2002 Growth of nanocrystalline diamond protective coatings on quartz glass *J. Appl. Phys.* **91** 10068–73

[174] Sillero E, Williams O A, Lebedev V, Cimalla V, Röhlig C C, Nebel C E and Calle F 2009 Static and dynamic determination of the mechanical properties of nanocrystalline diamond micromachined structures *J. Micromechan. Microeng.* **19** 115016

[175] Gaidarzhy A, Imboden M, Mohanty P, Rankin J and Sheldon B W 2007 High quality factor gigahertz frequencies in nanomechanical diamond resonators *Appl. Phys. Lett.* **91** 203503

[176] Krauss A R *et al* 2001 Ultrananocrystalline diamond thin films for MEMS and moving mechanical assembly devices *Diam. Relat. Mater.* **10** 1952–61

[177] Martin H B, Argoitia A, Landau U, Anderson A B and Angus J C 1996 Hydrogen and oxygen evolution on boron-doped diamond electrodes *J. Electrochem. Soc.* **143** L133–6

[178] Stotter J, Zak J, Behler Z, Show Y and Swain G M 2002 Optical and electrochemical properties of optically transparent, boron-doped diamond thin films deposited on quartz *Anal. Chem.* **74** 5924–30

[179] Stotter J, Show Y, Wang S and Swain G 2005 Comparison of the electrical, optical, and electrochemical properties of diamond and indium tin oxide thin-film electrodes *Chem. Mater.* **17** 4880–8

[180] Silva E, Bastos A C, Neto M, Fernandes A J, Silva R, Ferreira M G S, Zheludkevich M and Oliveira F 2014 New fluorinated diamond microelectrodes for localized detection of dissolved oxygen *Sensors Actuators* B **204** 544–51

[181] Silva E L, Gouvêa C P, Quevedo M C, Neto M A, Archanjo B S, Fernandes A J S, Achete C A, Silva R F, Zheludkevich M L and Oliveira F J 2015 All-diamond microelectrodes as solid state probes for localized electrochemical sensing *Anal. Chem.* **87** 6487–92

[182] Williams O A 2011 Nanocrystalline diamond *Diam. Relat. Mater.* **20** 621–40

[183] Williams O A, Curat S, Gerbi J E, Gruen D M and Jackman R B 2004 N-type conductivity in ultrananocrystalline diamond films *Appl. Phys. Lett.* **85** 1680–2

[184] Gajewski W, Achatz P, Williams O A, Haenen K, Bustarret E, Stutzmann M and Garrido J A 2009 Electronic and optical properties of boron-doped nanocrystalline diamond films *Phys. Rev.* B **79** 045206

[185] Gracio J J, Fan Q H and Madaleno J C 2010 Diamond growth by chemical vapour deposition *J. Phys. D: Appl. Phys.* **43** 374017

[186] Ashcheulov P *et al* 2013 Conductivity of boron-doped polycrystalline diamond films: influence of specific boron defects *Eur. Phys. J.* B **86** 443

[187] Neto M A, Pato G, Bundaleski N, Teodoro O M N D, Fernandes A J S, Oliveira F J and Silva R F 2016 Surface modifications on as-grown boron doped CVD diamond films induced by the B_2O_3–ethanol–Ar system *Diam. Relat. Mater.* **64** 89–96

[188] Wang W, Hu C, Li F N, Li S Y, Liu Z C, Wang F, Fu J and Wang H X 2015 Palladium Ohmic contact on hydrogen-terminated single crystal diamond film *Diam. Relat. Mater.* **59** 90–4

[189] van Enckevort W J P, Janssen G, Vollenberg W, Schermer J J, Giling L J and Seal M 1993 CVD diamond growth mechanisms as identified by surface topography *Diam. Relat. Mater.* **2** 997–1003

[190] May P W, Ashfold M N R and Mankelevich Y A 2007 Microcrystalline, nanocrystalline, and ultrananocrystalline diamond chemical vapor deposition: experiment and modeling of the factors controlling growth rate, nucleation, and crystal size *J. Appl. Phys.* **101** 1–9

[191] May P W and Mankelevich Y A 2006 Experiment and modeling of the deposition of ultrananocrystalline diamond films using hot filament chemical vapor deposition and Ar/CH$_4$/H$_2$ gas mixtures: a generalized mechanism for ultrananocrystalline diamond growth *J. Appl. Phys.* **100** 1–9

www.ingramcontent.com/pod-product-compliance
Lightning Source LLC
Chambersburg PA
CBHW082117210326
41599CB00031B/5792